X-RAY AND INNER-SHELL PROCESSES

X-RAY AND INNER-SHELL PROCESSES

17th International Conference

Hamburg, Germany   September 1996

EDITORS
R. L. Johnson
University of Hamburg, Germany

H. Schmidt-Böcking
University of Frankfurt, Germany

B. F. Sonntag
University of Hamburg, Germany

AIP CONFERENCE PROCEEDINGS 389

American Institute of Physics   Woodbury, New York

Authorization to photocopy items for internal or personal use, beyond the free copying permitted under the 1978 U.S. Copyright Law (see statement below), is granted by the American Institute of Physics for users registered with the Copyright Clearance Center (CCC) Transactional Reporting Service, provided that the base fee of $6.00 per copy is paid directly to CCC, 222 Rosewood Drive, Danvers, MA 01923. For those organizations that have been granted a photocopy license by CCC, a separate system of payment has been arranged. The fee code for users of the Transactional Reporting Service is: 1-56396-563-1/ 97 /$6.00.

© 1997 American Institute of Physics

Individual readers of this volume and nonprofit libraries, acting for them, are permitted to make fair use of the material in it, such as copying an article for use in teaching or research. Permission is granted to quote from this volume in scientific work with the customary acknowledgment of the source. To reprint a figure, table, or other excerpt requires the consent of one of the original authors and notification to AIP. Republication or systematic or multiple reproduction of any material in this volume is permitted only under license from AIP. Address inquiries to Office of Rights and Permissions, 500 Sunnyside Boulevard, Woodbury, NY 11797-2999; phone: 516-576-2268; fax: 516-576-2499; e-mail: rights@aip.org.

L.C. Catalog Card No. 96-80388
ISBN 1-56396-563-1
ISSN 0094-243X
DOE CONF- 9609283

Printed in the United States of America

Contents

Preface ix
Acknowledgments x

I. HISTORICAL INTRODUCTION

A Century of X Rays in Atomic Physics 3
    B. Crasemann

II. RADIATION SOURCES

New Lamps for Old: Synchrotron Radiation and Free-Electron
Laser X-Ray Sources 21
    R. P. Walker
The Ultimate Storage Ring X-Ray Source 43
    P. Elleaume
Transition Radiation in the X-Ray Region from a Low-Emittance
855 MeV Electron Beam 57
    H. Backe, K.-H. Brenzinger, F. Buskirk, S. Dambach, Th. Doerk,
    N. Eftekhari, H. Euteneuer, F. Görgen, C. Herberg, F. Hagenbuck,
    K. Johann, K.-H. Kaiser, O. Kettig, G. Knies, G. Kube, W. Lauth,
    B. Limburg, J. Lind, H. Schöpe, G. Stephan, Th. Walcher, Th. Tonn,
    and R. Zahn
Properties of Channeling and Parametric X-Radiation 73
    J. Freudenberger

III. HIGHLY CHARGED IONS

Highly Charged Ions in Storage Rings 93
    D. Liesen
High-Resolution X-Ray Spectra from Low-Temperature, Highly
Charged Ions 121
    P. Beiersdorfer
Complete Characterization of Final States in Double Electron Capture
by Slow Multiply Charged Ions 137
    H. Khemliche and M. H. Prior
The Dynamics of Target Single and Double Ionization Induced
by the Virtual Photon Field of Fast Heavy Ions 153
    R. Moshammer, J. Ullrich, H. Kollmus, W. Schmitt, M. Unverzagt,
    H. Schmidt-Böcking, and R. E. Olson
Electron Emission for He Double and Single Ionization after Heavy
Ion Impact 169
    S. Hagmann

IV. INSTRUMENTATION AND METHODS

Bent Crystal Optics for High-Energy Synchrotron Radiation 175
    P. Suortti, U. Lienert, and C. Schulze

Soft X-Ray Optics for Spectromicroscopy at the Advanced Light Source 193
    H. A. Padmore

Techniques and Applications of Spectromicroscopy with Soft X Rays 209
    J. Voss

Total Reflection X-Ray Fluorescence Analysis with Synchrotron
Radiation and Other Sources for Trace Element Determination 233
    P. Wobrauschek and C. Streli

Structural and Phase Transition Studies of Layered Materials
by X-Ray Standing Waves. 249
    B. N. Dev

Single-Pulse Laue Diffraction, Stroboscopic Data Collection,
and Femtosecond Flash Photolysis on Macromolecules 267
    M. Wulff, F. Schotte, G. Naylor, D. Bourgeois, K. Moffat,
    and G. Mourou

Atomic Holography with Electrons and X Rays 295
    P. M. Len, C. S. Fadley, and G. Materlik

V. NUCLEAR SCATTERING

Nuclear Scattering of Synchrotron X Rays. 323
    G. V. Smirnov

Applications of X-Ray Nuclear Resonant Scattering with the Use
of Synchrotron Radiation 351
    S. Kikuta, Y. Yoda, I. Koyama, T. Shimizu, H. Igarashi, K. Izumi,
    Y. Kunimune, M. Seto, T. Mitsui, T. Harami, X. Zhang, and M. Ando

VI. ELECTRON MOMENTUM SPECTROSCOPY

3D Electron Momentum Densities of Solids 369
    F. F. Kurp, M. Vos, Th. Tschentscher, A. S. Kheifets,
    H. Schulte-Schrepping, J. R. Schneider, E. Weigold, and F. Bell

Electron-Momentum Spectroscopy of Solids by the *(e, 2e)* Reaction 383
    A. S. Kheifets, M. Vos, S. A. Canney, X. Guo, I. E. McCarthy,
    and E. Weigold

Momentum-Space Magnetism Studied by Magnetic Compton Scattering. 399
    N. Sakai, A. Koizumi, N. Miyamoto, and Y. Tanaka

VII. THEORETICAL ASPECTS

Inner-Shell Ionization and Excitation 415
    M. Ya. Amusia
X-Ray Natural Widths, Level Widths, and Coster-Kronig Transition
Probabilities... 431
    T. Papp, J. L. Campbell, and D. Varga
Giant Resonances in Photon Emission Spectra of Atoms, Clusters,
and Solids 447
    L. G. Gerchikov, A. V. Korol, A. G. Lyalin, and A. V. Solov'yov
Double Ionization of a Helium-Like Atom or Ion by a High
Energy Photon 465
    T. Surić, K. Pisk, and R. H. Pratt
Double Ionization of Helium by Photons and Charged Particles 475
    J. Burgdörfer, Y. Qiu, J. Wang, and J. H. McGuire

VIII. X-RAY ABSORPTION SPECTROSCOPY

X-Ray Absorption and Dichroism of Transition Metal Compounds........... 497
    F. de Groot
New Applications of X-Ray Magnetic Circular Dichroism.................. 521
    G. Schütz, P. Fischer, K. Attenkofer, and D. Ahlers
Achievements and Prospects in X-Ray Absorption Spectroscopy 535
    Y. Yacoby and E. A. Stern
Linear Polarization Effects in X-Ray Emission and Absorption Spectra....... 557
    G. Dräger
From the S_2 Molecule to Its Condensed Molecular and Polymerized Phases:
An X-Ray Absorption Study at the S K Edge 569
    R. C. Karnatak
EXAFS and Diffraction Studies on High T_c Superconductors:
Mesoscopic Stripes in the CuO_2 Plane 585
    N. L. Saini, A. Lanzara, and A. Bianconi

IX. ELECTRON AND X-RAY EMISSION SPECTROSCOPY

Auger Electron and X-Ray Spectroscopy of Hollow Atoms.................. 603
    R. Morgenstern
Photoexcitation and Decay of Hollow Lithium States...................... 625
    F. J. Wuilleumier, S. Diehl, D. Cubaynes, and J.-M. Bizau
Extending Synchrotron-Based Atomic Physics Experiments
into the Hard X-Ray Region.. 647
    T. LeBrun
Nondipolar Photoelectron Angular Distributions......................... 659
    B. Krässig, M. Jung, D. S. Gemmell, E. P. Kanter, T. LeBrun,
    S. H. Southworth, and L. Young

Inelastic X-Ray Scattering Including Resonance Phenomena 671
    K. Hämäläinen, S. Manninen, W. Caliebe, C.-C. Kao, and J. B. Hastings

Angular Correlations in Auger and Fluorescence Cascades. 689
    N. M. Kabachnik

High Resolution Core-Level Electron Spectroscopy on Free Molecules and Atoms 703
    S. Svensson

Selection Rules in Resonant X-Ray Emission of Free Molecules. 723
    P. Glans, P. Skytt, K. Gunnelin, J.-H. Guo, and J. Nordgren

Free Clusters Studied by X-Ray Spectroscopy. 737
    A. V. Soldatov and T. Ivanchenko

Soft X-Ray Emission Excited Resonantly and Nonresonantly by Synchrotron Radiation 749
    D. L. Ederer, J. A. Carlisle, J. Jiminez, J. J. Jia, Ling Zhou, T. A. Callcott,
    R. C. C. Perera, A. Moewes, L. J. Terminello, E. Shirley, A. Asfaw,
    J. van Ek, E. Morikawa, and F. J. Himpsel

Electronic Structure of Advanced Materials Studied by X-Ray Emission Spectroscopy 771
    E. Z. Kurmaev, V. R. Galakhov, Yu. M. Yarmoshenko, V. A. Trofimova,
    S. N. Shamin, V. M. Cherkashenko, A. I. Poteryaev, and V. I. Anisimov

Author Index 787

Preface

This book contains the written versions of the invited talks and progress reports presented at the 17th International Conference on X-Ray and Inner-Shell Processes (X-96), which took place in Hamburg, Germany, September 9–13, 1996. Previous conferences in this series were held in Sendai, Japan (1978), Stirling, Scotland (1980), Leipzig, Germany (1984), Paris, France (1987), Knoxville, USA (1990), Eugene, USA (1992), and Debrecen, Hungary (1993). The present series combines the earlier series of conferences on X rays, which started in 1965 (held both in Ithaca and Leipzig) and continued until 1976 (Washington), and the series on Inner-Shell Ionization (held 1972 in Atlanta and 1976 in Freiburg).

The 1996 conference was attended by 305 scientists from 33 countries. The program covered recent advances and developments in both experimental and theoretical studies of X-ray and inner-shell processes and their applications in other disciplines and technologies. The invited speakers reviewed the latest developments in their fields and were chosen from the many suggestions offered by the members of the International Scientific Committee and the Organizing Committee. A plenary lecture by Professor B. Crasemann and a public evening lecture by Professor W. Martienssen celebrated the centennial of the discovery of X rays by Wilhelm Conrad Roentgen.

In total, 316 contributed papers were presented at the conference in three poster sessions. The wide variety of contributed papers demonstrated the breadth and the vitality of the field of X-ray and inner-shell phenomena in both theory and experiment.

Abstracts of the nine plenary lectures, 32 progress reports, three hot-topic reports, and 316 contributed papers were distributed as a Book of Abstracts to all participants at the start of the conference.

This conference proceedings volume contains all the plenary lectures, progress reports, and hot-topic reports and thus provides an overview of the present state-of-the-art in the fields of X-ray and inner-shell processes.

R. L. Johnson
Editor

B. F. Sonntag
Conference Chair

H. Schmidt-Böcking
Conference Co-Chair

NEXT CONFERENCE

The International Scientific Committee decided that the next conference in this series, the 18th International Conference on X-Ray and Inner-Shell Processes (X-99), will be held in Chicago, USA, in 1999 and Dr. Donald S. Gemmell will be the Conference Chairman.

ACKNOWLEDGMENTS

The X-96 Conference organizers are especially indebted to the following laboratories for their financial support and their assistance during the preparation of the conference:

Hamburger Synchrotronstrahlungslabor HASYLAB at
Deutsches Elektronen-Synchrotron DESY
Universität Hamburg
Gesellschaft für Schwerionenforschung GSI Darmstadt

The financial support by the following institutions is gratefully acknowledged:

Freie und Hansestadt Hamburg
Deutsche Forschungsgemeinschaft DFG
International Union of Pure and Applied Physics IUPAP
Alexander von Humbold-Stiftung
WE-Heraeus-Stiftung

The conference organizers acknowledge with special thanks the efforts of the staff of HASYLAB and the University of Hamburg, who were instrumental in making this conference a success.

I. HISTORICAL INTRODUCTION

A Century of X Rays in Atomic Physics

Bernd Crasemann

Physics Department, University of Oregon, Eugene, Oregon 97403, USA

Abstract. Photon-atom interactions have played an essential role in leading to present understanding of atomic structure and dynamics. The discovery of x rays, of characteristic radiation and of Auger transitions opened inner shells to exploration and greatly advanced the field. More recently, synchrotron-radiation sources and increasingly sophisticated instrumentation are bringing about a remarkable enhancement of experimental capability. Tempting opportunities arise for new investigations of many-body effects, fundamental processes, and relativistic and quantum electrodynamic phenomena. Some aspects of the history and present status of the subject are outlined.

INTRODUCTION

Near the end of the "classical" era, 1896 was an eventful year in the history of physics. In Hermann von Helmholtz's laboratory, Wilhelm Carl Werner Otto Fritz Franz Wien formulated his equation relating the wavelength of heat radiation to the temperature [1], thus helping to set the scene for Planck's epochal quantum hypothesis. In Leiden, Pieter Zeeman verified Hendrik Antoon Lorentz' prediction of the splitting of spectral lines in magnetic fields [2]. Guglielmo Marconi moved his invention of the wireless telegraph to London [3]. Albert Einstein was admitted to the *Eidgenössische Technische Hochschule* in Zürich, after having failed the entrance examination the previous year, and in Paris, Henri Becquerel detected the radiation from a uranium compound that became the subject of Marie Curie's doctoral dissertation [2]. None of these developments, however, penetrated the public's consciousness to the extent that Wilhelm Conrad Röntgen's discovery of x rays [4] did, which pervaded newspapers all over the world [5] and found applications almost immediately, first in medicine [6]. From a broader perspective, Röntgen's discovery proved to be a major factor in the evolution of 19th-century classical physics into our present-day view of the natural world [2]. Of the pervasive spin-off that Röntgen's discovery engendered in science and technology [7], the present paper is devoted to some aspects of the role that x rays have played in the development of atomic physics, a contribution that still continues at a most vigorous pace [8].

CHARACTERISTIC RADIATION

Following the announcement in December, 1895 of Röntgen's discovery [4], research started in many laboratories on the absorption and scattering of x rays and the ionization of gases by them. Already in April, 1896, M. I. Pupin reported to the New York Academy of Sciences on improvements that he had made in x-ray equipment in his Columbia University laboratory, exalting "It was in this manner only that I succeeded in photographing ... the whole chest, shoulders and neck of my assistant, with an exposure of seventy minutes and at a distance of three feet between the plate and the tube. The collar button and the buttons and clasps of the trousers and the vest show very strongly through the ribs and the spinal column ..." Continuing to describe work on "reflection and refraction" of x rays, Pupin concluded that his experiments "prove beyond all reasonable doubt that the Röntgen radiance is diffusely scattered through bodies, gases not excepted" [9]. This, in fact, had already been noted by Röntgen himself.

Extensive studies of the scattering of x rays were performed by Charles Glover Barkla (1877-1944) who, in his third year at Cambridge in J. J. Thomson's laboratory, initiated investigations of the secondary x rays emitted by substances in the path of an x-ray beam, a topic of research that he pursued during the following forty years [10]. Measuring hardness of radiation in terms of absorbability in aluminum foil, and intensity in terms of the rate at which the x rays caused a gold-foil electroscope to discharge, Barkla in 1903 reported that "all gases subject to x rays are a source of secondary radiation" and "the absorbability of the secondary radiation is ... the same as that of the primary radiation producing it" [11]—contrasting his results with those of Sagnac [12] who had found that secondary radiation from metals was distinctly more absorbable than the primary radiation. In fact, "all gases" studied by Barkla did not include heavier atoms than sulfur in "sulphuretted hydrogen" so that the fluorescence radiation would have been too soft for him to detect.

Further work, however, soon led to the crucial discovery of characteristic fluorescence radiation [13] that would earn Barkla the Nobel Prize for 1917. In addition to scattered x radiation, of the same penetrating power as the primary, the investigators found a "second type of secondary X-radiation emitted by many elements, and probably by all, ... a homogeneous radiation characteristic simply of the element emitting it" shown "to be emitted only when the exciting primary radiation was of more penetrating type" [13].

Measurements soon showed that the characteristic radiation contained two components, "a very easily absorbed radiation and a very penetrating homogeneous radiation superposed" [13], which Barkla denoted by A and B, respectively. In 1911, Barkla tabulated the absorbability of fluorescence radiation *vs.* atomic weight for a number of elements from Ca through Bi, but he had second thoughts with regard to notation and denoted the two groups by "Series K" and "Series L" explaining, in a footnote:

"Previously denoted by letters B and A (Proc. Camb. Phil. Soc. May 1909). The letters K and L are, however, preferable, as it is highly probable that series of radiations both more absorbable and more penetrating exist" [14].

Thus the puzzling notation was established that led to the innermost atomic shell in the Bohr model being called the K shell. The M and N x-ray series were soon found by Manne Siegbahn, but no J series was ever proven to exist although Barkla, in a tragic turn of his scientific career, pursued this chimera for the remainder of his life. Barkla had become a world leader in his field. He had not only discovered characteristic x rays but also the polarization of x rays and, on the basis of J. J. Thomson's theory of scattering, calculated the number of electrons per atom—at least approximately: he found it to be 15 for aluminum and concluded that "the number of scattering electrons per atom is about half the atomic weight . . ." [15]. After 1916, however, in a development that has been described as "pathological," Barkla gradually "passed out of physics," became oblivious to the work and results of others, engaged in involved and confused attempts to express his own results, and committed himself for the ensuing decades to a search, by absorption measurements, of the J series of radiation—more penetrating than the K series—which later grew in his mind to a more ubiquitous "J-Phenomenon" [10].

X-RAY SPECTROMETRY AND THE BOHR MODEL

Quantitative x-ray spectroscopy has its origins in Max von Laue's brilliant idea, in the spring of 1912, to send x rays through crystals. The crucial experiment by Friedrich, Knipping, and Laue and its explanation [16] not only established the nature of x rays as very short electromagnetic waves but also made precise measurements of their wavelengths possible and led the way toward a most powerful approach to studying the structure of atoms and of matter. Albert Einstein called Laue's discovery one of the most beautiful in physics.

Laue's interest in "great, general principles" however led him to stay with fundamental theory. The practical side of Laue's discovery was advanced forthwith by William Lawrence Bragg, whose father William Henry had persuasively deduced the particulate nature of x rays. In November, 1912, the younger Bragg showed how the Laue phenomenon could be seen as a reflection of electromagnetic waves from atom-rich crystal planes and derived *Bragg's law*. In January, 1913, he succeeded in constructing the first x-ray spectrometer [18].

The Braggs had been joined by H. G. J. Moseley, from Rutherford's laboratory, who brought the use of ionization chambers as detectors to the group and then also employed photographic plates. Moseley performed a systematic study of the characteristic radiations emitted by various targets, resolved components of the K and L series, and found the characteristic x-ray spectra of the elements to be homologous. Moseley discovered that the square roots of the $K\alpha$ frequencies of the ten elements from Ca to Zn fall on a straight line when plotted against *atomic number Z*, to within 0.5 percent precision: thus he established that atomic number is a much more fundamental quantity than atomic weight, as far as frequencies of characteristic lines are concerned. These frequencies, he found [19], follow the astonishingly simple relation that became known as "Moseley's Law,"

$$\frac{\nu_{K\alpha}}{R} = \frac{3}{4}(Z-1)^2, \qquad (1)$$

where R is the Rydberg frequency. Emboldened by this formula, particularly by the appearance of the Bohr-Balmer factor $3R/4$, Moseley inferred an analogous formula for the $L\alpha$ frequencies [19] "on very little evidence" [20]. These formulas allowed Moseley to test the periodic table for completeness and to spot the then missing elements $_{43}$Tc, $_{61}$Pm, and $_{75}$Re.

It was also in 1913 that James Chadwick, a young collaborator of Ernest Rutherford's, succeeded in observing the production of characteristic x rays from targets which had been exposed to alpha rays from radioactive sources [21]. These first inelastic atomic collision experiments with energetic ions mark the beginning of highly productive investigations at the interface between atomic and nuclear physics, still an active field at present [22,23].

Harry Moseley, with well-nigh prophetic intuition and extraordinary imagination, built upon the discovery of his "law" to forge a link with the Rutherford-Bohr atom. He argued, albeit qualitatively, that Bohr's quantization rules appeared to apply also to the deep-lying electrons of heavier atoms and suggested fruitful alterations to the theory, *viz.*, that electrons should not be restricted to move in a single plane through the nucleus, that the population of the innermost orbits should not be allowed to increase with Z, and that every electron should not be limited to $1\hbar$ of angular momentum. In the early days of the First World War, Sommerfeld's associate Walter Kossel povided the basis for further work that would confirm these guesses [24], although Moseley's formulas could never be derived from Bohr's theory.

At the outbreak of the First World War, Moseley joined the Royal Engineers, disregarding pleas of colleagues and family members. In August of 1915 he was killed, at age 27, in a furious attack by Kemal Ataturk's soldiers during an attempt by his unit to reach the ridge of Sari Bair on Turkey's Gallipoli Peninsula.

In his brief life Moseley, "the most promising of all English physicists of his generation" [24], accomplished so much so quickly thanks to his native genius and energy, according to his biographer J. L. Heilbron, but also thanks to luck which brought him to his research precisely at the right moment to exploit the seminal discovery of Laue and the atomic theories of Rutherford and Bohr: "It was also luck that the research proved 'so very easy,' as [Moseley] said, and gave 'so rich a return for a minimum of work'" [24]. Niels Bohr is quoted as having stated in 1962: "You see actually the Rutherford work [the nuclear atom] was not taken seriously. We cannot understand today, but it was not taken seriously at all. There was no mention of it in any place. The great change came from Moseley" [24].

The extraordinarily fruitful subsequent development of x-ray physics, symbiotically interlinked with that of quantum mechanics, is well-known and described in familiar textbooks and monographs (see, e.g., [25,26]). The names of many scientists come to mind whose skillful, patient, and insightful work was crucial in bringing the field to maturity, among them Parratt, Ewald, Kramers, Compton, Cauchois, Bearden, DuMond, and Deslattes.

The role that x-ray spectrometry would play in leading to a fundamental understanding of atomic structure was rather prophetically foreseen by Arnold Sommerfeld who, in the preface to the first edition of his famous *Atombau und Spectrallinien*, wrote [27] in 1919:

> "Seit der Entdeckung der Spektralanalyse konnte kein Kundiger zweifeln, dass das Problem des Atoms gelöst sein würde, wenn man gelernt hätte, die Sprache der Spektren zu verstehen. Das ungeheurige Material, welches sechzig Jahre spektroskopischer Praxis aufgehäuft haben, schien allerdings in seiner Mannigfaltigkeit zunächst unentwirrbar. Fast mehr haben die sieben Jahre Röntgenspektroskopie zur Klärung beigetragen, indem hier das Problem des Atoms an seiner Wurzel erfasst und das Innere des Atoms beleuchtet wird. Was wir heutzutage aus der Sprache der Spektren heraus hören, ist eine wirkliche Sphärenmusik des Atoms, ein Zusammenklingen ganzzahliger Verhältnisse, eine bei aller Mannigfaltigkeit zunehmende Ordnung und Harmonie. Für alle Zeiten wird die Theorie der Spektrallinien den Namen Bohrs tragen. Aber noch ein anderer Name wird dauernd mit ihr verknüpft sein, der Name Plancks. Alle ganzzahlige Gesetze der Spektrallinien und der Atomistik fliessen letzten Endes aus der Quantentheorie. Sie ist das geheimnisvolle Organon, auf dem die Natur die Spektralmusik spielt und nach dessen Rhytmus sie den Bau der Atome und der Kerne regelt."[1]

---

[1] Since the discovery of spectral analysis, no knowledgeable person could doubt that the problem of the atom would be solved once one had learned to understand the language of the spectra. The immense material amassed through sixty years of spectroscopy did, to be sure, at first appear uninterpretable because of its diversity. Almost more have the seven

RADIATIONLESSS TRANSITIONS

The role of x rays in the elucidation of atomic transitions cannot be adequately discussed without considering the complementary, competing process of nonradiative transitions, or *strahlungslose Quantensprünge*, as Wentzel called them [28]. It is rather surprising, in retrospect, that the discovery of the Auger effect lagged over a decade behind that of characteristic radiative transitions and drew far less acclaim. Yet, in the deexcitation of atomic vacancies (except for K holes in heavier elements), Auger transitions preponderate greatly—by up to six orders of magnitude for M-shell holes, for example [29].

To be sure, already at the time of discovery of characteristic x rays in 1909, there was suspicion that in ionization by x rays another process must take place in addition to fluorescence; Sadler wrote, in a footnote [13]: "Experiments now in progress indicate that when the exciting beam is more penetrating than the radiation characteristic of the tertiary radiator, part of the energy absorbed reappears as an easily absorbed corpuscular radiation..." Beatty (in 1911) and Barkla and Philpot (in 1913) measured total ionization of gases by x rays and found that, as the frequency of the radiation was increased through the K absorption limit, the total ionization increased sharply by two or three times, an increase that could not be due to the K series fluorescence radiation, little of which was absorbed in the gas [30]. In 1918, Barkla introduced the concept of *fluorescence yield* as the ratio of the energy carried by fluorescence radiation to the energy carried by the radiation absorbed in a sample, close to the present-day definition of fluorescence yield as the ratio of the radiative width of a hole state to its total width [29].

Only in the 1920's, however, physicists were closing in on the discovery of radiationless transitions. On the theoretical front, inspired by the experiments of Franck and Hertz, Klein and Rosseland considered statistical equilibrium between free electrons and atoms, including collisional excitation and deexcitation of the atoms whose discrete energy states were taken to be governed by Bohr's quantum theory. These authors pointed out that radiationless atomic deexcitation by collisions with initially slow electrons would suppress the emission of spectral lines [31].

In 1923, Rosseland extended this idea and pointed out that radiationless transitions could occur not only in collisions between atoms and free electrons,

---

years of x-ray spectroscopy contributed to clarification, by seizing the problem of the atom by its root and illuminating the innermost regions of the atom. What we can now hear out of the speech of the spectra is a veritable music of the spheres of the atom, a consonance of integer ratios, an order and harmony that is increasing despite its multiplicity. For all times, the theory of spectral lines will bear the name of Bohr. But still another name will be permanently tied to it, that of Planck. All integer laws of spectral lines and of atomic physics flow ultimately from quantum theory. This is the secretive organ on which Nature plays the spectral music and according to the rythm of which she regulates the structure of atoms and nuclei.

but also by emission of previously bound atomic electrons. He concluded that an excited atom possesses two channels for spontaneous decay: (1) emission of radiation, and (2) emission of one of its constituent particles ("corpuscular radiation"), but noted that nothing definite could be said as yet about the relative probabilities of the two competing processes [32].

Also in 1923, two experimental reports gave evidence, albeit fleetingly, of radiationless transitions. In a voluminous paper on observations of ionization in the cloud chamber he had invented, C. T. R. Wilson described almost in passing: "One or two examples of β-ray tracks with two branches springing from the same point have been found among the photographs... It is ... likely that the two electrons have been ejected simultaneously from one atom" [33]. Yet, Wilson appears to have given little importance to this phenomenon and did not even mention it in the summary of his paper.

Far more specific was the first observation of K Auger electrons by Lise Meitner, but it too appears to have been attributed little significance by the author who buried it in a paper on the β-ray spectrum of UX_1 and its interpretation [34]. The article is devoted to the presentation of apparent evidence for the hypothesis that nuclei emit β rays in discrete lines, the observed continuous spectrum or "band" being caused by energy-degrading "secondary influences," such as radiation by the emerging β particles.

Meitner studied UX, consisting of UX_1 (24-day ^{234}Th) and its decay products, notably UX_2 (1.2-minute isomeric ^{234}Pa^m), in secular equilibrium, and photographically recorded the β spectrum resolved in a magnetic spectrometer. In addition to the β-ray "band," Meitner identified three sharp lines and found "surprisingly" that within the accuracy of her measurements the energy of these lines equalled the Th $K\alpha$ x-ray energy less the L_2-, M_3-, and N_4-electron binding energies, respectively. This astute observation led her to the interpretation that (loosely translated)

> "The primary β rays of approximately 0.59 times the speed of light eject K electrons; thereby $K\alpha$ radiation is excited, which in turn releases L, M, or N electrons, where the whole process from emission of a primary β ray related to nuclear decay up to release of an L, M, or N electron takes place in one and the same atom..."

Only the first part of this interpretation is erroneous; the K vacancies arise primarily from γ-ray internal conversion. But the K vacancies' radiationless decay was correctly inferred.

Meitner apparently did not recognize the full significance of this first detailed identification of radiationless decay of an atomic inner-shell vacancy and treated it merely as part of the evidence that seemed to support the thesis that nuclei undergoing β decay emit monoenergetic "nuclear electrons." Had she changed the emphasis and pursued her observation, the Auger effect might well have become associated with Meitner's name, as speculated by Ruth Sime in her fascinating new biography of the eminent scientist [35].

Pierre Auger, with whose name we now identify radiationless transitions in atoms and the electrons emitted thereby, stayed on in Jean Perrin's laboratory in the Ecole Normale Supérieure in Paris after having completed his studies to become a science teacher. He constructed the first Wilson cloud chamber in France and, at Perrin's urging, varied experimental conditions in search of "the place where the truffels are" ("*la région des truffes*") [36]. At age 90, Auger recalled:

> "I ... noticed, in the cloud chamber snapshots, that photoelectron trajectories created along a beam of x rays contained a tiny group of small droplets at the start of each trajectory ... I replaced the gas in the chamber by wet hydrogen in order to lengthen the electron tracks, and observed in my photographs that the little group of drops lengthened into a new, very short trajectory, quite well visible, and which originated at the same point as that of the photoelectron ... By adding to the hydrogen atmosphere heavy atoms of such elements as krypton or xenon, I saw the additional tracks lengthening, indicating that the new electron had an energy that was characteristic of the atom absorbing the x rays, and increased with atomic number" [36].

These experiments provided unequivocal evidence for the nature of radiationless transitions [37]; for Auger they led to his doctorate, followed by a distinguished career in university teaching, cosmic ray research, science policy making and administration. The subject he had pioneered boomed; much work on the identification of different types of radiationless transitions in atoms and the measurement of their rates ensued [29,30]. Particularly noteworthy is the discovery of *Coster-Kronig transitions* from an analysis of L x-ray satellites [38]. These radiationless transitions, in which a vacancy "bubbles up" between the subshells of one principal shell, are among the fastest atomic transitions known; they can fill a vacancy in less than a femtosecond, resulting in level widths of tens of eV [39].

Gregor Wentzel first showed in 1927 how Auger transition probabilities can be calculated quantum mechanically on the basis of time-dependent perturbation theory [28]. The wide range of applicability of the formula that Wentzel derived imbued Enrico Fermi with such enthusiasm that in his lectures he would refer to it as the "Golden Rule No. 2" [40]. So great was his influence on his students that the formula is now commonly referred to as "Fermi's Golden Rule"; Fermi in his great modesty would have been revulsed by this apocryphalness. Extended to encompass many-electron and relativistic effects, calculations of radiationless transition probabilities have now grown from Wentzel's *ansatz* to reach a high degree of sophistication; comparison of theoretical results with measurements of Auger rates and level widths provide one of the most sensitive means to test models of atomic structure [41].

THE ERA OF SYNCHROTRON RADIATION

Photon-atom interactions have played a central role in generating present understanding of atoms and molecules. In this, photoexcitation is complemented by charged-particle collisions as a research tool but has the well-known advantages for interpretation that the interaction is relatively weak, one photon interacts with only one electron in first order, and perturbative approaches are applicable in most cases. Probably no technical advance since the development of Bragg spectrometers has had greater impact on the experimental study of atomic structure and dynamics through photoexcitation than the advent of synchrotron radiation sources.

It is interesting to trace progress on just one single problem that appears deceptively simple, yet is of central importance for the understanding of electron-electron Coulomb correlations: that of He, seemingly the most elementary atomic system next to hydrogen. Just to calculate the ground-state energy of He called for heroic variational efforts in the early days of quantum mechanics [42]; to understand He double ionization remains a tauntingly elusive problem to this day.

The potential of synchrotron radiation in atomic physics was first revealed strikingly through a series of classic experiments conducted on the National Bureau of Standards storage ring SURF in the early 1960's, which led to the discovery of the autoionizing doubly excited $^1P_0$ states of He [43,44]. This seminal finding induced the formulation of new correlation quantum numbers [45] and over the years was extended by ensuing experiments that led to new insights, as more sophisticated sources and instrumentation became available. In 1991, Domke et al. performed an analogous high-resolution photoionization measurement at BESSY in Berlin, observing more than 50 states as narrow as 0.1 meV, with evidence for interchannel interference that could be interpreted in terms of multichannel quantum-defect theory [46]. Most recently, Menzel et al., working on Berkeley's Advanced Light Source, measured partial photoionization cross sections and photoelectron angular distributions of He in the region of interfering Rydberg series below the $n = 5$ threshold; comparison with hyperspherical close-coupling calculations led to the most critical assessment of the dynamics of the two-electron excitations to date [47].

But the picture is not yet complete, and it is indeed a lesson in humility that the removal by a photon of both He electrons into the continuum is not fully understood. Experimental progress has been intimately tied to advances in sources and instrumentation [48], including development of an ingenious technique dubbed "cold target recoil ion momentum spectroscopy" by H. Schmidt-Böcking's group [49], going hand in hand with profound theoretical efforts [50]. Finally, perhaps, we can now see the light at the end of the tunnel.

The development of synchrotron-radiation sources, particularly of third-generation facilities, places us now at the threshold of significant new experiments that will become possible with the greater brilliance and higher energy of radiation from these machines. Progress can be expected to ensue on major frontiers of present understanding, in addition to various aspects of the ubiquitous many-electron problem epitomized by the He case just discussed [51].

Much challenging work lies ahead in the area of fundamental photon-atom interactions. As to photoionization, a good basis is provided by the fact that the theory rests on firm ground [52]. For some time, the *"complete" photoionization measurement* was considered the ultimate aim for experimenters, i.e., the determination of observables that contain, in some linearly independent combination, the five parameters that completely characterize the process in the low-energy, nonretarded dipole approximation [51]. Strictly speaking, there is *no* "complete" experiment, as U. Heinzmann has explained [53]. Yet, a wealth of relevant physics can be extracted when photon polarization, target orientation, photoelectron spin and angular distributions are resolved, as in current experiments by Becker and Heinzmann at ESRF in Grenoble [8].

A *breakdown of the dipole approximation* in photoionization has long been anticipated at higher energies, where it is necessary to consider the full retarded multipole expansion of the photon field [54]. In an elegant experiment, a group from the Argonne National Laboratory has recently measured the nondipolar asymmetries of Ar $1s$ photoelectrons produced by 3-5 keV x rays [55]. The results agree exceedingly well with calculations that include interference between electric-dipole and electric-quadrupole photoionization amplitudes in a nonrelativistic central-field model. These interference terms give a clear indication of angular-momentum transfer to core electrons. High-brilliance x-ray sources of higher energies will make it possible to further explore theoretical predictions for higher-order multipole and relativistic effects as well as photon-photoelectron polarization correlations, and even nondipolar effects in the angular distributions of Auger electrons [56], and to relate these measurements to features of atomic inner-shell dynamics.

Threshold resonances in photoexcited atomic transitions [57,58] are only now becoming accessible to detailed experimental studies, even though they were first observed well over a decade ago and are theoretically rather well-understood. *Resonant Raman transitions* arise from the **A**·**p** term in the operator that describes the interaction between atomic electrons and the radiation field (where **A** is the vector potential of the photon field and **p** is the electron-momentum operator) *in second order*. This can be seen from the generalized Kramers-Heisenberg cross section derived in terms of Åberg's generalization of time-dependent resonant scattering theory [59] for x-ray scattering from $|\mathbf{k}_1, \mathbf{e}_1\rangle$ to $|\mathbf{k}_2, \mathbf{e}_2\rangle$, where $\mathbf{k}_i$ is the wave vector and $\mathbf{e}_i$, the (complex) polarization vector of the radiation:

$$\frac{d^2\sigma}{d\omega_2 d\Omega} = r_0^2 \frac{\omega_2}{\omega_1} \left| \langle \phi^-_{\beta\varepsilon} \left| e^{i(\mathbf{k}_1-\mathbf{k}_2)\cdot\mathbf{r}} \right| \phi_\alpha \rangle (\mathbf{e}_1 \cdot \mathbf{e}_2^*) \right.$$

$$-\frac{1}{m}\sum_{\nu}\int d\tau \left[\frac{\langle \phi^-_{\beta\varepsilon}|e^{i\mathbf{k}_1\cdot\mathbf{r}}(\mathbf{e}_1\cdot\mathbf{p})|\phi_{\tau\nu}\rangle\langle\phi_{\tau\nu}|e^{-i\mathbf{k}_2\cdot\mathbf{r}}(\mathbf{e}_2^*\cdot\mathbf{p})|\phi_\alpha\rangle}{E_\nu^{(+)} - E_\alpha + \tau + \hbar\omega_1} \right.$$

$$\left. + \frac{\langle \phi^-_{\beta\varepsilon}|e^{-i\mathbf{k}_2\cdot\mathbf{r}_2}(\mathbf{e}_2^*\cdot\mathbf{p})|\phi_{\tau\nu}\rangle\langle\phi_{\tau\nu}|e^{i\mathbf{k}_1\cdot\mathbf{r}}(\mathbf{e}_1\cdot\mathbf{p})|\phi_\alpha\rangle}{E_\nu^{(+)} - E_\alpha + \tau - \hbar\omega_1 - i\frac{\Gamma_\nu}{2}} \right]\Bigg|^2$$

$$\cdot \delta\left(\omega_1 - \omega_2 + \frac{E_\alpha - E_\beta^{1+} - \varepsilon}{\hbar}\right). \quad (2)$$

The nonresonant term is seen to be of order of the square of the classical electron radius, $r_0 \cong 2.82 \times 10^{-13}$ cm, while near resonances the cross section increases typically by a factor of 10^6 and dominates overwhelmingly. Here, the intermediate states are virtual and their "natural lifetime widths" are not reflected in the energy width of the scattered radiation.

Not only is the physics of resonant Raman transitions of interest in itself, but these phenomena are rapidly finding applications in many fields. The narrow line width of the scattered radiation makes it useful for high-resolution studies of complicated atomic and molecular spectra [60], and resonant Raman spectrometry is becoming a good tool for the measurement of extraatomic effects [61], such as the momentum-resolved electronic structure of complex materials [62] and the density of states in the conduction band of metals, revealed in subthreshold excitation [63].

Other aspects of *photon-atom scattering* also offer very interesting opportunities for exploration with state-of-the-art sources as theory has progressed beyond form-factor approximation. Substantial progress in the understanding of elastic photon-atom scattering was made with the advent of numerical calculations of the amplitudes based on the second-order S matrix of quantum electrodynamics and relativistic wave functions [64]. Precise measurements of Rayleigh scattering are rather scarce to date [65]. Furthermore, progress in the calculation of Rayleigh scattering has made it possible, in principle, to measure the Delbrück amplitude for photon scattering by virtual electron-positron pairs created in the screened nuclear Coulomb potential [66]. This process is interesting because it belongs to the nonlinear effects of quantum electrodynamics that have no classical analog; although it has been known since 1933, only few experiments have been performed. Such measurements may, however, have to await the availability of yet higher-energy radiation sources.

Compton scattering from bound electrons has extensive applications in the study of electron momentum distributions in atoms and solids. An exact second-order S-matrix code for the relativistic numerical calculation of such cross sections has been developed within the independent-particle approximation [67]; detailed experimental tests would be useful.

We cannot conclude without alluding to the tempting possibilities that arise for measuring some *relativistic and quantum electrodynamic effects*. Since it is the Coulomb force that holds atoms, molecules, and solids together, it clearly is of interest to look at manifestations of the lowest-order relativistic correction to the static Coulomb potential, *viz.*, the Breit energy which accounts for the exchange of a single transverse photon and contains the effect of both the current-current interaction and of retardation:

$$H_{Breit}(\omega) = -\frac{1}{r_{ij}}[\alpha_i \cdot \alpha_j \cos \omega r_{ij} + (1 - \cos \omega r_{ij})], \tag{3}$$

where the α_n are Dirac matrices and r_{ij} is the distance between the two interacting charges; ω is the energy of the virtual photon. In the Pauli limit, the Breit operator corresponds to the orbit-orbit, spin-spin, and spin-other-orbit interactions between two electrons. Ordinarily, the effects of the Breit interaction in atomic structure are rather subtle, as in $\leq 1\%$ shifts of the 1s level energies, but they can become pronounced in cases where the static Coulomb interaction cancels out, as in j-splittings of double-hole states [68].

As a single instance, we mention one effect that arises exclusively from the relativistic Breit-Coulomb Hamiltonian, *viz.*, the splitting of excitations that comprise two vacancies in s states into distinct ^1S and ^3S levels. This splitting contains a dominant contribution from the spin dependence of the relativistic Coulomb interaction and a smaller contribution from the Breit operator [69]. Other splittings are greatly modified by the Breit-Coulomb Hamiltonian and in some cases even their energy ordering is reversed [68]. An unusual opportunity for relevant experiments arises here.

These few examples may serve to illustrate some of the range of fascinating physics that awaits ingenious experimenters on the new light sources, continuing in the venerable tradition of illuminating "the problem of the atom" with x rays.

The author is indebted to Sue Mandeville for indefatigable assistance, to Isabel A. Stirling, Head of the University of Oregon Science Library, for generous help in assembling historical material, and to Guy T. Emery for a constructive conversation.

REFERENCES

[1] W. Wien, Ann. Phys. (Leipzig) **294**, 662 (1896).
[2] A. Hermann, *Grosse Physiker* (Battenberg, Stuttgart, 1959).
[3] R. A. Chipman, in *Dictionary of Scientific Biography*, edited by C. C. Gillispie (Scribners, New York, 1976), vol. IX.
[4] W. C. Röntgen, Sitzungsber. Würzb. Physik-med. Gesellsch. **35** (1895), **45** (1896).
[5] Anonymous, CERN Courier **36**, No. 4, 9 (1996).

[6] O. Glasser, *Wilhelm Conrad Röntgen und die Geschichte der Röntgenstrahlen* (Springer, Berlin, 1931); *Forschung mit Röntgenstrahlen*, edited by F. H. W. Heuck and E. Macherauch (Springer, Berlin, 1995).

[7] W. R. Nitske, *Wilhelm Conrad Röntgen Discoverer of the X Ray* (University of Arizona Press, Tucson, 1971).

[8] See, e.g., *Proceedings of the Workshop on Atomic Physics with Hard X-Rays from High-Brilliance Synchrotron Light Sources*, edited by L. Young and E. Kanter (Physics Division, Argonne National Laboratory, Argonne, 1996); T. W. LeBrun, this Conference.

[9] M. I. Pupin, Science **3**, 538 (1896).

[10] P. Forman, in Ref. 3, vol. I.

[11] C. G. Barkla, Phil. Mag. **5**, 685 (1903).

[12] G. Sagnac, Compt. Rend. **126**, 521 (1898).

[13] C. G. Barkla and C. A. Sadler, Phil. Mag. **17**, 739 (1909); C. A. Sadler, Phil. Mag. **18**, 107 (1909); C. G. Barkla, Proc. Camb.Phil. Soc. **15**, 257 (1909).

[14] C. G. Barkla, Phil. Mag. **22**, 396 (1911).

[15] R. J. Stephenson, Am. J. Phys. **35**, 140 (1967).

[16] W. Friedrich, P. Knipping, and M. v. Laue, Ber. bayer. Akad. Wiss. 303 (1912); see also A. Hermann in Ref. 2 and Ref. 3, vol. VIII.

[17] W. L. Bragg, Proc. Camb. Phil. Soc. **17**, 43 (1913).

[18] P. Forman, in Ref. 3, vol. II.

[19] H. G. J. Moseley, Phil. Mag. **26**, 1024 (1913); **27**, 703 (1914).

[20] J. L. Heilbron, in Ref. 3, vol. IX.

[21] E. Merzbacher, in *Electronic and Atomic Collisions*, edited by J. Eichler, I. V. Hertel, and N. Stolterfoht (North-Holland, Amsterdam, 1984), p. 1.

[22] E. Merzbacher and J. M. Feagin, Comments Nucl. Part. Phys. **11**, 139 (1983).

[23] R. Anholt, in *Atomic Inner-Shell Physics*, edited by B. Crasemann (Plenum, New York, 1985), p. 581; P. Vincent, *ibid.*, p. 669.

[24] J. L. Heilbron, *H. G. J. Moseley: The Life and Letters of an English Physicist, 1887-1915* (University of California Press, Berkeley, 1974).

[25] A. H. Compton and S. K. Allison, *X-Rays in Theory and Experiment*, 2nd edition (Van Nostrand, New York, 1935); M. Siegbahn, *Spektroskopie der Röntgenstrahlen* (Springer, Berlin, 1931).

[26] F. K. Richtmyer and E. H. Kennard, *Introduction to Modern Physics*, 4th edition (McGraw-Hill, New York, 1947). (In later editions of this venerable text, much of the historical material is deleted).

[27] A. Sommerfeld, *Atombau und Spektrallinien*, 1st edition, 1919; 8th edition, 1960 (Vieweg, Braunschweig).

[28] G. Wentzel, Z. Phys. **43**, 524 (1927).

[29] W. Bambynek *et al.*, Rev. Mod. Phys. **44**, 716 (1972).

[30] E. H. S. Burhop, *The Auger Effect and Other Radiationless Transitions* (Cambridge U.P., Cambridge, 1952).

[31] O. Klein and S. Rosseland, Z. Phys. **4**, 46 (1921).

[32] S. Rosseland, Z. Phys. **14**, 173 (1923).

[33] C. T. R. Wilson, Proc. Roy. Soc. London **104**, 192 (1923).

[34] L. Meitner, Z. Phys. **17**, 54 (1923).
[35] R. L. Sime, *Lise Meitner: A Life in Physics* (U. of California Press, Berkeley, 1996).
[36] P. Auger, *La Recherche et la Creation du Nouveau*, address at the Symposium on the Auger Effect, l'Université Pierre et Marie Curie, Paris, 1989 (unpublished).
[37] P. Auger, C. R. Acad. Sci. **177**, 169 (1923), **180**, 65 (1925); J. Phys. Radium **6**, 205 (1925); Ann. Phys. (Paris) **6**, 183 (1926).
[38] D. Coster and R. de L. Kronig, Physica **2**, 13 (1935).
[39] See, e.g., M. H. Chen, B. Crasemann, and H. Mark, At. Data Nucl. Data Tables **24**, 113 (1979); Phys. Rev. A **27**, 2989 (1983); T. Papp, this Conference.
[40] *Nuclear Physics, a Course Given by Enrico Fermi*, notes compiled by J. Orear, A. H. Rosenfeld, and R. A. Schluter (U. of Chicago Press, 1950), revised edition, p. 142.
[41] See, e.g., M. H. Chen, in Ref. 23, p. 31; W. Mehlhorn, *ibid.*, p. 119.
[42] E. Hylleraas, Z. Phys. **54**, 347 (1929); **65**, 209 (1930).
[43] R. P. Madden and K. Codling, Phys. Rev. Lett. **10**, 516 (1963).
[44] J. W. Cooper, U. Fano, and F. Prats, Phys. Rev. Lett. **10**, 518 (1963); U. Fano, Rep. Prog. Phys. **46**, 97 (1983).
[45] D. R. Herrick and O. Sinanoğlu, Phys. Rev. A **11**, 97 (1975); C. D. Lin, Phys. Rev. A **29**, 1019 (1984).
[46] M. Domke *et al.*, Phys. Rev. Lett. **66**, 1306 (1991).
[47] A. Menzel *et al.*, Phys. Rev. Lett. **75**, 1479 (1995).
[48] J. C. Levin, G. B. Armen, and I. A. Sellin, Phys. Rev. Lett. **76**, 1220 (1996).
[49] R. Dörner *et al.*, Phys. Rev. Lett. **76**, 2654 (1996).
[50] L. R. Andersson and J. Burgdörfer, Phys. Rev. A **50**, R2810 (1994); P. M. Bergstrom, Jr., K.-i. Hino, and J. H. Macek, Phys. Rev. A **51**, 3044 (1995); see also the presentations by J. Burgdörfer and by T. Surić in this Conference.
[51] A splendid review of atomic photoionization studies with synchrotron radiation is given by V. Schmidt, Rep. Prog. Phys. **55**, 1483 (1992).
[52] R. H. Pratt, A. Ron, and H. K. Tseng, Rev. Mod. Phys. **45**, 273 (1973); A. F. Starace, in *Handbuch der Physik* **31**, edited by W. Mehlhorn (Springer, Berlin, 1982), p. 1; M. Ya. Amusia, *Atomic Photoeffect*, edited by K. T. Taylor (Plenum, New York, 1990).
[53] U. Heinzmann, in *International Workshop on Photoionization–WP94*, San Francisco, 1994 (unpublished).
[54] M. Peshkin, Adv. Chem. Phys. **18**, 1 (1970); A. Bechler and R. H. Pratt, Phys. Rev. A **39**, 1774 (1989); **42**, 6400 (1990); J. H. Scofield, Phys. Rev. A **40**, 3054 (1989); J. W. Cooper, Phys. Rev. A **42**, 6942 (1990); **45**, 3362 (1992); **47**, 1841 (1993).
[55] B. Krässig *et al.*, Phys. Rev. Lett. **75**, 4736 (1995).
[56] N. M. Kabachnik, this Conference.
[57] P. Cowan, in *Resonant Anomalous X-Ray Scattering: Theory and Experiment*, edited by G. Materlik, C. J. Sparks, and K. Fischer (North-Holland, Amsterdam, 1994), p. 449.

[58] T. Åberg and B. Crasemann, in Ref. 56, p. 431.
[59] T. Åberg and J. Tulkki, in *Atomic Inner-Shell Physics*, edited by B. Crasemann (Plenum, New York, 1985), p. 419.
[60] B. Langer *et al.*, Phys. Rev. A **53**, R1946 (1996); M. A. MacDonald *et al.*, Phys. Rev. A **51**, 3598 (1995); Z. F. Liu *et al.*, Phys. Rev. Lett. **72**, 621 (1994).
[61] H. Wang *et al.*, Phys. Rev. A **50**, 1359 (1994).
[62] Y. Ma *et al.*, Phys. Rev. Lett. **69**, 2598 (1992).
[63] W. Drube, R. Treusch, and G. Materlik, Phys. Rev. Lett. **74**, 42 (1995).
[64] L. Kissel, R. H. Pratt, and S. C. Roy, Phys. Rev. A **22**, 1970 (1980); L. Kissel *et al.*, Acta Crystallogr. **A51**, No. 3 (1995).
[65] P. P. Kane *et al.*, Phys. Repts. **140**, 75 (1986).
[66] A. I. Milstein and M. Schumacher, Phys. Repts. **243**, 183 (1994); R. Solberg, K. Mork, and I. Øverbø, Phys. Rev. A **51**, 359 (1995).
[67] P. M. Bergstrom *et al.*, Phys. Rev. A **48**, 1134 (1993).
[68] M. H. Chen, B. Crasemann, and H. Mark, Phys. Rev. A **25**, 391 (1982); B. Crasemann, M. H. Chen, and H. Mark, J. Opt. Soc. Am. B **1**, 224 (1984).
[69] S. J. Schaphorst *et al.*, Phys. Rev. A **47**, 1953 (1993).

II. RADIATION SOURCES

New Lamps for Old : Synchrotron Radiation and Free-Electron Laser X-ray Sources

R.P. Walker

Sincrotrone Trieste, Padriciano 99, 34012 Trieste, Italy
r.walker@elettra.trieste.it

Abstract. An overview of the field of synchrotron radiation and free-electron lasers is presented, including a basic introduction to the physical principles, the present state-of-the-art and the future possibilities, particularly with regard to the next generation of X-ray sources.

INTRODUCTION

Whenever a light charged particle (such as an electron or positron) is moving relativistically along a curved trajectory, under the influence of a transverse magnetic field, it emits electromagnetic radiation. Under normal conditions the radiation is emitted <u>incoherently</u>, that is the electrons emit with random phases and so the total intensity is proportional to the number of electrons. Such radiation is generally known as *synchrotron radiation* (SR), after its first observation in 1947 in the General Electric 70 MeV synchrotron, Schenectady, New York (1). Following various investigations of the properties of SR in the 50's, the first experimental programs began in the 60's firstly at the NBS synchrotron (USA) and later at many other laboratories in Europe, Japan, the USA and USSR, mainly using SR "parasitically" from rings built for high energy physics experiments. A second generation of dedicated SR sources were then built in the 70's and 80's, followed by the present third generation of high performance storage rings. In all about 54 rings either presently support research using SR or are under construction.

A much greater intensity of radiation can be produced if the electrons can be made to emit <u>coherently</u> i.e. all with the same phase, in which case the intensity is proportional to the square of the number of electrons. This can occur in either of two ways. In the first case it occurs when the length of the electron "bunch" is small with respect to the radiation wavelength. The possibility of generating high power millimetre waves using a bunched electron beam from a linear accelerator was first raised by Motz in the 1950's (2), who later succeeded in demonstrating coherent emission of 8 mm waves (3). Similar experiments carried out recently with much more highly compressed bunches have succeeded in producing coherent infrared radiation at 47 μm . The main aim of these experiments is to produce very short pulse lengths (less than 100 fs) in the infrared and far infrared region (4). The possibility of generating intense millimetre and sub-mm waves in a special electron storage ring has also been discussed (5). Such schemes are however of little interest in the present context, because of the impossibility of creating sufficiently small bunch lengths of the order of X-ray wavelengths.

A second possibility for obtaining coherent emission is when the electron bunch is modulated in intensity along its length. In this case the electrons emit in phase at a wavelength corresponding to the periodicity of the spatial modulation (as well as its harmonics). A device in which an initially un-bunched relativistic electron beam becomes bunched is called a *free-electron laser* (FEL). The idea of amplifying a radiation source by means of an interaction with an electron beam traversing an undulator was suggested as long ago as 1958 by Motz (6), but the key concept of feeding-back the emitted radiation by means of an optical cavity (as in a conventional laser) was suggested in 1971, by Madey (7). The first operation of a FEL oscillator working in the infrared region of the spectrum was later demonstrated in 1977 (8). Since then many more FELs have been operated, several of which have become user facilities, mainly in the infrared spectral region.

In the future, more facilities will be needed to meet the increasing demand for SR and FEL sources. Most new storage ring proposals (e.g. Diamond (UK), SOLEIL (France), SLS (Switzerland)) include some improvements compared to existing rings such as lower emittance and the possibility to include longer insertion devices, and also FELs. FELs have also been proposed at existing synchrotron radiation facilities (e.g. ELETTRA (Italy), MAX-lab (Sweden)) to offer the possibility of novel "pump-probe" experiments using a combination of infrared FEL radiation and high brightness synchrotron radiation.

Beyond these medium-term developments there is much activity directed towards future "fourth generation light sources" (9,10). Both soft X-ray and X-ray users at a recent workshop concluded that from the scientific point of view achieving higher brightness and spatial coherence, and shorter pulses, compared to existing sources was of great importance (11). A range of different approaches are involved, such as very low emittance storage rings for diffraction limited performance in the X-ray domain, short wavelength FELs based on linacs and storage rings, as well as other schemes for generating radiation with extremely short pulse length.

A progressive development of radiation sources can therefore be foreseen. The aim of the present article is to provide a brief introduction to this activity, including the basic principles involved, the present state-of-the-art and future possibilities.

SYNCHROTRON RADIATION

Introduction

The properties of synchrotron radiation that make it such a valuable and unique research tool may be summarised as follows; further information can be found in various books and review articles (12) :
- very wide spectral range, from γ-rays through to infrared, therefore covering many spectral regions where no other sources exist
- high intensity
- small angular divergence in the vertical plane
- small emitting source size, namely the electron beam
- polarization : SR is linearly polarized in the horizontal plane, and is elliptically polarized above or below the horizontal plane

$$\lambda_i = \frac{1}{i}\frac{\lambda_o}{2\gamma^2}(1+\frac{K^2}{2}+\gamma^2\theta^2) \tag{1}$$

where i is the harmonic number, and γ the usual relativistic factor. The intrinsic width of the lines is determined by the number of undulator periods, $\Delta\lambda/\lambda = 1/iN$, however the actual linewidth is broadened by the energy spread and divergence of the electron beam, as well as by the range of collection angles. Tuning of the output wavelength is achieved by varying the undulator field strength and hence K.

The total flux available at the on-axis wavelength (i.e. with $\theta = 0$ in Eq. 1) in the odd harmonics is given by the following, in standard units of photons/s/0.1% bandwidth :

$$\text{Flux} = 0.72 \; 10^{14} \; \frac{F_i(K)(1+K^2/2)}{i} \; \frac{L}{\lambda_o} I_b$$

where $F_i(K)$ is an expression involving Bessel functions and I_b is the beam current in Amps. This is a factor of two smaller than the usually quoted formula, since we are considering the situation where the intensity is peaked on-axis. For a given undulator length and wavelength the expression above maximizes with the shortest possible period length and the largest possible K value. A technological limit however is imposed by the fact that shorter periods result in a reduction in field strength and hence K value. In fact, for the permanent magnet systems that are usually employed in insertion device construction the field amplitude varies as $\exp(-\pi g/\lambda_o)$ where g is the gap between the magnet poles. The main limiting factor is therefore the minimum gap that allows sufficient aperture to contain the vacuum chamber and electron beam.

Figure 2 shows the maximum total flux achievable at a photon energy of 10 keV as a function of ring energy, for a fixed undulator length of 5 m. The importance of the gap can be clearly seen. For example, roughly the same flux at 10 keV obtainable in a 6 GeV ring with an undulator gap of 20 mm could be obtained at 4 GeV with a 10 mm gap, or at 3 GeV with a 5 mm gap. In practice however another factor has to be taken into account, namely the degree of tunability, since undulators are usually designed to operate over a range of wavelengths, not just at a fixed value. Taking this into account favours a somewhat higher energy than indicated in the figure. At lower energies it becomes beneficial to use higher harmonics, which will be discussed further below. The same graph can be applied to any photon energy by appropriate scaling of the horizontal axis e.g. for 1 keV, the horizontal scale should be divided by a factor $\sqrt{10}$. It is also worth noting that the flux is significantly higher than that of a bending magnet source, which has a peak value of only $4 \; 10^{12}$ photons/s/0.1%bandwidth per mrad of horizontal angle, per GeV, for the same beam current.

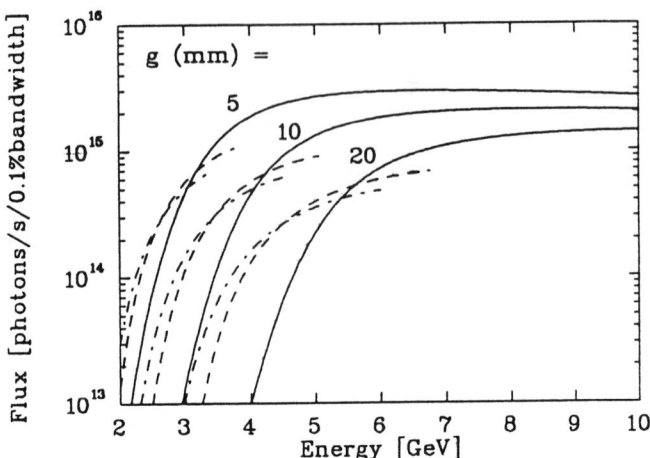

FIGURE 2. Maximum flux at a photon energy of 10 keV as a function of storage ring energy and undulator gap (g) in the first (solid lines), third (dashed lines) and fifth harmonics (dash-dotted lines). Undulator length = 5 m, beam current = 0.2 A. For clarity the curves for the third and fifth harmonics have been removed where the flux is less than that achievable on the first harmonic.

Brightness and Coherence

Source Size, Divergence and Brightness

We are, however, more interested in the brightness of the radiation than simply the total flux, and by this we mean the flux per unit solid angle, per unit source area. It will be sufficient here to use an approximate definition of the brightness as follows :

$$B = \frac{Flux}{4\pi^2 \Sigma_x \Sigma'_x \Sigma_y \Sigma'_y}$$

where the effective source sizes and divergences are given by the following :

$$\Sigma_x = (\sigma_x^2 + \sigma_R^2)^{1/2}, \; \Sigma_y = (\sigma_y^2 + \sigma_R^2)^{1/2}, \; \Sigma'_x = (\sigma'^2_x + \sigma'^2_R)^{1/2}, \; \Sigma'_y = (\sigma'^2_y + \sigma'^2_R)^{1/2}$$

i.e. a convolution between the electron beam sizes (σ_x, σ_y) and divergences, (σ'_x, σ'_y) in the two planes x, y and the radiation size (σ_R) and divergence (σ'_R). The conventional unit for brightness is photons/s/mrad^2/mm^2/0.1%bandwidth.

In the usual case the electron beam has a Gaussian distribution in both position and angular variables. At the location of the insertion device the rms widths are given by :

$$\sigma_x = \sqrt{\varepsilon_x \beta_x}, \quad \sigma'_x = \sqrt{\varepsilon_x/\beta_x}$$

where ε_x is the beam emittance and β_x is the value of "beta function" at the insertion device location, and similarly for the vertical plane (y). The product of size and divergence in each plane is thus equal to the electron beam emittance.

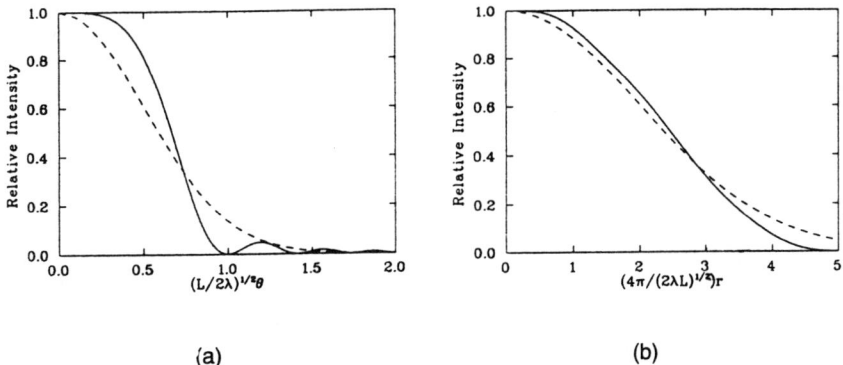

FIGURE 3. Angular (a) and spatial (b) distributions of undulator radiation, dashed lines - Gaussian approximations.

We now consider the radiation divergence and size. The link in Equation (1) between wavelength and angle implies that just as the interference results (at a given angle) in a squeezing of the spectrum into a series of lines, at a given wavelength the angular distribution is compressed into a series of narrow angular rings. Close to the axis the situation is complicated by the fact that the angular distribution varies significantly depending on the actual wavelength being considered. Since we are interested in examining the situation with very small electron beam emittance, it is appropriate to consider the wavelength which results in an angular distribution that is peaked on-axis (i.e. Eq. 1 with $\theta = 0$). As shown in Fig. 3(a), a Gaussian that gives the best fit to this distribution has a divergence $\sigma'_R = \sqrt{\lambda/2L}$.

The source size results from the finite length of the device combined with the above angular spread. More precisely, the spatial distribution at the centre of the undulator can be calculated by means of a diffraction integral, summing the amplitude at each angle, with an appropriate phase term :

$$I(r) = \left| \frac{1}{\lambda} \int_0^\infty \int_0^{2\pi} A(\theta,\phi) e^{-ik(\vec{n}.\vec{r})} \theta \, d\theta \, d\phi \right|^2$$

Carrying out this calculation one finds that the resulting spatial intensity distribution, shown in Fig. 3(b), is well approximated by a Gaussian with width $\sigma_R = \sqrt{2\lambda L}/2\pi$. The product of size and divergence in this case is therefore $\sigma_R \sigma'_R = \lambda/2\pi$ i.e. twice the minimum possible value that is obtained with the fundamental Gaussian mode of a laser resonator, often referred to as the "diffraction limit".

Temporal and Spatial Coherence

The degree of coherence of the radiation beam is an increasingly important parameter for various synchrotron radiation experiments. Basically, there are two measures of coherence - temporal (or longitudinal) coherence and spatial (or transverse) coherence. Temporal coherence determines the extent to which two parts of a beam, one delayed with respect to the other, can mutually interfere. Spatial coherence on the other hand determines the extent to which two spatially separated (but not delayed) parts can interfere. Temporal coherence is determined by the degree of monochromaticity of the radiation, and is described by the longitudinal coherence length over which waves of different wavelength remain in phase :

$$l_s = \frac{\lambda^2}{2\,\Delta\lambda}$$

In most cases this is determined by the monochromator bandwidth, which is usually considerably smaller than the linewidth of the undulator radiation.

Spatial coherence is determined by the phase-space area of the source i.e. the product of the size and divergence, and can be described by a transverse coherence length at a distance R from the source :

$$l_{x,y} = \frac{R}{d_{x,y}}\frac{\lambda}{4\pi}$$

where $d_{x,y}$ is the half-size of the source. In other words, only photons from within a phase-space area defined by $d_{x,y}(l_{x,y}/R) = \lambda/4\pi$ are spatially coherent. The fundamental mode of a laser resonator satisfies this condition and therefore has full spatial coherence. According to this simple definition even with zero emittance an undulator radiation beam would not be fully spatially coherent, which however is not correct. In fact, the single electron radiation pattern is always fully spatially coherent; partial coherence results only when a summation is taken over many electrons in an electron beam. To avoid this difficulty we will in the following consider a zero emittance undulator radiation beam to be diffraction limited and fully spatially coherent, i.e. neglecting the difference with respect to a Gaussian laser mode. The ratio of brightness to the zero emittance case therefore represents the fraction of spatially coherent flux.

Examples

Figure 4 shows brightness curves for several machines, in standard units. The solid lines indicate the peak brightness achievable at each photon energy with optimized undulator parameters (first harmonic only). The dramatic improvement between second generation (SRS) and third generation (ELETTRA, ESRF) machines can be clearly seen. The dashed lines indicate the performance that would be achieved for a zero emittance electron beam. At low photon energies up to the soft X-ray range (few 100 eV) machines like ELETTRA are close to the diffraction limit, and hence have a high degree of spatial coherence. In the X-ray region a machine of similar emittance, such as the ESRF, is much further from the corresponding limit. The brightness of a bending magnet (BM) and multipole wiggler (W) source in ELETTRA are also shown, confirming the much reduced brightness of these sources compared to undulators.

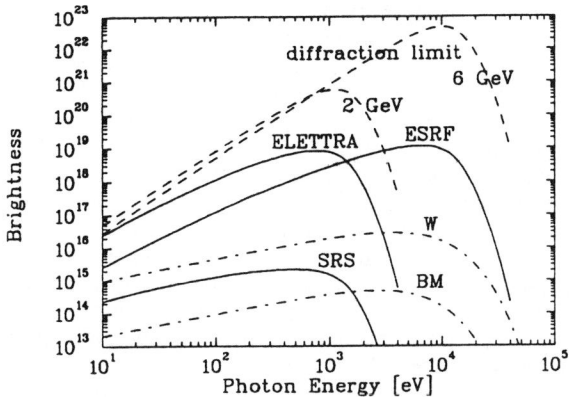

FIGURE 4. SR brightness (photons/s/mm^2/mrad^2/0.1%bandwidth) for various rings, Beam current = 0.2 A, undulator length = 1 m (SRS), 4.5 m (ELETTRA), 1.6 m (ESRF), 5 m (diffraction limit); undulator gap = 20 mm; see text for further details.

Extending the Limits of Insertion Device Performance

In general one wants to achieve the highest possible photon energies from an undulator in order to extend operation into the range that would otherwise only be accessible using lower brightness multipole wiggler sources. Higher photon energies also mean that the required energy (and hence cost) for a new storage ring facility can be minimized. According to Eq. 1 this can be achieved either by reducing the undulator period, or alternatively, by making use of the higher harmonics in the radiation spectrum. There are difficulties however in both approaches. As mentioned above, in order to maintain a useful field strength while reducing the period length requires a similar reduction in the gap between the magnet poles, and hence in the vacuum chamber aperture, which has an adverse effect on the beam lifetime and may also introduce instabilities. The limitation to the use of higher harmonics is the magnetic field quality, since errors cause a progressive deterioration of the strength of the higher harmonics. In recent years significant progress has been made in both of these areas.

When the first Insertion Device vacuum chambers are installed in a new ring, a conservative approach is usually taken when specifying the aperture. Later, with the benefit of a fuller understanding of the machine characteristics, there is a tendency to allow smaller and smaller apertures, thereby permitting reduced undulator gaps to be used. For example at the ESRF initial chambers had internal/external dimensions of 15/19 mm; subsequently chambers have been installed with 11/15 mm and also 9/13 mm, while chambers of 8/10 mm are under consideration (13). A chamber with 5 mm internal aperture and 8 mm external dimension will also be tested soon in the APS (14). In order to reach even smaller gaps variable gap vacuum vessels have been developed at various laboratories, for example at DESY and SSRL and more recently on small undulator period devices at ESRF, NSLS and MAX-lab. Experience with these devices shows that in general they are complicated and expensive, and it is difficult to obtain a very small difference between the minimum magnet gap and the vacuum chamber inner aperture. As a result there has recently been an increased interest in the alternative approach of placing the undulator in-vacuum, in which case the magnet gap is essentially the same as the vacuum gap. Early devices of this type have been operating successfully for some time at BESSY and in the TRISTAN accumulator ring. Of the initial eleven IDs being constructed for Spring-8, eight will be in-vacuum with an 8 mm minimum gap (15). A 24 mm period in-vacuum device, constructed by Spring-8, has recently been tested at the ESRF, while an 11 mm period device is due to be tested later this year at NSLS.

The performance specification for the insertion devices of the present 3rd generation light sources was generally based on the use of the 1st, 3rd and 5th harmonics only, limited by the field errors that were thought achievable at that time. Since then however dramatic progress has been made in reducing errors essentially to zero. The key to this development was the optimization of the phase of the radiation emission at each magnet pole, which has a direct relationship with the intensity of the radiation harmonics, rather than the underlying field amplitude errors, which are not well correlated with the radiation quality. The two main techniques used in the optimization are sorting of the permanent magnet blocks and magnetic "shimming" - placing thin ferromagnetic sheets on the magnet surface (16,17,18).

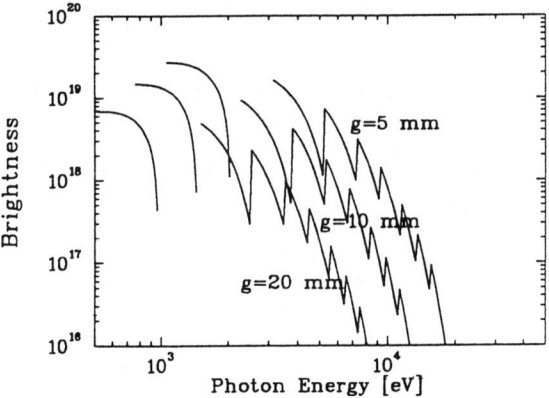

FIGURE 5. Calculated performance of various mini-gap devices in ELETTRA. Undulator length = 3 m, beam current = 0.4 A.

As a result of these two developments, undulator performance can now be pushed significantly beyond previous expectations. For example, Figure 5 shows the calculated performance of possible mini-gap devices in ELETTRA. At the ESRF, use of up the 13th harmonic of the installed mini-period undulator at 7 mm gap allows photon energies of up to 70 keV to be reached, compared to 40 keV on a conventional 20 mm gap device (17).

Insertion Devices for Circularly Polarized Radiation

Standard insertion devices generate only linearly polarized radiation because of the destructive interference of the circularly polarized component emitted by the positive and negative poles. Various special devices have therefore been devised and constructed in recent years to generate circularly polarized radiation with higher flux and brightness compared to alternative bending magnet sources (19).

In the first class of device combined horizontal and vertical magnetic field components force the electrons to follow an elliptical trajectory, rather than a sinusoid, thus producing elliptically polarized radiation on-axis. Helical current windings can be used to generate the necessary fields, however most devices use an arrangement of permanent magnets. One approach is to combine a horizontal field device with a conventional vertical field device, offset in position by one-quarter period length. Depending on the field strength such devices may function either as undulators or wigglers, and both kinds are in operation in Japan, in the TERAS, Photon Factory and TRISTAN AR rings.

A practical disadvantage of these initial designs is the need for magnets on all four sides of the electron beam tube and so various kinds of planar permanent magnet arrangements have now been devised that overcome this restriction. Planar devices of this kind are in operation at ESRF and SSRL, and are under construction for ALS, BESSY II and TLS. A second type of device is the asymmetric wiggler which has non-equal positive and negative field strengths, so avoiding the cancellation of the circularly polarized component. It thus provides circularly polarized radiation vertically off-axis, as in a bending magnet. Asymmetric wigglers are operational in the ESRF, DORIS and SUPERACO storage rings. A third type is the crossed-undulator scheme. In this case the radiation is generated by the interference of horizontal and vertically polarized radiation from two successive undulators. One device of this kind is in operation in BESSY I.

Various experiments using circularly polarized radiation involve the measurement of small differences in absorption or scattering between right- and left-handed polarizations. A convenient, and fast, way of switching between polarization states is therefore an advantage in order to increase the sensitivity of the measurement and several different methods have been devised :

- In the case of various permanent magnet geometries a mechanical translation can be used to alter the relative phase of the two field components and so change the polarization, however such a scheme is necessarily quite slow, and is limited to about 5 Hz.

- Electromagnets can be used for a faster switching, for example by combining a strong vertical magnetic field produced by permanent magnets with a weaker horizontal field (shifted by a quarter period) produced by an electromagnet. The first device of this kind was built by an APS/NPI/NSLS collaboration and has recently been tested in the NSLS X-ray ring at up to 100 Hz switching frequency (20). A second similar device is nearing completion at APS (14). More recently, alternative double-electromagnet schemes with open-sided structures have been developed at APS (21) and ELETTRA. Figure 6 shows a part of the structure that is being developed at ELETTRA. Electromagnets could also be employed in the crossed-undulator or asymmetric wiggler schemes, but have not been tried so far.

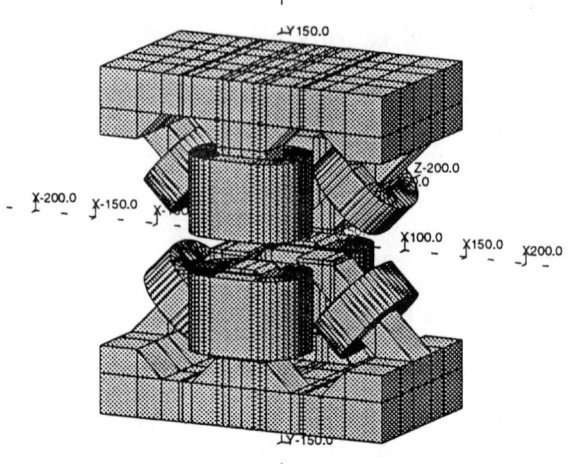

FIGURE 6. Section of double-electromagnetic elliptical wiggler.

- An alternative scheme is the one proposed for Spring-8 involving 2 helical devices of opposite handedness and a series of 5 kicker magnets which are used to deflect the electron beam in such a way as to switch between right and left-handed radiation on the same axis. A switching frequency of 10 Hz is envisaged (15).

- The fastest switching speeds can be obtained by generating two beams of radiation with opposite handedness at slightly different angles which are then fed into the same monochromator. A fast chopper can then be used to switch between the two beams. Such a method is employed at the ESRF and will be implemented also for the ALS and BESSY II.

The Next Generation of Synchrotron Radiation Source ?

Encouraged by the success in commissioning several "third-generation" storage ring light sources in recent years (see Table 1) the problems facing designers and builders of a possible "fourth-generation" are now actively being discussed (10). As shown in Fig. 2, the potential goal is a further 10^4 increase in brightness at 10 keV compared to existing machines such as the ESRF, i.e. an equivalent increase in brightness to that achieved in going from a second-generation to a third-generation machine. It remains unclear at this early stage however how close in reality one can get to the ultimate diffraction limit. The main limitation is the process of scattering of electrons off other electrons in each highly compressed bunch (intra beam scattering) which causes the emittance to increase above its natural value, as well as a reduction in beam lifetime.

Figure 7 shows the performance and Table 2 presents the main parameters for some existing and possible future light sources discussed at the recent Fourth Generation Light Sources Workshop (22, 23, 24). In the table the brightness values B_1 and B_{10} are calculated for optimized undulator designs at 1 keV and 10 keV respectively, using the formulae above. For simplicity, the beam current has been fixed at 0.2 A in all cases. The quantity R gives the improvement

TABLE 2. Parameters of Various High Brightness Synchrotron Radiation Sources. Units : Energy, E (GeV), Circumference, C (km), horizontal emittance, ε_x (nm rad), coupling factor, κ (%), beta functions, β_x, β_y (m); undulator length, L (m), gap, g (mm), brightness, B (photons/s/mm^2/mrad^2/0.1%bandwidth).

Ring	E	C	ε_x	κ	β_x, β_y	L	g	R_1	B_1	R_{10}	B_{10}
ESRF	6	0.85	4	1	27, 13	1.6	20	100	3 10^{18}	1700	9 10^{18}
ESRF-mod	6	0.85	3	0.2	27, 2.5	4.8	10	25	4 10^{19}	250	5 10^{20}
DE	3	0.40	0.56	1	5, 3	5.0	10	4.5	3 10^{20}	40	3 10^{20}
YC	3	1.1	0.31	1	5, 5	5.0	10	3.0	6 10^{20}	22	5 10^{20}
AR1	3	0.85	1	1	27, 2.5	5.0	10	11	1 10^{20}	90	1 10^{20}
AR2	6	1.7	0.5	0.2	27, 2.5	5.0	10	7.1	1 10^{20}	35	3 10^{21}

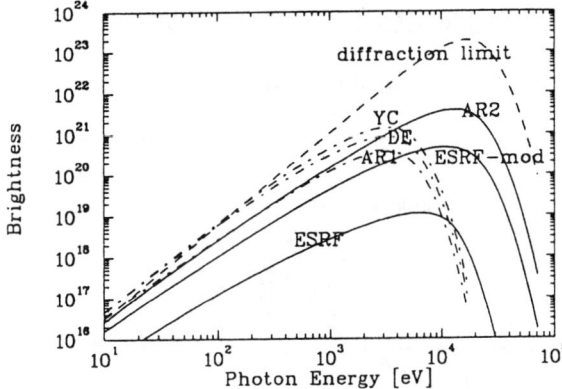

FIGURE 7. Undulator performance (first harmonic only) for various low emittance storage rings with the parameters given in Table 2. Beam current = 0.2 A.

in brightness that would be obtained in the zero-emittance case i.e. :

$$R = \frac{\Sigma_x \Sigma_x' \Sigma_y \Sigma_y'}{(\lambda/2\pi)^2}$$

which is therefore a function only of the electron beam parameters, the radiation wavelength and undulator length, independent of the flux, or gap. The first row presents data relevant to the present ESRF. The next row is the ESRF medium-term objective (25), with reduced emittance, lower coupling, increased undulator length and reduced vertical beta function. After this upgrade the brightness will be a factor of 25(250) below the diffraction-limit for a machine of the same energy and current at 1(10) keV respectively. Several new proposals (DE, YC, AR1) are based on a 3 GeV ring. It can be seen that such rings could provide performance that is within a factor 3-10 of the diffraction limit at 1 keV, and 20-90 at 10 keV. Despite the fact that the ring performance is closer to the diffraction limit, the actual brightness values do not exceed that of the ESRF upgrade, since the ring energy results in significantly lower flux at 10 keV. Figure 4 shows the brightness in the first harmonic, which is about 2 10^{19} at 10 keV.

Using the 7th harmonic gives the values in the table i.e. an order of magnitude improvement. A further improvement would be obtained with a smaller gap : for example, a 5 mm gap would allow an increase in brightness of a further factor of 3.2. The final row (AR2) is a 6 GeV machine with a circumference twice that of the ESRF that could arrive within a factor of 35 of the diffraction-limit at 10 keV.

In conclusion, extrapolation from operating machines and current lattice designs indicate that it should be possible to approach the diffraction limit to within a factor of about 20-50 in the X-ray region. Much work remains to be done however to optimize the designs, and to consider other aspects such as beam stability, diagnostics etc., as well as costs. A smaller, lower energy machine could be competitive, provided small undulator gaps can be used.

FREE ELECTRON LASERS

Introduction

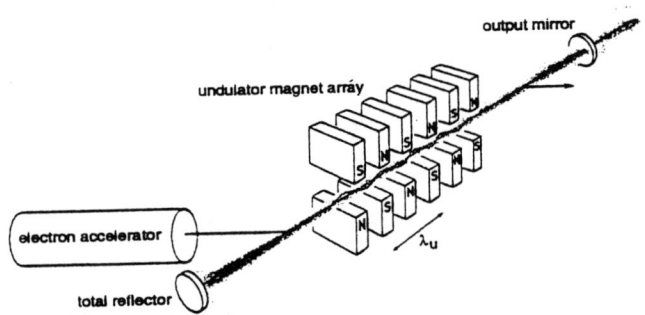

FIGURE 8. Schematic of a free electron laser oscillator.

In a FEL an electron beam passing through an undulator magnet interacts with a co-propagating radiation beam resulting in the stimulated emission of coherent radiation (see Fig. 8). Classically, it may be understood as an interaction between the radiation and electron beams in the undulator which causes the electrons to become bunched on the scale of the optical wavelength, and hence emit coherently. For the interaction to take place the electrons and photons must remain in step along the undulator length. Since the electrons travel slower than the velocity of light, such a resonance condition is obtained when the photons gain on the electrons by one radiation wavelength every undulator magnet period, which is therefore identical to the equation for the spontaneous radiation, Eq. 1. Figure 9 illustrates this resonance condition. It can be seen that for electrons at positions A the transverse electric field of the radiation acts always in the same direction as the electrons' transverse motion whereas for electrons at positions B the transverse field and velocity are in opposite directions. The A electrons are therefore accelerated at each undulator half-period, whereas the B electrons are decelerated, causing them to become bunched at the stationary points C, separated by one radiation wavelength. (Actually, although bunching would eventually occur the mechanism described above is not very efficient, since to first order as many electrons are decelerated as accelerated and hence there is no net energy transfer. A more detailed analysis shows that maximum electron energy loss, and hence amplification of the radiation field, occurs for a slight detuning away from the resonance condition.}

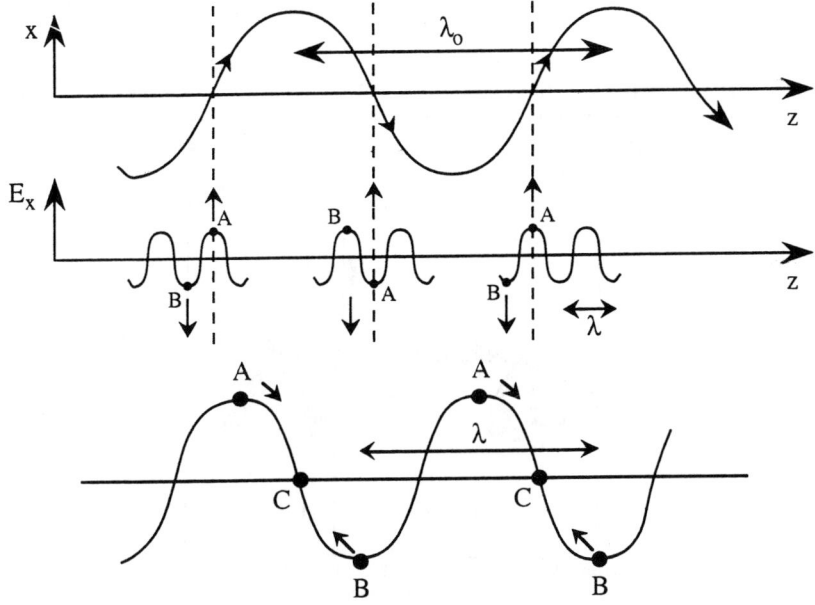

FIGURE 9. Illustration of the FEL resonance condition.

Since the radiation wavelength is not linked to particular atomic or molecular excitation levels the FEL has the advantage compared to conventional lasers that it can be tuned easily according to Eq. 1 using either magnetic field strength (i.e. K-value) or electron energy. Continuous and rapid tunability over a broad range is therefore possible. A second great advantage is that in a FEL there is no material lasing medium and hence no power limitation due to material breakdown. Very high peak and average power radiation can therefore be produced. Further important characteristics are as follows :
- short pulse length : picosecond and sub-picosecond pulses can be produced, much shorter than achievable with SR
- flexibility : variable pulse length, linewidth, and pulse format can be obtained
- regular time structure : since the FEL is driven by the radiofrequency system of the linear accelerator or storage ring, it can be synchronized to conventional lasers and other synchrotron radiation sources
- full spatial coherence i.e. diffraction limited
- variable polarization : all presently operating FELs generate linearly polarized radiation, however elliptical or circular polarization could be created using any of the undulator configurations developed as sources of SR.

The most common FEL scheme, adopted by all presently operating FELs, is the oscillator configuration (Fig. 8) where an optical cavity is placed around the undulator so that the spontaneously emitted radiation can be made to interact with subsequent electron bunches. Given a sufficient amplification per pass which exceeds the mirror and other losses, together with a sufficiently long train of electron bunches, the radiation intensity can build up to a saturation level. The radiation wavelength is then tunable according to Eq. 1, within the limits imposed by the mirror reflectivity etc. The choice of electron beam source is determined mainly by the radiation wavelength required. At the longest wavelengths, from the far infrared to the mm range, electrostatic accelerators or microtrons can be used. In the infrared the usual choice is a conventional linear accelerator (linac). For shorter wavelengths either linacs or storage rings can be used. Table 3 presents a list of currently operating FELs, as well as those under construction or commissioning.

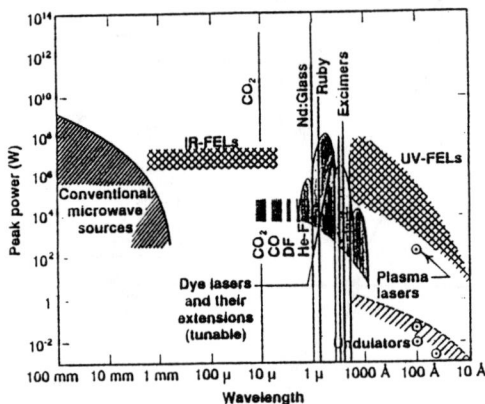

FIGURE 10. Peak power of present (IR) and future (UV) FELs and other competing sources (from K.J. Kim and A. Sessler (26))

The main role for the FEL is generally considered to lie in the IR/FIR region, from a few μm to a few mm wavelength, and in the VUV/soft X-ray below 100 nm (see Fig. 10). So far FELs have operated over the range from a few mm to a lower limit of 240 nm, which was achieved in 1988 using the VEPP3 storage ring. As shown in Table 3, there are presently about 19 operational FELs, about 16 under construction or commissioning, and more are planned. A number of user facilities are in operation. Further information on FELs can be obtained from several books and articles (26) as well as the proceedings of the annual International Free Electron Laser Conferences (27).

Short Wavelength FELs

There are difficulties in extending the oscillator scheme to shorter wavelengths due to the combined effect of reduced gain and the lower reflectivity (or complete lack) of mirror materials. The limit is being explored with a number of storage ring FEL projects. The UVSOR FEL has reached 260 nm and is hoping to lase below 200 nm using a new helical undulator (optical klystron) to reduce the mirror damage. Two new storage ring based FEL projects are also underway

TABLE 3. Free Electron Lasers in Operation and Under Construction.
Accelerator : E(electrostatic), L(linac), M(microtron), S(storage ring)
Status : C (construction), O(operational)

Country	FEL/Lab.	Wavelength	Accel., Status	Comments
China	IHEP (Beijing)	9 - 11 µm	L, O	
	CIAE	100 - 240 µm	L, C	
France	CLIO	3 - 50 µm	L, O	User facility
	SuperACO	350 nm	S, O	User facility
	ELSA	10 -20 µm	L, O	
Germany	Darmstadt	2.6 - 7.5 µm	L, C	
	DELTA (FELICITA I)	200 - 470 nm	S, C	Short wavelength
Holland	FELIX	5 - 110 µm	L, O	User facility
	TEUFEL	190 - 650 µm	M, O	
Japan	FELI	1 - 20 µm	L, O	User facility
	ILE/ILT (Osaka)	≥10 µm	L, O	
	ISIR (Osaka)	32 - 40 µm	L, O	
	JAERI	20 - 80 µm	L, C	High power
	Nihon Univ.	300 nm	L, C	
	NIJI-IV	350 - 600 nm	S, O	
	Sumitomo (Harima)	14 µm	L, O	
	UT-FEL (Tokyo)	43 µm	L, O	
	UVSOR	260 - 480 nm	S, O	
Russia	BINP (Novosibirk)	4 - 13 µm	M, C	Industrial use
USA	BNL-ATF (HGFEL)	3.5 µm	L, C	HGHG test
	BNL-ATF/MIT	0.53 µm	L, C	
	BNL-SDL	0.9 - 1 µm (1st phase)	L, C	SASE test
	Boeing (Seattle)	0.85 µm	L, C	High power test
	CIRFEL (Princeton)	12 - 20 µm	L, O	poss. user facility
	CREOL	225 - 800 µm	E, C	
	Duke Univ. (MK. III)	1.7 - 9.1 µm	L, O	User facility
	Duke Univ.	50 - 400 nm	S, C	Short wavelength
	LANL-AFEL	4 - 6 µm	L, O	
	Rutgers Univ.	160 - 280 µm	M, C	
	Stanford	3 - 65 µm	L, O	User facility
	TJNAF	(IR) 2-28 µm	L, C	Industrial use
		(UV) 150 nm -1 µm	L, C	Industrial use
	UCLA	10.6 µm	L, C	SASE test
	UCSB	30 µm - 2.5 mm	E, O	User facility
	Vanderbilt Univ.	2 - 9.1 µm	L, O	User facility

which should eventually go below this limit. The Duke University (USA) storage ring FEL is now in an advanced stage of commissioning, and should soon lase below 200 nm. Conventional mirrors will be used to extend to 120-140 nm. Thereafter a special ring optical cavity with in-vacuo coated Al mirrors may allow operation down to 50 nm. Ultimately, higher harmonics may allow operation as low as 4 nm in the more distant future. Shortly behind Duke is a similar storage ring project at Dortmund University (Germany), called DELTA. The first FEL project (FELICITA I) should allow the range between the visible and 200 nm to be reached. This may then be followed by a second experiment using a long undulator to reach the 25-100 nm range.

To go to very short wavelengths many believe that a single-pass approach must be used i.e. without an optical cavity. Such schemes require a very high gain, and so demand very intense, bright, electron beams, which has only become possible through the development of new electron beam sources, using laser driven r.f. cavity guns. Two main configurations have been proposed. The first involves a two-step approach : producing a density modulation in a first undulator followed by amplification at a higher harmonic frequency in the second. The first device may be either an oscillator, or a high-gain amplifier, using an input "seed" from a conventional laser. The latter, called the High Gain Harmonic Generation (HGHG) scheme, forms the basis of the proposed DUV (Deep UV) facility at Brookhaven for the 75-300 nm range. The most promising scheme for reaching even shorter wavelengths is that of self-amplified spontaneous emission (SASE). Here there are neither mirrors or an input seed and so to build up to high power requires a very long undulator (30-70 m) in addition to a very high brightness electron beam source. This is the basis of the proposed Linac Coherent Light Source at Stanford, as well as the TESLA Test Facility FEL at DESY.

Following the 1st Workshop on Fourth Generation Light Sources (9), a group of U.S. laboratories began in 1992 to consider the possibility of using the SLAC linac to drive a short wavelength FEL, called the Linac Coherent Light Source (LCLS), based on the SASE principle. Based on the current state of technology of r.f. photocathode injectors and undulators, and also on the understanding of linac beam dynamics and pulse compression it appears possible to use existing technology to construct an FEL that would operate at wavelengths down to about 30 Å. With improvements in components the operating wavelength could be reduced towards 1.5 Å. Recently, a more detailed study has been initiated (28) with the objective of producing a design report by August 1st 1997. Present timescales envisage implementation starting in the Autumn of 1998, operation at 3-5 nm in 1999-2000 with a 3 GeV beam and 1.5 Å in 2000-2001 with a 15 GeV beam.

Stimulated by this work, the DESY laboratory developed a proposal to use the superconducting linac being built as part of the TESLA Test Facility (TTF) to drive a FEL in the VUV region (29). A first experiment is funded which will use a beam energy of up to 380 MeV and a 15 m undulator to generate photons in the range 50-100 nm. No user operation is foreseen at this stage. Later, with additional funding to extend the linac energy to 1 GeV, and the undulator length to 30 m, a user facility is planned to make use of radiation in the 6-60 nm range. If funds become available in 1997, commissioning could start in the year 2000. The ultimate goal however is an Xray FEL integrated into the design of a next generation linear collider (30). Initial discussions on the scientific case for this ambitious project took place at DESY in February 1996. This will be followed, directly after the present Conference, by a further workshop that should formulate the scientific case for the XFEL.

Table 4 compares the main parameters of the TTF FEL, the two X-ray FELs under consideration at DESY based on linear colliders with either S-band (SBLC) or superconducting (TESLA) technology, and the Stanford LCLS. It can be seen that the peak brightness is very high. By comparison the peak brightness obtainable in a storage ring is approximately $2 \cdot 10^{22}$ at present (ESRF with 15 mA

TABLE 4. Main Parameters of Proposed Linac Based X-ray FELs.

	TTF	SBLC	TESLA	LCLS
Electron beam energy, GeV	1	20	20	15
Photon energy, keV	0.02-0.2	2-20	2-20	0.3-8.2
Pulse length rms, fs	170	80	80	100
Max. av. brightness†	$6\ 10^{21}$	$6 10^{25}$	10^{26}	10^{23}
Max. pk. brightness†	$5\ 10^{29}$	10^{34}	10^{34}	$4\ 10^{33}$
Bunch spacing, ns	111	16	93	12
Macropulse length, μs	800	2	1050	120
Repetition rate, Hz	10	50	5	120

† units photons/s/mm^2/mrad^2/0.1%bandwidth

single bunch of 45 ps length), 10^{24} in the future, and ultimately of the order 10^{25} i.e. many orders of magnitude less than an X-ray FEL. The average brightness can also be high, particularly in the case of the SBLC and TESLA FELs. This could allow the beam to be switched between a number of different undulators for multiple user operation, each with a time-averaged brightness exceeding that of a storage ring source.

There are however many open questions to be answered and therefore much R&D to be carried out before X-ray FELs become a reality. For example, high gain amplification has only been fully tested as low as 10 μm at the Lawrence Livermore National Laboratory in the USA. In particular the SASE scheme has only been tested at mm and sub-mm wavelengths (LLNL and MIT), although recently it has been observed at 47 μm (at the SUNSHINE experiment, Stanford) and at 5 μm (at the CLIO FEL facility). Tests of SASE are planned at UCLA (10-15 μm) and LANL (16 μm) in 1996-97. This should be followed in 1997-8 by tests at around 1 μm, at the Brookhaven Source Development Laboratory. The TTF-FEL first phase should then reduce the wavelength to 50-100 nm in 1998. Spring-8 are also proposing a 20 nm test, which should be ready in 1999. After this should come the TTF-FEL at 6 nm in 2000 followed by the LCLS at 0.15 nm in 2000-2001. Other questions relate to the problems of compression and transport of ultra-short bunches because of strong wake-field effects; the construction of very long (30-100 m), precise undulators with superimposed strong focusing fields; the construction of very low emittance, high peak current r.f. guns as well as photon beam handling and diagnostics.

OTHER DEVELOPMENTS : FEMTOSECOND X-RAY PULSES

As mentioned earlier, higher time-averaged brightness and hence higher spatial coherence is only one aspect of the definition of a next-generation source. There is a strong interest also in achieving very short, sub-picosecond, X-ray pulses for performing time-resolved studies on a range of ultra-fast phenomena in condensed matter physics, chemistry and bio-chemistry. In storage rings the radiation pulse length is determined by the length of the electron bunch, which is usually in the range 50-100 ps. Experiments have been carried out at ALS, ESRF and SUPERACO to try to reduce the bunch length as much as possible, however at both ALS and ESRF only very small beam currents could be stored with bunch lengths of 8-10 ps. This experience, backed up by theoretical calculations, leads to the conclusion that sub-picosecond pulses with reasonable intensity will be impossible to achieve in storage rings (31). On the other hand, the X-ray FELs described above should produce 100-200 fs pulses. X-ray FELs should therefore give the ultimate performance demanded by users, however, in the meanwhile there are several other approaches being developed for producing short X-ray pulses that could be competitive for experiments that do not need very high peak intensity. One possibility is the scattering of an intense laser beam at 90° off a relativistic electron beam. In this configuration, the shortness of the X-ray pulse is limited not by the length of the electron pulse, but rather by the length of the laser pulse or the transit time of the laser pulse across the waist of the focused electron beam. An experiment is under-way at LBL's Center for Beam Physics to scatter an intense 200 fs 800 nm laser pulse off a 10-30 MeV electron beam (32). The scattered photons are up-shifted in frequency by a factor $2\gamma^2$ giving between 1 and 10 Å in this case. Between 10^5 and 10^6 photons should be able to be produced in 100-300 fs pulses.

Another possibility is the backscattering of photons off a short electron pulse. Since in this case the photons produced are increased in frequency by a factor $4\gamma^2$ the X-ray domain can be reached with infrared radiation with a low energy electron beam. A high laser intensity is needed, which therefore suggests the use of a FEL to produce the radiation. A successful experiment has already been carried out at the CLIO FEL to generate 7-14 keV photons by interaction of a 50 MeV electron beam with 3.5-7 µm FEL radiation within the optical cavity (33). With achievable electron and laser pulse lengths as small as 3-4 ps and 0.2-0.5 ps respectively, a 2-ps X-ray source is produced.

Finally, an ingenious scheme has been suggested at the ALS to produce 100-300 fs FWHM X-ray pulses (34). The scheme works by inducing a large energy modulation on a 100 fs slice of a much longer electron bunch by an FEL-type interaction between an 800 nm laser and the electron beam passing through an undulator magnet, resonant on the first harmonic. Converting the energy modulation into a spatial modulation by means of a dispersion section then provides an offset between the photons emitted in the narrow slice and the rest of the bunch.

REFERENCES

1. F.R. Elder et al., Phys. Rev. 71, 829 (1947)
2. H. Motz, J. Appl. Phys. 22, 527 (1951)
3. H. Motz and D. Walsh, J. App. Phys. 33, 978 (1962)
4. H. Wiedemann, P. Kung and H.C. Lihn, Nucl. Instr. Meth. A319, 1 (1992)
5. J.B. Murphy and S. Krinsky, Nucl. Instr. Meth. A346, 571 (1994)
6. H. Motz and M. Nakamura, Proc. Symposium Millimeter Waves, Brooklyn, 1959, p. 155.
7. J.M.J. Madey, J. Appl. Phys. 42, 1906 (1971)
8. D.A.G. Deacon et al., Phys. Rev. Lett. 38, 892 (1977)
9. Proc. Workshop on Fourth Generation Light Sources, Stanford, Feb. 1992, SSRL 92/02.
10. Proc. 10th ICFA Beam Dynamics Panel Workshop on 4th Generation Light Sources, Grenoble, Jan. 1996, available from the ESRF.

11. I. Lindau ref. 10 p. 23; J. Als-Nielsen ref. 10, p. 33.
12. H. Winick, Scientific American, Nov. 1987, p. 72; G. Margaritondo, "Introduction to Synchrotron Radiation" (Oxford, New York, NY, 1988); Handbook on Synchrotron Radiation, Vols. 1-4 (North Holland).
13. P. Elleaume, private communication.
14. E. Gluskin, private communication.
15. H. Kitamura, private communication.
16. B. Diviacco, Proc. 1993 Particle Accelerator Conference, p. 1590.
17. P. Elleaume, Proc. 4th European Particle Accelerator Conference, World Scientific, (1994) p. 654.
18. B. Diviacco and R.P. Walker, Nucl. Instr. Meth. Phys. Res. A368 (1996) 522.
19. R.P. Walker, Proc. 4th European Particle Accelerator Conference, World Scientific, (1994) p. 310, and reference therein.
20. E. Gluskin et al., Proc. 1995 Particle Accelerator Conference, p. 1426.
21. E. Gluskin, ref. 10 p. WG1-56.
22. D. Einfeld, ref. 10 p. WG3-53.
23. Y. Cho, ref. 10 p. WG3-42.
24. J.L. Laclare, A. Ropert and and U. Weinrich ref. 10 p. WG3-7.
25. J.L. Laclare, ref. 10 p. 5.
26. T.C. Marshall, "Free Electron Lasers", MacMillan (1985); H.P. Freund and R.K. Parker, Scientific American, Apr. 1989, p. 56; C. Brau, "Free Electron Lasers", Academic Press (1990); P. Luchini and H. Motz, "Undulators and Free Electron Lasers", Clarendon Press (1990); C. Brau, Science 239, 1115 (1988); K.J. Kim and A. Sessler, Science 250, 88 (1990); W.B. Colson, C. Pellegrini and A. Renieri, Laser Handbook Vol. 6, Elsevier (1990);
27. Proc. 17th International Free Electron Laser Conference, Nucl. Instr. Meth. A375 (1996); and previous Conferences published in Nucl. Instr. Meth. Vols. A358, A341, A331, A318, and A304.
28. M. Cornacchia, private communication.
29. "A VUV Free Electron Laser at the TESLA Test Facility at DESY - Conceptual Design Report", DESY 95-03, June 1995.
30. J. Rossbach, unpublished.
31. A. Hoffman, ref. 10 p. 49.
32. S. Chattopadhyay, ref. 10 p. WG2-17.
33. J.M. Ortega, Proc. 5th European Particle Accelerator Conference, June, 1996, to be published.
34. S. Chattopadhyay et al., ref. 10 p. WG2-29.

The Ultimate Storage Ring X-ray Source

Pascal Elleaume
European Synchrotron Radiation Facility, B.P. 220, F-38043 GRENOBLE France

INTRODUCTION

Electron/positron storage rings were first developed as electron-positron colliders for high energy physics. They were later very successfully used as sources of synchrotron radiation. Nearly all storage rings built in the last 20 years are dedicated sources of synchrotron radiation. This success originates from several reasons.

- The small emittance of the electron beam combined with the strong collimation of synchrotron emission has resulted in photon beams covering a wide range of energies from 10 eV to 100 keV with laser-like collimated beams of extremely high brilliance and/or power (up to 20 kW cw in a single beamline).

- The stability of the photon beams and reliability of the sources has proven to be remarkable. It originates from the highly stable and reliable dc power supplies driving the dipoles, quadrupoles, sextupoles and corrector magnets.

- A storage ring can be organised very naturally as a multi-user facility with a great number of beamlines regularly spaced along the circumference operating simultaneously and independently from each other.

Storage rings are based on the principle of the recirculation of an electron beam with compensation of the energy loss due to synchrotron radiation by means of accelerating RF cavities. More precisely, let us consider the ESRF which is operated with a 200 mA 6 GeV electron beam corresponding to a circulating power of 1200 MW ($\approx$ a nuclear power plant). Every turn only 0.1% is extracted and converted into synchrotron radiation inside all bending magnets and insertion devices (IDs). This energy loss is compensated by a small acceleration of the electrons in the RF cavities. Any insertion device only extracts a typical 10^{-6} to 10^{-5} fraction of the circulating power resulting in a 1.2 to 12 kW cw X-ray beam. Due to the recirculation, the 1200 MW beam can be maintained for months during which the overall consumption of electricity is less than 5 MW.

Competitors to the storage ring X-ray sources are the X-ray tube and more recently the free electron laser (FEL) [1]. In an X-ray tube one extracts a typical 10^{-3} fraction of the electron beam power. The electron beam power is not recirculated and is dumped into the tube's anode. The cooling of the anode and the above mentioned extraction efficiency limits the power of the X ray beam to a few watts with a rather large angular divergence resulting in poor brilliance. The main advantage of X-ray tubes are their low size and cost. The free electron laser is another source of radiation which receives considerable interest nowadays. By analogy with conventional lasers, it can be understood as the result of the stimulated emission of radiation in an undulator, the synchrotron radiation being the equivalent of the spontaneous emission of conventional lasers. By this process, one can extract a much higher fraction of the circulating power (compared to simple synchrotron radiation emission). The tuning of the device is nevertheless much more complex and will require many years of R & D before being applied on the same large scale as storage rings. One distinguishes the oscillator FEL which has been routinely operated with average brilliances in the 10^{28} ph/sec/.1%/mm2/mr2 range at energies below 6 eV from the superradiant FEL also called self amplified spontaneous emission (SASE) which can, in principle, be operated at shorter wavelength down to 10 keV[1]. The SASE FEL is expected to give similar average brilliance as a storage ring undulator but a much higher peak brilliance because of the much shorter electron beam. The SASE operation requires a state-of-the-art ultra-short pulse low emittance linear accelerator of a class currently under development. The SASE principle while well established theoretically has yet to be demonstrated experimentally at short wavelength.

In the course of this paper, I shall describe the performance of the newly commissioned third generation synchrotron sources. I shall distinguish:

- The hard X-ray sources with high electron energies above 6 GeV optimized for the production of photon beams in the 1 to 100 keV energy range. They include APS, ESRF, PETRA and SPRING8.

- The medium energy X-ray sources with electron energies between 1.5 and 2.5 GeV optimized for the production of soft X-rays in the 0.1 to 10 keV range. They include ALS, BESSY II, ELETTRA and SRRC.

- The low energy sources with energies below 1 GeV optimized for the VUV range of the spectrum.

Some low energy X-ray sources are called compact synchrotron sources. They are of interest for industrial applications such as lithography. They are much cheaper than the most recent third generation medium energy and hard X-ray sources but generate much less flux and brilliance. They are out of the scope of this paper.

PERFORMANCE OF THIRD GENERATION SOURCES

The ring lattice of third generation synchrotron sources is essentially of the Chasman-Green type. The triple bend achromat type is also used on small circumference rings. All facilities accommodate a number of insertion devices each 4 to 5 m long. The total number of IDs depends on the size of the ring and ranges from 10 to 40. The stored current in the multibunch modes reaches 400 mA at the ALS[2] (1.5 GeV) and 200 mA at the ESRF (6 GeV). This performance is either obtained in the so-called uniform filling mode with a large number of bunches nearly equally filled or in a 80% (ELETTRA[3]) or 30% (ESRF) filling mode where only a part of the circumference is filled. The electron beam lifetime in this mode of operation ranges from 12 hours (ALS @ 400 mA),18 hours (ELETTRA @ 250 mA) to 30 hours (ESRF @ 200 mA). Single bunch currents of 20 mA with 2 hours lifetime (ALS) or 8 mA with 10 hours lifetime (ESRF) are routinely delivered to the users interested in time-resolved measurements. A number of facilities are also sometimes operated in a so-called multi single bunch mode of operation with a few bunches of high current equally spaced along the circumference of the ring both satisfying the beamlines interested in high flux or brilliance and those doing time resolved studies. Another exotic mode has been developed at the ESRF which consists in combining a 1/3 filling mode with a single bunch mode spaced 180 degrees from the other bunches. The rms longitudinal bunch length in these facilities is of the order of 10 to 50 ps with a lengthening by a factor 2 to 3 as the current per bunch increases. The electron emittance has also reached design goals ranging from 4 nm-rad in the horizontal plane (ALS, ESRF) to 7 nm-rad (ELETTRA). The ultimate vertical emittance depends on the effort spent on the coupling correction. As described below, a perfect coupling correction is only of interest for hard X-ray sources. A record vertical emittance as low as 20 pm-rad is routinely operated at the ESRF. All facilities report extremely stable beams with rms motion of the beam smaller than 10% of the rms dimension of the beam within the 10 to 20 hours separating two refills. Perturbation of the closed orbit induced by the gap change of an individual undulator or wiggler is either very low or fully compensated using correction coils. As a result, high brilliance of the undulators is reported with 10^{19} (ALS, ELETTRA) or 10^{20} (ESRF) photons/sec/.1%/mr^2/mm^2, three to four orders of magnitude higher than the previous generation of storage rings.

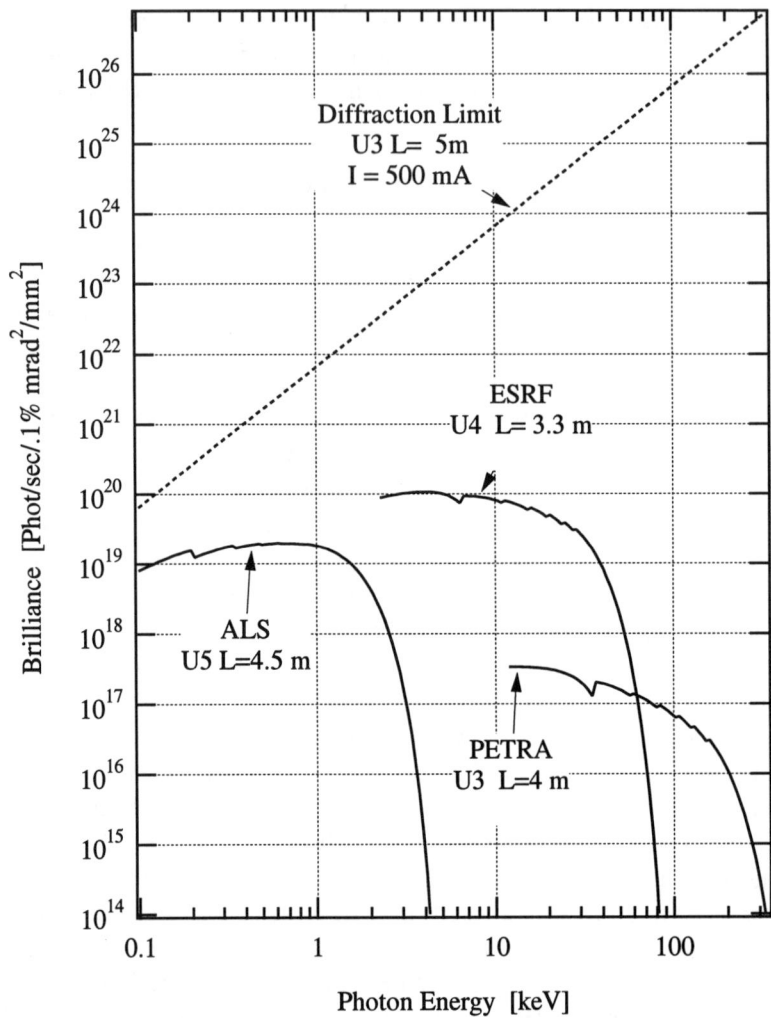

FIGURE 1 : Brilliance reached by ALS, ESRF and PETRA undulators as a function of photon energy. The ALS ring is operated with a 400 mA current of 1.5 GeV energy with horizontal (vertical) emittance of 4 (0.06) nm-rad. The ESRF ring is operated with a 200 mA current of 6 GeV with horizontal (vertical) emittance of 4 (0.02) nm-rad. The PETRA[4] ring is operated with a current of 60 mA, an energy of 12 GeV and emittances of 54 and 1 nm-rad. The dotted line corresponds to the maximum brilliance reached for a 500 mA current and an infinitely small electron emittance.

One consequence of the small emittance of the source is that the monochromatic undulator beam delivered to samples downstream has transverse dimensions smaller than 1 mm without the need of any focusing mirror. In the early stages, some facilities experienced multibunch longitudinal instabilities which result in a growth of the natural electron energy spread from a typical $1.0\ 10^{-3}$ to a few 10^{-3} which is detrimental to the use of high harmonics of the undulator spectrum. Several methods have been developed to reduce this instability including feedback and fine tuning of the RF cavities to the point that it is no longer a serious difficulty. Finally, nearly all facilities are highly reliable with an extremely low number of failures resulting in availability to the users in the 90% to 95% range.

To summarize, the performance of the recently commissioned third generation synchrotron sources exceeds the design goals by 1 to 2 orders of magnitude. Figure 1 presents the brilliance reached by various representative undulators currently in operation at ALS, ESRF and PETRA. The dotted line corresponds to the optimum brilliance reached by a 3 cm period 5 m long undulator installed on a ring with 500 mA current of infinitely small emittance (see next paragraph). This line must be understood as a fundamental limit to the highest brilliance reachable on a storage ring X-ray source at any photon energy. It appears that a medium energy source such as the ALS already operates closer to this limit than a hard X-ray source such as the ESRF.

GENERALITIES ON THE UNDULATOR RADIATION

In third generation sources, the most useful IDs are undulators which produce highly collimated beams with high brilliance. Undulators must be distinguished from wigglers whose spectrum extends to higher energies with greatly reduced brilliance and larger divergence. In this section, I briefly review the main parameters defining the spectral flux and brilliance of an undulator. Note that there always exists a few applications which require photon beams of large transverse dimension and therefore prefer a wiggler source point. At the ESRF, this is the case of the Topography and Medical beamlines. They nevertheless represent a minority. One should also note that a number of beamlines on medium and small energy rings make more use of wigglers than undulators to simply shift the spectrum to higher photon energies which cannot be reached by undulators. Even though cost-wise such an optimization makes sense, these beamlines cannot compete in brilliance and flux with the undulator beamlines of a higher energy facility.

Undulator radiation presents a series of peaks at the harmonically related energies ε_n given by:

$$\varepsilon_n[\text{keV}] = n\varepsilon_1[\text{keV}] = n\frac{0.95 \ E^2[\text{GeV}]}{\lambda_0[\text{cm}]\left(1+\frac{K^2}{2}\right)} \quad (1)$$

where n is an integer specifying the harmonic, E is the electron energy, λ_0 is the spatial period of the magnetic field and K is the deflection parameter which is related to the peak magnetic field B and the undulator period by the following relation

$$K = 0.934 \ \lambda_0[\text{cm}] \ B[\text{T}] \quad (2)$$

It is clear from Eq. (1) and Eq. (2) that, by varying the peak field B of the undulator, one changes the energy of the peak. One can show that the odd harmonics 1,3 and 5 are the most brilliant. In the following, I shall restrict the discussion to the most useful undulators with K=2.2 which provide full tunability between harmonics 1 and 3 of the spectrum. Lower K values result in slightly higher brilliance but a reduced range of tunability of the photon energy.

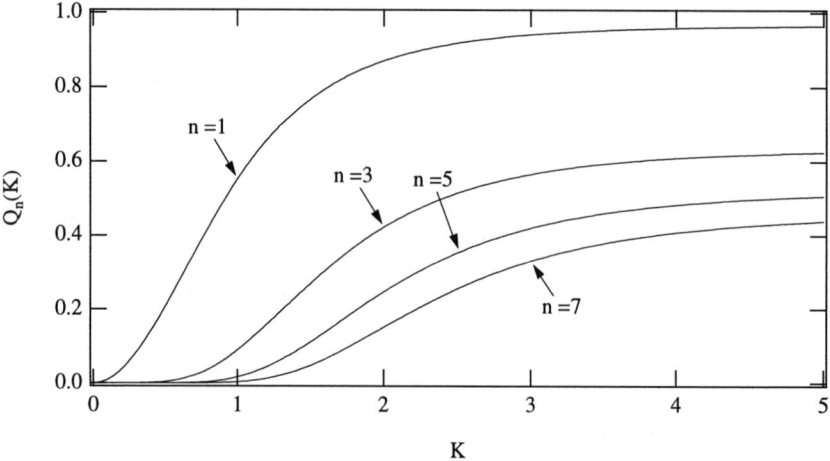

Q_n as a function of the deflection parameter K and harmonic number n.

The spectral flux of harmonic n is given by

$$F_n[\text{Phot./sec/.1\% bw}] = 1.431 \times 10^{14} \text{NIAmp} \ ] \ Q_n \quad (3)$$

where N is the number of period of the undulator and I is the electron beam current. Q_n is a dimensionless factor which depends on the deflection parameter K and the harmonic number n as shown in Fig. 2

The Brilliance B_n of the undulator radiation in harmonic n is:

$$B_n = \frac{F_n}{(2\pi)^2 \Xi_x \Xi_z} \qquad (4)$$

where Ξ_x and Ξ_z are the horizontal and vertical emittances of the photon beam which include contributions from the electron beam and from diffraction. Assuming an optimum beta functions of the ring lattice at the source point, Ξ_x and Ξ_z can be expressed as

$$\Xi_x \approx \varepsilon_x + \frac{\lambda}{2\pi} \qquad \Xi_z \approx \varepsilon_z + \frac{\lambda}{2\pi} \qquad (5)$$

where ε_x and ε_z are the horizontal and vertical emittances of the electron beam and λ is the wavelength of the radiation. $\frac{\lambda}{2\pi}$ is also called the diffraction limited emittance.

FUTURE IMPROVEMENTS

In this section, I shall present the various areas of potential improvement of storage ring X-ray sources. They include the use of larger rings, refinement of the insertion device technology and the optimization of the lattice functions at the source point.

Larger diameter Rings

From the point of view of the storage ring design, one can classify the methods for improving the undulator brilliance in two categories: Those resulting in a higher spectral flux and those resulting in a lower emittance of the photon beam.

Following Eq. (3), it is clear that to produce a higher spectral flux, one needs either a larger electron current in the ring or a larger number of undulator periods which implies a longer undulator. Present third generation rings operate between 200 and 500 mA and a doubling of the current is very unlikely to happen in the near future. The present limits on the current are the required RF power and RF cavities and the heatload on the absorbers which is already close to the limit imposed by the materials. The ESRF, for example, has already built a number of its photon absorbers out of "Glidcop", an alloy which, compared to copper, has a higher yield stress at elevated temperature. A doubling of the flux on all beamlines by doubling the length of the undulators is the maximum that one can reasonably envisage. A higher factor can be envisaged on a limited number of beamlines by using a few extremely long undulators[5].

The other method of increasing the brilliance is to decrease the photon beam emittance. The most efficient way to reduce the emittance is to reduce the field in the bending magnets which results in longer dipoles and therefore a longer circumference. One must distinguish medium energy sources from hard X-ray sources. In medium energy sources, the photon beam emittance is already close to the diffraction limit. Moreover, a further reduction of the emittance would reduce the lifetime of the stored beam which is not in the interest of the users. On the other hand, on a high energy source the horizontal emittance is typically 500 times greater than the diffraction limit and operation of the ring with the smallest possible horizontal and vertical emittances does not significantly degrade the lifetime of the stored electron beam. For example, the ESRF is routinely operated for users with a 0.02 nm-rad vertical emittance with a penalty on the lifetime of the order of 10 to 20%.

An ICFA workshop[1] was held in January 1996 where the potential for reducing the electron emittance of a ring was reviewed in line with the experience at the existing facilities. The conclusions were the following. The Chasman-Green lattice (the most widely used) as well as the triple bend achromat are both good candidates for future rings, no new lattice type is currently foreseen that could outperform them. A soft X-ray ring optimized for 1 keV photons could be envisaged by extrapolating the experience with low energy operation of the ESRF at 2 GeV. A 500 mA beam with typical emittance of 0.3 nm-rad is expected resulting in brilliances in the $5 \: 10^{22}$ photons/sec/.1%/mr^2/mm^2 range. In the hard X-ray range (10 keV photons), a 1.7 km circumference ring with an electron energy of 7 GeV could achieve a 0.5 nm-rad emittance (50 times the diffraction limit). Present experience in third generation sources indicates that the stability of the stored beam should be adequate. The most important question mark is the expected lifetime of the stored beam and possible growth of the emittance due to intra-beam scattering. These difficulties arise due to the long damping times inherent in these large diameter rings which make them more sensitive to any instability. It was recognized that further investigations are needed on these topics. In this respect, Table 1 presents the maximum current reached when operating the ESRF ring at low energy[5]. The horizontal electron emittance was measured at 5 and 6 GeV while at 1 and 2 GeV only theoretical values are quoted. These are preliminary results and higher currents should be possible at low energy by a proper tuning of the lattice.

Energy [GeV]	Current [mA]	Hor. Emittance [nm-rad]
6	200	4
5	170	2.8
2	8	0.4
1	1	0.1

Table 1: Current injected at the ESRF as a function of the electron energy together with the horizontal emittance. Emittances at 6 and 5 GeV have been experimentally confirmed. Those at 1 and 2 GeV are the values expected theoretically.

Another method of reducing the emittance is to install damping wigglers in some straight sections without dispersion. Damping wigglers are efficient if they are capable of producing a total synchrotron radiation power of the order or larger than that produced by all dipole magnets of the ring. Occupancy of a significant part of the straight sections for both the damping wigglers and the extra RF cavities required to compensate for the radiation in the wigglers sets a limit to this method. Taking the ESRF as an example, a reduction of the horizontal emittance by a factor 2 using damping wigglers would require 11 straight sections occupied with 2 T wigglers and two other sections for the additional RF cavities reducing the number of straight sections available for beamlines from 30 to 17! Obviously 2 or 3 damping wigglers could be used as synchrotron sources, however, they could not compete with high brilliance undulator sources which are the most in demand. The use of damping wigglers is more favourable in a larger diameter ring where more space would be available and where the radiation in the bending magnets is significantly reduced. To conclude, larger diameter rings with damping wigglers is the direction to follow to achieve smaller emittance photon beam but R & D is needed to assess the current lifetime and emittance really achievable in these rings operated with rather long damping constants.

Insertion Device Technology

The performance (flux and brilliance) of insertion devices is dictated by technology (room temperature or superconducting electromagnets or permanent magnets) and by the minimum magnetic gap that can be tolerated for stable operation of the ring. In this respect, one must distinguish high field devices built to produce the highest field from undulators. High field IDs use superconducting electromagnets, nearly all sources in the world have one or two such devices with fields in the range of 5 to 10 T. Higher fields require the development of superconducting materials with higher critical currents. The recent developments with high Tc superconducting materials have not yet resulted in practical applications in this field. Another way to shift the spectrum is to increase the electron energy. One can view a superconducting device on a medium energy ring (1.5 GeV) as a device whose radiation imitates that of a bending magnet or a few pole low field wiggler on a high energy ring (6 to 8 GeV). Similarly, the installation of a superconducting device on a high energy ring such as the 4 T three pole wiggler on the ESRF shifts the critical energy to 100 keV with high flux in the less demanded upper range of the spectrum from 100 to 400 keV. As a general rule, for the same photon energy, the performance of a super-conducting wiggler on a low energy ring is always lower than that of an undulator on a high energy ring.

Let us now consider the undulators. Nearly all undulators built in the world use permanent magnet technology which is the most economical and mature technology. One figure of merit for undulators is the peak magnetic field reached for a given ratio of magnet gap to spatial period. At this time, the use of NdFeB magnets allows peak fields as high as 0.7 T for a gap/period ratio of 0.35. In this respect the difference in field is not more than 5 or 10% between the pure permanent magnet technology or the most sophisticated wedged pole hybrid technology. The permanent magnet technology for the manufacture of undulators has reached a very advanced level. Undulators with several hundred periods and nearly ideal fields can now be built using sophisticated shimming techniques [7] and segmenting [8]. Another possibility consists in using superconducting electromagnet undulators. It has been shown that a 50% higher magnetic field[9] can be obtained from superconducting undulators with a cold bore. Little effort has been spent in this direction because of the high costs involved in both the manufacture and the operation of these devices. Unlike permanent magnets, the peak magnetic field of superconducting undulators decreases with the period at a fixed ratio of gap/period which makes them even less attractive compared to permanent magnet undulators for very short period devices. Another method to reach high field is to use a pulsed electromagnet device. Two to three times higher fields should be possible[10]. The low duty cycle of these undulators makes them unsuitable for the majority of experiments and little technological development has been made in this direction so far. Another important figure of merit of the undulators is the minimum gap achievable. A reduction in the gap allows both a reduction of the period (for the same deflection parameter K) resulting in a higher number of periods (for a fixed length) and most importantly a shift in the spectrum towards higher energies. This is illustrated in the next table which presents the photon energy range covered by the fundamental peak of a fully tunable undulator with K = 2.2 as a function of the minimum magnetic gap. The undulator is made of NdFeB magnets and the electron energy is 6 GeV.

Gap [mm]	Period [mm]	Field [T]	Energy Range of Fundamental [keV]
16	40	0.58	2.5 - 14
11	33	0.72	3 - 17
7	26	0.88	4 - 22

Table 2: Range of tunability of the fundamental peak of the spectrum as a function of the minimum magnetic gap assuming a 6 GeV electron energy and a NdFeB permanent magnet undulator.

The smallest gap is determined by the minimum aperture required by the electron beam. The minimum aperture results as a compromise with the lifetime of the stored beam. Most hard X-ray sources require a clear aperture of 7 to 8 mm for minimal reduction of the lifetime. Generally, one places the magnet blocks in the air outside a vacuum chamber. The thickness of the chamber combined with the alignment errors typically adds 3 to 5 mm to the minimum aperture resulting in a minimum magnetic gap between 10 and 13 mm. It is therefore clear from Table 2 that an undulator with magnet blocks in vacuum can be built with a 7 mm gap and would have a 26 mm period. An undulator with magnet blocks in the air and a minimum gap of 11 mm would have a period of 33 mm. The brilliance ratio and the shift in the energy of the harmonics between both devices is 33/26 = 1.3. One must not forget the significantly higher cost and lack of flexibility of the in-vacuum undulator technology. At present, nearly all undulators are built with magnet blocks in the air with the exception of the Spring8 project which is presently manufacturing a large number of in-vacuum undulators.

To conclude, one can say that, unless a real breakthrough takes place in magnet technology, one does not expect any dramatic improvement of the undulator brilliance beyond the present situation. A factor 1.3 in brilliance and energy of the spectrum can be reached using in-vacuum undulators of the same length and a factor 2 to 4 (depending on the photon energy) can be reached if the undulator length is increased from 5 to 10 m.

Optimization of the beta functions in the IDs.

In hard X-ray rings where the electron beam emittance dominates the diffractive emittance, it is possible to achieve an extremely small spot size by using a focused electron beam. In a beamline, the rms photon beam size Σ and divergence Σ' in a beamline can be expressed as:

$$\Sigma^2 \approx \varepsilon\left(\beta + 2\alpha d + \frac{1+\alpha^2}{\beta}d^2\right) + \frac{\lambda}{2L}d^2$$
$$(\Sigma')^2 \approx \varepsilon\frac{1+\alpha^2}{\beta} + \frac{\lambda}{2L} \quad (6)$$

α and β are the lattice functions of the electron beam in the middle of the undulator, ε is the emittance of the electron beam, d is the distance between the observation point and the middle of the undulator, λ is the wavelength of the radiation and L is the length of the undulator. The photon beam size and divergence of Eq. (6) can be understood as the sum of two contributions, one from the electron beam proportional to ε and one arising from single electron emission which is also called the diffractive contribution and is proportional to λ. In low and medium energy rings, the diffractive contribution dominates the electron beam contributions and the size and divergence of the photon beam is essentially determined by the wavelength and the undulator length. On high energy rings, the horizontal emittance is much greater than the diffractive emittance and the electron beam contribution to the beam size dominates. However, by using non-zero α and large β, which is physically equivalent to focusing the electron to a virtual point located downstream in the beamline one can achieve a very small size of the photon beam. The smallest spot size is set by diffraction:

$$\Sigma^2 \approx \frac{\lambda}{2L}d^2 \quad (7)$$

To illustrate this discussion, Table 3 presents the rms beam size and divergence for different choices of the alpha and beta functions.

α	β [m]	Σ [μm]	Σ' [μr]
0	2.5	1200	40
0	30	500	12
-3.3	100	200	22
-20	600	110	52

Table 3: Horizontal photon beam sizes and divergences for various choices of α and β functions. The photon energy is 20 keV and the horizontal emittance is 4nm-rad, the undulator length is 5m and the beam size is computed at a distance of 30 m from the undulator.

The first line with a β of 2.5 m corresponds to a perfect matching with the natural (or diffractive) beta function of the undulator resulting in a maximum brilliance. The second line corresponds to a slightly lower brilliance but a significantly smaller spot size and divergence at the point of experiment. Finally, the last two lines correspond to a focusing of the electron beam downstream. Significantly smaller beam sizes can be obtained at the price of a high horizontal beta function in the undulator straight section. This method of beam focusing is of particular interest at very high photon energies where the diffraction limited beam size given by Eq. (7) is small and focusing by mirrors becomes inefficient due to the grazing angles of reflection and the slope errors of the mirror surface. Obviously, a small spot size requires a large β function at the source point which can cause a number of complications in the operation of the storage ring. Expected difficulties are the correction of chromaticity, the sensitivity to the quadrupole positioning and undulator field errors. Nevertheless, no storage ring lattice has ever been designed with this criteria in mind and one may anticipate that a β function of 100 m is realistic. Note that a number of users would benefit from such a beam but not everyone. For example microfocus experiments may prefer a small size at the source (small β) and to demagnify it with a normal incidence zone plate which is aberration free[11].

To conclude, besides presenting small electron emittances, a new hard X-ray source should have several choices of lattice functions in the undulator straight sections optimized for each beamline. To ensure a large dynamic aperture and a long lifetime, the ID straight sections with equal lattice functions should probably be spaced symmetrically along the circumference of the ring.

CONCLUSION

The succession of generations of storage rings with an average brilliance increase of 1000 in brilliance every 10 years is probably over. There is limited room for higher brilliance in the soft X-ray range while a factor 100 can be expected in the hard x-ray range by means of very large perimeter (2 km) rings equipped with damping wigglers and very long undulators. A higher flexibility should be implemented in the optical functions of the source point for a better matching to the samples and optical components of the beamlines.

References

[1] 10th ICFA Workshop on 4th Generation Light Sources, Grenoble January 22-25, 1996.

[2] A. Jackson, Operational Experiences at the Advanced Light Source, Synchrotron Radiation Instrumentation Conference 1995, Argonne, October 17-20, 1995.

[3] C.J. Bocchetta, Operational Experience with ELETTRA, presented at the EPAC-96, Barcelona June 10-14, 1996.

[4] U. Hahn et al., Measurements of Emittance and Absolute spectral flux of the PETRA undulator at DESY Hamburg, submitted to the Journal of Synchrotron radiation.

[5] Spring8 International Workshop on 30 m Long Straight Sections, April 17-19, 1996.

[6] L. Farvacque et al. , same as [5].

[7] J. Chavanne, P. Elleaume, Undulator and Wiggler Shimming, Synchrotron Radiation News, Vol. 8, No. 1, p 18 (1995).

[8] J. Chavanne, P. Elleaume, P. VanVaerenbergh, J. Synchrotron Radiation (1996), Vol. 3, p 93-96.

[9] I. Ben-Zvi, R. Fernow, J. Gallardo, G. Ingold, W. Sampson, M. Woodle, Nuclear Instruments and Methods A318, 781 (1992).

[10] R. Warren, C. Fortgang, Nuclear Instruments and Methods A341, 436 (1994).

[11] E. Tarazona et al., Rev. Sci. Instrum., 65 (6), 1959 (1994).

Transition Radiation in the X-Ray Region from a low Emittance 855 MeV Electron Beam

H. Backe, K.-H. Brenzinger, F. Buskirk[*], S. Dambach, Th. Doerk,
N. Eftekhari, H. Euteneuer, F. Görgen, C. Herberg,
F. Hagenbuck, K. Johann, K.-H. Kaiser, O. Kettig, G. Knies,
G. Kube, W. Lauth, B. Limburg, J. Lind, H. Schöpe, G. Stephan,
Th. Walcher, Th. Tonn, and R. Zahn

Institut für Kernphysik, D-55099 Mainz, Germany, and
[*]*Naval Postgraduate School, Monterey, CA 93943, USA*

Abstract. A quasi-monochromatic hard x-ray beam with a photon energy of 33 keV has been produced from transition radiation (TR) at the Mainz Microtron MAMI. The radiator was a stack of 30 polyimide foils of 25 μm thickness and 75 μm separation and the monochromator a highly-oriented pyrolytic graphite crystal. The intrinsic bandwidth was measured with a critical absorption technique to be 100 eV. On the basis of these experiments a photon flux of $4 \cdot 10^9$/mm²s over an illuminated area of 5.7x125 mm² can be expected from an optimized beryllium radiator at a beam current of 100 μA. At the K-absorption edge of titanium at 5 keV narrow band transition radiation has been observed from a stack of four foils of 6 μm thickness and 294 μm separation originating from constructive interference in the region of anomalous dispersion. A new method is being developed with which the complex refractive index of thin foils in the x-ray region can be studied using such interference phenomena in the generation of transition radiation.

INTRODUCTION

Modern electron accelerators delivering high current low emittance electron beams with energies of about one GeV may present in the future attractive sources of high brilliance x-ray beams for applications in various fields of physics, material science, medicine and biology. Attention has been focused recently on the production of brilliant soft x-ray flashes in a single pass of high current electron bunches through an undulator by the process of self-amplified spontaneous emission (SASE) [see e.g. 1]. However, various other processes are also of interest for the production of soft and hard x-ray beams. The most important are: transition radiation (TR), channeling radiation (CR), parametric x-ray radiation (PXR), undulator radiation (UR), and Smith-Purcell radiation (SPR).

There are potential advantages of such new x-ray radiation sources over synchrotron radiation sources. First of all, since accelerators may become relatively inexpensive in the future, they could meet the radiation requirements of research laboratories or hospitals on the spot. Secondly, the x-ray beam can be triggered and its time structure adapted to nearly any experimental requirement. In particular, the electron beam can easily be turned off if the x-ray beam is not used minimizing power consumption and radiation production in the beam dump. Finally, materials can be brought into the electron beam and can be studied with new interferometric methods.

A research program has been started at the Mainz Microtron MAMI to demonstrate the feasibility and to explore the features of such new radiation sources. The spectral range covered by TR, PXR, UR and SPR extends from the hard x-ray region, about 40 keV for TR and PXR, down to the optical region, for UR and SPR. The brilliance of these radiation sources relies in all cases on the very good electron beam emittance of MAMI (horizontally 7 π nm rad (1 σ) and vertically a factor of 7 better at 855 MeV and 100 µA beam current).

In this contribution the present stage of the investigation of TR at MAMI is presented with special emphasis on the production of hard x-ray beams, the influence of anomalous dispersion at K-absorption edges on single and multi-foil interference, and the development of new interferometric methods with which the dispersion term of the refractive index of thin foils can be studied. The results of our investigations on the production mechanism of PXR will be published elsewhere [2]. For a survey about CR and PXR see the article of Freudenberger [3] in this volume.

BASIC THEORETICAL BACKGROUND

If a charged particle passes a single interface between a medium and vacuum broad band electromagnetic radiation is produced. The frequency spectrum emitted by a particle with charge e upon perpendicular traversal through a single interface between two media with different dielectric constants has been calculated by Ginsburg and Frank [4] and Garibian [5]. In the extreme relativistic case and for x-ray emission the number of photons of energy $\hbar\omega$ emitted from a single interface is given in direction θ with respect to the electron velocity vector, per electron, per relative photon energy interval $d(\hbar\omega)/(\hbar\omega)$, and per solid angle $d\Omega$ by [6]

$$\frac{d^2 N_0}{(d\hbar\omega/\hbar\omega)d\Omega} = \frac{\alpha\theta^2}{\pi^2}\left(\frac{1}{1/\gamma^2+\theta^2+2\cdot\delta(\omega)}-\frac{1}{1/\gamma^2+\theta^2}\right)^2 \quad (1)$$

where $\alpha = 1/137$ is the fine-structure constant, $\gamma = (1-(v/c)^2)^{-1/2}$ the Lorentz factor, v the speed of the electron, and c the speed of light. The quantity $\delta(\omega)$ is the deviation of the real part of the complex refractive index $n(\omega)$ from unity. Dispersion $\delta(\omega)$ and absorption $\beta(\omega)$ of the medium can be expressed by the angle-independent (forward) dipole atomic scattering factor $f(0,\omega) = f_1(0,\omega) + i f_2(0,\omega)$ [7] as

$$n(\omega) = 1 - \delta(\omega) - i \cdot \beta(\omega) = 1 - \tfrac{1}{2}(\omega_1/\omega)^2 \cdot [f_1(0,\omega) + i \cdot f_2(0,\omega)]/Z \qquad (2)$$

The quantity $\omega_1 = (4\pi r_e c^2 n Z)^{1/2}$ is the plasma frequency, with r_e the classical electron radius, n the atom number density of charge number Z. For simplicity it has been assumed that the medium consists of only one element. Far off atomic resonances the term $f_1(0,\omega)$ approaches the atomic number Z and $\delta(\omega) = -\tfrac{1}{2}(\omega_1/\omega)^2$, the high frequency approximation.

The radiation distribution resembles a Lorentz transformed dipole pattern with maximum intensity at an angle $\theta \cong 1/\gamma$. The energy spectrum extends up to a gradual cut-off energy $\gamma\hbar\omega_1$.

A radiator may be chosen in which the electron traverses many interfaces, e.g. a periodic stack of thin foils. In this case interference effects of radiation from the individual interfaces must be taken into account leading to a modulation of the angular and energy distribution obtained from a single interface. The differential energy spectrum of resonant XTR from M foils with thickness l_1, and spacing l_2 can be written as [6]

$$\frac{d^2 N_M}{(d\hbar\omega/d\hbar\omega)d\Omega} = \frac{d^2 N_0}{(d\hbar\omega/d\hbar\omega)d\Omega} \cdot F_2 \cdot F_3. \qquad (3)$$

The factor

$$F_2 = 1 + \exp(-\sigma) - 2 \cdot \exp(-\sigma/2) \cdot \cos(2l_1/Z_1) \qquad (4)$$

takes into account the interference of the radiation emitted from the two surfaces of a single foil of thickness l_1. The second factor

$$F_3 = \frac{1 + \exp(-M\sigma) - 2 \cdot \exp(-M\sigma/2) \cdot \cos(2MX)}{1 + \exp(-\sigma) - 2 \cdot \exp(-\sigma/2) \cdot \cos(2X)}, \qquad (5)$$

with $X = (l_1/Z_1) + (l_2/Z_2)$, describes the coherent summation of contributions from several foils in the stack. The formation lengths Z_1 and Z_2 are defined as

$$Z_i = \frac{4v/\omega}{1/\gamma^2 + \theta^2 + 2\cdot\delta_i(\omega)}, \quad i = 1,2. \tag{6}$$

Self absorption has been included in F_2 and F_3 [8] with $\sigma = \mu_1 \cdot l_1$, and μ_1 the linear absorption coefficient,. If self absorption can be neglected the well known expressions $F_2 = 4\sin^2(l_1/Z_1)$ and $F_3 = \sin^2(MX)/\sin^2 X$ are obtained from equations (4) and (5), respectively. If the resonance conditions $l_1/Z_1 = (N' - 1/2)\,\pi$, and $l_1/Z_1 + l_2/Z_2 = N\,\pi$ are satisfied simultaneously, with N' and N any integer greater or equal to one, the spectral intensity should increase by a factor $4M^2$. The decisive feature of periodic foil stacks is that the foil separation scales with the factor γ^2. The foil separation l_2 for a coherent interference superposition at a wavelength λ and an angle $\theta = 1/\gamma$ is given by $l_2 = \gamma^2 \cdot \lambda/2$. For $\gamma = 1673$ and a photon energy of 10 keV, i.e. $\lambda = 1.24$ Å, $l_2 = 173$ μm which is a mechanically realizable distance.

TRANSITION RADIATION AS A HARD X-RAY SOURCE

The possibility to exploit TR as a hard x-ray source has already been discussed in a large number of publications [see 9 and references cited therein]. Such a source would be indeed very attractive. The photon yield on the per electron basis, as given by equation (1), is large, the cut-off photon energy $\gamma\hbar\omega_1$ is high and photon energies in excess of 35 keV can easily be obtained with an electron beam energy of only about 1 GeV ($\gamma \approx 2000$). In addition, the x-ray beam is intrinsically highly directional with $\delta\theta \cong 1/\gamma$, which actually can be achieved at modern accelerators with beams of low emittance. Because of these reasons we are investigating the properties of such a hard x-ray source at MAMI.

Optimization of the foil stack

As foil materials low Z elements have to be used in order to keep deterioration of the electron beam quality by multiple scattering in the radiator foils and production of high energy bremsstrahlung to a minimum. In addition, the foils have to withstand the heat load of a 100 μA electron beam focused to a small spot size of 50 μm diameter or even less. This requires foil materials with a high melting point and good thermal conductivity. Since the temperatures are in general low radiation cooling can be neglected. Good candidates as radiator materials are beryllium, carbon, aluminum and titanium. The number of foils is limited by self absorption of the photons produced in the stack, and by multiple scattering of the electron beam which directly couples into the divergence of the photon beam. The former should not exceed about $1/e = 0.36$, the latter not $1/\gamma$ (= 0.6 mrad at MAMI). Optimization calculations for a photon energy of 33 keV have been performed taking into

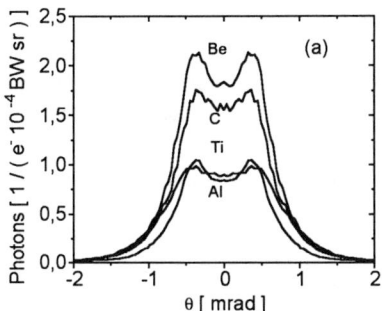 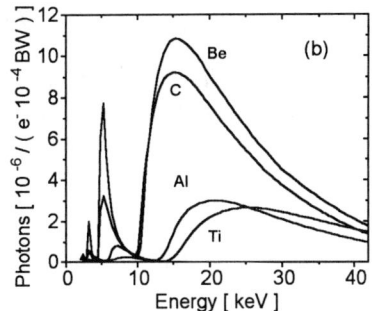

FIGURE 1. (a) Emission characteristics of TR as function of the angle for various foil stacks optimized for a photon energy of 33 keV. (b) Solid angle integrated photon number per electron and 10^{-4} bandwidth as function of the photon energy. The integration extends to an angle $\theta = 1$ mrad.

account the above requirements. The results are presented in Fig. 1 and the relevant parameters in Table 1 [10].

Calculations show that the maximum temperature of the beryllium and aluminum foils does not exceed 450 K at 100 µA beam current and a Gaussian beam profile with variances $\sigma_x = 34$ µm and $\sigma_y = 4.5$ µm. These calculations have been confirmed experimentally. A 14 MeV electron beam of 140 µA current and variances $\sigma_x = 11$ µm and $\sigma_y = 18$ µm has been focused on a stack of 10 aluminum foils of 25 µm thickness. After 60 s irradiation, which is a much longer time than the few ms required to reach thermal equilibrium by heat conductivity, the foils were inspected visually with the aid of a microscope. No indication of a deterioration of the foil material was observed [10].

TABLE 1. Collection of foils with various atomic number Z, plasma frequency ω_1, melting point T_{melt}. In addition the number of foils N, thickness of foils l_1, and separation of foils l_2 are given, as obtained from the optimization procedure.

	Z	ω_1 [eV]	T_{melt} [K]	N	l_1 [µm]	l_2 [µm]
Be	4	26.2	1551	30	32	69
C	6	24.9	> 3000	30	34	70
Al	13	32.9	933	10	25	68
Ti	22	41.5	1948	10	17	64

Experiments

A hard x-ray beam has been produced at MAMI with a test stack of 30 polyimide foils of 25 µm thickness separated by 75 µm. The calculated emission characteristics resembles that of carbon shown in Fig.1. The forward characteristics has been measured with a Ge(i) detector and is in fair agreement with the theory outlined above.

A highly oriented pyrolytic graphite (HOPG) crystal has been used to prepare a quasi-monochromatic 33 keV photon beam [11]. The experimental setup is shown in Fig. 2. The vertically focusing HOPG crystal with dimensions of 50x15x2 mm³ is bent with a radius of 250 mm at the 15 mm side. In a distance of 7.23 m from the radiator only part of the radiation cone was accepted by the crystal. The monochromized radiation was studied at a distance of 3.26 m from the HOPG crystal with a high resolution Ge(i) detector. A typical energy spectrum is shown in Fig. 3 taken with an aperture of 1.2 mm diameter in front of the Ge(i) detector. The x-ray spot has been scanned and found to be of 3.2x56 mm² size. The photon energy spread at the focal spot amounts to 5 keV, the measured photon flux to $2.8(3) \cdot 10^{-6}$ photons/(electron mm²), all in fair agreement with calculations using the mosaic spread of 0.49° as quoted by the manufacturer, and a calculated reflectivity R =

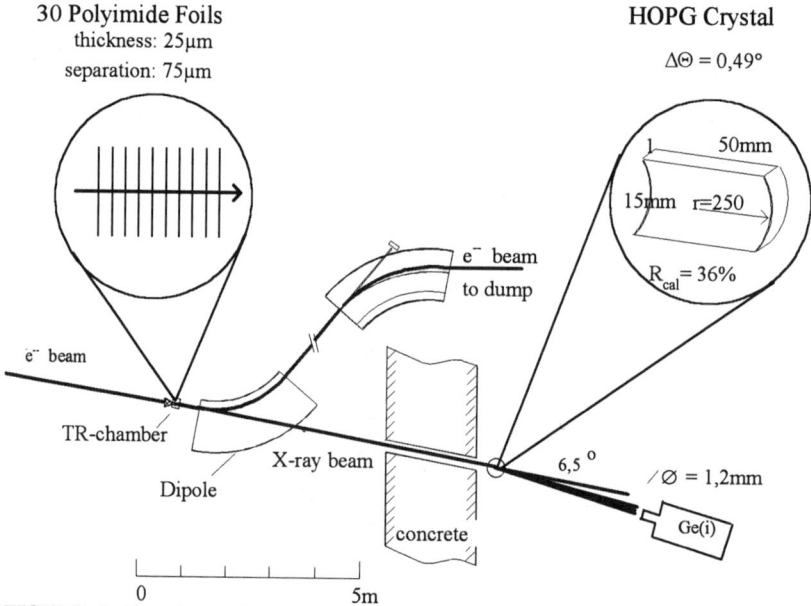

FIGURE 2. Experimental setup at MAMI for the investigation of TR characteristics from a polyimide foil stack. The electron beam is bent by dipole magnets and guided into the beam dump. The photon beam exits the dipole vacuum chamber through a 75 µm thick polyimide window. The experiment with the HOPG crystal monochromator could be performed in air since the absorption length at 33 keV photons is 28 m.

0.36 of the HOPG crystal [12]. An intrinsic band width of (94 ± 10) eV has been measured with a critical absorption technique at the K-absorption edge of tin at 29.2 keV. This value is in fair agreement with calculations on the basis of the about 300 crystal planes of a HOPG microcrystallite with 0.1 μm size. It is also important to notice that the background in the photon spectrum, displayed in Fig. 3, is very low.

Discussion

These test experiments show that a brilliant hard x-ray beam with photon energies of up to about 40 keV can be produced with a stack of 30 beryllium foils. The photon flux in the energy range between 20 and 40 keV would be rather high, see Fig. 4, although the electron beam energy is only 855 MeV and the beam current is lower by about three orders of magnitude than with synchrotron radiation sources. If the full 1 mrad radiation cone is covered by a vertically focusing HOPG crystal, of e.g. 0.49° mosaic spread, and optimum symmetrical distances between radiator, HOPG and detector are chosen [14], e.g. 7 m each, the quasi-monochromatic x-

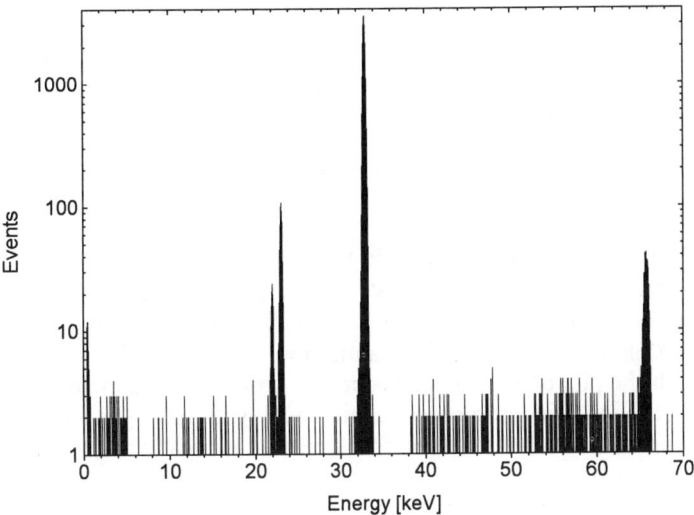

FIGURE 3. Typical energy spectrum taken with the Ge(i) crystal in at distance of 3.26 m from the HOPG crystal. A Bragg angle of 3.26° was chosen. The half width of the central line at 32.9 keV amounts to 302 eV a large fraction of which results from the detector resolution (265 eV) and geometry. The weak lines at 22 and 23.3 keV are $K_{\alpha,\beta}$ - escape peaks of germanium. The line at 66 keV is the second order (002) reflection.

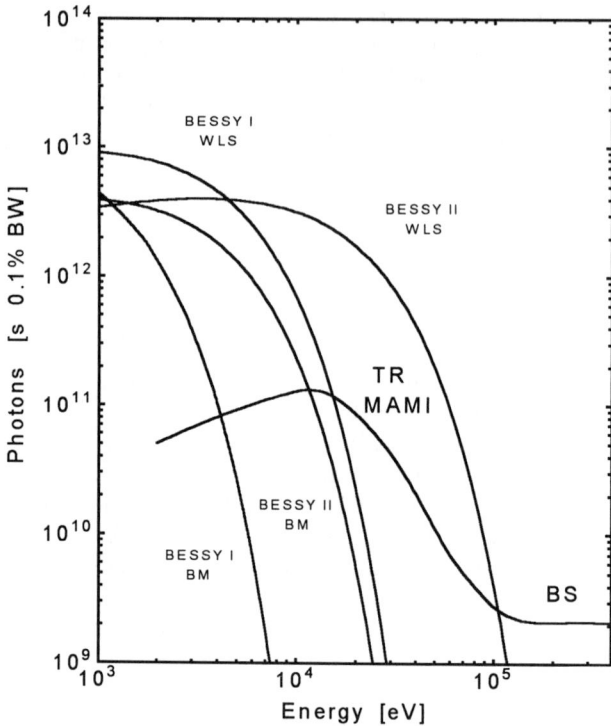

FIGURE 4. Solid angle integrated photon flux of the projected TR radiation source at MAMI at the maximum current of 100 μA in front of the HOPG monochromator. For comparison is shown the vertically integrated photon flux at a horizontal acceptance of 1 mrad of the synchrotron radiation sources BESSY I and BESSY II [13] which have comparable electron beam energies as MAMI but about a factor of 1000 higher beam current. BS: bremsstrahlung, BM: bending magnet, WLS: wave length shifter.

ray spot at position of the Ge-detector in Fig. 2 would have dimensions of 5.7x125 mm² at a photon energy around 33 keV. At this spot a photon flux of $4 \cdot 10^9$/mm²s can be expected at a beam current of 100 μA. The photon energy spread at the focal spot would amount to 5 keV at an intrinsic energy resolution of about 100 eV. Although the photon flux is still about two orders of magnitude below of that required for the intravenous coronary angiography with iodine as a contrast agent [15] such a beam would be of interest for numerous applications.

NARROW BAND RADIATION AT THE K-ABSORPTION EDGE OF TITANIUM

Particular interesting interference phenomena occur in the vicinity of absorption edges of the foil materials. Since the real and the imaginary part of the scattering factor $f(0,\omega)$ are connected by the Kramers - Kronig relation an anomaly in the linear attenuation coefficient $\mu = (4\pi /\lambda) \cdot \beta(\omega)$ is connected with an anomaly in the dispersion $\delta(\omega)$. Close to the K-absorption edge the real part of the scattering factor, $f_1(0,\omega)$, changes by many charge numbers e [7] resulting in a large relative change in the dispersion $\delta(\omega)$ and, consequently, also in the formation length Z_1. This anomaly can be exploited to produce narrow band transition radiation, as will be shown in the following.

Design of the foil stack

In order to determine the thickness and separation of the foils the dispersion term $\delta(\omega)$ must be known. It has been calculated numerically at the K-absorption edge of titanium from the total photoabsorption cross-section μ_a, as obtained from tables [16, 17], using the dispersion equation [7]

$$f_1(0,E) = Z + \frac{1}{\pi \cdot r_0 hc} \int_0^\infty \frac{\varepsilon^2 \mu_a(\varepsilon)(E^2 - \varepsilon^2)}{(E^2 - \varepsilon^2)^2 + (\eta \varepsilon)^2} d\varepsilon \qquad (7)$$

with $E = \hbar \omega$. A radiation damping constant $\eta = 0.7$ eV has been used. Details of the calculations are described in [18, 19]. Some results are depicted in Fig. 5 (a). Close to the K-absorption edge the phase $\phi(\omega) = -(\omega/c) \cdot \delta(\omega) \cdot l_1$ changes rapidly with respect to a wave traveling in vacuum, resulting in pronounced interference phenomena. This is reflected in the peak structure of the formation length, see Fig. 5 (b). The formation length is simply proportional to $1/\delta(\omega)$ since both, $1/\gamma^2$ and θ^2, are small in comparison to $2 \cdot \delta(\omega)$. We will return to this important feature in the last section of this paper.

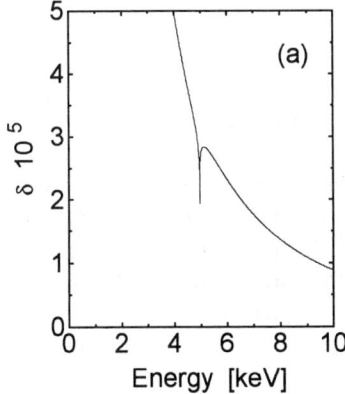
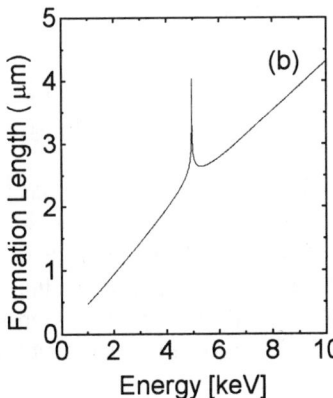

FIGURE 5. (a) Dispersion $\delta(\omega)$ for titanium ($Z = 22$) and (b) formation length Z_1 for an observation angle $\theta = 0.6$ mrad as function of the photon energy. The K-absorption edge is located at an energy of 4.966 keV.

Titanium with a plasma frequency of $\hbar\omega_1 = 41.5$ eV has been chosen as radiator material since, on the one hand, the K-absorption edge at an energy of 4.966 keV is high enough to be easily detectable with an existing experimental setup, equipped with a semiconductor detector of 850 eV resolution [20]. On the other hand, the charge number is low enough that multiple scattering of the electrons in the foil material does not spoil the coherent superposition of the TR amplitudes from the various interfaces. Finally, Ti foils can easily be handled and are commercially available in various thicknesses.

The optimization has been performed for the maximum beam energy of 855 MeV an observation angle $\theta = 1/\gamma = 0.6$ mrad (for which the TR amplitude from a single interface reaches its maximum) and the calculated minimum atomic scattering factor at the K-absorption edge $f_1(0,\omega_K) = 12.3$ which corresponds to $\delta(\omega_K) = 2\cdot 10^{-5}$, see Fig. 5 (a). The condition for constructive interference, i.e. $l_1/Z_1 = l_2/Z_2 = \pi/2$ (see chapter "Basic Theoretical Background"), yields an optimum foil thickness $l_1 = 6.3$ μm and an optimum separation $l_2 = 350$ μm results with these parameters. The number of foils $M = 4$ has been determined by the requirement that just below the K-absorption edge the attenuation of TR from the first interface due to self absorption in the succeeding foils should not be larger than $1/e = 0.36$. An actual foil thickness $l_1 = 5.89 \pm 0.09$ μm was chosen, since this foil was available, and the separation was reduced to $l_2 = 294 \pm 3$ μm. With these experimental parameters the line width increased from about 15 eV for the optimal parameters to about 140 eV. As a consequence, the intensity also increased which was necessary because of the modest resolution of the photo-diode detector. At poor resolution a line with too weak an intensity would not easily be detectable if superimposed by an intense continuous spectrum.

For this foil stack the spectral intensity was calculated taking into account self absorption of the photons and multiple scattering of the electrons in the foil material, the latter with a Monte-Carlo simulation. Details of the calculation procedure have been described in ref. [9, 18, 21]. Some results of these calculations are presented in Fig. 6 (b) and 7 (b).

Experimental results and discussion

The four foil TR interference pattern has been studied two-dimensionally with the experimental setup and procedures as described in ref. [9], in the energy range between 3 and 8 keV and for angles $\theta \leq 4$ mrad. A three dimensional representation of the experimental data as a function of observation angle and photon energy is shown in Fig. 6 (a). The spectral intensity agrees qualitatively with the calculations shown in Fig. 6 (b) for which multiple scattering, self absorption and the detector response function were taken into account. A more quantitative comparison with the theoretical calculations is shown in Fig. 7 which shows slices through the 3 D plots in Fig. 6 (a) and (b).

The good agreement with theoretical calculations indicates that indeed a narrow line at the K-absorption edge exists. The modest energy resolution of the silicon detector limits the investigations of details in the interference structures. In principle, the line may exhibits fine structure pattern originating from interference if the formation length differs from the theoretical values on the basis of which the foil stack has been calculated. Therefore, it would be worth reinvestigating the line structure using a crystal spectrometer with a resolution of the order of 1 eV. With such a measurement it should be possible to obtain detailed insight into near edge structures of the dispersion $\delta(\omega)$.

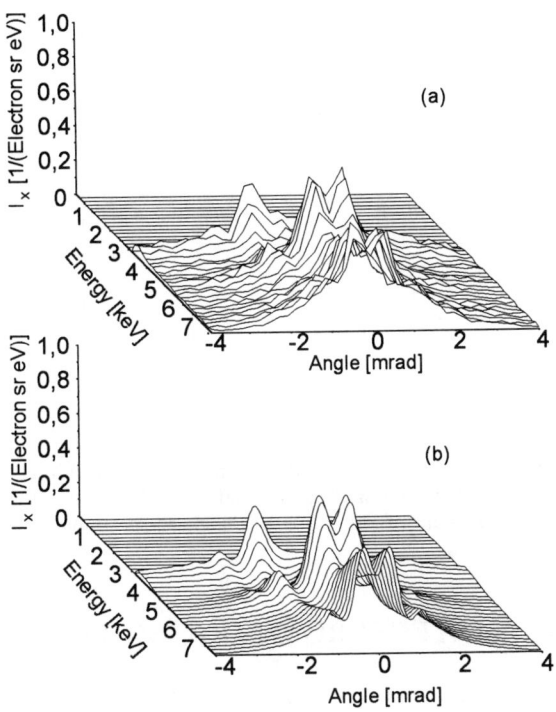

FIGURE 6. Three dimensional plots of (a) experimental and (b) calculated spectral x-ray intensities for a titanium foil stack at $\gamma=1673$. In the calculated spectra effects due to multiple scattering, self absorption and the detector response function were taken into account.

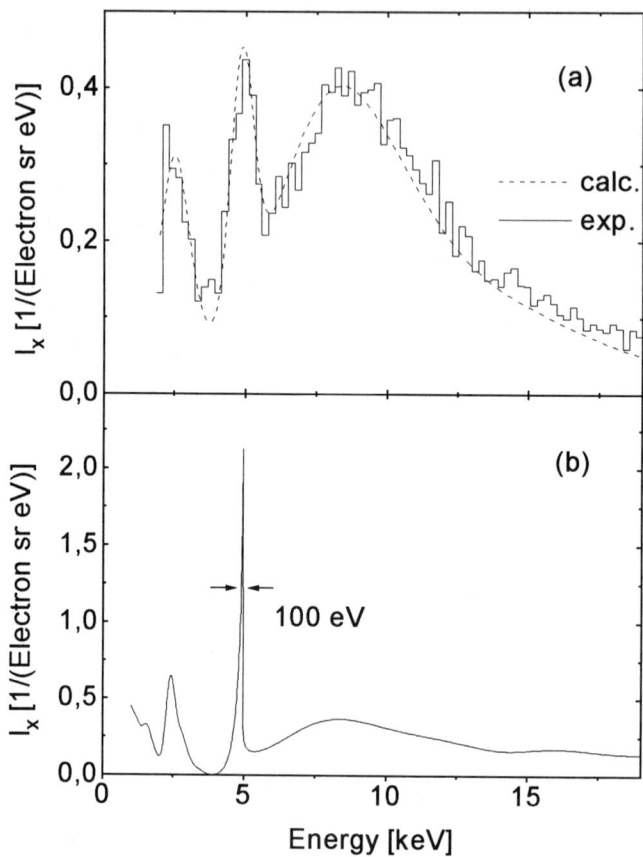

FIGURE 7. (a) Spectral intensity as function of the photon energy. The full line represents a cut at a fixed angle of ±0.55 mrad through the experimental data of Fig. 6 (a), the dashed line the same cut through the calculated one of Fig. 6 (b). (b) calculated spectrum as (a) without folding of the detector response function.

A NEW INTERFEROMETRIC METHOD

It has been demonstrated in the last section that anomalous dispersion has a large impact on the TR interference structures. This raises the question whether interference effects in the TR production can be exploited for the measurement of $\delta(\omega)$ [see also 22]. Inspecting equation (6) it can be seen that the formation length at MAMI beam energies is nearly independent on the kinematical variables θ and γ, since $\delta(\omega) \gg 1/\gamma^2 = 3.6 \cdot 10^{-7}$, provided the observation angle is chosen to be $\theta \approx 1/\gamma$. This offers a unique possibility to study the real part of the refractive

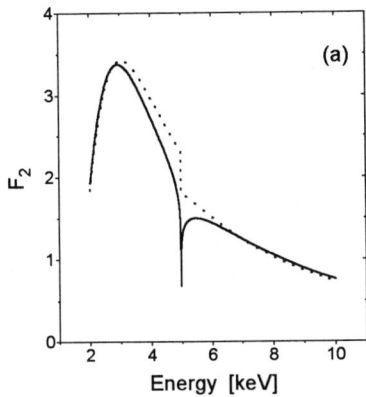 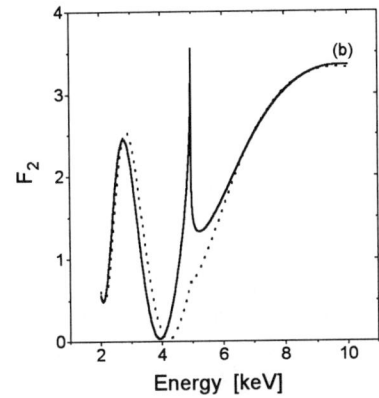

FIGURE 8. The interference term F_2 according to equation (4) as function of the photon energy for (a) 2 μm, (b) 6 μm thicknesses of the titanium foil. Note, that the interference effect changes sign if the foil thickness is increased. The dotted lines represent the high frequency approximation.

index $\delta(\omega)$ in the x-ray region which is not possible at low beam energies, i.e. if $\delta(\omega) \ll 1/\gamma^2$. Basically, there are three possibilities to study the interference effects: (i) on the basis of the single foil interference, (ii) on the basis of two foil interference in which case the relative distance between the foils must be varied, and (iii) interference between two identical stacks of thin foils of a light material such as beryllium between which the foil to be studied is inserted. In the following only one foil interference will be discussed in more detail. For the other possibilities we refer to [9].

Interference from a single foil is described by the term $F_2 = 4 \cdot \sin^2(l_1/Z_1)$, if self absorption can be neglected. It is clear that the intensity of the TR is essentially controlled by the phase

$$\phi_1 = \frac{l_1}{Z_1} = \frac{\omega}{4 \cdot v}(1/\gamma^2 + \theta^2 + 2 \cdot \delta(\omega)) \cdot l_1 \tag{8}$$

which we will discuss in the following in more detail. Since $\delta(\omega) \gg 1/\gamma^2$ and θ^2, an 30 to 50% change of $\delta(\omega)$ at the K-absorption edge of titanium, see Fig. 5, results in a phase change of the same relative magnitude. In Fig. 8 examples of the behavior of the interference term F_2 are depicted. To observe the effect associated with such a phase change it must be assured that the spreads $\Delta\gamma$, $\Delta\theta$, Δl_1, and $\Delta\omega$ have comparatively little influence on the phase. From equation (8) one obtains

$$\frac{\Delta\phi_1}{\phi_1} = \sqrt{\left(\frac{1}{\gamma^2\delta(\omega)}\cdot\frac{\Delta\gamma}{\gamma}\right)^2 + \left(\frac{(\theta\gamma)^2}{\gamma^2\delta(\omega)}\cdot\frac{\Delta\theta}{\theta}\right)^2 + \left(\frac{\Delta l_1}{l_1}\right)^2 + \left(\frac{\Delta\omega}{\omega} + \frac{\delta'(\omega)\cdot\Delta\omega}{\delta(\omega)}\right)^2}. \quad (9)$$

The relative phase changes $\Delta\phi_1/\phi_1$ due to the energy spread $\Delta\gamma/\gamma = 3\cdot 10^{-5}$ (1σ) of the electron beam can be neglected. Also, even a spread $\Delta\theta/\theta \approx 0.5$ due to beam angular divergence, multiple scattering, etc., can be tolerated if $\delta(\omega) = 2\cdot 10^{-5}$ and $\theta\gamma \approx 1$. The foil thickness must be homogeneous to a level of about a few % since $\Delta l_1/l_1$ is directly proportional to the relative phase change. The last term means that the energy resolution of the monochromator must be good enough in order to get meaningful results on $\delta(\omega)$ in an energy region where $\delta(\omega)$ varies rapidly. With an estimated value $\delta'(\omega)/\hbar = 5\cdot 10^{-6}$/eV and $\delta(\omega) = 2\cdot 10^{-5}$ for titanium at the K-absorption edge an energy resolution of the monochromator of about $2\cdot 10^{-4}$ is required.

It should be mentioned that the attenuation due to self absorption is superimposed on the intensity variations from the interference effect, see equation (4). To isolate both contributions it might be necessary to determine the self absorption of the foil in an additional experiment, or to perform measurements with various foil thicknesses.

CONCLUSION

Features of transition radiation from a low emittance 855 MeV electron beam have been reviewed. It has been demonstrated that TR from a multi-foil stack of light Z materials represents a potentially useful source of hard x rays for applications. Interference phenomena of TR from a stack of titanium foils at the K-absorption edge indicate the possibility of measuring the refractive index by new interferometric methods. For the measurement of the dispersion $\delta(\omega)$ a monochromator with a relative energy resolution of about $2\cdot 10^{-4}$ is required. The construction of such a monochromator and a dedicated beam line, making possible the mentioned measurements, are in progress at the Mainz Microtron MAMI.

ACKNOWLEDGEMENT

This work has been supported by DFG (SFB 201) and BMFT under contract 06 MZ 566.

REFERENCES

[1] Roßbach, J., Phys. Blätter **51**, 283 (1995).

[2] Brenzinger, K.-H., et al., Investigation of the Production Mechanism of Parametric X-ray Radiation, to be published in Z. Physik.

[3] Freudenberger, J., Properties of channeling- and parametric X-radiation, contribution to this conference.

[4] Ginsburg, V.L., and Franck, I.M., J. Phys. (Moscow) **IX**, 353 (1945).

[5] Garibian, G.M., Sov. Phys. JETP **6**, 1079 (1958); Garibian, G.M., Sov. Phys. JETP **12**, 237 (1961).

[6] Cherry, M.L., Hartmann, G., Müller, D., and Price, Th.A., Phys. Rev. **D 10**, 3594 (1974).

[7] Henke, B.L., Gullikson, E.M., and Davis, J.C., Atomic Data and Nuclear Data Tables **54**, 180 (1993).

[8] Fabjan, C.W., and Struczinski, W., Phys. Lett. **B 57**, 483 (1975).

[9] Backe, H., Gampert, S., Grendel, A., Hartmann, H.-J., Lauth, W., Weinheimer, Ch., Zahn, R., Buskirk, F.R., Euteneuer, H., Kaiser, K.-H., Stephan, G., and Walcher, Th., Z. Physik **A 348**, 87 (1994).

[10] Knies, G., Planung und Testmessungen zur Erzeugung eines 33 keV Röntgenstrahles hoher Brillanz am Mainzer Mikrotron, Diplomarbeit, Institut für Physik, Mainz, 1995, unpublished.

[11] Johann, K., Aufbau eines Monochromators für 33 keV Röntgenstrahlung am 855 MeV Elektronenbeschleuniger MAMI, Diplomarbeit, Institut für Kernphysik, Mainz, 1995, unpublished.

[12] Chabot, M., Nikolai, P., Wohrer, K., Rozet, J.P., Touati, A., Chetioui, A., Vernhet, D.,and Politis, M.F., Nucl. Inst. Meth. Phys. Res. **B61**, 377 (1991).

[13] Gaupp, A, Koch, E.E., Maier, R., and Peatman, W., Eine optimierte Undulator/Wiggler-Speicherring Lichtquelle für den VUV- und XUV-Spektralbereich, BESSY II Report, Berlin, 1986.

[14] Antonov, A.A., Baryshev, V.B., Grigoryeva, I.G., Kulipanov, G.N., and Shchipkov, N.N., Nucl. Inst. Meth. Phys. Res. **A 308**, 442 (1991).

[15] Dix, W.R., Progress in Biophysics and Molecular Biology, **63**, 159 (1995).

[16] Storm, E., and Israel, H.I., Atomic Data and Nuclear Data Tables, **A7**, 565 (1970).

[17] Henke, B.L., Lee, P., Tanaka, T.J., Shimabukuro, R.L., and Fujikawa, B.K., Atomic Data and Nuclear Data Tables, **27**, 1 (1982).

[18] Gampert, A.S., Messung resonant erzeugter Übergangstrahlung an der K-Absorptionskante von Titan, Diplomarbeit, Institut für Physik, Mainz, 1993, unpublished.

[19] Zahn, R., Messung resonanter Übergangstrahlung im Röntgenbereich mit einem 855 MeV Elektronenstrahl geringer Emittanz, Dissertation, Fach bereich Physik, Universität Mainz, 1994.

[20] Backe, H., Brenzinger, K.-H., Buskirk, F., Gampert, S., Euteneuer, H., Kaiser, K.-H., Kube, G., Lauth, W., Schöpe, H., Walcher, Th., and Zahn, R., to be published in Z. Physik.

[21] Hartmann, H.-J., Aufbau eines Detektorsystems zum Nachweis von Übergangsstrahlung, Diplomarbeit, Institut für Physik, Mainz, 1992, unpublished.

[22] Fiorito, R.B., Rule, D.W., Piestrup, M.A., Maruyama, X.K., Silzer, R.M., Skopik, D.M., and Shagin, A.V., Phys. Rev. **E51**, R 2759 (1995).

Properties of Channeling and Parametric X radiation

J. Freudenberger

Institut für Kernphysik, Technische Hochschule Darmstadt,
Schloßgartenstraße 9, 64289 Darmstadt, Germany

Abstract. The main properties of Channeling- and parametric X radiation - photon energy, number of photons, linewidth, angular distribution and their dependence on the electron energy, orientation and properties of the crystal - are summarized. Recent progress achieved with respect to the experimental investigation and the theoretical understanding of the radiation processes is presented. The possibility of applying the X radiation for medical and scientific purpose is discussed.

INTRODUCTION

Channeling radiation (CR) and parametric X radiation (PXR) are generated due to the interaction of relativistic charged particles passing through crystals. In contrast to transition radiation and coherent Bremsstrahlung (CBS), already moderate electron energies in the MeV range are sufficient to produce hard X radiation with energies of several keV. Furthermore, CR and PXR have been discussed as possible candidates for an intense, tunable, polarized and coherent radiation source based on an electron accelerator using moderate electron energies for medical purpose (1-3), especially for digital subtraction angiography (DSA, ref. (4)).
The present paper briefly summarizes the basic properties and recent developments of CR and PXR, i.e. the photon energy, number of photons, linewidth, angular distribution and their dependence on the electron energy, orientation and properties of the crystal. The following section describes the basic processes of CR and PXR. Subsequently, the experimental procedures are explained. The section 'radiation features' is a survey of CR and PXR properties and their theoretical description. Finally, the range of applicability of CR and PXR as radiation sources is discussed and an outlook to possible future developments of the fields is given.

RADIATION PROCESSES

For comparison of the experimental observations described below some simple relations valid for the most important observables - photon energy and number of photons - are of interest. They will be summarized in this section. It is far beyond the scope of the present paper to give full treatment of the theory of CR and PXR, which can be found in (5-7) for CR and in (8-13) for PXR.

Channeling radiation

Channeling radiation is generated when relativistic electrons penetrate a single crystal along one of its major axes or planes. In a quantum mechanical picture the part of the electron motion which is perpendicular to the axes or plane can be described by a crystal potential. This potential is due to the Coulomb interaction of the relativistic electron with the crystal atoms. The electrons can occupy bound and unbound states of the crystal potential and due to spontaneous transitions between these states CR is emitted. The radiation has been named planar CR if the electron moves along a plane and axial CR if the electron travels a crystal axis. Radiation occurs only if the electron enters the crystal at an angle smaller than the so-called critical Lindhard angle, which is typically below 0.5°. Thus, CR shows a strong dependence on the orientation of the crystal. Similar to Bremsstrahlung (BS) it is emitted into a narrow cone with an opening angle $\varphi \sim \gamma^{-1}$ in the direction of the electron trajectory, where $\gamma = 1 + E_o/m_o c^2$ with E_o being the kinetic electron energy and $m_o c^2$ its rest energy.

Since the radiation is emitted by the electron in its rest frame moving with respect to the laboratory coordinate system, the observed energy of the radiation, E_{CR}, is Doppler shifted according to

$$E_{CR} = \frac{1}{1-\beta\cos\psi}\Delta E \approx \frac{2\gamma^2}{1+\gamma^2\psi^2}\Delta E. \qquad (1)$$

Here ΔE denotes the energy difference between the two states involved in the transition, $\beta = v/c$ with v representing the electron velocity and ψ the angle between the electron trajectory and the direction of observation. As will be shown below there is a possibility to change the channeling radiation energy E_{CR} due to this Doppler-shift.

Depending on the character of the states involved in the transitions the CR spectrum will show single lines from bound-to-bound transitions on top of a broad distribution originating from free-to-bound and free-to-free transitions. For the calculation of the number of CR photons and the energy differences ΔE at low electron energies in the MeV region a quantum mechanical treatment is required, which is usually solved numerically. A complete description of the basic relations is given by (5-7).

Parametric X radiation

In contrast to CR, where the radiation is caused by the transition of electrons between discrete states, in the case of PXR the crystal itself is the source of the X rays. A relativistic electron moving through a crystal polarizes the crystal atoms located in the vicinity of its trajectory. Due to constructive interference of radiation emitted in connection with this polarization intense X radiation occurs. This process can be described as Bragg diffraction of the virtual photons associated with the electromagnetic field of the electron at the crystal planes (11,14), or as de-excitation of crystal atoms which were virtually polarized by the relativistic electron (10). Using the condition for constructive interference of radiation emitted by neighboring crystal planes it follows (11) that the PXR photon energy E_{PXR} is given by

$$E_{PXR} = \hbar c \frac{\vec{g}\vec{v}}{c - n\vec{v}\hat{k}} = \hbar g v \frac{\sin\phi \cos\alpha}{1 - \beta n \cos\theta}. \qquad (2)$$

Here $\vec{g}$ is the reciprocal lattice vector of the planes reflecting the virtual photons, $\vec{v}$ the electron velocity, the crystal's index of refraction is denoted by n, while θ is the angle between $\vec{v}$ and the direction of PXR observation, i.e. the direction $\hat{k} = \vec{k}/|\vec{k}|$ of the radiation wave vector $\vec{k}$. These two directions - the electron beam and the observation direction - define the radiation plane (see insert of Fig. 6b). The angles ϕ and α describe the orientation of the crystal and thus also of the reciprocal lattice vector. The first is located in the radiation plane and is zero if the reciprocal lattice vector is perpendicular to the electron beam axis. The second, i.e. α, is located in a plane perpendicular to the radiation plane. The Bragg condition for the diffraction of virtual photons is given by $\phi \approx \theta/2$, $\alpha = 0$. It is of interest to note that the PXR photon energy depends strongly on the orientation of the crystal but is almost independent from the electron energy.

The number of emitted PXR photons can be determined using the double differential PXR cross section for the emission of one photon into an energy interval dE and the solid angle $d\Omega$ per elementary crystal cell (10,15) which is given by the expression

$$\frac{d^2\sigma_{atom}^{PXR}}{dEd\Omega} = 8\pi\alpha^3\lambda^2 V_c \sum_{\vec{g},\sigma} \frac{|S_N F|^2 f^2}{(c - n\vec{v}\hat{k})\omega n} \left| \frac{\vec{e}_{\vec{k},\sigma}(n\omega\vec{v} - \vec{g}c^2)c}{c^2(\vec{k} + \vec{g})^2 - n^2\omega^2} \right|^2 \delta(E - E_{PXR}), \qquad (3)$$

where S_N, F and f represent crystal properties. The symbol S_N denotes the structure factor normalized to the volume V_c of one elementary crystal cell, F the atomic form factor and $f^2 = \exp(-u_t^2 g^2)$ the Debye-Waller factor with the transverse vibrational amplitude u_t of the crystal atoms, respectively. The polarization vectors $\vec{e}_{\vec{k},\sigma}$ are perpendicular to $\vec{k}$. The photon energy is denoted by E. It defines the radiation frequency ω through $E = \hbar\omega$. To get the number of PXR photons emitted the cross section has to be multiplied by the density of crystal cells and in the case of negligible X-ray absorption inside the crystal by the length of the electron path inside the crystal.

It is worth noting that CBS has the same energy as PXR. Since CBS appears due to the acceleration of the electron at the crystal planes it is directed into a narrow cone in forward direction of the electron trajectory. Thus, if PXR is observed under a small angle with respect to the electron trajectory, Eq. (3) has to be modified due to interference of PXR and CBS (9, 15,16).

Recently, Nitta (10) has pointed out that under certain conditions Eq. (3) needs to be modified in order to take into account the so-called dynamical effect. Nevertheless, for electron energies below about 100 MeV the dynamical effect is negligible. Furthermore, it has been claimed (10) that the theoretical model (13,17) describing PXR as some kind of Čerenkov radiation is misleading.

In our earlier work (18) we have already stated that for an accurate description of the experimental results obtained for electron bombarding energies in the MeV region multiple electron scattering (19) plays an important role. For comparison of experimental results with theoretical predictions, this influence can be taken into account by means of a Monte-Carlo simulation.

EXPERIMENTAL PROCEDURES

While CR was predicted by Kumakhov (20) and first observed by Alguard et al. (21), PXR was predicted by Ter-Mikaelian (14) and detected for the first time by Vorob'ev et al. (22). Since then CR and PXR have extensively been investigated (23). Most of the early experiments regarding PXR were conducted at fairly high electron energies (24-30), but the possibility of decreasing the electron energy was soon realized (31-34).

Experimental setup. CR and PXR can be produced simultaneously with the same experimental set-up. As a typical example of an experimental setup, the low energy channeling and PXR site at the superconducting Darmstadt Linear Accelerator S-DALINAC (35) is presented in Fig. 1. In this particular case a continuous wave (cw) electron beam provided by the accelerator interacts with a crystal, which can be oriented by means of a three axis goniometer. Behind the crystal the electron beam is deflected using a dipole magnet and collected by a Faraday cup. The magnet also allows to determine the electron energy and the Faraday cup is used to measure the beam charge. The beam position and size can be monitored by several scintillating targets using TV cameras. A Si(Li) detector located in forward direction ($\theta = 0°$) detects CR and a second one placed under an angle of $\theta = 43°$ with respect to the beam is used for the observation of PXR.

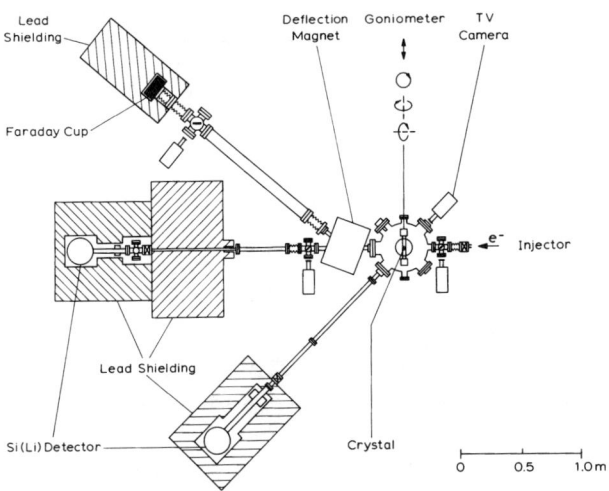

FIGURE 1. Low energy channeling and PXR site at the S-DALINAC as an example of a typical experimental setup for the observation of CR in direction of the electron beam and PXR under any direction, here 43°.

Spectra. A typical CR spectrum (36) which has been obtained using a diamond crystal of 55 μm thickness at 6.8 MeV electron energy is displayed in Fig. 2. The upper curve labeled (110) plane represents the spectral distribution observed if the crystal is oriented to meet the condition for planar channeling of electrons in this case along the (110) plane. The lower curve displays the background of BS at crystal orientations where the channeling condition is not fulfilled for any crystal plane or axis, i.e. a random orientation. Obviously, the difference of the two curves equals the radiation enhancement due to CR. The prominent line at 5 keV labeled 1-0 is caused by the transition of channeling electrons between the first excited bound level ($n = 1$) and the ground state ($n = 0$) of the crystal potential, respectively.

A typical PXR spectrum obtained under the same experimental conditions as the described ones for the CR spectrum but an observation angle of $\theta = 43°$ is displayed in Fig. 3. A prominent line appears at a photon energy of 8.2 keV due to the reflection of virtual photons at the (111) planes of diamond. The FWHM of the observed line amounts to about 300 eV and is mainly caused by the detector resolution, (264 ± 10) eV at this photon energy. The extremely small background is most likely caused by Bremsstrahlung (BS). If the crystal is turned, the photon energy of the prominent line changes as will be shown below.

From a comparison of the two spectra it becomes evident that CR is about 1000 times more intense than PXR. On the other hand the main advantage of PXR in relation to CR is the low background, the extremely narrow linewidth (see below) and the large range within which the PXR photon energy can be varied smoothly.

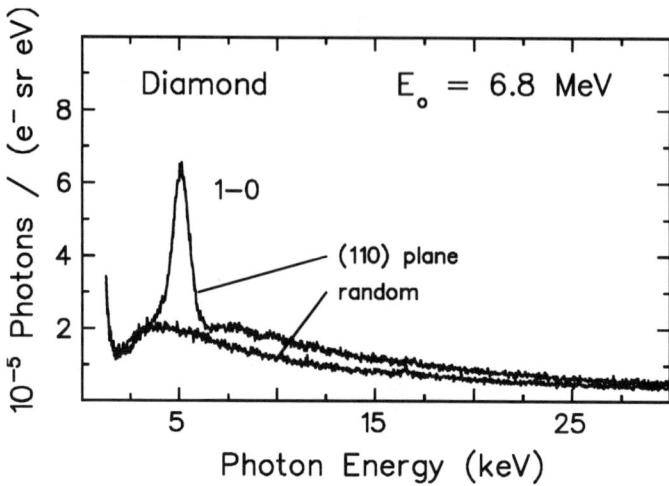

FIGURE 2. Typical planar CR and BS spectra obtained at 6.8 MeV electron energy and recorded with a Si(Li) detector placed under 0° with respect to the electron beam axis. The radiation occurring due to the 1-0 transition of planar channeling electrons along the (110) plane exceed the background of BS about 3 times at this energy. The upper spectrum was obtained with the diamond crystal oriented along the (110) plane, while the lower spectrum shows the contribution of BS obtained with the crystal in random position.

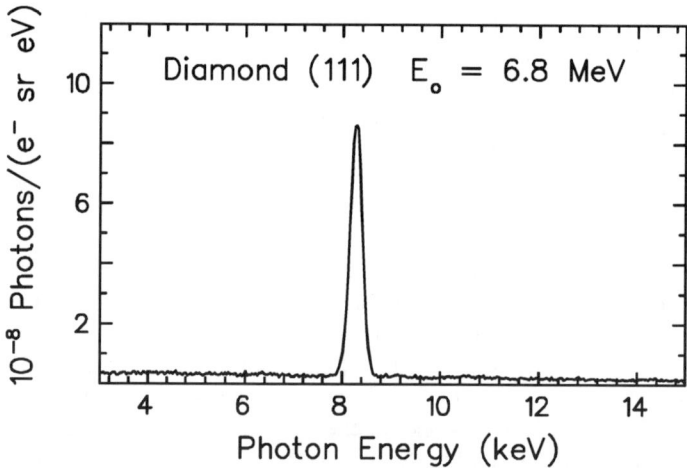

FIGURE 3. Typical PXR spectrum obtained at 6.8 MeV electron energy obtained by a Si(Li) detector placed under 43° with respect to the electron beam axis. Note that the FWHM of the prominent line of about 300 eV is mainly caused by the detector resolution of 264 eV.

For the evaluation of the data the background is subtracted to deduce the number of PXR or CR photons. The resulting spectra originate only from PXR or CR, respectively. In the case of CR the spectra are fitted taking into account the detector response function and the Lorentzian shape of the CR lines to deduce the observables, i.e. position, width and number of photons, of a single transition of channeling electrons in the crystal potential. In the case of PXR the photon energy and the linewidth of the peaks are given by the center of gravity and the variance of the distribution of PXR photons, respectively, which can be obtained after the background subtraction.

RADIATION FEATURES

In this section, the main features of CR and PXR are described based on experimental results obtained at low electron energies. The recent progress in both fields is presented.

Channeling Radiation

In recent years major progress has been achieved regarding the theoretical calculation of several CR properties (7). It is now possible to calculate the entire CR spectrum - consisting of free-to-free, free-to-bound and bound-to-bound transitions - very accurately for electron energies of about 9 MeV and probably also above (37). For lower energies the agreement is not quite as good. A comparison of the CR spectrum detected at 6.8 MeV (Fig. 2) and the theoretical prediction shows satisfactory agreement over the whole energy range, but the intensity of the prominent 1-0 transition is somewhat overestimated (Fig. 4). Here the solid line represents the CR spectrum with BS subtracted, the shaded area is the theoretical result. The difference between experiment and theory might be related to electron scattering at impurities in the diamond crystal. A detailed description of CR including the absolute calculation of CR properties (37) has recently been published.

It has been demonstrated (38) that the photon energy of CR can be changed by variation of the electron energy. The photon energy was found to scale according to $\gamma^{3/2}$. Another possibility to vary the photon energy is related to the Doppler shift (Eq. (1)). It has recently been observed that a swift change of the crystal orientation results in a change of the photon energy while the electron beam parameters are kept constant (Fig. 5). The photon energy has its largest value for electron entering the crystal along the (110) plane, i.e. the tilt angle $\psi = 0°$, and it decreases by about 500 eV if the crystal is turned out of this condition. The solid line in Fig. 5 represents a calculation using Eq. (1). The agreement of the theoretical dependence with the experimental findings is again quite satisfying.

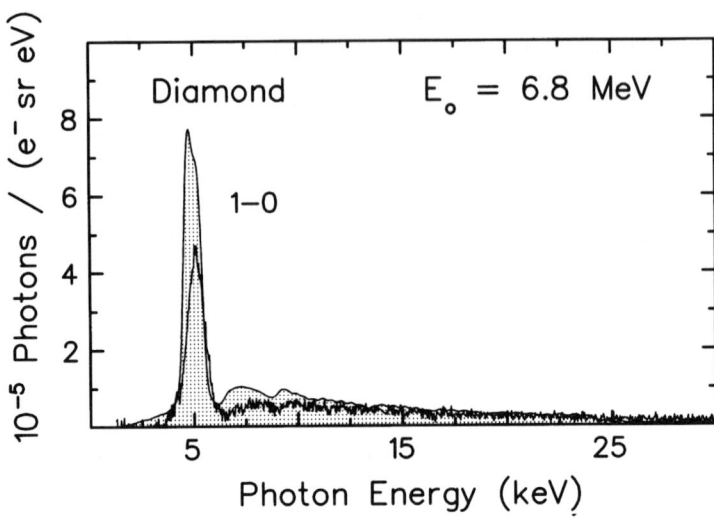

FIGURE 4. Comparison of experimental and theoretical CR spectrum. The shaded area indicates the calculated intensity summing up some 300 transitions, i.e. beside the bound 1-0 transition a large number of free-to-bound and free-to-free transitions.

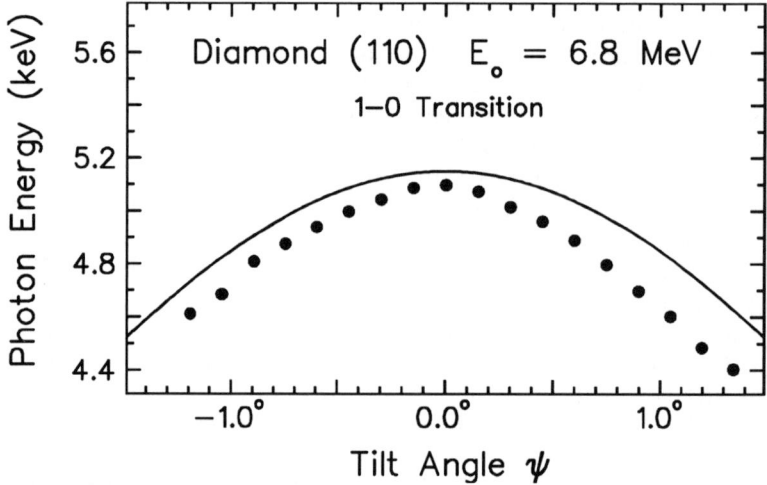

FIGURE 5. Variation of the CR photon energy as function of the tilt angle ψ which is the angle between electron beam direction and the (110) plane. The solid line represents the theoretical prediction of the variation caused by the angle dependence of the Doppler-shift according to Eq. (1).

However, it should be pointed out that the change of the photon energy is accompanied by a change of the number of photons emitted. At one degree away from $\psi = 0°$ the number of photons is about 10 times smaller than at the central position (36-37). Nevertheless, with respect to applications it might be useful to vary the photon energy by taking advantage of the orientational dependence due to the Doppler shift instead of the $\gamma^{3/2}$ dependence on the electron energy. Changing the electron energy usually involves adjusting the magnets in the accelerator which can be a time-consuming task.

For completeness it should be mentioned that the linewidth of CR is about 700 eV at 6.8 MeV electron energy (36-37) and varies according to γ^2. The number of photons per electron and steradian amounts to 1.7×10^{-2} photons/e$^-$/sr at this electron energy (36). At 9.0 MeV the maximum number of photons obtained so far, $7.7 \cdot 10^{-2}$ photons/e$^-$/sr, has been found. The number of photons scales, however, according to $\gamma^{5/2}$ (37, 39). Finally, it is worth to mention that it could be shown recently for the first time that planar CR is 100% linearly polarized (40).

Parametric X radiation

The most promising feature of PXR with respect to applications is the simplicity of changing the photon energy within a large range. A change of the crystal orientation leads to a variation of the energy as shown in Fig. 6. It varies according to $\sin\phi$ and $\cos\alpha$ as predicted by Eq. (2) and as indicated by the solid lines.

The energy range within the PXR energy can be varied depends on the observation angle and the reciprocal lattice vector, i.e. the crystallographic plane used to produce PXR. In the case of the (111) plane of diamond at $\theta = 43°$ the photon energy can be adjusted continuously in the range from 5 keV to 12 keV, for the (220) plane it increases from 8 keV to 19 keV. At different observation angles photon energies between 3 keV and 400 keV have been observed (31).

Large progress has been achieved regarding the knowledge about the lineshape and -width of PXR. The shape of the PXR line at low electron energies has been determined recently using an absorption technique (41,42) and is presented in Fig. 7. The experimental findings and their uncertainties are represented by the shaded area. It becomes apparent that the spectral line is asymmetric and its variance amounts to (49 ± 5) eV. The result of a Monte-Carlo simulation of the lineshape is represented by the solid line gives excellent agreement.

Under the experimental conditions of Fig. 7 the linewidth and -shape are mainly caused by multiple electron scattering, the mosaic structure of the crystal and the finite circular detector opening of $\Delta\Omega = 8.3 \cdot 10^{-6}$ sr. At higher electron energies the influence of multiple scattering decreases and at 855 MeV a linewidth, which is of the order of 1 eV, has been detected (43).

Furthermore, the linewidth depends also on the orientation of the crystal (41-42). The dependence of the linewidth on the tilt angle α shows a minimum near the

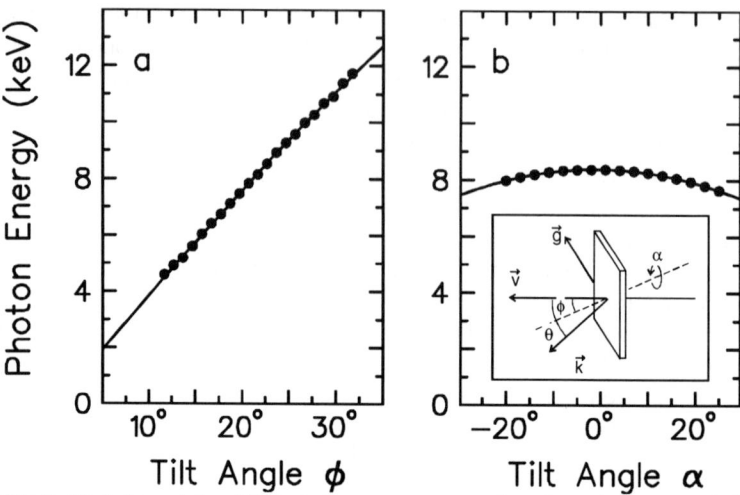

FIGURE 6. Variation of the PXR photon energy as a function of the crystal tilt angles ϕ and α at an electron energy of 6.8 MeV for diamond. The insert on the right side represents schematically the geometry including electron beam direction and crystal orientation. The reciprocal lattice vector $\vec{g}$ and the dashed line are perpendicular to each other. The dashed line lies in the plane defined by the electron velocity $\vec{v}$ and the wave vector $\vec{k}$.

center of the PXR reflection ($\phi \approx \theta/2$, $\alpha = 0°$) and it increases with increasing angular distance from this position due to the rising influence of multiple scattering of electrons inside the crystal (Fig. 8). The PXR linewidth σ can be deduced from the variance σ_{obs} extracted from the spectrum using the relation $\sigma = (\sigma_{obs}^2 - \sigma_{det}^2)^{1/2}$, where σ_{det} is the variance of the detector response function, i.e. the resolution divided by 2.35. Again, the result of a Monte-Carlo simulation (solid line in Fig. 8) is in perfect agreement with the experimental findings.

Since the first experimental results yielding absolute numbers of PXR photons produced at low electron energies became available (18,34) the different approaches of theoretical predictions could be tested. For this purpose the angular distribution of PXR will be compared to theoretical predictions.

A typical angular distribution is presented in Fig. 9. It exhibits two maxima of different height and a minimum at its center. The solid line represents the result of a Monte-Carlo simulation based on Eq. (3) taking into account multiple electron scattering, mosaic crystal structure, electron beam characteristics and the finite detector opening. With regard to the fact that the theoretical curve is absolute and not fitted to the data points, the agreement between experiment and theory can be considered very good. Thus, the kinematical theory (8-11,14) describes the experimental findings well, if multiple electron scattering is taken into account. A similar result has already been found at higher electron energies (29-30).

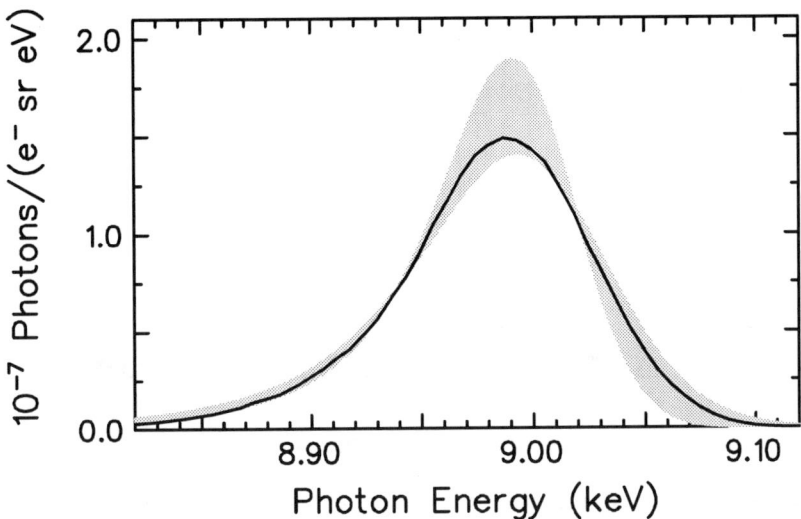

FIGURE 7. PXR lineshape obtained by bombarding a diamond crystal with electrons of 6.8 MeV. The solid line represents the result of a Monte-Carlo simulation based on Eq. (3) and taking into account multiple scattering of electrons, mosaic structure of the crystal, finite detector solid angle and various beam characteristics, the shaded area the experimental results.

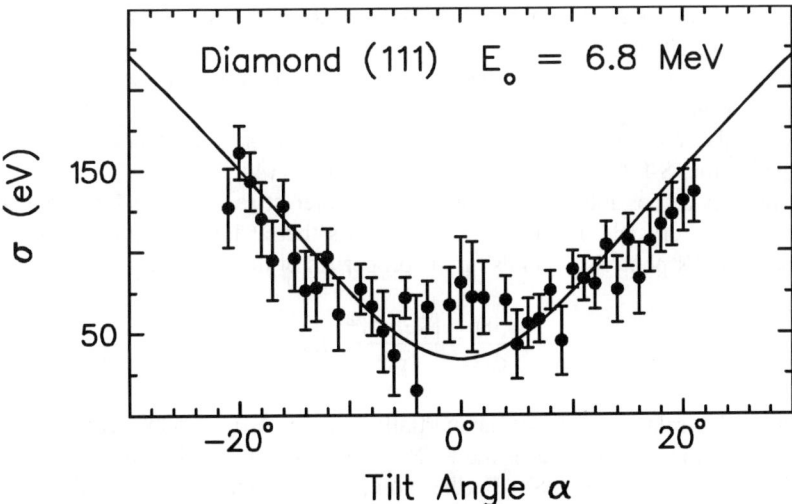

FIGURE 8. Variance σ of the PXR line as a function of the tilt angle α. The solid line represents the result of a Monte-Carlo simulation.

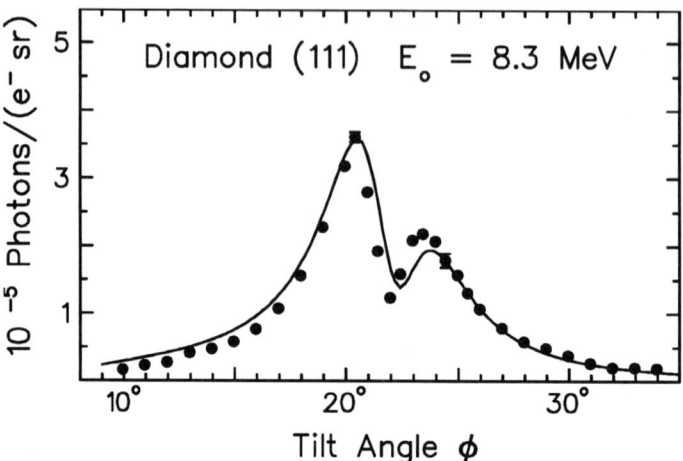

FIGURE 9. Number of PXR photons per electron and steradian as a function of the tilt angle. The solid line represents the result of a Monte-Carlo simulation.

For low and intermediate electron energies up to 30 MeV it was demonstrated (18, 34) that the height of the angular distribution increases according to γ^2 and the distance of the two maxima, i.e. the angular width, scales with $1/\gamma$. In the case of increasing electron energies the angular width decreases. The height as well as the angular width reaches a constant value which depends only on the dielectric susceptibility of the crystal and the possible influence of multiple scattering of the electrons (24, 29-30)).

Furthermore, PXR is linearly polarized. The polarization direction has an angular dependence and a degree of linear polarization up to 82 % has been found (26). Theoretically an even higher degree of linear polarization is expected and new data obtained at the S-DALINAC are currently evaluated (44).

With respect to possible applications it is of interest to compare PXR radiation from different crystals. From Eq. (3) it follows that at low electron energies the number of PXR photons depends on the properties of the crystal according to

$$\frac{dN}{d\Omega} \sim \frac{|S_N F|^2 f^2}{g^3} \cdot L. \qquad (4)$$

Recently, this relation (Eq.(4)) has partially been investigated in different works. The temperature dependence of the PXR intensity due to the Debye-Waller factor has been examined in (28). Furthermore, the proportionality to the crystal thickness L is only valid for thin crystals. With increasing crystal thickness the influence of absorption becomes important and the crystal thickness L in Eq. (4) must be replaced (10, 12, 45) by $L_a \cdot (1 - \exp(-L/L_a))$ with L_a being the absorption length of the crystalline medium.

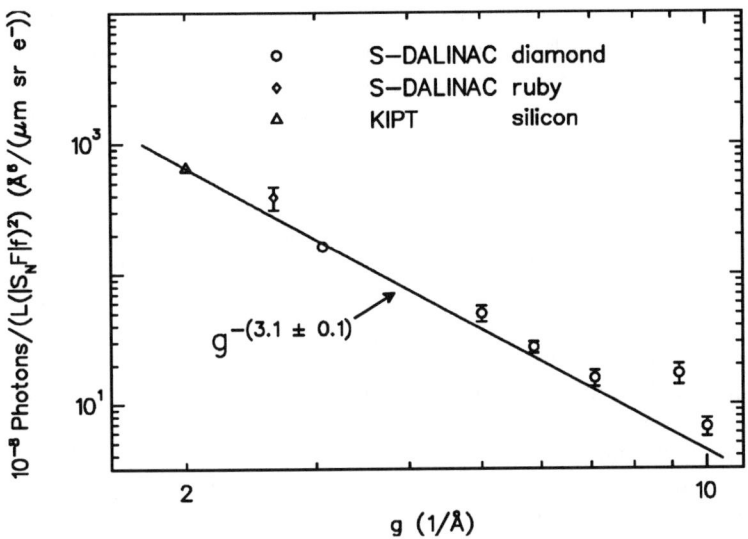

FIGURE 10. Dependence of the number of PXR photons divided by the structure factor, the atomic form factor, the crystal thickness and the Debye-Waller factor on the size of the reciprocal lattice vector. The point for silicon has been taken from (34). The solid line represents a fit of a function $A \cdot g^{-B}$ to the experimental data points.

From Eq. (4) it becomes apparent that $dN/(d\Omega |S_N F|^2 f^2 L) \sim g^{-3}$. Recently, this relation has been investigated for all absolute data points of PXR observed with electron energies below 30 MeV (Fig. 10). The results of (34) have been scaled to $E_o = 6.8$ MeV using the γ^2 scaling law. A fit of a function $A \cdot g^{-B}$ to the various data points, were A and B are fitting parameters, yields $B = 3.1 \pm 0.1$ which is in good agreement with the theoretical prediction of Eq. (4).

In a frequently quoted paper (46) dealing with approximations regarding radiation intensities for various processes simplified expressions were deduced which predicted that PXR under certain conditions might reach a photon flux of 10^{-5} Photons/e^- similar to synchrotrons. Using our experimental results scaled to the conditions of (46) we obtained a value of about 10^{-7} photons/e^-, which is two orders of magnitude less. Thus, our findings do not confirm the result of (46) and apparently the PXR intensity is not comparable to that of synchrotron radiation.

DISCUSSION AND OUTLOOK

CR and PXR have been investigated extensively in recent years. Both processes can now be regarded as basically understood and only a few questions concerning special conditions, i.e. parametric surface radiation (47), the crystal thickness limiting the CR intensity (37) and the Bragg-reflection of CR (48), remain. The theoretical descriptions have been improved and allow to estimate quite accurately the expected radiation properties for applications.

With regard to the most frequently discussed application digital subtraction angiography (DSA) it became apparent (49) that PXR and CR at low electron energies do not yet produce enough photons for the currently used method (4), which requires $3 \cdot 10^{11}$ photons/s/mm^2 within an energy band of 300 eV close to a photon energy of 33 keV. For comparison, CR produced by 21 MeV electrons in a diamond crystal of 55 μm thickness using 60 μA beam current, which are typical parameters at linear electron accelerators, yields about $2 \cdot 10^8$ photons/s/mm^2 at 1 m distance of the crystal. The PXR photon flux is about 1000 times smaller than the CR flux.

Nevertheless, there might be possibilities to increase the intensity of CR and PXR. In the first case, the spectral distribution of CR depends strongly on the shape of the crystal potential. If an almost harmonic crystal potential could be found, all transitions should contribute to the same photon energy leading to an intensity enhancement in a small energy region. In the case of PXR instead of one crystal a stack of well aligned crystals might be used to produce PXR. In this case, if an appropriate geometry is applied, the effective thickness of the crystal should not be limited by the absorption length of the crystal medium but by the range of the electron in the material. In addition, a pulsed accelerator might be used to achieve a larger average beam current for a time when, for example, an image for DSA is taken.

However, if a possibility of increasing the CR or PXR intensity is found in the future additional problems occur with respect to DSA. As shown above, CR is always accompanied by BS. Thus, absorbers or monochromators are necessary to reject photons with undesired energy which lead to a reduction of intensity in the desired energy band. This problem might possibly be overcome by the use of advanced X-ray optical devices with extremely high transmission.

A candidate for such a device might be a Kumakhov lens (50), as has recently been suggested (51). These lenses effects only the soft part of a spectrum. If a bent lens is applied the soft part can be guided to a large angle with respect to the incident photon beam, while the hard part remains unaffected. In combination with an appropriate absorber the system could work as a band path filter. Therefore, investigations using Kumakhov lenses have recently being initiated at the S-DALINAC.

In the case of PXR the observation angle must be of the order of 10° or less depending of the crystal to obtain a photon energy of 33 keV or larger. At this angle in the low electron energy range BS cannot be avoided resulting in a continuous background due to incoherent BS. Due to this fact, a larger electron energy in the order of 100 MeV must be chosen to separate PXR and BS.

Furthermore, applications of scientific and medical interest appear to be possible. The CR and PXR intensities seem to be sufficient to realize a new X-ray source for mammography (1). Additionally, as has recently been pointed out (52), the influence of the electron beam characteristics on the angular distribution of PXR might be used as a tool to get information on the electron beam divergence. The rather high spectral density of PXR (41-42) might be useful whenever a large amount of radiation in a small energy band is necessary and synchrotron radiation with crystal monochromators is not available.

Recently, an investigation of PXR under an observation angle of 180° has been started at the S-DALINAC. Under this angle PXR may possibly be applied to produce X rays with sufficient intensity in the water window, i.e. the energy range between about 280 eV and 530 eV. Within the water window it is possible to observe living cells in water (53), what is important for biological research. In a first experiment the existence of PXR of energies below 3 keV produced at 180° by silicon and beryllium has been demonstrated. The results encourages the hope that it should be possible to achieve the required photon energies if a crystal with a lattice spacing between 12 – 22 Å is used.

CONCLUSION

Our results and those of other groups (1-51) demonstrate that the mechanisms leading to CR and PXR are well understood and can be used to produce coherent, monoenergetic X radiation of tunable photon energy. Even an electron energy of less than 10 MeV is sufficient to produce these types of radiation. So far, the number of photons generated by one crystal is not sufficient to compete with synchrotron sources. Future investigations have to solve the questions about special conditions, e.g. the detailed investigation of parametric surface radiation, the interference of PXR and CBS or the crystal thickness limiting the CR intensity, and also possible applications including mammography should be realized.

ACKNOWLEDGMENTS

The results described here were obtained by a cooperation of many people and groups. I am particularly thankful to L. Groening, P. Hoffmann-Stascheck, and U. Nething for providing data partially prior to publication. The Monte-Carlo simulations have been performed by V.V. Morokhovskii. The CR calculation originates from M. Weber. Furthermore, I want to express my gratitude for numerous and helpful discussion with H. Genz, H.-D. Gräf, V.L. Morokhovskii, H. Nitta, A. Richter and R. Zahn. The collaboration with members of the groups from Kharkov, München and Johannesburg has been extremely rewarding. The support by BMBF under contract no. 06DA665I, DFG contract no. 436UKR113-19, and the FAZIT foundation is gratefully acknowledged.

REFERENCES

1. Knüpfer, W., *Nucl. Instrum. Methods B* **87**, 98-103 (1994).
2. Fiorito, R.B., Rule, D. W., Piestrup, M. A., Qiang Li, Ho, A. H., Maruyama, X. K., *Nucl. Instrum. Methods B* **79**, 758-761 (1993).
3. Genz, H., Gräf, H.-D., Hoffmann, P., Lotz, W., Nething, U., Richter, A., Kohl, H., Weickenmeier, A., Knüpfer, W., and Sellschop, J. P. F., *Appl. Phys. Lett.* **57**, 2956-2958 (1990).
4. Besch, H.-J., *Nucl. Instrum. Methods A* **360**, 277-282 (1995).
5. Hau, L. V. and Andersen, J. U., *Phys. Rev. A* **47**, 4007-4032 (1993).
6. Andersen, J. U., Bonderup, E., and Pantell, R. H., *Ann. Rev. Nucl. Part. Sci.* **33**, 453-504 (1983).
7. Weber, M., Dissertation, Universität Erlangen, Theoretische Physik II, 1995.
8. Nitta, H., *Phys. Lett. A* **158**, 270-274 (1991).
9. Nitta, H., *Phys. Rev. B.* **45**, 7621-7626 (1992).
10. Nitta, H., *Nucl. Instrum. Methods B* **115**, 401-404 (1996).
11. Morokhovskii, V. L., *Coherent X-rays of Relativistic Electrons in Crystals*, CSRI atominform, Moscow, **39** (1989). A Translation is available on request by the author.
12. Feranchuk, I. D., and Ivashin, A. V., *J. Physique* **46**, 1981-1986 (1985).
13. Baryshevskii, V. G., Grubich, A. O., and Le Tien Hai, *JETP* **67**, 895-902 (1988).
14. Ter - Mikaelian, M. L., *High Energy Electromagnetic Processes in Condensed Media*, New York: Wiley, 1972, ch. 5, sec. 28, pp. 332-334.
15. Adejshvily, D. I., Gavrikov, V. B., Morokhovskii, V. L., *About Interference between Parametric X-ray Radiation of Type B and Coherent Bremsstrahlung of a Fast Particle in a Crystal*, Kharkov(Kharkov Institute of Physics and Technology): Preprint KFTI 96-2, 1996, pp. 1-9.
16. Blazhevich, S. V., Bochek, G. L., Gavrikov, V. B., Kulibaba, V. I., Maslov, N. I., Nasonov, N. N., Pirogov, V. N., Safronov, A. G., and Torgovkin, A.V., *JETP Lett.* **59**, 524-526 (1994).
17. Fainberg, IA. B., and Khizhniak, N. A., *JETP* **5**, 720-729 (1957).
18. Freudenberger, J., Gavrikov, V. B., Galemann, M., Genz, H., Groening, L., Morokhovskii, V. L., Morokhovskii, V. V., Nething, U., Richter, A., Sellschop, J. P. F., and Shul'ga, N. F., *Phys. Rev. Lett.* **74**, 2487-2490 (1995).
19. Particle Data Group, *Phys. Rev. D* **45/II**, III.15 (1992).
20. Kumakhov, M. A., *Phys. Lett. A* **57**, 17-18 (1976).
21. Alguard, M. J., Swent, R. L., Pantell, R. H., Berman, B. L., Bloom, S. D., and Datz, S., *Phys. Rev. Lett.* **42**, 1148-1151 (1979).
22. Vorob'ev, S. A., Kalinin, B. N., Pak, S., and Potylitsyn, A. P., *JETP Lett.* **41**, 1-4 (1985).
23. Kumakhov, M. A., Komarov, F. F., *Radiation From Charged Particles in Solids*, New York: AIP, 1989, ch. 2.
24. Adishchev, Yu. N., Didenko, A. N., Mun, V. V., Pleshkov, G. A., Potylitsin, A. P., Tomchakov, V.K., Uglov, S. R., and Vorobiev, S. A., *Nucl. Instrum. Methods B* **21**, 49-55 (1987).
25. Avakyan, R.O., Avetisyan, A. E., Adishchev, Yu. N., Garibyan, G. M., Danagulyan, S. S., Kizogyan, O. S., Potylitsyn, A. P., Taroyan, S. P., Elbakyan, G. M., and Yan Shi, *JETP Lett.* **29**, 396-399 (1987).
26. Adishchev, Yu. N., Verzilov, V. A., Vorob'ev, S. A., Potylitsyn, A. P., and Uglov, S. R., *JETP Lett.* **48**, 342-346 (1988).
27. Afanasenko, V. P., Baryshevskii, V. G., Zuevskii, R. F., Lobko, A. S., Moskatel'nikov, A. A., Panov, V. V., Potsuiluiko, V. P., Skorokhod, S. V., and Shvarkov, D. S., *JETP Lett.* **54**, 494-497 (1991).

28. Amosov, K. Yu., Kalinin, B. N., Potylitsin, A. P., Sarychev, V. P., Uglov, S. R., Verzilov, V. A., Vorobiev, S. A., Endo, I. and Kobayashi, T., *Phys. Rev. E* **47**, 2207-2209 (1993).
29. Asano, S., Endo, I., Harada, M., Ishii, S., Kobayashi, T., Nagata, T., Muto, M., Yoshida, K., and Nitta, H., *Phys. Rev. Lett.* **70**, 3247-3250 (1993).
30. Amosov, K. Yu., Kalinin, B. N., Kustov, D. V., Naumenko, G. A., Potylitsin, A. P., Vnukov, I. E., Verzilov, V. A., Endo, I. and Yoshida, K., Characteristics of parametric X-ray radiation near threshold, in *Proceedings of the International Symposium on Radiation of Relativistic Electrons in Periodical Structures*, Tomsk, Russia, 1993, pp. 53-61.
31. Morokhovskii, V. L., and Shchagin, A. V., *Sov. Phys. Tech. Phys.* **35**, 623-624 (1990).
32. Shchagin, A. V., Pristupa, V. I., and Khizhnyak, N. A., *Phys. Lett. A* **148**, 485-488 (1990).
33. Fiorito, R. B., Rule, D. W., Maruyama, X. K., DiNova, K. L., Evertson, S. J., Osborne, M. J., Snyder, D., Rietdyk, H., Piestrup, M. A., Ho, A. H., *Phys. Rev. Lett.* **71**, 704-707 (1993).
34. Shchagin, A. V., Pristupa, V. I., and Khizhnyak, N. A., Absolute Differential Yield of Parametric X-Ray Radiation, in *Proceedings of the International Symposium on Radiation of Relativistic Electrons in Periodical Structures*, Tomsk, Russia, 1993, pp. 62-75.
35. Auerhammer, J., Genz, H., Gräf, H.-D., Hahn, R., Hoffmann-Stascheck, P., Lüttge, C., Nething, U., Rühl, K., Richter, A., Rietdorf, T., Schardt, P., Spamer, E., Thomas, F., Titze, O., Töpper, J., and Weise, H., *Nucl. Phys. A* **553**, 841c-844c (1993).
36. Groening, L., Diplomarbeit, Technische Hochschule Darmstadt, Institut für Kernphysik, Darmstadt, 1995, ch. 6, p. 40-41.
37. Genz, H., Groening, L., Hoffmann-Stascheck, P., Richter, A., Höfer, M., Hormes, J., Nething, U., Sellschop, J. P. F., Toepffer, C., and Weber, M., *Phys. Rev. B* **53**, 8922-8936 (1996).
38. Lotz, W., Genz, H., Hoffmann, P., Nething, U., Richter, A., Weickenmeier, A., Kohl, H., Knüpfer, W., Sellschop, J. P. F., *Nucl. Instrum. Methods B* **48**, 256-259 (1990).
39. Nething, U., Galemann, M., Genz, H., Höfer, M., Hoffmann-Stascheck, P., Hormes, J., Richter, A., and Sellschop, J. P. F., *Phys. Rev. Lett.* **72**, 2411-2413 (1994).
40. Rzepka, M., Buschhorn, G., Dietrich, E., Kotthaus, R., Kufner, W., Roessl, W., Schmidt, K. H., Hoffmann-Stascheck, P. Genz, H., Nething, U., Richter, A., and Sellschop, J. P. F., *Phys. Rev. B* **52**, 771-777 (1995).
41. Freudenberger, J., Galemann, M., Genz, H., Groening, L., Hoffmann-Stascheck, P., Morokhovskii, V. L., Morokhovskii, V. V., Nething, U., Prade, H., Richter, A., Sellschop, J. P. F., and Zahn, R., *Nucl. Instrum. Methods B* **115**, 408-410 (1996).
42. Freudenberger, J., Morokhovskii, V. L., Morokhovskii, V. V., Nething, U., Richter, A., Sellschop, J. P. F., and Zahn, R., *First Determination of the Shape and Width of Parametric X-Radiation Observed at Low Electron Energy*, submitted to *Appl. Phys. Lett.*.
43. Backe, H., Brenzinger, K.-H., Dambach, S., Euteneuer, H., Herberg, C., Kaiser, K. H., Knies, G., Kube, G., Lauth, W., Limburg, B., Schöpe, H., Walcher, Th., und die X1-Kollaboration, Verhandl. DPG (VI) 31, 1030 (1996).
44. Schmidt, K. H., private communication.
45. Endo, I., Harada, M., Kobayashi, T., Lee, Y. S., Ohgaki, T., Takahashi, T., Muto, M., Yoshida, K., Nitta, H., Potylitsin, A. P., Zabaev, V. N., and Ohba, T., *Phys. Rev. E* **51**, 6305-6308 (1995).
46. Baryshevsky, V.G., and Feranchuk, I.D., *Nucl. Instrum. Methods* **228**, 480-495 (1985).
47. Andriyanchik, A. A., Baryshevsky, V. G., Kaminsky, A. N., *Nucl. Instrum. Methods B* **83**, 482-494 (1993).
48. Matsuda, Y., Ikeda, T., Nitta, H., Minowa, H., and Ohtsuki, Y. H., *Nucl. Instrum. Methods B* **115**, 396-400 (1996).
49. Freudenberger, J., Genz, H., Groening, L., Hoffmann-Stascheck, P., Knüpfer, W., Morokhovskii, V. L., Morokhovskii, V. V., Nething, U., Richter, A., and Sellschop, J. P. F., Channeling Radiation and Parametric X-Radiation at Electron Energies below 10 MeV, presented at the Workshop on CHANNELING AND OTHER COHERENT CRYSTAL

EFFECTS AT RELATIVISTIC ENERGY, Aarhus, Denmark, 1995, expected to be published in *Nucl. Instrum. Methods B* (1996).
50. Kumakhov, M. A., and Komarov, F. F., *Phys. Rep.* **191**, 289-350 (1990).
51. Kumakhov, M.A., Poturaev, S.V., Stirin, A.I., and Khatkov, T.A., Possibilities of using a new wide-band X-ray capillary optics in channeling radiation experiments, in *Optics of Beams* edited by M.A. Kumakhov, 1993, pp. 86-93.
52. Kalinin, B. N., Potylitsin, A. P., Verzilov, V. A., Vnukov, I. E., Endo, I., Harada, M., Kobayashi, T., Kuwamoto, T., Yoshida, K., Muto, M., and Nitta, H., *Nucl. Instrum. Methods A* **350**, 601-604 (1994).
53. Rymell, L., Berglund, M., and Hertz, H. M., *Appl. Phys. Lett.* **66**, 2625-2627 (1995).

III. HIGHLY CHARGED IONS

Highly Charged Ions in Storage Rings

D. Liesen

GSI - Darmstadt
Planckstr. 1, D-64291 Darmstadt, Germany

Abstract. The possibilities for experiments with highly-charged ions have decisively been enlarged by the new generation of heavy-ion storage rings employing cooling of the stored beams. Some of the basic operation principles of heavy-ion storage rings and of electron cooling as the most commonly used method will be discussed. In combination with appropriate in-beam targets, the cooled, brilliant beams in the heavy-ion storage ring ESR offer a unique access to the determination of the binding energies of hydrogen-like, very heavy ions by the precise measurement of the energies of corresponding x-ray transitions. These investigations provide an experimental access to a special domain of quantum electrodynamics (QED), namely the radiative corrections to the binding energies of states bound in very strong nuclear Coulomb fields. In contrast to light ions, the electrons are relativistic and nonperturbative methods have to be applied for the calculations of the QED effects in order to properly treat the interaction between the electron and the nucleus. Since these corrections are largest for electrons in the ground states, special emphasis is given to measurements of x-ray transitions into the ground states of hydrogen-like ions up to uranium. The present status of the experimental and theoretical results will be discussed together with further experimental perspectives.

INTRODUCTION

The interest in the study of multiply charged ions was certainly one of the major driving forces for the development of ion storage rings, because such ions have always been in the focus of plasma physics, astrophysics, atomic physics, and theoretical physics. This is demonstrated in the following examples. Ions in high charge states occur in high-temperature plasmas in a laboratory as well as in hot stellar plasmas. From the distribution of ionic charge states many parameters can be derived which are important for the characterisation of the plasma as temperature, composition, densities of the charge carriers, reactivities of the components etc.. The nucleosynthesis in hot stellar plasmas is influenced by the degree of ionisation of the constituents. Depending on the Q-value of the nuclear reaction and on the charge state, some of the constituents may undergo a beta decay, in which the emitted electron ends up in a bound state. Thereby, stable

© 1997 American Institute of Physics

neutral elements can decay, if they are highly ionised (1). Ion impact on living cells can lead to lethal effects due to DNA breaking. Recent investigations indicate that this breaking is a consequence of the decay of inner-shell vacancies which are produced with high probabilities by ion impact (2).

Unfortunately, in most cases the environments are by far too complicated and inaccessible for a detailed study of the many individual processes occurring and their sequence. A consistent overall explanation of the phenomena observed is obviously only possible, if the underlying elementary processes are understood, at least the essential ones. In many cases it is even difficult to decide which processes are the essential ones. At this point at the latest atomic and theoretical physics come into play, because most of these elementary processes have been and are subject of extensive experimental and theoretical investigations in these fields. It is probably fair to state that the overwhelming part of this research originates from pure scientific interest, not looking to possible applications. The study of inner-shell vacancy production may serve as a typical example: The solving of the mechanisms of inner-shell vacancy production in ion-atom collisions was a fundamental problem for experimental and theoretical atomic physics, not having a presentiment of its importance for the process of DNA breaking.

The essential prerequisite for an experiment with highly charged ions attacking basic questions is a machine which provides ions of sufficient intensity in the desired charge states. Depending on the experiment to be performed, the degree of ionisation may vary from singly ionised to bare ions and from low nuclear charges Z up to uranium with $Z = 92$. In addition, the velocity of the ions as well as their velocity distribution has to be known accurately in order to be able to correct for possible kinematic effects like the Doppler effect and the transformation of solid angles. The accuracy of those corrections generally increases with decreasing velocity. Therefore, a low velocity of the ions is desired in many experiments, especially when high precision is aimed at. High precision implies in most cases a low efficiency of the detecting system. In order to get sufficient statistics in reasonable times, the luminosity should be as large as possible.

Nowadays two different types of such machines are available for experiments covering the whole range of nuclear charges and ionic charge states, each of which having its own figures of merit: Electron beam ion traps (EBIT) (3) and heavy ion storage rings. In an EBIT, highly charged ions are produced and trapped in electron beams of high current densities. The ions are at rest in the laboratory system and kinematic effects are therefore negligibly small. However, the charge state definition is not unique, because usually several charge states are confined in the trap. In storage rings, the velocity of the highly charged ions is large, but the charge state of the ions is well defined. The EBIT and its use for the study of highly charged ions is addressed in the talk of Peter Beiersdorfer at this conference (4) and a comprehensive review of the subject may be found in (5). Therefore, we will discuss in the following only heavy ion storage rings and related experiments.

Storage rings have been widely used in nuclear and particle physics. Although their potential for investigations with highly charged ions has been recognised only a few years ago, the amount of data obtained so far is impressive and far beyond the scope of this article. For an overview of the current status the reader is referred to the recent papers by M. Larsson (6) and P. H. Mokler and Th. Stöhlker (7). We have chosen out of the broad spectrum a few topics and will demonstrate that still 100 years after their discovery by C. W. Röntgen x rays are a tool to attack fundamental problems in modern fields of research like Quantum Electrodynamics.

HEAVY-ION STORAGE RINGS

In the following chapter we will shortly present the basics of heavy-ion storage rings, the principles and the results of cooling, and the usefulness of these rings as a source for x rays.

Some Basics of Heavy-Ion Storage Rings

In contrast to an EBIT, the ions to be stored in a ring are produced not in the device itself, but are extracted from an external ion source. In order to obtain the desired charge state, they are accelerated in an injector and passed through a thin solid foil, usually a copper foil. According to Bohr's criterion, bound electrons with orbital velocities smaller than the ion velocity will be stripped in the foil. This means an acceleration to a velocity of about $\beta = 0.5$ (in units of the speed of light) to strip also the last electron from a uranium nucleus bound to it with an energy of about 130 keV. Out of the distribution of charge states behind the foil, the desired one has to be chosen by means of a system of magnetic deflectors and injected into the storage ring.

In the ring, the injected ions have to be stacked and accumulated by suitable kickers, and confined to stable, closed orbits by an array of bending and focusing magnets. This array is called the lattice of the ring. Eventually, rf-cavities are build in for a further acceleration or a deceleration of the beam. As an example, Fig. 1 shows schematically the main components of the storage ring ESR (Experimentier-Speicher-Ring) of the GSI Darmstadt having a circumference of about 108 m and a magnetic rigidity of 10 Tm. The ions are injected either from the fragment separator FRS or from the heavy ion synchrotron SIS; the stripper foil (not shown in Fig. 1) is mounted in the transfer line to the ESR. The ESR will be our guideline for the further discussion of heavy-ion storage rings, because it is world-wide the only ring in which ions of all nuclear charges Z in all charge states can be stored, even bare uranium ions. Besides the ESR four other storage rings for heavy ions are operating with atomic and molecular physics forming a major part of the experimental program: TSR in Heidelberg, Germany, TARN II in Tokyo, Japan, ASTRID in Aarhus, Denmark, and CRYRING in Stockholm, Sweden. All the heavy-ion storage rings are more or less the children of father LEAR (Low-Energy-Antiproton-Ring) at CERN which came into operation in

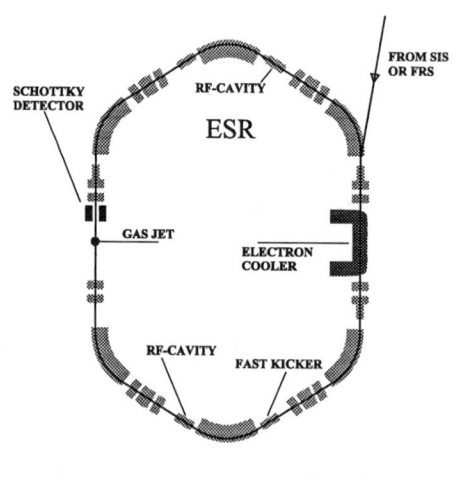

FIGURE 1: The storage ring ESR with its main components

1984.

The lattice of a ring has to be designed carefully in order to fulfil two essential conditions: To guarantee the stability of the motion of the ions and to provide appropriate places for in-ring experiments. Neither of these conditions is trivial. Stability of the orbits of the ions implies control of the overall focusing properties of the lattice, avoid of the blowing up of destructive resonances of the betatron oscillations, and of collisions with residual gas target atoms leading to a change of the ionic charge state. The latter process can strongly be reduced by maintaining a very good vacuum of 10^{-11} mbar or better in the ring. This may become rather difficult, especially if the ring is equipped with an internal gas jet target for experiments as it is the case for the ESR (see Fig. 1). The sine-like betatron oscillations originate from a deviation of the injected ions from the central orbit and an angular distribution around it and from a finite momentum distribution, in other words, from the transverse and longitudinal emittance of the injected beam. It can be shown that these emittances correspond to ellipses in the two-dimensional phase space (8). The shape of the ellipse changes along the position in the ring according to its focusing properties; the area of the ellipses stays constant unless the velocity of the particles changes. If the velocity changes, the area varies proportional to 1/p, where p is the momentum of the particles.

Unavoidable geometrical and magnetic field imperfections lead to an increase of the amplitude of these oscillations and thereby to a loss of stored particles, if the number of horizontal or vertical betatron oscillations per turn is an integer. The lattice has therefore to be designed in a way that the sum of these numbers (and integer multiples) which are called the horizontal and the vertical tune of the machine, differs from an integer. Shifts of the tune caused by the space charge of the stored beam have to be kept small for a stable operation of the machine, which in turn limits the number of particles that can be stored. This number is about proportional to the energy of the beam and proportional to the ratio between the ion mass and the square of the charge state. As an example we give numbers for the ESR (9): At 50 MeV/u beam energy ($\beta = 0.315$) $4.4 \cdot 10^8$, at 200 MeV/u ($\beta = 0.568$) $2.2 \cdot 10^9$, and at 556 MeV/u ($\beta = 0.78$) $9.3 \cdot 10^9$ bare uranium ions can be stored at maximum, respectively; the corresponding revolution frequency is in the order of MHz.

From an experimenter's point of view, the advantage of a storage ring consists of the possibility that those ions which did not react with the target, are not lost but can be recycled for a next passage through the target. This is the case, provided the lifetime of the beam in the ring is large enough, the velocity stays constant, the charge state doesn't change, and the quality with regard to the transverse and longitudinal emittance is preserved. At this point, cooling of the beam comes into play.

Electron Cooling of Ion Beams

The transverse and longitudinal emittance of the beam appear to an observer in a reference frame moving with the ion velocity as a random, three dimensional motion of the particles which he could describe by a temperature. An increase of the width of the velocity distribution or of the betatron amplitude for example due to reactions with target atoms, rest gas and intra-beam scattering means rising temperatures. In order to preserve or even increase the beam quality, the stored ions have to be cooled. G. Budker proposed 1966 (10) to cool heavy charged particles by overlaying a cold electron beam of the same velocity on a straight section of the ring. The Coulomb interaction between ions and electrons leads to a non-conservative friction force, so that the ions will be cooled and the electrons heated. Although the length of the cooling section is typically only (1 - 2) m, the circulating ions pass the cooler about 10^6 times per second and interact always with fresh cooling electrons. Thereby a thermodynamic equilibrium will be reached after a time of the order of seconds or below (11), mainly determined by the nuclear and the ionic charge. The final equilibrium is determined by the balance between cooling and the dominating heating process, namely intra-beam scattering. For a detailed description of electron cooling the reader is referred to the review by H. Poth (11). Budker's proposal which originally aimed at a luminosity increase in proton-antiproton colliders, was first successfully realised in 1974 at the NAP-M proton storage ring (12). LEAR was the first device which was equipped with an electron cooler as a fixed part of the ring.

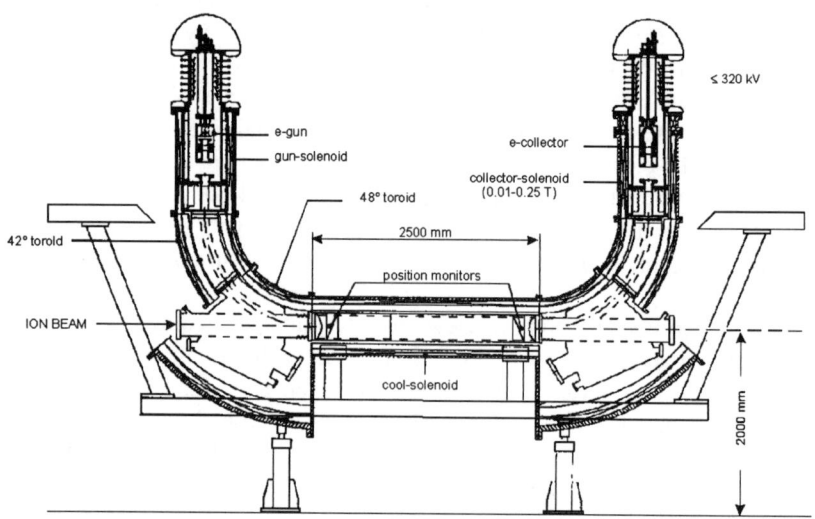

FIGURE 2: The electron cooler of the ESR

Electron cooling has proved to be a very effective method to cool heavy ions of all nuclear charges and in all charge states. This is the reason, why an electron cooler is installed in all of the above mentioned rings. Since the construction of the different coolers is quite similar, we show in Fig. 2 only the ESR cooler. The electrons are extracted from a cathode heated to about 1300 K and accelerated in a single gap. The electron beam is guided in a longitudinal magnet field provided by a solenoid in order to prevent blowing up of the beam due to its own space charge. Two toroids bend the beam onto the axis of the ion beam which is then imbedded in the electron beam on a length of 2.5 m. In the effective cooling section, the electron beam is again guided in the longitudinal magnetic field of the cooler solenoid. A second pair of toroids deflects the electrons towards the collector. Before hitting the collector, the beam is decelerated and afterwards widened up in order the reduce the thermal load of the collector. Tab. 1 gives the values of some of the most important parameters which are typical for the beam times up to now (13).

Table 1: Typical Operation Parameters of the ESR Electron Cooler

Diameter	50 mm
Current	(0.05 - 0.5) A
Energy	(27 - 200) keV
Transverse temperature	≈ 0.1 eV
Longitudinal temperature	< 0.1 meV
Magnetic field	(0.08 - 0.12) T

The ions are injected from the SIS into the ring with a relative momentum spread in the order of $\Delta p/p \approx 10^{-3}$ and a transverse emittance of 5π mm·mrad (14). Fig. 3

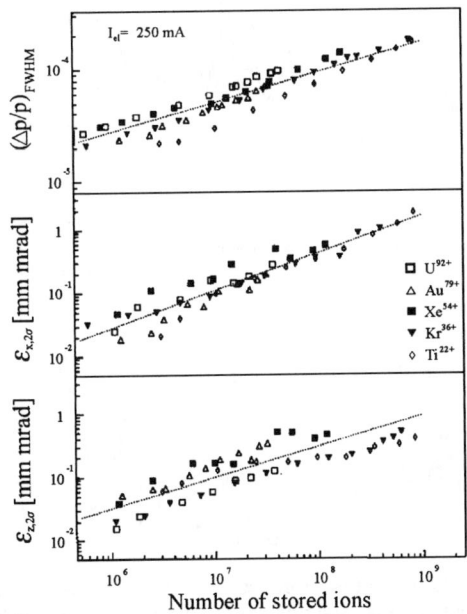

FIGURE 3: Relative momentum spread and transverse emittances of the stored beam after cooling as function of the number of stored ions. The electron current was 250 mA

shows the corresponding values after cooling for a large range of bare ions as function of the number of stored particles (15). The data were taken at an energy of 250 MeV/u and an electron current of 250 mA. For high numbers of stored particles the initial emittances and momentum spreads are reduced by almost one order of magnitude and for low numbers by almost two orders of magnitude. The reduction is nearly independent of the nuclear charge and its dependence on the particle number is reasonably well described by power laws (dashed lines). The longitudinal cooling times are smaller than 100 msec in accordance with theoretical estimates (11,13,16). These estimates take into account two effects which strongly reduce the cooling time: The flattening of the longitudinal velocity distribution of the electrons after acceleration to the final velocity (cf. Tab. 1) and the influence of the guiding longitudinal magnetic field. First measurements of the

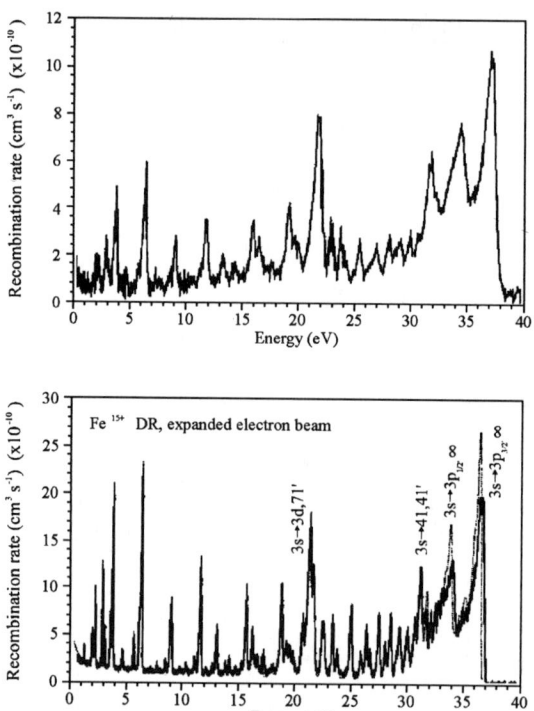

FIGURE 4: Measurements of dielectronic $\Delta n = 0$ resonances of Fe^{15+} with a conventional electron cooler (upper part) and with an expanded electron beam (lower part). The theoretical calculation (dotted line) is convoluted with a velocity distribution characterised by a longitudinal temperature of (0.15 ± 0.05) meV and a transverse temperature of (15 ± 5) meV (from (21)).

transverse cooling time with 350 MeV/u bare argon ions gave a surprisingly large value of about 100 sec which is roughly a factor of three larger than expected (17).

An important step towards an improvement of electron cooling efficiency was achieved by the suggestion of H. Danared to reduce the transverse electron temperature (18) and the first prove of the success of this concept (19). It uses the fact that the ratio of the transverse temperature to the magnetic field is an invariant, if charged particles are adiabatically expanded into a longitudinally varying magnetic field (20). Thus, the decrease of the longitudinal guiding field for the electrons by a certain factor results in a reduction of the transverse electron temperature by the same factor. The consequence of an improved cooling for experiments can excellently be demonstrated by dielectronic recombination (DR) because it is a resonance process. In DR a free electron is resonantly captured by a non-naked ion, if the energy released is equal to the excitation energy of at least of one of the bound electrons into an excited state. The energy released is given by the sum of the energy of the final bound state and the relative kinetic energy between the ion and the captured electron. The width of the observed resonance depends therefore besides the intrinsic resolution of the apparatus on the width of the velocity distribution of the captured electron. The smaller this width is, the smaller are the widths of the observed resonances and the more resonances can be resolved. By tuning the relative velocity between stored ions and cooling electrons for short times, the electrons in the cooler serve as a target for such experiments. As an example, Fig. 4 shows the rate for dielectronic recombination of cooling electrons with Fe^{15+} ions in the range of center-of-mass energies (0 - 40) eV measured at the TSR in Heidelberg (21). The measurement was performed in a cycle of cooling for 20 msec, detuning the voltage of the cooler for 20 msec, cooling for 20 msec, detuning for 20 msec for the next voltage and so forth. The energy range covered in the co-moving frame of reference the range of 3s ↦ 3p, 3d, and 4l transitions of the 3s core electron. The striking difference in resolution between the upper part of the data which was obtained with a conventional cooler, and the lower part obtained with a cooler making use of the adiabatic expansion is obvious. Along the axis of the beam, the magnetic field was lowered by a factor of 7.4. The dotted line is the result of a theoretical calculation which is convoluted with a longitudinal and transverse velocity distribution characterised by the temperatures $kT_{\parallel} = (0.15 \pm 0.05)$ meV and $kT_{\perp} = (15 \pm 5)$ meV, respectively.

The condition of adiabaticity requires that the longitudinal field strength varies slowly on the distance which is travelled by the particles during one period of the cyclotron motion around the field lines (20). Therefore, the technical realisation of a suitable magnetic field array provides more and more problems with increasing kinetic energy of the particles. For this reason, up to now only the low-energy rings are equipped with an expanding magnetic field configuration, while the ESR operating at high energies has still a conventional cooler. Nevertheless, ions stored and cooled in the ESR provide in combination with suitable targets a powerful source of x rays for fundamental research as will be shown in the following.

The ESR as a Source for X Rays

Heavy-ion accelerators have a long tradition for x-ray studies, reaching from beam-foil spectroscopy (for a review see for example (22)) to the investigation of superheavy collision systems (23 and rfs. therein). It should be noted that these experiments contributed a lot to our present knowledge of atomic physics. From the intensity, angular distribution, and impact parameter dependence of x rays emitted with the decay of exited inner-shell projectile and/or target vacancies, lifetimes, excitation mechanisms and in certain cases binding energies of electrons in superheavy atoms could be derived. Common to all of these experiments was the use of an ion beam in well defined charge states which was extracted out of an accelerator and directed on an external target. In most experiments, solid targets with a thickness in the order of 100 µg/cm^2 were used which were bombarded with typically 10^{10} ions /sec giving a luminosity around $6 \cdot 10^{29}$ /cm^2 sec.

Experiments at a storage ring have to use internal targets in order to profit from the revolution frequency of the stored ions. However, the thickness of the target (mostly gaseous targets) has to be kept so low, that the energy loss and the angular straggling of the ions after one passage through the target can be compensated for by cooling. A proper designed gasjet target containing around 10^{13} atoms/cm^2 fulfils the condition and guarantees the pressure gradient necessary for maintaining the ultra-high vacuum outside the target region (24). At 10^8 stored particles, a revolution frequency of 10^6 Hz and the above mentioned target thickness the luminosity amounts to 10^{27} /cm^2 sec which is considerably lower than the luminosities for external beam experiments. However, there is one distinct difference between the two cases which can be very important especially for experiments aiming at high precision. The internal target is a thin gas target which guarantees single collision conditions even for outer-shell electrons and a very small energy loss of the ions during the passage through the target. In contrast to solid targets, multiple excitation and capture processes due to peripheral collisions are negligibly small and the ion velocity stays practically constant. The latter is important for experiments measuring quantities which strongly depend on the velocity, e.g. resonances or precise energies of transitions in the ions. In order to achieve corresponding conditions in an extracted beam experiment, the solid target has to be replaced by a gas target resulting in a loss of luminosity of about 8 orders of magnitude.

An additional and peculiar source for x rays is inherent to an electron-cooler ring, namely the electron cooler itself which enables the investigation of the interaction between ions and electron at zero relative velocity. Ions passing the cooler may capture cooling electrons into bound states. Figure 5 shows schematically the possible recombination processes. For bare ions, radiative recombination (RR or REC) dominates which is a process where the energy released is taken over by a photon carrying the energy E_b of the bound state and of the relative kinetic energy E_{kin} between ion and cooling electron (upper part of Fig. 5):

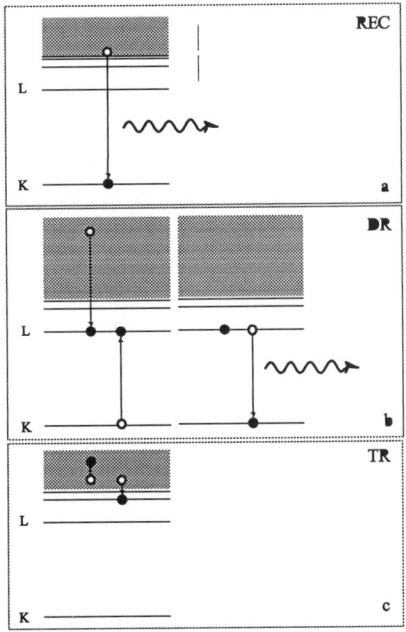

FIGURE 5: Electron capture processes in the cooler: a) radiative recombination, b) dielectronic recombination, c) three-body recombination

$$\hbar\omega = E_b + E_{kin} \qquad (1)$$

Since the average relative velocity between ions and cooling electrons is zero, the relevant kinetic energy is only determined by the temperatures of the beams which is small compared to the binding energies especially for high-Z ions. Thus, the energy of the emitted photon which is in the x-ray region for not too light elements, equals the energy of the bound state and from precise measurement of the photon energy the binding energy can be extracted. RR is the time-reversed process of photoionsation which has been treated theoretically by Stobbe (25). Based on Stobbe's work, radiative recombination coefficients for cooling electrons have been calculated (26) which are roughly proportional to $Z^{2.2}/(kT)^{1/2}$ $\cdot 1/n$, where kT is the electron beam temperature and n is the principal quantum number of the final bound state. Thus, capture into the ground states dominates, but the sum of all capture processes into higher states with subsequent cascading into the n = 2 state leads also to the emission of intense Lyman lines (n=2 → n=1) as will be seen later.

Due to the low density of only (10^8 - 10^9) electrons/cm^2, the luminosity of this x-ray source is rather small seen from an experimenter's point of view. In addition, the competing process of three-body recombination (see Fig. 5c)) which populates bound states with energies near kT and would finally lead to the emission of x rays from bound-bound transitions, is negligibly small (27). On the other side, low capture rates of cooling electrons lead to rather long 1/e-lifetimes of the stored ion beams of typically (200 - 1000) sec. In order to optimise the conditions for a planned experiment, one has to weigh the advantages and disadvantages of a higher luminosity by increasing the electron density vs. a shorter lifetime of the beam.

A further recombination channel for not fully stripped ions is the above mentioned resonance process of dielectronic recombination (see Fig. 5b)). The intermediate doubly excited state will especially in high-Z ions mostly decay by x-ray emission. It is however much easier to detect the down-charged ions than to measure the x rays for an investigation of this process. Therefore, dielectronic recombination plays only a minor role as an x-ray source at the time being.

Performing experiments either at the internal gas target or at the electron cooler of the ESR or of any other storage ring, one is faced with the Doppler effect. In order to correct for this effect, the results obtained in the lab system have to transformed into the moving frame of reference. Since the ion velocity $\beta \approx$ (0.3 - 0.5), the proper transformation is the relativistic Lorentz transformation. Photons emitted with an energy E_0 in the ion rest frame are observed in the lab system with an energy E given by

$$E = E_0 \cdot \{\gamma(1 - \beta \cos\Theta)\}^{-1/2} \tag{2}$$

with $\gamma = (1 - \beta^2)^{-1/2}$ and Θ the observation angle in the lab system. For a proper determination of E_0, the absolute values of the beam velocity and of the observation angle have to be known. Any uncertainty of $\Delta\beta$ and $\Delta\Theta$ will introduce an uncertainty ΔE_0 which can easily be calculated from the derivative of Eq. (2):

$$\left(\frac{\Delta E}{E}\right)^2 = \left(\frac{\beta \sin\Theta}{1 - \beta \cos\Theta}\Delta\Theta\right)^2 + \left(\frac{\cos\Theta - \beta}{1 - \beta \cos\Theta}\gamma^2 \Delta\beta\right)^2 + \left(\frac{\Delta E_0}{E_0}\right)^2. \tag{3}$$

An inspection of Eq. (3) reveals that three observation angles are very attractive for experiments. At the angles $\Theta = 0^0$ and 180^0 the uncertainty due to $\Delta\Theta$ is eliminated and at the magic angle $\Theta_{mag} = \cos^{-1}(\beta)$ the uncertainty due to $\Delta\beta$ plays no role. However, one has to pay a price: Under parallel and antiparallel observation the contribution from $\Delta\beta$, and at the magic angle the contribution from $\Delta\Theta$ to the uncertainty ΔE_0 are largest. Thus, in order to profit from a minimisation of the angular uncertainties, the absolute value of the ion velocity has to be known precisely and vice versa.

The size of the uncertainties strongly decreases with decreasing ion velocity. Deceleration of the beam is therefore an appropriate procedure to reduce the problems of the Doppler effect.

In the following chapter we will demonstrate the potential of brilliant cooled heavy ion beams in combination with the x-ray sources for precision experiments with one-electron ions and discuss methods in order to overcome the problems of the Doppler effect. The experiments which will be presented, aim at a fundamental test of QED calculations for one-electron, high-Z ions.

X-RAY SPECTROSCOPY AT THE ESR

The Lamb Shift of High-Z, Hydrogen-like Ions

The investigation of the spectrum of hydrogen as the simplest bound atomic system had a decisive influence on the development of quantum mechanics. In 1928 Dirac (28) proposed his famous relativistic equation of the electron which can be solved analytically for one electron bound to a point-like nucleus. It takes into account the spin of the electron and describes correctly the fine structure of the level scheme. An important consequence of Dirac's theory is the degeneracy of levels with the same principal quantum number n and with the same total angular momentum number j. Thus, the $2s_{1/2}$ state in hydrogen or hydrogen-like ions should have the same energy as the $2p_{1/2}$ state. However, in a series of famous experiments with hydrogen, Lamb and Retherford (29) showed that this is not the case. Instead, the states are split by about 1057 Mhz (the most precise experimental value for this splitting is 1057.845(9) Mhz (30)). The discovery of this splitting which is known as the "Lamb shift", had an enormous impact on the development and formulation of Quantum Electrodynamics. QED provides not only an explanation for the splitting, but a description of the interaction between a charged particle and its own (virtual) electromagnetic field. The latter interaction which is missing in the Dirac theory, turns out to be responsible for the observed splitting, and, more general, for a shift of all atomic levels due to QED corrections usually denoted as radiative corrections. The finite size of the nucleus also leads to a level shift. Therefore, the expression "Lamb shift" has been extended to shifts of any atomic level due to radiative corrections and finite nuclear size effects with respect to the eigenenergies of the Dirac equation for point-like nuclei.

Two effects represent the dominant contribution to the Lamb shift:

1. The self energy of the electron which is described in QED as the emission and reabsorption of a virtual photon by an electron. The Feynman diagram for this process which is the dominant part of the radiative corrections, is shown in the left part of Fig. 6.
2. The vacuum polarisation which may be thought of as a modification of the nuclear Coulomb potential that the electron feels in a region close to the nucleus (31). The modification arises from a virtual electron-positron pair creation described by the Feynman diagram in the right part of Fig. 6.

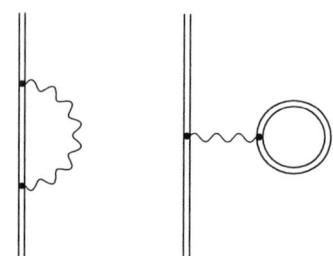

FIGURE 6: Feyman diagrams for the self energy (left) and the vacuum polarisation (right)

The importance of radiative corrections increases with increasing external field strength or in other words with decreasing distance of the electron from the nucleus and with increasing nuclear charge. Therefore, the level shifts are largest for wavefunctions with high radial densities at the origin such as s-states. For a given n, the shift of the s-states is much larger than the shift of the p- and higher l-states. As an example we note that the $2s_{1/2}$ - $2p_{1/2}$ splitting in hydrogen is caused to 98% by the shift of the $2s_{1/2}$ level and only to 2% by the shift of the $2p_{1/2}$ level (32).

Theoretically, the Lamb shift E_{LS} of a level of an electron with principal quantum number n bound to an extended nuclear charge Z is commonly described by (32, 33)

$$E_{LS}(n, Z) = \frac{\alpha}{\pi} \frac{(Z\alpha)^4}{n^3} m_0 c^2 F(Z\alpha) \qquad (4)$$

where α is the fine structure constant and $m_0 c^2$ the rest mass of the electron. $F(Z\alpha)$ is a slowly varying function of Z which comprises all the QED corrections including the effects of the finite nuclear size. Eq. (4) clearly reflects the fact that the shift is largest for the ground states of electrons in the fields of high-Z nuclei.

For low-Z ions, $F(Z\alpha)$ can be calculated in a perturbative $Z\alpha$ expansion. $Z\alpha$ is the characteristic coupling parameter for the interaction between the electron and the nucleus. For high-Z ions, nonperturbative methods have to be applied, because the electrons are fully relativistic in the very strong Coulomb field of the nucleus and bound-state propagators must be used for the calculations. The method to calculate for example the vacuum polarisation nonperturbatively is based on the potential expansion shown in Fig. 7. The self energy of the strongly bound electron is decomposed into a series, where the first term represents the zero-potential, the second term the one-potential and the third term the many-potential contribution. In Fig. 7, single solid lines represent free electron propagators,

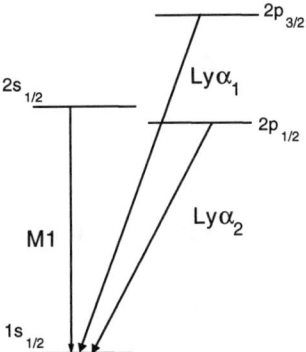

FIGURE 7: Potential expansion of the vacuum polarisation

double solid lines bound electron propagators, and the cross with the dotted line

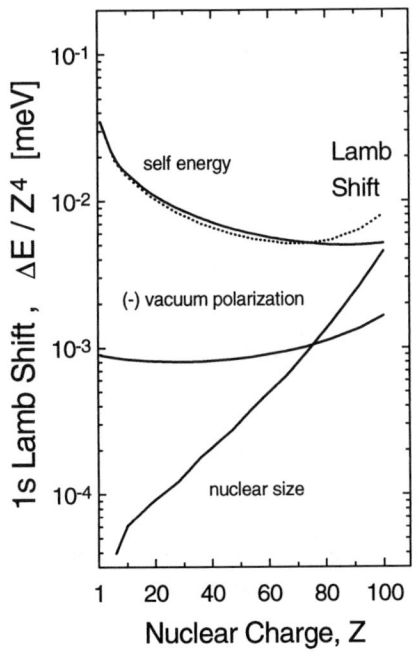

FIGURE 8: The various contributions to the 1s Lamb shift in hydrogen-like ions as function of the nuclear charge Z according to (32); in addition, the lowest lying transitions are indicated.

the instantaneous Coulomb interaction between electron and nucleus. Note that in the calculations the finite size of the nucleus has to be taken into account. Fig. 8 gives the various contributions to the Lamb shift of the ground state for hydrogen-like ions as function of the nuclear charge (32). The influence of the many-potential part becomes more and more important with increasing Z and can only be investigated experimentally by using the heaviest species available such as hydrogen-like uranium.

For light one-electron systems like atomic hydrogen, the results of the QED calculations have been confirmed with an extraordinary precision (34); first precise experiments in the strong-field region have been performed and are pursued at the ESR. In these experiments, the ground-state Lamb shift is mostly inferred from precise measurements of the energies of the Lyman lines lying in the region of hard x rays around 100 keV in the emitting frame of reference. The difference between the measured line energy and the transition energy calculated from the Dirac equation for a point-like nucleus is essentially the 1s Lamb shift because of the smallness of the shift for the higher levels.

Experiments at the Cooler

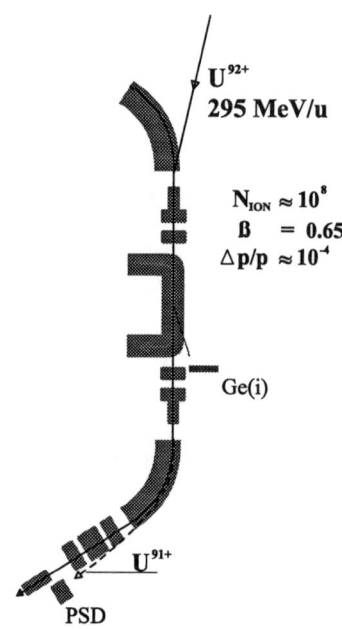

FIGURE 9: Experimental set-up at the cooler

The experimental set up at the electron cooler is shown in Fig. 9. Bare heavy ions (gold and uranium) are stored and cooled in the ESR at an energy around 300 MeV/u. X rays emitted with the capture of cooling electrons are detected in a Ge(i) detector mounted 4.2 m downstream of the middle of the cooler at an observation angle of 0.55^0 thereby taking full advantage of the above mentioned 0^0 geometry for the Doppler correction. The solid angle of the detector is $4.3 \cdot 10^{-5}$ sr and the energy resolution about 700 eV at 122 keV.

The x rays were detected in coincidence with hydrogen-like uranium ions as produced by the capture of one electron in the cooler. The ions were detected in a position-sensitive detector (PSD) which was installed in a pocket behind the first dipole magnet downstream of the cooler. The essential advantage of the coincidence set-up is the strong reduction of the x-ray background and a unambiguous determination of the charge state of the projectile.

An important aspect of the experiments is the precise calibration of the x-ray energy. Although the intrinsic resolution of the Ge(i) detector is rather low, small energy differences between adjacent lines can be determined with high accuracy (35). In order to take advantage of this property, the projectile energy was chosen

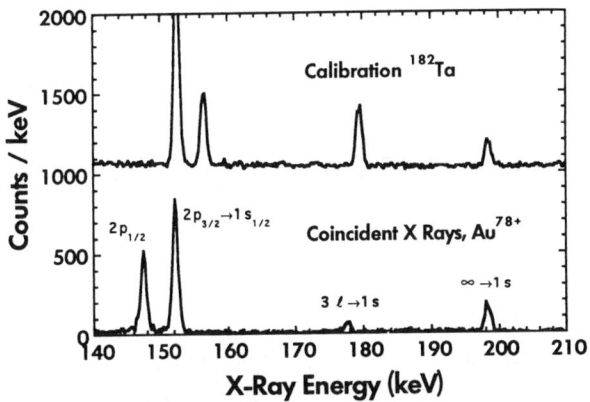

FIGURE 10: Coincident x-ray spectrum of Au^{78+} at a projectile energy of 266 MeV/u. For comparison, also the ^{182}Ta calibration spectrum is shown.

such that the Doppler-shifted x-ray lines were as close as possible to suitable γ lines used as standards. These γ lines (from ^{182}Ta and ^{192}Ir sources) have been measured previously on an absolute scale with an uncertainty below 0.3 eV (36). Figs. 10 and 11 show coincident x-ray spectra of Au^{78+} and U^{91+} recorded at projectile energies of 266 MeV/u and 321 MeV/u, respectively (37,38). The peaks are identified as the characteristic $Ly_{\alpha 1}$ ($2p_{3/2} \rightarrow 1s$), $Ly_{\alpha 2}$ ($2p_{1/2} \rightarrow 1s$), Ly_{β} and higher ($3l, 4l \rightarrow 1s$) transitions and as a line from the radiative recombination of cooling electrons into the ground state ($\infty \rightarrow 1s$). Due to the coincidence condition, the spectra are almost free of background. The relative intensities of the lines can shown to be characteristic for the population of the states by RR at small relative velocities and subsequent cascading (39). As a consequence of these processes, the population of the $2s_{1/2}$ state amounts to less than 5% of the total population of the $n = 2$ states; therefore, the $Ly_{\alpha 2}$ line contains only a very small contribution from the $2s_{1/2} \rightarrow 1s_{1/2}$ M1 transitions and is marked as $2p_{1/2} \rightarrow 1s_{1/2}$ transition. The tails on the low energy side of the Ly lines in the uranium spectrum come from

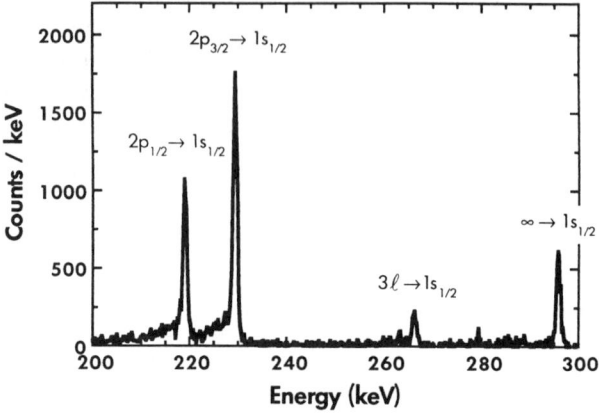

FIGURE 11: Coincident x-ray spectrum of U^{91+} recorded at a projectile energy of 321 MeV/u.

delayed emission on the way between the cooler and the detector giving rise to a Doppler shift towards lower energy; the shape and the intensity of the tails can properly be accounted for (38).

In order to extract the binding energy in the ion rest frame from the observed lines, the x-ray energies measured in the lab system have to be transformed correspondingly. For the geometry at the cooler, the absolute value of the beam velocity is the decisive quantity. Three methods have been used so far for the determination of β:

(a) the measurement of the revolution frequency of the circulating beam;
(b) the determination of energy differences between adjacent lines in the recorded x-ray spectrum;
(c) the high voltage of the cooler.

The first method suffers from the fact that although the frequency can be determined with a precision of $7 \cdot 10^{-6}$, the length of the closed orbit is not known to better than (108.36 ± 0.02) m resulting in an absolute precision of the projectile velocity of only ~ 10^{-4}. A similar precision is obtained from method (b) by selecting two lines whose energy difference is independent of the 1s binding energy such as the line from RR into the ground state and the $Ly_{\alpha 1}$ line. According to Eq. (3) this precision leads to an uncertainty of (30 - 40) eV for the $1s_{1/2}$ binding energy. Method (c) is based on an absolute calibration of the high-voltage generator of the cooler by the PTB (Physikalisch Technische Bundesanstalt) to an uncertainty of ± 25 V or better. From this value, the projectile velocity is determined with an absolute precision of $4 \cdot 10^{-5}$ which is almost one order of magnitude better than the results of the other two methods; the absolute value is lying within those error bars. It is satisfactory that the projectile velocity determined from the cooler voltage agrees within one standard deviation with the velocity obtained from laser-induced recombination of Ar^{18+} ions (40).

Table 2: Experimental Results at the Cooler for Hydrogen-like Gold and Uranium Ions

	1s binding energy (eV)	1s Lamb shift (eV)	statistical uncertainty (eV)	uncertainty from beam velocity (eV)
Z=79	93257.4 ± 7.9	202.3 ± 7.9	± 4.5	± 3.4
Z=92	131810 ± 16	470 ± 16	± 6.9	± 9.1

The results of the experiments with gold and uranium at the cooler are summarised in Table 2. The 1s Lamb shift was obtained by subtracting the Dirac energy for point-like nuclei corrected for the reduced mass effect form the ground-state binding energy. The errors from statistics and from the beam velocity are given separately and added up linearly for the final uncertainty. The results are up to now the most precise experimental values for the ground-state Lamb shift in hydrogen-like, high-Z ions.

Experiments at the Internal Target

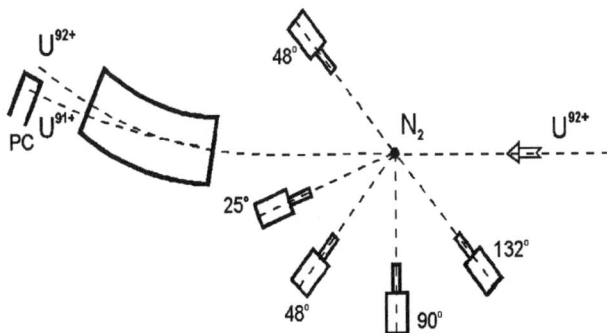

FIGURE 12: Experimental set-up at the internal gas target of the ESR. The direction of the gas jet is perpendicular to the plain of drawing. X rays registered in one of the detectors are measured in coincidence with hydrogen-like uranium ions detected in the particle detector PC

In contrast to the set-up at the cooler, the x-ray observation under angles near 0^0 or 180^0 is not possible at the internal gas target of the ESR for geometrical reasons. In order to get control over the Doppler corrections, several x-ray detectors are mounted around the gas target at well defined angles measured by laser assisted trigonometry providing a highly redundant set of spectra (41,42). Fig. 12 shows the arrangement of the detectors for the measurement of the Ly$_\alpha$ radiation of high-Z ions at the internal target. The target (usually a jet of N_2 molecules) is surrounded by four Ge(i) detectors mounted at observation angles of 48^0, 90^0, and 132^0 with respect to the predicted ion-optical beam axis. The two detectors at 48^0 are close to the magic angle, where the uncertainty of the beam velocity plays no role, but the uncertainties of the observation angle and the Doppler broadening of the lines due to the finite observation angle are largest. One of the 48^0 detectors is a conventional Ge(i) detector with a collimator in front for the reduction of the broadening and the other one is built up of seven precisely measured, equidistant, parallel segments. Since each of the segment is provided with a separate read-out, a sum spectrum with a small Doppler broadening can be generated from the known geometry of each of the segments. A similar detector is installed at 90^0, whereas the 132^0 detector is a conventional one without a collimator. For calibration purposes, γ lines from ^{179}Yb and ^{182}Ta sources were used as standards. The emitted x rays were measured in coincidence with projectiles having captured one electron in the gas target. The down-charged particles were registered in a fast plastic scintillator mounted behind the next following dipole magnet.

A typical coincident x-ray spectrum obtained from collisions between bare uranium ions at an energy of 358 MeV/u with nitrogen molecules is shown in the upper part of Fig. 13 (43). Note that the spectrum has already been transformed into the emitting frame of reference. Besides the well-resolved narrow Ly

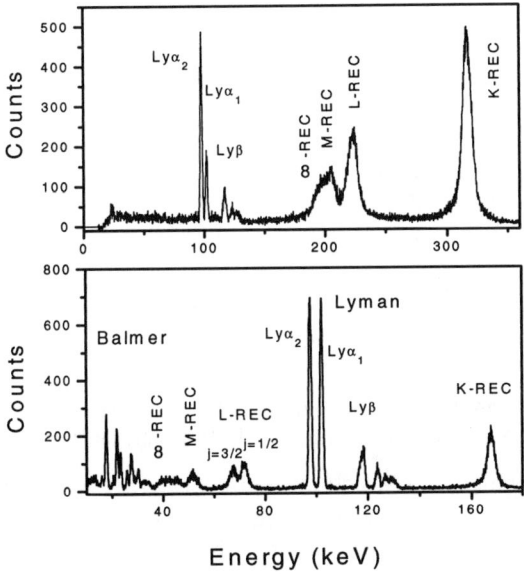

FIGURE 13: Coincident x-ray spectra of hydrogen-like uranium measured at projectile energies of 358 MeV/u (top) and 68 MeV/u (bottom) (42,43). The X-ray energy is transformed into the emitting system.

transitions, broader lines at higher energies dominate the spectrum. These lines are emitted with the radiative capture (REC) of bound target electrons into different levels of the ions and appear at energies given by the sum of the binding energy and of the kinetic energy of an electron moving with the projectile velocity. In contrast to the capture of free cooling electrons giving narrow lines (see Figs. 10,11), the momentum distribution of the initial bound target states leads to the observed line broadening (the so-called Compton profile). The Compton profile can be modelled quite well as can be seen from Fig. 14 which shows the REC-spectrum measured in 292 MeV/u Dy^{66+} + Ar collisions and the calculated profile (44). However, a determination of the binding energy of the final state in the ion is possible only to a precision of the order of 20 eV due to the asymmetry of the line.

The x-ray spectrum changes drastically when the projectile velocity is reduced by decelerating the ions in the ring. The bottom part of Fig. 13 shows the hydrogen-like uranium spectrum recorded at an energy of 68 MeV/u (42, 43). At

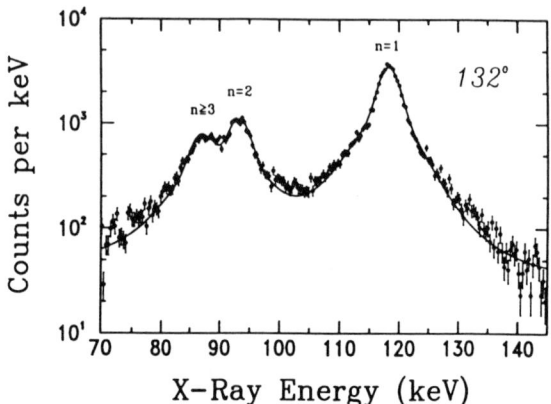

FIGURE 14: REC spectrum of Dy^{66+}; the points with the error bars are the experimental results and the solid line is the calculated profile.

low projectile velocity the spectrum is dominated by a variety of characteristic Balmer ($n = 3,4,... \to n=2$) and Lyman transitions, while the REC lines appear only rather weakly; note that the L-REC line is even split up into the j = 3/2 and j = 1/2 components. The increase of the intensity of the characteristic lines and the decrease of the REC lines is caused by the fact that at low beam energies the target electrons are captured mainly nonradiatively into high lying states and subsequent cascades lead to the emission of Balmer and Lyman lines. Since in addition the uncertainties introduced by the Doppler corrections reduce with decreasing velocity, experiments with decelerated beams have an enormous potential for high-precision x-ray spectroscopy of highly charged heavy ions. At the time being, the evaluation of the data is still in progress and a value of the 1s Lamb shift can not yet be given.

In a first U^{92+} experiment at the gas target only one segmented and one conventional detector was used (41). Although at that time the number of stored ions was at maximum only 10^7 and the projectile velocity could be determined to a precision of only $4 \cdot 10^{-4}$, a reasonable accuracy for the 1s Lamb shift could already be achieved. The results are summarised in Tab. 3 which clearly shows that the final result was essentially determined by the counting statistics. The individual uncertainties are summed up quadratically for the final uncertainty.

Table 3: Experimental Results at the Gas Target for Hydrogen-like Uranium Ions

	1s binding energy (eV)	1s Lamb shift (eV)	uncertainty in beam velocity (eV)	statistical uncertainty (eV)	calibration (eV)
Z=92	131769±63	429±63	±32	±46	±30

Comparison of the Experimental Results with Theory

A comprehensive comparison between experimental and theoretical results for the Lamb shift of the ground state can be made by plotting the function $F(Z\alpha)$ vs. the nuclear charge Z. This function is obtained acc. to Eq. (3) from a division of the results by $(\alpha/\pi)(Z\alpha)^4 m_0 c^2$. The plot is shown in Fig. 15, where the solid symbols represent the results obtained at the facilities of the GSI; for an overview of all the relevant references see for example (45). The solid line represents the theoretical predictions (32). Over the whole range of nuclear charges, an excellent agreement between experiment and theory is observed. For the high-Z region most of the experimental data provide a test of the ground-state Lamb shift on a level of only 30% with the exception of the recent experiments at the ESR: The uranium data from the gastarget have a sensitivity of 15% and the gold and uranium data from the cooler have a sensitivity of 4% and 3%, respectively.

Most of the experimental efforts have been concentrated on U^{91+} as the heaviest available hydrogen-like ion. Therefore, the inset in Fig. 15 separately presents the results in comparison to theory obtained within the last few years; the solid symbols are the data obtained at the GSI. The size of the error bars of the experimental results clearly demonstrates the substantial improvement in precision compared to former experiments performed at the BEVALAC accelerator (47, 48). The solid line shows the theoretical development over the years (32, 46). The present theoretical value for the Lamb shift is (465.5 ± 2) eV (46) which has to be compared to the most precise experimental value of (470 ± 16) eV. The largest contribution to the theoretical uncertainty comes from higher-order QED terms which are not yet calculated. Note that for uranium the finite size of the nucleus contributes to more than 40% to the ground-state Lamb shift. The uncertainty introduced by this effect is 0.3 eV (46) and still small compared to the present experimental accuracy, but it may finally prevent an experimental test of the QED calculations for the highest-Z ion with a precision below 1 eV. For such

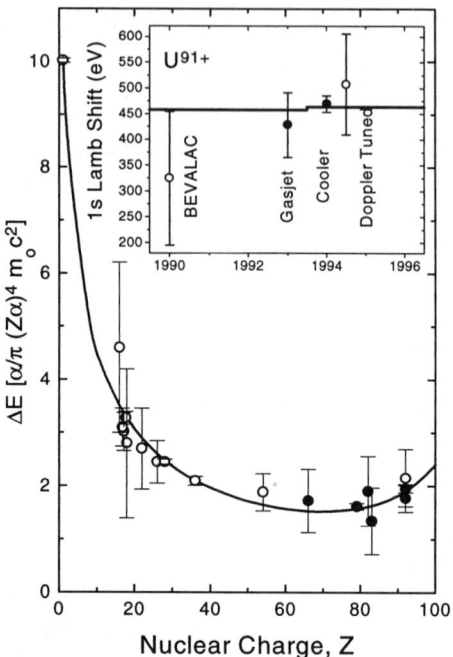

FIGURE 15: Comparison between experimental (dots with error bars) and theoretical (solid line taken from (32)) results of the ground-state Lamb shift in hydrogen-like ions. The inset shows the four available data points for uranium; the solid points refer to the experiments at the ESR. The solid lines shows the recent theoretical results (32, 46).

a purpose, a measurement using the much better known double-magic nucleus ^{208}Pb seems to be appropriate. In order to reach the accuracy of ± 2 eV of the present calculation, an improvement of the experimental precision by one order of magnitude is required which means a 10^{-5} precision in the measurement of x-ray energies around 100 keV.

CONCLUSIONS AND OUTLOOK

In summary, we have discussed some of the important features of heavy-ion storage rings with electron cooling which provide beams of excellent quality for experiments. The electron cooler and the internal gas target at the ESR represent suitable x-ray sources for experiments aiming at a high precision. Each of the targets has its own advantages and disadvantages regarding the luminosity and the Doppler effect. The latter effect introduces unavoidable uncertainties into the measurements the energy of x rays emitted by fast ions. Due to the favourable conditions for controlling the Doppler effect at the cooler, in a first series of measurements of the ground-state Lamb shift in hydrogen-like heavy ions, an accuracy of ± 16 eV for uranium could be achieved. This is a substantial improvement by almost one order of magnitude compared to the former experiments at the BEVALAC (47,48) using extracted, uncooled ion beams.

The next step will consist of an increase of the spectral resolution of the set-up at the cooler without loss of efficiency using the Doppler-tuned techniques (49). For this purpose, an additional Ge(i) will be mounted close to 180^0 with an absorber foil placed in front of it having a well known energy and shape of the K-absorption edge. By a variation of the beam energy, the red shifted characteristic x-ray lines are moved across the edge and from the dependence of the x-ray transmission on the velocity, the x-ray energy may be determined to about 5 eV.

It has been demonstrated that the deceleration of the ion beam in the storage ring leads to a very efficient production of characteristic x rays from the projectiles at the internal target. Since in addition the problems connected with the Doppler effect decrease with decreasing projectile velocity, experiments with decelerated beams at the internal target will pave the way for an accuracy of about 1 eV. This accuracy will provide a meaningful test of the recent QED calculations of the ground-state Lamb shift. In order to obtain such a precision, the Ge(i) detectors at the target have to be replaced by high-resolution instruments like transmission-type crystal spectrometers with position sensitive x-ray detectors (50) and by calorimetric low-temperature detectors (51); both schemes are presently under construction.

ACKNOWLEDGMENTS

The experiments at GSI have been performed in a pleasant and fruitful collaboration with the guests and co-workers of the Atomic Physics group. Special thanks are due to H. Beyer, F. Bosch, R. Deslattes, A. Gallus, P. Indelicato, G. Menzel, P. Mokler, and Th. Stöhlker who form the "hard core" of the x-ray spectroscopy group and to the members of the ESR team: K. Beckert, H. Eickhoff, B. Franzke, F. Nolden, and M. Steck. H.-J. Kluge, H. Persson, and G. Soff have always been stimulating partners in many discussions. I would like to express my gratitude particularly to L. Labzowsky for his steady willingness and patience, to answer my questions about QED

REFERENCES

1. Jung, M., Bosch. F., Beckert, K., Eickhoff, H., Folger, H., Franzke, B., Gruber, A., Kienle, P., Klepper, O., Koenig, W., Kozhuharov, C., Mann, R., Moshammer, R., Nolden, F., Schaaf, U., Soff, G., Spädtke, P., Steck, M., Stöhlker, Th., and Sümmerer, K., Phys. Rev. Lett. **69**, 2164 (1992)
2. Chetoui, A., Guiraud, L., Despiney, I., and Sabatier, L., in *Physics with Multiply Charged Ions*, ed. D. Liesen, NATO ASI Series B, Vol. 348, Plenum Press New York, 1995, p. 357
3. Marrs, R. E., Comm. At. Mol. Phys. **27**, 57 (1991)
4. Beiersdorfer, P., Proceedings of this Conference
5. Knapp, D. A., in ref. 2., p. 143
6. Larsson, M., Rep. Prog. Phys. **58**, 1267-1319 (1995)
7. Mokler, P. H. and Stöhlker, Th., The Physics of Highly-Charged Heavy Ions Revealed by Storage/Cooler Rings, to be publ. in *Advances in Atomic, Molecular and Optical Physics*
8. Bruck, H., Accélérateurs Circulaires de Particules, Presse Universitaires de France, Paris, 1966
9. Franzke, B., *Information about ESR Parameters*, GSI-ESR-TN/86-01, 1986 (Internal Report)
10. Budker, G. I., Sov. J. Atom. Energy **22**, 438 (1967)
11. Poth, H., Phys. Rep. **196**, 135 (1990)
12. Budker, G. I., Dikansky, N. S., Kudelainen, V. I., Meshkov, I. N., Parkhomchuk, V. V., Pestrikov, D. V., Skrinsky, A. N., and Sukhina, B. N., Part. Acc. **7**, 197 (1976)
13. Winkler, Th., Thesis, Univ. of Heidelberg, 1996
14. Langenbeck, B., Private Communication
15. Steck, M., Beckert, K., Bosch, F., Eickhoff, H., Franzke, B., Klepper, o., Moshammer, R., Nolden, F., Spädtke, P., and Winkler, Th., Proc. 4th Europ. Part. Accel. Conf., London 1994, ed.V. Suller and Ch. Petit-Jean-Genaz, World Scientific Singapore 1994, p. 1197
16. Parkhomchuk, V. V., Skrinsky, A. N., Rep. Prog. Phys. **54**, 919 (1991)
17. Winkler, Th., Bourgeois, W., Franzke, B., Steck, M., Unverzagt, M., Schmidt-Böcking, H., Haberle, K., GSI Scientific Report 1995, ISSN 0174-0814, ed. U. Grundinger, p. 162
18. Danared, H., Nucl. Instrum. Methods Phys. Res. **A335**, 397 (1993)
19. Danared, H., Andler, G., Bagge, L., Herrlander, C. J., Hilke, J., Jeansson, J., Källberg, A., Nielsson, A., Paal, A., Rensfelt, K.-G., Rosengård, U., Starker, J., and af Ugglas, M., Phys. Rev. Lett. **72**, 3775 (1994)
20. Jackson, J. D., *Classical Electrodynamics*, John Wiley & Sons, Inc., Sixth Printing 1967, p. 419
21. Linkemann, J., Kenntner, J., Müller, A., Wolf, A., Habs, D., Schwalm, D., Spies, W., Uwira, O., Frank, A., Liedtke, A., Hofmann, G., Salzborn, E., Badnell, N. R., and Pindzola, M. S., Nucl. Instr. Meth. **B98** (III), 1995
22. *Beam-Foil Spectroscopy* in Topics in Current Physics, Vol. I, ed. S. Bashkin, Springer Verlag, Berlin, Heidelberg, New York, 1976; Andrä., H. J., in *Progress in Atomic Spectroscopy*, Part B, ed. W. Hanle and H. Kleinpoppen, Plenum Press, New York and London, 1978, p. 829
23. Mokler, P. H.and Liesen, D., in *Progress in Atomic Spectroscopy*, Part C, ed. H. J. Beyer and H. Kleinpoppen, Plenum Press, New York and London, 1984, p. 321
24. Gruber, A., Bourgeois, W., Franzke, B., Kritzer, A., and Treffert, C., Nucl. Instr. and Meth. **A282**, 87 (1989)
25. Stobbe, M., Ann. Phys. (Leipzig) **7**, 661 (1930)
26. Bell, M., and Bell, J. S., Part. Accel. **12**, 49 (1982)
27. Beyer, H. F., Liesen, D., and Guzman, O., Part. Accel. **24**, 163 (1989)
28. Dirac, P. A. M., Proc. Roy. Soc. **A117**, 610 (1928)
29. Lamb, W. E., and Retherford, R. C., Phys. Rev. **72**, 241 (1947); Lamb, W. E., Rep. Prog. Phys **14**, 23 (1951)

30. Lundeen, S. R., and Pipkin, F. M., Phys. Rev. Lett. **46**, 232 (1981)
31. Feynman, R. P., *The Theory of Fundamental Processes*, W. A. Benjamin, Inc., Reading MA, 1962
32. Johnson, W. R., and Soff, G., At. Data Nucl. Data Tabl. **33**, 405 (1988)
33. Mohr, P. J., At. Data Nucl. Data Tabl. **29**, 453 (1983)
34. Weitz, M., Huber, A., Schmidt-Kaler, F., Leibfried, D., and Hänsch, T. W., Phys. Rev. Lett. **72**, 328 (1994); Bourzeix, S., de Beauvoir, B., Nez, F., Plimmer, M. D., de Tomasi, F., Julien, L., Biraben, F., and Stacey, D. N., Phys. Rev. Lett. **76**, 384 (1996)
35. Helmer, R. G., Greenwood, R. C., and Gehrke, R. C., Nucl. Instrm. Meth. **155**, 189 (1978)
36. Kessler, e. G., Deslattes, R. D., Henins, A., and Sauder, W. C., Phys. Rev. Lett. **40**, 171 (1978), Borchert, G. L., Scheck, W., and Schult, O. W. B., Nucl. Instrum. Meth. **124**, 107 (1975)
37. Beyer, H. F., Liesen, D., Bosch, F., Finlyson, K. D., Jung, M., Klepper, O., Moshammer, R., Beckert, K., Eickhoff, H., Franzke, B., Nolden, F., Spädtke, P., and Steck, M., Phys. Lett. **A184**, 435 (1994)
38. Beyer, H. F., Menzel, G., Liesen, D., Gallus, A., Bosch, F., Deslattes, R. D., Indelicato, P., Stöhlker, Th., Klepper, O., Moshammer, R., Nolden, F., Eickhoff, H., Franzke, B., and Steck, M., Z. Phys. **D35**, 169 (1995)
39. Liesen, D., Beyer, H. F., Finlayson, K. D., Bosch, F., Jung, M., Klepper, O., Moshammer, R., Beckert, K., Eickhoff, H., Franzke, B., Nolden, F., Spädtke, P., Steck, M., Menzel, G., and Deslattes, R. D., Z. Phys. **D30**, 307 (1994)
40. Borneis, S., Becker, St., Engel, T., Klaft, I., Klepper, O., Kohl, A., Kühl, T., Marx, D., Meier, K., Neumann, R., Schmitt, F., Seelig, P., and Völker, L., in Proc. Res. Ion. Spectrosc., Bernkastel-Kues, 1994, ed. H. J. Kluge, J. E. Parks, and K. Wendt, AIP 1995
41. Stöhlker, Th., Mokler, P. H., Beckert, K., Bosch, F., Eickhoff, H., Franzke, B., Jung, M., Kandler, T., Klepper, O., Kozhuharov, C., Moshammer, R., Nolden, F., Reich, H., Rymuza, P., Spädtke, P., and Steck, M., Phys. Rev. Lett. **71**, 2184 (1993)
42. Mokler, P. H., Stöhlker, Th., Dunford, R. W., Gallus, A., Kandler, T., Menzel, G., Prinz, H.-T., Rymuza, P., Stachura, Z., Swiat, P., and Warczak, A., Z. Phys. **D35**, 77 (1995)
43. Stöhlker, Th., GSI-Nachrichten, 05-95 (1995)
44. Beyer, H. F., Finlayson, K. D., Liesen, D., Indelicato, P., Chantler, C. T., Deslattes R. D., Schweppe, J., Bosch, F., Jung, M., Klepper, O., König, W., Moshammer, R., Beckert, K., Eickhoff, H., Franzke, B., Gruber, A., Nolden, F., Spädtke, P., and Steck, M., J. Phys. **B26**, 1557 (1993)
45. Liesen, D., Beyer, H. F., and Menzel, G., Comments At. Mol. Phys. **32**, 23 (1995)
46. Persson, H., Salomonson, S., Sunnergren, P., Lindgren, I., Gustavsson, M. G. H., subm. to Hyperfine Interaction (1996)
47. Briand, J. P., Chevallier, P., Indelicato, P., Ziock, K. P., and Dietrich, D., Phys. Rev. Lett. **65**, 2761 (1990)
48. Lupton, J. H., Dietrich, D. D., Hailey, C. J., Stewart, R. E., and Ziock, K. P., Phys. Rev. **A50**, 2150 (1994)
49. Schmieder, R. W., and Marrus, R., Nucl. Instr. Meth. **110**, 459 (1973)
50. Beyer, H. F., in ref. 2., p. 31
51. Egelhof, P., Beyer, H. F., McCammon, D., v. Feilitzsch, F., v. Kienlin, A., Kluge, H.-J., Liesen, D., Meier, J., Moseley, S. H., and Stöhlker, Th., Nucl. Instr. Meth. **A370**, 26 (1996)

High-Resolution X-Ray Spectra from Low-Temperature, Highly Charged Ions

Peter Beiersdorfer

Department of Physics and Space Technology
Lawrence Livermore National Laboratory, Livermore, CA 94550, USA

Abstract. The electron beam ion traps (EBIT) at Livermore were designed for studying the x-ray emission of highly charged ions produced and excited by a monoenergetic electron beam. The precision with which the x-ray emission can be analyzed has recently been increased markedly when it became possible to decouple the temperature of the ions from the energy of the electron beam by several orders of magnitude. By adjusting the trap parameters, ion temperatures as low as 15.8 ± 4.4 eV for Ti^{20+} and 59.4 ± 9.9 eV for Cs^{45+} were achieved. These temperatures were more than two orders of magnitude lower than the energy of the multi-keV electron beam used for the production and excitation of the ions. A discussion of the techniques used to produce and study low-temperature highly charged ions is presented in this progress report. The low ion temperatures enabled measurements heretofore impossible. As an example, a direct observation of the natural line width of fast electric dipole allowed x-ray transitions is described. From the observed natural line width and by making use of the time-energy relations of the uncertainty principle we were able to determine a radiative transition rate of 1.65 fs for the $2p$-$3d$ resonance transition in neonlike Cs^{45+}. A brief discussion of other high-precision measurements enabled by our new technique is also given.

INTRODUCTION

A reduction in the velocity of an atom increases the spectroscopic precision and creates the opportunity for new classes of observations. This has now been shown extensively in the case of trapped atoms, molecules, and singly charged ions cooled by lasers [1,2]. Undoubtedly, the spectroscopy of *highly* charged ions can benefit from a reduction of the ion motion as well. The increase in the spectroscopic precision associated with the preparation of cool, highly charged ions enables measurements of the energy levels with greater accuracy for precise determinations of quantum electrodynamical effects, of nuclear parameters, or of line coincidences for photopumping of x-ray lasers. It enables the study of line shapes and the application of laser spectroscopy.

We report on experiments performed on the electron beam ion trap (EBIT) facilities at the Lawrence Livermore National Laboratory that have systematically lowered the thermal motion of trapped, highly charged ions [3,4]. The reduction in thermal motion is carried out to the point where the observed x-ray line width of fast electric dipole transitions is limited by the Heisenberg uncertainty relations and the Lorentzian line shape is observed. In these experiments, a decoupling on the order of 100 times is achieved between the kinetic energy of the ions and the energy of the electron beam used for excitation of the observed x-ray lines. Such a marked decoupling is needed because the natural line width is a much smaller fraction of the transition energy for x-ray lines than for lines in the ultraviolet or visible, given the same lifetimes of the relevant levels. Observations of the Lorentzian line shape of x-ray lines from highly charged ions further enlargen the arsenal of precision spectroscopic observations available to atomic physicists, complementing, for example, such observations in neutral or few-times ionized ions excited by synchrotron sources [5].

Measuring the width of the Lorentzian line shape allows us to determine the radiative transition rates of fast, electric dipole allowed resonance transitions. Existing techniques for measuring fast transition rates fail to yield results for lifetimes shorter than a few picoseconds [6,7]. With our technique we measure lifetimes in the femtosecond range, i.e., in a regime that is 1000 times faster and has never before been open to experimental scrutiny. Resonance transitions with a radiative lifetime shorter than 10 femtoseconds are found, for example, among the K-shell transitions of heliumlike ions above argon (Z=18) and among the L-shell transitions of neonlike ions above krypton (Z=36). In other words, femtosecond radiative lifetimes are common for most resonance transitions in highly charged ions. These transitions form the dominant lines in a given x-ray spectrum and play an important role in the density and temperature diagnostics of high-temperature plasmas, such as those found in laser-produced, tokamak, and astrophysical plasmas [8–11]. Because fast transition rates correspond to large absorption oscillator strengths, these resonance transitions dominate the Planck mean opacity of a high-temperatures plasma, and accurate knowledge of their radiative rates is important for plasma opacity and line transfer [12,13]. Mesurements of the radiative lifetimes of these fast resonance transitions are needed to validate atomic physics calculations. Unlike transition energy measurements, they test the long-range behaviour of atomic wave functions and thus complement atomic structure measurements.

The electron beam ion trap devices are well suited for the spectroscopy of isolated spectral lines from highly charged ions. Because the devices employ a quasi monoenergetic electron beam for the production and excitation of highly charged ions, the charge state of interest and the excitation process can be selected by the appropriate choice of the electron beam energy [14].

Line blending can thus be avoided. Most importantly, line broadening caused by dielectronic satellite transitions with high-n spectator electrons, which is a common occurrence in most high-temperature plasma sources [15,16], can be avoided easily. Density and opacity broadening, which may broaden x-ray lines, for example, in laser-produced plasmas [17], are also not an issue because of the relatively low density of the electron beam ($\leq 5 \times 10^{12}$ cm^{-3}). Directional ion motion, as found in heavy-ion accelerators, does not occur in electron beam ion traps because the ions are confined in a narrow volume. Lines, therefore, are not Doppler shifted.

The lines generated in an EBIT device may be broadened by Doppler broadening as a result of their thermal motion. They may also be broadened by the response function of the instruments used in their observation. In order to ascertain the thermal motion of the ions, it is necessary to deploy spectrometers with very high resolving powers that do not obscure the x-ray line shape. This is especially the case when the ion temperature is low. Such instrumentation was recently implemented on our electron beam ion trap facilites [18]. Resolving powers exceeding $\lambda/\Delta\lambda=$ 60,000 were achieved, enabling the studies presented in this progress report.

This paper is organized as follows. We first present a brief description of how ions are produced and trapped in an EBIT device and describe the spectroscopic instrumentation employed in our measurements. We present measurements of the temperature of the trapped ions inferred from the Doppler-broadened line profiles and discuss our attempts to sytematically lower the temperature without degrading the signal to noise of the spectral measurements. We then apply our techniques to the $(2p^5_{3/2}3d_{5/2})_{J=1} \rightarrow (2p^6)_{J=0}$ transition in neonlike Cs^{45+} and record the Lorentzian shape of the line. From the time-energy relations of the Uncertainty Principle, we determine a 1.65-fs radiative lifetime of the excited level. In the conclusion we briefly discuss possible applications of the spectroscopy of cold ions in atomic and nuclear physics and give specific examples of future measurements.

EBIT OPERATION

The EBIT device employs a monoenergetic 60-μm-diameter electron beam to produce, trap, and excite a particular charge state of interest, as described in detail by Levine et al. [19,20]. The interaction between the ions and electrons takes place within the 2-cm-long cylindrical confinement region illustrated in Fig. 1. Confinement of the ions in the axial direction is accomplished by biasing the three drift tubes. In standard operation, an axial trapping potential $V_{ax} = 100$ V is applied. Confinement in the radial direction is accomplished by the space charge potential V_b of the electron beam, which depends on the beam current I and the beam energy E [19]:

$$V_b \approx 0.5(1-f)I/\sqrt{E}. \tag{1}$$

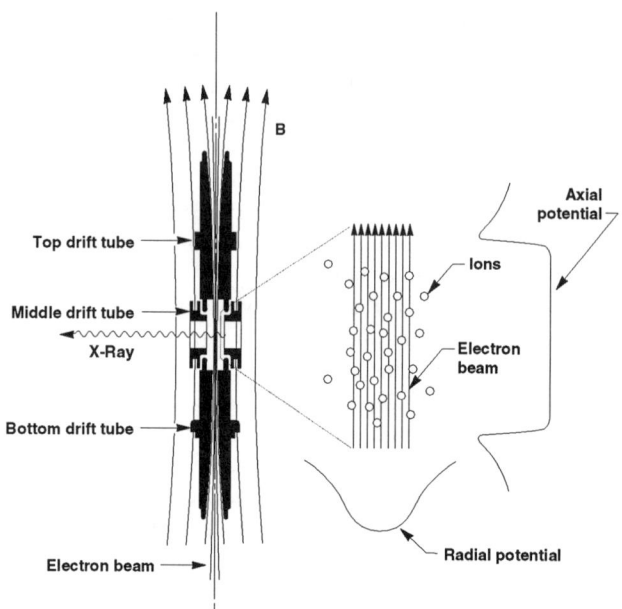

FIGURE 1. Schematic of the trapping geometry of the electron beam ion trap. The ions are trapped axially in the potential applied to the top and bottom drift tubes and radially in the space charge potential of the electron beam. The middle drift tube is slotted to allow a direct view of the trapping region and for recording emitted x rays.

Here, f is the fraction of the beam space charge neutralized by the presence of the ions. In standard operation, the radial potential at the beam edge relative to the beam center is about 10–20 V.

Elastic collisions with the electron beam heats the ions at a rate of several keV/s [19]. The energy gained from this interaction is shared among all ions, as the ion-ion collision frequency is much faster than that of ion-electron collisions [21]. The ions are heated until their kinetic energy is larger than the potential well and they are able to leave the trap. Because low-charge ions experience a potential well that is shallower than the well experienced by ions with higher charge, low-charge ions will leave the trap at a temperature significantly below that necessary for highly charged ions to leave. Loss of highly charged ions can, therefore, be prevented or greatly reduced by providing a constant source of low-charge ions, such as N^{7+} or Ne^{10+}, which carry with them the heat deposited by the electron beam [19,22,23].

The mechanisms of production and trapping suggests the ability to produce highly charged ions with "arbitrarily" low temperature. The temperature of the highly charged ions is limited by the temperature needed for low-charge ions to leave the trap. By reducing the potential well of the trap, this temperature is reduced, and we expect a drop in the overall ion temperature. The measurements described below confirm our expectation.

SPECTROSCOPIC INSTRUMENTATION

Our measurements concentrated on the ions Ti^{20+} and Cs^{45+}. The two ion species emit K-shell or L-shell x-ray lines, respectively, in the wavelength range 2.60–2.64 Å. The ion temperature T_i is determined from the Doppler broadening of the emitted line radiation $\Delta\lambda$ using the relation

$$T_i = \frac{m_i c^2}{8\ln 2}\left[\frac{\Delta\lambda}{\lambda}\right]^2, \qquad (2)$$

where m_i is the ion mass, c is the speed of light, and λ is the wavelength of the measured transition.

For such measurements to succeed it is necessary to employ spectrometers with very high resolving power. In our measurements, we employed a high-resolution crystal spectrometer based on the geometry proposed by von Hámos [24]. The spectrometer utilized a cylindrically bent analyzing crystal oriented such that the axis of the cylinder lies in the plane of dispersion. The instrument functions like a flat-crystal spectrometer in the plane of dispersion and provides focusing for rays perpendicular to the plane of dispersion. A schematic of the focusing properties of the von Hámos geometry is shown in Fig. 2. The von Hámos geometry is well suited for measurements on an EBIT, because EBIT represents a slit-like line source whose dimensions are determined by the 60-μm-diameter electron beam and the 2-cm-long trap length.

As illustrated in Fig. 2, x rays with different wavelengths λ are focused at different positions along the axis of curvature of the crystal. Aberrations are

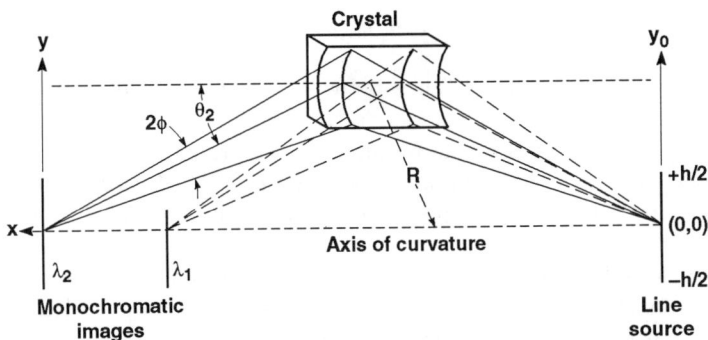

FIGURE 2. Focusing geometry of the EBIT von Hámos-type high-resolution crystal spectrometer. A cylindrically bent crystal generates monochromatic images of the EBIT source at different location along its axis of curvature. The images are recorded with a position sensitive proportional counter.

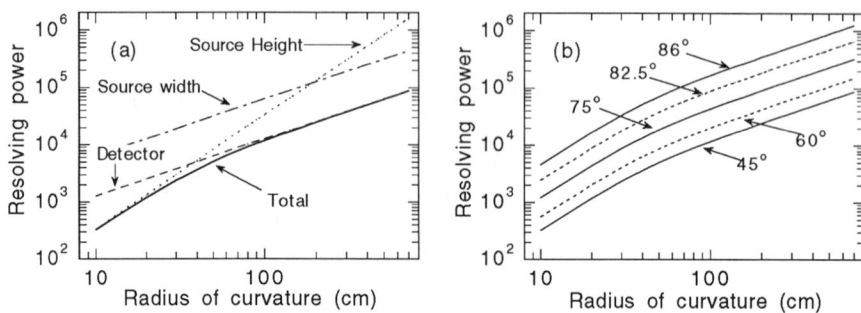

FIGURE 3. Estimated resolving power achievable with a crystal of different radii of curvature. Limitations from a 300-μm spatial resolution of the detector, a 1.2-cm source height, and a 60-μm source width are listed separately in (a). Here, the assumed Bragg angle is 45°. The variation of the resolving power for different Bragg angles is shown in (b).

introduced by the finite height h of the source, the finite diameter of the EBIT source Δs, and the finite spatial resolution of the detector Δx. Using a crystal with a radius of curvature R and an opening angle ϕ less than the Bragg angle θ, the limitation in the resolving power caused by the source height is given by

$$\lambda/\Delta\lambda \approx 4R^2/h^2. \tag{3}$$

The limitation caused by the source diameter is

$$\lambda/\Delta\lambda = 2D\tan\theta/\Delta x. \tag{4}$$

That caused by the finite detector resolution is

$$\lambda/\Delta\lambda = 2D\tan\theta/\Delta s. \tag{5}$$

Here, $D = R/\sin\theta$ is the distance between crystal and detector as well as between crystal and source. The quadratic sum of these aberrations is shown in Fig. 3(a) as a function of the radius of curvature of the crystal. Here, we took $h = 1.2$ cm, $\Delta x = 60$ μm, and $\Delta s = 300$ μm, which were typical values for our setup. The resolving power also varies with $\tan\theta$, as illustrated in Fig. 3(b).

In order to obtain the highest resolving power, we employed a crystal with a large radius of curvature and operated at a large Bragg angle. The layout of our very high-resolution spectrometer on the EBIT device is shown in Fig. 4. Our instrument employs a large, $120 \times 50 \times 0.25$ mm^3 quartz($20\bar{2}3$) crystal, which has a lattice spacing $2d = 2.750$ Å. The crystal was bent to a radius of curvature $R = 240$ cm. The nominal resolving power attainable with this crystal was $\lambda/\Delta\lambda = 22,000$ at a Bragg angle $\theta = 45°$. However, the Bragg angle to observe the Ti^{20+} and Cs^{45+} lines was about $\theta = 72°$. Because of the

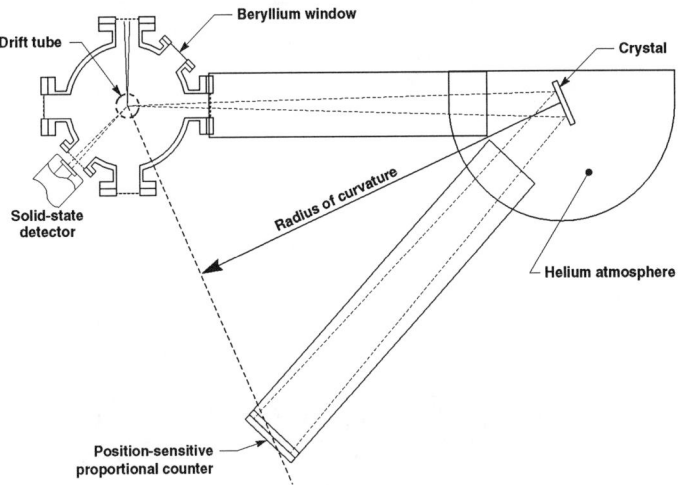

FIGURE 4. Layout of the high-resolution crystal spectrometer on EBIT. The electron beam is out of the page. The spectrometer operates in a helium atmosphere to reduce x-ray absorption by air. The vauccum interface is provided by a 125-μm beryllium widow.

$\tan \theta$ dependence the nominal resolving power thus was much higher, i.e., it was increased to $\lambda/\Delta\lambda \geq 66,000$. This value may be reduced by the intrinsic resolving power of crystal. Quartz($20\bar{2}3$), however, has an intrinsic resolving power of about 200,000 [25]. This is in excess of the nominal resolving power of the spectrometer and does not significantly reduce our estimate.

ION TEMPERATURE MEASUREMENTS

Spectra of the heliumlike Ti^{20+} intercombination lines $1s2p\ ^3P_1 \rightarrow 1s^2\ ^1S_0$, labeled y, and $1s2p\ ^3P_2 \rightarrow 1s^2\ ^1S_0$, labeled x are shown in Fig. 5. These spectra were obtained for two different operating conditions. The first was obtained with a beam current $I = 131$ mA and an axial well potential $V_{ax} = 200$ V. The second was obtained with a beam current $I = 51$ mA and an axial well potential $V_{ax} = 0$ V. The beam energy in both cases was about 5 keV. This value equaled the energy to excite some of the KMM resonances. These resonances enhanced the line emission of lines x and y by nearly a factor of two over the emission produced by direct electron-impact excitation alone [26]. The electron-ion interaction energy was slightly different for the two spectra because of the change in the space charge of the electron beam as the beam current was changed. As a result, a somewhat different set of KMM resonances was excited, and the relative intensities of lines x and y varied between two spectra.

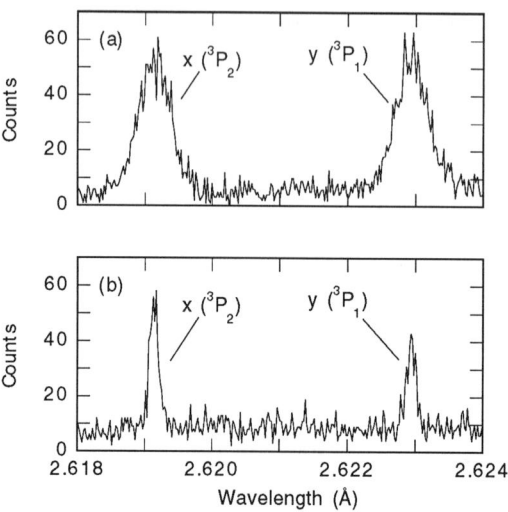

FIGURE 5. Spectra of the heliumlike Ti^{20+} intercombination lines $1s2p\ ^3P_2 \to 1s^2\ ^1S_0$ and $1s2p\ ^3P_1 \to 1s^2\ ^1S_0$, labeled x and y, respectively, obtained for two different EBIT operating conditions. The ion temperatures inferred from the line widths are 368 ± 22 and 31.2 ± 4.2 eV.

The measured line widths in Fig. 5(a) are 0.56 mÅ. These correspond to a temperature of 368±22 eV. The line widths in (b) are 0.16 mÅ and correspond to an ion temperature of 31.2 ± 4.2 eV. By contrast, the resolving power of the spectrometer corresponds to an effective ion temperature of 1.8 eV and does not contribute to the observed line widths.

The reduction in the temperature evident in Fig. 5 was achieved by reducing the axial potential and the beam current, i.e., the radial potential. The effect of each parameter separately on the ion temperature is seen in Fig. 6. In Fig. 6 (a) we plot the inferred ion temperature as a function of the applied axial potential V_{ax}. The current was kept constant at 130 mA. Reducing V_{ax} from 200 V to 0 V reduced the ion temperature from about 300 eV to slightly below 100 eV. Fixing the axial potential at $V_{ax} = 0$ V, the temperature can be lowered further by reducing the beam current. A reduction from 130 mA to 50 mA lowered T_i from about 100 to about 30 eV. In fact, we observed a temperature as low as 15.8 ± 4.4 eV under these conditions.

In Fig. 6 we also plot the dependence of the temperature of Cs^{45+} ions on the applied potential and electron beam current. The temperature was inferred from the observed Gaussian line width of the $(2p^5_{1/2}3s_{1/2})_{J=1} \to (2p^6)_{J=0}$ transition. This line is located at 2.6079 Å, and thus falls into the same wavelength region as the titanium lines. A description of the spectrum is given in the next Section. The Cs^{45+} ions were excited by a 7.5-keV electron beam, i.e., at an

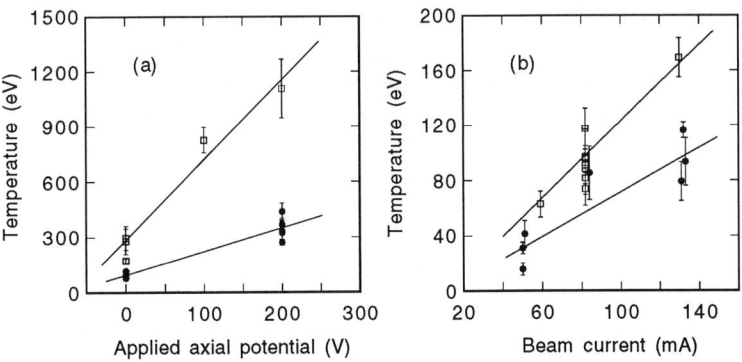

FIGURE 6. Dependence of the inferred ion temperature on (a) the applied axial potential and (b) electron beam current. Solid circles represent data from heliumlike Ti^{20+}, open squares from neonlike Cs^{45+}. The solid lines represent least-squares fits to the data and are shown as a guide to the eye only.

energy about 50% larger than that for the Ti^{20+} measurements.

The Cs^{45+} data show a similar depedence on the trap parameters as the titanium data. Lowering the potential well from 200 to 0 V reduced the temperature of the ion from well about 1200 eV to 300 eV. A further reduction was achieved by lowering the electron beam current. Choosing a beam current of 59 mA and an axial trap of 0 V, we measured an ion temperature of 59.4±9.9 eV.

The temperature of the Cs^{45+} ions is sytematically higher than that of the Ti^{20+} ions. In fact, a difference of about a factor of two to three is found. However, because the charge of the Cs^{45+} ions is higher than that of the titanium ions we expect them to be more deeply bound in the trap. Dividing the measured temperature by the charge of the ion provides a measure of the effective trapping potential. This value is about the same for both ion species for similar trapping parameters. Values less than 1 V/q are found for the shallowest trap conditions.

The low temperatures measured for shallow trap conditions contrast starkly with the temperatures measured in deep traps. The highest temperature we measured for Ti^{20+} was 439 ± 47 eV. The parameters for this measurement were $I = 135$ mA, $V_{ax} = 200$ V, and $E_{beam} = 6.5$ keV. The higest temperature for Cs^{45+} was 1145 ± 72 eV at a beam current of 145 mA, a beam energy of 7.5 keV, and an applied potential of 300 V.

OBSERVATION OF LORENTZIAN LINE WIDTHS

By the Heisenberg time-energy relations [27], electric dipole x-ray transitions in highly charged ions exhibit a large energy uncertainty. Based on the Dirac equation, Weisskopf and Wigner showed that this uncertainty results in a lifetime-limited, Lorentzian-shaped "natural" line width [28]. For an excited state decaying to the ground state, the line width ΔE is given by

$$\Delta E = \hbar/\Delta t, \qquad (6)$$

where Δt is the excited-state lifetime.

A line width $\Delta E = 0.658$ eV results for a lifetime of $\Delta t = 1$ fs. Such a width would be relatively easy to measure in a visible or UV transition, where ΔE represents a large fraction of the overall transition energy. We note, though, that the transition rates for such transitions are typically no more than 10^{13} s^{-1} reducing the natural line width accordingly. By contrast, it is much more difficult to measure a Lorentzian width in the x-ray regime, where it represents a rather small fraction of the x-ray energy.

In the following we concentrate on measuring the natural line width of the $(2p_{3/2}^5 3d_{5/2})_{J=1} \rightarrow (2p^6)_{J=0}$ x-ray transition in neonlike Cs^{45+}. This line is situated within 7.2 mÅ from the $(2p_{1/2}^5 3s_{1/2})_{J=1} \rightarrow (2p^6)_{J=0}$ transition, thus allowing observation of the two lines in a single spectrum with the same spectrometer arrangement described above. The radiative lifetime of its upper level is predicted in single-configuration calculations to be 1.39 fs, which corresponds to natural line width of 0.47 eV.

The natural line width of the $(2p_{3/2}^5 3d_{5/2})_{J=1} \rightarrow (2p^6)_{J=0}$ transition is easily masked by other broadening effects. To produce neonlike Cs^{45+} ions in the Princeton Large Tokamak required an electron temperture of about 5 keV [29]. Assuming an ion temperature of just half the electron temperature, the thermal broadening is more than three time the Lorentzian width of the $2p$-$3d$ transition, and the natural line width is obscured.

The natural line broadening of the $2p$-$3d$ transition should become obvious provided the ion temperature is less than about 200 eV, i.e., for temperatures where the Doppler-broadened line width is less than the natural width. By contrast, the predicted natural width of the $2p$-$3s$ transition is predicted to be an order of magnitude smaller than the instrumental broadening and cannot be observed. In our measurements, the $2p$-$3s$ line serves instead as an indicator of the temperature of the Cs^{45+} ions. The observed width was typically $\Delta E \geq 0.25$ eV with a Gausssian lineshape indicative of thermal Doppler broadening, from which the ion temperature T_i was determined.

Three spectra of the $2p$-$3d$ and $2p$-$3s$ transitions obtained under different operating conditions are shown in Fig. 7. The reduction in the width of both lines as the temperature is lowered is clearly seen. Unlike for the $2p$-$3s$ line, the reduction in width nearly ceases for the $2p$-$3d$ line at the lowest temperatures,

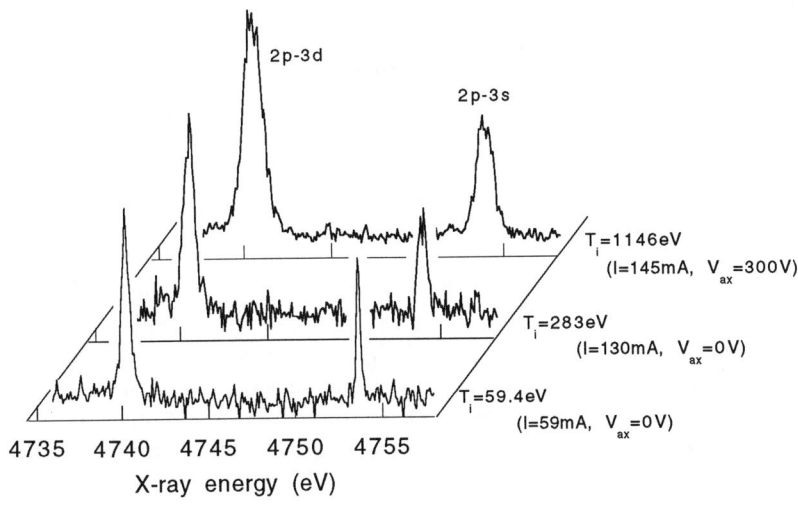

FIGURE 7. Spectra showing the two x-ray transitions from upper levels $(2p^5_{3/2}3d_{5/2})_{J=1}$ and $(2p^5_{1/2}3s_{1/2})_{J=1}$ to the $(2p^6)_{J=0}$ closed-shell neonlike ground state in Cs^{45+}. The ion temperature T_i inferred from the width of the 2p-3s line drops from 1146 eV to 59.9 eV as the electron beam current I and axial trapping potential V_{ax} are reduced to the values indicated.

and the Lorentzian line shape becomes evident. A detailed view of the line shape of the 2p-3d transition recorded at an ion temperature of 110 eV is given in Fig. 8. Fitting the line with a Lorentzian trial function provides an excellent fit. The goodness of the fit is indicated by the reduced residuals shown in Fig. 8(a), which are defined as the difference between data and fit normalized to the square root of the fit value at each point. By contrast, fitting the line with a Gaussian trial function provides a poor fit, as illustrated in Fig. 8(b).

Residual contributions of thermal broadening on the line shape are accounted for by fitting the 2p-3d line with a Voigt profile [30], which represents a Lorentzian convolved with a Gaussian profile. For this, the temperature parameter for the Gaussian profile is taken from the width of the 2p-3s line. A value of 0.398 eV averaged over all observations was obtained for the natural width of the 2p-3d line. The statistical uncertainty is ±0.012 eV. A systematic uncertainty is introduced by the possibility that natural broadening may also affect the width of the 2p-3s line. This would result in an inferred temperature that is too high. Because of the close vicinity of the upper level of each line, configuration interaction might decrease the lifetime of the $(2p^5_{3/2}3s_{1/2})_{J=1}$ level while decreasing that of the $(2p^5_{3/2}3d_{5/2})_{J=1}$ level. No lifetime-limited line broadening of the 2p-3s line was detected within the resolution of our measurements. However, the minimum amount detectable with our technique introduces a systematic error of -0.049 eV, which combines with the statistical uncertainty and yields a natural line width of $0.398^{+0.012}_{-0.050}$ eV for the $2p$-$3d$ line. From this we infer a radiative lifetime of the excited level of 1.65 fs with an uncertainty of $+0.24$ fs and -0.05 fs.

The measured value differs from the 1.39 fs predicted by single-configuration calculations. Allowing for configuration interaction by including all 36 excited levels with a vacancy in the $n = 2$ shell and an excited electron in the $n = 3$ shell in a multi-configuration Dirac-Fock calculation in the extended average level (EAL) model as described by Grant *et al.* [31], we get 1.98 fs. This value also differs from the observed value, albeit in the opposite direction than the single-configuration result. This disagreement with theory as well as between theoretical approaches shows that measurements are needed even for very fast electric dipole transitions in order to guide atomic calculations, especially when configuration interactions play a dominating role.

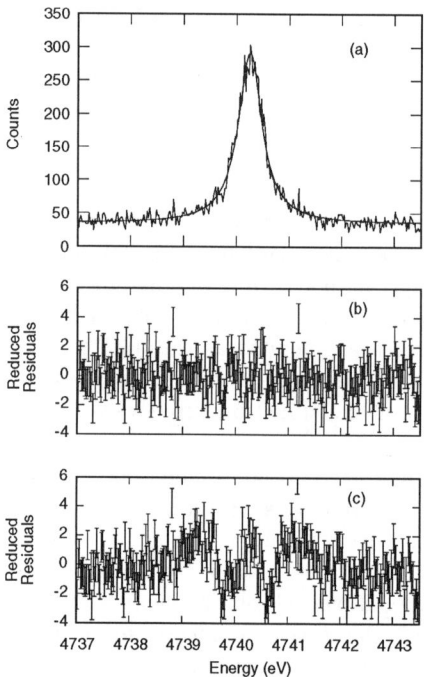

FIGURE 8. Observed lineshape of the $2p$-$3d$ transition at an ion temperature of 110 eV. The result of a least-squares fit of a Lorentzian trial function is superimposed for comparison. The reduced residuals of the fit are shown in (b). The reduced residuals of a least-squares fit of a Gaussian trial function are shown in (c).

DISCUSSION

Our measurements have demonstrated that the temperature of the ions in an electron beam ion trap can be reduced to values below 1 V/q despite continuous interactions with a multi-keV electron beam. A decoupling of the energy of the ions from that of the electrons employed for production and excitation by more than two orders of magnitude was achieved by reducing the potential well of the trap and by allowing the hottest ions to leave. Although the total number of ions in the trap dropped as the potential well was reduced, enough ions remained to prevent a significant degradation of the signal-to-noise ratio, as seen in Figs. 5 and 7. In fact, further decoupling appears possible. Because fewer photons are needed to map out a narrower line, the peak line intensity, and thus the signal-to-noise ratio, can be expected to remain nearly constant relative to the background level as the potential well and the ion energy are reduced further. Such a decoupling is crucial for precision measurements, since in the absence of x-ray or gamma-ray lasers, probing by energetic electrons is the method of necessity for studying the structure of highly charged ions.

Cooling, i.e., the reduction in the translational motion, has been applied effectively to atoms and singly charged ions to increase the precision and range of fundamental atomic and nuclear physics measurements [1]. In our progress report we showed that a controlled reduction in the translational energy is now possible for highly charged ions as well.

The production of low-temperature highly charged ions opens up a multitude of high precision studies. We have already shown that it is possible to perform measurements of the radiative lifetimes in the femtosecond regime. In this regime, no such measurements have been possible before. The fastest radiative lifetimes measured by other techniques are three orders of magnitude lower, in the ten-picosecond range [6,7]. Theoretical calculations of femtosecond radiative lifetimes, thus, have never before been validated by a measurement.

The production of low-temperature highly charged ions also allows measurements of energy levels with unprecedented accuracy. In the present measurements, relative x-ray line widths as low as $\Delta\lambda/\lambda = 5 \times 10^{-5}$ have been achieved. Because the location of a line can typically be determined to within better than a tenth of the value of its width, such small line widths, in principle, enable determinations of x-ray transition energies with a precision of one part per million with relative ease. Very precise tests of relativistic energy calculations and of QED contributions are thus possible.

Precise energy level determinations also play an important role in x-ray laser research. Here, coincidences between two x-ray lines could be used for resonant photopumping of lasing transitions. Examples are given in Refs. [32–34]. A precise experimental determination of the degree of overlap between two lines is essential for validating the feasibility of a proposed photopumping scheme.

The production of low-temperature highly charged ions also enables possible observations of x-ray lines split by the hyperfine interaction. Two obvious candidates are the $2s_{1/2}$–$2p_{1/2}$ and the $2s_{1/2}$–$2p_{3/2}$ transitions in lithiumlike Bi^{80+}. These transitions are split by about 0.8 eV [35]. This splitting requires a minimum resolving power of $\lambda/\Delta\lambda = 3000$ for the $2s_{1/2}$–$2p_{3/2}$ transition. This can only be achieved, if the ion temperature is decoupled from the estimated minimum electron energy of 30–50 keV needed for the production of such a highly charged ion.

As a final example of the possibilities created by the production of low-temperature highly charged ions we mention measurements of the isotopic variation of nuclear charge radii. The feasability of such measurements using highly charged ions has been demonstrated just recently [36] in a measurement of ^{233}U and ^{238}U. The uncertainties associated with this measurement were comparable to the uncertainties associated with the more standard methods of determining variations in nuclear charge radii. Using low-temperature ions for such a measurement could reduce the uncertainty by nearly an order of magnitude and thus make this method the method of choice for the measurement of such fundamental nuclear parameters.

ACKNOWLEDGMENTS

This work was performed at the Lawrence Livermore National Laboratory under the aupices of the Department of Energy under Contract No. W-7405-Eng-48. Support from the Office of Basic Energy Sciences is gratefully acknowledged.

REFERENCES

1. *Atomic Physics 14*, AIP Conference Proceedings No. 323, ed. by D. J. Wineland, C. E. Wieman, and S. J. Smith (AIP, New York, 1995), p. 193-275.
2. W. M. Itano, J. C. Bergquist, J. J. Bollinger, and D. L. Wineland, Phys. Scripta **T59**, 106 (1995).
3. P. Beiersdorfer, V. Decaux, and K. Widmann, Nucl. Instrum. Methods **B98**, 566 (1995).
4. P. Beiersdorfer, A. L. Osterheld, V. Decaux, and K. Widmann, (submitted to Phys. Rev. Lett.) (1996).
5. S. Diehl *et al.*, Phys. Rev. Lett. **76**, 3915 (1996).
6. L. Engström and P. Bengtsson, Phys. Scripta **43**, 480 (1991).
7. S. Cheng *et al.*, Phys. Rev. A **50**, 2197 (1994).
8. B. A. Hammel *et al.*, Phys. Rev. Lett. **70**, 1263 (1993).
9. C. J. Keane, B. A. Hammel, A. L. Osterheld, and D. R. Kania, Phys. Rev. Lett. **72**, 3029 (1994).
10. M. Bitter *et al.*, Phys. Rev. Lett. **42**, 304 (1979).

11. J. L. Culhane et al., Astrophys. J. **244**, L141 (1981).
12. T. S. Perry et al., Phys. Rev. Lett. **67**, 3784 (1991).
13. M. J. Seaton, Yu Yan, D. Mihalas, and A. K. Pradhan, Mon. Not. R. Astron. Soc. **266**, 805 (1994).
14. P. Beiersdorfer et al., in *UV and X-Ray Spectroscopy of Astrophysical and Laboratory Plasmas*, ed. by E. Silver and S. Kahn (Cambridge University Press, Cambridge, 1993), p. 59.
15. F. Bely-Dubau, A. H. Gabriel, and S. Volonté, Mon. Not. R. Astron. Soc. **189**, 801 (1979).
16. M. Bitter et al., Phys. Rev. Lett. **47**, 921 (1981).
17. A. A. Hauer, N. D. Delameter, and Z. M. Koenig, Laser and Particle Beams **9**, 3 (19921).
18. P. Beiersdorfer, V. Decaux, S. Elliott, K. Widmann, and K. Wong, Rev. Sci. Instrum. **66**, 303 (1995).
19. M. A. Levine, R. E. Marrs, J. R. Henderson, D. A. Knapp, and M. B. Schneider, Phys. Scripta **T22**, 157 (1988).
20. M. A. Levine et al., Nucl. Instrum. Methods **B43**, 431 (1989).
21. L. Spitzer, The Physics of Ionized Gases (J. Wiley & Sons, New York, 1962).
22. M. B. Schneider, M. A. Levine, C. L. Bennett, J. R. Henderson, D. A. Knapp, and R. E. Marrs, in *International Symposium on Electron Beam Ion Sources and their Applications - Upton, NY 1988*, AIP Conference Proceedings No. 188, edited by A. Hershcovitch (AIP, New York, 1989), p. 158.
23. B. M. Penetrante, J. N. Bardsley, M. A. Levine, D. A. Knapp, and R. Marrs, Phys. Rev. A **43**, 4873 (1991).
24. L. v. Hámos, Ann. der Physik **17**, 716 (1933).
25. A. Burek, Space Sci. Instrum. **2**, 53 (1976).
26. S. Chantrenne, P. Beiersdorfer, R. Cauble, and M. B. Schneider, Phys. Rev. Lett. **69**, 265 (1992).
27. W. Heisenberg, Z. Phys. **43**, 172 (1927).
28. V. Weisskopf and E. Wigner, Z. Phys. **63**, 54 (1930).
29. P. Beiersdorfer et al., Phys. Rev. A **37**, 4153 (1988).
30. A. Unsöld, *Physik der Sternenatmosphären*, 2nd ed. (Springer-Verlag, Berlin, 1955).
31. I. P. Grant, B. J. McKenzie, P. H. Norrington, D. F. Mayers, and N. C. Pyper, Comput. Phys. Commun. **21**, 207 (1980).
32. R. C. Elton, Phys. Rev. A **38**, 5426 (1988).
33. J. Nilsen, Phys. Rev. Lett. **66**, 305 (1991).
34. P. Beiersdorfer et al., Phys. Rev. A **46**, R25 (1992).
35. V. M. Shabaev, M. B. Shabaev, and I. I. Tupitsyn, Phys. Rev. A **52**, 3686 (1995).
36. S. R. Elliott, P. Beiersdorfer, and M. H. Chen, Phys. Rev. Lett. **76**, 1031 (1996).

Complete characterization of final states in double electron capture by slow multiply charged ions

Hocine Khemliche[1] and M. H. Prior

University of California, Lawrence Berkeley National Laboratory
Chemical Sciences Division
Berkeley, California 97420 U.S.A.

Abstract. We have determined the scattering angle dependence of the complex amplitudes for populating the magnetic substates of the doubly excited states (1s2l2l') ^{2}L of C^{3+} and B^{2+} following double electron capture at low energies. This is achieved by measuring the three-dimensional Auger anisotropy through detection of the energy-analyzed Auger electron in coincidence with the scattered projectile. These measurements represent the most detailed study of double electron transfer. One of the most striking results is the tendency for the angular momentum transferred to the projectile to reach its maximum value, i.e. |L|=1, 2 for ^{2}P, ^{2}D states respectively as has been observed for excitation and single capture processes.

INTRODUCTION

Double electron capture by slow multiply charged ions from neutral targets has been the subject of extensive studies (see for example 1-5); these follow upon the large body of literature devoted to single electron capture which has reached a state of maturity. Two electron capture is the simplest example of multiple electron transfer, and, as compared to the case of single capture, includes any dynamic effects of the electron-electron interaction between the active electrons. Since normally the initial state of the projectile and target are known, description of the collision process follows from determination of the final internal states of the products and their momenta.

The internal states are often revealed by spectroscopic techniques, which, for double capture, often means Auger electron spectrometry. This follows because Auger electron emission is the predominant decay path for the doubly excited states formed by the two electron transfer. In the case of a He target, the interpretation of the Auger spectra becomes unambiguous since no more than two

---

[1] Present address: KVI, Atomic Physics, Zernikelaan 25, 9747 AA Groningen, The Netherlands

electrons can be captured. Thus in principle, relative cross sections for the population of levels of (n, l)(n', l') configurations can be extracted from the relative intensities of the different Auger electron lines (1, 2, 3). However because of the alignment that occurs in these collisions, the Auger electron emission is not isotropic (except when the emission is from an S level) with respect to the polar angle, the emission angle with respect to the beam direction. For a P state, the Auger emission intensity at the so called "magic angle" (54.7 degrees) is a direct measure of the cross section; however no such angle exists for a D state or any level with angular momentum greater than one. The distribution of the Auger emission with polar angle depends upon the relative populations of the magnetic substates (6). Therefore, the measurement of that distribution allows one to extract relative cross sections for the population of the substates (7). This last step represents the most complete information that can be derived from a single particle detection. Further insight can only be gained by refined measurement of the anisotropy in the electron emission, i.e. determination of the azimuthal distribution with respect to the collision plane. This requires use of coincidence techniques, and leads to nearly full access to the final quantum state.

We present here such a measurement made by detecting the energy analyzed Auger electron in coincidence with the position resolved scattered projectile. The results provide the most detailed study available of double electron capture. However the principle of this experiment has been previously applied to study the double excitation mechanism during the collision of single charged ions with neutral targets. The authors, Kessel et al. (8) and Oud et al. (9), have measured the polar and azimuthal angular distribution of Auger electrons from the decay of D states but in coincidence with projectiles scattered at a specific angle (few degrees), whereas in the experiment presented here, all projectile scattering angles are recorded.

We have studied two systems, namely C^{5+} + He and B^{4+} + He at energies 15, 25, 50 keV and 20 keV respectively, through the reactions:

$$A^{q+}(1s) + He(1s^2) \rightarrow A^{(q-2)+}(1s2l2l')^2L + He^{2+}$$
$$A^{(q-2)+}(1s2l2l')^2L \rightarrow A^{(q-1)+}(1s^2) + e^-(E, \theta_e, \varphi_e)$$
(1)

We focus on the main states, $(1s2s2p)^2P$ and $(1s2p^2)^2D$. Since both initial and final states of the collision system are S states, the measurement of the 3 dimensional distribution of auger emission leads to the determination of the complex population amplitudes (i.e. magnitudes and relative phases) for all substates and their dependence upon the projectile scattering angle.

It is worth noting that, in the case of single capture, such a full characterization of the excited final state can be achieved through the study of the angular distribution and/or polarization of the emitted photon in coincidence with the scattered projectile (10). However this method can only be applied to dipole transitions, i.e. mainly between an excited P level decaying to a lower S level.

EXPERIMENTAL METHOD

Fig. 1 shows a schematic of the experimental arrangement with a simplified diagram of the data acquisition system. The multiply charged ions are produced by the LBNL Electron Cyclotron Resonance ion source at energies between 15 and 50 keV. After collimation by two apertures of size 1.0x1.0 mm at 50 keV and 1.5x1.5 at lower energies, they enter the collision chamber and intercept a He gas target. The Auger electrons emitted at polar angle θ_e, with energies around 160 eV for B^{2+} and 240 eV for C^{3+}, are energy analyzed by a movable parallel plate electron spectrometer with a resolution better than 1%.

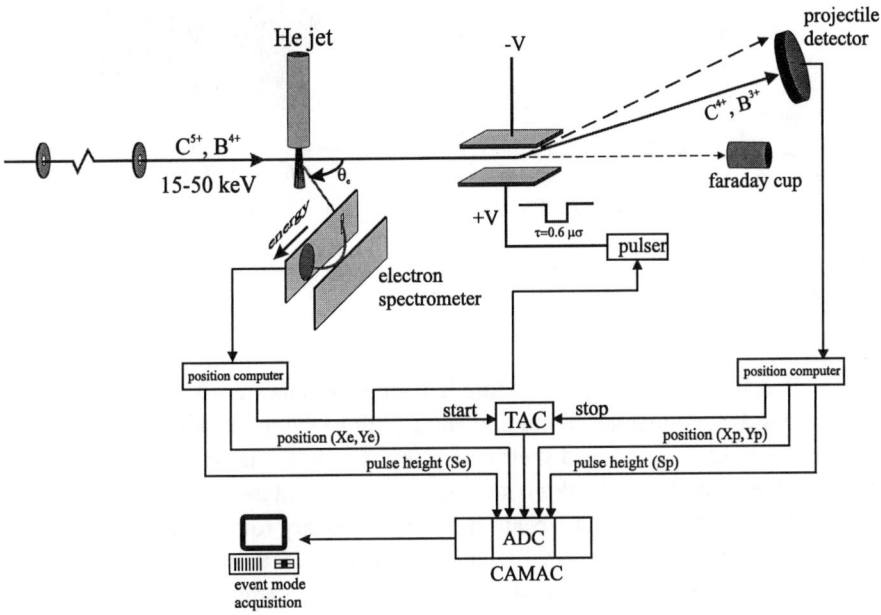

FIGURE 1. Experimental setup and data acquisition diagram.

The spectrometer is fitted with a position sensitive microchannel plate detector on the exit plane and a deceleration lens at the entrance. The electron pass-energy, typically 50-80 eV, is adjusted to resolve the Auger lines (see Fig. 2) and spread the spectrum across the position detector active area. The energy range that can be recorded simultaneously is around 50% of the pass-energy, defined as the energy measured at the detector center. Following capture and Auger decay, the projectile scattered at angle θ_s is deflected 17° onto a second position sensitive microchannel plate detector. The arrival position of the projectile determines both the collision plane and the scattering angle.

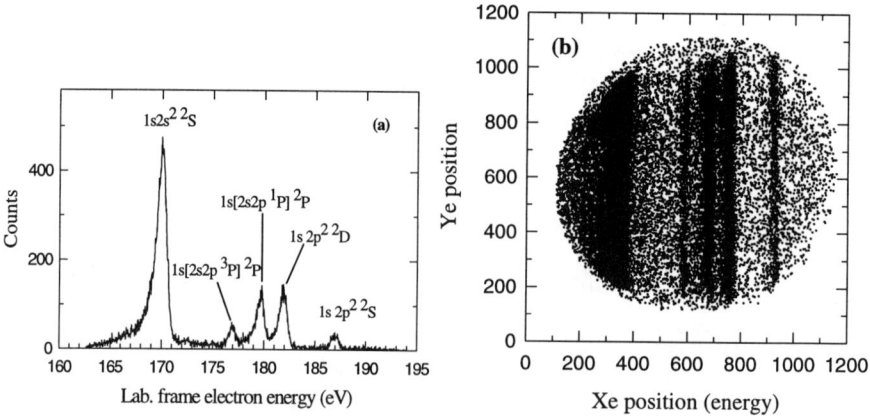

FIGURE 2. (a): Auger spectrum from B^{4+} + He at 20 keV, (b): 2D image of the spectrometer detector. (a) is the projection of (b) onto the Xe axis.

Because single electron capture has a much higher cross section and may occur from the residual gas along the whole path to the deflection plates, the rate of projectiles having changed their charge by one unit- via one electron capture compared to that from autoionizing two electron capture from the He jet- is extremely high if one uses the beam intensity necessary for a reasonable electron signal (at least few tens of pA for an electron rate of about 1-3 s^{-1}). Thus the projectile detector is easily saturated with ions that are unrelated to electrons seen by the spectrometer. To circumvent this major problem, the ions are over-deflected so they do not strike the projectile detector until an electron has been detected. The electron signal is then used to trigger a pulser that reduces the deflection voltage to direct the ions towards the projectile detector center. This pulse is applied before the ions reach the deflection plates. Following the delay needed for the ions to cross the deflection region, the voltage is brought back to its initial value for over-deflection. In this way higher beam intensities resulting in feasible electron rates can be used without saturating the projectile detector. The scattering angles are usually less than 10 milliradian, and the flight distance from the collision center is adjusted in order to optimize the spot size on the projectile detector.

Both microchannel plate detectors are mounted with resistive anodes to determine the particle impact position; the output pulses are fed into position computers which calculate the actual X and Y coordinates and also provide the signal pulse height. The latter information is exploited to discriminate one and two-particle impact on the projectile detector. For every event, seven parameters are recorded simultaneously: the electron and ion position coordinates (Xe, Ye), (Xp, Yp), their respective pulse height Se, Sp and the output of a Time to Amplitude Converter (TAC), which is started by the electron and stopped by the ion. These quantities are stored into an event-mode file for off-line analysis. For a typical current a 20 pA, the true coincidence rate is in the range 10-35 per minute.

Fig. 2a shows an Auger spectrum from B^{4+} + He at 20 keV, one notices the peak shape characteristic of PCI (Post-Collision Interaction) effect, especially for the low energy S state which has a shorter lifetime (see 7 and references herein). The spectrum is the projection of the 2D image of the detector surface (Fig. 2b) on the Xe axis. The spectrometer entrance slit is parallel to Ye, thus the Auger lines normally appear as parabolas in the 2D picture. The scatter plot of Fig. 2b has been corrected for this effect in order to provide a better resolution in the projected spectrum.

DATA ANALYSIS

The mathematics of the data analysis has already been described elsewhere (11,12), so only a summary will be given here. During the collision, each substate M within the state ^2L is populated with the complex amplitude:

$$a_M(\theta_s) = |a_M(\theta_s)| e^{i\beta_M(\theta_s)} \tag{2}$$

where θ_s represents the projectile scattering angle. When using the so called right handed laboratory frame, with Z axis along the beam and Y axis pointing from the target to the impact direction (Fig. 3a), the intensity of Auger emission in the direction (θ_e, φ_e) is described by:

$$I_L(\theta_e, \varphi_e, \theta_s) = K_L \frac{d\sigma_L}{d\theta_s} \left| \sum_M a_M(\theta_s) Y_L^M(\theta_e, \varphi_e) \right|^2 \tag{3}$$

where K_L is a geometrical factor that includes the beam intensity, the target density, etc., and Y_L^M are the spherical harmonics. One has to keep in mind that this expression holds only in the case where both target and projectile initial and final states following Auger decay are S states, otherwise expression (3) has to be summed over all initial and final substates weighted by their respective population.

The electron emission polar angle θ_e is defined by the spectrometer position, and Fig. 3b shows how the azimuthal angle φ_e is determined. The projectile arrival position (X_p, Y_p) on the ion detector fixes the collision plane (Y, Z) and, since the electron is emitted in the horizontal plane (Xp, Z), φ_e is therefore the angle between X and Xp. However it is often more convenient to describe the collision in the so called natural frame, obtained from the lab. frame by a rotation of 90° around the Y axis (Fig. 3a). In this coordinate system, the conservation of the reflection symmetry allows only population of M=±1 for the P state and M=0,±2 for the D state, which in the laboratory frame leads to the equality: $a_{+M}=a_{-M}$ for both states.

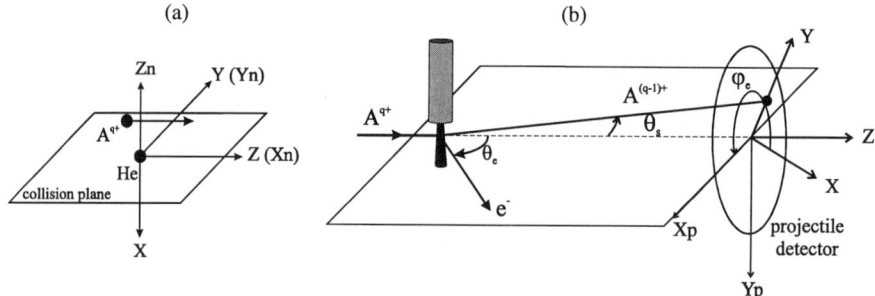

FIGURE 3. (a): Frame definition. (X,Y,Z) and (Xn,Yn,Zn) represent the laboratory and natural frames respectively. The latter is obtained from the former by a rotation of 90° around Y. (b): φ_e is determined from the projectile impact position on the ion detector.

The population amplitudes in expression (2) are first determined in the laboratory frame, they can be easily transformed into the natural frame by using the appropriate rotation matrices.

The Auger anisotropy, i.e. the left-hand side of expression (3), is obtained by gating the projectile position on both the electron line of interest (S, P or D) in the Auger spectrum and the TAC peak. This leads a distribution (Xp,Yp), which after transformation to polar coordinates becomes $Ip_L(\theta_e, \varphi_e, \theta_s)$. A similar procedure is carried by gating on the random part of the TAC spectrum, which leads to the distribution of random events $Ir_L(\theta_e, \varphi_e, \theta_s)$. Further both sets of events are divided into regions of θ_s and the difference of the subsequent φp_e and φr_e distributions represents the net azimuthal Auger anisotropy. As an example, Fig. 4 shows 2D images of the projectile detector for events coincident with the low energy ^2S and the ^2D Auger lines (Fig. 4a and 4b respectively). As expected the pattern for the former case is isotropic, whereas the latter shows a fourfold structure.

The basis of the data analysis is to fit the net φ_e distribution for each region θ_s with the right-hand side of expression (3) to extract the quantum quantities, which are for the P state $|a_1|$, $|a_0|$ and the relative phase $\Delta\beta_{01} = \beta_0 - \beta_1$,

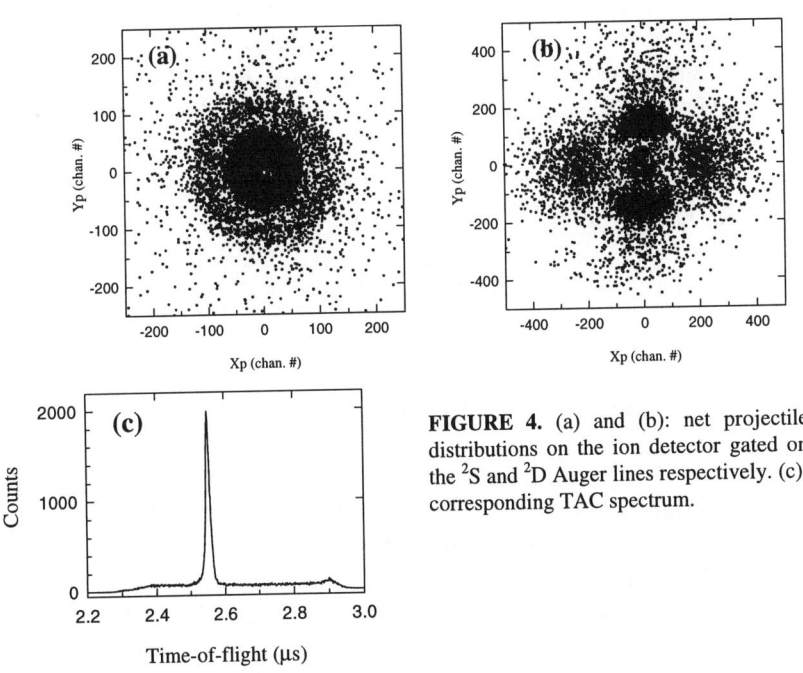

FIGURE 4. (a) and (b): net projectile distributions on the ion detector gated on the ^2S and ^2D Auger lines respectively. (c): corresponding TAC spectrum.

and for the D state $|a_2|$, $|a_1|$, $|a_0|$, and two relative phases, e.g., $\Delta\beta_{12} = \beta_1 - \beta_2$ and $\Delta\beta_{02} = \beta_0 - \beta_2$. However, the spectrometer position, i.e. the choice of the polar angle, gives usually access to only some of these parameters. For instance, at $\theta_e = 90°$:

$$I_P(90,\varphi_e,\theta_s) = K_P \frac{d\sigma_P}{d\theta_s} \frac{3}{4\pi} |a_1|^2 (1 - \cos(2\varphi_e)) \quad (4)$$

$$I_D(90,\varphi_e,\theta_s) = K_D \frac{d\sigma_D}{d\theta_s} \frac{5}{16\pi} \cdot \left[6|a_2|^2 + |a_0|^2 - 2\cdot\sqrt{6}\ |a_0||a_2|\cos(\Delta\beta_{02})\cos(2\varphi_e)\right] \quad (5)$$

TABLE 1. Quantities accessible at the respective polar angles.

θ_e	^2P	^2D
90°	$|a_1|^2$	$|a_2|^2$, $|a_0|^2$, $\cos(\Delta\beta_{02})$
54.7°	$|a_1|^2$, $|a_0|^2$, $\sin(\Delta\beta_{01})$	$|a_2|^2$, $|a_1|^2$, $\sin(\Delta\beta_{12})$
45°	$|a_1|^2$, $|a_0|^2$, $\sin(\Delta\beta_{01})$	$|a_2|^2$, $|a_1|^2$, $|a_0|^2$, $\sin(\Delta\beta_{12})$, $\cos(\Delta\beta_{02})$, $\sin(\Delta\beta_{01})$

Thus, in order to make the fitting procedure as reliable as possible, we record several sets of data at specific polar angles where the number of fitting parameters is reduced. To further restrict the degrees of freedom in the fit, all data sets are fitted simultaneously. Table 1 gives a list of polar angles used and the quantities that are actually accessible for both P and D states. It is worth noting that at 45°, the relative phases of the D state are related through: $\Delta\beta_{12} - \Delta\beta_{02} + \Delta\beta_{01} = 0$, furthermore, the populations are normalized; that is:

$$\sum_M |a_M|^2 = 1 \quad (6)$$

RESULTS AND DISCUSSION

Part of the C^{5+} results have been presented elsewhere (11, 12). Here we compare some of these results to recently available calculations (13) and present new results from B^{4+} + He. In all the following figures, the experimental points are fitted with a spline curve to guide the eye.

Fig. 5 and Fig. 6 show the different quantities extracted from C^{5+} + He at 25 keV, compared to the close coupling calculation by C. D. Lin and W. Fritsch (13). The theoretical results have been convoluted with our experimental resolution of around 1.2 mrad., and the agreement is often satisfactory. There is no explanation of the relative shift seen in the differential cross-section. The interpretation of these quantities, and especially the relative phases, is not obvious but they can be exploited to draw the 3D Auger anisotropy, which exhibits the same symmetry as the actual charge cloud (14). Snapshots of this Auger anisotropy from C^{3+} are show in Fig. 7 for few scattering angles. One can notice the flat shape of the distribution, i.e. the capture takes place mainly in the scattering plane. This effect is more important for large impact parameters where this plane is well defined.

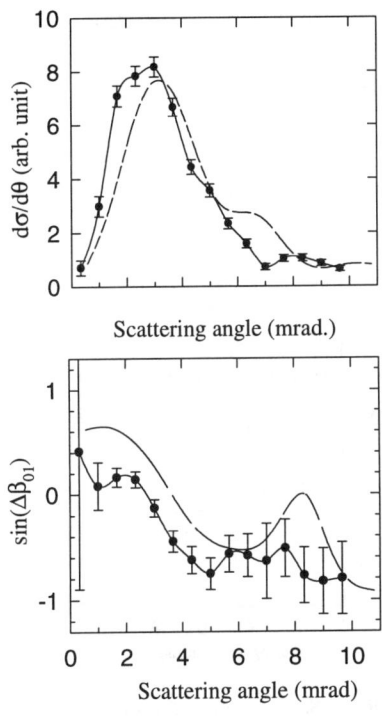
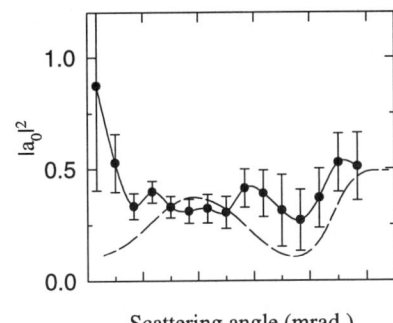

FIGURE 5. Results for C^{3+} ($1s2s2p$ ^2P) at 25 keV. Dashed curves are from close-coupling calculation convoluted with our experimental resolution.

These pictures are also a good tool for evaluating at first sight the strength of the rotational coupling which can have two effects: (i) during the rotation of the internuclear axis, it acts as a friction causing a lag in the rotation of the charge cloud, (ii) it allows transition between Σ and Π states (or Π and Δ states), that is, induces population of final substates with $|M| > 0$. Furthermore, quantities from Fig. 5 and 6 give access to a critical parameter: the angular momentum transferred to the projectile during the collision. It is perpendicular to the collision plane and is defined as:

$$L_\perp^P = \frac{|a_1|^2 - |a_{-1}|^2}{|a_1|^2 + |a_{-1}|^2} \qquad \text{for the P state} \qquad (7)$$

$$L_\perp^D = \frac{2(|a_2|^2 - |a_{-2}|^2)}{|a_2|^2 + |a_{-2}|^2 + |a_0|^2} \qquad \text{for the D state} \qquad (8)$$

when the population amplitudes are expressed in the natural frame. Fig. 8a and Fig. 8c show its behavior for C^{5+} (15 keV) on He as a function of scattering angle, also shown is the so called alignment angle γ between the beam direction and the ^{2}P state major axis.

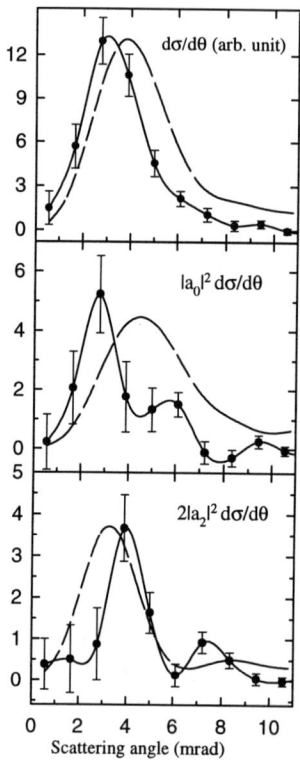
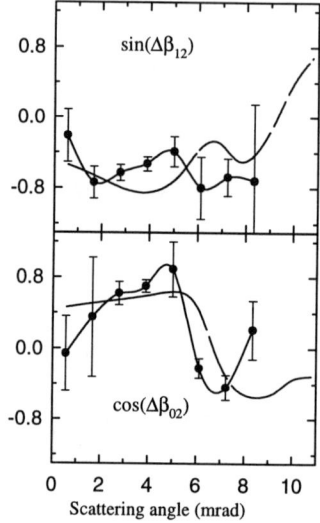

FIGURE 6. Same as in Fig. 5 for C^{3+} ($1s2p^2$ ^{2}D)

The behavior of γ, illustrated in Fig. 8b, clearly shows in this case the development of the rotational coupling induced lag for decreasing impact parameters, i.e. increasing scattering angles. Regarding the behavior of $L_\perp^{P,D}$, they both nearly reach their maximum value around the maximum of the cross-sections, namely ±1 and ±2 in units of $\hbar$ for P and D respectively. As can be seen in table 1, we do not have direct access to the relative phases but only to their sine or cosine. This leaves ambiguity on the absolute sign of $L_\perp$ as determined from the experiment, however theory is expected to clarify this uncertainty. In Fig. 8a and 8c, the label $\pm L_\perp$ corresponds to the experimental points, whereas the theoretical curves display $+L_\perp$.

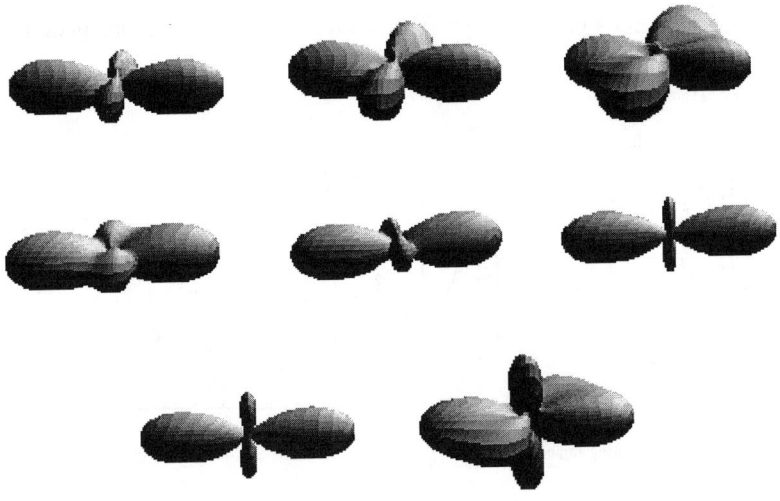

FIGURE 7. Three-dimensional images of the Auger anisotropy from C^{3+} ($1s2p^2$ ^2D) at 15 keV. The corresponding scattering angles are from left to right, starting from the top: 2.8, 3.6, 4.4, 5.2, 6.0, 6.8, 7.5 and 8.3 mrad. The charge clouds exhibit the same symmetry.

In the case of the P state, the value of angular momentum gives full information on the relative populations (M=±1) in the natural frame, but for the D state the situation is more complicated because of the possible population of M=0 substate in the natural frame. Thus in the latter case, it is fruitful to calculate these fractions from the ones derived in the laboratory frame (see previous chapter), they are shown in Fig. 8d.

We notice that population of M=0 is mostly responsible for the loss of angular momentum at larger scattering angles, whereas the population of M=±2 (depending on the sign of $L_\perp^D$) remains very weak. We notice that theory does not reproduce the behavior of $L_\perp^D$, whereas the agreement is acceptable for $L_\perp^P$ and γ. The experimental results from the second system investigated are displayed in Fig. 9. Both P and D levels have nearly the same energies, although their differential cross-sections peak at well separated scattering angles. This is an indication of rather different capture pathways in the molecular curve diagram (15). We notice also the logical increase of $|a_2|^2$ with scattering angle, since its population is made possible through the rotational coupling. Similarly, $|a_1|^2$ for the P state is weak for small scattering angles, and then increases for larger angles. However these simple intuitive features can be strongly perturbed by the interplay between the numerous possible capture pathways and the interference pattern that is often induced.

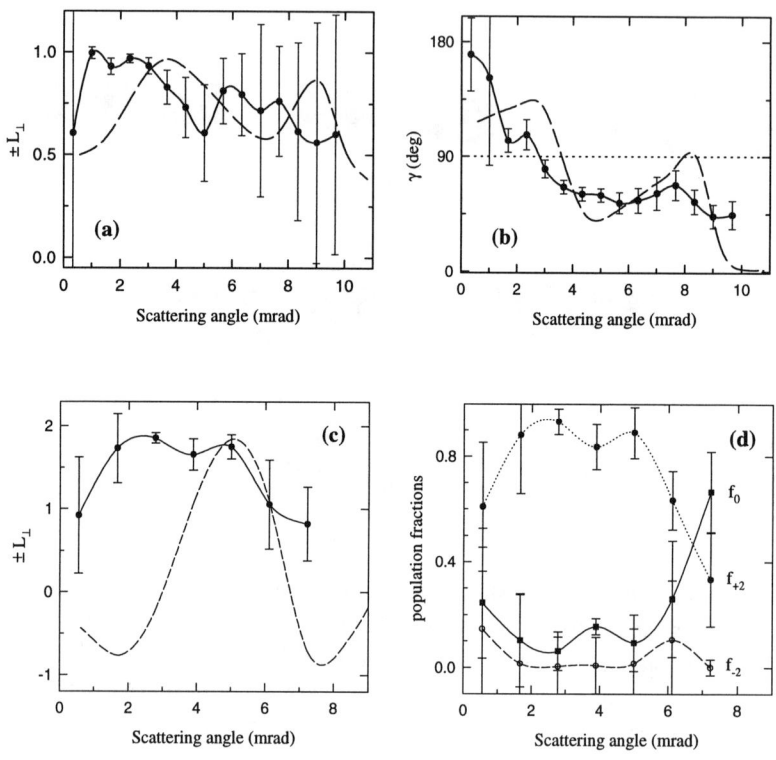

FIGURE 8. (a) and (c): angular momentum for C^{3+} ^2P and ^2D state respectively. (b): alignment angle from the P state. (d): population fractions in the natural frame for the D state. The dashed curves in (a, b, c) are from close-coupling calculations.

As in the case of C^{5+}, we can calculate the angular momentum transferred to the projectile, the fractional populations in the natural frame for the D state, and the alignment angle γ for the P state, these are shown in Fig. 10. The angular momenta display the same tendency as is seen in the case of C^{5+} at all energies: they reach or approach their maximum value around the maximum of the cross-section.

The tendency of the angular momentum to reach −1 for a P state and −2 for a D state, often referred to as a the propensity rule, has been previously calculated (16, 17) and observed (18, 19) in S→P and S→D excitation by atom and ion impact, and calculated (20, 21, 22) and observed (10) in single capture into P states. This behavior can be intuitively understood from the tendency of the electrons to follow the rotation of the internuclear axis during the collision.

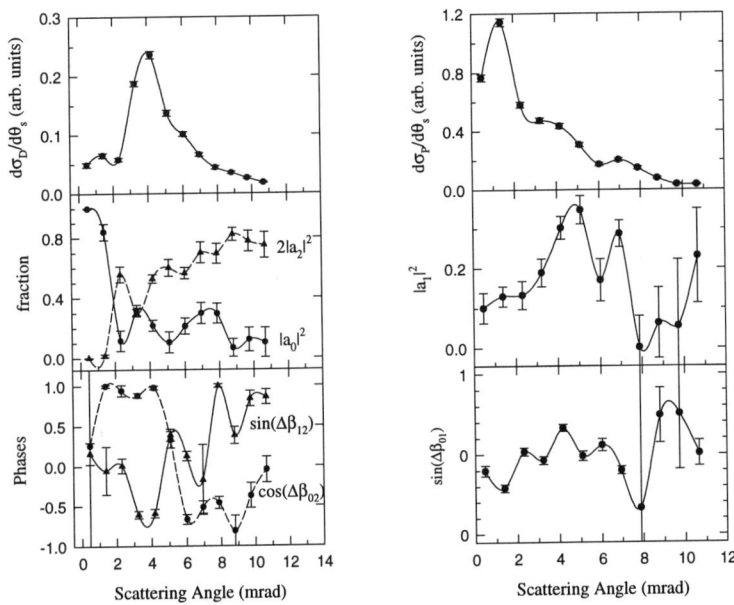

FIGURE 9. Results for $B^{2+}(1s2s2p)$ ^2P (left side) and $B^{2+}(1s2p^2)$ ^2D (right side) at 20 keV.

Our results for double capture are consistent with these observations, though the agreement is incomplete since the absolute sign is not determined. For the system C^{5+} + He, the calculations by C. D. Lin and W. Fritsch (13) show that this tendency holds for capture to $C^{4+}(1s3p)$, which is the dominant single capture channel. Furthermore, they show that double capture happens partially through the intermediate single capture channel. However, their calculated $L_\perp$ exhibit a sign opposite to that expected if the propensity rule is to be valid for double capture.

From the D state fractional populations in the natural frame (Fig. 8d and 10d), we observe that the loss of angular momentum is mostly due to the increasing population of M=0 substate, the population of one of the other two allowed substates (f_{+2} or f_{-2} depending on the sign of $L_\perp^D$) remains near zero for all scattering angles. This remarkable effect is in agreement with the angular momentum being induced by the rotation of the internuclear axis. In the natural frame, the velocity matching condition implies that f_{-2} is preferably populated, and for large scattering angles, i.e. for close collisions, the classical electron angular momentum with respect to the projectile is reduced in favor of f_0. Regrettably, the calculated $L_\perp$ do not fit this physical picture.

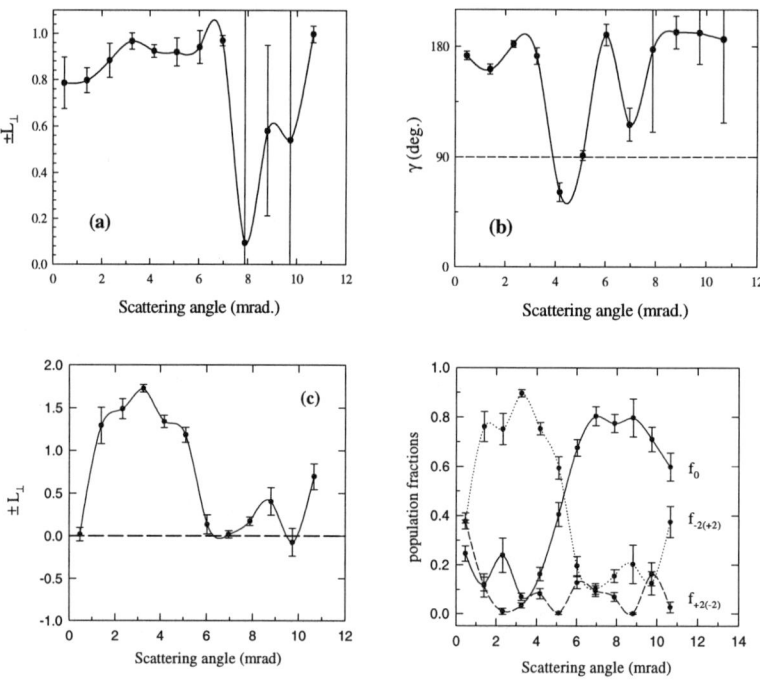

FIGURE 10. Same as Fig. 8 for B^{2+} ^2P and ^2D states at 20 keV.

In conclusion, we have developed a technique for measuring the scattering angle dependence of substate complex populations following double electron capture at low energies. The extremely detailed description derived from these measurements for C^{5+} and B^{4+} on He represents a challenge for theory which, to some extent, is able to reproduce many features of the experimental results. A better agreement is observed for the $(1s2s2p)^2P$ configuration for which the anisotropy is only due to one of the captured electrons, as opposed to the more complex $(1s2p^2)^2D$ configuration.

Such refined studies, both experimental and theoretical, provide a large amount of new information, which can form the basis for a better understanding of double electron transfer in slow collisions. This understanding might include simple intuitive mechanisms or models capable of accounting for the major features of these collisions. The experiments described here are the first of their kind. Further investigations, e.g., a systematic study of similar or, perhaps, simpler systems, together with advanced calculations are needed to determine whether intuitive physical models or only the full power of modern calculations are necessary to describe slow two electron transfer collisions.

ACKNOWLEDGMENTS

We gratefully thank C. Lyneis and Z. Xie of the LBNL Nuclear Science Division for valuable assistance and D. Schneider of the Lawrence Livermore National Laboratory for the loan of parts of the experimental setup. This work was supported by the Director, office of Energy research, Office of Basic Energy Sciences, U.S. Dept. Of Energy under contract No. DE-AC03-76SF00098.

REFERENCES

1. E. M. Mack, Thesis (1987)
2. M. Mack and A. Niehaus, *Nucl. Inst. Meth. B* **23**, 109-115 (1987)
3. R. Mann, *Phys. Rev. A* **35**, 4988-5004 (1987)
4. P. Roncin et al., *J. Phys. B* **22**, 509-524 (1989)
5. C. Harel and H. Jouin, *J. Phys. B* **25**, 221-237 (1992)
6. E.G. Berezhko and N. M. Kabachnik, *J. Phys. B* **10**, 2467-2477 (1977)
7. M. H. Prior, R. A. Holt, D. Schneider, K. L. Randall, and R. Hutton, *Phys. Rev. A* **48**, 1964-1974 (1993)
8. Q. C. Kessel, R. Morgenstern, B. Muller, and A. Niehaus, *Phys. Rev. A* **20**, 804-812 (1979)
9. M. Oud, S. F. te Pas, W. B. Westerveld, and A. Niehaus, *J. Phys. B* **26**, 1641-1653 (1993)
10. P. Roncin, C. Adjouri, N. Andersen, M. Barat, A. Dubois, M. N. Gaboriaud, J. P. Hansen, S. E. Nielsen, and S. Z. Szilagyi, *J. Phys. B* 27, 3079-3091 (1994)
11. H. Khemliche, M. H. Prior, and D. Schneider, *Phys. Rev. Lett.* **74**, 5013-5016 (1995)
12. M. H. Prior and H. Khemliche, "Double electron capture: complex amplitudes from Auger anisotropy measurements", in *Proceedings of the Conference on the Physics of Electronic and Atomic Collisions*, 1995, pp. 557-566.
13. W. Fritsch and C. D. Lin, to be published in *Phys. Rev. A*, (1996)
14. P. Van der Straten and R. Morgenstern, *Com. At. Mol. Phys.* **17**, 243-260 (1986)
15. W. Fritsch and C. D. Lin, *Phys. Rev. A* **45**, 6411-6416 (1992)
16. N. Andersen and S. E. Nielsen, *Z. Phys. D* **5**, 309-319 (1987)
17. S. E. Nielsen and N. Andersen, *Z. Phys. D* **5**, 321 (1987)
18. G. S. Panev, N. Andersen, T. Andersen and P. Dalby, *Z. Phys. D* **5**, 331(1987)
19. N. Andersen, T. Andersen, P. Dalby and T. Royer, *Z. Phys. D* **9**, 315 (1988)
20. N. Toshima and C. D. Lin, *Phys. Rev. A* **49**, 397-401 (1994)
21. N. Toshima and C. D. Lin, *Phys. Rev. A* **47**, 4831-4836 (1993)
22. J. P. Hansen, L. Kocbach, A. Dubois, and S. E. Nielsen, *Phys. Rev. Lett.* **64**, 2491-2494 (1990)

The dynamics of target single and double ionization induced by the virtual photon field of fast heavy ions

R. Moshammer [1], J. Ullrich [2], H. Kollmus [1], W. Schmitt [2], M. Unverzagt [1], H. Schmidt-Böcking [1] and R.E. Olson [3]

[1] *Institut für Kernphysik, August-Euler-Strasse 6, D-60486 Frankfurt, Germany*
[2] *GSI, Planckstrasse 1, D-64291 Darmstadt, Germany*
[3] *Department of Physics, University of Missouri-Rolla, Missouri 65401, USA*

Abstract

The collision dynamics of helium single and double ionization after impact of 3.6 MeV/u Se^{28+} projectiles has been explored using the high resolution electron - recoil-ion momentum spectrometer at GSI. The complete three particle final state in momentum space was determined with a resolution of $\Delta p \approx \pm 0.1$ a.u. in the case of single ionization by measuring the three momentum components of the emitted electron and the recoiling target-ion in coincidence. For double ionization the longitudinal momenta of both individual electrons were determined in coincidence with the recoil-ion. The final momenta of the electrons reflect the correlated motion in the ground state of helium (neon) during a time interval (collision time) which is short compared to the average revolution frequency in the bound state. The ionization of the target atom shows correspondence to photodisintegration by the equivalent photon field of the passing fast highly charged projectile.

INTRODUCTION

The dynamical response of a correlated quantum mechanical many-particle system under the action of a time dependent perturbation is one of the most fundamental subjects in atomic physics. In contrast to the enormous precision which is obtained in the investigation of the static structure of many-electron atoms, basic and unsolved problems are still present in the understanding of the simple dynamic reactions like double ionization of helium by charged particle [1] impact. Moreover, the dynamic correlation of the two electrons in the ground state of helium is not analytically calculable in closed form and approximations of the two-electron ground state wave function contain a large number of adjustable parameters. A correct and full quantum mechanical calculation for the evolution of the correlated motion of the two electrons from the ground state into the final many particle continuum state is presently beyond theoretical capabilities. There is only one quantum mechanical theory [2] available that treats double ionization of helium by ion impact with closed coupling methods including the (e-e) interaction which correctly predicts ratios of double to single ionization cross sections but, unfortunately, no differential cross sections were calculated. This is a pity because our data give clear indications that the momentum correlation between both electrons in the final state after double ionization by 3.6 MeV/u Se^{28+} reflect to a certain extent the initial two-electron ground state of helium. Thus, besides double photoionization of atoms (γ,2e experiments) [3, 4] which is known to be sensitively dependent on the (e-e) correlation, double ionization induced by ion impact delivers further information concerning the short time dynamic correlation of many electron systems in their ground state.

Concerning fast ion induced single ionization of atoms considerable progress has been achieved in the past in describing the ionization collision dynamics with quantum mechanical [5], semiclassical [6] and classical [7] treatments. In the regime of high velocities for small perturbations, where the Born approximation is applicable, methods to calculate the complete three body kinematics have been developed [8]. With increasing perturbation (for $q/v_p > 1$; q, v_p: projectile charge and velocity) the dipole approximation breaks down and three body interactions strongly influence the electron emission characteristics. Here it was found that almost all the electrons are emitted into the forward direction [9] due to the "post collision interaction (PCI)" with the outgoing projectile. The emitted electron is attracted while the recoiling target-ion is pushed backward by the receding projectile ion. Two-center effects on the electron emission have been included recently [5] and total ionization cross sections as well as the details of the electron emission were successfully calculated. Since, however, not the complete three-particle problem is solved, the transverse scattering of the recoil-ion and of the projectile is not accessible within these approaches. Thus, the very fundamental three-particle problem, single ionization of a target atom by an ion in the non-perturbative regime, is still out of the capabilities of present quantum mechanical calculations. On the other hand tremendous progress has been achieved using semiclassical or classical methods. Classical trajectory Monte Carlo (CTMC) calculations are the only treating the full 3-particle problem without any approximation beyond the classical approximation for the interaction of the particles. Quantum behavior is partly included statistically by using the correct quantum mechanical momentum distribution of the electron in the bound state of an hydrogen atom. Methods have been developed for treating many electron transitions (n-body CTMC) [7] accounting for the (e-e) interaction in the ground state. More sophisticated techniques were reported to treat the (e-e) interaction during the collision accounting for the monopole part of the interaction (dynamical screening: dCTMC) [10] as well as on its full implementation [11]. Since only classical calculations are available presently predicting differential cross sections for single and double ionization of helium, parts of our data are compared to results obtained with the nCTMC approach.

Experimentally, no kinematically complete data set for target ionization by ion impact has been published in literature up to now providing information on the electron emission, the recoil-ion scattering, the energy loss and the angular straggling of the projectile. Only two studies have been presented recently where the longitudinal momentum balance was completely determined for single [12] and double ionization [11] of helium. In contrast, single ionization by electron impact has been extensively investigated in kinematically complete so called (e,2e) experiments (for a review see [13]). Even for double photoionization a considerable amount of complete (γ,2e) studies have been performed (see [4] and references therein). The almost complete lack of such data for ion impact is mainly due to the enormous difficulties one is faced with in performing such experiments. Nearly 90% of the electrons are emitted with energies below 50 eV and the recoiling target-ions of interest have energies in the sub-meV regime. For most collision systems the change in projectile energy and transverse momentum (scattering angle) are not accessible directly by analyzing the outgoing projectile. The involved tiny relative projectile energy losses and scattering angles would require a beam quality which is beyond the capabilities of even the best available accelerators or ion storage rings. Thus, the only possible strategy leading to kinematically complete experiments is the coincident detection of the emitted electron and the recoiling target-ion where the final momenta of both have to be determined with sufficient resolution. During the last few years efficient recoil-ion detection techniques, based on ultracold supersonic jet-targets, have been developed which are sensitive to such small energy transfers to the target nucleus [14]. Efficient methods for the detection of low energy electrons, which have completely been missing, were developed at GSI [15] recently. By combining these techniques kinematically complete experiments on fast ion induced target ionization have become feasible for the first time.

EXPERIMENT

The reaction products after target ionization by fast ion impact, the emitted electrons and the scattered recoil-ion, were analyzed with high resolution and extremely high coincidence efficiency by means of the recently developed combined electron recoil-ion momentum spectroscopy [15, 16]. An internally cold (50 mK), dense (10^{12} cm^{-2}) and well defined ($\Delta x \approx 2$ mm) target is provided by a two stage supersonic gas jet. This atomic beam is crossed with a well collimated beam (1mm × 1mm) of 3.6 MeV/u Se^{28+} ions delivered from the UNILAC at GSI. The projectiles were charge state analyzed after the collision and Se^{28+} ions (no charge exchange) were recorded by a fast scintillation counter with a rate of up to 1 MHz. Recoil-ions and electrons produced in the reaction zone are extracted along the ion beam into opposite directions by a weak uniform electric field of 1-5 V/cm applied over a length of 22 cm (Fig.1). After extraction the recoil-ions and the electrons both drift over 22 cm before they are post accelerated (2000 V for recoil-ions and 200 V for electrons) and detected by two-dimensional position-sensitive channel-plate detectors. An additional homogeneous magnetic field of typical 10 - 20 Gauss, generated by two Helmholtz coils (1.5 m diameter), is applied almost parallel to the electric field forcing the electrons on cyclotron trajectories. This way the electrons are guided onto the detector guaranteeing a high detection efficiency. From the position of detection and the time of flight, measured in an electron - recoil-ion - projectile coincidence, the recoil-ion charge state and the three momentum components of both, the recoil-ion and the emitted electron, can be deduced. All recoil-ions of interest and all electrons ($\Delta \Omega = 4\pi$) with energies up to 50 eV are projected onto the detectors. Due to detector efficiencies and grid transmissions a total triple coincidence efficiency of about 20 % was achieved. When a double ionization event occurs the time of flight of both electrons is registered if their spacing in time is more than 8 ns. This allows us to extract the longitudinal momenta of both electrons in coincidence with the complete momentum vector of the He^{2+} recoil-ion.

Our new technique for low energy electron detection removes many of the tremendous experimental difficulties of conventional spectrometers. The target extension is well defined by the supersonic jet. Electrons from the residual gas are completely suppressed in the triple coincidence spectra. The influence of electric and magnetic fringe fields is drastically reduced by extracting the electrons, and the final charge state of the target and that of the projectile is well defined. Furthermore, since the inelasticity, or the Q-value, of the reaction is measured in addition, accompanied electronic excitation of the projectile is excluded.

In the near future fast multihit capable electron detectors with delay-line readout [17] will be implemented. This then enables kinematically complete

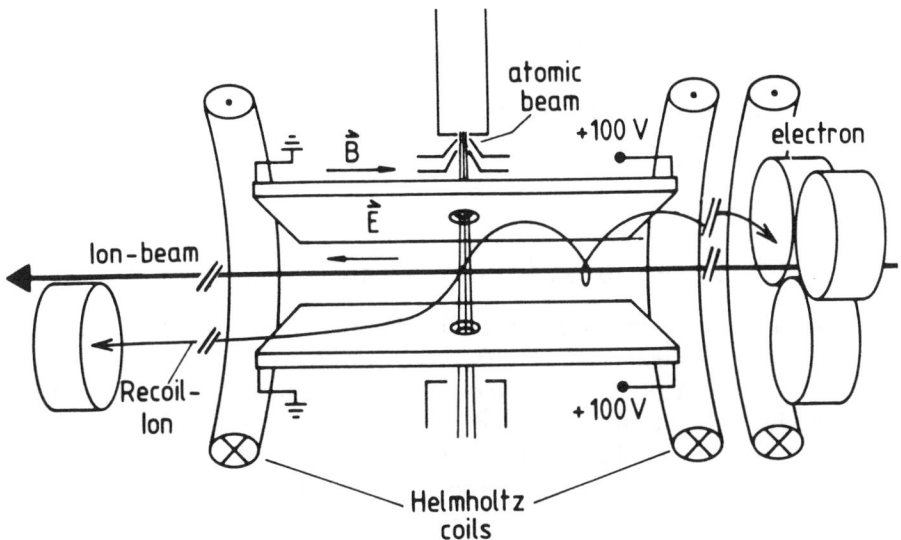

FIGURE 1: Schematic drawing of the combined recoil-ion many-electron momentum spectrometer

experiments for a large variety of collision induced reactions.

RESULTS

From the high-resolution measurement of the complete momentum vectors (with $\Delta P \leq \pm 0.1$ a.u.) of both, the recoil-ion and the emitted electron, the kinematics of single ionization is completely determined. The missing momentum components (the three components of the projectile momentum change in our case) of the three particles in the final state as well as the inelasticity (the Q-value) of the reaction can be deduced from momentum and energy conservation laws. This way the energy loss of the 0.28 GeV Se^{28+} projectile ion was determined with a resolution of $\Delta E_p/E_p \approx 10^{-7}$ and scattering angles as small as $\Delta\vartheta \approx \pm 60$ nrad became accessible. This would require the measurement of a deflection of only 0.6 mm on a detector placed 10^4 m behind the target region if the outgoing projectile would be analyzed directly. In the case of double ionization the complete longitudinal momentum balance was measured in coincidence with the full He^{2+} recoil-ion momentum vector.

Since all reaction products are detected simultaneously the absolute cross section is easily obtained by normalizing the sum of all events on the measured total single ionization cross section of $\sigma^{1+} = (3.3\pm 0.5)\times 10^{-15} \text{cm}^2$ [18]. In the case of double ionization the data are normalized on 50% of the total double ionization cross section to correct for the estimated loss of solid angle due to the limited electron energy acceptance of $E_e < 50$ eV. The corresponding experimental total cross section is $\sigma^{2+} = (4.9 \pm 0.5) \times 10^{-16} \text{cm}^2$ [18].

Longitudinal momentum balance

The longitudinal momentum balance (along the ion-beam direction) for helium single and double ionization are shown in Fig.2. As in previous measurements with 3.6 MeV/u Ni^{24+} [12] and with 5.9 MeV/u U^{65+} [19] projectiles the electrons are found to be emitted dominantly into the forward direction and their longitudinal (sum)momentum is almost completely balanced by the backscattered recoil-ion. The momentum transferred by the projectile is small compared to the measured final momenta of the recoil-ions and the electrons. Moreover, the width obtained for the projectile momentum loss distribution is mainly determined by our experimental resolution. For large projectile velocities v_p its momentum change $\Delta P_p = (Q + E_e)/v_p$ (Q: initial binding energy of the active electron, E_e: electron continuum energy) becomes extremely small and approaches the minimum possible momentum transfer for ionization by a photon in the limit of $v_p \to c$. At 3.6 MeV/u ($v_p = 12$ a.u. and $v_p/c \approx 0.1$) it follows that $\Delta P_p < 0.25$ a.u. for the overwhelming part of all ionization reactions with $E_e < 50$ eV. Thus, the observed much broader widths of the recoil-ion and the electron momentum distributions are not a result of direct momentum transfer but must have been "stored" in the bound state of the helium before the encounter. Therefore they closely reflect the single and two-electron Compton profile in the helium bound state prior to single and double ionization respectively. For negligible momentum transfer by the projectile it follows from momentum conservation that $\sum P_{e||} \approx -P_{R||}$ and, thus, that the recoil-ion mirrors the electron (sum)-momentum or, in other words, their center of mass motion. This feature is clearly visible in the data for single ionization as well as for double ionization. The passing projectile delivers energy but only very little momentum: The target atom dissociates in the strong and long ranging electric field of the projectile ion. Thus, the action of the fast heavy projectile reveals similarities to photoionization where the electron perfectly balances the recoil-ion momentum to the extend of the negligibly small momentum transferred by the absorbed photon. In fact, the field of a swift charged projectile can be described as an intense electromagnetic pulse containing a broad band of photon frequencies. In the framework of the Weizsäcker Williams method target ionization is described as absorption of those virtual photons by the atom [20]. Since almost no momentum is carried by the incident photon only those can interact whose energy corresponds to the momentum of the electron in the bound state at the instant of absorption. Hence the obtained momenta of the electrons and the recoil-ions are directly related to the internal momenta in the helium atom at the instant of collision with the projectile ion.

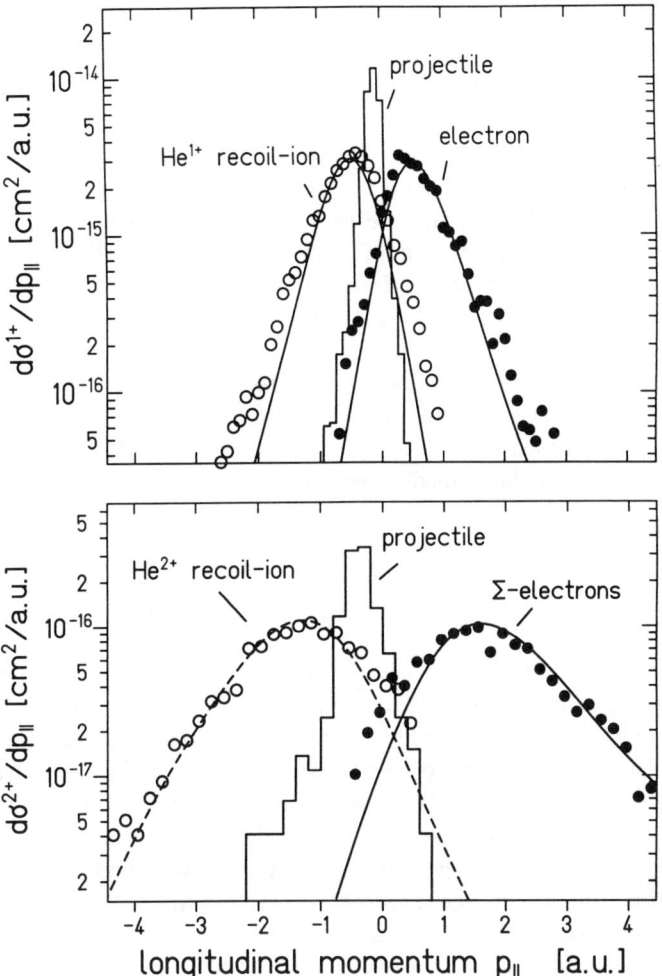

FIGURE 2: Longitudinal momentum distributions of the recoil-ions, the electrons (their sum-momentum is plotted in the case of double ionization) and the momentum change of the projectiles for single and double ionization of He with 3.6 MeV/u Se^{28+} (see text). Lines: Results of nCTMC calculations normalized on the experimental cross section.

The momentum distributions of the electrons and the recoil-ions are considerably shifted into the forward and backward direction respectively. This asymmetry can be ascribed to the influence of the strong and long ranging potential of the outgoing projectile leading to a "pulling behind" of the electron and a "pushing away" of the remaining target ion. This "post collision interaction" (PCI) is very well predicted by the classical calculation (nCTMC) demonstrating the importance of treating the 4-particle problem completely. Although nCTMC underestimates the total single ionization cross section the predicted shapes of the distributions are in excellent agreement with the experimental data. In this model the two independent and distinguishable "classical" electrons are set on different Kepler orbits bound with the sequential binding energies (nCTMC). This way the total binding energy of the initial state is correct and the initial state single electron momentum distribution as well as the two-electron sum-momentum distribution in the bound state is reproduced. Obviously these requirements have to be fulfilled to reproduce the experimental data. Explicit quantum mechanical features like the (e-e) correlation due to the symmetry of the many electron wave function and the electron-electron interaction are not essential to describe the recoil-ion momentum distributions or the longitudinal electron (sum)momentum. This changes when the electron-electron correlation in the final state is considered as will be discussed in one of the following sections.

Details on He single ionization

In contrast to the longitudinal direction where recoil-ion and electron momenta are linked via the energy conservation equation the transverse momentum balance (perpendicular to the ion-beam direction) reveals the full three body dynamics. Here, the transverse recoil-ion momentum reflects the transverse momentum of the ejected electron and in addition the internuclear momentum exchange between the target and the projectile. In Fig.3 the momentum distribution of the emitted electron and the momentum change vectors of the outgoing projectile are projected onto the azimuthal plane (the plane perpendicular to the ion-beam direction) and plotted with respect to the recoil-ion transverse momentum vector. For different conditions on the transverse recoil-ion momentum vectors, as indicated by the arrows in Fig.3, the corresponding electron and projectile momenta are displayed. We want to emphasize that the transverse projectile momentum change is directly connected to the scattering angle $\vartheta_p = P_{p\perp}/P_0$ ($P_{p\perp}$: transverse projectile momentum change, P_0: incoming projectile momentum) and that a $P_{p\perp}$ of 1 a.u. corresponds to a deflection angle of only 6×10^{-7} rad. Similar to the longitudinal direction only

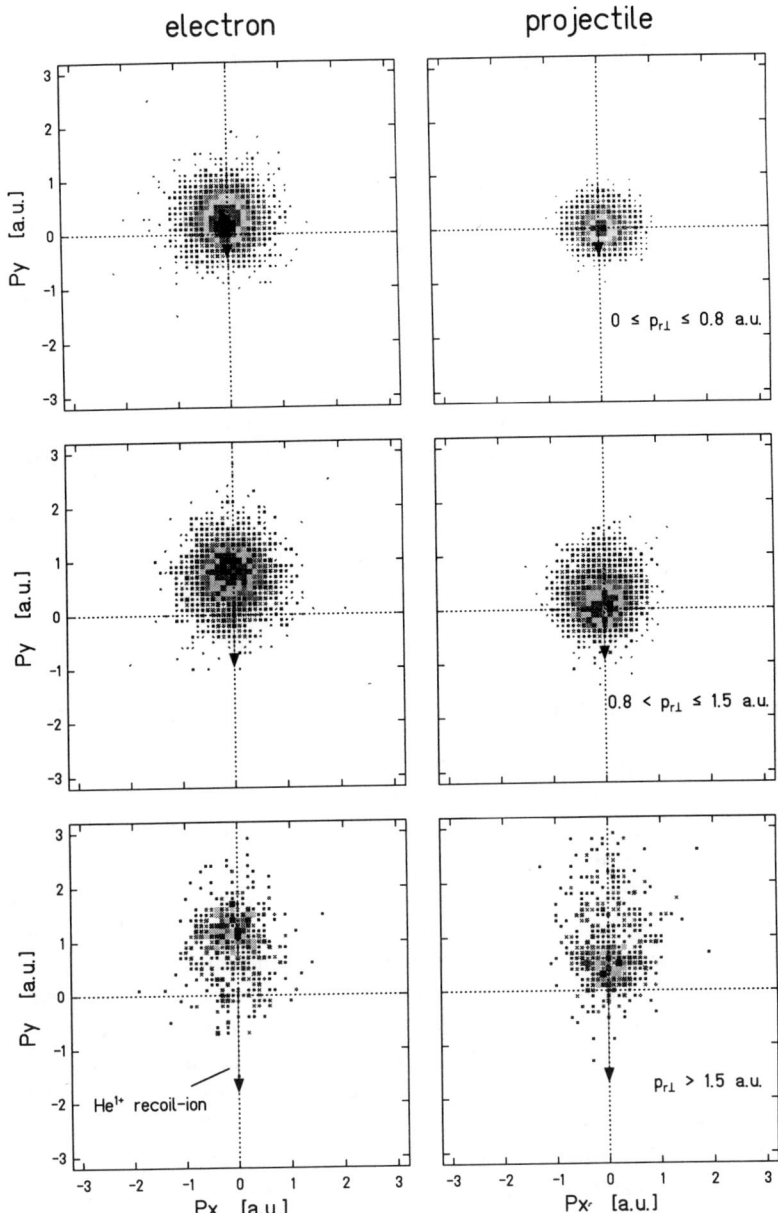

FIGURE 3: Projections of the projectile (right part) and the electron (left part) momenta onto the azimuthal plane (the plane perpendicular to the beam direction) for selected transverse recoil-ion momenta. The data are plotted on a logarithmic scale. The He^{1+} recoil-ion is scattered into the $-P_y$ direction with a transverse momentum as indicated by the length of the arrows.

a negligibly small amount of momentum is transferred in the transverse direction by the projectile. Its deflection is almost independent of the magnitude of the transverse recoil-ion momentum and only for large recoil-ion momenta the direct nucleus-nucleus interaction leads to an increase of projectile scattering angles. Obviously, for practically all collisions, the projectile scattering can not be described as a two-body collision with either the target nucleus or the electron. One important consequence is the general impossibility to extract the involved impact parameters for these soft collisions leading to single ionization of helium. The projectile interacts with the whole target atom delivering energy but only very little momentum demonstrating again the correspondence to photoionization. Moreover, the preferred back to back emission of recoil-ion and electron (see Fig.3) and the fact that their momenta compare well to the internal momenta of the target atom supports the interpretation that ionization is a result of the interaction with the equivalent photon field of the passing projectile ion. Although the Weizsäcker Williams formalism fails in reproducing i.e. the total cross section, since it is valid only in the limit of high velocities, the assumption that dipole excitations contribute mainly to ionization is still valid in the present regime of large perturbations.

Electron-electron correlation after double ionization

Up to now double photoionization has been considered the ideal reaction to study the correlated motion of electrons since the ejection of two electrons after absorption of a single photon is prohibited in the independent particle approximation and therefore sensitively depends on the details of the electron-electron interaction. In contrast, using fast heavy ions as projectiles in the regime of large perturbations, helium double ionization is dominated by direct and independent interactions of the projectile with both individual electrons during one encounter and no momentum or energy exchange between the electrons is required to obtain double ionization. This is well known from the velocity dependence of the double to single ionization cross section ratio (R=15% for the present collision system) considerably exceeding the velocity independent "high velocity limit" for charged particle impact of R$\approx$ 0.25% [21] as well as the maximum value obtained by photoionization of R$\approx$ 3% [4]. Thus, one might argue that results obtained with highly charged projectiles are well described within the framework of the independent electron approximation and hence are not an adequate probe to study the electron-electron correlation. Instead, however, the results demonstrate that only negligibly small momentum is transferred to the atom during the ionization and that the final momenta of the ejected electrons after single ionization closely reflect their momentum distribution in the bound state. The same is true for double ionization.

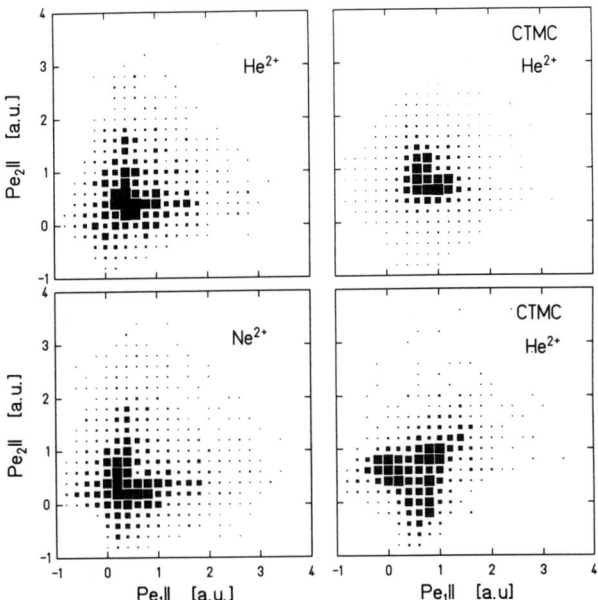

FIGURE 4: $P_{e1\parallel}$ versus $P_{e2\parallel}$ of the two electrons emitted in He and Ne double ionization by 3.6 MeV/u Se^{28+} impact (left part). Right part: results of CTMC calculations using two different simplified models for the correlated ground state of helium (see text).

Here, the observed longitudinal momentum distributions display the correlated sum-momentum distribution in the bound state, i.e. the two-electron Compton profile. The electrons correlated motion can be explored in more detail when the final longitudinal momenta of both electrons are plotted in a 2-dimensional representation. In Fig.4 the experimental $P_{e1\parallel}$-versus-$P_{e2\parallel}$ spectra for helium and neon double ionization are compared to simplified classical calculations. This representation can be considered as plotting the electrons longitudinal center of mass momentum $(P_{e1\parallel} + P_{e2\parallel})$ versus their momentum difference $(P_{e1\parallel} - P_{e2\parallel})$ if the figure is rotated by 45°. Two distinct features of the experimental data are remarkable. First, both electrons are mainly emitted into the forward hemisphere which has been identified before to be a result of the PCI. Second, if one electron is fast the other electron most probably has a small longitudinal momentum or, in other words, with increasing electron sum-momentum their momentum difference also increases. (Surprisingly, this feature is even more pronounced for double ionization of neon.) This cannot be explained with any independent particle model indicating that our data are sensitive to the (e-e) correlation caused by the direct electron-electron interaction due to the $1/r_{12}$-potential and the symmetry of the two-electron wave function.

We have performed various CTMC calculations to elucidate the influence of any electronic initial state correlation on the electrons final momentum correlation. Within a classical calculation it is impossible to correctly describe the bound initial two-electron ground state since the explicit inclusion of the $1/r_{12}$-interaction leads to autoionization of the unperturbed helium atom. Sophisticated methods have been developed to circumvent this problem partly by switching on the $1/r_{12}$-potential at a time when one electron is in a continuum state [11]. Already the results obtained with the most simple classical model for the initial state, the Bohr-model, are illustrative and helpful for the interpretation of our data. Here, the two-electron helium atom is approximated by two independent electrons moving on identical circular orbits around the target nucleus [23]. The orbits are selected by the condition that each electron is bound with one half of the total electronic binding energy and both orbits are in the same plane (a 2-dimensional atom). Two distinctively different configurations with specific correlations between the electrons are possible: First (upper right part of Fig.4), both electrons are orbiting in the same direction of rotation staying always on opposite sides of the nucleus with ($\mathbf{v}_{e1} = -\mathbf{v}_{e2}$). Second (lower right part of Fig.4), they circulate with opposite direction of rotation and thus zero total angular momentum ($\mathbf{L}_{e1} = -\mathbf{L}_{e2}$). It seems not surprising that a qualitative agreement with the experimental data is achieved only with the model where both electrons move in a highly correlated manner on opposite sides of the target nucleus. The completely different results of these calculations demonstrate that the final electron momenta sensitively depend on the facets of the initial state correlation. Furthermore, a direct comparison of the data with the quantum mechanical two-electron momentum distribution (see [11]) of the ground state of helium revealed qualitative similarities and led to the suggestion that the initial state two-electron wave function is mapped in a direct and unperturbed way.

This interpretation might also be understood considering the time dependent electromagnetic field of the projectile as an extremely short ($\Delta t < 10^{-17}$ sec) and intense (I$\approx 10^{18}$ W/cm^2) pulse of virtual photons covering a broad band of photon energies up to hundreds of eV. Both target electrons are independently "photoionized" by the absorption of a photon whose energy corresponds to the individual electron momentum at the instant of absorption. Two essential requirements have to be fulfilled to extract the initial state correlation from the final momenta: First, the collision should be soft or, equivalently, the momenta imparted by the absorbed photons should be small compared to the momentum of each bound electron. Under this condition the initial correlation of the many-electron system is not significantly perturbed. Second, the collision time should be short compared to the electron revolution time in the

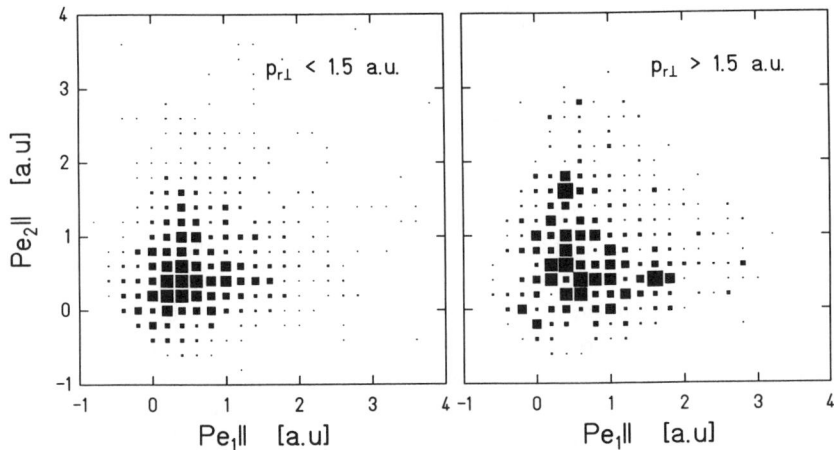

FIGURE 5: $P_{e1\parallel}$ versus $P_{e2\parallel}$ after He double ionization under the condition of large ($P_{r\perp} > 1.5$ a.u.) and small ($P_{r\perp} < 1.5$ a.u.) transverse momenta for the He^{2+} recoil-ion.

bound state. Then the positions of the electrons do not change significantly during the photoabsorption and no momentum exchange between the electrons or between each electron and the nucleus occurs.

Both requirements are fulfilled. As demonstrated by our results, the total momentum transfer from the projectile is small and the back to back emission of recoil-ion and electron indicates the importance of dipole transitions. For the present collision system typical impact parameters of about $b = 1.5$ a.u. are deduced from classical calculations and with a beam velocity of $v_p = 12$ a.u. the collision time can be estimated to be $\Delta t \approx 0.1$ a.u.. The orbital time period of a bound helium electron is about 0.25 a.u..

The collision time is further a direct measure of the photon energies involved which are high for close collisions. Assuming that in the case of double ionization the recoil-ion transverse momentum is large for small impact parameters and vice versa then the experimental data allow a selection of impact parameters or photon frequencies. The $P_{e1\parallel} - P_{e2\parallel}$ pattern should depend on the transverse recoil-ion momentum. This dependence is shown in Fig.5 for helium double ionization where the final electron momenta are plotted for large ($P_{r\perp} > 1.5$ a.u) and small ($P_{r\perp} < 1.5$ a.u) recoil-ion momenta. Obviously the emitted electrons are faster for close collisions (large $P_{r\perp}$) or, in other words, for small collision times and therefore high equivalent photon energies. Moreover, a more pronounced pattern is observed compared to that of Fig.4 (upper left part).

CONCLUSIONS

It has been demonstrated in a kinematically complete experiment on helium single ionization that a fast highly charged projectile is an extremely soft tool to ionize a target atom on a time scale which is short compared to the ground state electron revolution time. The interaction of the outgoing projectile with all participating particles causes a considerable forward-backward asymmetry of recoil-ion and electrons. These final state interactions between all particles in the continuum deliver an ideal situation to investigate the three and four-body Coulomb continuum. In the case of single ionization it has been shown that the attempt to separate the many particle problem into two body interactions fails. For the present collision system the strongest correlation occurs between the recoil-ion and the electron and not, as might be expected, between the recoil-ion and the projectile via nuclear Coulomb deflection. Thus, and very important, it is definitely impossible for the major part of all collisions resulting in single ionization to extract the impact parameter from any observable quantity. The projectile merely transfers momentum to the target electrons, acting very much like a photon field. Since no momentum or energy exchange between the two electrons is required to obtain double ionization the final two electron momentum distributions have been demonstrated to be a sensitive probe of the initial (e-e) correlation. The same method, termed as "Heisenberg microscope", has recently been used to explore the correlated motion of the two halo neutrons in ^{11}Li [22]. In a kinematically complete experiment fast ^{11}Li projectiles were photo disintegrated with only little momentum transfer in the virtual photon field of heavy target nuclei and the angular correlation between the neutrons in the bound state were deduced.

In the near future experiments will be performed using 1 GeV/u U^{92+} projectiles from the SIS at GSI extending the investigations to the highest possible perturbation in the high velocity limit. Then the modification of the electron spectra by the final state interaction with the receding projectile will be decreased by about a factor of 5 substantially reducing the influence of the PCI. The collision time will be less than 1% of typical revolution frequencies of outer-shell electrons in bound states of atoms, molecules or clusters. In combination with further substantial improvements to detect up to five electrons simultaneously, which seems to be feasible, this then allows to take even better "snapshots" of the short time correlation of many electron systems in their ground state. Fast GeV/u highly charged ions represent a unique source of extremely intense, attosecond (10^{-18} sec) and broadband pulses of equivalent photons. One might envision that kinematically complete experiments on double (multiple) ionization will become a standard "attosecond microscope" for the investigation of the bound state many-electron correlation in atoms, molecules, clusters or even solids.

REFERENCES

[1] J.H. McGuire, J. Phys. B **28**, 913 (1995).

[2] A.L. Ford and J.F. Reading, J. Phys. B **21**, L685 (1988).

[3] M. Pont and R. Shakeshaft, Phys. Rev. A **51** R2676 (1995).

[4] R. Dörner et al., Phys. Rev. Lett. **76**, 2654 (1996).

[5] P.D. Fainstein, V.H. Ponce and R.D Rivarola, J. Phys. B. **24**, 3091 (1991).

[6] M. Horbatsch, Phys. Rev. Lett. A **137**, 466 (1989).

[7] R.E. Olson, J. Ullrich and H. Schmidt-Böcking, Phys. Rev. A **39**, 5572 (1989).

[8] H. Fukuda, I. Shimamura, L. Vegh and T. Watanabe, Phys. Rev. A **44**, 1565 (1991).

[9] S. Suárez, C. Garibotti, W. Meckbach, and G. Bernardi, Phys. Rev. Lett. **70**, 418 (1993).

[10] V.J. Montemayor and G. Schiwietz, Phys. Rev. A **40**, 6223 (1989).

[11] R. Moshammer, J. Ullrich, H.Kollmus, W. Schmitt, M. Unverzagt, O. Jagutzki, V. Mergel, H. Schmidt-Böcking, C.J. Woods and R.E. Olson, Phys. Rev. Lett. **77**, 1242 (1996).

[12] R. Moshammer, J. Ullrich, M. Unverzagt, W. Schmitt, P. Jardin, R.E. Olson, R. Mann, R. Dörner, V. Mergel, U. Buck and H. Schmidt-Böcking,
Phys. Rev. Lett. **73**, 3371 (1994).

[13] I.E. McCarthy and E. Weigold, Rep. Prog. Phys. **54**, 789 (1991)

[14] J. Ullrich, R. Dörner, V. Mergel, O. Jagutzki, L. Spielberger and H. Schmidt-Böcking,
Comments At. Mol. Phys. **30**, 285 (1994).

[15] R. Moshammer, M. Unverzagt, W. Schmitt, J. Ullrich and H. Schmidt-Böcking,
Nucl. Instr. Meth. B **108**, 425 (1996).

[16] H. Kollmus, W. Schmitt, R. Moshammer, M. Unverzagt and J. Ullrich,
Nucl. Instr. Meth. B (to be submitted).

[17] S. Sobottka and M. Williams, IEEE Trans. Nucl. Sci. **35**, No.1 (1980).

[18] H. Berg et al., J. Phys. B **25**, 3655 (1992).

[19] M. Unverzagt, R. Moshammer, W. Schmitt, R.E. Olson, P. Jardin, V. Mergel, J. Ullrich and H. Schmidt-Böcking,
Phys. Rev. Lett. **76**, 1043 (1996).

[20] E.J. Williams, Phys. Rev. **45**, 729 (1934).
J.D. Jackson, Classical Electrodynamics (Wiley, New York, 1975).

[21] J. Ullrich, R. Moshammer, H. Berg, R. Mann, H. Tawara, R. Dörner, J. Euler, H. Schmidt-Böcking, S. Hagmann, C.L. Cocke, M. Unverzagt, S. Lencinas, V. Mergel,
Phys. Rev. Lett. **71**, 1697 (1993).

[22] K.Ieki et al., Michigan State University MSUCL-1036 (1996).

[23] K. Richter, G. Tanner and D. Wintgen, Phys. Rev. A **48**, 4182 (1993).

Electron Emission for He Double and Single Ionization After Heavy Ion Impact

Siegbert Hagmann

*J.R. Macdonald Laboratory, Department of Physics,
Kansas State University, Manhattan, KS 66506-2601*

The double ionization of He has been a central topic of interest in the effort to understand the general dynamics of multiple ionization processes. Substantial progress has been brought by extended studies, both experimentally and theoretically, of the ratio R of total double to single ionization cross sections for photoionization, Compton scattering and for charged particle impact; in the latter case collisions with weak to strong perturbation strength q/v were investigated by varying incident projectile charge state q and collision velocity v. Ratios R for double to single ionization have been found to be as high as 0.3 for low velocities and to monotonically decrease down to 2.6×10^{-3} for asymptotically high velocities and for photoionization to converge to 1.7×10^{-2}, both in good agreement with theory [1-4]. Whereas double ionization in the case of photoionization arises from electron correlation in the target atom, for charged particle impact this is only true for asymptotically high velocities with a small perturbation strength q/v.

The situation is more complex in slow collisions and strong perturbations and where the collision strength is high.

It has been argued that here the independent interaction of the highly charged projectile with both electrons ought to be the main channel contributing to double ionization of the target. This conclusion, however, cannot unambiguously be reached solely from the ratios of total recoil production cross sections R being large in comparison to the asymptotic value and from also rising steeply with increasing perturbation strength because all relevant dynamical parameters were integrated over.

A meaningful signature of double ionization via independent interaction of the projectile with the target electron can only be the spectral shape of the electron continua associated with single and double ionization, i.e. doubly differential cross sections coincident with He^{2+} and He^{1+}. For consecutive independent events leading to double ionization the shape of the continua coincident with He^{1+} and $2+$ recoils is then expected to differ only with respect to the role of increased binding energy of the electron to be ionized in the second step.

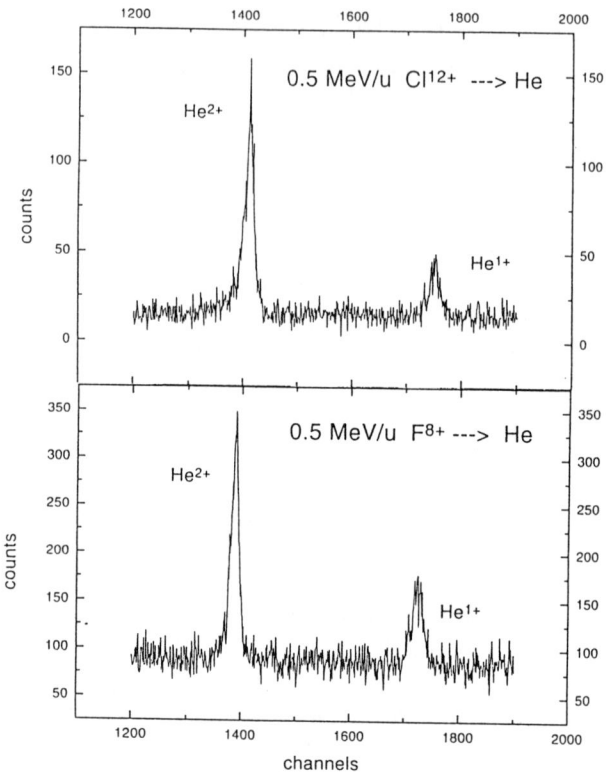

FIG. 1. Electron recoil time of flight spectrum.

A first step in this direction to create a deeper insight into the mechanism contributing to double ionization has been taken: for very low electron energies in the continuum where electrons can be energy analyzed using time of flight techniques COLTRIMS has recently provided a complete kinematical analysis for some selected cases [3].

The experiments reported here on the other hand focus on higher electron energies in the continuum between 100 eV and 1000 eV, i.e., electron velocities comparable to collision velocities. We have measured doubly differential cross sections DDCS for electron emission for single and for double ionization of He in the range of strong perturbation q/v between 0.2 and 3.

The experiment was performed at the EN tandem of the J.R. Macdonald Laboratory using projectiles ranging from F to Cl in charge states 2^+ to 12^+ for characteristic energies between 0.5 and 2 MeV/u, i.e., a perturbation strength between .2 and 3.

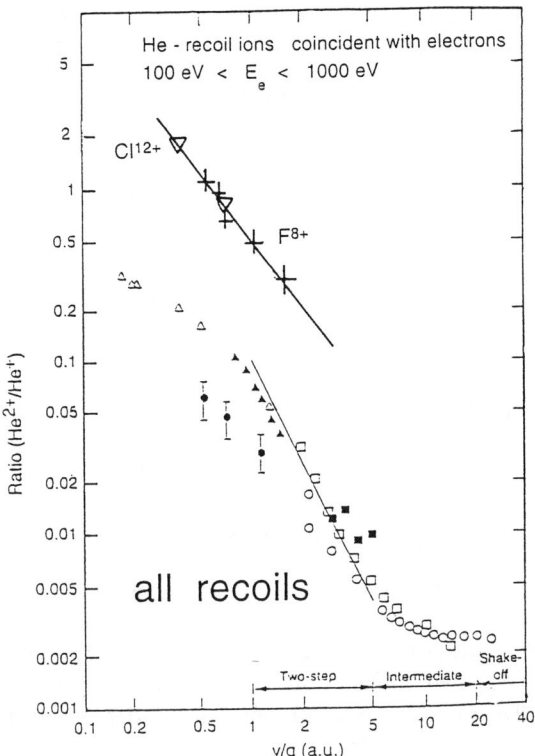

FIG. 2. Cross Section ratios He^{2+}/He^{1+} as function of inverse perturbation strength (F^{8+}, Cl^{12+}=present data, other data are from Ref. 1).

Electrons were energy analyzed using a toroidal electrostatic electron spectrometer which allows for electron emission angles between 0° and 180° simultaneous detection on a channelplate detector equipped with a 2D position sensitive anode. Electrons with energies between 100 eV and 1000 eV were detected in coincidence with recoil ions which were charge state analyzed using a time of flight spectrometer (Fig. 1).

We find that ratios for recoil production cross sections R where a condition on the energy of the emitted electron is imposed are significantly higher than those for cross section ratios where all recoil ions produced are counted (Fig. 2). This is true for all collision systems investigated. R increases strongly as a function of electron energy between 100 eV and 400 eV from R=1.1 to R=3.8 and remains constant for electron energies above 400 eV (Fig. 3).

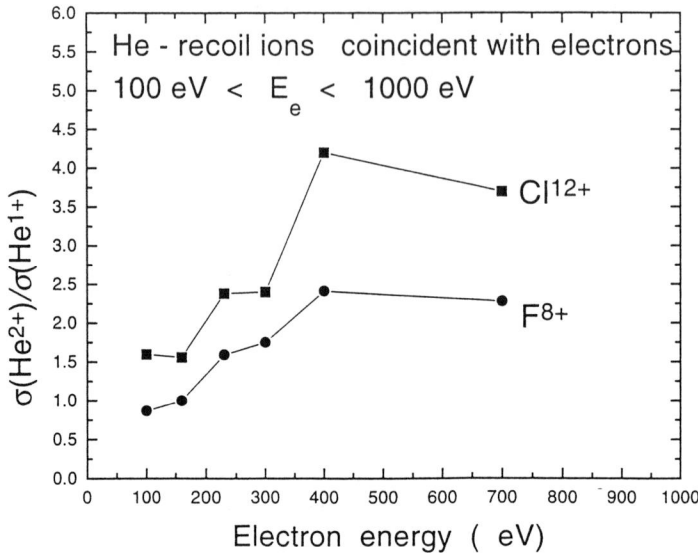

FIG. 3. Cross section ratio He^{2+}/He^{1+} as function of electron energy.

In the present result electron emission angles have been integrated between 0 and 180°. More detailed information on the dynamics of the collision mechanism lies in the DDCS for electron emission for well defined recoil charge states He(2+) and He(1+), reflecting SI and DI respectively. Preliminary results for electron emission angles between 45 and 90° show that for single ionization the cross section decreases monotonically with electron energy. For DI, however, the cross section steeply rises with electron energy up to 340 eV, i.e 5 a.u velocity and then decreases again with further increase of E.

The spectral shapes associated with DI and SI differ significantly; it is thus apparent that in the strong perturbation region the model of two independent interactions of the projectile with the target electrons cannot be the dominant channel as had been assumed previously. At present work on complete angular distributions, also for systems over a wider range of perturbation strength is in progress.

This work was supported by the Division of Chemical Sciences, Office of Basic Energy Sciences, Office of Energy Research, U.S. Department of Energy.

1. J. Ullrich, et al., Nucl. Inst. and Meth. B**87**, 70 (1994).

2. R. Doerner, et al., Phys. Rev. Lett. **76**, 2654 (1996).

3. R. Moshammer, et al., Phys. Rev. Lett. **77**, 1241 (1996).

4. J. Burgdoerfer, et al. Invited talk, X-96, 17th Conf. on x-ray and inner-shell phenomena, Hamburg, Sept. 1996.

IV. INSTRUMENTATION AND METHODS

Bent Crystal Optics for High Energy Synchrotron Radiation

P. Suortti*, U. Lienert* and C. Schulze[†]

*European Synchrotron Radiation Facility, B.P. 220, F-38043 Grenoble, France
[†]Swiss Light Source Project Paul Scherrer Institute, CH-5232 Villigen PSI, Switzerland

Abstract. The use of elastically bent perfect crystals as optical elements of high energy synchrotron radiation beamlines is reviewed. The geometrical principles of focusing are described, and formulas for focal lengths and energy band-passes are given for transmission (Laue) and reflection (Bragg) cases. The effects of bending on the reflectivity of the crystal are discussed within models that combine trajectories of the beam inside the crystal with the conditions of dynamical diffraction. It is shown that the reflectivity of the crystal can be tailored for a given application by changing the bending radius, asymmetric cut, and thickness. Slightly different x-ray energies are reflected at different depths, so that the reflectivity curve is broadened. At high energies, where the crystals become almost transparent, large gains of intensity are achieved. This is important and very useful for x-ray spectroscopy of weak scattering, where the resolution and efficiency of the crystal spectrometer must be optimized. The integrated reflectivity can reach the kinematical limit.
In the Laue case the reflectivity curve is almost flat-topped, and the maximum reflectivity may be close to unity. Calculations based on the Penning-Polder model agree well with the measured reflectivity curves. In the Bragg case the reflectivity curve is calculated using a layer-crystal model, and also these results are substantiated by experimental results.
The above ideas are used in several constructions of crystal monochromators and analyzers at the ESRF. These include focusing single-bounce Bragg-type monochromators for inelastic scattering experiments, thick asymmetrically cut Laue-type monochromators for scattering studies above 200 keV, and tuneable Laue-Bragg monochromators with fixed exit beam and focal length for energies between 50 keV and 120 keV. A Laue-type monochromator has been built for the beamline dedicated for dispersive EXAFS, and a scanning spectrometer with Rowland circle focusing is used for high-resolution Compton profile measurements. The solutions for dynamical bending and cooling of the crystals are described and performances of the various instruments are given.

© 1997 American Institute of Physics

INTRODUCTION

The principles of focusing by curved mirrors have been known for centuries, and it was realized very soon after the discovery of diffraction of x-rays by crystals that the same principles could be used to improve the efficiency of x-ray spectrometers (1). The first x-ray spectrometers employing large cylindrically bent (almost) perfect crystals were built in early 1930's (2, 3, 4), and it is remarkable that these works include thorough analyses of geometrical aberrations. Double focusing for a fixed wavelength by toroidally bent crystals has turned out to be difficult, but an approximate solution has been introduced (5). No fundamentally new geometrical solutions can be expected, and the improvements will be due to optimization and matching of the optical components.

The fact that the reflectivity of a perfect crystal is profoundly changed by bending was realized much later (6). The orientation and lattice spacing of the crystal change with distance from the surface, and the reflectivity curve broadens accordingly. Several authors have treated diffraction in deformed crystals starting from dynamical theory (7, 8, 9, 10, 11), but application of these results in practical cases is cumbersome, and simplified models have been introduced. The penetration depth of a ray is controlled by absorption and extinction, and the effects due to bending remain rather small for strong reflections at x-ray wavelengths of the order of 1 Å, and changes of reflectivity were usually ignored. On the other hand, the effects of bending are large for neutrons, and the so-called lamellar crystal model was introduced for calculation of the reflectivities (12). There is a rich literature in this field, and it has been summarized recently (13). The integrated reflectivity for x-rays has been interpreted using models for extinction (14) and interpolation between the limiting cases of dynamical and kinematical diffraction (15). The lamellar model was introduced in the x-ray case for calculation of the reflectivity curves in synchrotron radiation applications (16).

One of the present authors initiated a study where the geometrical focusing and reflectivities of bent crystals were combined for the optimal performance of an x-ray spectrometer (17). This spectrometer was designed to operate with a conventional x-ray source, but it was soon realized that the most important applications were those where high energy synchrotron radiation is used (18). This paper will review these developments, and the emphasis is put on applications in x-ray spectroscopy.

FOCUSING BY BENT CRYSTALS

There is an important difference between the reflection of light from a mirror and reflection of x-rays from a crystal. The direction of the ray reflected from a crystal is determined by the orientation of the Bragg planes, which may not coincide with the surface of the crystal. Furthermore, only a narrow energy band is reflected by the Bragg planes, and when the ray penetrates into a bent crystal, this band shifts, the direction of the reflected ray changes, and there is a lateral displacement of the reflected ray. For a complete description of diffraction by a bent crystal the trajectories of the incident and reflected rays inside the crystal must be calculated and then integrated over the incident beam and the active volume of the crystal in the position-angle space. In idealized cases the geometry of the crystal and the incident beam can be treated separately from the trajectory of an individual ray by considering very thin crystals.

Geometrical Focusing

Focusing by a very thin crystal is perfect in the plane of diffraction when the Bragg planes are curved cylindrically about an axis normal to the focusing circle and when the surface of the crystal is ground to coincide with the circle. This is the geometry of the Johansson-type monochromators used with conventional x-ray sources (4). The limitation is the fixed radius of the focusing circle, which limits the possibilities of tuning particularly in the synchrotron radiation applications. In the following only crystals which are flat when unbent are considered. In these crystals the angle between the surface and the relevant Bragg planes is constant.

Perfect point-to-line focusing is obtained with an elliptically bent symmetric crystal when the source is at one of the focii. Different wavelengths are reflected along the crystal, and the correlation between the direction of the reflected ray and wavelength can be utilized in instruments for dispersive EXAFS (19). On the other hand, when the crystal is bent to the shape of a logarithmic spiral, and the source is placed at the focus of the spiral, there is no wavelength dispersion, but the reflected rays form a pseudo-focus on the caustic. As a compromise solution, this was used in the spectrometer mentioned above (17).

In the synchrotron radiation applications the aberrations inherent in the Johann geometry are usually not important. In this geometry the crystal is bent cylindrically with a radius equal to the diameter of the focusing (Rowland) circle, so that the crystal surface does not coincide with the focusing circle, but the displacement is negligible when the beam divergences are small. At high energies the penetration of the beam into the crystal is far more important.

Four cases of focusing by a cylindrically bent crystal are shown in Fig. 1. The reflected beam can be convergent or divergent, monochromatic or polychromatic, depending on the location of the source and image. When the source is on the focusing circle the crystal acts as a monochromator, while a polychromatic beam is reflected in the other cases. The beam reflected by a Bragg monochromator is focused back to the Rowland circle, but the beam from a Laue monochromator

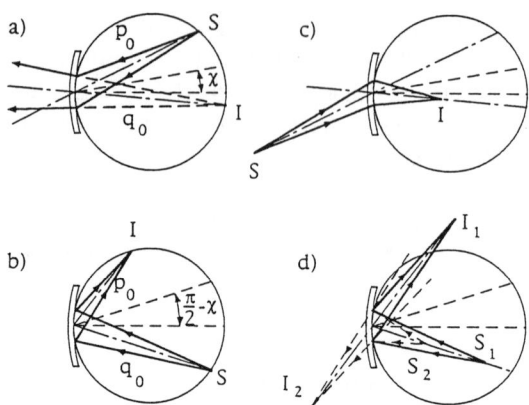

FIGURE 1. Focusing by an ideal, cylindrically bent crystal. Here (a) is the transmission or Laue case when the source S (real) and image I (virtual) are on the Rowland circle, (b) is the reflection or Bragg case for monochromatic focusing, (c) polychromatic focusing by a Laue crystal, and (d) polychromatic focusing by a Bragg crystal. In (c) and (d) the image is real or vital, depending on the location of the source.

diverges. The focal distances p (source-to-crystal) and q (crystal-to-image) are related through

$$q = q_0/(2-p_0/p), \qquad (1a)$$

where

$$p_0 = p\gamma_0 = \rho \cos(\chi \pm \theta), \qquad (1b)$$

$$q_0 = p\gamma_H = \rho \cos(\chi \pm \theta), \qquad (1c)$$

are those for monochromatic focusing (18). Here γ_0 and γ_H are the direction cosines of the incident and reflected beams, respectively, and χ is the angle between the Bragg planes and the surface normal of the crystal. The signs are chosen such that p is positive for a real source, q is of the same sign as p when on the same side of the crystal, and p is positive when the incident beam is on the concave side. A band of energies, ΔE, is reflected due to the equatorial divergence of the incident beam, $\Delta \psi = h_0/p$, unless the source is on the focusing (Rowland) circle. The relative width of the band is

$$\Delta E/E = \cot\theta\, h_0[1/p_0 - 1/p]. \tag{2}$$

The geometries shown in Fig. 1 have been used in different constructions including monochromators for coronary angiography, inelastic scattering at very high photon energies, dispersive EXAFS, single crystal diffraction, and x-ray spectrometers. Examples are shown in subsequent chapters.

Beam Trajectories in the Crystal

The width of the reflectivity curve of a flat thick crystal is the Darwin width

$$w_D = \sqrt{M}\,(4r_e d^2/\pi V_c)\,CF'\tan\theta = \sqrt{M}\,2\lambda/\pi\Lambda \sin 2\theta, \tag{3}$$

where $M = \cos(\chi - \theta)/\cos(\chi + \theta)$ is the magnification factor, $r_e = e^2/mc^2$ the classical electron radius, d the spacing of lattice planes, V_c the volume of the unit cell, C the polarization factor (1 for σ polarization and $\cos 2\theta$ for π polarization), F' the real part of the structure factor, $\Lambda = V_c/r_e\lambda CF'$ the extinction length, and λ the x-ray wavelength. In the energy scale the relative width of the reflectivity curve is independent of energy (wavelength),

$$(\Delta E/E)_D = \cot\theta\, w_D. \tag{4}$$

When the crystal is bent the Bragg planes are curved and their spacing changes due to elastic compliance. In the lamellar model the effect on a polychromatic pencil beam is described by a term arising from the change of orientation of the Bragg planes and by two terms resulting from local deformation of the crystal. Within this model, the relative energy band reflected by an elastically isotropic crystal of thickness T is

$$(\delta E/E)_L = \cot\theta\,(T/\rho)\times\{\tan(\chi+\theta) + 1/2\,(1+\nu)\sin 2\chi - \tan\theta\,[\cos^2\chi - \nu\sin^2\theta]\}, \tag{5}$$

where ν is the Poisson ratio (20, 21, 22).

The Penning-Polder theory applies in the Laue case where the reflected beam exits by the backside of the crystal (7). The propagation of a ray in a crystal where the reciprocal lattice vector changes slowly is treated in analogy to a light beam passing through a medium of inhomogeneous index of refraction. It is assumed that the x-ray wavefield adjusts itself to the slowly varying lattice parameter, and its propagation is described locally in terms of dynamical theory of x-ray diffraction. In technical terms, the dispersion surface moves through the crystal along the beam trajectory, or the tie-point moves within certain limits on the dispersion surface.

This corresponds to an angular range of diffraction, and in the case of cylindrical bending the reflected energy band is

$$(\delta E/E)_{PP} = \cot\theta\,(T/\rho)\,\{\sin\chi/[\cos(\chi+\theta)\cos\theta]\}$$
$$\{1 + 1/2\,(\cos 2\theta + \cos 2\chi)\,[1 - (s_{23} + s_{34}\cot\chi)/s_{33}\},\quad(6)$$

where s_{23}, s_{34}, and s_{33} are elastic compliances (23). The isotropic case is obtained by putting $s_{23}/s_{33} = -\nu$, and $s_{34} = 0$, and then Eqs. (5) and (6) yield very closely the same results, although the functional form of $(\delta E/E)_{PP}$ differs from that of $(\delta E/E)_L$. The reflectivity curves can be calculated from the Penning-Polder model, and the results have been substantiated by measurements (24).

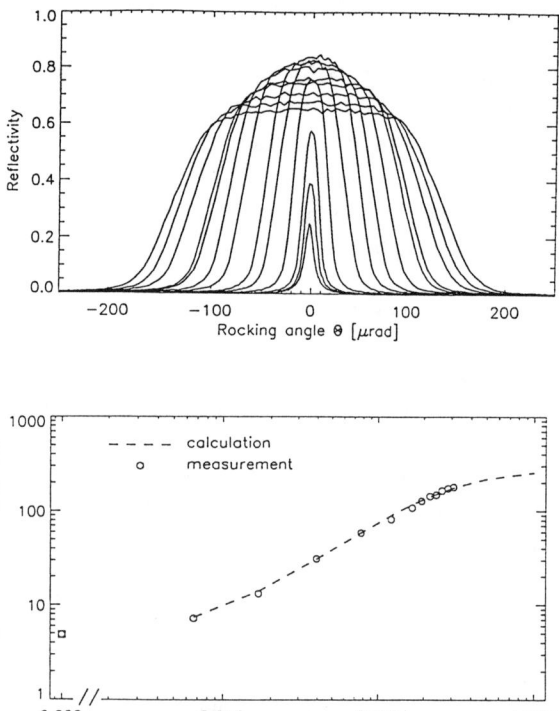

FIGURE 2. Measured rocking curves (a) of bent Laue crystal monochromator at different bending radii (from infinity to 3 m). The reflection is Si(111) at 33.17 keV, the thickness of the crystal is 0.7 mm and the asymmetry angle 26.22°. Integrated reflectivity (b) as a function of the inverse bending radius; the calculation is based on extended Penning-Polder theory (24).

The lamellar model has been used with success for calculation of the reflectivity curves of cylindrically bent crystals in the Bragg geometry. The total width of the reflectivity curve is given by Eq. (5), but the FWHM may be much less because of absorption. Typical reflectivity curves are shown in Figs. 2a and 3 for Laue and Bragg cases, respectively. There are several remarkable features in these curves. First, the width may be one or two orders of magnitude larger than that for a flat crystal. The energy band reflected by the crystal increases in the same proportion, and this is very important in applications where maximum flux is needed. The curves are almost symmetrical and flat-topped in the Laue case, and the maximum value may be close to unity instead of the value of 0.5 for a thick flat crystal. The calculated results have been corroborated by several measurements, and some of those results are shown in Fig. 2b.

Polychromatic Focusing

When a pencil beam enters the crystal the so-called Borrmann fan is formed inside the crystal between the incident and reflected beam directions. The reflected energies are spread over this fan, and the position-angle coupling produces a polychromatic focus, which is either real or virtual. Most applications require demagnification, and in general the polychromatic focus does not coincide with the geometrical one, limiting the achievable focus size.

The polychromatic divergence of the diffracted beam is described by

$$\Delta\theta_p = (1 + M^{-1}) \Delta\theta_o + B/\rho, \tag{7}$$

where $\Delta\theta_o$ is the rocking curve width and B the size of the Borrmann fan on the exit surface of the crystal. The first term is also present in the flat crystal case. It originates from the boundary condition for wave vectors at an interface, namely that the tangential components are continuous (25). The second term is due to the change of direction of the normal to the crystal surface within the Borrmann fan. It is seen that by adequate choices of the asymmetry angle and bending radius $\Delta\theta_p$ can be adjusted to bring the geometrical and polychromatic focii together, so that the resulting focus size is limited only by the source size.

APPLICATIONS IN HIGH ENERGY SYNCHROTRON RADIATION RESEARCH

Combination of the focusing properties of bent crystals with crystal reflectivity offers most possibilities at high x-ray energies, where absorption is small and the crystals almost transparent to radiation. From the practical point of view, it is important that small absorption reduces the heat load of the synchrotron beam, so that the cooling problem can be solved even in cases which require dynamical bending of the crystal.

Broad-Band Monochromators in Laue Geometry

A broad energy band is reflected by a Laue crystal when the source is outside the focusing circle and the focus is formed inside the circle. This geometry has been used with success in coronary angiography with synchrotron radiation (20, 22, 26, 27). In this application the horizontal fan beam is focused vertically to a line. The height of the line is typically 0.5 mm, and it is due to a difference between geometrical and polychromatic focal lengths. In one construction the center of the beam is blocked, so that two beams of slightly different energies cross at the geometrical focus (26). The difference in energy is about 300 eV, and the monochromator is tuned so that the energies of the two beams bracket an absorption edge of the contrast agent (usually iodine with K-edge at 33.17 keV). The contrast agent is injected intravenously to the patient who is seated in a chair and scanned through the double beam. The transmitted beams are recorded by two linear position-sensitive detectors at a frequency of at least 1 kHz, so that two-dimensional images are acquired. When the low-energy image is subtracted logarithmically from the high-energy image the distribution of the contrast agent in the coronary arteries is obtained, and this reveals possible decrease of the arterial lumen.

The focusing Laue monochromator has been used recently in the dispersive EXAFS method (28). When the energy is 10 keV or larger the Laue geometry offers several advantages over the usual Bragg geometry. At these energies the beam penetration into the crystal becomes important. In the Bragg case the angle of incidence is small and together with beam penetration to the crystal this causes substantial loss of resolution. In the Laue case the effects of geometrical aberrations are smaller, and the thickness and asymmetric cut of the crystal can be optimized for flux and energy resolution. For instance, a 0.15 mm thick Si(400) crystal with $\chi = 6°$ and bending radius of 1.38 m gives at the Pd K-edge an overall resolution of 1.6 eV, which is mostly due to the detector. At the same time, the integrated reflectivity is 4 times higher than that of a symmetrically cut crystal. The footprint of the beam is much smaller in the Laue case than in the Bragg case, which makes the construction of the monochromator simpler. The first version of the Laue monochromator for dispersive EXAFS was uncooled, but a cooled monochromator has been built and is being tested at the ESRF (29).

The possibility that the geometrical and polychromatic foci can be brought together by an appropriate choice of the asymmetry angle χ was used in the monochromator for high pressure measurements (30). In this application the focus size should be below 20 μm to reduce scatter from the diamond anvil cell. On the other hand, a rather wide energy band can be tolerated at small scattering angles. In this experiment an 1.5 mm wide beam was focused down to 8.4 μm, when the demagnification ratio p/q was 52:1. The width of the focus is due to the source size, demonstrating ideal focusing.

Horizontally Focusing Bragg Type Monochromators

The standard solutions of monochromatization and focusing do not work at high photon energies. The mirrors would be prohibitively long due to the small critical angle, and sagittal focusing in the horizontal direction becomes impossible due to the small width of the perfect crystal rocking curves. Furthermore, the energy resolution of the flat crystal monochromators is dominated by the beam divergences, as seen from Eq. 2. These difficulties are overcome by horizontally focusing monochromators in Bragg geometry, shown in Fig. 1. The source, monochromator and the focus are on the Rowland circle, and the focal lengths are adjusted by the asymmetric cut of the crystal.

The High Energy Inelastic X-ray Scattering Beamline (BL 25) at the ESRF is designed to work at energies above 30 keV (31). The source is a 7-period permanent magnet wiggler with critical energy of 45 keV at minimum gap of 20 mm. 0.8 mrad of the horizontal fan is intercepted by a 400 mm long monochromator, which is placed at 42 m from the source. The scattering angle is 7.5º and the asymmetry angle 2.5º to yield 5:1 demagnification. The calculated monochromatic flux is shown in Fig. 3. Reflections with odd indices are used in most cases to avoid the first harmonic energy. The reflectivity curve in the insert demonstrates the intensity gain achieved by bending. The integral width of the bent Si(331) crystal is about 10 μrad, while the Darwin width is only 0.4 μrad. The relative energy band reflected by the monochromator is typically less than 5×10^{-4}, which is ideal for Compton scattering experiments.

Combination of cylindrical bending and efficient cooling of a long Bragg-type crystal is difficult. The width of the focus of an ideally bent 1 mm thick crystal is about 0.5 mm, as seen in Fig. 4, so that the angular deviations from the ideal shape should stay below 10 μrad over the whole length of the crystal. The best results so far have been obtained with a watercooled bender, which has an opening in the middle allowing the direct beam to pass through the crystal. The crystal is curved to cylindrical shape by opposite moments acting on the ends of the crystal, and a thermal contact with the cooling frame is obtained by In-Ga eutectic.

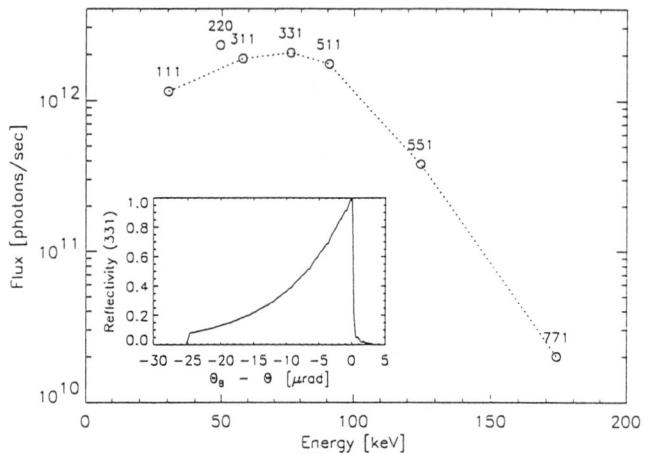

FIGURE 3. Calculated vertically integrated monochromatic flux from horizontally focusing Bragg-type Si monochromators at BL 25 of the ESRF. The crystal thickness is 1 mm, and the asymmetry angle 2.5° to yield 5:1 demagnification. The reflectivity curve of the 331 reflection is shown in the insert. The Darwin width of the reflection from a flat crystal is 0.4 μrad.

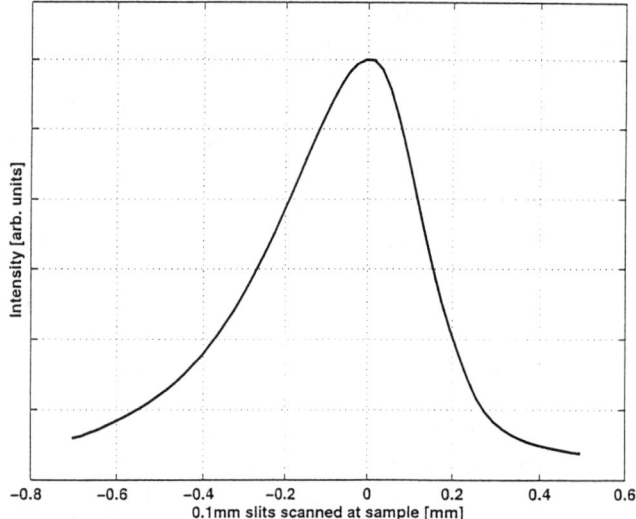

FIGURE 4. Horizontal profile of the focused beam at sample position of BL 25. Si (311) monochromator is used, and the energy is 58 keV. The measured integrated flux is 2×10^{12} photons/sec/100 mA, and the energy bandpass 9 eV.

Simple bending mechanisms can be used at very high energies, where the part of the spectrum below 100 keV, say, can be removed by suitable filters (1 to 2 mm of Cu). If the crystal is operated in air or inert gas environment, no cooling is needed. A triangular crystal that is clamped by the base curves to cylindrical shape when the tip is pushed (32). This construction can be used also in two-crystal combinations for the second crystal, which is not heated by the direct beam.

Double Crystal Laue-Laue Monochromators

Bent two-crystal monochromators have been discussed recently by the present authors (33). Different non-dispersive settings of two Laue crystals are shown in Fig. 5. These include two broad-band (a and c), and one narrow-band monochromator (b). For a tunable fixed-exit monochromator the second crystal is placed on a longitudinal translation stage. There are many possible applications including medical imaging and resonant scattering at high energies. Tests for medical applications at the iodine K-edge (33.17 keV) demonstrated that bending stabilizes the crystals, and the increased width of the reflectivity curves makes tuning easy (34). The two-crystal rocking curve is the convolution of the individual rocking curves when the bending radii are exactly matched, and the measured width agreed closely with the calculated value. The peak reflectivity of the monochromator with Si(111) crystals exceeded 60%.

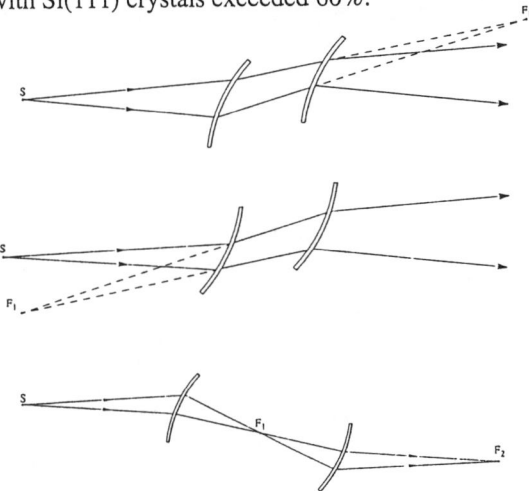

FIGURE 5. Non-dispersive setting of two Laue-type crystals. In each case, the focus of the first crystal is the source point for the second crystal: (a) corresponds to polychromatic and (b) monochromatic focusing, where the source for the second crystal is virtual, and the exit beam diverges. In the polychromatic case (c), the source for the second crystal is real, and the exit beam then converges.

It is seen from Eqs. 4 and 5 that when χ is zero, the bent crystal reflectivity curve is that of the flat crystal. Small asymmetry can be used to increase this small intrinsic width to match the needs of the experiment. Even high-order reflections have high reflectivity when χ is small, and this can be used in construction of efficient narrow-band monochromators for high photon energies. This is demonstrated in Fig. 6, where the reflectivities and rocking curve widths of Si(111) and Si(771) reflections are compared at 100 keV. The bending radius is determined by geometrical constrains, but the peak reflectivity and bandwidth can be optimized by crystal thickness and asymmetry angle. These possibilities were utilized in an experimental set-up used at the High Energy Diffraction Beamline of the ESRF. The bandpass with a 2 mm thick Si(511) crystal with $\chi=6.0°$ is 14 eV at 88 keV (Pb K-edge). The corresponding values for Si(711) with $\chi=1.6°$ are 3 eV and 80%. It is important that the effect of the beam divergence is eliminated when the source is on the Rowland circle, so that the energy resolution is determined by the width of the reflectivity curve and the size of the source. The applications of such a monochromator are in studies of resonant scattering at the K-edges of heavy elements and in experiments involving nuclear excitations.

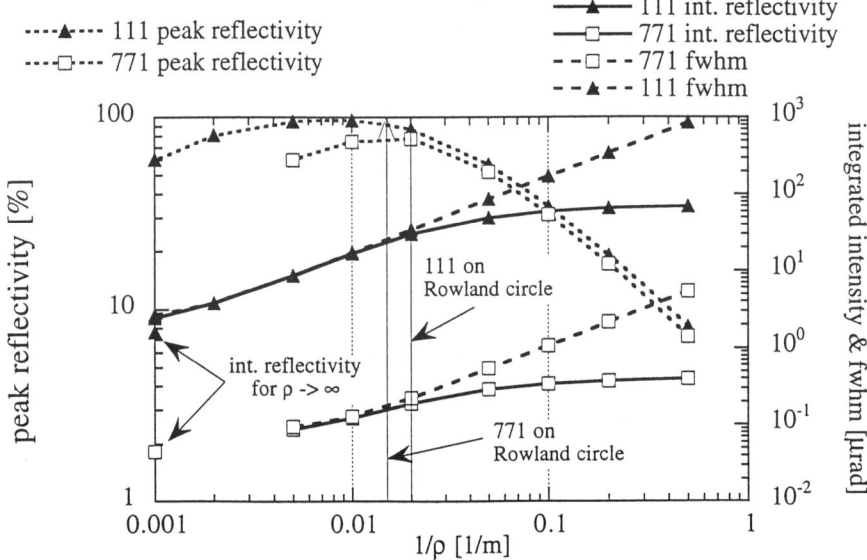

FIGURE 6. Calculated reflectivities of asymmetric reflections from Si(111) and Si(771) crystals in Laue geometry. The distance to the source is 50 m, and the photon energy 100 keV. The bending radii which correspond to monochromatic focusing are marked by arrows. The kinematical limit is reached in both cases when $\rho = 10$ m.

Double Crystal Laue-Bragg Monochromators

The Laue-Bragg combination of two flat crystals is a well-known construction for a fixed exit monochromator (35). However, when the crystals are bent many new possibilities appear (36). It is obvious on geometrical grounds that a tunable fixed-exit Laue-Bragg or Bragg-Laue monochromator can be constructed, but it turns out that with suitable choices of the focal lengths the focus is also practically fixed for large energy ranges (33). The broadening of the reflectivity curves due to the effective thickness of the crystals offers many possibilities of matching and fine tuning of the crystal pair.

The geometry of the Laue-Bragg monochromator is shown in Fig. 7. The source point for the Laue crystal is on the Rowland circle and it forms an virtual image, which is in turn source for the Bragg crystal at the intersection of the Rowland circles of the two crystals. The beam is focused on the Rowland circle of the Bragg crystal. The focal distances are obtained from Eqs. 1. In a typical case the beam is demagnified at both crystals, so that the distance from the source to the monochromator may be 10 times larger than the distance from the monochromator to the focus. It is important that the contribution of the beam divergence to the energy passband is eliminated. Compared to the standard combination of flat and sagittally bent crystals, the photon flux is enhanced and the total band-pass is reduced (30).

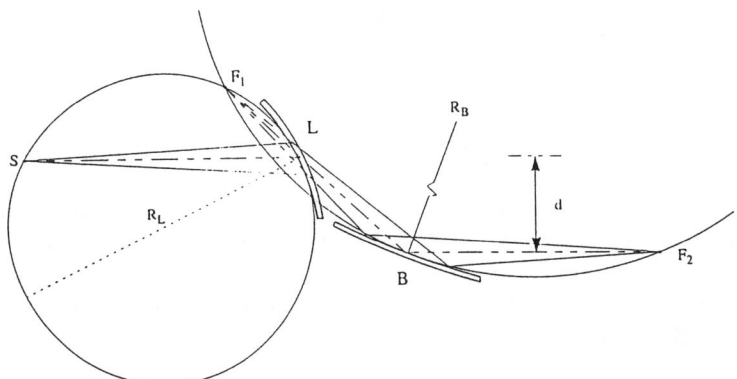

FIGURE 7. Focusing Laue-Bragg monochromator. The source S is on the Rowland circle of the Laue crystal L (bending radius R_L), the virtual focus of the reflected beam is at F_1, which is on the Rowland circle of the Bragg crystal B (bending radius R_B). The exit beam is displaced by the distance d, and a real focus is at F_2. With a suitable choice of the asymmetry angles of the crystals the focus F_2 is almost fixed.

Prototype monochromators using Si(511), Si(531) and Si(551) reflections were built and tested at the High Energy Diffraction Beamline and Optics Beamline (bending magnet source) of the ESRF at energies between 50 and 125 keV. Up to 400 mm long crystals were used to accept 50 mm of the horizontal fan beam. A focus size of 0.5 mm was obtained, which is in good agreement with the expected broadening due to the polychromatic divergences. The relative energy bandpass is typically 2×10^{-4}, and the vertically integrated flux can be up to 10^{12} photons/sec/100 mA on a wiggler source. It was verified experimentally that there is a constant energy gradient across the focus and that the monochromatic focus size is due to the source size. The linear distribution of energy in the focus may be used for energy compensation when a crystal analyzer is used in energy loss spectroscopy. In that case the slope error of the crystals must be smaller than the angular width of the source (few µrad). Slope errors of the order of 10 µrad may be tolerated without substantial degradation of the total focus size or flux, but the energy distribution at the focus would be smeared out.

The Laue crystal of the prototype monochromator was not cooled, and a filter was used to eliminate the low-energy part of the spectrum. This is adequate at energies above 70 keV, say, but a cooled crystal will be needed at lower energies. A prototype of such a crystal has been built and tested successfully at the ESRF (37). A long rectangular crystal is clamped by one end and bent by applying a moment at the free end. Also a transverse displacement can be introduced for compensation of third order terms in the crystal figure, which may arise from friction forces and uneven thickness of the crystal. Cooling is achieved by immersing the lower part of the crystal in a narrow vessel filled with In-Ga eutectic.

Scanning X-Ray Spectrometer

The scanning x-ray spectrometer, which is operated at BL 25 of the ESRF, takes advantage of several properties of bent crystals. The main use of the spectrometer is measurement of Compton profile, which is the projection of the electron momentum density onto the scattering vector along the z-axis. The recorded energy spectrum is converted to the momentum scale by the relation

$$p_z = mc \{r - 1 + E_2(1-\cos 2\theta)/mc^2\}/\{1 + r^2 - 2r \cos 2\theta\}^{1/2}, \tag{8}$$

where m is the electron rest mass, c the velocity of light, E_1 the energy of the incident photon, E_2 that of the scattered photon, $r = E_2/E_1$, and 2θ the scattering angle (38). In atomic units mc = 137.

The layout of the scanning spectrometer is shown in Fig. 8. The incident beam is focused on the sample by a bent Bragg monochromator, and it is limited by a horizontal slit which is typically 0.15 mm to 0.5 mm. The incident energy is determined by the available low-order reflections of the Si monochromator, because only a small tunability of 10% to 15% is achieved by a change of the monochromator angle and focal lengths (see Fig. 5). The sample is the source for the spectrometer, which works in the Rowland circle geometry. The Bragg type analyzer crystal is bent cylindrically to the radius equal to the diameter of the Rowland circle, and typically the crystal is cut asymmetrically. The scattering angle is fixed, and the analyzer and the receiving slit in front of a scintillation detector move on the Rowland circle by synchronized translations and rotations. The distances between the sample, analyzer, and receiving slit are given by Eq. 1.

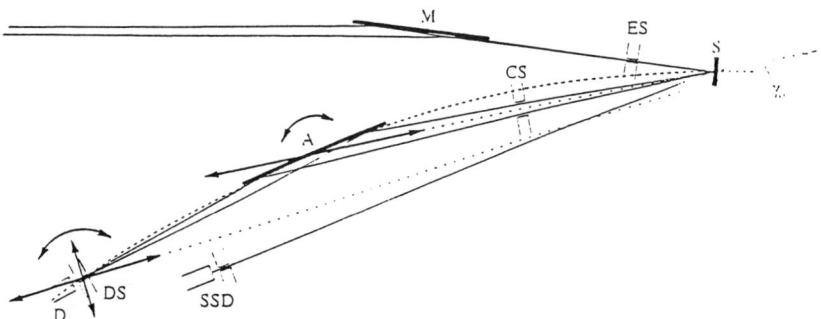

FIGURE 8. Schematic construction of the scanning x-ray spectrometer. A 0.8 mrad horizontal fan from the wiggler source is focused on the sample S by an asymmetrically cut bent Bragg crystal M. The cylindrically bent analyzer crystal A is translated and rotated during a scan, and the receiving slit DS and detector D track the focus of the reflected beam by translations and rotation, so that the sample, analyzer and receiving slit stay on the Rowland circle (broken line).

The spectrometer has been used so far at incident energies of 49 keV and 58 keV. The analyzer is Ge(440) crystal, which is bent to a radius of 6.1 m. The asymmetry angle is 2° to increase the distance between the sample and analyzer, because the dominant factor in the energy resolution is the angular width of the active sample volume seen by the analyzer. If the sample is a disc of thickness T, and its normal is along the scattering vector, this width is

$$\Delta\theta_A = w_s/p = \{w_b^2 + [2T\sin(\pi/2-\theta)]^2\}^{1/2}/p, \qquad (9)$$

where w_b is the width of the incident beam. It is seen that a backscattering geometry is essential for good energy resolution. The resolution of the spectrometer is between 0.1 a.u. and 0.2 a.u. depending on the effective width of the sample.

The spectrum of the scattered radiation is recorded as a function of the analyzer angle; an example is shown in Fig. 9. Due to the focusing geometry, which allows the use of a narrow receiving slit, the background is low and easy to subtract. Weak parasitic reflections are subtracted, and the spectrum is corrected for the effects depending on the energy and geometry. The contribution of multiple scattering in the sample is subtracted, and the spectrum is converted to the momentum scale using Eq. 8. A full report on the spectrometer is being prepared (39).

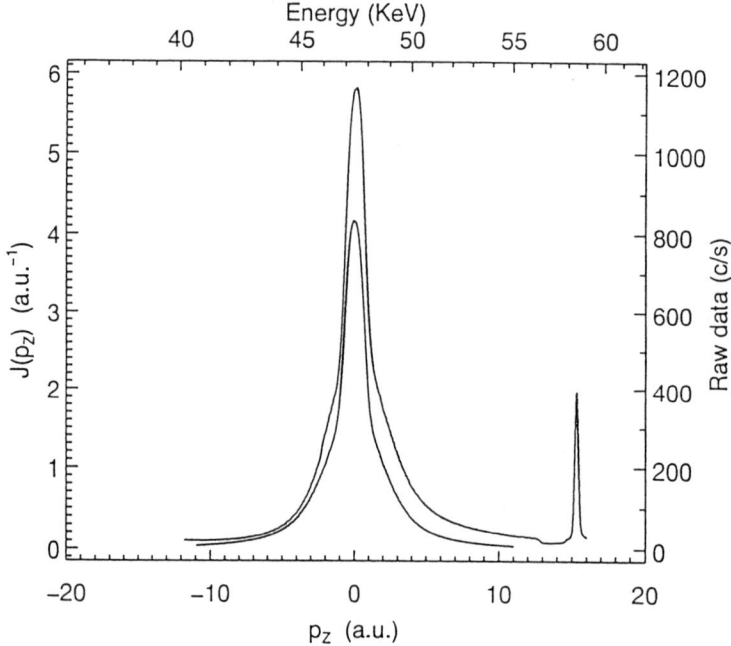

FIGURE 9. Spectrum of the radiation scattered to 161° from a 0.76 mm thick Si crystal (upper curve, energy scale). The absolute Compton profile is obtained after correction for instrumental factors, background and multiple scattering (lower curve, momentum scale). The resolution at the Compton peak is 0.15 a.u., which is mostly due to the effective width of the sample volume.

SUMMARY AND CONCLUSIONS

The use of bent crystal optics has provided many breakthroughs in research with high energy synchrotron radiation. Intravenous coronary angiography suffered from beam instabilities and low flux before the use of bent Laue crystals. Energy resolution of flat crystal monochromators is dominated by angular divergence at high energies, and small Darwin widths make monochromators inefficient and horizontal focusing by sagittally bent crystals difficult or impossible. On the contrary, horizontally focusing bent Bragg or Laue-Bragg configurations become more efficient with increase of energy, and their bandpass can be tailored to the needs of the experiment. Upon Bragg reflection, and x-ray beam travels only a distance of the extinction length in a flat crystal, while the whole thickness of a bent crystal may be effective. This results in spectacular gains of reflecting power, which has made possible Compton profile measurements at energies of 1 MeV.

Many of the x-ray optical systems using bent crystals are still in prototype state. Best performance can only be achieved by experienced users at present, but this situation is expected to change in the near future when the instruments are used more routinely. Continuous tunability of monochromators is needed in many applications, and this will require development of sophisticated encoding and feedback systems.

REFERENCES

1. Wagner, E., *Phys. Zeitschrift* **17**, 407 (1916).
2. Johann, H.H., *Z. Phys.* **69**, 185 (1931).
3. Cauchois, Y., *J. de Phys.* **3**, 320 (1932).
4. Johansson, T., *Z. Phys.* **82**, 507, (1933).
5. Berreman, D.W., *Rev. Sci. Instrum.* **26**, 1048 (1955).
6. White, J.E., *J. Appl. Phys.* **21**, 855 (1950).
7. Penning, P., and Polder, D., *Philips Res. Rep.* **16**, 419 (1961).
8. Kato, N., *J. Phys. Soc. Japan* **19**, 971 (1964).
9. Takagi, S., *Acta Cryst.* **15**, 1311 (1962); *J. Phys. Soc. Jpn.* **26**, 1239 (1969).
10. Taupin, D., *Bull Soc. Fr. Minéral. Cristallogr.* **87**, 469 (1964).
11. Chukhovskii, F.N., and Petrashen, P.V., *Acta Cryst* **A33**, 311 (1977).
12. Egert, G., and Dachs, H., *J. Appl. Cryst.* **3**, 214 (1970).
13. Popovici, M., and Yelon, W.B., *J. Neutron Res.* **3**, 1 (1995).
14. Brown, D.B., and Fatemi, M., *J. Appl. Phys.* **45**, 1544 (1974).
15. Kalman, Z.H., and Weissmann, S., *J. Appl. Cryst.* **16**, 295 (1983).
16. Boeuf, A., Lagomarsino, S., Mazkedian, S., Melone, S., Puliti, P., and Rustichelli, F., *J. Appl. Cryst.* **11**, 442 (1978).
17. Suortti, P., Pattison, P., and Weyrich, W., *J. Appl. Cryst.* **19**, 336 (1986).
    Suortti, P., Pattison, P., and Weyrich, W., *J. Appl. Cryst.* **19**, 343 (1986).
    Pattison, P., Suortti, P., and Weyrich, W., *J. Appl. Cryst.* **19**, 353 (1986).
18. Suortti, P., *Rev. Sci. Instrum.* **63**, 942 (1992).
19. Tolentino, H., Dartyge, E., Fontaine, A., and Tourillon, G., *J. Appl. Cryst.* **21**, 15 (1988).
20. Suortti, P., and Thomlinson, W., *Nucl. Instrum. Meth.* **A269**, 639 (1988).
21. Erola, E., Eteläniemi, V., Suortti, P., Pattison, P., and Thomlinson, W., *J. Appl. Cryst.* **23**, 35 (1990).
22. Suortti, P., Thomlinson, W., Chapman, D., Gmür, N., Greene, R., and Lazarz, N., *Nucl. Instrum. Meth.* **A297**, 268 (1990).
23. Schulze, C., Dissertation, Univ. Hamburg, 1994.

24. Schulze, C., and Chapman, D., *Rev. Sci. Instrum.* **66**, 2220 (1995).
25. Matsushita, T., and Kaminaga, U., *J. Appl. Cryst.* **13**, 472 (1980).
26. Suortti, P., Thomlinson, W., Chapman, D., Gmür, N., Siddons, D.P., and Schulze, C., *Nucl. Instrum. Meth.* **A336**, 304 (1993).
27. Illing, G., Heuer, J., Reime, B., Lohmann, M., Menk, R.H., Schildwächter, L., Dix, W.R., and Graeff, W., *Rev. Sc. Inst.* **66** (1995).
28. Hagelstein, M., Ferrero, C., Hatje, U., Ressler, T., and Metz, W., *J. Synchrotron Rad.* **2**, 174 (1995).
29. Hagelstein, M., private communication.
30. Schulze, C., and Lienert, U., *ESRF Newsletter* **25**, 38 (1996).
31. Suortti, P., and Tschentscher, Th., *Rev. Sci. Instrum.* **66**, 1798 (1995).
32. Lemonnier, M., Fourme, R., Rousseaux, F., and Kahn, R., *Nucl. Instrum. Meth.* **152**, 173 (1978).
33. Suortti, P., and Schulze, C., *J. Synchrotron Rad.* **2**, 6 (1995).
34. Schulze, C., Suortti, P., and Chapman, D., *Synchrot. Rad. News* **7**, No.3, 8 (1994).
35. Mills, D.M., *Nucl. Instru. Meth.* **208**, 355 (1983).
36. Sparks, C.J., *BNL Report* 26740 (1979).
37. Honkimäki, V., private communication.
38. Manninen, S., Paakkari, T., and Kajantie, K., *Phil. Mag* **29**, 167 (1974).
39. Suortti, P., Buslaps, Th., Fajardo, P., Honkimäki, V., Kretzschmer, M., Lienert, U., McCarthy, J.E., Renier, M., Shukla, A., Tschentscher, Th., and Meinander, T., in preparation.

Soft X-ray Optics for Spectromicroscopy at the Advanced Light Source

H. A. Padmore

Advanced Light Source, Lawrence Berkeley National Labboratory, Berkeley, CA 94720, USA

Abstract. A variety of systems for performing spectromicroscopy, spatially resolved spectroscopy, are in operation or under construction at the Advanced Light Source (ALS). For example, part of our program is centered around the surface analysis problems of local semiconductor industries, and this has required the construction of a microscope with wafer handling, fiducialization, optical microscopy, coordinated ion beam etching, and X-ray Photoelectron Spectroscopy (XPS) integrated in this case with Kirkpatrick-Baez (K-B) grazing incidence micro-focusing optics. The microscope is to be used in conjunction with a highly efficient entrance slitless Spherical Grating Monochromator (SGM). The design and expected performance of this instrument will be described, with emphasis on the production of the elliptically curved surfaces of the K-B mirrors by elastic bending of flat mirror substrates. For higher resolution, zone-plate (Z-P) focusing optics are used and one instrument, a Scanning Transmission X-ray Microscope (STXM) is in routine operation on undulator beamline 7.0. A second Z-P based system is being commissioned on the same beamline, and differs from the STXM in that it will operate at Ultra-High Vacuum (UHV) and will be able to perform XPS at 0.1 µm spatial resolution. Spatially resolved X-ray Absorption Spectroscopy (XAS) can be performed by imaging electrons photoemitted from a material with a Photo-Emission Electron Microscope (PEEM). The optical requirements of a beamline designed for PEEM are very different to those of micro-focus systems and we give examples of bending magnet and undulator based instruments.

INTRODUCTION

The development of spatially resolved spectroscopy, spectromicroscopy, and its use in the analysis of problems in surface and materials science is in its infancy. Although several microscopes have been developed on second generation synchrotron radiation sources, third generation sources such as the ALS were specifically designed to give extremely high brightness, and hence give optimized performance for x-ray microscopy. This is now opening the way for the use of spectromicroscopy as a widely used analytical tool in materials science.

There are two distinct methods of image formation, scanning and full field imaging. In the scanning method, a monochromatic object is demagnified by an x-ray optical system to a small image at a sample surface, and an image is acquired by recording a response of the sample as a function of the beam position. The demagnification can be done in a number of ways, for example using a mirror,

© 1997 American Institute of Physics

tapered capillary or zone plate (Z-P). Each of these have very different optical characteristics; the Z-P is a circular diffraction grating whose focal length is a function of wavelength, while the other two are achromatic. They also have different acceptances and efficiencies; the reflection systems have a high reflection efficiency as well as a large acceptance, while the Z-P has a small efficiency (although phase shifting soft x-ray Z-Ps can deliver a peak efficiency of 20%) and small acceptance. The Z-P has however the best spatial resolution, approximately equal to the outer most zone width, and a resolution of around 50 nm is becoming routine. With the advent of new electron beam machines dedicated to Z-P writing, we can expect significant improvements in resolution in the next few years, ultimately approaching the limit imposed by the recording medium of photoresist, of around 15 nm.

The full field imaging method requires that a field of view is illuminated, and that either the transmitted or fluorescent photons or the photoemitted electrons are imaged. In the case described later, we are concerned with Photo-Electron Emission Microscopy (PEEM). Unlike the scanning technique described above, the full field methods acquire data in a parallel process. In addition, as only a condensed illumination field has to be produced, we can have an optical system with a large acceptance. Together, this implies that the parallel imaging techniques are intrinsically fast. Moreover, the resolution of the electron imaging technique, PEEM, depends only on the quality of the electron optics. We routinely achieve 200 nm resolution with secondary electron yield x-ray absorption spectroscopic imaging, and recent developments in this field have shown a clear path to achieving < 5 nm resolution.

In this paper, we describe both a bending magnet and an undulator beamline optical system optimized for PEEM and for XPS. In the latter case, the bending magnet beamline uses K-B mirror micro-focusing, whereas for the undulator beamline zone plate focusing is used. For PEEM, in each case mirror condensors are used. These choices are examined in this paper, with emphasis on the use of K-B mirror systems to produce micro-focii on bending magnet beamlines, and on the selection of systems for scanning and full field techniques on an undulator. This discussion is with reference to two new microscope beamlines currently under construction at the ALS, beamline magnet beamline 7.3, and elliptically polarized undulator beamline 4.0.

FOCUSING SYSTEMS

In this section, we restrict discussion to zone plate and grazing incidence mirror focusing systems. Tapered capillary optics however are becoming useful for condensing systems, but although excellent spatial resolution has been achieved (< 1μm), throughput is limited due to roughness of the reflecting glass surfaces. In addition, straight capillaries reduce the brightness of the source, and although not important in terms of flux density, the effect is to produce an image with much greater divergence than a mirror system of equivalent image size. This usually requires that the tip of the capillary be close to the sample to achieve the expected resolution. In addition to the grazing incidence mirror systems discussed later, normal incidence optics can be used, usually arranged in a Schwartzchild configuration of a convex - concave spherical mirror pair. Coated with a metal, the maximum photon energy is set by the reflectivity cut-off at normal incidence of around 25 eV (1). They can also be coated with a synthetic Bragg crystal (multilayer) and by choosing the appropriate materials and periodicity, high reflectivity within a narrow energy range can be achieved (2). Mo-Si multilayers are used for photon energies just lower than the Si $L_{2,3}$ edge (typically 90 - 95 eV), although using boron nitride instead of silicon as the 'spacer' element has recently allowed operation at 130 eV. Using both these types of system, a resolution of around 100 nm has been achieved (1,2).

Although zone plate micro-focusing systems are commonly used, the use of grazing incidence mirrors has been relatively little used due to the problems of fabrication of the optical surface to the required figure tolerance. Mirror systems although not challenging zone plates in terms of resolution, have a role to play where modest resolution at very high throughput is required, and have the advantage of being achromatic. They have been used for soft x-ray spectromicroscopy and achieved μm resolution (3). Recent advances in the production of aspheric mirrors by elastic bending of flat substrates have opened up the possibility of achieving near perfect elliptical surfaces, thus providing a way of producing a high efficiency achromatic system for sub-micron focusing and this methodology is described in detail in a later section (4).

Zone Plate Focusing

In order to obtain a diffraction limited image size, the Z-P can only accept the fraction of the radiation that is spatially coherent, and for practical purposes this means that the product of the object size and divergence (phase space) that can be used is approximately equal to the wavelength. In the absence of electron beam divergence, undulators produce radiation that is transversely spatially coherent and hence are well matched to the requirements of Z-Ps. Third generation sources attempt to have an electron beam emittance less than the photon beams produced by the undulators, so that the fraction of spatially coherent versus incoherent radiation will be high. In the case of the ALS, the vertical emittance is set to around 1×10^{-10} m.rads by skew quadrupoles in the accelerator lattice (to couple a small fraction of the horizontal emitance into the vertical plane to increase the lifetime to an acceptable value), and the horizontal emittance is 4×10^{-9} m.rads. For these values we can expect a high fraction of the light to be spatially coherent for wavelengths greater than 0.6 nm and 24 nm in the vertical and horizontal planes respectively. In general we are concerned with wavelengths from 10 to 1 nm (covering for example the K edges of the light elements and the $L_{2,3}$ edges of the 3d transition metals), and so we can see that over the whole range the ALS will produce spatially coherent radiation in the vertical plane, but in the horizontal plane, except at the longest wavelengths only a small fraction will be coherent. In actuality, the fraction of spatially coherent light is less than is indicated by the simple analysis above as it assumes that the phase space ellipses of the photon beam generated by a single electron passing through the undulator and the real electron beam are concentric. In fact the betatron functions of the ALS are too large to give this optimum matching, and the coherent fraction of the light is less than 1% at 1 nm wavelength. However, the total monochromatized flux can be high and this to some extent offsets the low diffraction efficiency of the Z-P and the low coupling efficiency of the source to the Z-P. For example, the spectromicroscopy beamline BL 7.0 has an 89 period undulator with a period of 5 cm, and together with an SGM beamline produces 10^{12} to 10^{13} photons/sec in a bandpass of 10^{-4} over the wavelength range described above (5,6,7). After spatial filtering, and focusing with an amplitude zone plate, the flux in the center of the wavelength range is typically 10^7-10^8 photons/sec, presently focused into a spot size of around 100 nm.

This flux allows Scanning Transmission X-ray Microscopy (STXM) at high spatial resolution and with fast frame speeds. The ALS beamline 7.0 STXM uses a piezo driven x-y stage for fast scanning, a flexure z stage to allow the focus position to be tracked as the photon energy is changed during local area spectroscopic measurements, and a selection of pulse counting and analog detectors. The system is operated at 1 atmosphere of helium or at rough vacuum. It has been mostly used for studies on polymeric systems, and a review of recent work can be found in (8).

A second system that operates at UHV and uses an XPS detection system has been constructed and is presently being commissioned. This is far more demanding in terms of flux, as the XPS analyzer only integrates over a small fraction of solid angle and over a very small fraction of the whole photo-electron energy spectrum. For many practical experiments, the available flux will in fact limit the achievable resolution, set by the maximum time allowable to acquire an image. In a later section, we describe a system in which the optical system is designed to optimally couple the source to the zone plate focusing system, in this way maximizing the coherent flux, and allowing the full high resolution capabilities of the zone plate to be utilized for XPS imaging.

Grazing Incidence Mirror Focusing

The two main types of grazing incidence mirror arrangement are the K-B system consisting of two crossed mirrors, and systems in which only a single reflection is used. Usually, in order to avoid or minimize aberrations, an elliptical or ellipsoidal surface is used, or a close approximation to this surface. In the case of single mirror systems, complete surfaces of revolution of an ellipse (2) or sections of a complete surface of revolution (9) have been used and good resolution has been achieved, albeit at modest collection aperture.

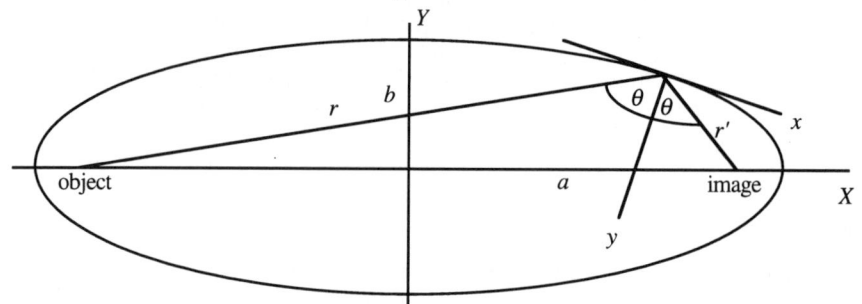

FIGURE 1. Geometry of the Ellipse

As it is extremely difficult to produce aspheric surfaces at the required figure tolerances, we have adopted an approach in which one starts with a 'perfect' flat substrate, and produce the desired elliptical shape by elastic bending, and this approach is described in detail later. The bending action is produced by the application of unequal couples to the ends of the beam in a method originally devised for the production of collimators (9). The main advantages of a grazing incidence system is that it is achromatic, and can collect a large solid angle. The maximum collection aperture is essentially limited by the critical angle of reflection for the maximum photon energy to be used. This can be seem in Fig. 1, in which the geometry of a grazing incidence ellipse is shown. If a beam of angular width ϕ from the object is collected and focused by the ellipse, the convergence at the image is $\phi'=\phi r'/r$. If this convergence angle becomes similar to the critical angle, it implies that at one end of the mirror light is propagating almost parallel to the surface, and at the other is becoming close to the critical angle, and hence reflectivity efficiency is being lost. A further decrease in the image distance will simply result in only the upstream end of the mirror reflecting light. As a general rule, the maximum convergence is restricted to half of the critical angle of reflection for the maximum photon energy in the design range. It can also be seen that as we demagnify more, this implies that less angular aperture can be accepted, and so we can say that the product of image size and divergence is constant. For example, in the case of a system we are building for μ-XPS on a bending magnet, we are aiming for an image size of 1 μm, and with a grazing incidence angle of 1.6° dictated by the upper working photon energy of 1.2 KeV, we can say that the convergence angle is limited to around 14 mrads. The acceptance aperture is then set by the required demagnification which is set from the source size and required image size. In the case of the ALS which has a vertical source size of around 30 μm FWHM, it can be seen that for the 30:1 demagnification needed to produce a 1 μm focus size, 14/30 or approximately 0.5 mrads vertical aperture can be collected from the source. This is a little larger than the vertical angular distribution (fwhm), and so we can say that it is possible to demagnify the ALS source for 1.2 KeV photons to a 1 μm spot size without geometrical loss. In the horizontal direction however, the ALS source is approximately 7 times larger than in the vertical direction, and hence only 1/7 of the angular aperture can be taken if the objective is to demagnify to a 1 μm image. In this case therefore, a horizontal aperture of only around 70 μradians can be accepted. This emphasizes the importance of the brightness of a source, and the fact that even with a 3rd generation bending magnet source, only a relatively small aperture can or needs to be taken.

PRODUCTION OF ELLIPTICAL MIRRORS BY CONTROLLED BENDING OF FLAT SUBSTRATES

One of the key technologies that we are developing for a range of applications, including μ-XPS, is the production of highly accurate elliptical surfaces by the controlled deformation of flat mirror substrates. The reason for developing this approach over conventional grinding and polishing methods is that the latter are difficult to make accurate enough particularly in the case of highly aspheric surfaces, while the production of flat surfaces to the required tolerances is routine. The remaining problem is to produce a mirror bender capable of exerting the right combination of couples and forces to produce the desired shape. If we take a mirror substrate of length L and apply couples C_1 and C_2 respectively at the two ends, the bending of the substrate is given by the Bernouilli-Euler equation (10),

$$EI_0 \frac{d^2y}{dx^2} = \frac{C_1+C_2}{2} - \frac{C_1-C_2}{L}x \qquad (1)$$

where E is Young's modulus, I is the section modulus which is equal to $bh^3/12$, where b is the substrate width and h is the substrate thickness. The second derivative is effectively the curvature of the ellipse, and as this is a function of displacement along the beam from the center (x=0), a variable in (1) has to be controlled to give the required value. C_1 and C_2 are independent of x, and so clearly the section modulus I has to be varied, either by changing the width b(x) or the thickness h(x). In the present arrangement we have chosen to vary the width. The individual couples C_1 and C_2 are chosen assuming a rectangular beam of width b_0 and this ensures that the smallest variation of b(x) has to be made to give true elliptical bending. These couples are given by (4),

$$C_1 + C_2 = 4EI_0 a_2 = \frac{2EI_0}{R_0} \qquad (2)$$

$$C_1 - C_2 = -6EI_0 L a_3 = \frac{3EI_0 L}{R_0} \frac{\sin\theta}{2}\left(\frac{1}{r} - \frac{1}{r'}\right) \qquad (3)$$

where R_0 is the radius of curvature at x=0, and θ is the angle of incidence.

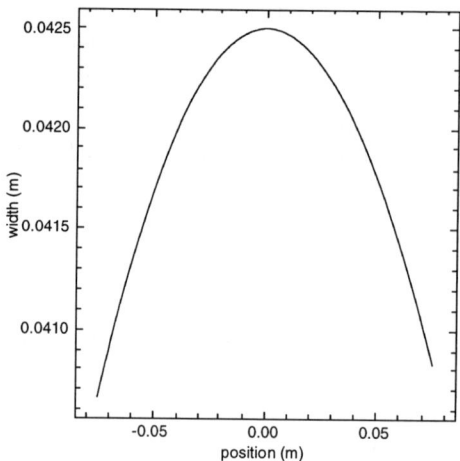

FIGURE 2. Width of a mirror substrate as a function of displacement from the center; r= 31 m, r' = 0.5 m, θ = 89.67°, b_0 = 42.5 mm, L = 150 mm.

The design process first fixes the width and thickness of the substrate at the center, the couples are calculated from (2) and (3) and then these values are used in (1) together with the known curvature of the required ellipse to calculate the variation of width with displacement. As an example, Fig. 2 shows the substrate width b(x) as a function of displacement x, for a mirror to be used for hard x-ray focusing, with r= 31m, r' = 0.5 m and θ = 89.67°. This edge profile is produced by computer controlled grinding. The most important aspect of this approach is the method of applying couples. The method we have been developing for both hard and soft x-ray micro-focusing is shown schematically in Fig. 3.

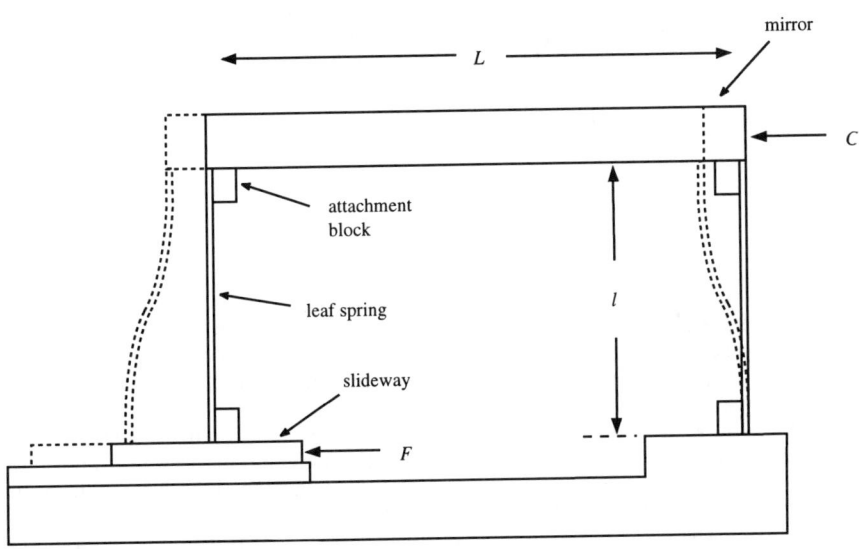

FIGURE 3. 's' spring bender shown in the relaxed and extended positions.

The mirror substrate is attached to two blocks which are joined to leaf springs at either end. At the right hand end, the other end of the leaf spring is rigidly attached to a base, and at the left hand end is attached to a slideway. If a force F is exerted on the slideway to move it to the left, the springs are deformed in a symmetric manner and exert equal and opposite couples on the ends of the beam. If the mirror substrate is pushed left by the application of a force G, the extension of the right hand spring increases, and the extension of the left hand spring decreases, resulting in a linear change in moment as required along the beam. The system is highly controllable and insensitive to temperature due to the large extension of the springs. Typically we use an extension of around 1/5 of the length of the spring, and in many cases this amounts to around 1 cm. The mirror substrates are made of glass or steel, and polished by conventional techniques to sub-microradian flatness, and a few Å rms roughness.

In order to adjust the couples to the correct values, and to measure the deviation of the surface from the required elliptical shape, we use a slope measuring device known as a Long Trace Profiler (11). The results of this measurement for a mirror with parameters given in Fig. 2 is shown in Fig. 4.

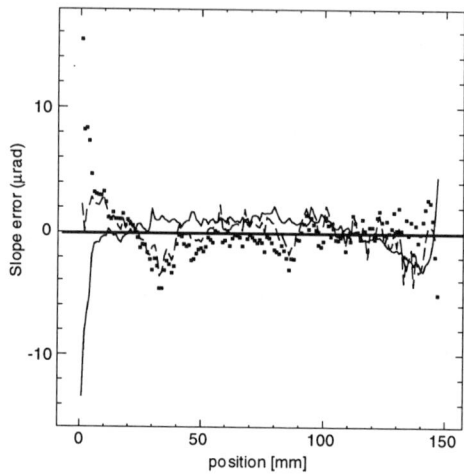

FIGURE 4. Solid line gives the slope error of the unbent surface from a flat, the dotted line shows the error from the defined ellipse, and the dots give the difference.

It can be seen that the slope error deviation from the required ellipse is small, and in fact has a standard deviation of only 1.2 µradians. The details of production of this class of mirror are given in reference (4). In using the same approach in the production of mirrors for µ-XPS, the main difference is that the radius of curvature has to be much less, due to the increased grazing angle that is used. Of course, the maximum grazing angle is used commensurate with obtaining good reflectivity in order to collect the maximum angular aperture. In the case of a Kirkpatrick-Baez system we are building for µ-XPS on a bending magnet source, the horizontally focusing mirror has the following parameters; r= 4m, r'=0.1m, θ = 88.4° and an optical length of 50 mm, giving a central radius of curvature of 7 m. This contrasts to the case described above for hard x-ray micro-focusing in which the radius of curvature is around 150 m. In that case the beam thickness was 9.5 mm, and therefore if the stress in the µ-XPS mirror substrate were to be the same, the thickness would have to be 0.45 mm. Clearly this is impractical, and so the thickness has to be increased to a level where it is stiff enough to be polished flat to µradian slope tolerance, and still have a stress that is reasonable. For metals, it is relatively easy to decide what is a reasonable stress based on the yield strength, micro-yield point and creep. For brittle materials such as glass and silicon, the behavior of the material depends on such things as micro-crack density and length, as well as in the case of glass materials, moisture content. Based on this information we are using a thickness of 2 mm of glass for the horizontal focusing µ-XPS mirror. In order to reduce the crack density, the machined edges are optically polished, and the back surface is acid etched (12). Using this approach we have successfully obtained micro-radian flatness and have bent the material to the desired radius without failure. We are currently performing bending tests to assess residual slope errors.

An alternative to glass is to use a metal substrate material. The advantage is that they can withstand much higher stress, and so can be made substantially thicker. This in turn makes them easier than glass to fabricate to the required flatness, but more care is required to produce the required micro-finish. In fact, using a precipitation hardened steel, we have obtained mirrors for this application 4mm thick, polished to around 1 µradian flatness and with a micro-roughness as measured with an optical profiler (Micromap) of < 3Å rms (13). Bending tests of these mirrors is underway.

BEAMLINE 7.3.1 for µ-XPS and PEEM

Beamline 7.3.1 was specifically designed to be optimized for Photo-Electron Emission Microscopy (PEEM), with emphasis on the application of the technique to the study of magnetic materials using Magnetic Circular Dichroism (MCD) spectroscopy (14, 15). The beamline layout is shown in Figure 5.

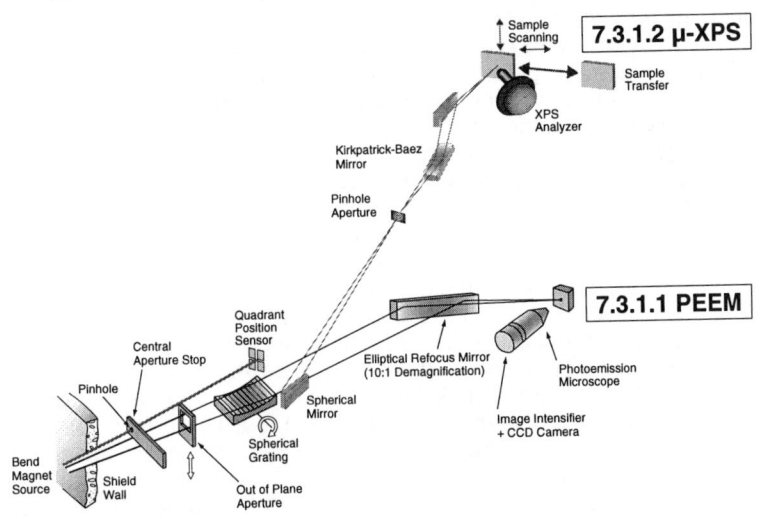

FIGURE 5. Beamline 7.3.1 for µ-XPS and PEEM.

The monochromator is entrance slitless, and produces a monochromatic image of the source at the position of the sample in the case of the PEEM beamline and at a pinhole aperture in the case of the µ-XPS branch. The system is designed to cover from 280 eV to around 1200 eV. The vertical source size at the center of the central bending magnet of the triple bend achromat structure of the ALS is less than 30 µm (FWHM). The required field of view of the PEEM imaging microscope is around the same value, and so it allows us to directly image the source to the sample. In addition, as the source size is small, this allows us to use a low line density grating, and this in turn results in a high diffraction efficiency (16). It also leads to a very slow variation of focal length with wavelength, and therefore it is not necessary to track the focus with the sample position. In the horizontal direction, a 1m long elliptical mirror is used to collect 2 mrads of horizontal aperture at a grazing angle of 2°, and to demagnify it by 10:1 onto the sample. As before, this demagnification and aperture are simply dictated by the critical angle of reflection for the highest photon energy within the design range, in this case 1200 eV. The design is simple, and by optimizing each component for a specific experimental requirement, extremely high performance can be achieved. In a design bandpass of 1 eV at 1 KeV, the predicted photon flux is expected to be 3×10^{12} ph/sec, with the storage ring running at 1.9 GeV 400 mA, and this flux will be contained in a monochromatic spot size of around 30 μm diameter. This not only will allow rapid spectroscopic PEEM imaging, but should also allow us to undertake time-resolved studies at standard video frame rates.

In the case of μ-XPS, the optical requirements are significantly different. In this case, the grating together with a horizontally deflecting spherical mirror produce a monochromatic image of the source at an adjustable pinhole aperture. In the vertical direction the grating produces a 1:1 image, and in the horizontal direction, the mirror produces a 2:1 demagnified image. The monochromatic image size at this point is therefore 30 by 120 μm (FWHM) in the vertical and horizontal direction respectively. The slit is set to 20 μm by 40 μm (v, h) and the following Kirkpatrick-Baez (K-B) mirror pair image at 20 and 40 :1 respectively. The K-B system uses a pair of elliptically bent mirrors, as described in the previous section. From consideration of the critical angle for our upper photon energy (1200 eV), we can determine as previously described that the convergence onto the image should be a maximum of 14 mrads. If we assume that the K-B mirrors are at 100 and 200 mm from the sample, and we assume a 1.6° grazing angle, we can see that the mirror lengths are approximately 50 and 100 mm (horizontal and vertical respectively). If we set the object aperture as above to 40 by 20 μm (h, v), then we need 40:1 and 20:1 demagnification to achieve our aim of a 1 μm image, and hence the angular acceptance from the object aperture is 0.35 and 0.7 mrads respectively (h, v). As the monochromator operates at almost 1:1 magnification, the 0.7 mrads in the vertical direction corresponds to the acceptance from the source. This compares to the divergence from an ALS bending magnet source at 1.2 KeV of 0.5 mrads (FWHM), and so we can say that at this energy, our K-B mirror system can demagnify the source to a 1 μm image with only the geometrical loss caused by reducing the vertical monochromatic image size by a factor of 2 to 20 μm. In the horizontal case, the 0.35 mrads divergence from the object slit corresponds to 0.175 mrads from the source due to the 2:1 demagnification of the horizontally focusing and deflecting mirror. This mirror is at 11 m from the source at a grazing angle of 2°, and is therefore only 55 mm long. The acceptance from the source is 0.175 mrads, and as the image size at the object slit is 120 μm in comparison to the horizontal slit size of 40 μm, only around 1/3 of the focused beam is transmitted. The photon flux in this case is expected to be around 2×10^{10} ph/sec in a focused spot of 1 μm diameter. This flux is commensurate with that necessary for X-ray Photoelectron Spectroscopy (XPS), and we are presently completing a microscope with XPS detection specifically designed to meet the needs of the semiconductor microstructure community. Images are created by scanning the sample in the fixed micro-focused beam, and recording some response function of the sample. In order to get topographic information, or to measure x-ray absorption related features, we will record the total electron yield as a function of position. Once the area of interest has been identified, we will measure high resolution XPS data from that position, or create local area maps by choosing a particular photoemission feature and recording its intensity as a function of position. In order to investigate microstructured surfaces that have been examined in other microscopes, the system has a rapid sample introduction system, optical registration with a high resolution in-situ optical microscope, and a laser interferometer to record the sample position. In this way, we will be able to go directly to features previously identified to micron precision. This system is in the final stages of construction and will be commissioned from November 1996.

The optimized design of 7.3.1, together with the high brightness of an ALS bending magnet source allows us to achieve high enough flux density to permit PEEM imaging at < 1000Å resolution at normal CCD frame rates (30 frames/sec), and to permit XPS imaging at 1 μm spatial resolution. The optical systems are simple and relatively inexpensive, and this should in turn allow us to duplicate the apparatus when demand requires. This opens the way to solving one of the problems that has emerged in synchrotron radiation microscopy, that of having sufficient time to solve real world materials science problems. The traditional approach of either sharing beamlines with a multiplicity of other users, or bringing a microscope to a beamline for specific run are not practical, and we have been driven to conclude that microscopes should be developed for specific purposes on dedicated beamlines. There will be some systems where the higher resolution afforded for zone plate microscopes, or aberration corrected PEEMs require the far higher brightness of undulators and one new system we are developing is described in the following section.

ELLIPTICALLY POLARIZING UNDULATOR BEAMLINE 4.0

Undulators produce quasi-harmonic radiation, and radiate over a narrow range of angles. If one observes the undulator at the energy of the peak of an odd harmonic, the angular emission cone has a width (1 sigma) of approximately $(\lambda/L)^{0.5}$ where λ is the energy of the harmonic, and L is the length of the undulator. For example, at 1 KeV, a 2.5 m long undulator will radiate in a cone width of 22 μradians (1 sigma). This narrow angular range, combined with the addition of flux from each pole results in extremely high brightness when used in conjunction with the small electron beam emittance of a third generation source such as the ALS. For the case of any micro-focusing system, brightness translates directly into the flux available in the focus.

Because of the small angular width of undulator radiation, traditionally a single monochromator has been used to provide light to different end stations for both spectroscopy and microscopy. An example is beamline 7.0 at the ALS, in which a watercooled SGM is used in conjunction with a 5 cm period length, 89 period undulator to provide light to a range of experimental stations including high spectral resolution XPS, PEEM, and zone plate based microscopes (5,6,7). Inevitably in designing an optical system to satisfy several different requirements, many compromises in performance have to be made.

We are presently constructing undulators and optical systems for both spectroscopy and microscopy on a new line at the ALS, beamline 4.0. The layout of the system is shown in Figure 6. The main difference in comparison to existing systems is that we have designed two beamlines, one for spectroscopy, one for microscopy, both with highly specific design constraints. In addition, we are using two undulators in the straight section that can supply beam to both of the beamlines simultaneously. As we only have 10 straight sections available, doubling the number of available undulator beamlines is a significant benefit at the cost of a modest reduction in the already extremely high brightness of full length 4.5 m undulators.

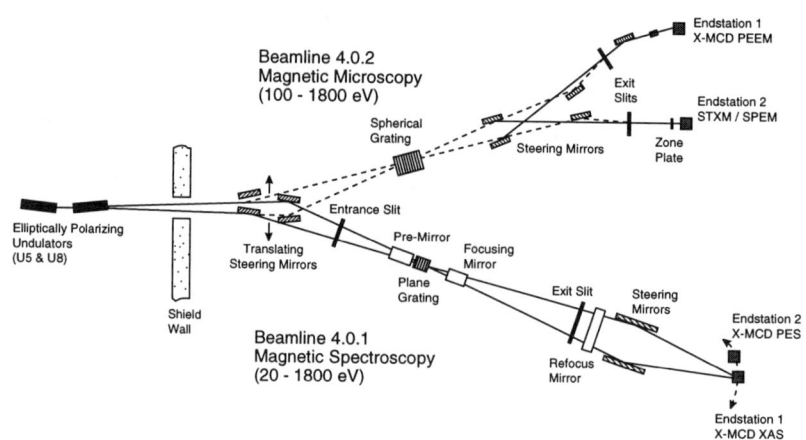

FIGURE 6. Elliptically polarizing undulator beamlines

The electron beam in the straight section is deflected before, between and after the undulators so that the two beams will be separated by an angle of 2.5 milliradians. The undulators are elliptically polarizing (19,20) and by correct phasing can produce radiation that is linearly polarized in-plane, out of plane, and circularly polarized. Initially we are building a 5 cm period device to cover up to 1.8 KeV, followed by an 8 cm period device to cover down to 15 eV. A third 5 cm period device will be built at a later date and installed in the upstream end of the straight. This will be used for experiments in which the two 5 cm undulators will be tuned to the same photon energy but opposite helicity, and the two beams will then be directed by a mirror switchyard into one beamline. The two beams will be alternately chopped by a mechanical disc, thus causing the helicity of the light at the sample to reverse. This will be used for magnetic circular dichroism experiments in which the dichroism to be measured is very small and by changing the beam helicity at high rates (KHz), noise introduced by motion of the beam or the optical system can be minimized.

The mirror switchyard is designed to direct light from either undulator to either beamline, or both undulator beams to one beamline. The 4.0.1 beamline is designed for high resolution circular dichroism spectroscopy, and a design requirement was that the photon energy range should cover 20 eV - 1.8 KeV. This very wide energy range can only be covered if the monochromator uses a grating arrangement in which the deviation angle can be changed as a function of photon energy (16). We have chosen to base our design on the SX 700 (21) design in which a plane mirror and plane grating establish the variable deviation angle, but in which a spherical mirror at low demagnification is used to produce a monochromatic astigmatic image (22). The 4.0.1 switchyard mirrors (one for each of the two beams) are horizontally focusing toroids, and focus in the saggital direction onto the entrance slit of the monochromator, and in the tangential direction onto the sample. Although the beams focus in the horizontal direction at the sample, they are directed to cross at the grating, and are redirected by plane mirrors after the exit slit to cross at the sample focus. The reason for the crossing at the grating is that we will then illuminate the same optical area for each of the two beams for the resolution determining optics. This should help to eliminate the small energy shifts that afflict other beamlines in which 2 beam mode chopping has been attempted. A further innovation is that the grating is designed to have multiple stripes with differing groove depth. In this way, a diffraction efficiency of greater than 15% is predicted throughout the whole energy range from 20 - 1.8 KeV. This possibility is allowed by the fact that we will be employing a lateral grating interchange mechanism, so that changing areas on the grating is straightforward, and that the undulator beam size is typically only millimeters wide compared to the grating width of many centimeters.

The mirror switchyard can also direct beams into the 4.0.2 beamline. This optical system is specifically designed for PEEM and zone plate based microscopy. The energy range is more restricted, 100 - 1.5 KeV, and so a less complex optical arrangement can be used. This system is still being studied, but either a multi-grating SGM will be used, or a plane grating arrangement as in 4.0.1 with only a single variable line spaced (VLS) grating. The VLS arrangement has the advantage in this case that it allows an arbitrary choice of incidence and diffraction angles, and thus allows a choice of dispersion to be made. In many cases in microscopy, we wish to image a fixed pinhole, but to vary the energy bandpass, and this VLS plane grating design should allow us to do so. The system will be entrance slitless as small electron beam induced wavelength shifts are less important in this case than on 4.0.1, as almost all the magnetic systems to be studied are concentrated, but of small area. The switchyard mirrors will focus only in the horizontal direction, and will form an image at the exit slit of the monochromator. Another switchyard will then relay light either to a zone plate based microscope, or to a PEEM microscope. The choice of magnification of the relay optics is not so important in the case of the zone plate microscope, but they must preserve the beam brightness and so near unity magnification will be used. In the case of the PEEM system however, it is very important to produce the highest flux density on the sample. With the advent of a practical scheme for chromatic and spherical aberration correction in emission microscopes (23), in order to utilize the extremely high spatial resolution that such systems will achieve (<5 nm), and to obtain statistically significant data in a reasonable time (seconds / frame), we need to achieve very high flux density at the sample. We will do this by using a Kirkpatrick-Baez mirror pair as described previously to demagnify to a point that maximizes flux density. Our initial design shows that it should be possible to achieve a spot size of less than 1 μm without geometrical loss in both the vertical and horizontal directions, if the system is optimized for 1 KeV operation (ie., between the 3d transition metal $L_{2,3}$ edges and the rare earth $M_{4,5}$ edges). To view wider regions for initial survey work, the mirror system will be slightly defocused, thus achieving the coordinated condensing and magnification action used in a transmission electron microscope.

We have attempted to show that by optimized optical design, bending magnet sources can be used for a wide range of demanding microscopy, such as 1 µm scanning XPS, and 0.1 µm resolution PEEM. The use of such systems will help to solve the problem of providing an adequate number of microscopes at modest cost. In addition, where the ultra-high brightness of undulator sources is needed, we have also shown that the use of two undulators in one straight section and separate optimized beamlines for spectroscopy and microscopy results in a highly effective overall solution. In addition, in order to achieve optimum optical coupling, the refocus systems for zone plate and electron emission microscopes need to be separated and individually optimized.

ACKNOWLEDGEMENTS

The work reported here is an ongoing effort by many members of the ALS Experimental Systems Group. Among the contributors to this effort are S. Anders, A. Cossy, R. Duarte, K. Franck, F. Gozzo, M. Howells, Z. Hussain, S. Irick, T. Lauritzen, V. Martynov, J. Meng, G. Morrison, T. Renner, R. Sandler, R. Steele, T. Warwick and T. Young.

This work was supported by the Director, Office of Energy Research, Office of Basic Energy Sciences, Materials Sciences Division of the U.S. Department of Energy, under Contract no. DE-AC03-76SF00098.

REFERENCES

1. Voss, J., Fornefett. M., Kunz, C., Moewes. A., Pretorius, M., Ranck. A., Schroeder, M., Wedemeier, V., *J. Elect. Spect. Relat. Phenom.* **80**, 329-335 (1996)

2. Cerrina, F., Ray-Caudhuri, A. K., Ng, W., Liang, S., Singh, S., Welnak, J. T., Wallace, J. P., Capasso, C., Underwood, J. H., Kortright, J. B., Perera, R.C.C., and Margaritondo, G., . *Appl. Phys. Lett.* **63**(1), 63-65 (1993)

3. J. Voss, this conference.

4. Padmore, H. A., Howells, M. R., Irick, S., Renner, T., Sandler, R., and Koo. Y-M., "Some new schemes for producing high-accuracy elliptical x-ray mirrors by elastic bending," in Proceedings of SPIE conf. 2856 (1996), *"Optics for high brightness synchrotron radiation beamlines II"*, Denver Aug. 1996.

5. Warwick, A., Heimann, P., Mossessian, D., McKinney, W., and Padmore, H. A., *Rev. Sci. Instrum.* **66**, 2037 (1995)

6. Padmore, H. A. and Warwick, A., *J. Synch. Rad.* **1**, 27-36 (1994)

7. Padmore, H. A. and Warwick, A., J. Elect. Spect. Relat. Phenom. **75**, 9-22 (1995)

8. Warwick, T., Ade, H., Hitchcock, A. P., Padmore, H. A., and Tonner, B. P., "Soft X-ray spectromicroscopy development for materials science at the Advanced Light Source" Special issue of *J. Elect. Spect. Relat. Phenom.*, Ed. H. Ade, in press

9. Underwood, J. H., *Space Sci. Instrum.*, 3, 259-270 (1977)

10. Ugural, A. C., S. K. Fenster, *Advanced Strength and Applied Elasticity*, Prentice Hall, Englewood Cliffs, 1995.

11. Irick, S., "Improved measurement accuracy in a long trace profiler", *Nucl. Inst. Meth.*, **A347**, 226-230 (1994).

12. Rockwell Power Systems, 2511C Broadbent Parkway NE, Albuquerque, New Mexico 87107

13. Howells, M., Lawrence Berkeley National Laboratory (private communication)

14. Stohr, J., *J. Elect. Spect.* **75**, 253-272 (1995)

15. Smith, N. V., and Padmore, H. A., *Materials Research Society Bulletin* **XX**, 41-44 (1995)

16. Padmore, H. A., Martynov, V., and Hollis, K., *Nucl. Inst. Meth.* **A 347**, 206-215 (1994)

17. Martynov, V. V., Young, A. T., and Padmore, H. A., "Elliptically polarizing undulator beamline 4.0.1 for magnetic spectroscopy at the Advanced Light Source", in Proceedings of SPIE conf. 2856 (1996), *"Optics for high brightness synchrotron radiation beamlines II"*, Denver Aug. 1996.

18. Young, A. T., Padmore, H. A. and Smith, N. V., *J. Vac. Sci. Technol.* **B 14(4)** 3119 - 3125 (1996)

19. Sasaki, S., *Nucl. Inst. Meth.*, **A 347**, 83 (1994)

20. Carr, R. and Lydia, S., *Nucl. Inst. Meth.*, **A 347**, 77 (1994)

21. Petersen, H., *Opt. Commun.*, **40**, 402 (1982)

22. Padmore, H. A., *Rev. Sci. Instrum.*, **60(7)** 1608-1615 (1989)

Techniques and Applications of Spectromicroscopy with Soft X-rays

J. Voss

*II. Institut für Experimentalphysik, Universität Hamburg,
Luruper Chaussee 149, 22769 Hamburg, Germany*

Abstract. Spectromicroscopy is developing rapidly and is being applied in diverse fields. In this article different technological approaches are outlined and their characteristics discussed.
The scanning soft X-ray microscope at Hasylab has the capability to use photoelectrons, luminescence, photodesorbed ions, reflected, scattered and transmitted photons as signals for imaging and spectroscopy. Mirror optics for grazing and normal incidence provide a lateral resolution in the micron and submicron range for photon energies of 15 to 1500 eV. Experimental results on lithium niobate, porous silicon, barium fluoride and gadolinium oxysulfide are presented. Peliminary results on linear magnetic dichroism in VUV-reflectivity and defects in multilayer structures are discussed.

INTRODUCTION

Spectromicroscopy and microspectroscopy, the combinations of conventional soft X-ray spectroscopy techniques with lateral resolution, have developed into powerful analytical methods to characterize structured and inhomogeneous samples. There is growing interest in material and surface science in laterally resolved information about elemental, chemical, electronic, magnetic and topographic features The availability of synchrotron radiation sources (1-3) has provided the possibility of developing instruments with dedicated characteristics. The different requirements and priorities concerning lateral resolution, tunability of the exciting radiation, number of the utilizable signals, spectral resolution, exposure time, and the properties of the optics lead to different technological approaches with characteristic advantages and disadvantages. The developments in the field of soft X-ray microscopy and details of more than 50 soft X-ray microscopes operating worldwide are described in a series of conference proceedings (4-8).

In principle the methods can be divided into two categories: full-field imaging microscopes and scanning instruments, as shown schematically in Figure 1. In the imaging type microscope the sample is illuminated over a large area and the signal which originates from this area is imaged by an optical system. This configuration

© 1997 American Institute of Physics

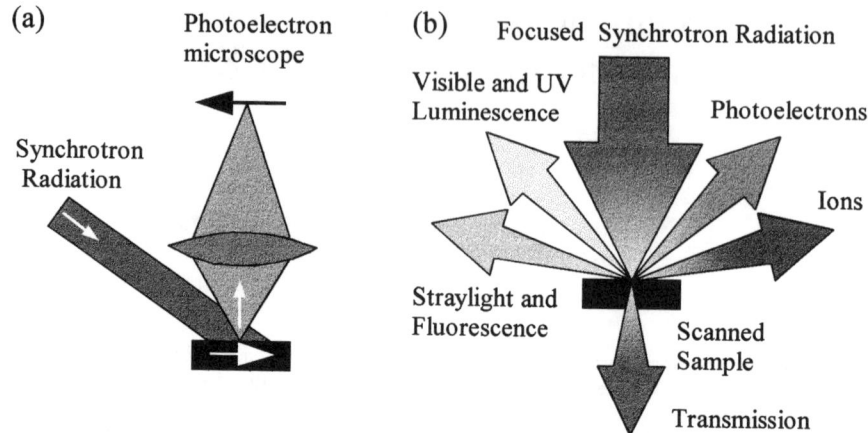

FIGURE 1. Techniques to combine lateral and spectral information. (a) The imaging type microscope. (b) The scanning type microscope.

is used in different types of photoelectron emission microscopes with electrostatic (9-14) or magnetostatic lenses (15,16). Present activities try to combine high lateral and spectral resolution in one instrument (17). Transmittive samples can be investigated with imaging X-ray optics, Fresnel zone plates (18-20) and Wolter mirrors (21).

In scanning microscopes the lateral resolution is achieved by focusing the incident radiation. Images are acquired one pixel at a time while rastering the specimen. Compared to full-field imaging this method is significantly slower, but it is very versatile in the choice of the signals. Imaging and spectral analysis of several signals becomes possible, standard spectrometers can be used, and if the optical arrangement permits, different signals can be detected at the same time.

The optical elements for focusing of soft X-rays which are used in the scanning type instruments are single and multiple reflection grazing incidence mirrors (22-24), systems of normal incidence mirrors with single layer (25) or multilayer coatings (26,27) and zone plates (28-34). These elements are discussed more in detail in section 2.

Figure 2 shows with an example of a silicon/silicon nitride test pattern the different information one can obtain if lateral as well as spectral resolution is available. In spectromicroscopy (Figure 2a) a certain spectral signal is used for image formation. This signal can depend on the exciting photon energy and/or its kinetic energy. Microspectroscopy describes the spectroscopy of signals originating only from small areas of the sample. Furthermore in microspectroscopy two different classes are distinguished: in primary microspectroscopy (Figure 2c) the signal is measured as a function of the excitation photon energy, in secondary microspectroscopy the signal itself is energy analysed (Figure 2b).

The example shown in Figure 2 describes the capabilities of these methods. The absorption edges in the excitation spectra as well as the peaks in the photoelectron energy distribution curves can be used to obtain element contrast. The changes of the chemical states of an atom due to its different surroundings leads to a chemical shift in the kinetic energy of the photoelectrons. In Figure 2(b) this is demonstrated for the Si 2p electrons of Si and Si_3N_4. This chemical contrast mechanism is used in Figure 2(a) to image the pure Si-areas of the sample.

This example demonstrates the importance of a variable photon energy for the primary excitation. Besides lateral and spectral resolution, tunability is one of the most important parameters and has to be taken into account in the design of a microscope.

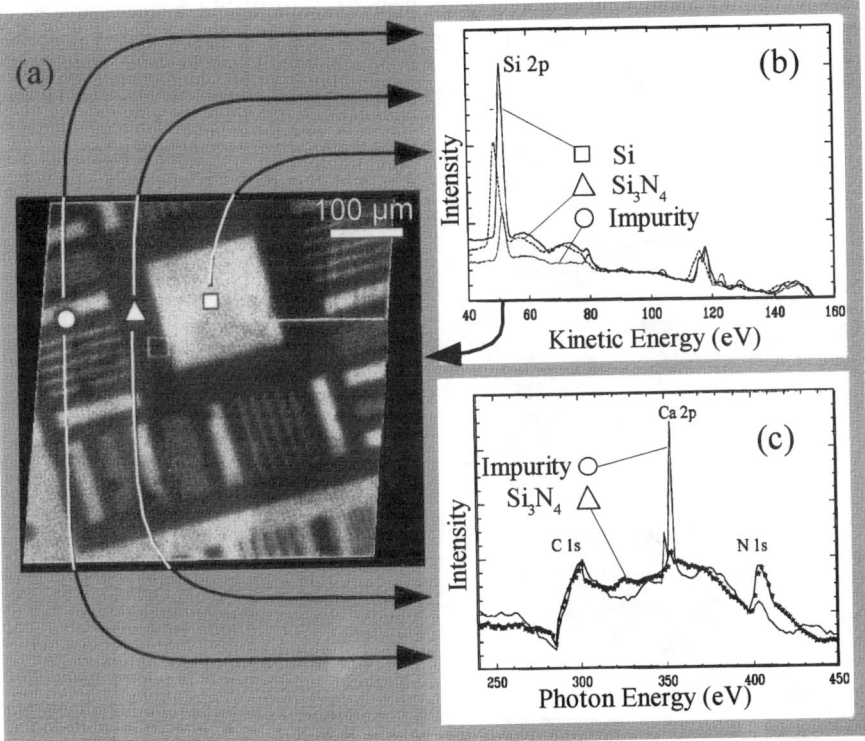

FIGURE 2. Different methods for obtaining lateral and spectral information: the sample is a silicon/silicon nitride test pattern (35). (a) *Spectromicroscopy*: micrograph with Si 2p electrons from pure Si, hν=160 eV. (b) *Secondary microspectroscopy*: photoelectron secondary spectra recorded at the highlighted points of (a). (c) *Primary microspectroscopy*: photoelectron excitation spectra (total electron yield) taken at Si_3N_4 and the impurity.

OPTICS FOR SOFT X-RAY SPECTROMICROSCOPY AND MICROSPECTROSCOPY

Depending on the requirements of different applications the operating soft X-ray microscopes have been realized as imaging or scanning type instruments. Full field imaging microscopes use either electron or X-ray optics. Investigations of transmissive samples in most cases have been made with Fresnel zone plates (18-20) (Figure 3d), but some attempts have also been made with Wolter-objectives (21), i.e. combinations of grazing-incidence ellipsoid and hyperboloid mirrors (Figure 3b). Fresnel zone plates achieve lateral resolutions down to 30 nm, but due to strong chromatic aberrations, tunability is inseparably connected with simultaneous variation of the working distance. The problem of chromaticity is absent in case of Wolter mirrors, but as for all aspherical elements surface figure errors preclude achieving the highest lateral resolution.

Conductive samples can be imaged with different types of photoelectron-

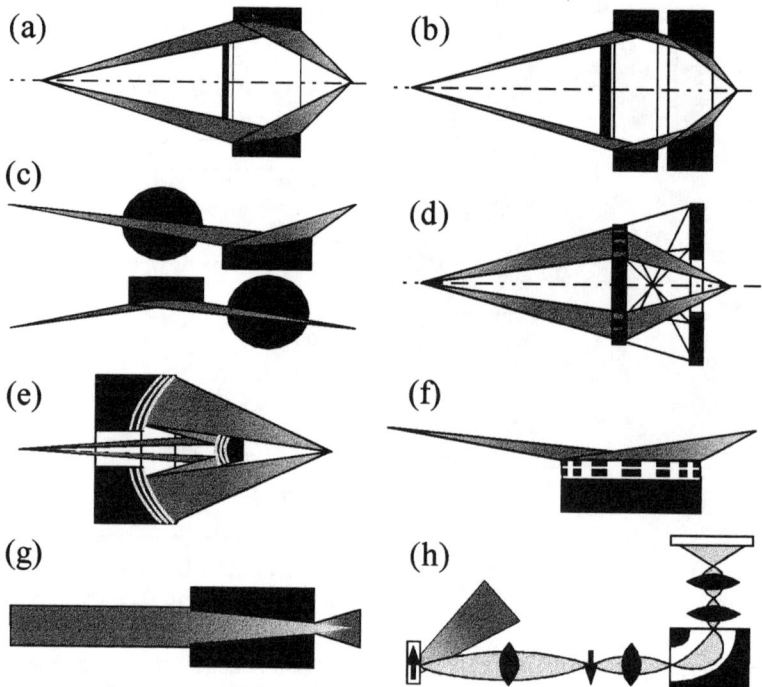

FIGURE 3. Optics for soft X–ray microscopy and microspectroscopy. (a) Ellipsoid mirror, a concentrating but nonimaging element. (b) Wolter objective, imaging combination of grazing incidence ellipsoid and hyperboloid mirrors. (c) Kirkpatrick-Baez arrangement of two grazing incidence mirrors, each focusing in one direction (vertical and horizontal view). (d) Fresnel zoneplate, diffractive element for imaging and focusing. (e) Schwarz-schild objective, imaging combination of two spherical mirrors, for soft X–rays coated with multilayers. (f) Bragg-Fresnel optics. (g) Capillary, multiple reflection concentrator. (h) Photoelectron emission spectromicroscope, electron optics both for imaging and spectral analysis.

emission-microscopes (PEEM, Figure 3h). With these instruments the photon energy can be tuned easily and no high quality X-ray optics is necessary. The main disadvantage of most of the existing emission microscopes is the incompatibility of lateral resolution with spectral analysis of the kinetic energy of the photoelectrons. The PEEM, using electrostatic lenses, has a lateral resolution of about 0.1 μm, but no spectral resolution (9-11). The Escascope, an imaging hemispherical analyzer, provides spectral information but the lateral resolution is in the order of 5 μm (12-14). A similar resolution has been reported for a number of microscopes based on divergent magnetic fields (15,16). Recently reported results demonstrate simultaneous energy resolution and high lateral resolution of 40 nm (17). Several projects are under development aiming at a resolution of a few nanometers.

Unlike imaging microscopes, scanning instruments only need an optical system which is capable of concentrating soft X-rays into a small focus area. Grazing incidence mirrors seem to be suitable for this purpose, they have the advantages of full tunability and large working distances, which simplifies the detection of surface signals, but in spite of efforts in manufacturing technology, the attainable quality of aspherical surfaces prevents utilizable lateral resolutions below the limit of visible light microscopes (22-24).

Kirkpatrick-Baez systems, consisting of two crossed, single focusing mirrors (Figure 3e), have been used mainly for harder X-rays, but there is a new attempt to use elliptically bent plane mirrors to produce a soft X-rays probe in the submicron range (36). Grazing incidence mirrors with an axis of rotation such as ellipsoid mirrors (Figure 3a) or Wolter systems (Figure 3b), have large numerical apertures, but due to the necessary central obscuration diffraction reduces the contrast compared to rectangular or circular apertures.

Spherical mirrors, which can be produced with the smallest figuring errors, can be combined to form imaging Schwarzschild objectives (Figure 3e). The normal incidence geometry limits the maximum utilizable photon energy to about 30 eV if single layer coatings are used. In the VUV range a diffraction-limited resolution of 0.1 μm could be achieved with a comparatively small technical and financial effort (25). For higher photon energies these objectives can be multilayer coated to reflect at a predefined, fixed wavelength. For photon energies of 70 to 95 eV, reflectivities of about 0.6 and a lateral resolution of 100 nm were obtained with multilayers consisting of alternating layers of Si and Mo (26,27). The use of these optics with other material combinations at higher energies still fails due to the low reflectivity caused by the interface roughness.

The restricted tunability of zone plates (Figure 3d) due to their chromaticity, makes excitation spectroscopy over large energy ranges more complicated. If a small energy window is required, which is the case for applications investigating near edge X-ray absorption fine structures (NEXAFS), this problem can be overcome by simultaneous changing of the working distance (28-30).

Bragg-Fresnel (39) and capillary optics (37,38) which are also shown in Figure 3(f) and (g) at present are of less importance in the soft energy range, but there is increasing interest for applications with harder X-rays.

THE SCANNING MICROSCOPE AT HASYLAB

The microscope at Hasylab was designed to investigate the different signals emitted from the surface of conducting and insulating solid state samples following primary excitation with soft X-rays. To meet the requirements of wide range tunability (15-1500 eV) and the variety of signals it was decided to build a scanning instrument with a grazing-incidence ellipsoid mirror. In a further development high resolution normal-incidence Schwarzschild optics have been added to optimize the imaging properties in the vacuum ultraviolet (VUV, 15-30 eV).

The essential features of the microscope are summarized in Table I and Figure 4, detailed descriptions have been published elsewhere (22,40-44).

TABLE I. The parameters of the microscopy beamline at Hasylab

Beamline	DORIS wiggler/undulator W1	
Polarisation	Linear	
Monochromator	Plane grating	
Energy	15 eV< hν<1500 eV	
Energy resolution	E/ΔE~200-400	
Optics	Ellipsoidal ring mirror	Schwarzschild objective
Working distance	30 mm	10 mm
Energy range	15 -1500 eV	15 - 30 eV
Intensity	10^6 - 10^9 photons/(s µm² 100 mA)	10^6 - 10^7 photons/(s µm² 100 mA)
Minimum probe Ø	1 µm	100 nm
Scanned volume	(3 mm)³	
Scan resolution	40 nm	
Sample orientation	arbitrary	
Signals (S: spectrometer)	Photoelectrons (S) Photoluminescence (S) Photodesorbed ions (S) Transmitted, reflected and scattered photons	

The light source for the microscope is a 32 period wiggler/undulator operated at the W1 beamline on the Doris storage ring. The Flipper plane grating monochromator supplies the microscope with linearly polarised synchrotron radiation in the range of 15 to 1500 eV with an energy resolution of $E/\Delta E=200\text{-}400$. The exit pinhole of the monochromator defines the object of the microscope optics. The minimum probe diameter is 100 nm for the Schwarzschild objective and 1 μm for the grazing incidence ellipsoid ring mirror. The intensity in a 1 μm focal spot varies from 10^6 to 10^9 photons/s/100mA depending on the photon energy.

The alignment is controlled by a high precision autocollimation system with an angular resolution of 1 μrad. Scanning of the sample and focusing is managed by three actuators installed outside the ultra high vacuum chamber. The usable image volume has a size of $(140\ \mu m)^3$ using piezo elements and of $(3\ mm)^3$ with stepping

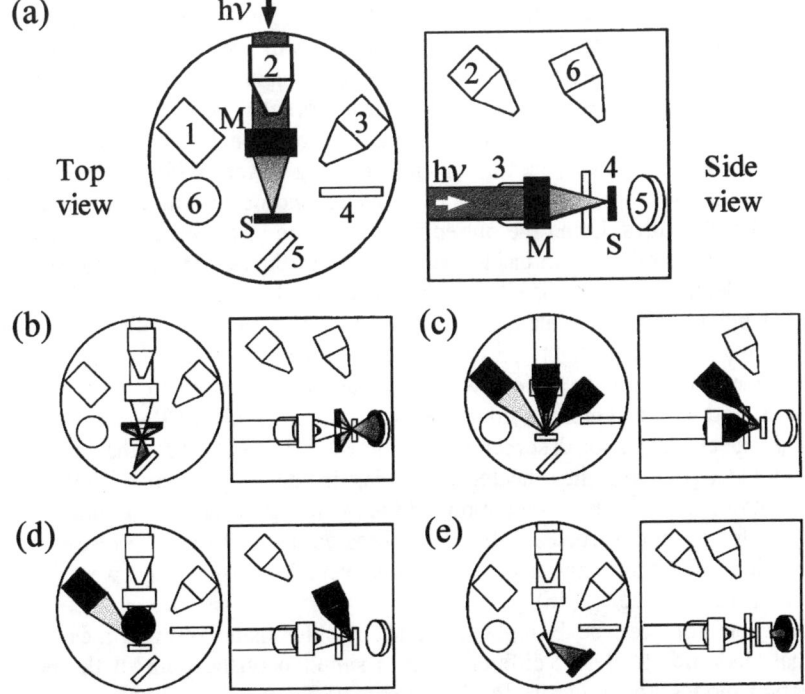

FIGURE 4. The operating modes of the microscope at Hasylab. (a) Schematic representation of the optics and the detector arrangement in the microscope UHV–chamber. /M/ focusing mirror, /S/ sample. /1/ Spectrometer for visible and UV luminescence, /2/ time of flight (TOF) electron analyser, /3/ hemispherical electron analyser, /4/ multichannel-plate (MCP) with a central hole for detection of desorbed ions, scattered, emitted and reflected soft X-rays, /5/ transmission MCP, /6/ TOF analyser for photodesorbed ions. (a) All detectors deactivated. (b) Center hole MCP and transmission detector activated. (c) Luminescence spectrometer and electron analysers activated. (d) TOF ion analyser and luminescence spectrometer activated. (e) Detector for reflected light in grazing incidence activated, the scanning plane can be oriented arbitrarily.

motors. The position of the sample can be monitored with a resolution of 40 nm. The microscope is equipped with a separate preparation chamber to sputter, anneal, and put in samples. The samples are transfered to the analysis chamber by means of a magnetically coupled transfer rod.

Due to the large working distance of the mirror objectives used in the microscope, the free solid angle of the illuminated area of the sample is large enough to arrange detectors and analyzers for all conceivable types of signals emitted from the surface. A schematic diagram of the optics and the different detectors mounted permanently in the microscope vacuum chamber is shown in Figure 4.

In Figure 4(a) /1/ is the spectrometer for visible and ultraviolet luminescence. The self-luminous or scattering area of the illuminated sample area is imaged to the entrance pinhole of the spectrometer by a Schwarzschild preoptics with a numerical aperture of 0.3. The spectrometer attached outside the vacuum consists of a spherical flat-field grating, a position sensitive detector for readout of complete spectra and three photomultipliers for different detection modes. Light of wavelengths from 190 to 770 nm can be analyzed with a spectral resolution of 1.3 nm. In addition to this, detection of scattered visible light can be used to determine topographic structures of the sample. For this darkfield microscopy the sample is illuminated with zero order light from the monochromator. Detector /2/ consists of a multichannelplate (MCP) and a conical screening cover. With different voltage polarities this detector can be used to measure photoelectrons or reflected and scattered soft X-rays. Using the pulsed time structure of the synchrotron radiation the drift distance of 11.5 cm inside the cone allows time-of-flight spectroscopy of the photoelectrons (45). /3/ is a hemispherical photoelectron analyzer with a radius of 5 cm and a conical multielement electron lens. /4/ is an MCP with a central hole for the cone of focused light. This detector is mounted between the objective and the sample with the active side towards the sample. It is used to detect reflected and scattered light and emitted fluorescence. In addition this MCP works as a time-of-flight spectrometer for desorbed positive ions. /5/ is an MCP for the detection of transmitted light and light reflected in grazing incidence geometry. The scanning stage allows an arbitrary orientation of the scanning plane. /6/ a time-of-flight analyzer for photodesorbed positive ions. Masses up to 20 can be resolved with this device. With this analyzer desorbed ions excited by soft X-rays were used for the first time for imaging (46).

The detectors can be brought close to the illuminated sample area without mutual hindrance to detect different signals simultaneously. Four of the possible detection modes are shown in the lower part of Figure 4. In (b) the center hole MCP and the transmission MCP are activated, in (c) the electron analyzers and the luminescence spectrometer are activated, in (d) the time-of-flight ion analyzer and the luminescence spectrometer are activated. (e) represents the case of detection of the reflectivity in grazing incidence geometry.

APPLICATIONS OF THE HASYLAB MICROSCOPE

Over the past few years the microscope has been used increasingly for scientific investigations. To give a summary of the activities, examples which demonstrate the capabilities of the microscopy beamline have been selected. Imaging and spectroscopy results of different samples, using photoluminescence, photoelectrons, reflected, scattered and emitted soft X-rays and desorbed ions, are described.

Lithium Niobate

Lithium niobate ($LiNbO_3$) is a very interesting material for optical applications. It is photorefractive, used as frequency multiplier and for lightwave coupling. The crystal investigated with the microscope was pure bulk $LiNbO_3$ doped with stripes of Ti and Er. The width of the Ti lines was 10 μm spacing at about 100 μm, the corresponding values for Er were 200 μm and 1 mm. Thin layers of these materials were evaporated on the substrate (Ti: 10 nm, Er: 1.8 nm) and thermally diffused into the bulk. As a first step it was demonstrated that the microscope can image the Ti and Er doping pattern.

Figure 5 shows a compilation of some results obtained with different types of signals. In Figure 5(a) a large area of the sample is shown, using backscattered VUV light of 40 eV. At this photon energy only the presence of Ti can be detected, which can be explained by the high normal incidence reflectivity of Ti compared to lithium niobate. The high contrast of 10:1 can be achieved, if a Ti layer with a thickness of more than 5 nm remained at the surface. The normal incidence reflectivity of lithium niobate at 40 eV is 0.23% and that of a 5nm Ti-layer 2.1% (calculated with optical constants from (47)). Further investigations showed that the Ti stripes are nonconducting, which indicated that the Ti is diffused incompletely into the bulk during preparation and metallic islands, smaller than the microscope resolution, remained at the surface. This assumption is supported by the results shown in the Figures 5(c) to (g) representing a magnified section of (a).

Figure 5(c) was measured with scattered 40 eV photons. The main structures consist of two Ti stripes of different widths, separated by about 100 μm. Image (d) resulted from detection of the visible and UV light, scattered and emitted by the sample if illuminated with primary zero order light. It is the sum of the scattered visUV part of the exciting spectrum and the luminescence yield. Expectedly the assumed metallic Ti stripes appear to be the brightest regions, but furthermore there emerges an additional bright area in the upper half. This can be attributed to contributions of luminescence from Er doped lithium niobate with the help of the results of image (e). In (e) the total luminescence yield was detected at an excitation photon energy of 120 eV. The metallic Ti on the $LiNbO_3$:Er reduces the luminescence signal, which cannot be explained by the penetration depth of the incident radiation (300 nm for bulk Ti). The reason for this can only be found out, if wavelength-resolved luminescence data are available. The lower Ti stripe shows a weak luminescence due to the fraction of Ti diffused into lithium niobate.

The Ti also can be made visible with the total photoelectron yield at the Ti 2p absorption edge. This is shown in micrograph (f) and the corresponding excitation spectra in (a). Additional structures resolved in the TEY image above the O 1s edge corresponding to those in the VUV micrograph (c) may be caused by surface topography and/or impurities.

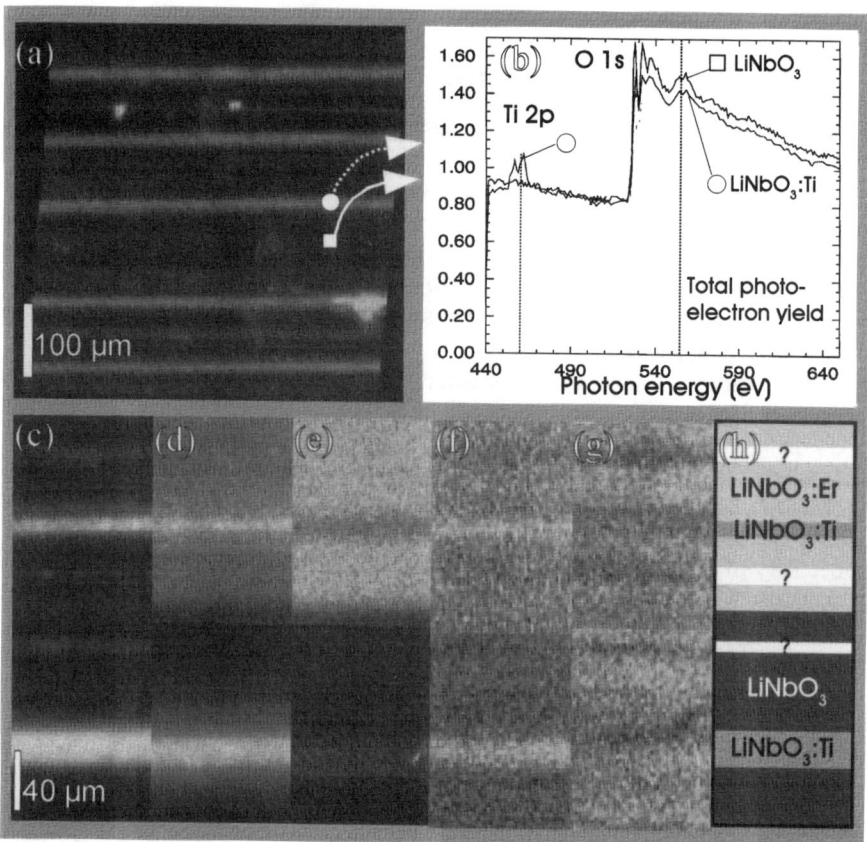

FIGURE 5 Micrographs and microspectra of a lithium niobate crystal with linetype Ti and Er dopings, probe ⌀ ~ 3 μm. (a) Micrograph with backscattered soft X-rays, hν=40 eV. (b) Total photoelectron yield excitation spectra (TEY) of $LiNbO_3$ (□) and $LiNbO_3$:Ti(O). (c)–(g) Micrographs of a magnified area of (a) taken with different signals. (c) Scattered soft X-rays, hν=40 eV. (d) Total visible and UV scatter and luminescence yield, illuminated with primary 0. order. (e) Total visible and UV luminescence yield, hν=120 eV. (f) TEY at Ti 2p, hν=462 eV. (g) TEY at O 1s, hν=555 eV. (h) Model of the structure of the examined area.

Porous Silicon

Investigations have been made on porous silicon produced by electrochemical etching of p-doped Si-wafers. The origin of the observed efficient visible luminescence, which makes this material so interesting for applications, is connected with the surface nanostructures as represented in Figure 6 and interpreted as a quantum confinement effect. The role of different associated factors is still not completely understood and is discussed intensively and controversially in the literature (48-50).

The measurements with the microscope revealed that the spectra of the emitted photoluminescence were independent of the porosity of the samples examined. In addition to the well-known emission band in the red part of the spectrum, the emission of UV-luminescence with wavelengths from 250 to 500 nm was clearly identified.

Imaging of porous silicon with the microscope turned out to be very difficult, due to the strong degradation of the photoluminescence signal during illumination of the sample with soft X-rays (51) depicted in Figure 7(a). Dots and rectangular areas only appear dark due to the previous acquisition of spectra and images in these regions. Calculations of the thermal behavior of the irradiated parts of the sample demonstrate that other effects must be responsible for the degradation.

To analyse the influence of the hydrogen passivation of the surface, investigations of the photon-stimulated desorption of positive ions have been performed. Figure 7(b) shows one of the time of flight spectra observed with primary zero order illumination. In addition to the dominant H^+-peak, H_2^+, OH^+, O^+, F^+ and some higher masses could be detected. Simultaneous detection of luminescence and H^+-desorption shows a remarkable agreement in time response as presented in Figure 7(c) and (d). This leads to the conclusion that the presence of a hydrogen passivated surface is one of the necessary factors which leads to the observed visible luminescence (52).

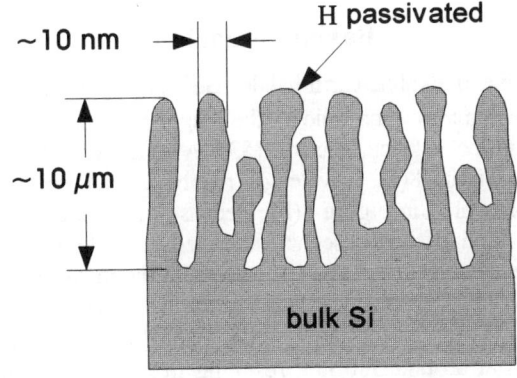

FIGURE 6. Model of the porous silicon layer on bulk silicon.

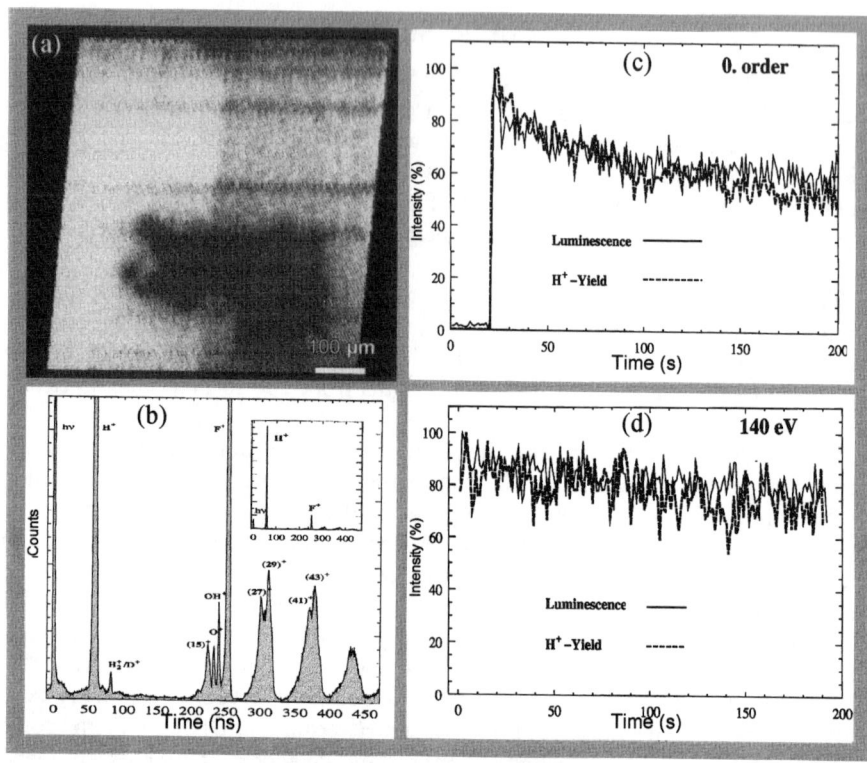

FIGURE 7. Some results measured on porous silicon. (a) Luminescence micrograph, $h\nu$=140 eV, all dark areas due to previous illumination. (b) Time of flight spectrum of photodesorbed positive ions, illuminated with primary 0. order, probe $\varnothing \sim 3$ µm. (c) Time dependence of the simultaneously measured luminescence and H$^+$ signals by illumination with primary zero order light, probe $\varnothing \sim 1$ µm. (d) As in (c) but $h\nu$=140 eV.

Barium Fluoride

Like other alkali and alkaline earth halides BaF$_2$ shows a fast intrinsic ultraviolet luminescence. This luminescence follows the Auger-free recombination of a Ba 5p core hole by a fluorine valence electron (53). A schematic representation of this process is given in Figure 8(a), wavelength resolved luminescence spectra around the 5p edge are presented in Figure 8(b). Because of the interatomic character of this process, this kind of luminescence is called cross-luminescence (CL). The subnanosecond decay constant of the CL, shown in the measured time spectrum in Figure 8(b), made BaF$_2$ a promising candidate for use as a scintillator material (54). Unfortunately, fast UV-CL at 230 nm is accompanied by an interfering slow visible luminescence band at around 300 nm, resulting from the decay of excitons. This luminescence has a decay constant in microsecond range and is visible in the slowly decreasing component of the time resolved spectrum in Figure 8(c).

Worldwide a number of groups are trying to suppress the visible component by various dopants and dopant concentrations. We have investigated a cleaved, undoped single crystal with regard to its homogeneity and obtained data on the characteristics of luminescence in the region of the F 1s and Ba 3d core excitations.

Figure 8(b) shows luminescence spectra recorded from different sample areas around the Ba 5p threshold. Microspectra and the corresponding micrographs revealed inhomogeneities of the crystal, where the ratio of the two luminescence components varies strongly with the position of the microprobe (23).

At higher excitation energies the surface of the crystal also shows topographic structures and inhomogeneities. Particularly the range from 650 to 820 eV, containing the F 1s and Ba 3d absorption edges, has been studied intensively. As an example a compilation of micrographs and microspectra is shown in Figure 9. Figure 9(a) is an overview of the investigated area measured with the total luminescence yield. This area contains a cleavage step and some impurities, leading to a reduced luminescence signal. The highlighted area is magnified in the images (b) to (d). The white points in in Figure 9(b) to (d) indicate the position of the microprobe recording the spectra in (e) and (f).

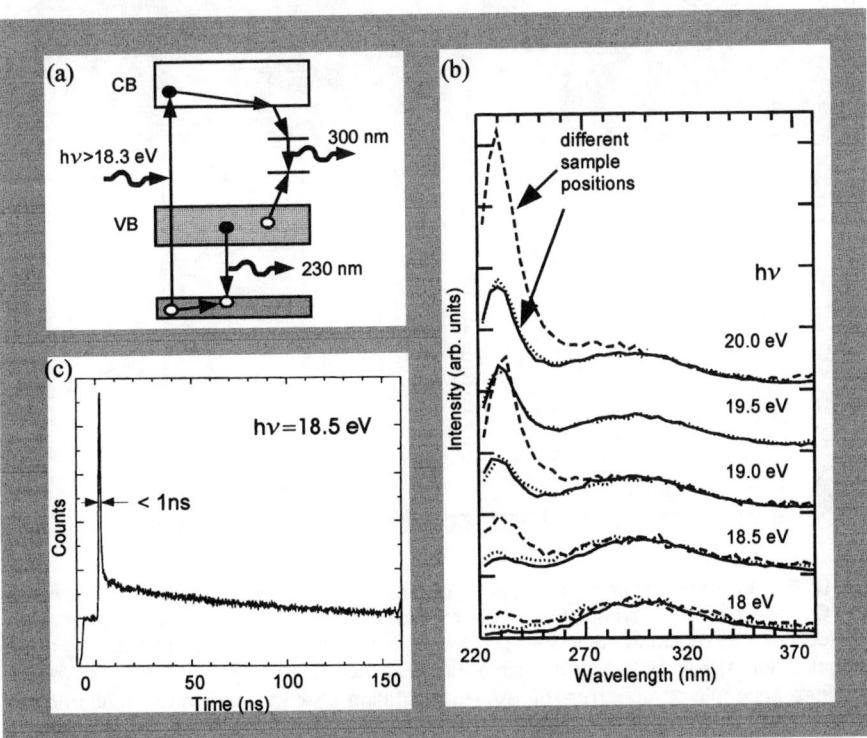

FIGURE 8. Photoluminescence of barium fluoride. (a) Diagram of the luminescence process. (b) Photoluminescence microspectra of a cleaved BaF_2 single crystal. (c) The total luminescence time decay spectrum, measured with zero order of the spectrometer.

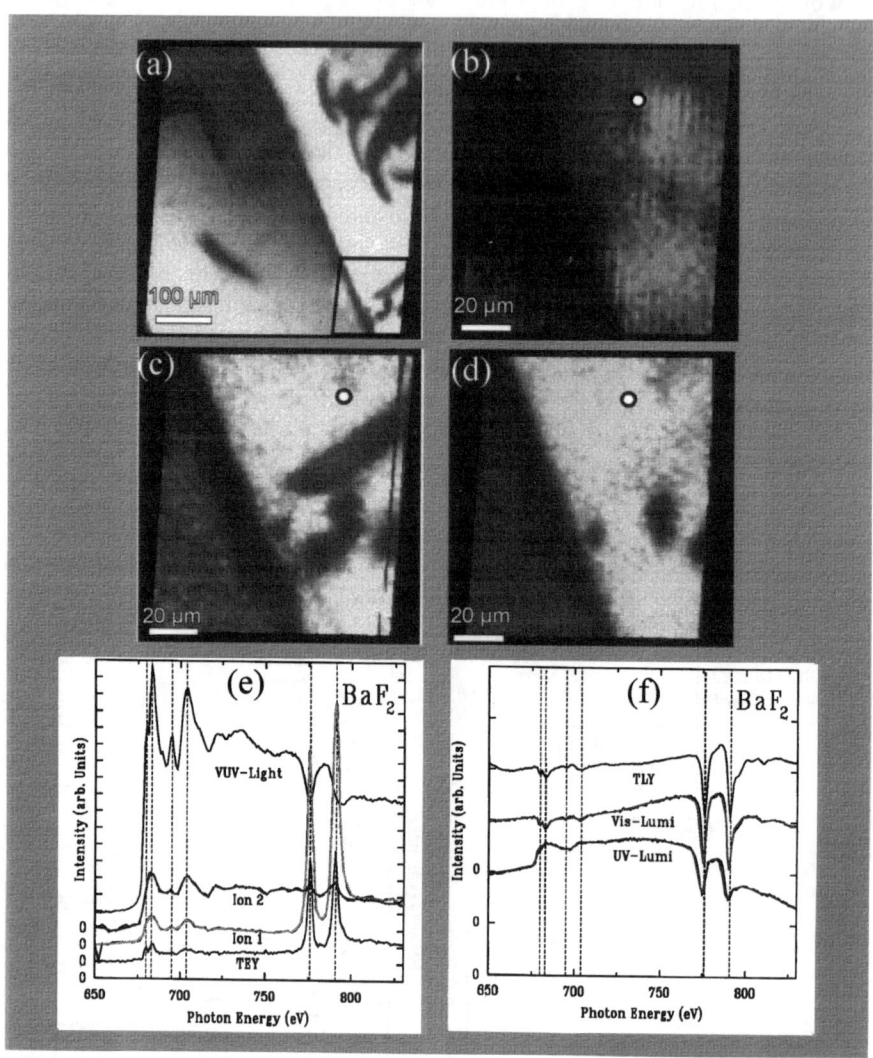

FIGURE 9. Micrographs and microspectra of a cleaved, undoped barium fluoride crystal. (a) Overview of the investigated area measured with the total luminescence yield, $h\nu$=683 eV, the region in the lower right corner is magnified in (b)-(d). (b) Mapping of the signal of ion 1, $h\nu$=790 eV. (c) Visible luminescence micrograph, $h\nu$=750 eV. (d) UV luminescence micrograph, $h\nu$=750 eV. (e) Excitation spectra taken in O : scattered soft X-rays (VUV-light), different ions and total electron yield (TEY). (f) Excitation spectra taken in O : total luminescence yield (TLY), visible and UV luminescence.

The H⁺ ion micrograph in (b) was measured with the central hole MCP at 790 eV photon energy (Ba $3d_{3/2}$ peak). The yield is strongly correlated to the Ba 3d edge as shown in Figure 9(e). A second ion species was found to be correlated exclusively to the F 1s threshold which were identified as F⁺ ions.

The wavelength resolved luminescence micrographs in Figure 9(c) and (d) reveal that some of the dark regions shown in the total luminescence yield in (a) are due to a reduced efficiency of the visible luminescence. In the UV-micrograph (d) a 20 μm wide region appears bright, which was dark in the visible component of the spectrum (c). In further investigations no radiation damage effects of this UV efficient area have been observed, which was the case for crystals doped to suppress the visible component. Another interesting detail appeared in the excitation spectra of the wavelength resolved luminescence presented in (f). In contrast to the visible luminescence the UV luminescence shows an increase at the F 1s threshold. This is surprising since the UV signal follows exclusively a Ba 5p excitation. It was expected to see a decrease at F1s and Ba 3d comparable to the characteristics of the visible luminescence (57).

Bragg Microscopy

As shown in Figure 4, a multichannelplate detector with a central hole can be put between the sample and the mirror objective to measure light, reflected or scattered by the sample. If the wavelength of the incident radiation is shorter than twice the lattice constant, under certain angles Bragg reflection can occur from single crystals or polycrystalline samples as shown schematically in Figure 10.

Gadolinium Oxysulfide

Bragg reflections were measured for the first time on a polycrystalline Gd_2O_2S sample as shown in Figure 11. The different spectra in Figure 11(a) were recorded at different positions of the sample. A detailed analysis of these data revealed, that Bragg reflection of planes perpendicular to the c-axis of the hexagonal structure

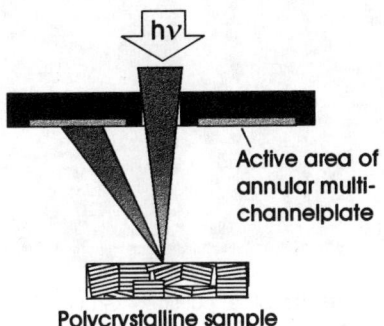

FIGURE 10. Geometry of the multichannelplate with a central hole and the sample in Bragg reflection.

can explain these results (55). Figure 11(c) shows a micrograph of the Gd_2O_2S ceramics taken at a photon energy of 930 eV. Only crystallites matching the Bragg condition at this energy, appear bright. Reflecting crystallites could be found in this small area of the sample at all photon energies between 930 and 990 eV. In Figure 11(d) the same part as in (c) is illuminated with photons of 970 eV. These results are very important for the interpretation of structures obtained in investigations of the photoluminescence efficiency (56,57).

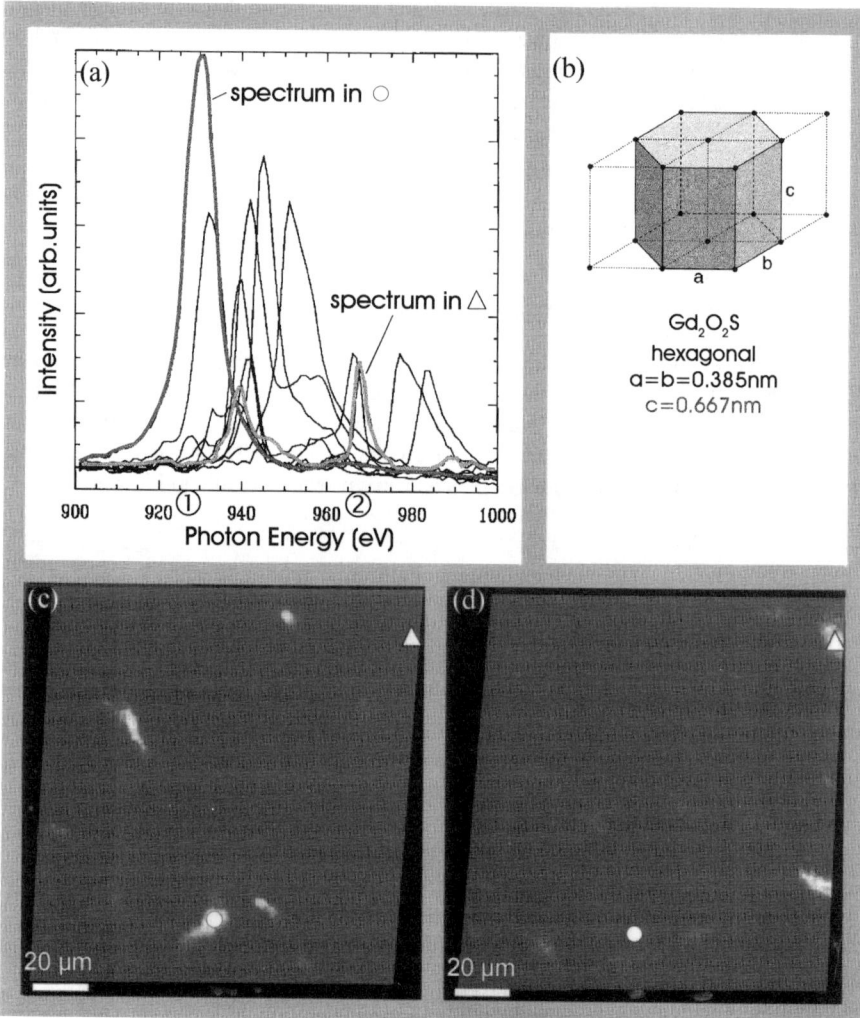

FIGURE 11. Experimental results of a polished Gd_2O_2S:Pr ceramics. (a) Excitation spectra of the backscattered X-rays originating of different sample areas. (b) The unit cell of Gd_2O_2S:Pr. (c) Micrograph of the sample at hν=931 eV(①). (d) Same area as in (c), hν=969 eV (②)

Defects in Multilayer Structures

There is increasing interest in multilayer coated optics for X-ray focusing and imaging, especially in EUV projection lithography (58,59). With this technique reflection masks are demagnified by imaging mirror optics to produce resist structures of sizes below 100 nm. To obtain sufficiently large image fields and high resolution, Si/Mo multilayer coated reflection masks and normal incidence mirror optics are used, optimized for photon energies around 95 eV. To avoid errors in the resist pattern, the reflection mask has to be defect-free. With our microscope the homogeneity and reflectivity of multilayer mirrors as well as multilayer mask structures can be analysed over a broad band of design wavelengths. In addition, the different signals have different sensitivities to surface, bulk and substrate defects (60). A compilation of some results from a Si/Mo multilayer fabricated with a multilayer-free area is given in Figure 12.

Figure 12(a) shows total electron yield (TEY) excitation spectra recorded from different positions in the defect-multilayer transition zone. This signal is a measure of the amplitude of the standing wave field, formed by the superposition of the incident and reflected waves around the Bragg peak. At 79 eV the TEY of the uncoated Si substrate is higher than from the multilayer, which has its interferene minimum at this energy. This strong contrast was used in the TEY-micrograph at 79 eV in Figure 12(c), the multilayer appears dark, the uncoated substrate bright. Tuning the photon energy to 83 eV reverses the contrast as shown in image 12(d). Analysis of the TEY interference structure provides the possibility of an indirect determination of the multilayer reflectivity.

The straylight spectra in Figure 12(b) were measured with the central hole MCP detector. In the normal incidence geometry, the specular reflected light is reflected back through the hole. Interpreting the spectra needs to consider incident light scattered by the focusing mirror. If the Bragg condition is fulfilled for the primary scattered halo surrounding the focal spot it can dominate the signal. This leads to the non-vanishing Bragg peak in the spectra taken from the multilayer-free area. Therefore contrast is reduced but it is sufficient to image the structure of the sample as shown in Figure 12(e) and (f). The higher signal on the multilayer is caused by the roughness of the enhanced number of contributing interfaces.

Further investigations have been performed with visible darkfield microscopy and with specularly reflected VUV light (not presented here). In visible darkfield microscopy the sample is illuminated with zero order light and scattered light is detected with the luminescence spectrometer. This method gives information about the topography and impurities on the surface. The specular VUV reflection can be detected if the sample is inclined with respect to the incident beam. This signal is a direct measure of the quality of the multilayer structure. Micrographs of the specular reflectivity can be measured at different angles of incidence since the scanning stage allows arbitrary orientations of the scanning plane.

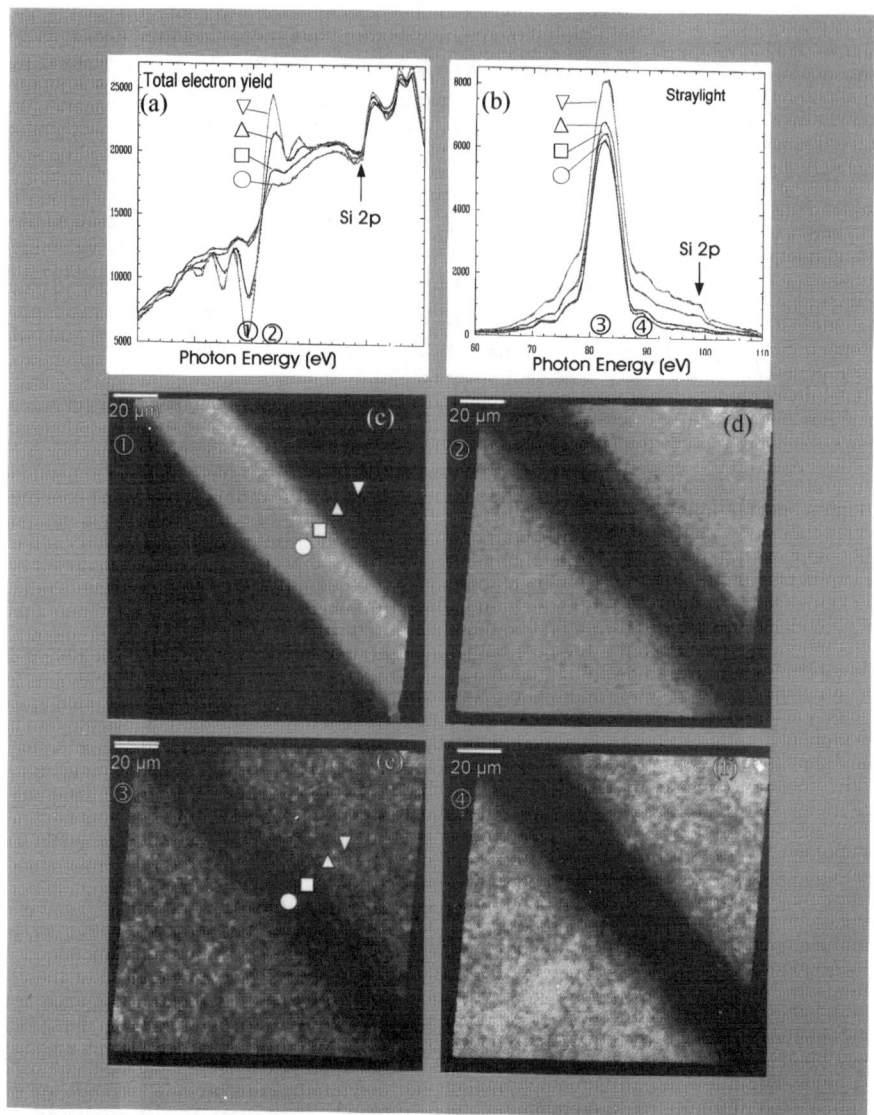

FIGURE 12. Investigations of an artificial defect in a silicon/molybdenum multilayer (60). The excitation spectra were recorded from the highlighted areas of (c) and (e). (a) Total electron yield excitation spectra at the highlighted points in (c). (b) Straylight excitation spectra at the highlighted points in (e). (c) Electron micrograph of the defect, hν=79 eV(①). (d) Electron micrograph, hν=83 eV(②). (e) Straylight mapping of the same area, hν=81 eV(③). (f) Straylight micrograph, hν=90 eV(④).

Linear Magnetic Dichroism

At the absorption edges of the transition metals different signals show a dependence on the polarization of the exciting radiation and the direction of the magnetization of the sample. At the Fe 2p threshold dichroism in the photoyield was observed using circularly polarized synchrotron radiation. Circular dichroism was used to provide contrast when imaging structures of magnetic storage media with a photoelectron emission microscope (9). Circular dichroism in transmission has been used to investigate thin magnetic films with an imaging X-ray microscope (62). Magnetic dichroism was also detected with linearly polarized radiation, both in photoyield (61) and reflectivity (63,64), but without lateral resolution. Strength of linear dichroism in reflectivity (XLMDR) is expressed by the asymmetry parameter

$$A = \frac{I_{up} - I_{down}}{I_{up} + I_{down}},$$

with the reflectivity signals I_{up} and I_{down} corresponding to magnetization directions parallel and antiparallel to the y-axis (Figure 13). An asymmetry of 17 % has been observed at the 3p edge of a Fe single crystal, illuminated with p-polarized light at an incidence angle θ of 60.5° (65). Investigations of magnetic structures using XLMDR at the microscope at Hasylab are in preparation.

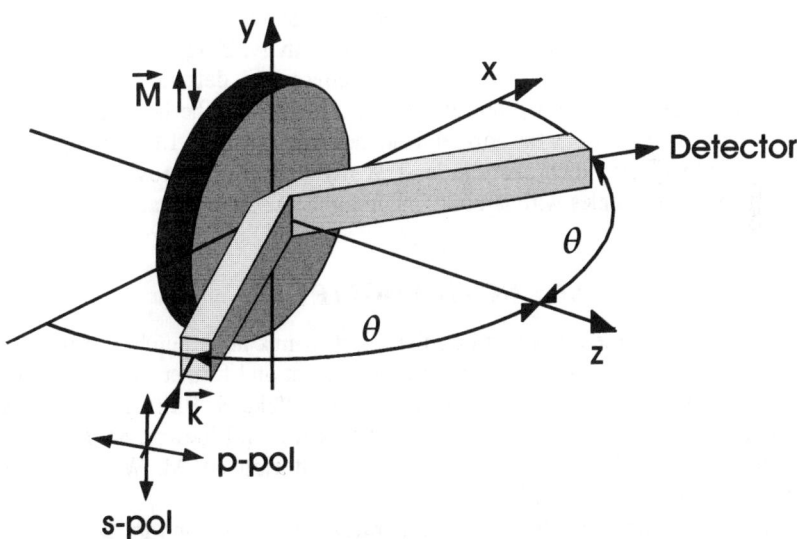

FIGURE 13. Geometry to detect the linear magnetic dichroism in soft X-ray reflectivity.

SUMMARY AND OUTLOOK

Our scanning soft X-ray microscope at Hasylab uses mirror optics for focusing of monochromatized synchrotron radiation to a microspot. The achromaticity of the grazing incidence optics provides full tunability for photon energies from 15 eV to 1500 eV with a lateral resolution in the micron range. An optionally usable normal incidence optics, restricted to photon energies below 30 eV, reaches a diffraction limited resolution of 100 nm.

A number of different signals can be used for spectromicroscopy and microspectroscopy. Photoelectrons can be detected in the total yield mode and spectrally analysed with time-of-flight and hemispherical analysers. Photodesorbed positve ions can be detected and analysed with multichannelplates and a time-of-flight spectrometer. Photoluminescence in the visible and UV is detected with several photomultipliers and spectrally resolved with a flat-field grating spectrometer. Specularly reflected, scattered and transmitted radiation is measured with different MCP detectors. The scanning plane is rotatable, the maximum scanning range has a volume of $(3mm)^3$ with a spatial resolution of 40 nm. In addition the defined pulse structure of the storage ring allows investigations with high temporal resolution.

Results of a variety of different samples have been presented to give an impression of the capabilities of our microscope. Low concentration dopants of lithium niobate are imaged with photoluminescence, scattered VUV-radiation and photoelectrons. Simultaneous detection of photoluminescence and photodesorbed hydrogen ions from porous silicon prove the influence of H-passivation of nanostructured Si. Besides a wealth of spectral information the investigation of barium fluoride shows the possibility of suppressing the slow, visible contribution to photoluminescence without increasing radiation sensitivity. Bragg reflection is used to image microcrystallites in luminescence ceramics and defects of multilayers. Total photoelectron yield is a promising method to analyse the homogeneity and quality of multilayer coatings for normal incidence optics used in EUV-lithography. Finally linear magnetic dichroism in soft X-ray reflectivity provides element-specific imaging of samples with laterally inhomogeneous magnetization.

ACKNOWLEDGMENTS

The microscopy project has required the effort from a large number of persons. Amongst the contributors to this work are the present and former members of the microscopy group at Hasylab: K. Berens von Rautenfeld, S. Büttner, H. Dadras, A. Föhlisch, M. Fornefett, J. Friedrich, H. Lübberstedt, A. Moewes, M. Pretorius, A. Ranck, G. Roy, M. Schroeder, H. Sievers, I. Storjohann, M. Wachmuth, V. Wedemeier, H. Wongel and C. Kunz.

We thank the following persons for preparation of samples and for discussions: J. Becker, University of Hamburg (barium fluoride); S. Eisebitt, Institut für Schicht- und Ionentechnik Jülich (porous silicon); W. H. Brünger, Fraunhofer Institut für Mikrostrukturtechnik Berlin (silicon nitride); E. Louis, N. B. Koster, H.-J. Voorma, and F. Bijkerk, FOM-Institute for Plasma Physics "Rijnhuizen", (multilayers); W. Sohler and coworkers, Universität Paderborn (lithium niobate).

This project is supported by the German Federal Minister of Education and Research (BMBF) under contract number 05 644 GUA 9.

REFERENCES

1. C. Kunz (ed), *Synchrotron radiation – techniques and applications*, Topics in current physics 10. Springer, Berlin Heidelberg New York , (1979).
2. E.E. Koch (ed), *Handbook on synchrotron radiation*, North Holland, Amsterdam New York Oxford, vol la, 1b, 2 (1983).
3. H. Winick, *Synchrotron radiation sources*, Series on synchrotron radiation techniques and applications 1. World Scientific, Singapore New Jersey London Hong Kong Bangalore, (1994).
4. D. Parsons (ed), *Ultrasoft X-ray microscopy, its application to biological and physical sciences*. Annals of the New York Academy of Sciences 342:148, (1980).
5. G. Schmahl, D. Rudolph (eds), *X-ray microscopy*. Springer series in optical sciences, vol 43. Springer, Berlin Heidelberg New York, (1984).
6. D. Sayre, M. Howells, J. Kirz, H. Rarback (eds), *X-ray microscopy II*. Springer series in optical sciences, vol 56. Springer, Berlin Heidelberg New York, (1988).
7. A.G. Michette, G.R. Morrison, C.J. Buckley (eds), *X-ray microscopy III*. Springer series in optical sciences, vol 67. Springer, Berlin Heidelberg New Yor, (1992).
8. V.V Aristov, A.I. Erko (eds), *X-ray microscopy IV*. International Conference on X-ray Microscopy, Chernogolovka/Russia, 1993 Bogorodski Pechatnik Publ. Comp., Chernogolovka Moscow Region, (1995).
9. J. Stöhr, Y. Wu, B.D. Hermsmeier, M.G. Samant, G.R. Harp, S. Koranda, D. Dunham and B. Tonner, *Science* **259**, (1993) 658 – 661.
10. G. de Stasio et. al., *Synchr. Rad. News* **7 (5)**, (1994) 18 – 21.
11. Y. Hwu, C.Y. Tung, J.Y. Pieh, S.D. Lee, P. Almeras, F. Gozzo, H. Berger, G. Margaritondo, G. De Stasio, D. Mercanti and T. Ciotti, *Nucl. Instrum. Meth.* **A 361**, (1995) 349 – 353.
12. T. Kachel, K. Holldack, W. Gudat, M. Neuber and C. Wilde, *Journal de Physique* **4**, (1994) 439 – 444.
13. J.D. Denlinger et.al., *Rev. Sci. Instrum.* **66 (2)**, (1995) 1342 – 1345.
14. M. Kinzler, M. Grunze, N. Blank, H. Schenkel and I. Scheffler, *J. Vac. Sci. Technol. A* **10(4)**, (1992) 2691 – 2697.
15. T. Komeda, G.D. Wadill, P.J. Benning and J.H. Weaver, *Phys. Rev. B* **43(10)**, (1991) 8713 – 8716.
16. C. Kim, P.L. King and P. Pianetta, *J. Vac. Sci. Technol. B* **10(4)**, (1992) 1944-1948.
17. G. Lilienkamp, C. Koziol, Th. Schmidt, and E. Bauer, *X-ray microscopy V*. Proceedings of the International Conference XRM 96, to be published.
18. G. Schmahl, D. Rudolph, P. Guttmann, G. Schneider, J. Thieme and B. Niemann, *Rev. Sci. Instrum.* **66 (2)**, (1995) 1282 – 1286.
19. W. Meyer – Ilse, D. Rudolph, G. Schmahl, H. Medecki, L. Jochum, E. Anderson, D. Attwood, C. Magowan, R. Balhorn and M. Moronne, *Synchr. Rad. News* **8(3)**, (1995) 29 – 33.

20 J.- D. Wang, Y. Kagoshima, T. Miyahara, M. Ando, S. Aoki, E. Anderson, D. Attwood and D. Kern, *Rev. Sci. Instnsm.* **66 (2),** (1995) 1401 – 1403.
21 S. Aoki, T. Ogata, S. Sudo and T. Onuki, *Jpn. J. Appl. Phys.* **31,** (1992) 3477-3481.
22 J. Voss, H. Dadras, C. Kunz, A. Moewes, G. Roy, H. Sievers, I. Storjohann and H. Wongel, *J. X-ray Sci. Techn.* **3,** (1992) 85.
23 J. Voss, M. Fornefett, C. Kunz, A. Moewes, M. Pretorius, A. Ranck, M. Schroeder and V. Wedemeier, *Journal of Electron Spectroscopy and Related Phenomena* **80,** (1996) 329-335.
24 U. Johansson, R. Nyholm, C. Törnevik and A. Flodström, *Rev. Sci. Instrum.* **66 (2),** (1995) 1398-1400.
25 M. Pretorius, M. Fornefett, J. Friedrich, A. Ranck, K. Berens von Rautenfeld, M. Schroeder, V. Wedemeier and J. Voss, *X-ray microscopy V.* Proceedings of the International Conference XRM 96, to be published.
26 M. Hasegawa and K. Ninomiya, *Rev. Sci. Instrum.* **66 (2),** (1995) 1361-1363.
27 F. Cerrina, A.K. Ray – Chaudhuri, W. Ng, S. Liang, S. Singh, J.T. Welnak, J.P. Wallace, C. Capasso, J.H. Underwood, J.B. Kortright, R.C.C. Perera and G. Margaritondo, *Appl. Phys. Lett.* **63(1)**, (1993) 63-65.
28 C.J. Buckley, *Rev. Sci. Instrum.* **66 (2),** (1995) 1318 – 1321.
29 S. Williams, C. Jacobsen, J. Kirz, H. Ade, M. Rivers, J. Maser, S. Wirick and X. Zhang, *Rev. Sci. Instrum.* **66 (2),** (1995) 1271 – 1275.
30 H. Ade, A.P. Smith, S. Cameron, R. Cieslinki, G. Mitchell, B. Hsiao, E. Rightor, *Polymer* **36(9),** (1995) 1843-1848.
31 J. Kirz, H. Ade, E. Anderson, C. Buckley, H. Chapman, M. Howells, C. Jacobsen, C.-H. Ko, S. Lindaas, D. Sayre, S. Williams, and X. Zhang, *Nucl. Instrum. Meth.* **B 87,** (1994) 92 – 96.
32 Y. Kagoshima, T. Miyahara, M. Ando, J.-D. Wang and S. Aoki, *Rev. Sci. Instrum.* **66 (2),** (1995) 1534 – 1536.
33 C. Ko, J. Kirz, H. Ade, E. Johnson, S. Hulbert, E. Anderson, *Rev. Sci. Instrum.* **66 (2),** (1995) 1416 – 1418.
34 G.R. Morrison and M.T. Browne, *Rev. Sci. Instrum.* **63,** (1992) 615 – 618.
35 I. Storjohann, C. Kunz, A. Moewes and J. Voss, in: X-ray *Optics and Microanalysis, Inst. Phys. Conf. Series, Bristol* **180,** (1993) 587 – 590.
36 H.A. Padmore, M.R. Howells, S. Irick, T. Renner, R. Sandler and Y.-M. Koo, *Optics for high Brightness SR Beamlines II,* Proceedings of the SPIE Conference, (1996), to be published.
37 P. Engström, S. Larsson, A. Rindby, A. Buttkewitz, S. Garbe, G. Gaul, A. Knöchel, and F. Lechtenberg, *Nucl. Instrum. Methods Phys. Res.* **A(302),** (1991) 547.
38 D.H. Bilderback, S.A. Hoffman, and D.J. Thiel, *Science* **263,** (1994) 201.
39 A. Snigirev, *Rev. Sci. Instrum.* **66,** (1995) 2053.
40 F. Senf, K. Berens v. Rautenfeld, S. Cramm, J. Lamp, J. Schmidt-May, J. Voss, C. Kunz, and V. Saile, *Nucl. Instrum. Methods* **A246,** (1986) 314.
41 J. Voss, I. Storjohann, C. Kunz, A. Moewes, M. Pretorius, A. Ranck, H. Sievers, V. Wedemeier, M. Wochnowski and H. Zhang, in Ref 8, (1995).
42 C. Kunz and J. Voss, *Rev. Sci. Instrum.* **66,** (1995) 2021-2029.
43 H. Zhang, A. Föhlisch, C. Kunz, A. Moewes, M. Pretorius, A. Ranck, H. Sievers, I. Storjohann, V. Wedemeier and J. Voss, *Rev. Sci. Instrum.* **66(6)**, (1995) 3513-3519.
44 C. Kunz and J. Voss, *Fresenius J. Anal. Chem.* **335,** (1995) 494-498.

45 A. Ranck, M. Fornefett, J. Friedrich, M. Pretorius, K. Berens von Rautenfeld, M. Schroeder, V. Wedemeier and J. Voss, *X-ray microscopy V*. Proceedings of the International Conference XRM 96, to be published.
46 M. Schroeder, M. Fornefett, J. Friedrich, M. Pretorius, A. Ranck, K. Berens von Rautenfeld, V. Wedemeier and J. Voss, *X-ray microscopy V*. Proceedings of the International Conference XRM 96, to be published.
47 B.L. Henke, E.M. Gullikson, and J.C. Davis, *Atomic Data and Nuclear Tables* **54**, (1993) 181-342.
48 A. Takazawa, T. Tamura and M. Yamada, *J. Appl. Phys.* **75(5)**, (1994) 2489-2495.
49 K.J. Nash, P.D. Calcott, L.T. Calham, M.J. Kane and D. Brumhead, *J.Lumin.* **60+61**, (1994) 297–301.
50 A. Kux and M.B. Chorin, *Phys. Rev. B* **51(24)**, (1995) 17535–17541.
51 A. Föhlisch, Diploma Thesis, Universität Hamburg, (1995).
52 M. Fornefett, Diploma Thesis, Universität Hamburg, (1996).
53 C.W.E. Van Eijk, *Nucl. Tracks Radiat. Meas.* **21(1)**, (1993) 5–10.
54 V.N. Makhov and N.M. Khaidukov, *Nucl. Instrum. Meth.* **A 308**, (1991) 205-207.
55 V. Wedemeier, K. Berens von Rautenfeld, M. Fornefett, J. Friedrich, M. Pretorius, A. Ranck, M. Schroeder, and J. Voss, *X-ray microscopy V*. Proceedings of the International Conference XRM 96, to be published.
56 A. Moewes, H. Zhang, C. Kunz, M. Pretorius, H. Sievers, I. Storjohann and J. Voss, in Ref (8), (1995).
57 A. Moewes, C. Kunz and J. Voss, *Nucl. Instrum. Meth.* **A 373**, (1996) 299-304.
58 H.J. Voorma, and F. Bijkerk, *Microelectron. Eng.* **17**, (1992) 145-148.
59 F. Bijkerk et .al., to be published in Proc. MNE 95, Aix en Provence, Sept. 26-28, (1995).
60 J. Friedrich, K. Behrens v. Rautenfeldt, M. Fornefett, M. Pretorius, A. Ranck, M. Schroeder, H. Sievers, J. Voss, V. Wedemeier, and E. Louis, N. B. Koster, H.-J. Voorma, and F. Bijkerk, *X-ray microscopy V*. Proceedings of the International Conference XRM 96, to be published.
61 T. Kinoshita, H.B. Rose, Ch. Roth, D. Spanke, F.U. Hillebrecht, and E. Kisker, *J. Magnet. Mat.* **148**, (1995) 64-65.
62 P. Fischer, G. Schütz, G. Schmahl, P. Guttmann, and D. Raasch., *X-ray microscopy V*. Proceedings of the International Conference XRM 96, to be published., *X-ray microscopy V*. Proceedings of the International Conference XRM 96, to be published.
63 F.U. Hillebrecht, T. Kinoshita, D. Spanke, J. Dresselhaus, Ch. Roth, H.B. Rose, and E. Kisker, *Phys. Rev. Lett.* **75(11)**, (1995) 2224-2227.
64 C. Kao, J.B. Hastings, D.P. Siddons, and G.C. Smith, *Phys. Rev. Lett.* **65(3)**, (1990) 373-376.
65 M. Pretorius, J. Friedrich, A. Ranck, M. Schroeder, V. Wedemeier, and J. Voss, to be published.

Total Reflection X-ray Fluorescence Analysis with Synchrotron Radiation and other Sources for Trace Element Determination

Peter Wobrauschek and Christina Streli

*Atominstitut der Österreichischen Universitäten,
Schüttelstr. 115, A-1020 Wien*

Abstract. Total reflection x-ray fluorescence analysis (TXRF) is an accepted powerful analytical tool for trace element determination in various kinds of samples. In typical applications like environmental, medical and technical sample analysis as well as for quality control during production processes, ultralow concentrations at the pg/g level, or femtogram masses, have to be determined. The combination of synchrotron radiation (SR) and multilayer monochromators together with TXRF is perfectly suited to meet the requirements. Best results can be expected from SR-TXRF though cost and accessability to SR sources limit the application. In some cases the additional inherent advantage of XRF as a nondestructive method is important. Another approach to reach such low detection limits is to increase the photon flux on the sample by means of high power x-ray tubes and multilayer focusing x-ray optics. With standard laboratory equipment the choice of appropriate anode materials for efficient excitation of specific elements and an optimal design of the energy dispersive spectrometer can also increase sensitivity. Various experimental setups used for EDXRF of the elements from B to U by K-shell excitation will be presented and discussed. The results from ultralow trace element analysis of surface impurities on Si wafers demonstrate the excellent potential of this method. With SR-TXRF the detection limits for medium Z elements can be below 20 femtogram.

INTRODUCTION

Analytical techniques for trace element determination can be quantified, following the IUPAC definition, by their lower limits of detection (LLD) given by:

$$LLD = \frac{3}{S} \cdot \sqrt{\frac{I_B}{t}} \qquad (1)$$

The sensitivity S (cps/ng) is the intensity of the fluorescence signal for a certain sample mass, the background intensity I_B and the counting time t. To improve the detection limit the sensitivity S must be increased and the background decreased. Increasing the measuring time is limited for practical reasons. In practice other aspects such as elemental range, simultaneous detection, speed of analysis, and sample throughput are equally important.

TXRF and Polarized Radiation

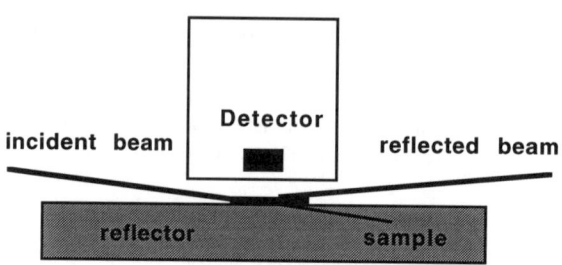

FIGURE 1. Experimental geometry for TXRF

Total reflection x-ray fluorescence analysis (TXRF) is a well established analytical technique for the detection of trace elements (1-9). The sample lies on top of a smooth polished reflector which is illuminated by a low-divergence x-ray beam incident below the critical angle for total external reflection. Reflectors made of quartz, fused silica, germanium, boron nitride and perspex have been used successfully. Regarding the surface quality, a flatness of $\lambda/20$ (589 nm) and a mean roughness of 1 nm are required. Excitation in the total reflection geometry significantly reduces the spectral background due to scattering from the substrate, as only a small fraction of the beam penetrates into the reflector. Another advantage is the additional excitation of the sample by the reflected beam, which can in principle double the fluorescence signal. Due to the grazing incidence angle (1 mrad or less) the detector can be brought very close to the sample, resulting in a large solid angle, leading to efficient detection (see Fig. 1). All these advantages make this method a sensitive analytical technique. Detection limits in the low pg range can be achieved for medium Z elements if a 2 kW x-ray tube is used for excitation.

By using polarized incident radiation the background from coherently scattered radiation can be reduced (10-17). Since the fluorescence is essentially isotropic (as indicated in Fig. 2) by placing the detector at a position which minimises the contribution from electronic dipole scattering the signal to background ratio can be improved.

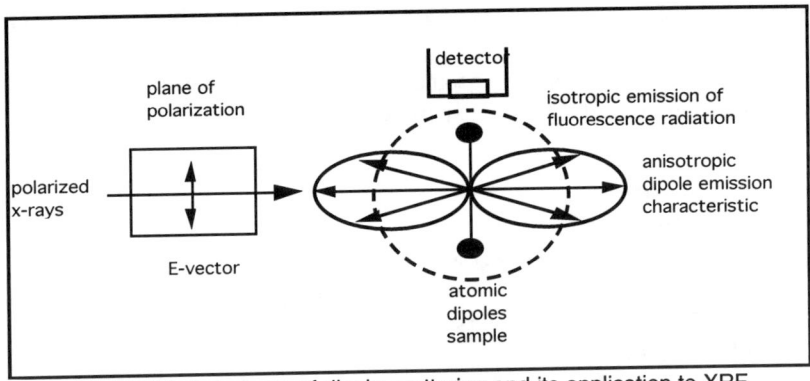

FIGURE 2. Anisotropy of dipole scattering and its application to XRF

The use of monochromatic primary radiation reduces the background because only photons with one energy are scattered. Another advantage of monochromatic radiation is the simple conversion of measured intensities into concentrations - or mass units - using the fundamental parameter method. With tunable radiation the sensitivity can be enhanced by adjusting the excitation energy to just above the absorption edge of the element of interest.

Synchrotron Radiation and Multilayer Monochromators

The outstanding properties of synchrotron radiation offer new possibilities for improving the analytical power of TXRF. The natural collimation and high degree of linear polarization of the intense continuum radiation are ideal for excitation in the total reflection geometry. Figure 3 shows the calculated spectral brightness of a typical bending magnet beamline suitable for TXRF - beamline L at HASYLAB (18). The usable continuum extends from eV and to about 100 keV and filters, mirrors, and monochromators can be used to tailor the spectral distribution for optimal sample excitation conditions.

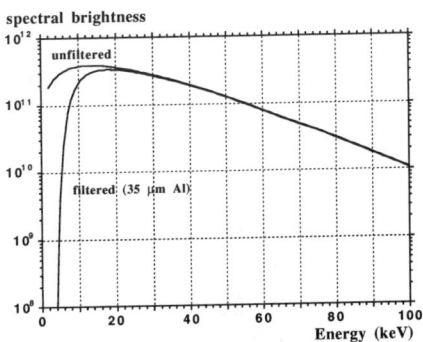

FIGURE 3. Calculated spectral brightness (ph/s•mA•mrad² 0.1%•ΔE/E).

Multilayer monochromators are particularly useful for SR-TXRF because compared to Si(111) crystal they offer a larger bandwidth, which leads to a much higher photon flux on the sample (19-26).

TXRF USING SYNCHROTRON RADIATION (SR-TXRF)

As shown in Fig. 4 various geometrical arrangements of the reflector and detector can be employed in TXRF experiments with synchrotron radiation.

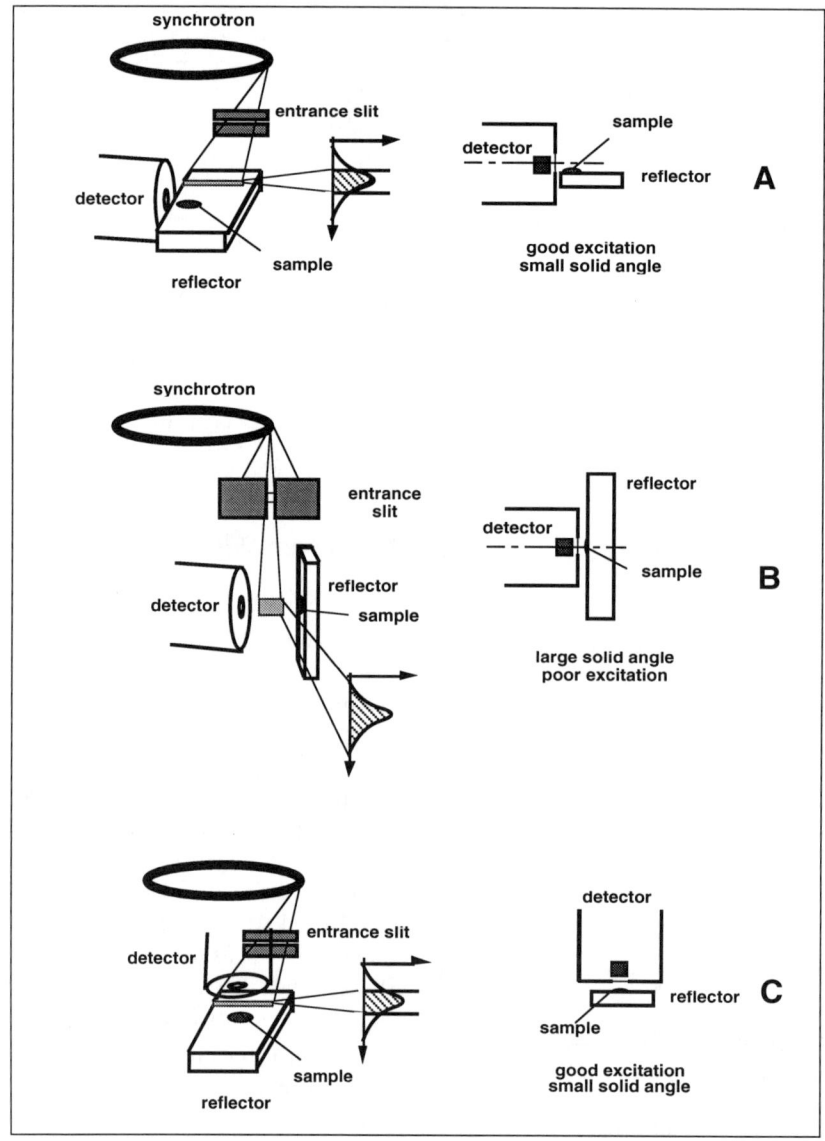

FIGURE 4. Various possible arrangements of sample reflector and detector in SR-TXRF.

In geometry A the polarization effect is fully utilized by positioning the detector axis in the plane of the orbit. Scattered radiation is not emitted in that direction. The sample is excited efficiently and illuminated uniformly by the wide beam in the horizontal plane. There are hardly any losses due to the collimators because the beam is naturally collimated in the vertical plane to 0.1 to 0.2 mrad depending on the energy as shown in Fig. 5. The detection of the fluorescence signal is not optimal because the detector must be side-looking to use the polarization effect. The fluorescent radiation has a long path in the sample before reaching the detector. In B the excitation conditions are poor because the slits restrict the beam in the horizontal direction and the illumination in the vertical direction is nonuniform. The sample area should be restricted to 2 - 4 mm diameter to ensure that the intensity and polarization distribution in the vertical plane is uniform. Figure 5 shows the intensity distribution of the p- and s-polarized components with respect to the vertical emission angle for different photon energies. Assuming that the experiment is about 20 m from the source, at about 2mm above the plane of orbit the intensity begins to drop significantly. However, the detection efficiency is good because of the large solid angle. Excellent excitation and detection can be achieved with arrangement C, however, the polarization advantage is lost. If the sample is small, which is the case in ultralow trace-element analysis, the scattering contribution from the sample itself is negligible.

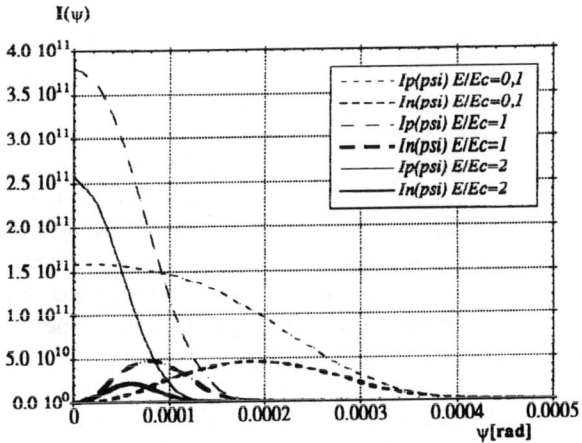

FIGURE 5. Brightness (ph/s•mA•mrad² 0.1%•ΔE/E) of parallel Ip and normal In components of synchrotron radiation as a function of vertical emission angle for different photon energies (E/Ec)

EXPERIMENTAL

Experiments were performed at the bending magnet beamline L in HASYLAB at DESY. The experimental geometry is shown in Fig. 6 and corresponds to arrangement B in Fig. 4 with a vertical reflector and a side-looking detector. All measurements were performed under vacuum conditions to achieve the lowest background. More details are given in (27).

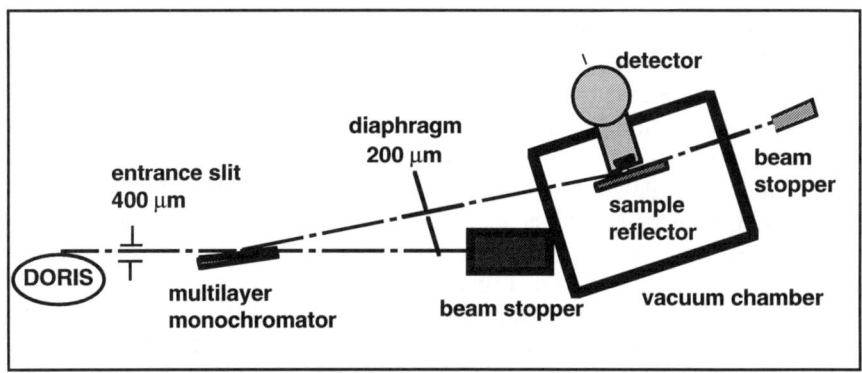

FIGURE 6. Experimental setup at HASYLAB beamline L

The beam is collimated by a primary slit system to about 400 μm and then impinges on a multilayer monochromator (W/C, 2d = 68 Å, 100 layer pairs) and is reflected. The primary beam which penetrates straight through the monochromator is absorbed in a beamstop made of high Z densimet material. In addition to the beam stop, the vacuum chamber is shielded with lead which is extremely important to avoid multiple scattered high energy photons which would penetrate into the measuring chamber through the 20mm Al wall. The monochromatic beam enters the chamber through an 8 μm kapton window and is totally reflected by a rectangular suprasil reflector of 100 mm length, with a 1 mm Ta plate in front to prevent the impinging beam from being scattered by the end of the mirror. The detector is also shielded with lead and the fluorescence radiation reaches the Be-window of the detector through an 8 mm aperture. The energy dispersive analyzer is a 30 mm^2 Si(Li) detector.

The sample is prepared by pipetting a few μl of the liquid sample for analysis on the reflector with the droplet deposited right opposite the detector entrance window. The solvent is evaporated either in a vacuum vessel or with an infrared lamp leaving a thin layer of irregularly formed microcrystals on the surface of the reflector. An internal standard element of known concentration is added to the sample before the analysis. A typical multielement spectrum is shown in Fig. 7. Assuming one element added to the sample as an internal standard the quantitative analysis of the unknown elemental concentrations is calculated using:

$$c^i = \frac{I^i}{I^{St}} \cdot \frac{S^{St}}{S^i} \cdot c^{St} \qquad (2)$$

where S^i and S^{St} are the sensitivities of the element i and the standard.

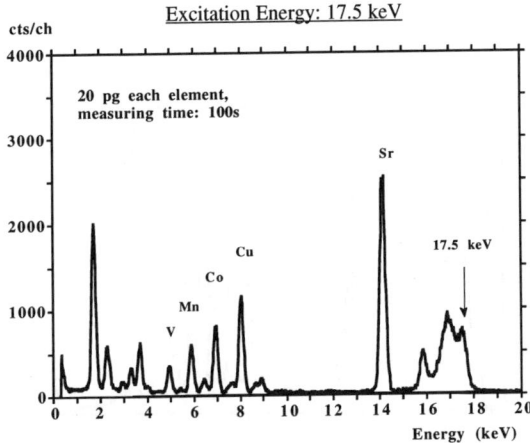

FIGURE 7. Spectrum of a multielement sample (HASYLAB beamline L)

The sensitivities can be found either by analysis of a number of samples of known concentrations, or by theoretical calculations if the excitation spectrum and all of the fundamental parameters and the detector efficiency are known. With continuum excitation integration of the intensity multiplied by the fluorescence cross sections over the energy range is necessary. No absorption corrections are required because the thin layer approximation is always valid for standard TXRF.

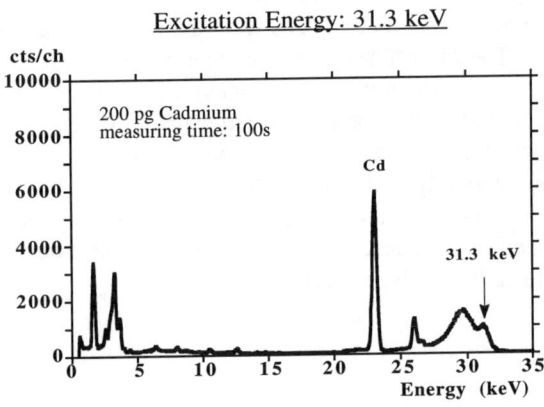

FIGURE 8. Spectrum of a sample containing 200 pg Cd (HASYLAB beamline L)

Comparison of different excitation modes

The spectra in Fig. 8 and Fig. 9 show measurements from various samples excited with monochromatic radiation at various primary energies provided by the W/C multilayer monochromator (see Ref. 28 for details)

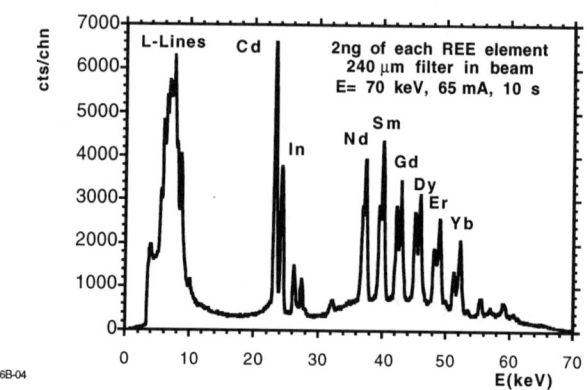

FIGURE 9. Spectrum of a sample containing 2 ng of REE elements

For the spectrum in Fig. 10 the primary beam was filtered with 6 mm Al and 1 mm Cu to remove the low energies from the white synchrotron beam. It shows the advantage of exciting the K-shells even for rare earth elements (REE) or uranium, because it avoids the overlap problems of the L-lines.

In another experiment a high energy cut-off reflector of polished glass was used to modify the primary spectrum. Figure 11 shows the spectrum of a sample containing 200 pg Ni and Sr. For comparison 20 pg of Ni and Sr were analyzed excited with a monochromatic beam. (The resulting intensities are different as can be seen in Fig. 7). The cut-off leaves a low energy continuum which excites Ni much better than Sr, whereas a decrease in intensities from Sr to Ni is observed with monochromatic radiation. The cut-off reflector, however, not only increases the intensity, but also the background which leads to worse detection limits. Using the full white beam with only a filter to excite the high Z elements results in higher intensities but also in higher backgrounds. Figure 12 shows a spectrum of REE excited with a white filtered beam. A comparison of the detection limits achieved is given in Table 1 (details are given in ref. 29).

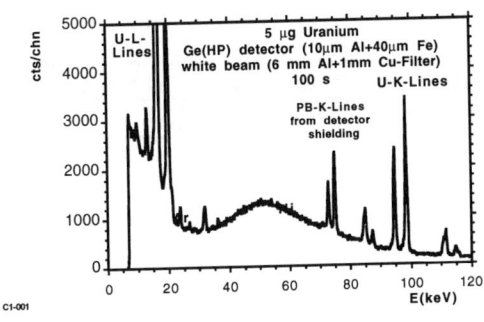

FIGURE 10. Spectrum of a sample containing 5 μg U excited with filtered white radiation (HASYLAB beamline L)

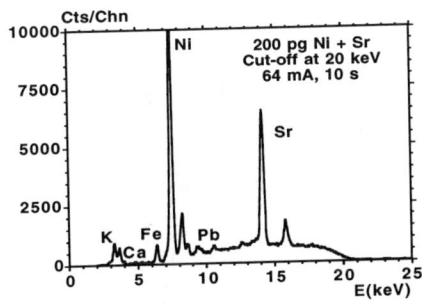

FIGURE 11. Spectrum of a sample containing 200 pg Ni and Sr , excited with a beam after a cut-off at 20 keV (HASYLAB beamline L)

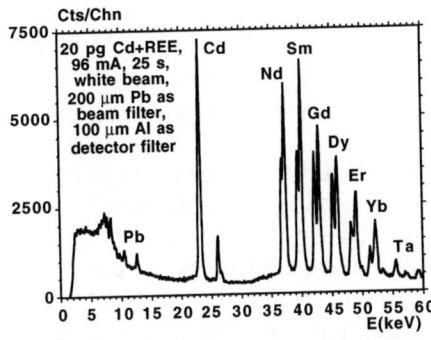

FIGURE 12. Spectrum of a REE multielement sample excited with the filtered white beam (HASYLAB beamline L)

TABLE 1 Comparison of Detection Limits obtained under different excitation conditions

Spectr.Mod	E(keV)	I(mA)	Mn	Ni	Sr	Cd	Nd	Pr
Multilayer	10	103	15	12				
	17	70			20*			
	31	70				150		
	50	57					1500	600
Cut-off	E < 20	64		40	73			
Cut-off	E < 40	86		121	155	563		
White/Filter		96				16000	15000	12000

Analysis of Ni on Si wafer surfaces

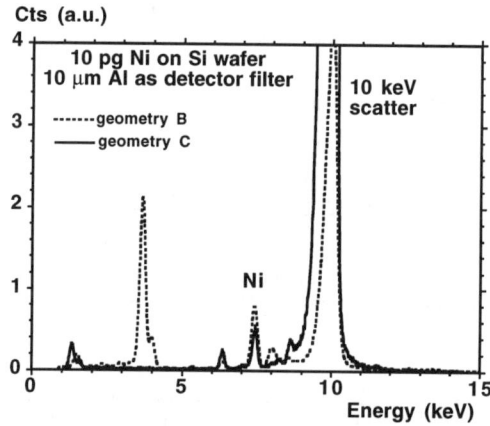

FIGURE 13. Comparison of the spectra from 10 pg Ni on a Si-wafer obtained with different arrangements according to Fig. 4, normalized to the same beam current.

The spectra obtained from measurements on a sample containing 10 pg of Ni in the different geometries corresponding to Fig. 4B and 4C are compared in Fig. 13 and more detailed data are given in Table 2. The limiting factor in geometry C (horizontal reflector, down-looking detector) is the high scattered intensity which although expected, is 10 times higher than in geometry B because the polarization of the primary beam is not used (more details are given in Ref. 30).

The analysis of Ni and other metallic contaminations on Si-wafer surfaces is extremely important for the semiconductor industry, necessitating detection limits in the low fg range. In a cooperation with Wacker Siltronic, Burghausen, Germany and HASYLAB, Hamburg, a series of investigations were performed and under the best circumstances detection limits as low as 13 fg were achieved as shown in Fig. 14. Assuming an inspected area of 1 cm^2 the Ni concentration corresponds to 1.3 x 10^8 atoms/cm^2 Ni on the surface of the wafer.

If the special sample preparation technique called vapor-phase decomposition (VPD) (32) is used, the collection area is the entire surface of the wafer. For a 200 mm wafer the expected detection limits are in the 10^5 atoms/cm^2 range.

TABLE 2 Comparison of results obtained with the configurations shown in Fig. 4

Arrangement	Ni (pg)	t (s)	I (mA)	N-netto	N-back	S (cps/ng/mA)	LLD (fg)	LLD (fg) (120 mA)
B	10	100	77	44066	3395	572	13	10
C	10	50	29	4737	1451	327	54	26

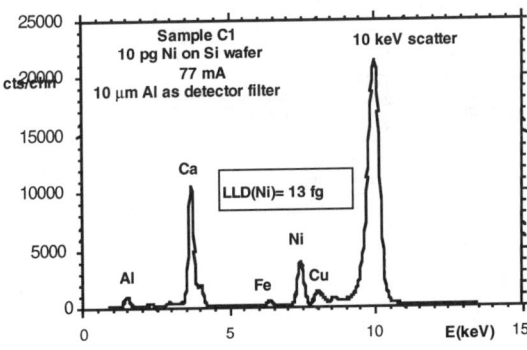

FIGURE 14. Spectrum from 10 pg Ni on a Si-wafer obtained in geometrie B.

SR-TXRF of Light Elements

Another interesting application of synchrotron radiation induced TXRF is the use of the lower energy spectrum for exciting low Z elements. With x-ray tubes it was found that provided a special detector is used TXRF is well suited for analyzing light elements (33-35).

A beamline originally designed for lithography (36) at SSRL (Stanford Synchrotron Radiation Laboratory, California), which is equipped with a Au-coated mirror which cuts off radiation above 3 keV, was used for these experiments. The experimental setup is shown in Fig. 15. The beamline is equipped with a differential pumping system which permits the measuring chamber to be directly coupled to the beamline allowing the complete low-energy spectrum to impinge on the sample. Filters and a double-multilayer monochromator can be inserted in the beam to modify the spectral distribtion. Due to the downlooking detector only geometry C shown in Fig. 4 could be used. Figure 16 shows the spectrum of a sample of 100 pg Mg on a Si wafer excited with the white beam filtered with 12 μm Si to produce a "quasi-monochromatic" beam with a FWHM of about 400 eV at 1.6 keV. Detection limits of 62 fg were obtained for Mg. At these low energies self-absorption corrections are large and can significantly effect quantitative determinations. The determination of Na and Al on Si wafer surfaces is potentially a very interesting application, but requires excellent clean room conditions.

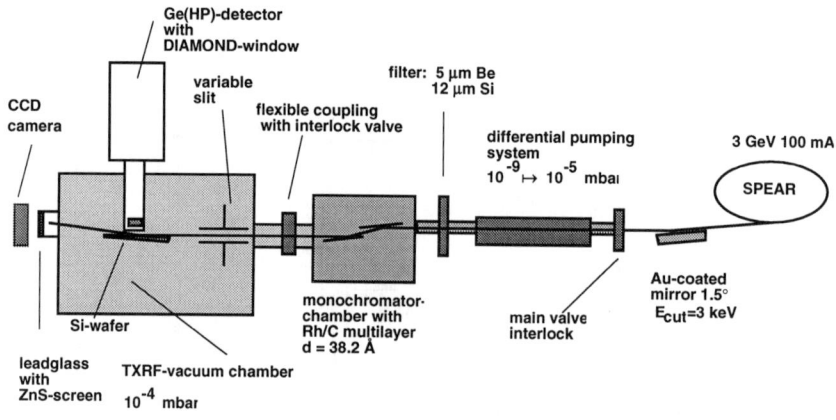

FIGURE 15. Schematic of the TXRF setup at Beamline III-4 at SSRL

Another challenge is the detection of boron, which was achieved by using the low energy part of the synchrotron spectrum which is totally reflected by the top layer of the multilayer acting as a mirror for these low energy photons. Figure 17 shows a spectrum from a wafer with about 10^{14} atoms/cm^2 boron implanted in the surface. The boron peak at 185 eV is separated from the carbon peak at 270 eV. Details are described in Ref. 37.

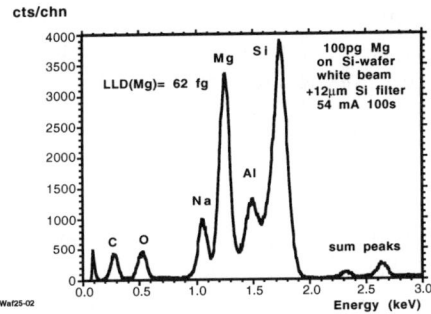

FIGURE 16. Spectrum of 100 pg Mg on a Si-wafer excited with filtered white radiation from Beamline III-4 at SSRL

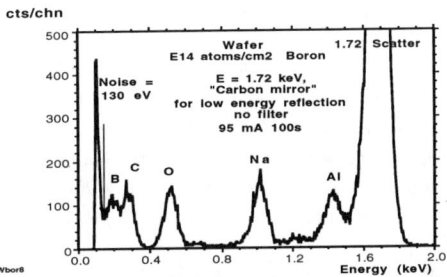

FIGURE 17. Spectrum of boron in the surface of a Si-wafer, excited with the low energy spectrum from Beamline III-4 at SSRL reflected by the multilayer

CONCLUSIONS

TXRF with synchrotron radiation excitation is potentially a powerful analytical tool for ultralow trace-element analysis. The multielement capability, rapid, and nondestructive analysis are unique features of EDXRF. The attainable detection limits for medium Z elements are in the low fg region. The range of detectable elements extends from boron to uranium with K-shell excitation. Higher sensitivities are feasible with the higher flux available at the new synchrotron sources, however, at present conventional detectors and the associated electronics limit the maximum integrated count rate to about 40 kHz. Efficient shielding is mandatory to achieve the lowest detection limits. In future it would be useful to be able to modify the polarization of the incident radiation. A permanent SR-TXRF station would enable the high sample throughput necessary for industrial applications, for interdisciplinary projects, and fundamental research. Good results have been achieved, but further improvements have to be made in the areas of x-ray optics, high-resolution detectors, and high count-rate processors.

ACKNOWLEDGMENTS

The authors acknowledge the valuable contributions of W. Drabek, R. Görgl, R. Koppitsch, P. Kregsamer, W. Ladisich and R. Rieder from the Atominstitut and the local teams at HASYLAB and SSRL (specially P. Pianettta, and P. Biltoft, R. Ryon from LLNL). The financial support of the Austrian National Sience Foundation for the projects P9460 and P11429 and from Planseewerke Lechbruck, Germany are gratefully acknowledged.

REFERENCES

1. Yoneda Y.and Horiuchi T., *Review of Sci.Inst.* **42** (7), 1069 (1971).
2. Aiginger H. and Wobrauschek P., *Nucl.Instr.Meth.* **114**, 157 (1974).
3. Wobrauschek P. and Aiginger H., *Anal. Chem.* **47**(6), 852 (1975).
4. Knoth J.and Schwenke H., *Fresenius Z. Anal. Chem.* **291**, 200 (1978).
5. Schwenke H. and Knoth J., *Nucl.Instr.Meth.* **193**, 239 (1982).
6. Von Bohlen A., Eller R., Klockenkämper R.and Tölg G. *Anal. Chem.* **59**, 2251 (1987).
7. Wobrauschek P., Kregsamer P., Streli C., Rieder R., *Adv.X-Ray Anal.* **35**, 925 (1992).
8. Klockenkämper R., Knoth J., Prange A., Schwenke H., *Anal. Chem.* **64** (23), 1115A (1992).
9. Prange A. and Schwenke H., *Adv. X-Ray Anal.* **35,** 899 (1992).
10. Dzubay T.G., Jarrett B.V., Jaklevic J.M., *Nucl. Instr. Meth.* **115,** 297 (1974).
11. Howell R.H. and Pickles W.L., *Nucl. Instr. Meth.* **120**, 187 (1974).
12. Aiginger H., Wobrauschek P. and Brauner C., *Nucl. Instr. Meth.* **120**, 541 (1974).
13. Ryon R.W., *Adv. X-Ray Anal.* **20**, 575 (1977).
14. Ryon R.W., Zahrt J.D., Wobrauschek P., Aiginger H., *Adv. X-Ray Anal.* **25,** 63 (1982)
15. Ryon R.W.and Zahrt J.D., "Polarized Beam X-ray Fluorescence", *Handbook of X-ray Spectrometry*, p.491, Marcel Dekker(1993).
16. Wobrauschek P., Aiginger H., *Adv. X-ray Anal.* **28**, 69 (1985).
17. Kanngießer B., Beckhoff B., Scheer, Swoboda J., *Adv. X-Ray Anal.*, **35**, 1001 (1992).
18. Rieder R., "Verbesserung der Nachweisgrenzen bei der TXRF durch den Einsatz von Synchrotronstrahlung und Bau einer neuen Meßkammer", *Doctoral Thesis,* Technical University Vienna (1994).
19. Bilderback, H., Lairson B. M., Barbee, T. W. Jr., Ice, G. E., and Sparks, C. J., Jr., *Nucl. Instr. Meth.* **208**, 251 (1983).
20. Pianetta P., and Barbee T., *Nucl. Instr. Meth.* **A266**, 441 (1988).
21. Dhez P., *Ann. Phys. Fr.* **15**, 493 (1990).
22. Day R., Grosso J., Bartlett R., *Nucl. Instr. Meth.* **208**, 245 (1983).
23. Biltoft P.J., *X-ray Spectrom.* **22**(4), 293 (1993).
24. Smith A., Riedel C., Edwards B., Savage D., Lai B., Ray-Chaudhuri A., Cerrina F., Lagally M., Underwood J., Falco C., *Rev.Sci.Instr.* **60**(7), 2003 (1989).
25. Kortright J.B., DiGennaro R.S., *Rev.Sci.Instr.* **60**(7), 1995 (1989).

26. Bilderback D., Hubbard S., *Nucl. Instr. Meth.* **195**, 85 (1982).
27. Rieder R., Wobrauschek P., Ladisich W., Streli C., Aiginger H., Garbe S., Gaul G., Knöchel A., Lechtenberg F., *Nucl. Instr. Meth.* **A355**, 648 (1995).
28. Wobrauschek P., Kregsamer P., Ladisich W., Rieder R., Streli Ch., Garbe St., Haller M., Knöchel A.and Radtke M., accepted for publication in *Adv.X-ray Anal.* **39**, (1996).
29. Görgl R., Wobrauschek P., Kregsamer P., Streli Ch., Haller M., Knöchel A. and Radtke M., Submitted for publication in *X-ray Spectrometry* (1997), Proceedings of the EDXRS Conference, Lisbon 1996.
30. Wobrauschek P., Görgl R., Kregsamer P., Streli Ch., Pahlke S., Fabry L., Haller M., Knöchel A. and Radtke M., submitted for publication in *Spectrochimica Acta Part B* (1997),proceedings of the 6th TXRF Conference, Eindhoven and Dortmund (1996)
31. Wobrauschek, P., Kregsamer, P., Ladisich, W., Streli, Ch., Pahlke, S., Fabry, L., Garbe S., Haller, M., Knöchel, A. and Radtke, M., *Nucl. Instr. Meth.* **A363**, 619 (1995).
32. L. Fabry, S. Pahlke, L. Kotz, *Adv. X-Ray Chem. Anal. Japan* **27**, 345 (1996).
33. Streli C., Aiginger H., Wobrauschek P., *Spectrochim. Acta* **48B**, 163 (1993).
34. Streli C., Wobrauschek P., Ladisich W., Rieder R., *X-ray Spectrometry*, **24**, 137 (1995).
35. Streli, C., Wobrauschek, P., Ladisich, W., Rieder, R., Ryon, R., Pianetta, P. and Aiginger, H., *Nucl.Instr. Meth.* **A345**, 399 (1994).
36. Pan, L., King, P.L., Pianetta, P., Seligson, D., and Barbee T.W., *Nucl.Instr.Meth.* **A266**, 287 (1988).
37. Streli, C., Wobrauschek, P., Bauer, V., Kregsamer, P., Görgl, R., Pianetta, P., Ryon, R., Pahlke, S.and Fabry, L., submitted to Spectrochim. Acta B, proceedings of the 6th TXRF Conference, Eindhoven and Dortmund (1996).

Structural and Phase Transition Studies of Layered Materials by X-Ray Standing Waves

B. N. Dev

Institute of Physics, Sachivalaya Marg, Bhubaneswar 751 005, India

Abstract. X-ray standing waves have been extensively used for the structural analysis and some phase transition studies of thin and ultrathin epitaxial layers grown on nearly perfect crystal substrates. Recently, superlattice structures have been analyzed by generating standing waves with Bragg reflections from the superlattice. Grazing incidence X-ray standing waves, generated under total external reflection conditions, and the associated resonance enhancement of X-ray intensity have found uses in the analysis of Langmuir-Blodgett and polymer layers. The developments have been discussed with a few examples.

INTRODUCTION

As we celebrate the centennial of the discovery of X-rays, the X-ray standing wave (XSW) technique is in its fourth decade. Although the theoretical basis of the XSW technique – the dynamical theory of X-ray diffraction – was worked out quite early, experiments had to wait for the development of appropriate materials, namely, nearly perfect single crystals. X-ray diffraction from such crystals are used to generate standing waves. The first XSW experiment by Batterman [1] in 1964 involved the observation of the variation of Ge K_α fluorescence yield while rocking a Ge single crystal through a diffraction peak. This observation demonstrated the movement of X-ray standing waves in the crystal due to a change in angle of incidence. In the first two decades following Batterman's work there were very slow developments with about five groups using the XSW technique and about 20 papers published. Later, the availability of synchrotron radiation facilities has brought a rapid change. Now there are about twenty groups using the XSW technique and the number of publications has reached around 300.

It is now appropriate to provide a historical perspective about the developments. Although Batterman was the first to detect the effect of X-ray standing waves on a secondary process, the existence of standing waves was proven in the Borrmann effect [2]. As diffraction from crystals is the basic element necessary for the formation of standing waves, besides X-rays, other kinds of radiation employed in diffraction studies could also form standing waves. Evidence of neutron standing waves and their effect on the inelastic yield from lattice atoms were obtained earlier by Knowles [3]. With electrons the formation of standing waves and inelastic scattering were shown by Duncumb and Hirsch et al. [4].

Having demonstrated the movement of standing waves in the crystal with the angle of incidence, Batterman showed in 1969 how the lattice position of impurity atoms in a crystalline substrate can be detected by monitoring the fluorescence yield from the impurity atoms as a function of incident angle over the reflecting region. In this experiment, fluorescence yield — a secondary process excited by the standing wave field — carried the information about the movements of the nodes and antinodes and the relative position(s) of the fluorescing atoms with respect to the nodes and antinodes. Other secondary processes excited by the X-ray standing wave field can also be detected for the same purpose. Brümmer and Stephanik [5] used photocurrent and Shchemelev and Kruglov [6] used electron emission to detect the effects of standing waves. Standing wave effects were also detected through luminescence [7].

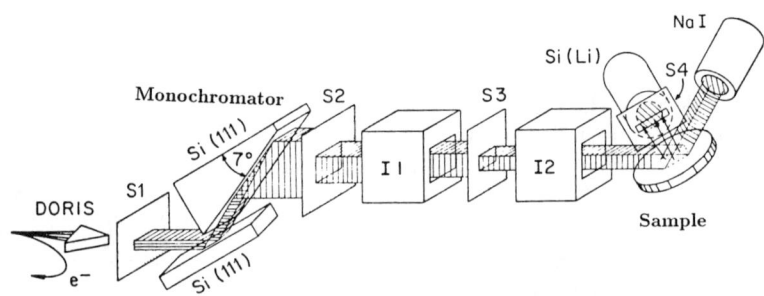

FIGURE 1. A schematic diagram of a typical X-ray standing wave experimental arrangement with a double crystal monochromator. DORIS - SR source; S1, S2, S3, S4 - slits; I1, I2 - ionization chambers; NaI - detector for reflectivity measurement; Si(Li) - energy dispersive fluorescence detector.

The first quantitative measurements of lattice position of impurity atoms were made by Golovchenko et al. [8] and Andersen et al. [9] avoiding the extinction effect by detecting fluorescence from impurities in a thin implanted layer. They demonstrated that the position of atoms can be determined with a precision of 0.02 Å. Cowan et al. [10] demonstrated that the standing waves extend outside the surface of the crystal in which they are generated. They determined the position of an adsorbed Br layer on a Si(110) surface by generating a standing wave field with the (220) Bragg reflection from the substrate. This opened up the application of XSW to surface physics.

The developments mentioned so far used conventional X-ray sources, such as sealed tubes or rotating anode X-ray sources. The advent of synchrotron radiation (SR) sources providing a high flux of photons, energy tunability and polarized radiation began to change the scenario; from the early eighties the number of XSW experiments took a quantum jump upwards. Materlik and Zegenhagen [11] advanced the measurement techniques with SR and demonstrated the advantages of SR for bulk and surface studies with XSW. Synchrotron radiation sources became the favorites, however, conventional sources remained in use.

Although the XSW technique has emerged as an important structural tool for surface and interface studies, it has been used in several other interesting areas. The coherent Compton effect was observed under the standing wave conditions [12,13]. Interference effects were observed in the inelastic scattering by phonons when the experiment was performed by generating standing waves. These interference effects were utilized to determine the phases of the phonon eigenvectors in a crystal [14]. The XSW technique has been applied to solve the phase problem in crystallography [15,16].

STRUCTURE OF AN OVERLAYER

Structural studies of thin layers on single crystal surfaces by the conventional XSW technique till the end of 1992 has been listed in Ref. [17]. The progress since 1993 will be summarized later in this section. The theoretical basis for the generation of X-ray standing waves is discussed in the review article by Batterman and Cole [18]. The formalisms for the structural analysis of ultrathin layers (submonolayer and monolayer regime), modified layers of the substrate surface and thin (≤ 1000 Å) overlayers may be found in references [17,19–21].

As an example of XSW studies on an ultrathin overlayer, the Pb/Ge(111) system will be presented. The choice of Pb on a Ge(111) surface has been made for several reasons. The Pb/Ge(111) system shows interesting behavior with two well-defined room temperature (RT) phases for $\frac{1}{3}$ monolayer (ML) and $\frac{4}{3}$ ML coverages of Pb. (A monolayer is equivalent to the number of atoms on an ideal substrate surface. For a Ge(111) surface 1 ML = 7.22×10^{14} atoms/cm^2). Both phases exhibit $(\sqrt{3} \times \sqrt{3})R30°$ reconstructions. The $\frac{1}{3}$ ML phase is known as the α phase and the $\frac{4}{3}$ ML phase the β phase. The α phase shows a transition to a (3×3) phase at lower temperature (LT) while the β phase shows a transition to a (1×1) phase at higher temperature (HT). Both the transitions are reversible. It appears that the transition of the α phase at LT is due to a charge density wave [22] and the transition of the β phase is of the order-disorder type. While no standing wave studies are available on

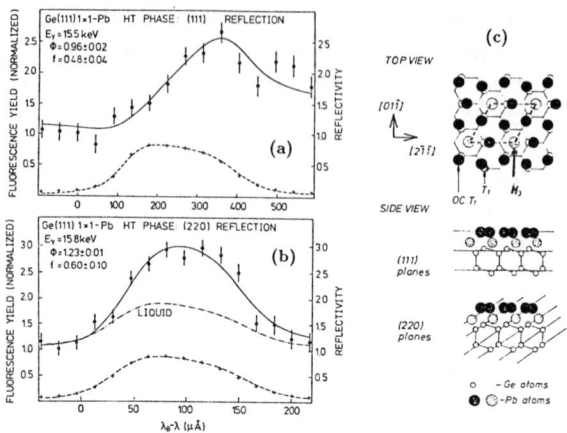

FIGURE 2. Measured reflectivity (+) and Pb L-fluorescence yield (•) as a function of the wavelength of the incident X-rays and the corresponding fits (- - -, ——) to the dynamical theory. λ_B is the wavelength that satisfies the Bragg condition: $2d\sin\theta = \lambda_B$, for a given θ. (a) Measurement with a Ge(111) reflection, (b) with a Ge(220) reflection. The Fourier components of the Pb atom density distribution for both the RT and the HT phases, obtained from this analysis, are presented in Table II (from Ref. [26]). (c) The structural model for the RT β phase, constructed from Ref. [26,27] with 1-ML Pb at OC T_1 site and $\frac{1}{3}$-ML at H_3 site at heights of 2.85 Å and 1.55 Å, respectively, above the top Ge surface.

both sides of the transition of the α phase, for the β phase they are available. Hence both the structural and the phase transition aspects can be discussed with the single example of the β phase of Pb/Ge(111).

Table I. Different models for the RT β phase

No.	Coverage (ML)	Pb-sites in the unit cell	Reference/ Technique
1	1	Pb-T_4	[23]/LEED
2	1	Pb Trimer- OC T_1	[24]/STM
3	$\frac{4}{3}$	Pb(111)- OC T_1, H_3	[25]/SXRD
4	$\frac{4}{3}$	OC T_1, H_3, bilayer	[26,27]/XSW
5	$\frac{4}{3}$	OC T_1, H_3, bilayer	[28,29]/LEED
6	$\frac{4}{3}$	OC T_4, H_3 bilayer	[30]/STM
7	$\frac{4}{3}$	Pb-chain : OC T_1, OC T_4	[31]/MD

In an XSW experiment one uses a Bragg reflection from the substrate below the layer. Simultaneously one measures the reflectivity and the yield of an inelastic signal (e.g., fluorescence, photoelectrons, Auger electrons etc.) from one (or more) kind(s) of atoms in the layer as a function of angle of incidence, or wavelength, of the incident radiation. The angular, or wavelength, variation of the inelastic signal yield contains the structural information. The reflectivity and the inelastic yield are fitted to the dynamical theory of X-ray diffraction to extract this information. What one determines are different Fourier components (FC's) of the atomic distribution. A typical experimental arrangement is shown in Fig. 1. An example of results of XSW measurements is shown in Fig. 2 for the Pb/Ge(111) β phase. The existing structural models of the RT β phase are presented in Table I. The Fourier components of the Pb atom distribution measured by the XSW technique and those obtained by a molecular dynamics (MD) simulation study are presented in Table II. The results shown in Fig. 2 are for the HT β phase. However, the results for the RT β phase are almost identical as can be seen from the measured FC's presented in Table II.

Table II. Different Fourier components of the Pb-atom distribution obtained from XSW experiments and an MD simulation: phase (coherent position), ϕ and amplitude (coherent fraction), f

Coverage (ML)		ϕ_{111}	f_{111}	ϕ_{220}	f_{220}	$\phi_{11\bar{1}}$	$f_{11\bar{1}}$
$\frac{4}{3}$ ^{a} (XSW)	RT	0.96 ±0.03	0.50 ±0.05	1.23 ±0.02	0.64 ±0.12	—	—
	HT	0.96 ±0.02	0.48 ±0.04	1.23 ±0.01	0.60 ±0.10	—	—
$1-\frac{4}{3}$ ^{b} (XSW)	RT	0.91 ±0.01	0.90 ±0.03	—	—	1.01 ±0.09	0.27 ±0.04
	HT	0.92 ±0.01	0.88 ±0.02	—	—	1.06 ±0.09	0.33 ±0.04
$\frac{4}{3}$ ^{c} (MD)	RT	0.96	0.96	—	—	1.22 ±0.02	0.25 ±0.10
	HT	1.0	0.86	—	—	0.93 ±0.02	0.26 ±0.06

a — from Ref. [26], b — from Ref. [32], c — from Ref. [31]

Let us first discuss the structure of the RT β phase in the light of the available results. The XSW results of Franklin et al. [32], presented in Table II have been claimed to agree best with the models 3 and 5 in Table I; but they could not distinguish between them. They did not compare their results with model 4 in Table I, which was earlier suggested by Dev et al. from XSW measurements [26,27]. In model 3 all the Pb atoms are in one plane and their height from the top Ge plane was not given. In model 4, the Pb atoms have the same lateral positions as in model 3. However, they form a bilayer with $\frac{1}{3}$ ML Pb at H_3 site and 1 ML Pb at off-centered (OC) T_1 site, with vertical distances of 1.55 Å and 2.85 Å above the top Ge layer, respectively. This was the first time the bilayer model was suggested. Model 5 agrees [28,29] with the bilayer model 4. In model 5 the height of the upper layer Pb, 2.70 Å, is in good agreement with that of model 4 within the error bars of both the values. However, there is a discrepancy between the heights of the Pb atoms at the H_3 site, which is 2.22 Å in model 5. The smaller weight at the H_3 site is probably partly responsible for a larger uncertainty in the determination of this position.

Table III. Layered systems studied by XSW since 1993

System	Ref.	System	Ref.
Layers on semiconductor surfaces:			
Alkali/Si(111)	[33]	Rb/Si(211)	[34]
Alkali/Si(110)	[35]	Rb/Si(001)	[36]
Ga/Si(001)	[37]	Ga/Si(111)	[38]
In/Si(111)	[39]	Ge/Si(100)	[40]
Pb/Ge(111)	[32]	As/Si(111)	[41]
As/Si(001)	[42]	Sb/Si(001)	[43]
Bi/Si(001)	[44]	Sb/Ge(111)	[45]
Bi/Si(111)	[46]	Br/Si(001)	[47]
Br/Si(211)	[48]	Au/Si(111)	[49]
Ag/Si(111)	[50]	RbBr/Si(111)	[51]
LiBr/Si(111)	[52]	GaSe/Si(111)	[53]
GaAs/Si(001)	[54]	Bi/GaP(110)	[55]
Si/GaAs(001)	[56]	Sb/GaAs(110)	[57]
Bi/GaAs(110)	[58]	Sb/GaAs(001)	[59]
S/GaAs(111) A and B	[60]	S/GaAs(001)	[61]
Al,S/GaAs(111)	[62]	Fluoride/GaAs(111)	[63]
H$_2$S/InP(110)	[64]	SrF$_2$/S/GaAs(111)A and B	[65]
Buried layers in semiconductors:			
Si/CoSi$_2$/Si(111)	[66]	Si/Ge/Si(111)	[67]
GaAs/InAs/GaAs(001)	[68]	Si/Ge/Si(001)	[69]
Layers on metal and insulator surfaces:			
Na,O/Al(111)	[70]	Rb/Al(111)	[71]
Rb/Cu(111)	[72]	Cl,Br/Cu(111)	[73]
Y-Ba-Cu-O/SrTiO$_3$(001)	[74]		
Superlattice:			
(AlAs)$_m$(GaAs)$_n$	[75]	Si/Ge	[76]
Modified layers:			
As in Si	[77]	Si in GaAs	[78]
Ti in LiNbO$_3$	[79]	Er in LiNbO$_3$	[80]
Fe in LiNbO$_3$	[81]	Fe in InP	[82]
Pb on and Mn in CaCO$_3$	[83]		

For a coverage slightly less than $\frac{4}{3}$ ML the α and β phases coexist. Therefore, it is very important to measure the coverage very accurately, when determining the atomic positions in the β phase of the Pb/Ge(111) system.

Structural analysis of thin and ultrathin layers by XSW has progressed considerably since the last review in 1993 [17]. An overview of recent work with XSW on surfaces and interfaces is summarized in Table III.

ORDER-DISORDER TRANSITION IN AN OVERLAYER

Phase transitions in thin physisorbed layers, which interact weakly with the substrate, has been a subject of extensive research for many years. Recently there has been an increasing interest in the study of phase transitions in chemisorbed layers, where the adlayer-substrate interaction is stronger. An ultrathin Pb layer on a Ge(111) substrate has been treated as a model system and studied by various techniques, such as, low energy electron diffraction (LEED), surface X-ray diffraction (SXRD), XSW, scanning tunneling microscopy (STM) and reflection high energy electron diffraction (RHEED) and theoretically by an MD simulation. A complete understanding of this system is yet to be achieved.

In order to study a temperature-driven phase transition in an overlayer on a substrate by the XSW technique, it is necessary first to characterize the behavior of the standing waves at higher substrate temperature, because diffraction from the substrate is used to generate the Pb standing waves. This was explored for a Ge(111) substrate for a substrate temperature about twice the Debye temperature of Ge. It was found that an atomic form factor modified by the appropriate Debye-Waller factor is adequate for the description of the standing wave fields in the hot substrate [26]. The high-temperature XSW studies were then applied to the reversible phase transition $(\sqrt{3} \times \sqrt{3})R30°$ $\longleftrightarrow$ (1x1), observed for the β phase of the Pb/Ge(111) system at $\sim$ 470 K [26].

The transition and the HT phase

The $(\sqrt{3} \times \sqrt{3})$ β phase undergoes an apparently continuous reversible phase transition as the temperature is raised. The HT phase is a (1×1) phase. The transition temperature T_c seems to vary with Pb coverage. Just below the completion of the bilayer T_c remains close to $\sim$ 440 K. Very close to the full coverage there is a sharp increase of T_c to $\sim$ 570 K. The transition temperature falls again to $\sim$ 470 K as the layer becomes defective [23,84]. The nature of this transition is not well characterized. The nature of the HT (1×1) phase is also controversial. Based on RHEED experiments Ichikawa [84] attributed the HT phase to a 2D isotropic liquid, essentially unperturbed by the underlying Ge substrate. Metois and Le Lay [23], based on their LEED studies, suggested that the HT (1×1) phase is an ordered Pb layer with a coverage of 1 ML. These are conflicting results. In XSW experiments above T_c, Dev et al. [26] found a nonzero value of the amplitude of the (220) FC of the Pb atom distribution [see Table II]. For a 2D isotropic Pb layer this value would be zero and the expected fluorescence yield curve would be the one marked "LIQUID" in Fig. 2(b). Thus the HT (1×1) phase was found to be inconsistent with the model of a 2D isotropic liquid Pb layer. The practically unchanged (111) and (220) FC's (see Table II) are also very unlikely to agree with an ordered (1×1) Pb layer, as this would involve a new structure.

From their observations, Dev et al. [26] suggested the existence of strong local order in the HT phase and that the HT phase is a disordered phase consisting of small domains of the RT ($\sqrt{3} \times \sqrt{3}$) phase. Thus the ($\sqrt{3} \times \sqrt{3}$) $\longrightarrow$ (1×1) transition was interpreted as an order-disorder transition. In a SXRD study, Grey et al. [85] observed a ring of diffuse scattering characteristic of a 2D liquid layer. However, the ring was found to be azimuthally anisotropic. This was attributed to the interaction between the Pb layer and the Ge substrate. HT STM studies concluded that in the HT phase thermal fluctuation disrupts the long range order of the RT β phase [24], while the Pb-Ge bonding remains basically the same as in the RT phase, indicating an order-disorder transition. The interpretation of the ($\sqrt{3} \times \sqrt{3}$) $\longrightarrow$ (1×1) transition as an order-disorder transition is also supported by later XSW results [32]. A recent MD simulation by Ancilotto et al. [31] apparently has some success in uniting different interpretations for the HT (1×1) phase, although the structure they obtained for the RT β phase does not quite agree with experiments. They find that large in-plane fluctuations of Pb atoms are already present in the RT ($\sqrt{3} \times \sqrt{3}$) structure. At T $\sim$ 800 K (although much higher than the observed transition temperature) they observed a (1×1) disordered structure showing in-plane diffusion of Pb atoms. A strong correlation with the Ge substrate is present leading to preferential residence sites and persistence of local order in the overlayer over medium-range distances. The qualitative character of the diffusion is neither clearly jumplike, as in the presence of significant activation barriers, nor continuous, as expected for a 2D liquid. Ancilotto et al. [31] used 800 K in order to have faster atomic motions during the limited time of their simulations. At T $\sim$ 470 K, they estimate, the residence time at sites is of the order of nanoseconds. In this situation, however, diffusion would not be very significant. Simulations at temperatures close to T_c will be useful.

The XSW results of Franklin et al. [32] for the RT and the HT phases (Table II) are the same within the experimental error as in those of Dev et al. [26]. However, they proposed a possible structural model for the HT (1×1) phase. In this model, with the assumption of all Pb atoms in a single layer (not a bilayer), OC T_1 atoms move to T_1 sites and the Pb atoms at the H_3 sites randomly occupy H_3 positions. In this structure the FC's would be different from their observed values. They attributed the absence of this difference in the measured FC's to an enhanced thermal vibration of Pb atoms.

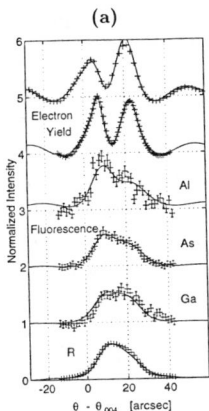
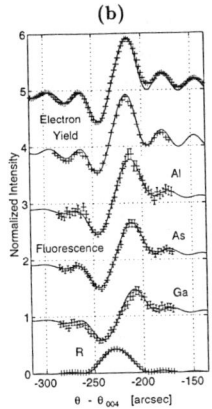

FIGURE 3. Reflectivity (R), fluorescence yield and electron yield for XSW measurements with (a) a substrate GaAs(004) reflection and (b) with the 0-th order superlattice satellite (AlAs)(GaAs)(004,0) reflection. Electron yields are shown for two slightly different samples. (From Ref. [75], with permission)

XSW WITH REFLECTIONS FROM OVERLAYERS AND SUPERLATTICES

Usually in an XSW experiment a Bragg reflection from the substrate is used to generate the standing waves. The overlayer simply lies in the generated standing wave pattern. Recently, XSW experiments using Bragg reflections from overlayers [86] and from superlattices [75] have also been performed. Some results using a superlattice reflection are presented in Fig.3. In these studies the superlattice sample consisted of 460 $(AlAs)_3(GaAs)_7$ layer pairs on a GaAs(001) substrate capped by a 100 Å GaAs layer. The analysis of the results led to the conclusion that the wavefield at the angular position of the (AlAs)(GaAs)(004,0) satellite has a periodicity equal to the average lattice plane spacing of the superlattice. Good crystalline order was found in the AlAs layers, while Ga and As atoms showed the occupation of additional positions. This was interpreted as an indication for a shift of the atomic planes at the interfaces between AlAs and GaAs.

GRAZING INCIDENCE X-RAY STANDING WAVES

For X-rays any solid or liquid medium has a refractive index slightly less than unity. Thus for grazing incidence of X-rays from air or vacuum onto such a medium total external reflection can be obtained below a critical angle of incidence. Cowan et al. demonstrated diffraction of evanescent X-rays during total external reflection, which combined XSW with evanescent wave studies [87]. However, even in the absence of any diffraction, standing waves are generated while total reflection takes place. Fresnel's theory can account for this phenomenon. This grazing incidence XSW (GIXSW) technique has been used by Bedzyk et al. [88] for the analysis of Langmuir-Blodgett multilayers on a gold mirror surface. The multilayer consisted of octadecyl thiol (ODT), cadmium arachadate (CdA), ω-tricosanoic acid (ωTA) and zinc arachadate (ZnA) layers. The periodicity of the grazing incidence standing waves varies

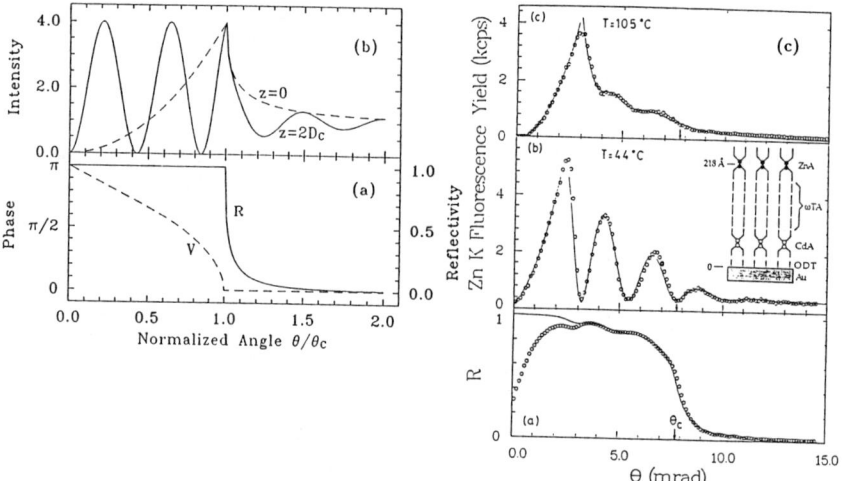

FIGURE 4. (a)The angular dependence of the reflectivity R and the relative phase v of the reflected plane wave X-ray beam. (b) The angular dependence of the E-field intensity at the surface ($z = 0$) of the mirror and at a distance ($z = 2D_c$, twice the critical period) above the surface. D_c is nearly energy independent, and for a gold mirror it is ~77 Å. The calculations ignore absorption. The results are normalized to the incident intensity. (c) The angular dependence of the reflectivity and Zn fluorescence yield: experiment (o), theory (—). (From Ref. [88], with permission).

with the angle of incidence in contrast to the conventional XSW and the period is also longer. Some features of this standing wave are shown in Fig. 4(a,b). Fig. 4(c) shows the results of their measurements, the sample is shown schematically in the inset. They measured reflectivity and the Zn K_α fluorescence yield as a function of angle of incidence. As the angle of incidence increases up to the critical angle (θ_c), for the sample at T = 44° C, the first, second and third antinodes pass through the zinc layer giving rise to three peaks in the zinc fluorescence. At T = 105° C the changed features in the fluorescence yield indicates a shift of the Zn layer towards the gold mirror surface and an enhancement in the spread in the vertical distance of the Zn atoms from the mirror surface. The transition has been found to be irreversible. Such XSW measurements can help understand the interaction between the molecular chains in the LB layer. Ion permeation through LB layers has also been studied by the GIXSW technique [89]. In some other

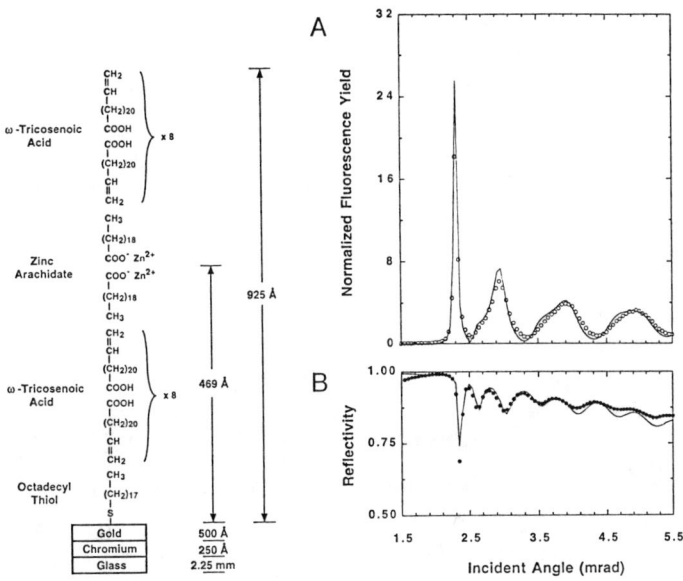

FIGURE 5. Resonance enhancement of X-ray intensity at the Zn positions, approximately at the middle of the organic layer. The enhancement is evident in the Zn fluorescence yield. The first order maximum occurs near an incident angle of 2.3 mrad. (From Ref. [93], with permission). Reprinted with permission from *Science* **258**, 775-778 (1992); ©1992 AAAS.

studies, periodic multilayers have been used to generate standing waves using both diffraction and total external reflection for analyzing overlayers and/or the multilayers themselves [90].

Under the grazing incidence condition a resonance enhancement of X-ray intensity in a thin-film waveguide [91] or in an overlayer [92] can be obtained under suitable conditions. Wang et al. [93] have detected this resonance enhancement by placing a marker layer approximately at the middle of an organic overlayer on a gold mirror surface and detecting fluorescence from the marker layer in the GIXSW experiment. The results are shown in Fig.5.

In order to observe the resonance enhancement of X-rays it is not really necessary to have a marker layer in the film at a well defined position. Dev et al. [94] have detected this resonance enhancement in GIXSW experiment by detecting fluorescence from atoms which are almost uniformly distributed in the overlayer. They used a thin layer of a blend system of polystyrene (PS) and the statistical copolymer poly(styrene-stat-para bromo styrene) (PBr$_x$S) on a gold-coated silicon substrate [PS-PBr$_{0.06}$S (350 Å)/Au(350 Å)/Si]. This polymer blend layer contained a very small amount of bromine. In the experiment, Br K_α fluorescence and the reflected beam were simultaneously detected as a function of angle of incidence. The raw data are presented in Fig. 6, which also shows the theoretical curves for reflectivity and X-ray field intensity integrated over the whole film. Resonance enhancement peaks in the Br fluorescence yield are evident from the figure. The shape of the in-

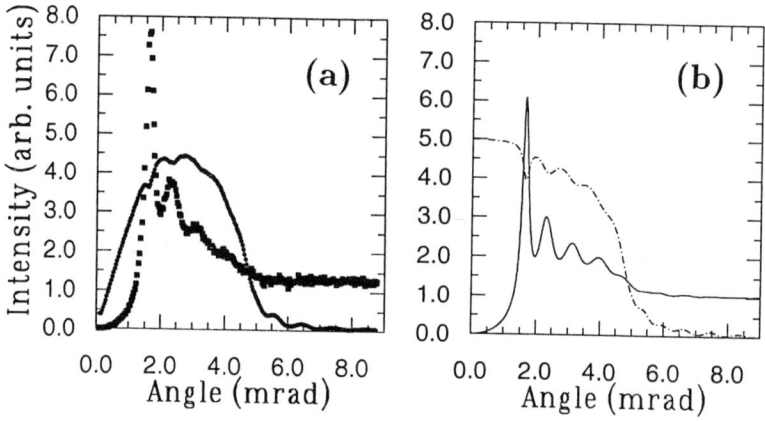

FIGURE 6. (a) Experimental reflectivity (•) and Br K_α fluorescence yield (■) (raw data) as a function of angle of incidence for a PS-PBr$_{0.06}$S (350 Å)/Au(350 Å)/Si sample annealed at 170° C for ten hours. (b) Theoretical reflectivity (×5) (- . . -) and the integrated fluorescence yield (–) for Br uniformly distributed in the polymer layer. The first, second and third order resonance enhancements of X-ray intensity in the polymer layer are evident in the data.

tegrated fluorescence signal is sensitive to the distribution of the fluorescing atoms in the overlayer. This could provide an effective tool to study segregation in polymer blends and interface broadening in multilayer films. Diffusion in polymers can be studied. The possibility of measuring ultrasmall diffusion coefficients ($\sim 10^{-21}$ cm^2/sec) in polymers with the GIXSW technique has been suggested earlier [95]. In all the measurements mentioned above, a gold-coated substrate was used. However, it is not necessary to have the layer on a gold-coated surface. Resonance enhancement has also been observed in a polymer blend layer on a silicon substrate [94]. The main requirement is to prepare thin films of these materials on substrates of higher electron density.

CONCLUSIONS

The status of the analysis of layered materials by X-ray standing waves under Bragg diffraction as well as total external reflection has been presented. The availability of synchrotron radiation played a very important role in these developments. Future X-ray sources, especially X-ray free electron lasers, are expected to broaden the applications of the XSW techniques with fewer constraints.

ACKNOWLEDGMENTS

I wish to acknowledge the helpful discussions with G. Materlik and thank him for his comments on the manuscript. I would like to thank R. L. Johnson and M. Nielsen for useful discussions. The financial supports under the Indo-German collaboration project (Physics 25) and from Hamburgersynchrotronstrahlungslabor HASYLAB are gratefully acknowledged.

REFERENCES

1. Batterman, B. W., *Phys. Rev.* **133**, A759 (1964).
2. Borrmann, G., *Phys. Z.* **42**, 157 (1941); *Z. Phys.* **127**, 297 (1950).
3. Knowles, S., *Acta Cryst.* **9**, 61 (1956).
4. Duncumb, P., *Philos. Mag.* **7**, 2101 (1962); Hirsch, P. B., Howie, A., and Whelan, M. J., *Philos. Mag.* **7**, 2095 (1962).
5. Bruemmer, O, and Stephanik, H., *Phys. Status Solidi(a)* **36**, 617 (1969).
6. Shchemelev, V. N., and Kruglov, M. V. *Soviet Phys. Solid State* **14**, 2988 (1973).
7. See *Dynamische Interferenz Theorie*, Leipzig: AVG (1976), Bruemmer, O., and Stephanik, H., (Eds.), p 185.
8. Golovchenko, J. A., Batterman, B. W., and Brown, W. L., *Phys. Rev.* **B10**, 4239 (1974).
9. Andersen, S. K., Golovchenko, J. A., and Mair, G., *Phys. Rev. Lett.* **37**, 1141 (1976).
10. Cowan, P. L., Golovchenko, J. A., and Robbins, M. F., *Phys. Rev. Lett.* **44**, 1680 (1980).
11. Materlik, G., and Zegenhagen, J., *Phys. Lett.* **A104**, 47 (1984); Zegenhagen, J., Materlik, G., and Uelhoff, W., *J. X-Ray Sci. Technol.* **2**, 214 (1990).
12. Golovchenko, J. A., et al., *Phys. Rev. Lett.* **46**, 1454 (1981).
13. Schuelke, W., Bonse, U., and Mourikis, S., *Phys. Rev. Lett.* **47**, 1209 (1981).
14. Spalt, H., Zounek, A., Dev, B. N., and Materlik, G., *Phys. Rev. Lett.* **60**, 1868 (1988); *Phys. Rev. Lett.* **61**, 2819 (1988).
15. Bedzyk, M. J., and Materlik, G., *Phys. Rev.* **B32**, 6456 (1985).
16. Dev, B. N., and Materlik, G., in *Resonance Anomalous X-ray Scattering: Theory and Applications*, Elsevier Science B. V., 1994, pp. 119-143.

17. Zegenhagen, J. *Surf. Sci. Rep.* **18**, 199 (1993).
18. Batterman, B. W., and Cole, H., *Rev. Mod. Phys.* **36**, 681 (1964).
19. Hertel, N., Materlik, G., and Zegenhagen, J., *Z. Phys.* **B58**, 199 (1985).
20. Dev, B. N., Aristov, V., Hertel, N., Thundat, T. and Gibson, W. M., *Surf. Sci.* **163**, 457 (1985).
21. Vlieg, E., Fischer, A. E. M. J., van der Veen, J. F. Dev, B. N., and Materlik, G., *Surf. Sci.* **178**, 36 (1986).
22. Carpinelli, J. M., Weitering, H. H., Plummer, E. W., and Stumph, R., *Nature* **381**, 398 (1996).
23. Metois, J. J., and Le Lay, G., *Surf. Sci.* **133**, 422 (1983) and *Appl. Surf. Sci.* **17**, 131 (1983); Le Lay, G., Peretti, J., Hanbucken, M., and Wang, W. S., *Surf. Sci.* **204**, 57 (1988); Le Lay, G., Hricovini, K., and Bonnet, J. E., *ibid.* **41/42**, 25 (1989).
24. Hwang, I.,-S., and Golovchenko, J. A., *Phys. Rev. Lett.* **71**, 255 (1993).
25. Feidenhans'l, R., Pedersen, J. S., Nielsen, M., Grey, F., and Johnson, R. L., *Surf. Sci.* **178**, 927 (1986); Pederson, J. S., Feidenhans'l, R., Nielsen, M., Grey, F., and Johnson, R. L., *ibid.* **189/190**, 1047 (1987).
26. Dev, B. N., Grey., F., Johnson, R. L., and Materlik, G., *Europhys. Lett.* **6**, 311 (1988).
27. Dev, B. N., *Phys. Rev. Lett.* **64**, 1182 (1990).
28. Huang, H., Wei, C. M., Li., H., Tonner, B. P., and Tong, S. Y., *Phys. Rev. Lett.* **62**, 559 (1989). A single layer model was earlier proposed by Tonner et al. (see *Phys. Rev.* **B34**, 4386 (1986)).
29. Huang, H., Wei, C. M., Tonner, B. P., and Tong, S. Y., *Phys. Rev. Lett.* **64**, 1183 (1990).
30. Seehofer, L., Falkenberg, G., and Johnson, R. L., *Surf. Sci* **290**, 15 (1993).
31. Ancilotto, F., Selloni, A., and Car, R., *Phys. Rev.* **B50**, 15158 (1994).
32. Franklin, G. E., Bedzyk, M. J., Woicik, J. C., Liu, C., Patel, J. R., and Golovchenko, J. A., *Phys. Rev.* **B51**, 2440 (1995).
33. Lagomarsino, S., Scarinci, F., Castrucci, P., Giannini, C., Fontes, E., and Patel, J. R., *Appl. Surf. Sci.* **56-58**, 402 (1992); Etelaniemi, V., Michel, E. G., and Materlik, G., *Phys. Rev.* **B48**, 12023 (1993).
34. Michel, E. G., Etelaniemi, V., and Materlik, G., *J. Vac. Sci. Technol.* **A 11**, 1812 (1993).
35. Michel, E. G., Etalaniemi, V., and Materlik, G., *Journal of Phys.: Condensed Matter* **5**, A85 (1993).
36. Castrucci, P., Lagomarsino, S., Scarinci, F., and Franklin, G. E., *Phys. Rev.* **B51**, 5043 (1995).
37. Quian, Y., Bedzyk, M. J., Tang, S., Freeman, A. J., and Franklin, G. E., *Phys. Rev. Lett.* **73**, 1521 (1994); Tang, S., Freeman, A. J., Quian, Y., Franklin, E., G., and Bedzyk, M., J., *Phys. Rev.* **B51**, 1593 (1995). Quian, Y., and Bedzyk, M, J., *J. Vac. Sci. Technol.* **A13**, 1613 (1995).
38. Zegenhagen, J., Artacho, E., Freeland, P. E., and Patel, J. E., *Phil. Mag.* **B70**, 731 (1994); Artacho, E., Molinas-Mata, P., Bohringer, M., Zegenhagen, J., Franklin, G. E., and Patel, J. R., *Phys. Rev.* **B51**, 9952 (1995).

39. Woicik, J. C., et al., *Phys. Rev. Lett.* **71**, 1204 (1993). Woicik, J. C, et al., *J. Vac. Sci. Technol.* **A11**, 2359 (1993).
40. Fontes, E., Patel, J. R., and Comin, F., *Phys. Rev. Lett.* **70**, 2790 (1993); Kruger, P., Pollman J., & Fontes, E., Patel, J.,R., and Comin, F., *Phys. Rev. Lett.* **72**, 1130 (1994).
41. Sakata, O., and Hashizume, H., *Acta Cryst.* **A51**, 375 (1995).
42. Franklin, G. E., Fontes, E., Quian, Y., Bedzyk, M., J., Golovchenko, J. A., and Patel, J., R., *Phys. Rev.* **B50**, 7483 (1994).
43. Izumi, K., Saito, A., Kikuta, S., and Zhang, X. W., *Jap. J. Appl. Phys. (Part 1)* **32**, 1772 (1993); Lyman, P. F., Quian, Y., and Bedzyk, M. J., *Surf. Sci.* **325**, L385 (1995) and *Applications of Synchrotron Radiation Techniques to Materials Science II.*, Terminello, L. J., Shin, N. D., Ice, G. E., D'Amico, K. L., and Perry, D. L. (Eds.): Pittsburgh, Mater. Res. Soc. (1995) p-121; and *Scanning Microscopy* **9**, 969 (1995).
44. Franklin, G. E., Tang, S., Woicik, J. C., Bedzyk, M. J., Freeman, A. J., and Golovchenko, J. A., *Phys. Rev.* **B52**, R5515 (1995); Quian, Y., Bedzyk, M. J., Lyman, P. F., Lee, T. L., Tang, S., and Freeman, A. J., *Phys. Rev.* **B54**, 4424 (1996).
45. Kendelewicz, T., et al., *J. Vac. Sci. Technol.* **B11**, 1449 (1993); Kendelewicz, T., Woicik, J. C., Miyano, K. E., Yoshikawa, S. A., Pianetta, P., and Spicer, W. E., *J. Vac. Sci. Technol.* **A12**, 1843 (1994).
46. Woicik, J. C., et al., *Phys. Rev.* **B50**, 12246 (1994).
47. Cao, P. -L., and Zhou, R. -H., *J. Phys.: Condens. Matter* **5**,2897 (1993).
48. Etelaniemi, V., Michel, E. G., and Materlik, G., *J. Vac. Sci. Technol.* **A13**, 1583 (1995).
49. Falta, J., et al., *Surf. Sci.* **330**, L673 (1995).
50. Woicik, J. C., et al., *Phys. Rev.* **B53**, 15425 (1996).
51. Etelaniemi, V., Michel, E. G., and Materlik, G., *Surf. Sci.* **287-288**, 288 (1993).
52. Gog, T., Follis, G. C., and Durbin, S. M., *Appl. Surf. Sci.* **81**, 485 (1994).
53. Zheng, Y., Koebel, A., Petroff, J. F., Boulliard, J. C., Capelle, B., and Eddrief, M., *J. Cryst. Growth* **162**, 135 (1996).
54. Kawamura, T., Takenaka, H., Hayashi, T., Tachikawa, M., and Mori, H., *Appl. Phys. Lett.* **68**, 1969 (1996).
55. Herrera-Gomez, A., et al. *J. Vac. Sci. Technol.* **12**, 2473 (1994).
56. Sugiyama, M., Maeyama, S., and Oshima, M., *Appl. Phys. Lett.* **68**, 3731 (1996).
57. Kendelewicz, T. et al. *J. Vac. Sci. Technol.* **A11**, 2351 (1993).
58. Herrera-Gomez, A., et al. *J. Vac. Sci. Technol.* **A11**, 2354 (1993).
59. Sugiyama, M., Maeyama, S., Maeda, F., and Oshima, M., *Phys. Rev.* **B52**, 2678 (1995).
60. Sugiyama, M., Maeyama, S., and Oshima, M., *Phys. Rev.* **B48**, 11037 (1993).
61. Sugiyama, M., Maeyama, S., and Oshima, M., *Phys. Rev. Lett.* **71**, 2611 (1993); *Phys. Rev.* **B50**, 4905 (1994).
62. Sugiyama, M., Maeyama, S., Scimeca, T., and Oshima, M., *Appl. Phys. Lett.* **63**, 2540 (1993).

63. Niwa, T., Sugiyama, M., Nakahata, T., Sakata, O., and Hashizume, H., *Surf. Sci.* **282**, 342 (1993).
64. Dudzik, E., et al., *J. Phys.: Condens. Matter* **8**, 15 (1996).
65. Sugiyama, M., Maeyama, S., Watanabe, Y., Heun, S., and Oshima, M., *J. Cryst. Growth* **150**, 1098 (1995).
66. Satayam, P. V., and Dev, B. N., *Indian J. Phys.* **68A**, 23 (1994).
67. Falta, J., Gog, T., Materlik, G., Muller, B. H., and Horn-von Hoegen, M., *Phys. Rev.* **B51**, 7598 (1995); see also the conference proceedings in the previous reference.
68. Giannini, C., et al., *Phys. Rev.* **B48**, 11496 (1993); Woicik, J. C., et al., *Phys. Rev.* **B52**, R2281 (1995); Lee, T. -L., et al., *Physica* **B** (in press).
69. Takahasi, M., et al., *Jap. J. Appl. Phys.* **34**, 2278 (1995); Falta, J., Gog, T., Materlik, G., Muller, B. H., Horn-von Hoegen, M., *Application of Synchrotron Radiation Techniques to Materials Science II.*, Terminello et al (Eds.): Pittsburgh, Mater. Res. Soc., 1995, p- 177; Falta, J., Bahr, D., Materlik, G., Muller, B. H., and Horn-von Hoegen, M., *Appl. Phys. Lett.* **68**, 1394 (1996).
70. Scragg, G., et al., *Surf. Sci.* **328**, L533 (1995).
71. Scragg, G., et al., *J. Phys.: Condensed Matter* **10**, 1869 (1994).
72. Shi, X., Su, C., Heskett, D., Berman, L., Kao, C. C., and Bedzyk, M. J., *Phys. Rev.* **4**, 14638 (1994); Heskett, D., Xu, P., Berman, L., Kao, C. -C., and Bedzyk, M. J., *Surf. Sci.* **344**, 267 (1995).
73. Kadodwala, M. F., et al., *Surf. Sci.* **324**, 122 (1995).
74. Nakanishi, M., et al., *Phys. Rev.* **B48**, 10524 (1993); Zegenhagen, J., Siegrist, T., Fontes, E., Berman, L. E., and Patel, J. R., *Solid State Commun.* **93**, 763 (1995).
75. Lessmann, A., Munkholm, A., Schuster, M., Brennan, S., Riechert, H., and Materlik, G. *Phys. Rev. (submitted) ; Rev. Sci. Instrum.* **66**, 1428 (1995).
76. Lagomarsino, S., Castrucci, P., calicchia, P., Cedola, A., and Kazimirov, A., *Strained Layer Epitaxy – Materials, Processing, and Device Applications.*, Fitzerald, E. A., Hoyt, J., Cheng, K., -Y., Bean, J. C. (Eds.): Pittsburgh, Mater. Res. Soc., 1995, p 243.
77. Herrera-Gomez, A., et al., *Appl. Phys. Lett.* **68**, 3090 (1996).
78. Shih, A., et al., *J. Appl. Phys.* **73**, 8161 (1993).
79. Gog, T., Harasimowicz, T., Dev, B. N., and Materlik, G., *Europhys. Lett.* **25**, 253 (1994); Gog., T., Griebenov, M., Harasimowicz, T., and Materlik, G., *Ferroelectrics* **153**, 249 (1994).
80. Gog, T., Griebenow, M. and Materlik, G., *Phys. Lett.* **A181**, 417 (1993).
81. Gog, T., Schotters, P., Falta, J., and Materlik, G., *J. Phys.: Condensed Matter* **35**, 6971 (1995).
82. Bocchi, C., et al., *J. Appl. Phys.* **76**, 7239 (1994).
83. Quian, Y., Sturchio, N. C., Chiarello, R. P., Lyman, P. F., Lee, T. -L., and Bedzyk, M. J., *Science* **265**, 1555 (1994).
84. Ichikawa, T., *Solid State Commun.* **46**, 827 (1983); **49**, 59 (1984).
85. Grey, F., Feidenhans'l, R., Pedersen, J. S., Nielsen, M., and Johnson, R. L., *Phys. Rev.* **B41**, 9519 (1990).

86. Dev, B. N., Satyam, P. V., Gog, T., and Materlik, G., *HASYLAB Annual Report*, (1994), p-349.
87. Cowan, P. L., Brennan, S., Terrence, J., Bedzyk, M. J., and Materlik, G., *Phys. Rev. Lett.* **57**, 2399 (1986).
88. Bedzyk, M. J., Bommarito, G. M., and Schildkraut, J. S., *Phys. Rev. Lett.* **62**, 1376 (1989).
89. Zheludeva, S. I., et al., *Materials Science and Engineering C: Biometic Materials, Sensors and Systems* **C3**, 211 (1995).
90. Bedzyk, M. J., et al., *Science* **241**, 1788 (1988); Chernov, V. A., et al., *Nucl. Instrum. and Meth.* **A359**, 175 (1995); Hayashi, K., et al., *Appl. Phys. Lett.* **68**, 1921 (1996).
91. Feng, Y. P., Sinha, S. K., Deckman, H. W., Hastings, J. B., and Siddons, D. P., *Phys. Rev. Lett.* **71**, 537 (1993).
92. de Boer, D. K. G., *Phys. Rev.* **B44**, 498 (1991).
93. Wang, J., Bedzyk, M. J., and Caffrey, M., *Science* **258**, 775 (1992).
94. Dev, B. N., Dev, S., Schubert, D. W., Stamm, M., and Materlik, G., (to be published).
95. Dev, B. N., and Dev, S., *Indian J. Phys.* **70A**, 189 (1996).

Single-Pulse Laue Diffraction, Stroboscopic Data Collection and Femtosecond Flash Photolysis on Macromolecules

Michael Wulff[*], Friedrich Schotte, Graham Naylor, Dominique Bourgeois, Keith Moffat[§] and Gerard Mourou[§§]

The European Synchrotron Radiation Facility, B.P. 220, 38043 Grenoble, France
[§] *Department of Biochemistry and Molecular Biology and the Consortium for Advanced Radiation Sources,*
University of Chicago, 920 East 58th Street, Chicago, IL 60637, USA
[§§] *Center for Ultrafast Optical Science, University of Michigan, 2200 Bonisteel Blvd, Ann Arbor, MI 48109-2099, USA*

Abstract. We review the time structure of synchrotron radiation and its use for fast time-resolved diffraction experiments in macromolecular photo-cycles using flash photolysis to initiate the reaction. The source parameters and optics for ID09 at ESRF are presented together with the phase-locked chopper and femtosecond laser. The chopper can set up a 900 Hz pulse train of 100 ps pulses from the hybrid bunch-mode and, in conjunction with a femtosecond laser, it can be used for stroboscopic data collection with both monochromatic and polychromatic beams.

Single-pulse Laue data from Cutinase, a 22 kD lipolic enzyme, are presented which show that the quality of single-pulse Laue patterns is sufficient to refine the excited state(s) in a reaction pathway from a known ground state. The flash photolysis technique is discussed and an example is given for heme proteins. The radiation damage from a laser pulse in the femto and picosecond range can be reduced by triggering at a wavelength where the interaction is strong. We propose the use of microcrystals between 25-50 µm for efficient photolysis with femto and picosecond pulses.

The performance of circular storage rings is compared with the predicted performance of an X-ray free electron laser(XFEL). The combination of micro beams, a gain of 10^5 photons per pulse and an ultrashort pulse length of 100 fs is likely to improve pulsed diffraction data very substantially. It may be used to image coherent nuclear motion at atomic resolution in ultrafast uni-molecular reactions.

1. INTRODUCTION

The aim of this article is to describe the time structure of the radiation from a third generation synchrotron such as the European Synchrotron Radiation Facility(ESRF), the Advanced Photon Source(APS) and SPring8 and how it can be used to determine the spatial organization of macromolecules down to ca. 100 ps when it is combined with laser flash photolysis and spectroscopy. The beamline for structural kinetics at the ESRF is ID09 and we will describe its insertion devices, focusing optics, phase-locked chopper and fs laser. The control of reaction initiation by flash photolysis is also discussed and illustrated by calculations of the release of O_2 in Myoglobin. Firstly, this information should enable researchers to evaluate new opportunities(and limitations) for time-resolved experiments using X-ray diffraction and spectroscopy. Secondly, the performance of circular storage rings for time-resolved experiments will be compared with the expected performance of a free electron laser(XFEL).

© 1997 American Institute of Physics

The instrumentation for time-resolved experiments on ID09 was developed initially for single-shot Laue diffraction. The key components are the synchronization electronics setting the delay between the photolysing laser and the probing X-ray pulse[1]. The first reversible macromolecular reaction in a crystal which was studied using single-pulse Laue diffraction was the kinetics of the CO-photocycle in carbonmonoxy myoglobin[2]. Using the focused polychromatic beam from a wiggler, Laue images before and after photolysis were recorded on a 170 ps single pulse from a 15 mA bunch of electrons in the storage ring, operated in the single bunch mode. Difference maps of the electron density were extracted for times between 4 ns to 2 ms after photolysis. They show the departure of CO from its original Fe-CO site, an ultrafast displacement of the iron out of the heme plane and more subtle relaxations in the heme pocket[2]. Apart from the biological results, the experiment is important since it shows that fast reaction initiation in a crystal is possible under carefully controlled conditions and that single-pulse Laue data can produce difference maps of atomic resolution with a potential time resolution of 100 ps. Time-resolved diffraction techniques may therefore be able to determine intermediate states in other biological reactions such as the production and storage of energy in the photo synthetic reaction center[3], the speed and pathway of ligand binding in heme proteins[4] and time-resolved recognition and discrimination.

The best static structure determinations of macromolecules are produced by X-ray diffraction from crystallized samples. In a majority of cases this gives physiologically meaningful results despite the constrains imposed by the intermolecular contacts in the crystal. It is less clear however to what extend these constraints modify the time course or even the nature of a reaction. Moreover to what extent does the laser pulse bias the results and which parameters can one use to reduce the side effects from a short pulse? Are there any fundamental limits for the obtainable time resolution imposed by the cross section of photolysis and the magnitude of the electric field in a short pulse? We will discuss these questions at the end of this article.

2. THE STORAGE RING AND INSERTION DEVICES

In a third generation synchrotron highly focused electrons or positrons are accelerated to an energy in the 6-8 GeV range and then sent into a low emittance storage ring. As the electrons arc through the bending magnets they emit synchrotron radiation along the direction of motion in a narrow cone with a polychromatic spectrum. The opening angle of the cone is $1/\gamma = mc^2/E$, where m is the mass of the electron, E is its energy and c is the speed of light. The radiation from a bending magnet is confined within a narrow cone of opening angle ca 0.1 mrad in the vertical direction but spread out horizontally over several degrees as the electrons sweep through the bending magnets. The advantage of new synchrotrons is the use of undulators and wigglers, which raise the intensity over that of bending magnets by several orders of magnitude[5]. The devices are inserted in straight sections between the bending magnets and consist of linear arrays of dipoles with alternating polarity. These devices, also called insertion devices, may be up to 5 m long and may contain hundreds of poles. As an electron traverses an insertion device, the Lorenz force directs the electrons into a slalom motion with photon emission near the turning points, see figure 1.

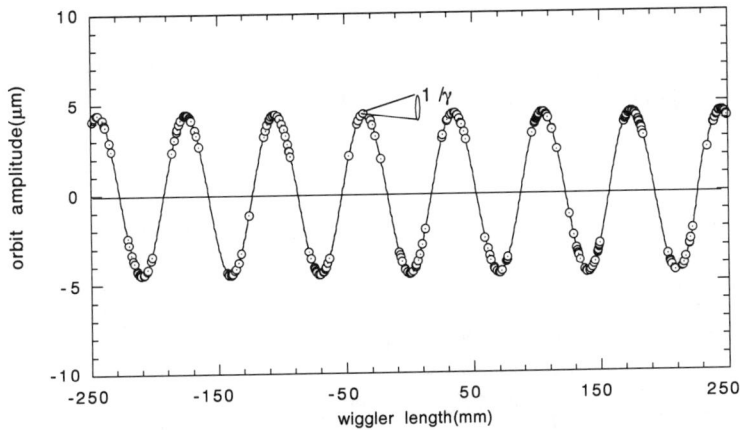

Figure 1. The trajectory of a single electron filament in a K=5 wiggler on a 70 mm period. The circles show the positions of spontaneous emission. Note that the emission is tangential and concentrated near the turning points where the acceleration is greatest. The ESRF straight sections may accommodate up to three insertion devices each 1.6 m long.

Undulators and wigglers have the same magnetic construction but produce different types of radiation. The difference is expressed by the deflection parameter K, which is a measure of the maximum deflection angle of the electron in units of $1/\gamma$

$$K = 0.0934 \, B_0(T) \lambda_m(mm), \qquad (1)$$

where B_0 is the peak magnetic field in the dipole array and λ_m is its period. The fundamental energy is:

$$E_f = 0.950 \, E^2(GeV) / (\lambda_m(mm) + K^2/2). \qquad (2)$$

When $K \geq 1$, the electron path swings outside the radiation cone, the radiation adds incoherently and the device is said to be a wiggler. The spectrum is an energy-shifted version of a bending magnet, but it is brighter by a factor roughly equal to the number of poles. By contrast an undulator is a device where $K < 1$. Here the electron does not move outside the radiation cone, so that the radiation from successive bends overlaps and interferes coherently. The radiation spectrum is characterized by a sharp peak at the fundamental wavelength, with less intense peaks at higher harmonics nof the fundamental. The fundamental wavelength is tunable by changing the size of the gap of the magnetic array. The radiation characteristics are summarized in table 1.
The third generation synchrotrons designed for hard X-rays are the 6 GeV ESRF in Grenoble, the 7 GeV APS at Argonne National Laboratory and the 8 GeV SPring-8 in Harima Science Garden City near Osaka. The ESRF is an electron ring while the APS and SPring-8 rings have a positron option which is expected to prolong the life time of the beam. Such an option was also considered at the

ESRF but later dropped due to the very good life time of ca 50 h with electrons. The circumference of the rings are 844, 1104 and 1436 m respectively and their emittance(transverse electron-positron phase space) is practically identical. In principle the higher energy rings have the advantage of a larger undulator tunability and higher angular power but recent advances in minigap undulators make the difference less important[6].

TABLE 1. Radiation formulae for undulators and wigglers. E_f is the undulator fundamental, E_c is the critical energy, the midpoint of the power distribution. The angular power P_{ang} is the power per unit space-angle in the center of the beam. The length of the insertion device is L and B_0 is the peak field which can vary from 0.2-1.8 T depending on the magnet technology and the gap.

K	$0.0934\ B_0(T)\ \lambda_m(mm)$
E_f(keV)	$0.95\ E^2(GeV)/\ (\lambda_m(mm)(1+K^2/2))$
E_c(keV)	$0.665\ E^2(GeV)\ B_0(T)$
P_{ang}(kW/mrad^2)	$10.84\ E^4(GeV)\ B_0(T)\ N\ I(A)$
P(kW)	$0.633\ E^2(GeV)\ B_0^2(T)\ L(m)\ I(A)$

From table 1 it is seen that the angular power P_{ang} of an insertion device increases as E^4. This relativistic focusing of the beam in high energy machines can only be fully exploited if the monochromators and focusing optics can handle this power density without geometrical distortion[7,8]. These heat load problems were a great challenge at the ESRF where wigglers may produce up to 10 kW on the monochromator crystal. They are now generally solved by the use of liquid nitrogen cooled silicon monochromators which take advantage of the very small thermal expansion of silicon around 100 K[9,10].

3. THE X-RAY TIME STRUCTURE AT THE ESRF

The storage ring is not filled uniformly but has a bunch structure which arises in the following way. An electron in orbit emits synchrotron radiation and the energy loss makes it drift towards the center of the ring. To compensate for the energy loss, radio frequency(RF) oscillator sets up a tangential field at a harmonic of the ring frequency f= c/l, where c is the speed of light and l is the circumference. The harmonic number determines the number of bunches which can be stored. The RF at the ESRF operates at 352.2 MHz and the machine can store up to 992 bunches evenly spaced along its 844 m circumference. The time separation of adjacent bunches is 2.84 ns. In the steady state storage ring mode, the electrons arrive in the RF cavity while the field is increasing. The electrons in the head of the bunch receive therefore a smaller enery than electrons in the tail. The head electrons will concentrate on the inner radius of the orbit and visa versa and the bunch is consequently compressed. The length of the bunch is 18-60 mm corresponding to a 60-200 ps X-ray pulse, see figure 2.

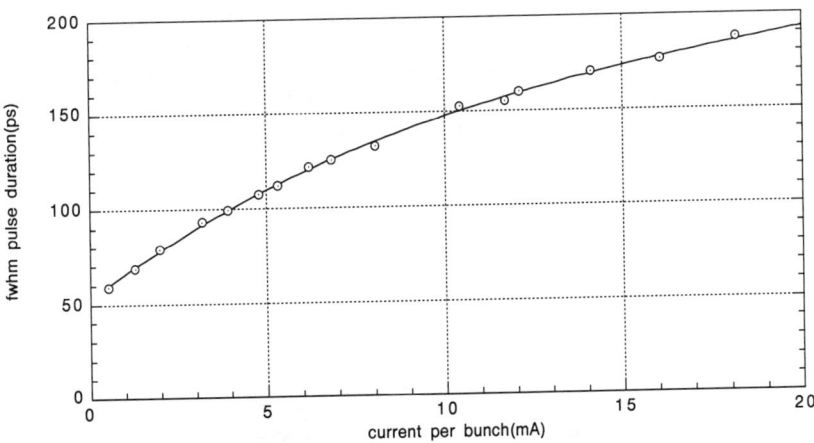

FIGURE 2. The pulse length of a bunch measured by an optical streak camera. In the single-bunch mode, the current is typically 10-20 mA and the pulse length 150-190 ps. In the multi-bunch mode, a higher total current, a longer life time and lower emittance are produced by distributing the current over 332 consecutive bunches in 1/3 of the ring. The stored current can here reach 200 mA with 0.6 mA per bunch(with permission from Kees Scheidt, ESRF).

The bunch structure is an advantage for fast time resolved experiments in the 100 ps to 1µs range where the time structure of the sample, the excitation pump, the shutter and detector may be optimized by a unique pulse structure. The most commonly used bunch structures are shown in figure 3. The exposure time of an n-pulse experiment is :

$$\tau = (n-1)\delta\tau_{ib} + \delta\tau_b \tag{3}$$

where $\delta\tau_{ib}$ is the inter-bunch separation and $\delta\tau_b$ is the bunch length. The intensity in one pulse is in most cases insufficient to obtain satisfactory data and the exposure time is determined by the long dark period between pulses. But for those experiments where there is sufficient intensity in one exposure, the exposure time is reduced to the pulse length itself. The exposure time increases thus from ca 0.1 ns to 2.9 ns in going from a one- to a two-pulse exposure. An interesting possibility exists in time-resolved experiments of reversible phenomena where the sample excitation can be repeated over and over until sufficient intensity is collected by the detector. Laue diffraction is the prime example of a single-pulse experiment performed in the single bunch mode. In this experiment the polychromatic beam is focused onto a stationary crystal and the diffracted intensities are recorded on an integrating area detector. The single-bunch revolution time of 2.82 µs is sufficiently long to allow isolation of a single pulse by a mechanical chopper.

The storage ring at the ESRF is normally operated by default in a 200 mA mode where 1/3(332) of the buckets are filled and 2/3 empty. Many experiments only use the average intensity and it turns out that distributing the current over many bunches has the advantage of lowering the electron "evaporation" caused by the repulsive electron-electron interaction, an effect called the Touschek effect [11-

13]. In the low density 1/3 filling, the lifetime is bremsstrahlung-limited to about 52 hours for a vacuum of 10^{-9} torr whereas a 10 mA bunch has a Touschek lifetime of 11 hours. A suitable compromise is made by placing a single bunch opposite the center of the 1/3 pattern which is called the hybrid mode. In this way a phase-locked chopper can isolate a 100 ps pulse while other beamlines take full advantage of the high average intensity from the 1/3 filling pattern.

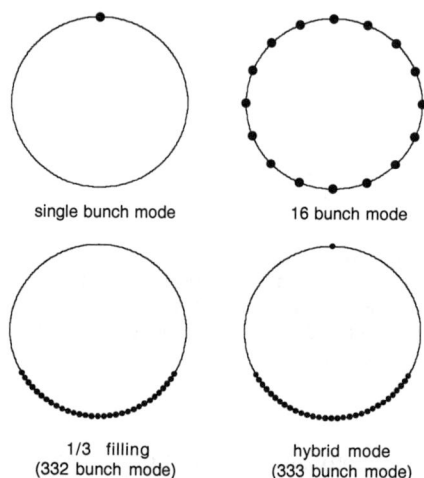

FIGURE 3. : The most frequent bunch structures at the ESRF. The default mode is the 1/3 filling which produces a high current(200mA), a long life time(50h) and a low emittance. The 16 bunch mode is used for nuclear diffraction and spectroscopy. The single-bunch and hybrid mode are used for pulsed experiments on the μs, ns and ps time scale.

The storage ring at the ESRF is normally operated by default in a 200 mA mode where 1/3(332) of the buckets are filled and 2/3 empty. Many experiments only use the average intensity and it turns out that distributing the current over many bunches has the advantage of lowering the electron "evaporation" caused by the repulsive electron-electron interaction, an effect called the Touschek effect [11-13]. In the low density 1/3 filling, the lifetime is bremsstrahlung-limited to about 52 hours for a vacuum of 10^{-9} torr whereas a 10 mA bunch has a Touschek lifetime of 11 hours. A suitable compromise is made by placing a single bunch opposite the center of the 1/3 pattern which is called the hybrid mode. In this way a phase-locked chopper can isolate a 100 ps pulse while other beamlines take full advantage of the high average intensity from the 1/3 filling pattern.

4. THE INSERTION DEVICES ON ID09

At the ESRF the straight sections between bending magnets can accommodate up to three insertion devices each 1.6 m long. In this way one beamline can do very different experiments. In a Laue experiment one sends a polychromatic beam on to a stationary crystal and records the diffraction pattern on an area detector. The number of spots is proportional to the bandwidth and the best insertion device is a wiggler with a smooth spectrum and high intensity. A much broader class of experiments needs monochromatic beams. In cases where the tunability range is limited, one can use an undulator with essentially only one harmonic.

To show how an insertion device can be tailor made for a specific experiment, we will briefly outline how the magnetic period for the Laue wiggler was determined. Ideally one would like to insert as many magnetic poles per unit length as possible but closely spaced poles weaken the field. Specifically, an array of alternating dipoles produces a sinusoidal vertical field with a peak field[14]:

$$B_0 = B_1 \exp(-\pi \frac{g}{\lambda_m}) \qquad (4)$$

where g is the gap of the insertion device and λ_m is the magnetic period. B_1 is a material parameter which is 2.31T for the NdFeB magnets used at the ESRF. It is seen that the field increases with increasing period; a strong field wiggler has therefore fewer poles than an undulator. The angular power in the forward direction per unit length is(see table 1):

$$P_{ang}(kW/mrad^2) = 21.68 \frac{B_0(T)}{\lambda_m(mm)} E^4(GeV) I(A) \qquad (5)$$

where the function $\exp(-\pi g/\lambda_m)/\lambda_m$ has a maximum at $\lambda_m = \pi g$ which we shall call the wundulator period[15-17]. The properties of wundulators are summarized in table 2. Note that the wundulator period is a geometrical property of the array. The line integral of the power collected in the forward direction is here optimal. The wundulator period is independent of the electron energy whereas the deflection parameter and critical energy depend on both E and B_1. Note that at 20 mm gap, NdFeB magnets on a 62.8 mm period produce a K= 5 wiggler with a critical energy of 20.6 keV. Such a device is ideal for Laue diffraction where it is important to have a broad and smooth spectrum. The gap was originally 20 mm at the ESRF. It is presently 16 mm with the aim of reaching 11 mm in 1997. At 10 mm gap, the angular power is doubled while the critical energy and total power are unchanged. Moreover the heat load on the mirror - measured by the critical energy - is unchanged. Note that wundulators become more undulator-like at lower gaps with a K =2.5 at 10 mm. The spectral smoothness is determined by the density of harmonics $1/E_f$ which is proportional to $\lambda_m(1+K^2/2)$. Mini gap wundulators therefore need a taper option or a sligthly varying period to produce a variation in K along the length of the device.

TABLE 2. Insertion devices for optimal power density optimized at 10 and 20 mm gap. The magnet technology is NdFeB(B_1=2.31 T) and the current is 100 mA.

	Wundulator Optimum	g=10mm	g=20mm
λ_m(mm)	π g(mm)	31.4	62.8
K	0.108 B_1(T) g(mm)	2.5	5.0
ε_c(keV)	0.245 E^2(GeV) B_1(T)	20.6	20.6
P_{ang}(kW/mrad^2)	1.269 E^4(GeV) B_1(T) L(m) I(A)/ g(mm)	62.4	31.2
P(W)	0.085 E^2(GeV) B_1^2(T) L(mm) I(A)	2648	2648

Presently we are using two wigglers(or wundulators) on a 46 mm(K =2.72) and 70 mm(K =5.40) period and a single-line undulator on a 26 mm(K =0.65) period. The source parameters are summarized in table 3 and the polychromatic spectra shown in figure 4. The source size of a wiggler is defined as the virtual image on a screen in the center of the wiggler waist. The image is broadened by the divergence of the beam over the half-length of the insertion device. In a coherent line source on the other hand, the photon tracks with the electron and the cone becomes gradually more collimated. The beam dimensions at the exit determine the source size and the distance to the focusing optics.

TABLE 3. The calculated source parameters of the U26 mono-harmonic undulator and the U46 and W70 wigglers on a low b site in the electron beam. The insertion devices have 129, 71 and 43 poles respectively. Note the sub mm source size which is ideal for protein crystallography, high pressure research and liquid scattering. The magnetic gaps are 20.1 mm for W70, 16.2 mm for U26 and U46. The powers are quoted at 100 mA. Source dimensions are fwhm.

ID	E_f(keV)	E_c(keV)	K	P(kW)	P_{ana}(kW/mr²)	S_x(µm)	S_z(µm)	S_x'(mr)	S_z'(mr)
U26	11.02	6.5	0.65	0.276	24.86	195	117	0.31	0.117
U46	1.60	15.4	2.72	1.516	32.50	241	117	0.62	0.117
W70	0.32	20.0	5.40	2.370	25.63	331	117	1.01	0.117

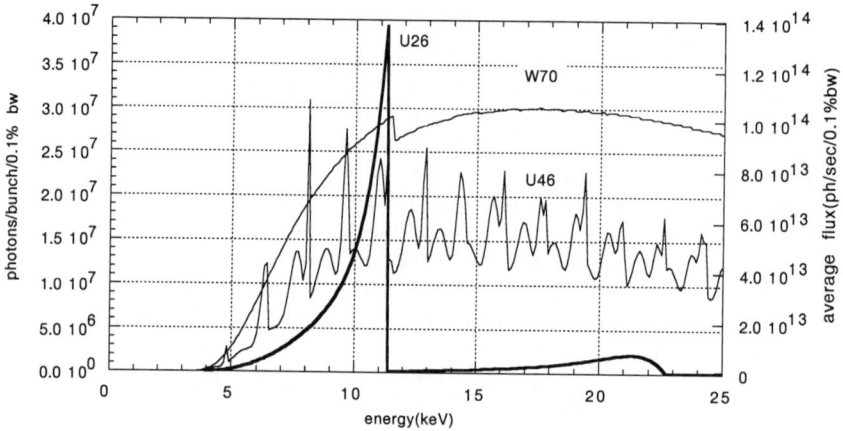

FIGURE 4. The pulsed flux in the focal spot from the three insertion devices from a single-bunch of 10 mA. The focus is formed by a Pt coated toroidal mirror placed 29.7 m from the source. The entrance slit is here 20 mmh and 1.0 mmv and the mirror is at 2.32 mrad glancing angle. The average flux is shown on the right at 100 mA. In the single bunch mode all three insertion devices can operate simultaneously.

5. OPTIMIZATION OF THE TOROIDAL AND PARABOLIC MIRROR

The large radius of the new storage rings makes it necessary to install the first optical components far from the source point in order to free the photon optics from the electron beam. Thus the first pair of slits at ESRF are installed 27.0 m from the source. The long distance combined with the high heat load make the design of focusing optics a challenge in order to preserve the brilliance and guarantee a stable beam.

The first mirror on ID09 is toroidal and the second is plane. Both mirrors are mounted in a bender for bending in the vertical plane. Both mirrors reflect vertically and are cooled to withstand the polychromatic beam. The toroidal mirror is used for focused Laue diffraction and the plane mirror for microfocusing in one direction. The toroidal mirror is also used as a premirror for the Laue-Bragg monochromator. It focuses the monochromatic beam sagittally whereas the vertical focusing is accomplished naturally by a Laue wafer in 1:1 focusing mode. The monochromator is protected by a graphite prefilter which lowers the heatload to a level suitable for water cooling. The trade-off is a poor throughput at low energies which limits the energy range to 12-40 keV. The mirrors are mounted in a three-point bender. The mirror rests on four piezoes at the corners which lift the mirror against four reaction points 100 mm away. In addition a piezo is mounted under the mirror to compensate the gravity sag. The water cooling is accomplished by spring-loaded Cu plates which float against the mirror sides using a layer of InGa to ensure good thermal contact. The mirror parameters are summarized in table 4.

TABLE 4. Toroidal and parabolic mirror parameters

item	mirror 1	mirror 2
shape	cylindrical(toroidal)	plane(parabolic)
sagittal radius(mm)	64.53	-
bender	three-point actuator	three-point actuator
dimensions : L x W x H(mm^3)	1000 x 130 x 100	1200 x 100 x 50
material	SiC on graphite	Si crystal
coating	Pt	Pt
rms surface roughness(Å)	1.8	1.0
rms longitudinal slope error(μrad)	8.4§	6.0
p(m)	31.3	35.5
demagnifications used	0.36, 0.57, 0.83 §§	0.2-0.5
incidence angles (mrad)	3.86, 2.83, 2.28 §§	1.25-5.0
energy range(keV)	5.0-21.5, 5.0-29.7, 5.0-37.1 §§	5.0-55.0
gravity sag(km)	4.7	6.9
R_m bending range(km)	1.8-4.7	1.5-6.9
R_m pushing range(km)	4.7-12.4	6.9-20.0

§4.1 μrad over 380 mm in the center
§§for the optical tables T1, T2 and T3

A mirror can be considered a low pass filter. It reflects and focuses an energy band $E < E_{max}$, where E_{max} is a function of the incidence angle. By displacing the beam out of the orbital plane, which is contaminated by highly penetrating 6 GeV bremsstrahlung, the background is reduced which is of importance in experiments using integrating detectors such as image plates and CCDs. The meridional (or in-plane) radius R_m and the sagittal (or out-of-plane) radius r_s are related to the distance p between the source and the mirror and the distance q between the mirror and the focal spot :

$$R_m = \frac{2pq}{(p+q)\sin\theta}, \quad r_s = \frac{2pq\sin\theta}{(p+q)} \tag{6}$$

where θ is the incidence angle. Since r_s is fixed, changing the incidence angle induces a change in q :

$$q = \frac{p\,r_s}{r_s - 2p\sin\theta}. \tag{7}$$

Raising the angle brings therefore the focus nearer while reducing the bandwidth. The toroidal and parabolic surfaces deviate significantly from the ideal elliptical shape at strong defocusing and produce aberrations and therefore loss of intensity. The stroke of the figure error is reduced on the other hand. In the meridional focusing case, which applies to the parabolic mirror here, we have found an expression for the fwhm vertical spot size[16] :

$$F_z = \sqrt{(\frac{3}{8}L^2\frac{\theta(1-M^2)}{p\,M} + S_z M + S_z(M+1)\frac{L}{4p})^2 + (2(2.235\,\alpha)M\,p)^2} \tag{8}$$

where L is the length of the mirror, S_z is the vertical source size and α is the rms longitudinal slope error. The optimal defocusing M_{opt} and its associated focal size is found by solving $dF_z/dM = 0$ giving the optimal defocusing :

$$M_{opt} \approx L\sqrt{\frac{3\theta}{64\alpha p^2\,4LS_z + 16pS_z - 3L^2\theta}} \tag{9}$$

which shows that greater α implies greater optimal defocusing. The focal spot size for the parabolic mirror on ID09 is shown in fig. 5 together with the intensity vs defocusing for a toroidal mirror at 2.0 mrad incidence angle. Note that in both cases one benefits from working in the range $0.15 \leq M \leq 0.7$ provided that the experiment can accept the larger divergence of the beam. In any case the advantage of working with great defocusing is that the beamline becomes shorter and therefore less expensive, and also less sensitive to thermal drifts.

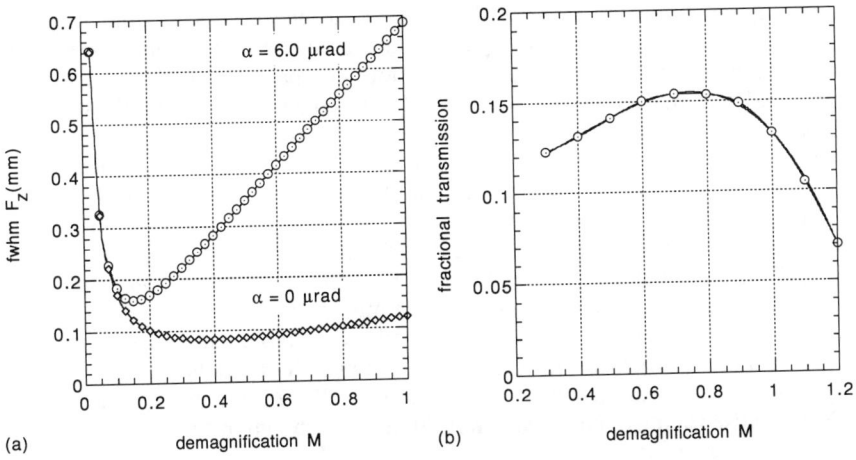

FIGURE 5. The optimal defocusing of an insertion device is a competition between aberrations and the figure error. (a) the optimal defocusing of a parabolic mirror is 0.15 for a slope error of 6.0 μrad rms. (b) the optimal defocusing of a toroidal mirror is non analytical and a ray tracing simulation of the flux through a 0.2 x 0.2 mm² pinhole is shown for a = 3.0 μrad rms. The incidence angle is in both cases 2.0 mrad.

6. THE CHOPPER AND PULSED MONOCHROMATOR

In order to select one or a set of X-ray single or super pulses, a fast shutter system has been designed, whose opening time and frequency match the time structure of the storage ring as well as the desired timing sequence of the experiment. The most efficient shutter design[1,17] consists in using a channel-cut rotating device, synchronized to the RF bunchclock, mounted in series with a magnetic preshutter. Rotation of the chopper about an axis perpendicular to the X-ray beam provides the best performance in terms of x-ray absorption and opening time. In such a configuration, the opening time τ and the opening profile (of trapezoidal shape) p(t) are given by, see figure 6 :

$$\tau = \frac{a}{2\pi R f \sqrt{1-(h/R)^2}} \quad (10)$$

and

$$p(t) = \begin{cases} 0 & \text{if } Abs[\theta(t)] \geq \theta_{min} \\ Min\left[1, \frac{2R}{S}\sin(\theta_{min} - \theta(t))\right] & \text{if } Abs[\theta(t)] \leq \theta_{min} \end{cases} \quad (11)$$

where a is the height of the channel, R the chopper radius, h the channel offset, f the rotation frequency,

$$\theta_{min} = Arc\sin(a/2R) \quad \text{and} \quad \theta(t) = 2\pi f t, \quad (12)$$

and S is the beam size. In principle, for a given τ, the best choice consists in retaining as high a rotation frequency as possible. Indeed, this allows at the same time, (i) a higher repetition rate for the case of stroboscopic selection of X-ray pulses, (ii) a larger channel section, which is desirable if a large beam size is used and to aid alignment procedures, and (iii) a more compact chopper size, which is more practical and makes accurate drilling of the channel (for example with the spark erosion technique) much easier. However, a high rotation frequency results in significant mechanical stress, and the material as well as the chopper shape must be carefully chosen without compromising the X-ray absorption capability of the device. A propeller-shaped titanium chopper is presently under construction, see figure 6. The parameters are : a = 0.8 mm, R = 97 mm, h = 25.5 mm, f_{max} = 900 Hz, τ_{min} = 1.47 µs. The velocity at the chopper periphery can reach 548.5 m/s thus exceeding the speed of sound so that the device must be run in vacuum. The overall repetition rate of the chopper can be reduced by suitable manipulation of the magnetic preshutter, down to the selection of a single X-ray pulse. The minimum opening time of the preshutter must be smaller than 2/f (2.2 ms in the ESRF case). The selected pulse train from the ESRF hybrid mode is shown in figure 7.

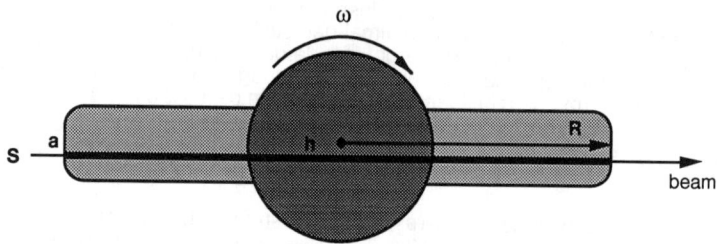

FIGURE 6. The phase-locked chopper rotates at supersonic speed and is kept in vacuum. It can rotate at 900 Hz. When it is used for single-bunch selection in the hybrid mode, it locked to the 391840 subharmonic of the RF at 898.84 Hz. The base-line of the opening window is here 1.47 µsec. The high repetition frequency matches the frequency of the femtosecond laser.

Chopper synchronization is necessary to avoid partial chopping of the X-ray beam and selection of inadvertent double pulses. It can be performed in two ways. An "electronic" synchronization can be obtained by monitoring the phase drift between an asynchronous chopper and the X-ray clock(RF clock). By assuming short term stability of the chopper rotation frequency (say, over a few turns), the conditions of phase matching can be predicted a few milliseconds ahead of time, which suffices to decide when the magnetic preshutter should be opened. This elegant solution, proposed by LeGrand et al has the advantage of requiring limited mechanical precision, therefore reducing the cost of the device[17]. However, this is achieved at the expense of repetition rate, which in practice is limited to a few Hz due to the rather rare occurrence of satisfactory phase matching. The best performance of a chopper system is obtained by mechanical phase locking. By making use of magnetic bearings and high precision speed control electronics, mechanical locking with less than 100 ns jitter relative to the X-ray clock can be achieved. This costly solution was adopted since it will allow to run stroboscopic experiments at a frequency of 900 Hz, thus dramatically improving the duty cycle of the device for those experiments which can run at high frequency.

The chopper was mainly designed for white beam experiments such as single-bunch Laue diffraction but we expect many applications in stroboscopic monochromatic data collection. A simple way of producing a pulsed monochromatic beam is accomplished by using the chopper in the focused white beam and sending the selected pulse train on to a channel-cut monochromator just before the sample. In the hybrid mode with a 150 mA super-bunch and a 15 mA single-bunch, one can extract the single bunch at 898.84 Hz. The chopper accepts thus 1:395 of the single bunches(and totally eliminates the super bunch). The heatload on the monochromator is consequently equivalent to an average current of 38µA. In addition, the heat pulse in a single 100 ps pulse is so short that the net planes in the monochromator have no time to react. In this way one has a 100 % efficient pulsed monochromator can be obtained without external cooling. This was illustrated by an experiment on the photo cycle of bacteriorhodopsin in which a chopper was run synchronously of the bunch clock at 10 Hz which reduced the heatload by a factor of 780 from 45 W to 58 mW. The chopper was run by a stepper motor at 10 Hz, which produced 115 µs polychromatic pulses which were loaded on to a Si(111) channel-cut monochromator. The monochromator gave its full theoretical flux over the 115 µs window as measured by a PIN diode(figure 8);. the focal spot is shown in figure 9.

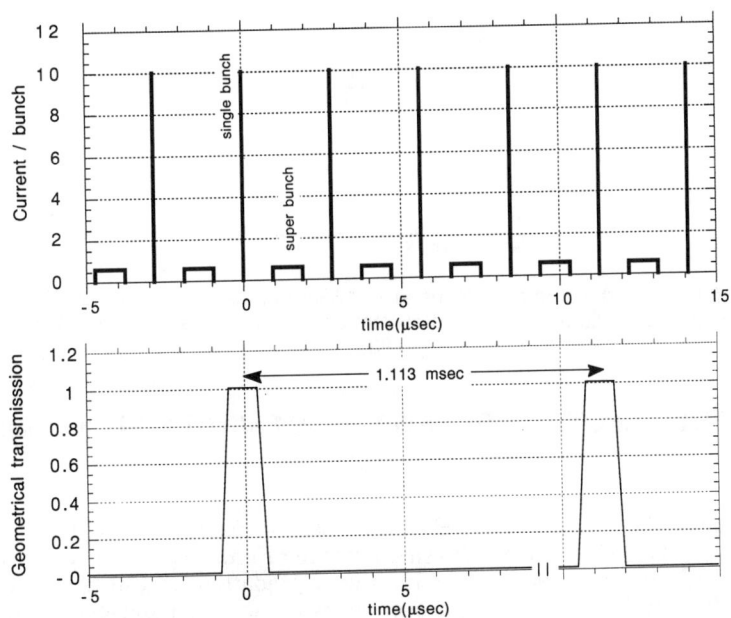

FIGURE 7. The time structure of the hybrid mode(above) and the geometrical acceptance of the chopper(below). The base-line of the opening window is 1.47 µs and the top-line is 0.94 µs for a vertical beam of 0.3 mm.

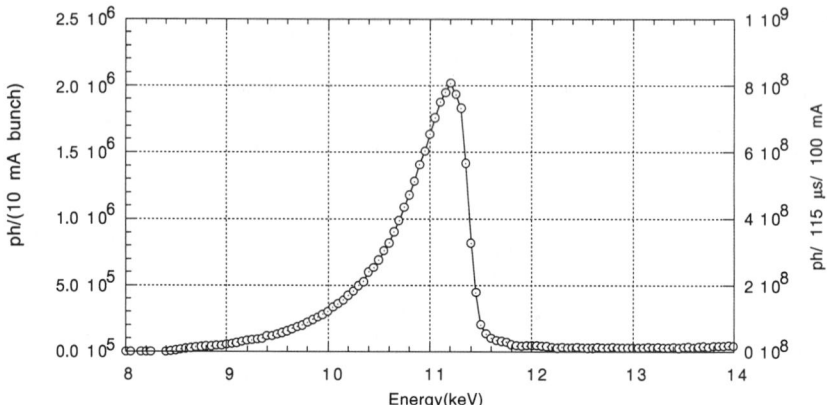

FIGURE 8. The flux in the Ø180 μm focal spot from the U26 fundamental in a single-pulse of 150 ps. The beam is monochromatized by a Si(111) channel-cut monochromator. In the experiment on the Bacteriorhodopsin photocycle, the polychromatic beam was chopped into 115 μs pulses at 10 Hz and the measured flux per pulse is shown on the right.

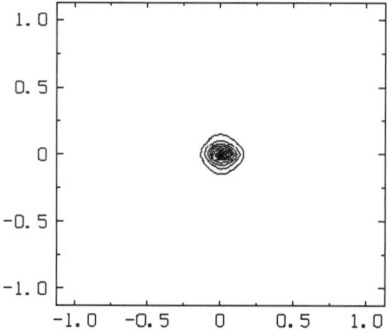

FIGURE 9. The monochromatic focus at 11.16 keV of the U26 undulator. The horizontal and vertical focal dimensions are 0.15 mm and 0.10 mm respectively. The incidence angle is 2.32 mrad and the defocusing 0.83.

7. THE ELECTRON DENSITY FROM SINGLE-BUNCH LAUE DATA

To examine the quality of Laue electron density maps from single-bunch patterns, we have collected data from the small enzyme *Fusarium solani* Cutinase, a 22 kD fungal enzyme responsible for the degradation of Cutin[18]. Cutinase crystallizes in a monoclinic space group with lattice parameters a =35.2 Å, b =67.3 Å, c =37.1 Å, $\alpha = \gamma = 90.0°$ and $\beta = 94.1°$. The ESRF was operated in single-bunch mode was operated at 13 mA which produces ca 150 ps X-ray pulses. The flux was increased by using two insertion devices in series, the 43-pole wiggler(K=5.4) and the 71-pole wiggler(K=2.7). In this configuration 3.3 x

10^{10} photons over 8-32 keV could be delivered to the crystal in a single pulse. The purity and intensity of the pulses were monitored with an avalanche diode placed close to the sample. The images were recorded with a Ø220 mm image intensifier coupled to a slow scan CCD camera[19]. This detector is particularly efficient at detecting very weak diffraction patterns(DQE ~ 50%). A Laue pattern based on three X-ray pulses accumulated on the CCD detector is shown in figure 10. The crystal was mounted on a horizontal spindle axis and 19 Laue images were recorded between 0-180°. The crystal was thus exposed to the beam for 8.5 ns and no direct sign of radiation damage was observed. Each image contains ca 5000 reflections and the completeness is 71.7 % up to 1.5 Å resolution. The Laue pattern have been indexed, integrated and merged and their intensity distribution vs resolution is shown in figure 11. Note that the strongest reflections, which are typically multiplets of the form (h, k, l) and (2h, 2k, 2l) etc. diffract up to 3×10^4 photons per pulse. With two insertion devices in series, the available flux was sufficient to refine a model of native Cutinase with R_{cryst} = 19.3% and R_{free} = 24.2%, the starting model for the refinement being the R196E mutant[18]. A similar situation is encountered in time-resolved experiments where the excited state of the protein has to be refined from a model of the ground state.

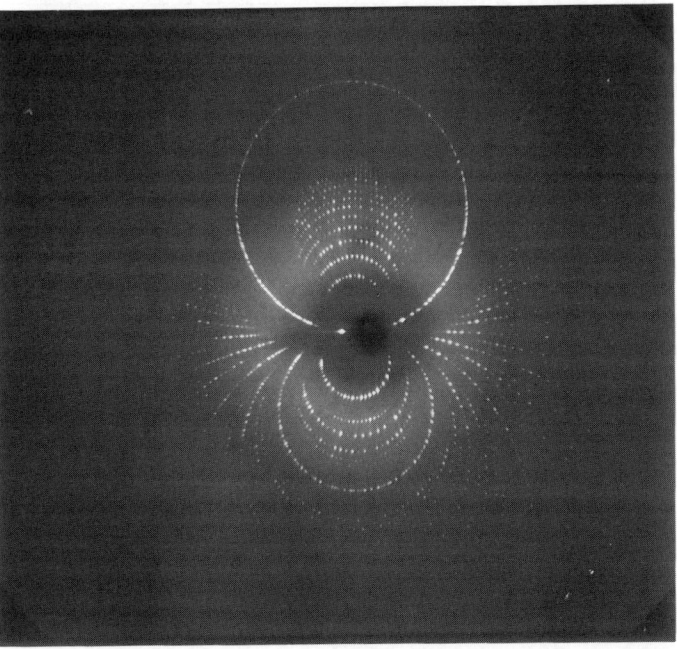

FIGURE 10. Laue pattern a Cutinase enzyme recorded on three 170 ps pulses accumulated on an image intensifier. The pattern was recorded with the two insertion devices U46 and W70 in series and 3.3×10^{10} photons per pulse could be delivered to the 0.3x0.4x0.4 mm3 sample at 13.6 mA. The crystal was supplied by Sonia Longhi and Christian Cambillau, CNRS Marseilles.

The first single-bunch Laue experiment at the ESRF was performed by Keith Moffat and his colleagues[2]. In a study of the photocycle of myoglobin carbonmonoxide(MbCO), the CO ligand was detached from the Fe atom by a 7 ns laser pulse at 635 nm and the movements of CO, Fe and the molecules in the heme pocket could be followed with ns resolution. Several effects were observed : a geminate recombination of the CO molecule where CO recombines with its parent Fe atom on the sub ns scale, a short-lived "docking site" some 4 Å from the initial position with a life time of 10-100 ns and a slower back diffusion of CO through the protein on the 100 µs scale. The Fe atom was also seen to move out of the heme plane by a small amount ~ 0.2 Å. For each time-point a new crystal was used and the power of the laser was carefully adjusted by monitoring the jump in the optical density to avoid unnecessary radiation damage. The crystal was rotated in steps of 4° over 0-184° with three exposures of the ground state and three exposures of the excited state for a given delay between the laser and the X-rays. The radiation damage during the 282 exposures seemed to be so small that complete time-points could be collected on one crystal. The synchronization of the laser, X-ray chopper and the CCD detector is described in[1].

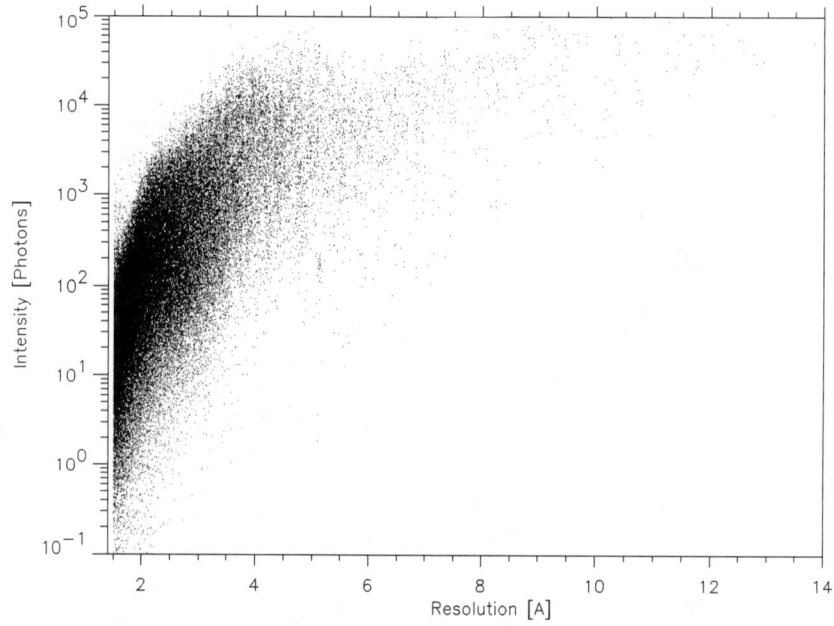

FIGURE 11. The intensity distribution of 19 Laue frames from the Cutinase enzyme integrated to 1.5 Å. The decrease in intensity at low d-spacing is due to the atomic form factor, finite temperature and disorder. The total exposure time for the 75000 reflections was 8.5 ns.

8. FLASH PHOTOLYSIS OF FAST MACROMOLECULAR REACTIONS

In a time-resolved experiment one needs a way of initiating the reaction which is faster than all subsequent reactions of interest and which is otherwise free of artefacts[20]. Some reactions may be initiated chemically in a flow cell by delivering an activator or inhibitor to the crystal. The time scale is here limited by the speed of diffusion into and through the crystal which typically requires tens of seconds. Primary reactions in biology such as electron or proton transfer and the formation and breakage of bonds take place on the fs to ms scale and here one is forced to work with either naturally photo-sensitive reactions or reactions which can be made photo-sensitive artificially. Reversible photo-reactions are convenient because they can be repeated and diffraction patterns accumulated to satisfy a given precision and resolution. By varying the delay between the laser and the X-rays, one can produce a series of three-dimensional stills of the electron density which has been thought of as a movie[21]. In this movie, the time scale of a given phenomenon is the convolution of the intrinsic time scale and the laser and X-ray pulse lengths :

$$\tau_{movie}^2 = \tau_{int}^2 + \delta\tau_{laser}^2 + \delta\tau_{x-ray}^2, \tag{13}$$

where we neglect the finite sample dimension(which becomes important on the sub ps scale). Consequently the time resolution of ultra fast phenomena can not be resolved beyond the X-ray pulse length of ca 100 ps with pump probe techniques on a third generation synchrotron. Will the use of picosecond or even femtosecond laser pulses make it possible to reach this 100 ps time-resolution with observable changes in the diffraction intensities? Since photolysis of macromolecular crystals with ultra short pulses has never been tried, we will look into this problem for liquids and crystals. The main difference between crystals and liquids is that the transmission of light is anisotropic or dichroic in crystals and has to be measured carefully.

In deciding the photolysis strategy, one may start by considering the available crystal and focal dimensions. In a single pulse experiment on radiation sensitive molecules, the X-ray intensity takes priority and the crystal should be smaller than the focal spot minimizing transverse gradients. Consider now a laser pulse of intensity I_0 impinging on a flat crystal. If the cross section for the event under consideration (eg. release of CO in MbCO) is σ, the inverse absorption length is :

$$\mu(\lambda, \tau) = n\, \sigma(\lambda, \tau), \tag{14}$$

where n is the concentration of scatterers, λ the laser wavelength and τ the time after photolysis. If the pulse is shorter than the time scale of the change in absorption and the photolysis is below saturation, the absorbed photon density (or photolysis density) is :

$$I_{abs} = I_0\,(1 - exp(-\mu t)), \tag{15}$$

the decay in a linear differential system. It is seen that if the crystal size is μ^{-1}, the longitudinal variation is $e^{-1} = 36.8\%$. A longitudinal photolysis gradient may be acceptable since non-photolysed unit cells cancel out in a difference map(while lowering δI/I). One may however split the laser beam and photolyse from both sides. This will not only boost the degree of photolysis but also lower the longitudinal gradient. In practice one should choose a crystal size between $0.25 \leq \mu t \leq 0.5$ to minimize the thermal bending and spot broadening from the longitudinal temperature profile. Moreover, it is important that the laser follows the crystal rotation to ensure that the photolysed volume is constant throughout a data collection.

The intensity I_0 needed to saturate the initial layer $0 \leq t \leq dt$ of the crystal is determined by :

$$I_0 \mu \, dt = n \, dt \qquad (16)$$

from which we get the *saturation limit* :

$$I_0 = \sigma^{-1}. \qquad (17)$$

It follows that the *energy saturation limit* is :

$$\varepsilon_{sat} = \frac{hc}{\sigma \lambda}, \qquad (18)$$

which numerically is $\varepsilon_{sat}(mJ/mm^2) = 1.987 \times 10^{-13}/(\sigma(mm^2) \lambda(nm))$. So apart from the energy term, ε_{sat} depends on the interaction σ only and not on the density of scatters. The wavelength dependence of ε_{sat} for heme containing proteins is therefore a generic function. In a crystal, σ is a function of the relative angle between the absorbing molecule and the polarization and finding the best angle between the crystal and the polarization could be important if one needs to work with larger crystals. Despite the simplified treatment, the above formula is a useful rule of thumb. Note that σ may differ substantialy from the value inferred from absorption measurements; the ratio is the quantum yield Q.

To illustrate the photolysis requirements, we will examine the release of O_2 in MbO_2. The absorption cross section σ_{abs}, derived from data in solution[22], is shown in figure 12 and the density of heme for the monoclinic($P2_1$) and hexagonal(P6) unit cells is listed in table 5. The absorption length μ^{-1} for monoclinic MbO_2 is shown in figure 13. It is seen that the only wavelengths which match the focal size on ID09 is 620-635 nm where $400 \leq \mu^{-1} \leq 600$ μm. But the price of the high penetration is a high ε_{sat} of 3.8 mJ/mm² at 620 nm, see figure 14. Alternatively, triggering at 500 nm brings ε_{sat} down to 0.50 mJ/mm² but an absorption length of $\mu^{-1} = 40$ μm requires micro crystals and focusing.

TABLE 5. Lattice parameters for MbCO in its monoclinic $P2_1$ and hexagonal P6 state. The concentration of Fe sites is also listed.

sample	a(Å)	b(Å)	c(Å)	a(deg)	b(deg)	g(deg)	V/Fe(Å³)	c(mMol)	N(Fe/mm³)
MbCO($P2_1$)	64.60	31.10	34.80	90.0	105.5	90.0	33686	49.3	2.97x 10¹⁶
MbCO(P6)	90.25	90.25	45.31	90.0	90.0	120.0	53268	31.2	1.88x 10¹⁶

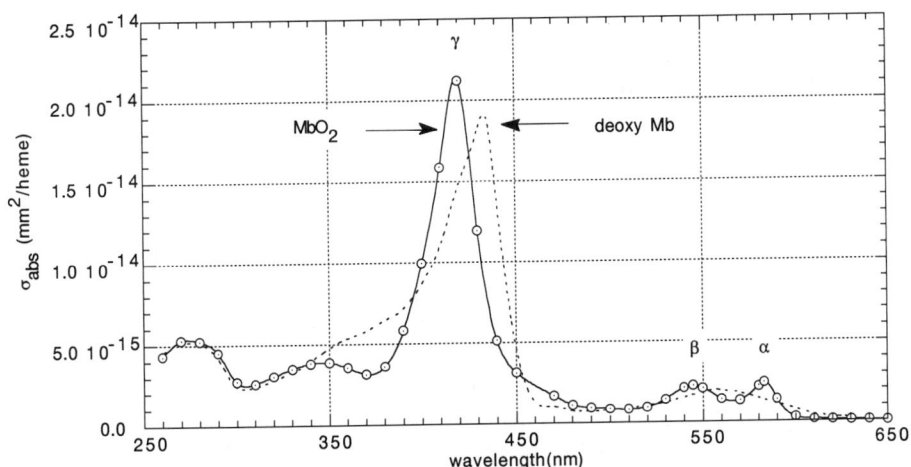

FIGURE 12. The absorption cross section of one heme unit in MbO$_2$ and deoxy Mb. The data are derived from measurements in solution and neglects crystalline anisotropy[22].

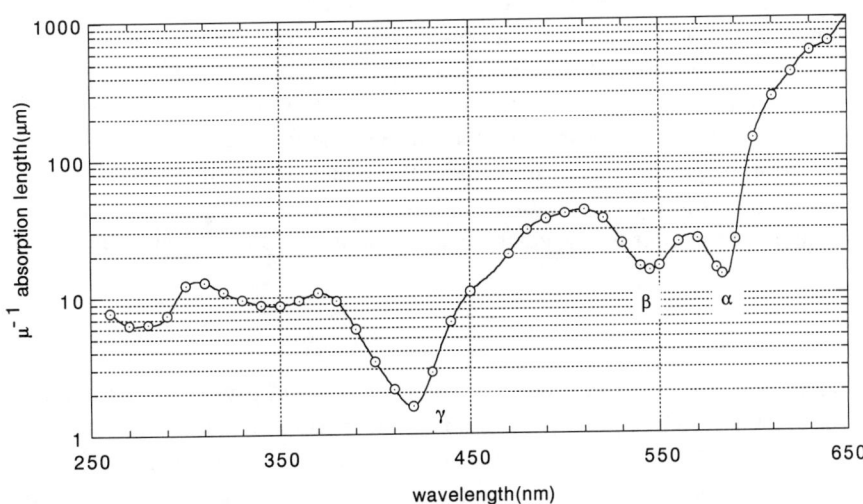

FIGURE 13. The absorption length for monoclinic MbO$_2$. To illuminate a volume comparable with the X-ray focal spot, one has to work in between 610- 635 nm, a range which can be covered by a Cr:fosterite femto second laser[23]. Alternatively one may use microcrystals at 500 nm.

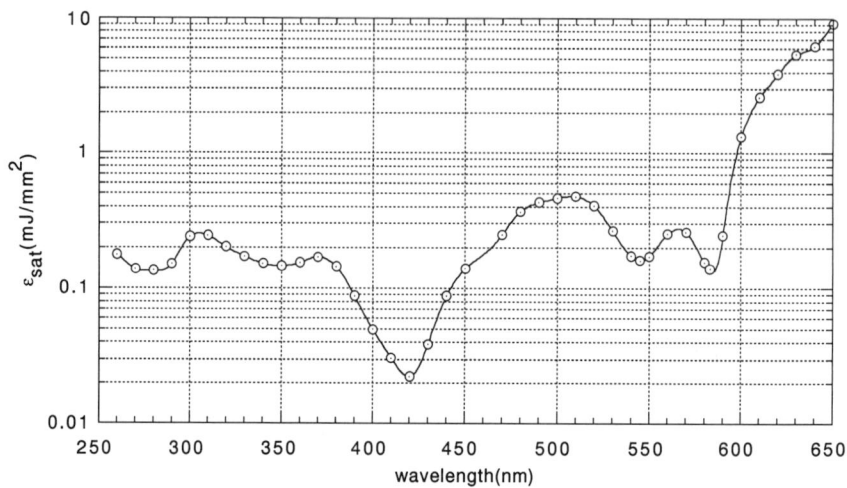

FIGURE 14. The energy density for complete photolysis of heme proteins in solution. The quantum yield is here assumed 1.

The behavior of the absorption spectrum during the pulse is an important issue. In figure 12 it is seen that the absorption in the deoxy state increases at 620 nm. It is therefore likely that the tail of the pulse is obstructed by the effect of the head, an effect called anti-bleaching. The photolysis and temperature gradient are thus greater than exponential. The homogeneity would be improved by choosing the α band where the absorption goes down in the photolysed state but a μ^{-1} of 15 μm is difficult to use. Precise information on the time dependence of the absorption spectrum is therefore vital for a carefully planned experiment. The photolysis parameters of a 200 μm thick MbO_2 crystal photolysed at 620 nm, are summarized in table 6. The adiabatic temperature jump is calculated by

$$\Delta T = \frac{E}{c_p M}, \tag{19}$$

where E is the absorbed energy, c_p the specific heat of water and M the mass of the crystal. Note the fairly modest temperature rise of less than 1 K. It should be emphasized that the above calculations assume that the O_2 release efficiency or quantum yield is 1 which is difficult to verify since the transient absorption spectra have not yet been measured. The energy densities mentioned in figure 14 are therefore lower limits.

TABLE 6. The photolysis of monoclinic and hexagonal MbO2 crystals. The adiabatic temperature rise is calculated using the heat capacity of water.

	MbCO($P2_1$)	MbCO(P6)
density(bonds/mm^3)	2.969 x 10^{16}	1.877 x 10^{16}
trigger wavelength(nm)	620	620
μ^{-1}(mm)	0.40	0.64
ε_{sat}(mJ/mm^2)	3.8	3.8
t(mm)	0.2	0.32
P_{abs}(mJ)	1.5	1.5
<ΔT>(K)	0.89	0.55

Having calculated ε_{sat}, one needs to compress this energy into a pulse compatible with the time resolution of the experiment without damaging the crystal. The electric field scales as $(\varepsilon_{sat}/\tau)^{1/2}$, where τ is the pulse length.

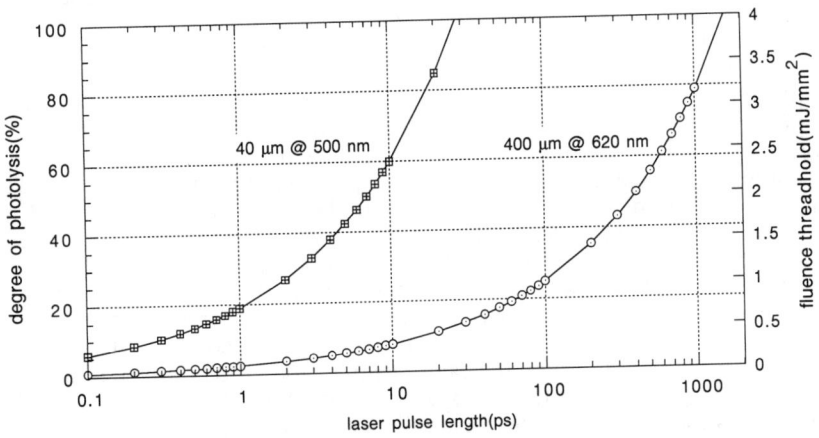

FIGURE 15. The maximum degree of photolysis in heme proteins as a function of the laser pulse length. A quantum yield of 1.0 is assumed. Two cases are shown: a microcrystal of 40 μm thickness triggered at 500 nm where ε_{sat} =0.50 mJ/mm^2 and a 400 μm thick crystal triggered at 620 nm where ε_{sat} =3.8 mJ/mm^2. Making optimal use of the finite electric field tolerated by the crystal, one should trigger at a strongly absorbing wavelength. In return the crystal has to be very thin.

So in going to shorter pulses, the electric field diverges and reaches eventually a breakdown point, sometimes called the fluence threshold, where ionization and continium creation takes place. The limit for myoglobin in solution is 30 μJ/mm^2 for 100 fs pulses[24] and it increases as $\sqrt{\tau}$ for longer pulses[25]. It is shown in figure 15 that the photolysis threshold is 0.8% at 100 fs and 100 % at 1.5 ns. The calculated $\delta I/I$ of a reflection of median intensity in MbCO has been shown to be ~ 10% at 100% photolysis[1] thus scaling down $\delta I/I$ to 8 x 10^{-4} at 100 fs, which is impossible to measure by Laue diffraction. At this level, one may use modulation techniques with lock-in amplification and measure reflections one by one. One way out of this dilemma may be to go to microcrystals combined with microfocusing on the 25-50 μm scale. Specifically, photolysing at 500 nm, μ^{-1} is

40 μm and ε_{sat} is 0.45 mJ/mm^2. At this wavelength the sample could be fully photolysed by a 20 ps pulse which is nicely matched the present X-ray pulse length 60-200 ps.

9. THE FEMTOSECOND LASER AND ITS SYNCHRONIZATION

The fs laser for ID09 is going to be used for the jitter-free streak camera[26] and stroboscopic photolysis in macromolecules. The photolysis of heme proteins has been used to guide the choice of laser technology. The laser must provide sufficient energy to trigger the photo-conductive switch in the streak camera and to excite the sample under study. Some 50μJ are required for the photoconductive switch and some 1-10 mJ for the sample. The repetition rate of the laser has been chosen to match the highest possible frequency of X-ray pulses after the chopper which is ~ 900 Hz. While the fundamental wavelength of 800 nm produced by commercially available Ti: Sapphire lasers is suitable for triggering the photo-conductive switch, the sample will often require some other wavelength; for example 500-640nm in heme proteins. This necessitates the use of an optical parametric amplifier(OPA) in order to produce 100 fs pulses in this region, see figure 16. Commercial systems can produce 10s of microjoules at these wavelengths. Higher energies in the 630 nm region could be achieved by amplifying 1.260 μm pulses from an OPA in a Cr:Forsterite regenerative amplifier and then frequency doubling the output. The two laser pulses, one triggering the photo-switch and one triggering the sample, are both exactly synchronised due to the nature of the optical parametric generation process. These pulses must in turn be synchronised to the 0-900 Hz X-ray pulses passed by the chopper. The synchronisation of the laser and X-ray pulse trains must be achieved with a jitter substantially less than the width of the X-ray pulse which means ca 1ps rms. The synchronisation is achieved either by adjusting the cavity length to be resonant with a sub-harmonic (4th or 5th) of the storage ring RF frequency and setting the correct phase either by using the sub-harmonic RF to mode-lock actively the cavity or to act on the cavity length in order to bring the timing of the output laser pulse (monitored with a fast photodiode) into a fixed phase synchronisation with the sub-harmonic RF. The synchronisation is summarised in the block diagram shown in figure 17.

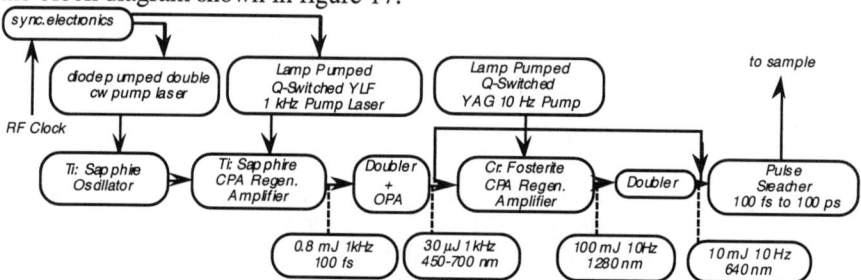

FIGURE 16. The Ti: Sapphire and Cr: Fosterite femtosecond lasers. The fundamental of the Ti: Sapphire can be tuned between 780-830 nm and delivers 0.8 mJ at 1 kHz in a 100 femtosecond pulse. The wavelength can be varied in theOPA between 450-700 nm and 900-2500 nm with 30 μJ per pulse. The Cr:Fosterite laser, seeded by the OPA can be frequency doubled to the important "heme band" 625-650 nm with ca 10 mJper pulse. The lasers can be focused to below Ø10 μm.

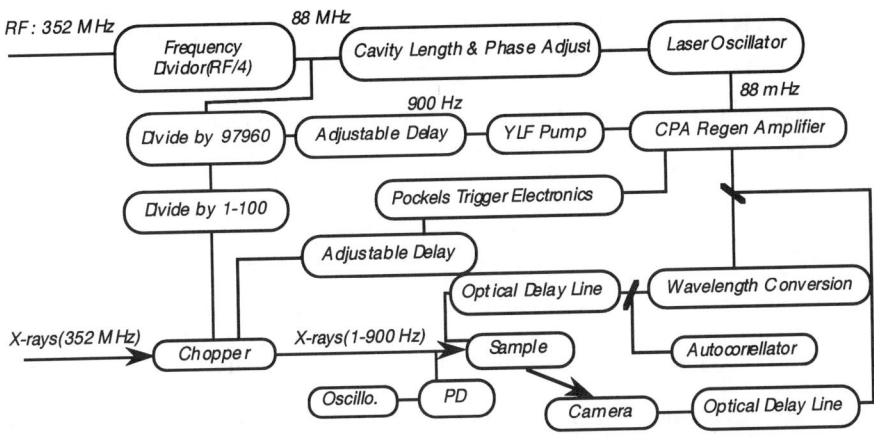

FIGURE 17. The synchronization of the Ti: Sapphire laser to the RF. The streak camera operation requires setting the delays between the the x-ray pulse, the photolysing pulse and the camera trigger.

The RF frequency used to accelerate the electron bunches in the storage ring is 352.2Mhz and defines the possible positions of the electron bunches in the ring. The revolution frequency of the storage ring is 355kHz. With a single bunch in the storage ring, the X-ray pulses produced by the insertion device on a particular straight section will therefore propagate down the beam line at a repetition frequency of 355kHz. The sample cannot however be excited by the laser at this frequency as it would overheat. The X-ray pulses are therefore chopped to a repetition frequency up to 900 Hz. A master laser oscillator is synchronised to the RF signal (which defines the timing of the electron bunches) at 88Mhz. There is in practice a drift of several picoseconds between the RF used to accelerate the electrons and the position of the electron bunch during the decay of the stored electron current. This is compensated by re-adjusting the RF phase by monitoring the exact position of the electron bunch with pick up electrodes. This compensated RF timing must then be distributed to the laser electronics some hundreds of meters away using a cable of low temperature coefficient dielectric in a temperature controlled environment. Single laser pulses are then selected from this train of pulses and amplified in a regenerative laser amplifier. The correct laser pulse corresponding to the X-ray pulse passed by the chopper is selected by opening and closing pockels cells in the regenerative amplifier cavity. The two pulse trains are brought into synchronisation on the 10s of picosecond time scale by adjusting the phase of the RF used to lock the laser oscillator while monitoring the relative timing with a fast photodiode. The synchronisation is then adjusted on the picosecond time scale by adjusting optical path lengths of the laser while monitoring using the streak camera itself. The relative timing of the optical excitation of the sample and the laser triggering of the streak tube is adjusted using a further optical delay line to the photo-conductor. The correct timing of the optical excitation of the sample will be verified by placing a doubling crystal in the excitation path of the sample and allowing a small amount of the scattered light to fall on the entrance slit of the camera.

10. OUTLOOK ON THE FUTURE FOR TIME-RESOLVED STRUCTURAL KINETICS

It appears that time-resolved crystallography can be conducted using existing techniques and apparatus[1] down to a time-resolution of a few hundred picoseconds, but that better time resolution require that the problem of the fluence threshold be separately confronted. Smaller crystals may offer some advantages(Figure 15) but they of course diffract more weakly and require a longer total X-ray exposure to accumulate a good diffraction pattern. New problems arise if even better time-resolution is required, of less than a typical bunch length and no clearly suitable, two dimensional detector exists. We here propose how a zero-dimensional streak camera could be used to identify the synchronization problems, but it will clearly require substantial effort to test these proposals.

The recent observation of coherent nuclear motion in bacterial reaction centers up to 3 ps after photolysis [27] and in smaller molecules such as HgI_2 up to 10 ps[28], makes it important to examine whether such phenomena can be resolved by diffraction or spectroscopy on existing X-ray sources. The shortest X-ray pulse length at the ESRF, obtained in a special low-α momentum compaction configuration, is 32 ps(fwhm) which is close to the theoretical limit for circular machines[29]. So knowing that there is interesting new physics on the 10 ps scale, we are developing a jitter-free streak camera in which the electron sweep is initiated by the same optical pulse used to trigger the sample[26]. Recent tests have demonstrated 1 ps jitter free accumulation[30] and it is possible that later versions with pulsed photo-cathodes may reach 200 fs. The low DQE of the photo cathode(1-4% @ 1 Å) is going to prolong the exposure time which may not always possible on protein samples. Nevertheless, the intensity distribution from the single-pulse Cutinase study in figure 11 looks very promising with low resolution reflections collection up to 3×10^4 ph/pulse. We expect that the streak camera will enable us to measure small oscillations in the diffracted intensity which are gradually damped out beyond 10 ps. It is of great importance to extend diffraction measurements to this regime which so far has only been accessible to laser spectroscopy which does not probe the nuclear positions directly. The reconstruction of a 3D potential energy landscape of coherent motion will not only be very important but also make a direct link to theoretical kinetics calculations of proteins. The quasi zero-dimensional nature of a streak camera is clearly a limitation but one should note that on the shortest time scales, only a few atoms have time to move. A small data set of say 100 time-resolved reflections may define a sufficient set of boundary conditions on the local environment near the reaction center. Guided by theory, one may be able to reconstruct the potential energy landscape in real time.

The radiation damage from a fs pulse indicates that one has to reduce the power and work with small perturbations $\delta I/I$. In fast reversible reactions one may take advantage of the low heat load per pulse and step up the frequency. It seems feasible for example to illuminate a crystal with $1 W/mm^2$ comprising 898.8 laser pulses per second. The X-ray pulse train is extracted by the phase-locked chopper and has a well-defined delay with respect to the laser pulse train. The sample is rotated through a small angle as in conventional monochromatic crystallography. But the exposure time is going to be long using conventional monochromators. Specifically, the reduction in average intensity associated with a 898.8 Hz extraction of the 15 mA single bunch opposite the 150 mA super bunch is 4345. The rotation speed for small proteins at the ESRF is 1° per second but this is due to the too narrow monochromaticity of $\delta E/E = 1.33 \times 10^{-4}$ for Si(111) monochromators. Increasing the bandwidth using a Be(002) mosaic monochromator($\delta E/E = 2 \times 10^{-3}$) or perhaps a 50-200 period multilayer with $\delta E/E = 1 \times 10^{-2}$ and a reflectivity of 50% may be the solution. It is thus straightforward to reduce the exposure time by a factor ca 35. The broadening of the diffraction spots arising from the rotation of the lattice planes through this quasi-white beam is smaller than 6.6 mrad for $\delta E/E = 1 \times 10^{-2}$ at 12.4 keV and a resolution limit of 1.5 Å. This seems acceptable for small unit cell proteins.

The Free Electron Laser(XFEL) proposed at DESY in Hamburg will in addition to short pulses of 100 fs, produce near optics-free beamlines with a beam size of 10 μm and a flux of 2×10^{12} ph/pulse over a 0.1% bandwidth[31], see table 7 for a comparison between ID09 and a XFEL. One can thus work with micro crystals and trigger heme proteins at 450 nm where the absorption length is 10 μm and ε_{sat} only 150 μJ/mm^2. Narrow bandwidth Laue and monochromatic data collection should become possible down to the bunch limit of 100 fs. The gain in intensity of 1×10^5 is likely to improve the data substantially and probably allow imaging of coherent motion in an enzyme directly. The structural kinetics community is still very small in Europe and in order to strengthen the scientific case for a free electron laser, we strongly encourage interested people to join the efforts at the ESRF, APS and SPring8. Exploring the picosecond range will lay the foundation for things to come.

TABLE 7. Comparison between the U26 undulator on a low beta position(s_x = 0.044 mm, s_z = 0.007 mm, s'_x = 0.087 mrad and s'_z = 0.003 mrad) and the predicted performance of a free electron laser(XFEL) based on a 250 GeV linac.

-	ESRF	XFEL
fwhm bunch length(s)	150×10^{-12}	200×10^{-15}
photons per bunch in 0.1% bw	4.0×10^7 §	$0.2\text{-}50 \times 10^{12}$
photons per bunch	1.4×10^{10} §	?
peak flux (ph/sec/0.1% bw)	2.7×10^{17} §	$1\text{-}270 \times 10^{24}$
average flux (ph/sec/0.1% bw)	1.4×10^{14} §§	$0.2\text{-}150 \times 10^{16}$
peak brilliance of the source (ph/s/mm^2/mr^2/0.1%bw)	2.3×10^{27} §§§	$1\text{-}100 \times 10^{32}$
average brilliance of the source (ph/s/mm^2/mr^2/0.1%bw)	3.9×10^{18} §§§	$0.1\text{-}50 \times 10^{24}$

§ measured in focal spot from a 10 mA single-bunch of 150 ps
§§ measured in focal spot at 100 mA
§§§ calculated values for the source

ACKNOWLEDGEMENTS

The authors wish to express their gratitude for informative discussion with V. Srajer, T.-Y. Teng, W. Schildkamp, C. Pradervand, J. Hajdu, R. Neutze, W. Bürmeister, J. Helliwell, P. Bösecke, M. Hanfland, P. A. Anfinrud, Y. Gauduel, M.-C. Bellissent-Funel, P. Trommsdorff, K. Scheidt and C Cambillau. We would also like to thank ESRF directors J.-L. Laclare, C.-I. Brändén, C. Kunz and Y. Petroff for moral and financial support.

REFERENCES

1. Bourgeois D, Ursby T, Wulff M, Pradervand P, Legrand A, Schildkamp W, Labouré S, Srajer V, Teng T.-Y, Roth M and Moffat K, "Feasibility and Realization of Single Pulse Laue Diffraction on Macromolecular Crystals at ESRF", *J. Synch. Rad*, **3**, 65 (1996)
2. Srajer V, Teng T.-Y, Ursby T, Pradervand C, Ren Z, Adachi S, Schildkamp W, Bourgeois D, Wulff M and Moffat K, "Nanosecond Time-Resolved Macromolecular Crystallography", Accepted by *Science* September 1996
3. Youvan D C and Marrs B L, "Photosynthetic Reaction Centers", Scientific American, **256**, 42 (1987)
4. Lim M, Jackson T A, Anfinrud P A, "Binding of CO to Myoglobin from Heme Pocket Docking Site to Form Linear Fe-C-O", *Science*, **269**, 962 (1995)
5. Elleaume P, "Insertion devices for the ESRF", *Rev. Sci. Instr*, **63(1)**, 321 (1992)
6. Elleaume P, "Spectral Shimming and Minigap Undulators", *The ESRF News Letter*, **23**, 10 (1995)
7. Batterman B, Bilderback D, "X-ray Monochromator and Mirrors", *Handbook on Synchrotron Radiation*, Vol edited by G. Brown and D. Moncton, Elsevier Science Publisher B.V. (1991)
8. Susini J, "Design Parameters for Hard x-ray Mirrors" : The European Synchrotron Radiation Facility Case, *Optical Engineering*, **34(2)**, 361 (1995)
9. Rogers C S, Mills D M, Lee W, Knapp G, Freund A, Wulff M, Holmberg J, Rossat M Hanfland M and Yamaoka H(1995), "High Heat Load Performance of a Liquid Nitrogen Cooled Thin Silicon Crystal Monochromator on a High-Power Focused Wiggler Synchrotron Beam", *Rev. Sci. Instr.*, **66**, 6 (1995)
10. Yamaoka H, Freund A K, Holmberg J, Rossat M, Wulff M, Hanfland M, Lee K W, and D.M. Mills, "Experience from a Cryogenically Cooled Diamond Crystal Monochromator", *Nuclear Instruments and Methods in Physics Research A*, **364**, 581 (1995)
11. Farvacque L, "Beam Life Time", *Internal Report ESRF*, (1995)
12. Piwinsky A, "Beam Losses and Life Times", *Proc of the CERN Accelerator School*, CERN, 19, 1984
13. Evans L R and Gareyte J, "Beam-beam Effects", *Proc CERN Accelerator School*, CERN, 87 (1985)
14. Halbach K, "The Magnetic Assembly of Insertion Devices", *Nucl. Instrum. Methods*, 187 (1981) 109
15. Wulff M, "The Optimization of Mirror Focusing of X-ray Sources : a Test Case at the ESRF", *Proceedings on High Heat Flux Engineering*, Spie, **1739**, 576, (1992)
16. Susini J and Wulff M (1993), "Study of the Design Parameters Governing the Performance of Synchrotron Mirrors", *Proceedings on High Heat Flux Engineering II*, Spie **1997**, 278 (1993)

17. LeGrand A D, Schildkamp W and Blank B, Nuclear Instruments and Methods in Physics Research A, **275**, 442 (1989), and LeGrand A D, Pradervand C and Schildkamp W, in preparation
18. Bourgeois D, Longhi S, Wulff M and Cambillau C, "Accuracy of Structural Information Obtained at the ESRF from very Rapid Laue Data Collection on Macromolecules", Accepted for publication in J. Appl. Cryst., August 1996
19. Moy J-P, Nuclear Instruments and Methods, **A348**, 641 (1994)
20. Moffat K(1995), "Macromolecular crystallography on the millisecond to nanosecond time domain", *SPIE Proceedings on Time-Resolved Electron and X-ray Diffraction*, 13-14 July, San Diego, **2521**, 182 (1995)
21. Hajdu J and Andersson I, "Fast X-ray Crystallography and Time-Resolved Structures", Annual Review of Biophysics and Biomolecular Structure, **22**, 467 (1993)
22. Antonini E and Brunori M, *Hemoglobin and Myoglobin in their Reactions with Ligands*, North Holland Publishing Company, 1971
23. Alfano M(1993), "The Cr:Fosterite Laser", *Optics Letters*, **18**, 891 (1993)
24. Anfinrud P A, private communication
25. Du D, Liu X, Korn G, Squier J and Mourou G, Appl. Phys. Lett. **54(23)**, 3071 (1994)
26. Mourou G, Naylor G, Scheidt K and Wulff M, "Jitter-free Accumulating Streak Camera with Femto-Second Time Resolution", *The ESRF Newsletter*, **26**, 32 (1996)
27. Vos M H, Rappaport F, Lambry J-C, Breton J and Martin J-L "Visualization of coherent nuclear motion in a membrane protein by femtosecond spectroscopy", *Nature*, **363**, 320 (1993)
28. "Relaxation of the product state coherence generated through the photolysis of HgI_2 in solution", Pugliano N, Szarka N K, Hochstrasser R M, *J. Chem. Phys.*, **104** (1996) in press
29. Limborg C and Laclare J-L, ESRF, private communication
30. Cote C Y, Kieffer J C, Gallant P, Rebuffie J C, Goulmy C, Maksimchuk A, Mourou G, Kaplan D and Bouvier M, "Development of a subpicosecond Large-Dynamic Range X-ray Streak Camera",22nd Int. Congress on High Speed Photography and Photonics, *SPIE Proceedings,* (1996)
31. Rossbach J, "Coherent X-Ray Sources as a Part of a Linear Collider Design", Internal report, DESY, Hamburg, (1996)

Atomic Holography with Electrons and X-rays

P. M. Len[1], C. S. Fadley[1,2], G. Materlik[3]

[1]*Department of Physics, University of California, Davis, CA 95616 USA*

[2]*Materials Sciences Division, Lawrence Berkeley National Laboratory, Berkeley, CA 94720 USA*

[3]*Hamburger Synchrotronstrahlungslabor (HASYLAB) am Deutsches Elektronen-Synchrotron (DESY), 22603 Hamburg, Germany*

Gabor first proposed holography in 1948 as a means to experimentally record the amplitude and phase of scattered wavefronts, relative to a direct unscattered wave, and to use such a "hologram" to directly image atomic structure. But imaging at atomic resolution has not yet been possible in the way he proposed. Much more recently, Szöke in 1986 noted that photoexcited atoms can emit photoelectron or fluorescent x-ray wavefronts that are scattered by neighboring atoms, thus yielding the direct and scattered wavefronts as detected in the far field that can then be interpreted as holographic in nature. By now, several algorithms for directly reconstructing three-dimensional atomic images from electron holograms have been proposed (*e.g.* by Barton) and successfully tested against experiment and theory. Very recently, Tegze and Faigel, and Gog *et al.* have recorded experimental x-ray fluorescence holograms, and these are found to yield atomic images that are more free of the kinds of aberrations caused by the non-ideal emission or scattering of electrons. The basic principles of these holographic atomic imaging methods are reviewed, including illustrative applications of the reconstruction algorithms to both theoretical and experimental electron and x-ray holograms. We also discuss the prospects and limitations of these newly emerging atomic structural probes.

INTRODUCTION

Historical Origin of Atomic Holography

Dennis Gabor first outlined in 1948 a direct experimental method of recording diffraction phases as well as intensities in an effort to surpass the then current resolution and lens aberration limits of electron microscopy and thus achieve atomic-scale image resolution (1). In Gabor's original scheme, an electron wavefront (of wavenumber k_0 and wavelength λ_0) diverging from a point focus illuminates an object as well as a detector (or image plate) directly. The interference pattern at this detector involves the wavefronts scattered by the object, and explicitly records the phases of these wavefronts relative to the direct or reference wavefront (Fig. 1(a)). This interference "hologram" thus contains spatial information about the scattering object, which can be retrieved as an image in several ways. Gabor suggested that the developed image plate could simply be re-illuminated by a visible light reference wavefront (of wavenumber k *and wavelength* λ), as shown in Fig. 1(b). The wavefronts thus diffracted by the image plate would create a virtual image of the original object visible to the naked eye, and magnified by a factor of k_0/k. But the image reconstruction can also be performed numerically using a Fourier-transform-like integral, as first pointed out by Wolf (2). Holography is now of course widespread in science and technology, with lasers at usually optical wavelengths providing the reference waves. Note that, since the three-dimensional information of the **r**-space object field $u(\mathbf{r})$ (shown in Fig. 1 as an optical mask of the letter "F") is "encoded" holographically into a single-wavenumber two-dimensional **k**-space diffraction pattern $\chi(\hat{\mathbf{k}})$, both a real and twin image of the optical mask are retrieved. This is due to the loss of spatial information perpendicular to the plane of the image plate recording the diffraction pattern, and is by now overcome in optical holography by recording a volume of holographic intensities by means of a thick recording medium (3).

Until recently, Gabor's goal of imaging at atomic resolution had not been attained, due to the lack of a source of sufficiently coherent radiation at such short wavelengths. However, in 1986, Szöke observed that there is an atomic-scale analog of Gabor's holographic scheme: photoexcited atoms produce outgoing photoelectron or fluorescent-x-ray wavefronts, which then reach a far field detector either directly, or after scattering by neighboring atoms surrounding the emitter (5). With a sub-Ångström source size and wavelength, scattered wavefront amplitudes and phases from atoms surrounding the emitter can thus be referenced to the directly emitted wavefront, as shown for the case of fluorescence in Fig. 2(a). It was also pointed out a little later by Barton (4a,4b) and subsequently by Tong *et al.* (4c) that, by measuring diffraction patterns at different wavenumbers, three-dimensional spatial information could be completely encoded into a three-dimensional **k**-space volume of diffraction intensities $\chi(\mathbf{k})$, from which atomic images free of twin-image effects and other aberrations should be directly obtainable.

Two other approaches for obtaining structural information at the atomic scale should also be mentioned, as illustrated in Fig. 3. First, atomic order and electron density maps can be determined by so-called direct methods from the kinematical (single-scattering) diffraction technique, which exploits the translational symmetry (Bragg planes) of a crystal (6) (Fig. 3(a)). The second is the use of multiple, dynamical scattering from single crystals to solve the phase

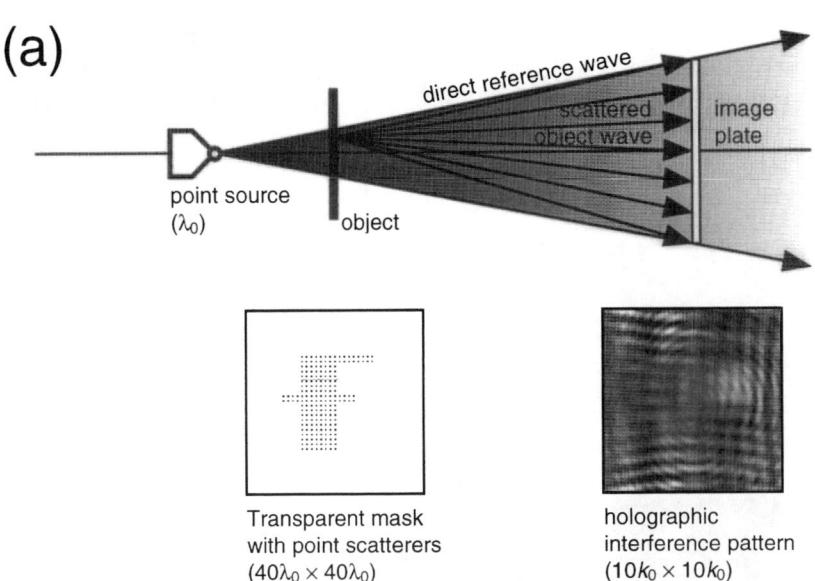

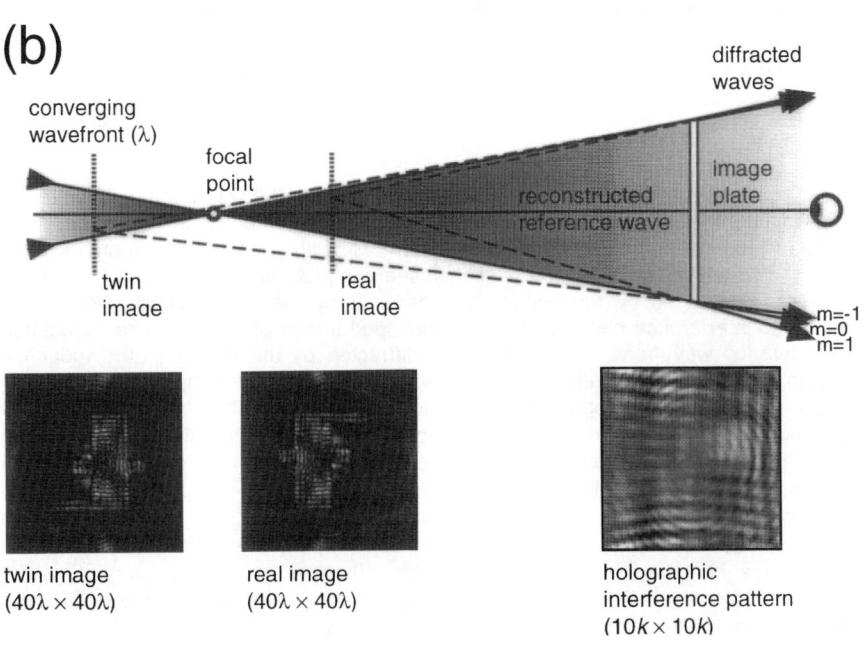

(c)

u(**r**): object field

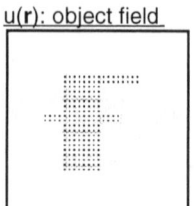

Transparent mask
with point scatterers
($40\lambda_0 \times 40\lambda_0$)

scattering
$K(\mathbf{k},\mathbf{r})$
⇨

χ(**k**): convolution

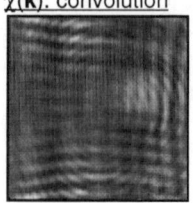

multiple-wavenumber
holographic
interference pattern
($10k_0 \times 10k_0 \times 0.3k_0$)

(d)

U(-**r**'): twin image

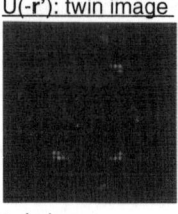

twin image
($40\lambda \times 40\lambda$)

U(**r**'): real image

real image
($40\lambda \times 40\lambda$)

reconstruct
$\kappa(\mathbf{k},\mathbf{r}')$
⇦

χ(**k**): convolution

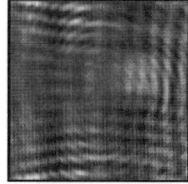

multiple-wavenumber
holographic
interference pattern
($10k_0 \times 10k_0 \times 0.3k_0$)

FIGURE 1. (a)-(b) An idealized numerical demonstration of the creation and inversion of single-wavenumber optical Gabor in-line holograms. (a) A point source of coherent radiation at the origin illuminates a transparent mask with point scatterers creating the letter "F" at **r**, as well as an image plate. This image plate is then exposed by a direct wavefront, as well as by the wavefronts scattered by the mask, which produces a holographic interference pattern. (b) The developed image plate is later re-illuminated by a reference wavefront. The wavefronts diffracted by the image plate produce a virtual (real) image of the mask at **r**, and a virtual conjugate twin image at the inverse position -**r**. (c)-(d) An analogous demonstration of the creation and inversion of optical multiple-wavenumber holograms. (c) A multiple-wavenumber normalized χ(**k**) hologram data set (of which one wavenumber is shown) is calculated from the object field u(**r**) by means of an **r**-space convolution, using a kernel K(**k**,**r**) that describes the emission and scattering physics involved (here, optical scattering in the far field regime). (d) The object field u(**r**) is recovered as an image intensity U(**r**') by a **k**-space deconvolution of χ(**k**), using a kernel κ(**k**,**r**') that is sufficiently orthogonal to K(**k**,**r**). Note that the conjugate twin image U(**r** = -**r**') has been suppressed, due to the volume of **k**-space enclosed in the multiple-wavenumber χ(**k**) considered here.

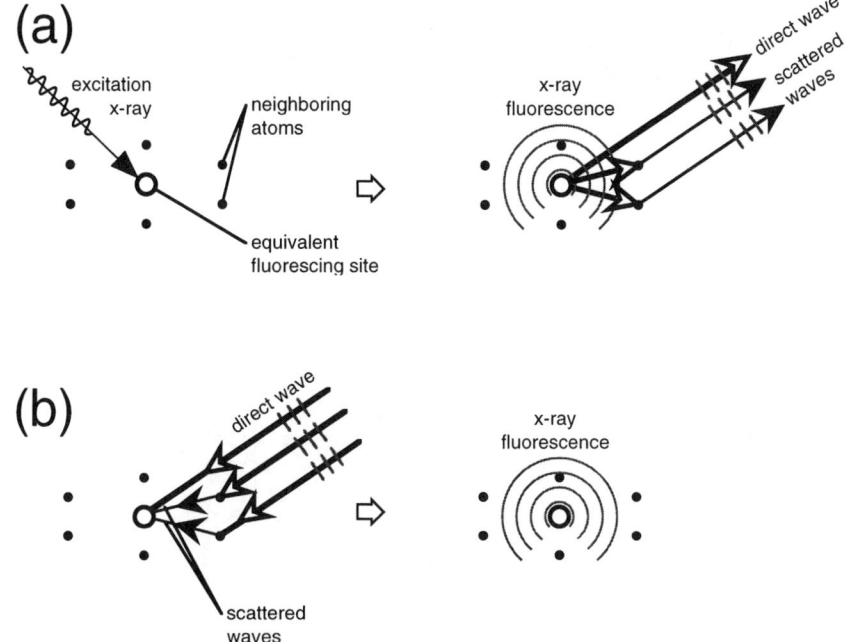

FIGURE 2. Atomic-scale analogs of Gabor holography. (a) The first scheme suggested by Szöke (5), in which an excitation x-ray first creates an inner-shell hole in one of many equivalent fluorescing atoms, and this atom then emits fluorescent x-ray (or electron) wavefronts that illuminate neighboring atoms, as well as a far field detector. This detector senses the interference between the direct wavefront and wavefronts scattered by the neighboring atoms. Moving the detector over a large solid-angle range builds up a holographic interference pattern. (b) The time-reversed case of (a) as suggested by Gog et al. (15), where a coherent far field excitation x-ray illuminates and photoexcites an emitter, and also illuminates and is scattered from atoms neighboring the emitter. The emitting atom senses the interference between the direct wavefront and wavefronts elastically scattered by the neighboring atoms. The net photoexcitation is then detected by a stationary, large solid-angle detector. Moving the far field source over a large solid-angle range builds up a holographic interference pattern. In both (a) and (b), atomic images can be reconstructed numerically.

problem of crystallography (7-9), either via Kossel lines (Fig. 3(b)) (10) or standing-wave methods (Fig. 3(c)) (11). The holographic approach is different from these two methods in that it uses the interference pattern which results from the direct unscattered wavefront emitted by a source atom, and the wavefronts which have been singly scattered from neighboring atoms. This does not require translational order (only rotational alignment) between the atomic neighborhoods to be imaged.

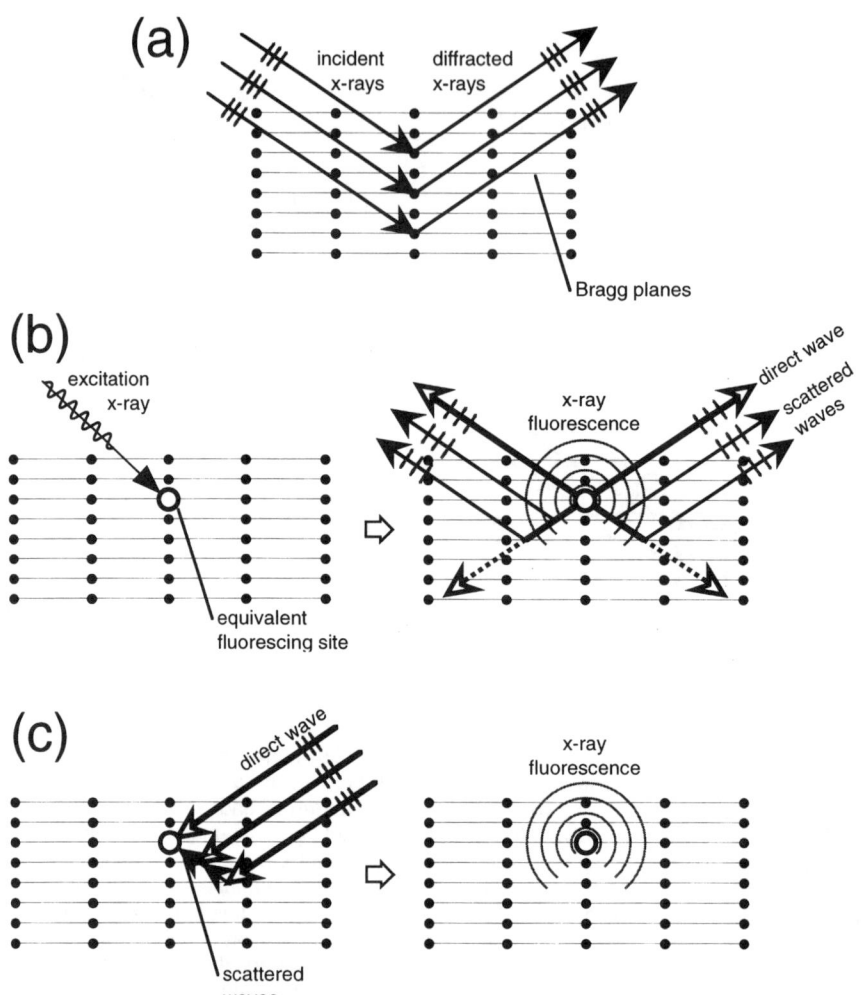

FIGURE 3. Diffraction probes of atomic structure related to atomic holography. (a) Conventional x-ray crystallography, where x-rays are diffracted by Bragg planes of atoms. Diffraction phases are determined by the simultaneous analysis of many Bragg intensities and other methods. (b) Kossel patterns (or Kikuchi bands, for the case of electrons). Fluorescent x-rays (or electrons) from a photoexcited emitter are diffracted by Bragg planes of atoms. Diffraction phases are thus here directly referenced to the unscattered portion of the fluorescence. (c) X-ray standing waves. This is the optical reciprocal of (b), where a coherent plane wave illuminates a fluorescing atom either directly, or after being scattered by Bragg planes of atoms. The interference between these wavefronts determines the amount of fluorescence by the emitter. Note that in all these above cases, the structure to be determined must have long-range atomic order, in contrast to the methods illustrated in Fig. 2.

Basic Principles of Atomic Holography

The process by which three-dimensional atomic image intensities are numerically reconstructed is to first measure the intensity $I(\mathbf{k})$ from a localized source over some range of directions $\hat{\mathbf{k}} = \mathbf{k}/k$ and perhaps also some range of wavenumbers k. Normalized holographic intensities $\chi(\mathbf{k})$ are then derived from either $[I(\mathbf{k}) - I_0(\mathbf{k})]/\sqrt{I_0(\mathbf{k})}$ or $[I(\mathbf{k}) - I_0(\mathbf{k})]/I_0(\mathbf{k})$, where $I(\mathbf{k})$ is the raw measured intensity, and $I_0(\mathbf{k})$ is the intensity that would be measured in the absence of atomic scattering; that is, $I_0(\mathbf{k})$ is the unperturbed intensity of the reference wave. The overall imaging process can be understood by first considering the hologram to be a convolution of the **r**-space object field $u(\mathbf{r})$:

$$\chi(\mathbf{k}) = \iiint_R d^3\mathbf{r} \cdot K(\mathbf{k},\mathbf{r})u(\mathbf{r}) + \iiint_R d^3\mathbf{r} \cdot K^*(\mathbf{k},\mathbf{r})u^*(\mathbf{r}), \qquad (1)$$

where the convolution kernel $K(\mathbf{k},\mathbf{r})$ somehow describes the physics of the emission and atomic scattering of the photoexcited wavefronts, and R denotes the volume in real space over which the object exists. This produces a three-dimensional $\chi(\mathbf{k})$ volume in **k**-space, so as to completely encode three-dimensional spatial information of the object field $u(\mathbf{r})$. The reconstruction algorithm is then most simply a **k**-space deconvolution of $\chi(\mathbf{k})$ to obtain a real-space $U(\mathbf{r}')$ image intensity:

$$U(\mathbf{r}') \equiv \iiint_K d^3\mathbf{k} \cdot \kappa^*(\mathbf{k},\mathbf{r}')\chi(\mathbf{k}), \qquad (2)$$

where the reconstruction kernel $\kappa(\mathbf{k},\mathbf{r}')$ has been chosen to be orthogonal to the scattering kernel $K(\mathbf{k},\mathbf{r})$, as integrated over a sufficiently large **k**-space volume, that is, so that:

$$\begin{aligned}\iiint_K d^3\mathbf{k} \cdot \kappa^*(\mathbf{k},\mathbf{r}')K(\mathbf{k},\mathbf{r}) &\propto \delta(\mathbf{r}-\mathbf{r}'), \\ \iiint_K d^3\mathbf{k} \cdot \kappa^*(\mathbf{k},\mathbf{r}')K^*(\mathbf{k},\mathbf{r}) &\approx 0 \end{aligned} \qquad (3)$$

If such a $\kappa(\mathbf{k},\mathbf{r}')$ can be found, then the object field $u(\mathbf{r})$ can thus be recovered as the image intensity $U(\mathbf{r}')$ from Eqs. (1)-(3):

$$\begin{aligned}U(\mathbf{r}') &= \iiint_R d^3\mathbf{r} \cdot \left[u(\mathbf{r})\iiint_K d^3\mathbf{k} \cdot \kappa^*(\mathbf{k},\mathbf{r}')K(\mathbf{k},\mathbf{r}) + u^*(\mathbf{r})\iiint_K d^3\mathbf{k} \cdot \kappa^*(\mathbf{k},\mathbf{r}')K^*(\mathbf{k},\mathbf{r}) \right] \\ &= \iiint_R d^3\mathbf{r}' u(\mathbf{r})\delta(\mathbf{r}-\mathbf{r}') \\ &= u(\mathbf{r}'). \end{aligned} \qquad (4)$$

So once the emission and scattering process that creates $\chi(\mathbf{k})$ can be sufficiently modeled by a $K(\mathbf{k},\mathbf{r})$ convolution kernel, then a deconvolution kernel $\kappa(\mathbf{k},\mathbf{r}')$ can in principle be formulated so as to directly reconstruct atomic images using Eq. (2).

Atomic Holography Reconstruction

The basic algorithms used in reconstructing atomic holographic images can be understood in the context of a single scattering (or kinematical) model of the scattering process. We consider e^{-ikr}/kr to represent the photoexcited electron or x-ray spherical wavefront that illuminates the (point-like) scattering atoms surrounding the emitter (with the emitted wave assumed to be isotropic for simplicity), $f(\Theta_r^k)$ to be the complex plane-wave atomic scattering factor ($\equiv |f(\Theta_r^k)|\exp[i\psi(\Theta_r^k)]$), where Θ_r^k is the scattering angle, and $\mathbf{k}\cdot\mathbf{r}$ is the phase of the scattered portion of this wavefront as it reaches the far field detector (Fig. 4). Thus the total geometrical path-length phase difference between the reference and scattered wavefronts is $(\mathbf{k}\cdot\mathbf{r} - kr)$. The convolution kernel for this scattering process can then be expressed as:

$$K(\mathbf{k},\mathbf{r}) = \frac{f(\Theta_r^k)}{kr} e^{i(\mathbf{k}\cdot\mathbf{r}-kr)}. \tag{5}$$

This choice for $K(\mathbf{k},\mathbf{r})$ does not include any allowance for anisotropy in magnitude or phase of the outgoing reference wave, which for the simple example of s-level photoemission, takes the form of an additional factor of $\varepsilon\cdot\mathbf{k}$, where ε is the polarization vector of the radiation (12). Thus, in photoemission, reference wave anisotropy is almost always present. However, for the case of Kα x-ray fluorescence to be considered below, the outgoing reference wave should be isotropic *and* randomly polarized, and thus be well described by Eq. (5).

Another advantage of x-rays lies in the nature of $f(\Theta_r^k)$. Figure 5 shows the magnitudes and phases of Ni atomic scattering factors for both x-rays and electrons with wavelength $\lambda = 0.79$Å (or wavenumber $k = 8.0$Å^{-1}). Note that the x-ray scattering factors (Fig. 5(a)) are much weaker (by ~1/2000) and more nearly constant in magnitude than those for electrons (Fig. 5(b)), and that the scattering phase shifts for x-rays are also much smaller (by ~1/100) and more nearly constant than those for electrons. Thus, for x-rays $|f(\Theta_r^k)| \approx$ constant $= f_0$, and $\psi(\Theta_r^k) \approx \psi_0 \approx 0$, such that the simplest possible optical scattering kernel results: $K_o(\mathbf{k},\mathbf{r}) \propto e^{i(\mathbf{k}\cdot\mathbf{r}-kr)}$. The reconstruction kernel that is most simply orthogonal to this optical scattering kernel is thus $\kappa_o(\mathbf{k},\mathbf{r}') \equiv e^{i(\mathbf{k}\cdot\mathbf{r}'-kr')}$, as first suggested by Barton and Terminello (4b). Thus for the scattering of fluorescent x-rays, the

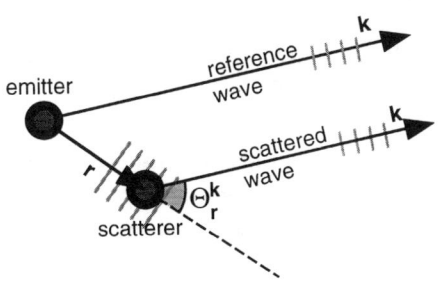

FIGURE 4. Scattering geometry between a photoemitter and a neighboring scattering atom. The photoemitter is placed at the origin, while the scatterer is located at the relative position **r**. The far field detector lies in the direction **k**. The portion of the direct wavefront that is scattered by the neighboring atom into the detector at **r** depends on the scattering angle Θ_r^k between **r** and **k** according to the complex phase factor $f(\Theta_r^k)$.

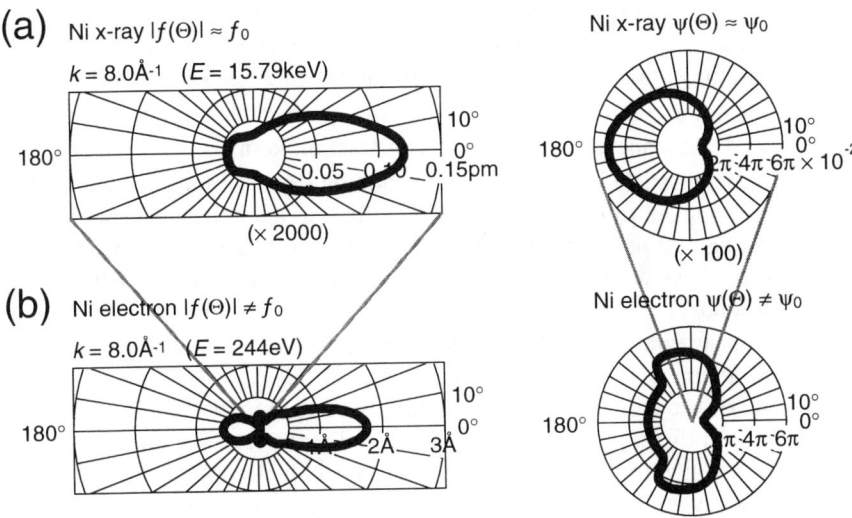

FIGURE 5. Ni scattering factor magnitudes ($|f(\Theta_r^k)|$) and phases ($\psi(\Theta_r^k)$), as a function of scattering angle Θ_r^k for (a) $k = 8.0$ Å^{-1} ($E = 15.79$keV) x-rays. (b) $k = 8.0$ Å^{-1} ($E = 244$eV) electrons. $\Theta_r^k = 0°$ is the forward scattering direction, $\Theta_r^k = 180°$ is the backscattering direction.

reconstruction algorithm of Eq. (2) becomes to a good approximation:

$$U(\mathbf{r'}) \equiv \iiint_K d^3\mathbf{k} \cdot e^{-i(\mathbf{k}\cdot\mathbf{r'}-kr')}\chi(\mathbf{k}). \quad (6)$$

This simple optical reconstruction algorithm has been used recently to obtain direct atomic images from experimental single-wavenumber (13,14) and multiple-wavenumber (15,16) x-ray holographic data sets, as discussed further below.

For example, Fig. 1(c) schematically shows the optical holographic $\chi(\mathbf{k})$ intensities that were numerically calculated from the transparent "F" mask using Eqs. (1) and (5), over a range of different directions ($\hat{\mathbf{k}}$) and wavenumbers (k). Figure 1(d) shows the numerically reconstructed real and twin images obtained from the volume $\chi(\mathbf{k})$ of Fig. 1(c). Due to the three-dimensional spatial information that was encoded in the k-space volume encompassed by $\chi(\mathbf{k})$, the reconstruction algorithm of Eq. (6) suppresses the spurious twin image, while increasing the fidelity of the desired real image (cf. Fig. 1(b)).

The optical reconstruction algorithm of Eq. (6) was in fact first used to reconstruct data from electron holographic data sets, e.g. from photoelectron diffraction (17-21). However, because of the generally anisotropic nature of the photoemitted source wave, and the strong, non-optical and often multiple nature of electron scattering, the single-scattering optical convolution kernel $K_o(\mathbf{k},\mathbf{r}') \propto e^{i(\mathbf{k}\cdot\mathbf{r}'-kr')}$ does not accurately describe the process by which electron holograms are produced, and consequently the optical reconstruction kernel $\kappa_o(\mathbf{k},\mathbf{r}') = e^{i(\mathbf{k}\cdot\mathbf{r}'-kr')}$ will not in general satisfy the orthogonality condition (Eq. (3)) for electrons. Thus Eq. (6), when applied to electron holograms, often results in images which suffer from aberrations and position shifts (22-24). Nonetheless, useful atomic structure information has been derived from electron holography, with various modifications to the basic optical reconstruction kernel $\kappa_o(\mathbf{k},\mathbf{r}')$, and to the definition of the reconstruction integral (Eq. (2)) itself being proposed (25-29), and comparative reviews of different methods appearing elsewhere (30,31).

In summary, the atomic scattering of x-rays is much more nearly ideal than that of electrons, and this suggests that a simple optical reconstruction kernel as in Eq. (6) can be straightforwardly used to directly obtain atomic images from holographic x-ray intensities. However, more sophisticated reconstruction kernels and deconvolution integrals will probably be necessary to account for the non-ideal nature of the propagation and scattering of electrons, in order to successfully obtain the most accurate atomic images from holographic electron intensities, as discussed elsewhere (25-31).

ATOMIC ELECTRON HOLOGRAPHY

In this section, we discuss the results of applying the imaging algorithm of Eq. (6) to experimental and theoretical photoelectron diffraction results for W $4f$ emission from the surface atoms of clean W(110), with the experimental data being obtained by Denlinger, Rotenberg, and co-workers at Beamline 7.0 of the Advanced Light Source at the Lawrence Berkeley National Laboratory (31). The $4f$ photoelectron peak (which contains d and g components due to the dipole selection rule) can be resolved into bulk and surface core-level-shifted components, of which atomic images reconstructed from only surface photoemission will be considered here. Photoelectron spectra were measured for kinetic energies of $E = 41\text{eV}$ to 197eV (wavenumbers $k = 3.3\text{Å}^{-1}$ to 7.2Å^{-1}), and collected over a polar takeoff angle range of $14° \leq \theta \leq 90° \equiv$ normal emission. These data points were measured at wavenumber intervals corresponding to $\delta k = 0.1\text{Å}^{-1}$, and angular intervals of $(\delta\theta, \delta\phi) = (3°, 3°\cos\theta)$ corresponding roughly to equal solid angle elements, making a total of 12,280 unique measurements in a symmetry-reduced 1/4th of the total solid-angle above the sample. For each different wavenumber and direction, the W $4f$ peak was resolved into bulk and surface emission components by integrating the areas under the lower and higher flanks of the bulk and surface W $4f$ peaks, respectively, as shown in Fig. 6(a). Figures 6(b)-(c) show the bulk and surface $I(\mathbf{k})$ data sets in k-space, respectively, as viewed down along $[\bar{1}\bar{1}0]$. Data points in the lower right quadrant have been cut away to reveal the intensities $I(\mathbf{k})$ for the minimum $k = 3.3\text{Å}^{-1}$; the other quadrants show the intensities $I(\mathbf{k})$ for the maximum $k = 7.2\text{Å}^{-1}$. The dark bands at the perimeter indicate the locations in k-space on these iso-wavenumber surfaces where data was not collected. Due to the strong atomic scattering of electrons, the anisotropy of the raw $I(\mathbf{k})$ data, which we measure as $\Delta I / I_0 \equiv (I_{max} - I_{min})/I_0$, is found to be $\approx 30\%$, and is easily discernible with this gray scale.

In order to determine the normalized $\chi(\mathbf{k})$ from the raw $I(\mathbf{k})$ intensities of Figs. 6(b)-(c), $I_0(\mathbf{k})$ was determined by fitting a low order polynomial in wavenumber k and polar angle θ to $I(\mathbf{k})$:

$$I_0(\mathbf{k}) = a_{00} + \sum_{m=1}^{3}\sum_{n=1}^{3} a_{mn} k^m \cos[(2n-1)\theta], \qquad (11)$$

where the coefficients a_{mn} are determined by a least-squares fit to $I(\mathbf{k})$. This is in contrast to previous more approximate methods for determining $I_0(\mathbf{k})$ where simple linear, low-order polynomial, or spline fits were separately made for each set of different wavenumbers along a given direction: $I_k(\hat{\mathbf{k}})$, or each set of different directions at a given wavenumber: $I_{\hat{k}}(k)$. Such separate normalizations within each scanned-wavenumber or scanned-angle set of data points in $I(\mathbf{k})$ arose from the historical development of electron holography, in which data tended to be collected with k-space resolution that was either fine-in-direction/coarse-in-wavenumber or coarse-in-direction/fine-in-wavenumber (30,31). There has in fact been a recent proposal to consider these k-space sampling choices as distinct atomic structure probes (17(e)), but these choices simply represent extremes of a continuous range of k-space sampling, of which the optimal choice has been shown to be in the intermediate range of roughly equally resolved direction and wavenumber data steps (30). Thus, this distinction (17(e)) is artificial, and not

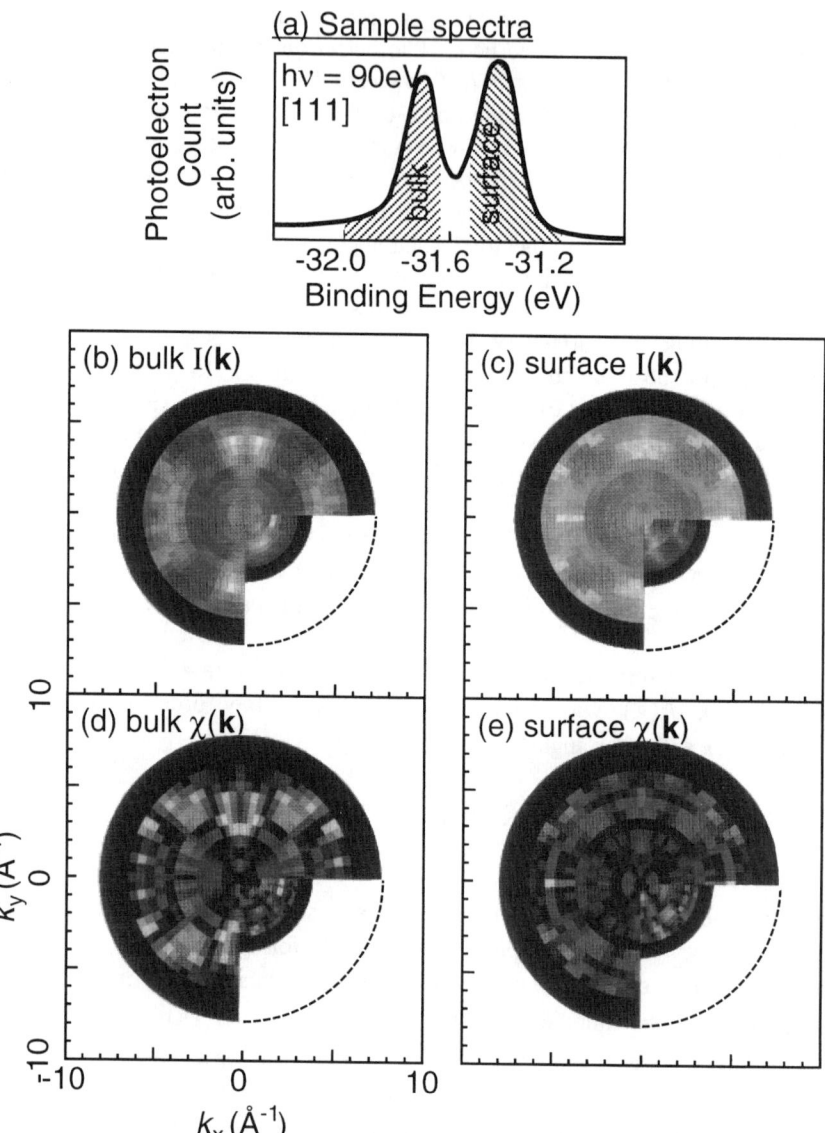

FIGURE 6. (a) Example W $4f_{7/2}$ photoelectron spectrum from clean W(110), with the bulk and surface emitter contributions used to generate the $I(\mathbf{k})$ intensity data points of (b)-(c) shaded in. (b)-(c) Schematic **k**-space representations of the raw $I(\mathbf{k})$ intensity data sets for bulk and surface W $4f_{7/2}$ emission, respectively. (d)-(e). Normalized bulk and surface emission $\chi(\mathbf{k})$ data sets, respectively, corrected for the unscattered intensity I_0 (as derived from Eq. (11)) and an inner potential of $V_0 = 14V$.

consistent with the optimal use of the holographic methodology. As a consequence, the normalization of $I(\mathbf{k})$ intensities should ideally be made via the determination of a general wavenumber and direction dependent $I_0(\mathbf{k})$ background, as done here, rather than determined separately for each wavenumber or direction in the $I(\mathbf{k})$ data set.

Figures 6(d)-(e) show the normalized bulk and surface $\chi(\mathbf{k})$ functions obtained from the raw $I(\mathbf{k})$ intensities of Figs. 7(b)-(c), using the wavenumber and angle fit $I_0(\mathbf{k})$ of Eq. (11), and after correcting for an inner potential of $V_0 = 14V$ (32) to yield electron wavenumbers and directions beneath the surface of the sample. These data points were then remapped to a $\delta k = 0.1$Å^{-1}, and $(\delta\theta, \delta\phi) = (5°, 5°)$ grid over the range $k = 3.85$Å^{-1} to 7.45Å^{-1} ($E = 56$eV to 211eV), and $40° \leq \theta \leq 90°$ range, for a final total of 6,697 unique intensities in the symmetry-reduced 1/4th of the solid angle above the sample.

For comparison, single-scattering and multiple-scattering models were used to calculate the surface emission $I(\mathbf{k})$ from a theoretical W(110) cluster (33). These theoretical photoemission intensities were then also normalized using Eq. (11).

Figure 7 shows the reconstructed images in the vertical $(\bar{1}1\bar{2})$ plane obtained from applying the optical reconstruction kernel of Eq. (6) to: (a) the experimental surface emission $\chi(\mathbf{k})$ of Fig. 6(e); (b) the theoretical single-scattering $\chi(\mathbf{k})$; and (c) the theoretical multiple-scattering $\chi(\mathbf{k})$. The expected atomic image resolution for this wavenumber and angular range of $\chi(\mathbf{k})$ in the horizontal $[\bar{1}11]$ direction is given by $\delta x \approx \pi / \Delta k_x \equiv \pi / (2k_{max} \sin(\theta_{max} - \theta_{min})) \approx 0.3$Å, and in the vertical $[110]$ direction is given by $\delta z \approx \pi / \Delta k_z \equiv \pi / (k_{max} - k_{min} \cos(\theta_{max} - \theta_{min})) \approx 0.6$Å (34), and these numbers are comparable to the actual atomic image resolutions in Fig. 7. As noted above, Eq. (6) makes no special effort to suppress aberrations due to the non-optical nature of the electron scattering process. In all of these images, the $\bar{1}\bar{1}0$ backscattering atom and the $\frac{1}{2}\frac{\bar{1}}{2}\frac{\bar{1}}{2}$ and $\frac{\bar{1}}{2}\frac{1}{2}\frac{1}{2}$ side scattering atoms are well-resolved, with experiment and the more accurate multiple-scattering theory showing the sharpest features for the backscattering atoms. In the experimental image of Fig. 7(a), the $\frac{1}{2}\frac{\bar{1}}{2}\frac{\bar{1}}{2}$ and $\frac{\bar{1}}{2}\frac{1}{2}\frac{1}{2}$ atoms are shifted in toward the emitter (by ≈ 0.7Å), and downward from the $z = 0$Å surface (by ≈ 0.2Å), this is probably primarily due to anisotropies in the photoemitted source wave and the atomic scattering factor for such side-scattering directions. As expected, the backscattering $\bar{1}\bar{1}0$ atom is better resolved due to the more ideal nature of electron backscattering (cf. Fig. 5), with no significant position shift. The experimental backscattering image is also less intense ($\approx 50\%$) than the side scattering atomic images; and image intensities above and below $z = -3.5$Å have been scaled accordingly. This difference in relative image intensity is qualitatively expected due to the longer inelastic attenuation path of the wavefront that illuminates, and is subsequently scattered by, the backscattering atom, as compared to the wavefront paths that involve the side scattering atoms. Despite these

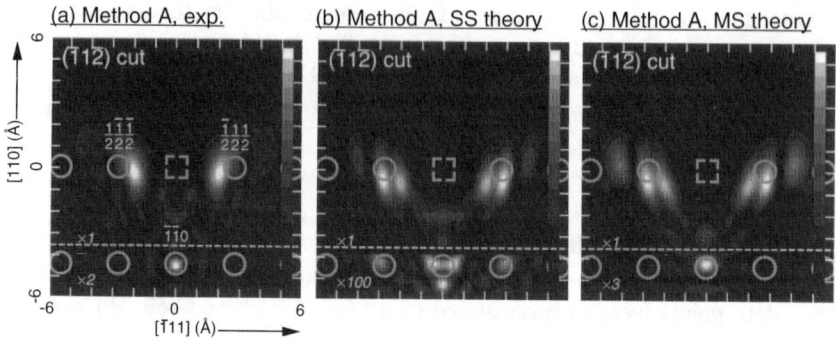

FIGURE 7. (a) W(110) atomic images obtained in the vertical $(\bar{1}1\bar{2})$ plane from experimental and theoretical W $4f_{7/2}$ surface emission $\chi(\mathbf{k})$ data sets, via (Eq. (6)). The surface emitter site is indicated by the dashed square, and the positions of the scatterers (assuming no surface relaxation) are indicated by circles. The nearest and next-nearest scattering positions have been labeled. Axes are marked off in 1Å units. Image intensities for $z \le -3.5\text{Å}$ have been rescaled, with the scale factors indicated on the figures. (a) Images reconstructed from the experimental $\chi(\mathbf{k})$ data set of Fig. 6(e). (b) Images reconstructed from a theoretical single scattering $\chi(\mathbf{k})$ data set. (c). Images reconstructed from a theoretical multiple scattering $\chi(\mathbf{k})$ data set.

position shifts and aberrations, this atomic image overall gives good *ab initio* estimates of the positions of the atoms surrounding the surface W(110) emitter, which could then be refined *e.g.*, using *R*-factor comparisons of experiment with model diffraction calculations for various structures.

The single and multiple scattering images of Figs. 7(b)-(c) are similar to experiment in that the $\frac{1}{2}\frac{\bar{1}}{2}\frac{\bar{1}}{2}$ and $\frac{\bar{1}}{2}\frac{1}{2}\frac{1}{2}$ side scatterers exhibit side lobes which are shifted in towards the emitter, and downward from the surface. However, the side scattering atomic images of Figs. 7(b)-(c) differ from those of Fig. 7(a) in that the theoretical image peaks are split. This splitting may be due to a number of reasons, among them the differences between the theoretical and actual wavenumber-dependent photoexcitation cross-sections, photoemitted source wave angular distributions, and atomic scattering factors. Still, these single- and multiple-scattering models produce other image features that rather closely match the experimental image of Fig. 7(a), even including the faint aberrations seen at $(x,z) \approx (\pm 4\text{Å}, 0\text{Å})$. The most marked difference between the experimental image of Fig. 7(a) and the single-scattering image of Fig. 7(b) is the triply-split backscattering $\bar{1}\bar{1}0$ atom in the latter, which is also very much weaker in intensity ($\approx 1\%$) relative to the $\frac{1}{2}\frac{\bar{1}}{2}\frac{\bar{1}}{2}$ and $\frac{\bar{1}}{2}\frac{1}{2}\frac{1}{2}$ image peaks. This is mainly due to the oversimplification of the single-scattering model, as seen by comparing Figs. 7(b) and (c). Note that in the multiple-scattering image of Fig. 7(c), the backscattering $\bar{1}\bar{1}0$ peak intensity relative to the side scattering $\frac{1}{2}\frac{\bar{1}}{2}\frac{\bar{1}}{2}$ and $\frac{\bar{1}}{2}\frac{1}{2}\frac{1}{2}$ image peaks ($\approx$ 33%) is more nearly that of Fig. 7(a) ($\approx$ 50%). This dramatic difference between single and multiple scattering can arise because each of the atoms in the multiple-scattering model becomes a secondary emitter, which can then illuminate the atoms surrounding them, especially the atom located at the $\bar{1}\bar{1}0$ relative position. In this way more scattering events contribute to the backscattering signal in the resulting holographic $\chi(\mathbf{k})$ intensities, and as such the reconstructed $\bar{1}\bar{1}0$ atomic intensity is much stronger for the image reconstructed from the multiple-scattering model than that from the single-scattering model. Thus, the closer match between Fig. 7(c) and the experimental image of Fig. 7(a) graphically illustrates that multiple-scattering more accurately describes the nature of the creation of the experimental holographic photoelectron intensities $I(\mathbf{k})$.

Atomic electron holography has been extensively tested on both bulk and surface structures, with some notable successes to date being the determination of structures of adsorbate overlayers (17c,19a,20a,27b-c,) and reconstructed surface structures (17e,19b). This technique is most useful in that initial atomic position estimates can be determined, which can then be refined using a more standard comparison of experiment and theory. Further improvements of image quality in atomic electron holography will lie primarily in the refinement of reconstruction kernels and algorithms that more accurately account for the non-ideal atomic scattering and propagation of electron wavefronts, as well as the wavenumber dependences and anisotropies in the source wave. Other holographic experiments that await implementation in the near future are the monitoring of temperature and coverage dependent structural phase changes; and spin-polarized photoelectron holography (SPPH) (35), where spin-specific photoemission (or detection) could be exploited to yield images of local atomic spin order.

ATOMIC X-RAY HOLOGRAPHY

In this section we review two experimental techniques for acquiring holographic x-ray data, and show the results of imaging experimental and theoretical x-ray holographic data sets involving both single and multiple wavenumbers.

The first atomic x-ray holographic images were recently obtained using what can be termed *x-ray fluorescence holography (XFH)*, as shown in Fig. 2(a). In this work, Tegze and Faigel (13) measured the hologram by monitoring the single-wavenumber Sr Kα emission ($k = 7.145 \text{Å}^{-1}$, $E = 14.10 \text{keV}$) from a single crystal of $SrTiO_3$. 2,402 intensities were measured over a full cone of 60° half angle above the surface. The final hologram was found to have anisotropies in intensity of $\Delta I / I_0 \approx 0.3\%$. These much smaller effects mean that more demanding detector counting statistics are required in x-ray holographic measurements than with comparable atomic electron holography measurements. The reconstruction of this hologram via the optical kernel algorithm of Eq. (6) yields images of the Sr atoms only, as the much weaker scattering strength of the Ti and O atoms renders their images invisible compared to those of the Sr atoms. Figure 8(a) shows the experimental image reconstructed in the (010) plane (36), and it is compared in Fig. 8(b) to an image reconstructed from a theoretical $\chi(\mathbf{k})$ for Sr Kα emission from a simple-cubic Sr cluster of 27 atoms (14). The expected atomic image resolutions at this hologram wavenumber and angular range are $\delta x \approx 0.3$Å in the horizontal [100] direction, and $\delta z \approx 0.9$Å in the vertical [001] direction (34), and are roughly comparable to the atomic images of Figs. 8(a)-(b).

Reconstructing three-dimensional atomic images from a single-wavenumber hologram yields twin images. In any structure with inversion symmetry, these twins can overlap with real atomic images so as to confuse structural interpretation (37,38). In addition, the real and twin atomic images for a particular wavenumber and system can overlap completely out of phase, leading to an artificial suppression of atomic image intensities (37,38). It is thus advantageous to reconstruct direct atomic images from multiple-wavenumber $\chi(\mathbf{k})$ data sets so as to avoid such real-twin image overlaps (4a,4b,38). However, such XFH holograms cannot be measured at arbitrary wavenumbers, with the latter being limited by the intensity and number of fluorescence lines of the photoemitting species (38,39).

Another method for obtaining x-ray holographic information at conveniently chosen multiple wavenumbers has also very recently been demonstrated for the first time by Gog *et al.* (15,16), and its basic principle is illustrated in Fig. 2(b). This method has been termed *multiple energy x-ray holography* (MEXH). MEXH is the time-reversed version of the conventional geometry of XFH (Fig. 2(a)), in that the wave motions are reversed, and the emitter and detector positions are interchanged (Fig. 2(b)) (15,16,39,40). The exciting external x-ray beam now produces the reference and object waves, and the fluorescent atom acts only to detect the interference between the direct and scattered wavefronts in the near field. The emitted x-rays are now collected by a distant detector with a large acceptance solid angle, in principle yielding much higher effective counting rates. The far field source wave can be set to any wavenumber (energy) above the fluorescence edge of the emitting species. thus permitting holograms at multiple wavenumbers and yielding in principle atomic images with no real-twin image overlaps (15,16,39). Specifically, multiple-wavenumber x-ray holograms have been measured to date for hematite ($\alpha - Fe_2O_3(001)$) (15,40), and for Ge(001) (16).

We show the results of applying the optical kernel algorithm of Eq. (6) to experimental and theoretical MEXH data for $\alpha - Fe_2O_3(001)$ as measured by Gog and co-workers on Beamline X-14A of the National Synchrotron Light

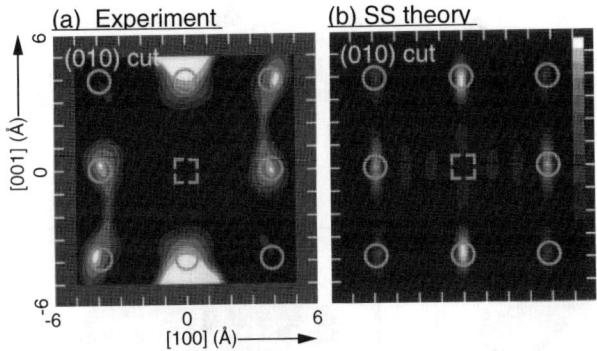

FIGURE 8. X-ray fluorescence holography atomic images of SrTiO$_3$ in the vertical (010) plane, obtained from (a) experimental (13) and (b) theoretical (14) Sr Kα χ(k) data sets, via Eq. (6). The Sr emitter site is indicated by the dashed square, and nearest-neighbor and next-nearest-neighbor Sr scatterers are indicated by circles. Axes are marked off in 1Å units.

Source at Brookhaven National Laboratory (15,40). Fe Kα fluorescence was excited by horizontally polarized radiation in the range k = 4.561Å^{-1} to 5.220Å^{-1} (E = 9.00keV to 10.30keV) that was incident on the α – Fe$_2$O$_3$(001) sample surface over a polar angle range of $60° \leq \theta \leq 90°$ = surface normal. These data points were measured at three wavenumbers with intervals of δk = 0.329Å^{-1} (δE = 650eV), and at angular intervals of $(\delta\theta, \delta\phi)$ = (5°,5°), making a total of 435 unique measurements in a symmetry-reduced 1/3rd of the total solid-angle above the sample. Figure 9(a) illustrates the orientation of the sample with respect to the horizontal ($\hat{e}_1$) polarization vector, with the vertical ($\hat{e}_2$) polarization vector shown also to permit discussing other possible experimental geometries. Figure 9(b) shows the raw measured $I(\mathbf{k})$ data set in k-space, as viewed down along [00$\bar{1}$], in the same format as Figs. 6(b)-(d). Data points in the fourth quadrant have been cut away to reveal the k = 4.561Å^{-1} $I(\mathbf{k})$ intensities, while the other quadrants show the k = 5.220Å^{-1} $I(\mathbf{k})$ intensities. Note that the much weaker atomic scattering of x-rays renders the anisotropy of the raw $I(\mathbf{k})$ data ($\Delta I / I_0 \approx 0.5\%$) barely discernible with this linear gray scale.

Due to the limited wavenumber range of this $I(\mathbf{k})$ data set, a separate $I_0(\mathbf{k})$ was determined for each of the three different wavenumber holograms via a low-pass filter (34), thereby including in $I_0(\mathbf{k})$ the reference wave, as well as corrections for the effects of x-ray absorption during both excitation and emission. Figure 9(c) shows the normalized $\chi(\mathbf{k})$ obtained by this method from the raw $I(\mathbf{k})$ intensities of Fig. 9(b).

For comparison to the experimental results, a single-scattering model (38,41)

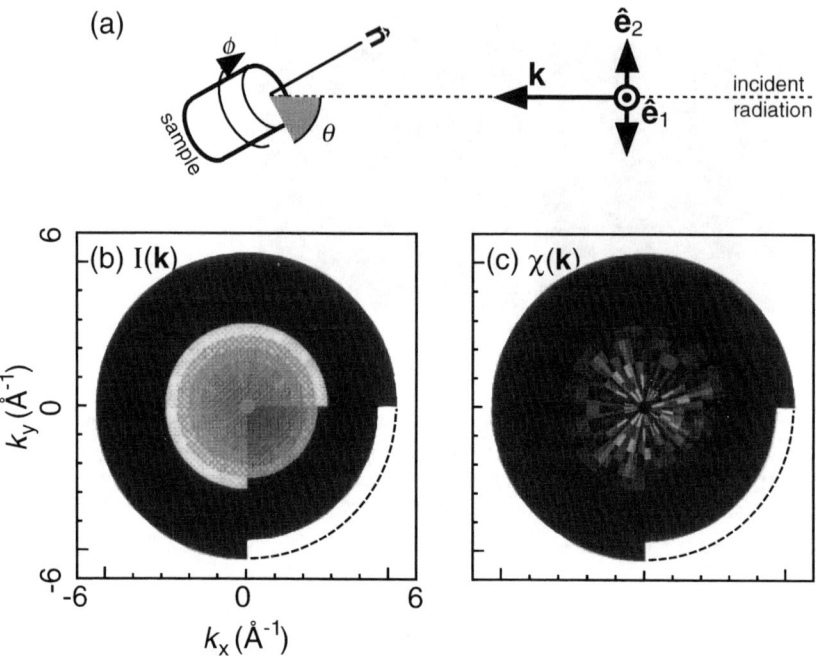

FIGURE 9. (a) Orientation of the sample (where $\hat{n}$ is the surface normal) with respect to the horizontal ($\hat{e}_1$) and vertical ($\hat{e}_2$) polarization vectors of the incident radiation **k**. (b) Schematic k-space representation of the raw measured $I(\mathbf{k})$ intensity data set for Fe Kα fluorescence from α-Fe$_2$O$_3$(001) excited by horizontally polarized radiation. (c) The normalized $\chi(\mathbf{k})$ data set. The format is the same as Figs. 6(b)-(d).

was used to calculate a theoretical $\chi(\mathbf{k})$ from an ideal $\alpha-\text{Fe}_2\text{O}_3(001)$ cluster containing 384 Fe atoms with two inequivalent Fe emitter sites as appropriate to the hematite lattice. The O atoms were not included due to their much smaller scattering power (15). The incident radiation in this model calculation is polarized horizontally with respect to the $\hat{\theta}$ and $\hat{\phi}$ rotation axes of the cluster (*cf.* Fig. 9(a)), as was the case in the measurement of the experimental $I(\mathbf{k})$ data set discussed above. Because the incident radiation is polarized, the x-ray scattering factor in Fig. 5(a) must be further multiplied by the Thomson scattering factor, which has the form $\sin^2\Theta_\varepsilon^{k'}$, where $\Theta_\varepsilon^{k'}$ is the angle between the polarization vector of the incident radiation ε, and the direction $\mathbf{k'}$ of the scattered radiation. Thus, there will be nodes in the incoming scattered object waves along the polarization direction, and emitter atoms near this direction will not be strongly influenced by x-ray scattering. For the present case, the use of horizontal polarization is therefore a

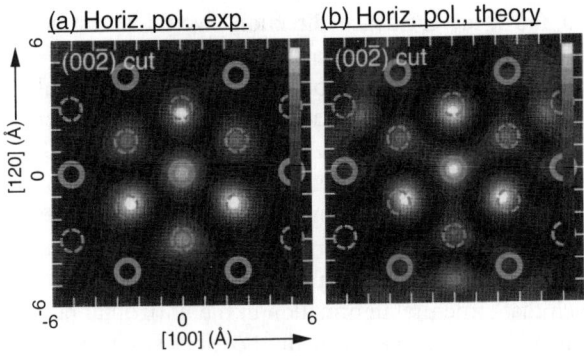

FIGURE 10. Multiple energy x-ray holography (MEXH) atomic images of α-Fe$_2$O$_3$(001) in the horizontal (00$\bar{2}$) plane situated 6.89Å below each of the two types of Fe emitters, obtained from (a) experimental and (b) theoretical Fe Kα $\chi(\mathbf{k})$ data sets, via Eq. (6). Fe scatterers in the bi-layer just above or below this plane are indicated by dashed circles, and Fe scatterers in relative positions common to both inequivalent Fe emitters are indicated by bold circles. Axes are marked off in 1Å units.

disadvantage in the imaging of horizontal planar structures such as those in $\alpha - \text{Fe}_2\text{O}_3(001)$, which is comprised of closely stacked horizontal Fe bi-layers with (001) orientation. The effect of such horizontally polarized incident radiation (via the Thomson cross section) is thus to strongly suppress atomic images in the basal (001) plane of the emitter, but to much less suppress images in horizontal planes farther above and below the emitter plane (40).

Figures 10(a) and (b) show the reconstructed atomic images in the (00$\bar{2}$) plane situated at $z = -6.89$Å below the emitter and obtained by applying Eq. (6) to the experimental and single-scattering theoretical $\chi(\mathbf{k})$ data sets, respectively. The expected image resolutions in the horizontal ([100] and [1$\bar{2}$0]) directions are $\delta x = \delta y \approx 0.6$Å (34). The experimental and theoretical images are very similar in that three of the Fe atoms from the neighboring upper bi-layer intrude into the (00$\bar{2}$) image plane. This intrusion is due to the limited wavenumber and angular range of the $\chi(\mathbf{k})$ data points in k-space (as compared to the larger wavenumber and angular range for the electron $\chi(\mathbf{k})$'s in the previous section), which results in atomic images much less resolved in the vertical [001] direction: $\delta z \approx 2.5$Å (34). Still, since these images are reconstructed from a multiple-wavenumber $\chi(\mathbf{k})$ data set, they should be freer of real-twin image overlaps (4,15,16,38-40).

As a future prospect, using unpolarized incident radiation in MEXH, or perhaps rotating the entire sample-detector complex by 90° so as to measure holograms with both horizontal polarization and vertical polarization (with the polarization vector in the plane formed by the azimuthal rotation axis (the normal of the sample surface) and the x-ray incidence direction, would allow atomic images to be reconstructed for atoms in all horizontal and vertical planes.

In fact, however, there also exist some classes of structures where it would be sufficient to utilize horizontally polarized incident radiation, and for which vertical structural information is more important than horizontal planar structure. These include some surface structure problems and buried epitaxial atomic layers. We specifically illustrate what might be learned for a buried atomic layer by considering theoretically a single Ge "δ-layer" buried in Si(001) (40). The Ge atoms in the δ-layer are assumed to lie in horizontal epitaxial sites with respect to the surrounding Si(001), such that structural information in the horizontal plane of a Ge emitter is relatively unimportant compared to the possibly strained vertical distances between the Ge δ-layer atoms and the Si neighbors above and below them (42). Thus using horizontally polarized incident radiation to record a MEXH Ge Kα $\chi(\mathbf{k})$ data set for this system may prove to be sufficient, and perhaps even advantageous.

As an example, Figs. 11(a)-(c) show the Thomson scattering factors for unpolarized, horizontally polarized, and vertically polarized incident radiation, respectively. Figs. 11(d)-(f) show the reconstructed atomic images in the vertical $(1\bar{1}0)$ plane obtained from applying Eq. (6) to a theoretical single-scattering $\chi(\mathbf{k})$ data set calculated for these polarization modes (unpolarized, horizontally polarized, and vertically polarized) for an ideal Ge δ-layer buried in a Si(001) cluster with no vertical strain. These MEXH $\chi(\mathbf{k})$ intensities were calculated at 7 wavenumbers (energies) for radiation of $k = 6.081$Å^{-1} to 9.122Å^{-1} ($E = 12.00$keV to 18.00keV) that was incident over a polar takeoff angle range of $10° \leq \theta \leq 90°$, and with wavenumber (energy) steps of $\delta k = 0.507$Å^{-1} ($\delta E = 1.00$keV) and angle steps of $(\delta\theta, \delta\phi) = (5°, 5°)$, yielding a total of 1,897 unique data points in the symmetry-reduced 1/4th of the total solid-angle above the cluster. The higher wavenumber and larger wavenumber and angular ranges of these MEXH $\chi(\mathbf{k})$ data sets ensure better resolved atomic images ($\delta x = \delta y \approx 0.2$Å; $\delta z \approx 0.4$Å) than those of Fig. 10 (34). The Ge δ-layer atoms are well-defined in the image obtained with unpolarized radiation (Fig. 11(d)), and the Si atoms in the layer directly above the Ge δ-layer are fairly well resolved, but the Si atoms in the top center of the image along the [001] direction are poorly resolved, being farther away from the emitter. In contrast, in the image obtained with horizontally polarized radiation (Fig. 11(e)), the Si atoms above and below the Ge δ-layer, including those at top center and bottom center of the image along the [001] direction, are clearly imaged compared to those in the basal plane of the Ge δ-layer. Thus, it appears that the strained vertical interlayer distances could be determined in an MEXH experiment on this system using horizontally polarized incident radiation. Figure 11(f) shows the image obtained with vertically polarized incident radiation, where in contrast to Fig. 11(e), the Ge δ-layer atoms are strongly evident, compared to the suppressed images of the Si atoms above and below. Should both vertical *and* horizontal

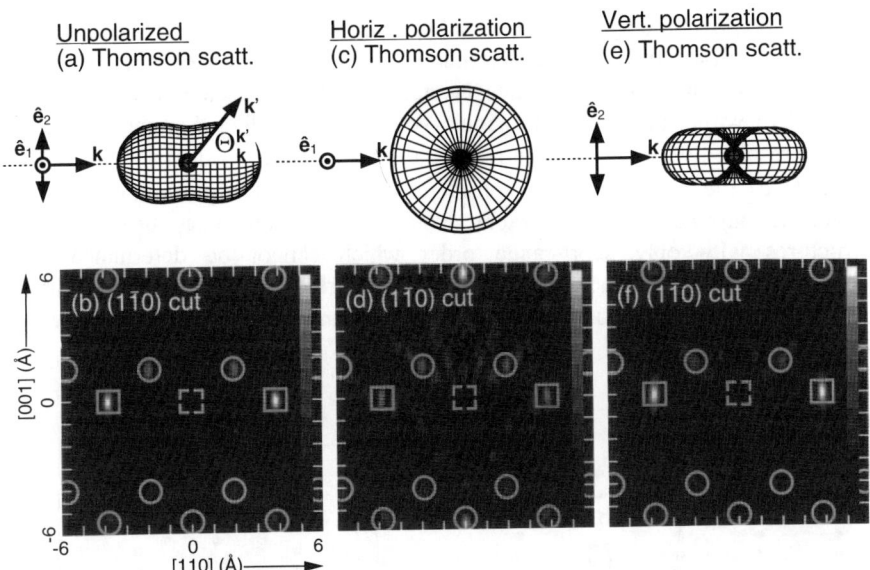

FIGURE 11. (a)-(c) Thomson scattering factors for unpolarized, horizontally polarized, and vertical polarized incident radiation, respectively. (d)-(f) Multiple energy x-ray holography images of a single Ge layer embedded in Si(001) (Si(001)/Ge-δ/Si(001)) in the vertical ($1\bar{1}0$) plane obtained from theoretical Ge Kα $\chi(\mathbf{k})$ data sets via Eq. (6), for: unpolarized, horizontally polarized, and vertically polarized incident radiation, respectively. The Ge emitter site is indicated by the dashed square, and the Ge δ-layer scatterers are indicated by solid squares. The Si atoms directly above and below the Ge δ-layer are indicated by circles. Axes are marked off in 1Å units.

structural information be desired for a given system with only linearly polarized incident radiation for excitation, then $\chi(\mathbf{k})$ intensities measured using horizontally and vertically polarized radiation separately could simply be added to determine the MEXH $\chi(\mathbf{k})$ intensities for most of the solid angle above the sample that one would measure using unpolarized incident radiation (40). The use of circularly polarized incident radiation should also be advantageous in this respect (40). In order to determine the vertical strain in this system (an effect of a few percent), increased spatial resolution of atomic images could be obtained by measuring holographic $\chi(\mathbf{k})$ intensities at higher wavenumbers (34).

Thus atomic x-ray holography holds much promise for the imaging of local atomic structure surrounding a specific emitter species of interest. The more ideal atomic scattering nature of x-rays produces reconstructed images that are relatively free of the aberrations, artifacts, and position shifts that are usually found in comparable electron atomic holographic images. XFH and MEXH also share the advantage of being element specific; thus the local structure around each atomic type in a sample can be determined. In addition, neither XFH or MEXH requires a sample with long range crystalline order; it need only be minimally ordered to within the potential imaging volume surrounding the emitter site that can be resolved with the **k**-space resolution of a given $\chi(\mathbf{k})$ data set (30,38). In contrast to the bulk structures considered in the initial implementation of this technique, atomic x-ray holography would be advantageously used to image structures with only short-range order which cannot be determined using conventional x-ray diffraction probes, such as surface and buried atomic layers; strained atomic lattice positions surrounding dopant sites (Fig. 12(a)); as well as

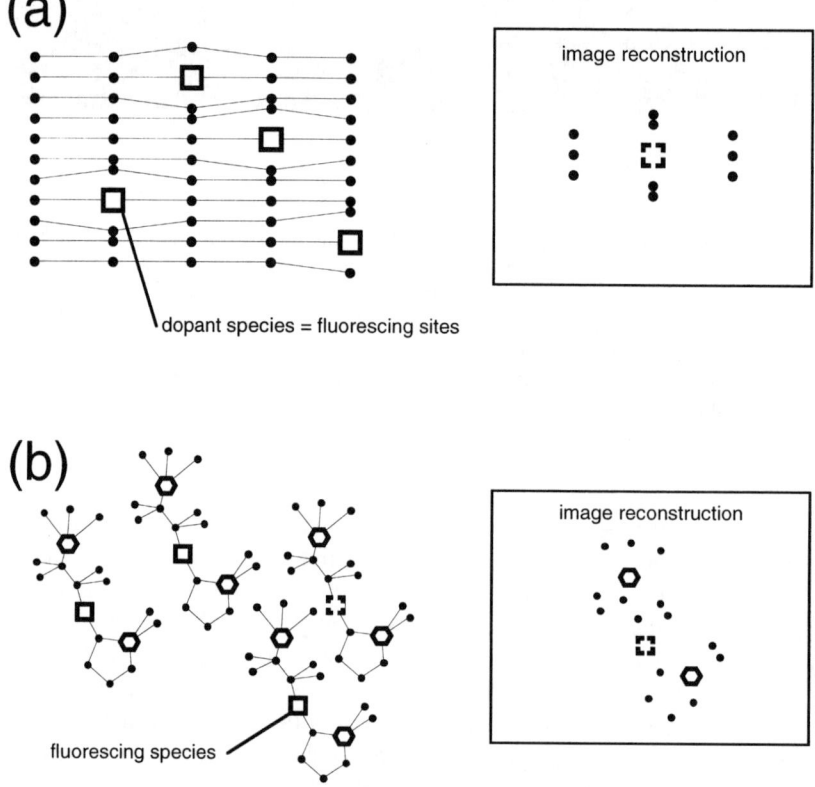

FIGURE 12. Schematic representation of two types of short-range-order atomic structures that could be fruitfully studied using atomic x-ray holography, together with their expected reconstructed images. (a) Strained lattice atoms surrounding dopant sites. (b) Rotationally aligned macromolecules with poor long-range translational symmetry.

the structure of macromolecules which do not exhibit perfect translational symmetry in crystal form (Fig. 12(b)). But one can also imagine using x-ray holography as a adjunct to conventional x-ray diffraction, with good estimates of local structures and phase relationships being derived to assist the diffraction analysis. Exploitation of linear and circularly polarized incident radiation in MEXH may also be utilized to emphasize horizontal and/or vertical structures of interest.

CONCLUDING REMARKS

In conclusion, holographic atomic imaging with localized single-atom sources of electrons or x-rays promises to become an important structural probe that will complement, or in some respects even surpass, conventional diffraction methods or other atomic structure probes. These holographic methods should be applicable to a wide variety of systems of practical and fundamental interest. X-ray holography of either the single-wavenumber fluorescence type or the multiple-wavenumber (inverse) type promises to yield more accurate images due to the more ideal scattering of x-rays, although the much weaker diffraction effects observed with x-rays also present challenges in measurement. However, with brighter sources of x-rays at next-generation synchrotron radiation facilities, and the development of faster detectors, these experimental problems should be surmountable. Thus, although much experimental and theoretical work lies ahead if we are to develop both the electron and x-ray techniques to their fullest potential, the final fulfillment of Gabor's dream for atomic-resolution holography seems well worth the effort.

ACKNOWLEDGEMENTS

Research at UC-Davis was supported in part by the Office of Naval Research (Contract Nos. N00014-90-5-1457 and N00014-94-1-0162), by the Director, Office of Energy Research, Office of Basic Energy Sciences, Materials Sciences Division of the U. S. Department of Energy (Contract No. DE-AC03-76SF00098), the National Energy Research Supercomputer Center, and by the International Centre for Diffraction Data. The authors would like to thank T. Gog, J. D. Denlinger, E. Rotenberg, F. Arfelli, R. A. Eisenhower, A. P. Kaduwela, R. H. Menk, D. Novikov, and S. Thevuthasan for their experimental and theoretical collaborative efforts, and M. A. Van Hove and R. L. Johnson for their useful comments during the preparation of this manuscript. Use of a preliminary version of a new multiple-scattering photoelectron diffraction program package was made possible by Y. Chen, H. Wu, and D. A. Shirley.

REFERENCES

1. Gabor, D. *Nature* (London) **161**, 777-778 (1948).
2. Wolf, E. *Optics Communications* **1**, 153-156 (1969), and *Optical Society of America* **60**, 18-20 (1970).
3. Syms, R. R. A., *Practical Volume Holography*, Oxford: Clarendon Press, 1990, ch. 1, pp. 21-28.
4. (a) Barton, J. J., *Physical Review Letters* **61**, 1356-1359 (1988), and *Physical Review Letters* **67**, 3106-3109 (1991). (b) Barton, J. J., and Terminello, L. J., paper presented at the Third International Conference on the Structure of Surfaces, Milwaukee, July 1990, and in *Structure of Surfaces III*, S. Y. Tong, M. A. Van Hove, X. Xide, and K. Takanayagi, eds., (Springer Verlag, Berlin, 1991) p. 107. (c) Tong, S. Y., Huang, H., and Wei, C. M., *Physical Review B* **46**, 2452-2459 (1992), and references therein.
5. Szöke, A., *Short Wavelength Coherent Radiation: Generation and Applications*, T. Attwood, J. Boker (eds.), AIP Conference Proceedings No. 147, (AIP, New York, 1986) pp. 361-367.
6. Karla, J., and Hauptmann, H. A., *Acta Crystallographica* **3**, 18 (1950).
7. Lipscomb, W. N., *Acta Crystallographica* **2**, 193 (1949).
8. Hümmer, K., Billy, H., *Acta Crystallographica A* **38**, 841 (1982).
9. Chang., S., L., *Physical Review Letters* **48**, 163 (1982).
10. Hutton, J. T., Trammell, G. T., and Hannon, J. P., *Physical Review B* **31**, 743 (1985), and *Physical Review B* **31**, 6420 (1985).
11. Bedyzk, M. J., and Materlik, G., *Physical Review B* **32**, 6456 (1985).
12. Saldin, D. K., Harp, G. R., Tonner, B. P., *Physical Review B* **45**, 9629 (1992).
13. Tegze, M., and Faigel, G., *Nature* **380**, 49-51 (1996).
14. Fadley, C. S., and Len, P. M., *Nature* **380**, 27-28 (1996), and unpublished results.
15. Gog, T., Len, P. M., Materlik, G., Bahr, D., Sanchez-Hanke, C., and Fadley, C. S., *Physical Review Letters* **76**, 3132-3135 (1996).
16. Gog., T., Menk, R.-H., Arfelli, F., Len, P. M., Fadley, C. S., and Materlik, G., *Synchrotron Radiation News* **9**, 30-35 (1996).
17. (a) Terminello, L. J., Petersen, B. L., and Barton, J. J., *Journal of Electron Spectroscopy and Related Phenomena* **75**, 229-308 (1995), and references therein. (b) Wu, H., Lapeyre, G. J. Huang, H., and Tong, S. Y., *Physical Review Letters* **71**, 251-254 (1993) and references therein. (c) Zharnikov, M., Weinelt, M., Zebisch, P., Stichler, M., Steinrück, H.-P., *Surface Science* **334**, 114-134 (1995), and references therein. (d) Denecke, R., Eckstein, R., Ley, L., Bocquet, A. E., Riley, J. D., Leckey, R. C. G., *Surface Science* **331-333**, 1085-1092 (1995). (e) Tobin, J. G., Waddill, G. D., Li, H., Tong, S. Y., *Surface Science* **334**, 263-275 (1995), and references therein.
18. (a) Li, H., Tong, S. Y., Naumovic, D., Stuck, A., and Osterwalder, J., *Physical Review B* **47**, 10036-10039 (1993). (b) Saldin, D. K., Harp, G. R. and Chen, X., *Physical Review B* **48**, 8234-8245 (1993), and references therein.
19. (a) Han, Z.-L., Hardcastle, S., Harp, G. R., Li, H., Wang, ,X.-D. Zhang, J., and Tonner, B. P., *Surface Science* **258**, 313-327 (1991), and references therein. (b) Hong, I. H., Shyu, S. C., Chou, Y. C., and Wei, C. M., *Physical Review B* **52**, 16884-16891 (1995), and references therein.
20. (a) Mendez, M. A., Glück, C., Guerrero, J., Andres, P. L., Heinz, K., Saldin, D. K., and Pendry, J. B., *Physical Review B* **45**, 9402-9405 (1992), and references therein. (b) Wei, C. M., Tong, S. Y., Wedler, H., Mendez, M. A., and Heinz, K., *Physical Review Letters* **72**, 2434-2437 (1994), and references therein.

21. Tong, S. Y., Huang, H., and Guo, X. Q., *Physical Review Letters* **69**, 3654-3657 (1992).
22. Saldin, D. K., Harp, G. R., Chen, B. L., and Tonner, B. P., *Physical Review B* **44**, 2480-2494 (1991).
23. (a) Thevuthasan, S., Herman, G. S., Kaduwela, A. P., Tran, T. T., Kim, Y. J., Saiki, R. S., and Fadley, C. S., *Journal of Vacuum Science and Technology A* **10**, 2261-2270 (1992). (b) Herman, G. S., Thevuthasan, S., Tran, T. T., Kim, Y. J., and Fadley, C. S., *Physical Review Letters* **68**, 650-653 (1992). (c) Tran, T. T., Thevuthasan, S., Kim, Y. J., Friedman, D. J., Fadley, C. S., *Physical Review B* **45**, 12106-12109 (1992), and *Surface Science* **281**, 270-284 (1993). (d) Thevuthasan, S., Ynzunza, R. X., Tober, E. D., Fadley, C. S., Kaduwela, A. P., and Van Hove, M. A., *Physical Review Letters* **70**, 595-598 (1993).
24. Hu, P. and King, D. A., *Physical Review B* **46**, 13615-13618 (1992).
25. Tonner, B. P., Han, Z.-L., Harp, G. R., and Saldin, D. K., *Physical Review B* **43**, 14423-14433 (1991).
26. Thevuthasan, S., Ynzunza, R. X., Tober, E. D., Fadley, C. S., Kaduwela, A. P., and Van Hove, M. A., *Physical Review Letters* **70**, 595 (1993).
27. (a) Tong, S. Y., Li, H., and Huang, H., *Physical Review B* **51**, 1850-1854 (1995). (b) Wu, H., and Lapeyre, G. J. *Physical Review B* **51**, 14549-14553 (1995). (c) Roesler, J. M., Sieger, M. T., and Chiang, T.-C., *Surface Science* **329**, L588-592 (1995), and references therein.
28. Rous, P. J. and Rubin, M. H., *Surface Science* **316**, L1068-1074 (1994).
29. Hofmann, P., Schindler, K.-M., Fritzsche, V., Bao, S., Bradshaw, A. M., and Woodruff, D. P., *Journal of Vacuum Science and Technology A* **12**, 2045-2050 (1994), *Surface Science* **337**, 169-176 (1995), and references therein.
30. (a) Len, P. M., Thevuthasan, S., Kaduwela, A. P., Van Hove, M. A., and Fadley, C. S., *Surface Science* **365**, 535-546 (1996). (b) Len, P. M., Zhang, F., Thevuthasan, S., Kaduwela, A. P., Van Hove, M. A., and Fadley, C. S., *Journal of Electron Spectroscopy and Related Phenomena* **76**, 351-357 (1995). (c) Len, P. M., Zhang, F., Thevuthasan, S., Kaduwela, A. P., Fadley, C. S., and Van Hove, M. A., submitted to the *Journal of Electron Spectroscopy and Related Phenomena*.
31. Denlinger, J. D., Rotenberg, E., Len, P. M., Kevan, S. D., Tonner, B. P., and Fadley, C. S., in preparation.
32. Ynzunza, R. X., private communication.
33. SCAT photoelectron diffraction program package, Chen, Y., Wu, H., and Shirley, D. A., private communication.
34. Harp, G. R., Saldin, D. K., Chen, X., Han, Z.-L., and Tonner, B. P., *Journal of Electron Spectroscopy and Related Phenomena* **57**, 331-355 (1991).
35. (a) Timmermans, E. M. E., Trammell, G. T., and Hannon, J. P., *Physical Review Letters* **72**, 832-835 (1994). (b) Kaduwela, A. P., Wang, Z., Thevuthasan, S., Van Hove, M. A., and Fadley, C. S., *Physical Review B* **50**, 9656-9659 (1994).
36. Faigel, G., private communication.
37. Tegze, M., Faigel, G., *Europhysics Letters* **16**, 41-46 (1991).
38. Len, P. M., Thevuthasan, S., Fadley, C. S., Kaduwela, A. P., and Van Hove, M. A., *Physical Review B* **50**, 11275-11278 (1994).
39. Len, P. M., Gog, T., Fadley, C. S., and Materlik, G., submitted to *Physical Review B*.
40. Len, P. M., Gog, T., Novikov, D., Eisenhower, R. A., Materlik, G., and Fadley, C. S., submitted to *Physical Review B*.
41. *International Tables for X-ray Crystallography*, edited by K. Lonsdale (Reidel, Dordrecht, 1968), Vol. III.
42. Falta, J., Gog, T., Materlik, G., Muller, B. H., and Horn-von Hoegen, M., *Physical Review B* **51** 7598-7602 (1995), and references therein.

V. NUCLEAR SCATTERING

Nuclear scattering of synchrotron X-rays

G. V. Smirnov

Russian Research Center "Kurchatov Institute", 123182 Moscow, Russia

The interaction of x-rays with the electrons in atoms is dominated by Thomson scattering. The resonant contributions in the scattering amplitude play the role of perturbations, which may be important in special applications. On the contrary, the interaction of x-rays and atomic nuclei is a case of pure resonance scattering. A Mössbauer nucleus presents a simple resonating two level system. Scattering of x-rays by a nuclear ensemble is characterized by the formation of collective coherent nuclear excitations - nuclear excitons.

The sharpness of the nuclear resonance correlates to a long lifetime of the nuclear excited state, which lies in the range $10^{-5} - 10^{-9}$s. Pulsed synchrotron radiation (SR) sources provide the best possibility to observe the time evolution of x-ray nuclear scattering. Since excitation by a SR pulse is almost instantaneous compared to the characteristic nuclear lifetimes, the two stages of the scattering process, excitation and decay are well separated in time: the nuclear excitations exhibit a free, delayed decay. Both the time differential and time integrated intensities of the delayed radiation deliver rich information about the states of nuclear ensembles.

Owing to specific properties of nuclear transitions unique information concerning crystalline magnetic and electric fields, magnetic and electric phase transitions, internal dynamics, such as thermal vibrations, diffusion, fluctuations, relaxation, etc., can be obtained by studying synchrotron resonant nuclear scattering. The physical principles of the new experimental technique are described and applications are illustrated by recent experiments.

INTRODUCTION

X rays are used widely and have proven to be very useful for studying the atomic structure of matter. The interaction between x rays and bound electrons in atoms have been primarily exploited for these purposes. In principle atomic nuclei can also be excited by x rays to their low energy levels. However, conventional sources of x-ray radiation do not have sufficient brilliance to permit measurements of the excitation of extremely narrow nuclear levels.

In 1958 Mössbauer discovered the possibility of recoilless emission and absorption of γ rays by nuclei. Due to this effect the spectral energy density of a radioactive source of γ radiation can exceed that of a powerful x-ray tube by four to five orders of magnitude. This allowed experimental studies of nuclear resonant scattering using γ rays from Mössbauer sources long before the advent of SR. The experiments with γ rays clearly showed that the process of resonant scattering of a Mössbauer quantum by a nuclear ensemble can be fully coherent over an extremely large interaction time. In case of the low energy nuclear states the interaction can last for 10^{-5} - 10^{-9}s (compared to 10^{-16} s in case of electronic scattering).

The inter-nuclear interference observed in scattering of a single photon is evidence of the formation of a delocalized, temporally and spatially coherent nuclear excitation [1,2]. A single excitation distributed coherently over a nuclear ensemble is called a *nuclear exciton* and is the central concept of the theory of nuclear resonant scattering [3-9] (see section 2.5). The subsequent radiative decay of a nuclear exciton is radically influenced by coherence. Due to the constructive contributions of all scattering paths the probability of γ-ray reemission is strongly enhanced. Under conditions where radiationless processes (internal electronic conversion) is dominant in the interaction with an isolated nucleus this *enhancement effect* is of great importance for the survival of a γ-ray in multiple scattering in a large nuclear ensemble.

The general (or dynamical) theory of nuclear resonant scattering, where the effects of multiple scattering are taken into account, was developed by Kagan and Afanas'ev [10-12] and by Hannon and Trammell [13,6]. Due to multiple scattering the γ-ray field traveling in a big nuclear ensemble is reconstructed and new effects appear. An important effect of the dynamical theory is the *effect of suppression of nuclear reaction* predicted by Kagan and Afanas'ev [10-12] . Internal electronic conversion can be completely suppressed under diffraction conditions in a regular nuclear ensemble. The nuclear excitation amplitude can vanish because of the formation inside a crystal of a standing wave-field pattern consistent with the crystalline structure (similar to the Borrmann effect).

Reviews of the kinematical and dynamical theories of nuclear resonant scattering are given by Afanas'ev and Kagan [3,12] and by van Bürck [15]. Experimental investigations of coherent nuclear scattering using a radioactive Mössbauer source were reviewed by Smirnov [16,17], by de Waard [18] and by Smirnov and Chumakov [19].

The studies of coherent phenomena with γ-rays over the last thirty years formed the basis of a new scientific direction. A new branch of experimental and theoretical optics, *γ-ray optics,* started to be developed. In 1974, Ruby [20] suggested that synchrotron radiation (SR) could be used for exciting nuclei. In 1985 Gerdau et al [21] made the first unambiguous observations of synchrotron x-rays resonantly scattered by ^{57}Fe nuclei. Since that pioneering work, many nuclear resonance fluorescence experiments have followed (see the reviews by

Gerdau et al [22], by Arthur et al [23], by Rüffer [24], by Gerdau and van Bürck [25], by Smirnov [26]).

The pulsed structure of SR is perfect for studying the time dependence of nuclear fluorescence. The free decay of a nuclear exciton created by a SR flash can be observed. The dynamical theory of nuclear resonant scattering in the time domain was developed by Kagan, Afanas'ev and Kohn [4,5] and by Hannon and Trammell [7,8]. For a recent review of the theory including the most interesting effects of coherence in the time domain see the article by Hannon and Trammell [9]. In the following sections we consider the physical principles and some recent applications of the new experimental technique.

2. Resonance scattering of x-rays by a nuclear ensemble

2.1 On the coherence of synchrotron radiation. An electron orbiting in a storage ring radiates when it is accelerated in a bending magnet, or a wiggler. An extremely short wave train of SR is emanated which lasts for about 10^{-18} sec and is of 10^{-8} cm in length. The relevant spectrum of electromagnetic radiation spreads from visible light to hard x-rays with the upper limit extending up to ~100 keV. The energy range thus covers the nuclear transitions of almost all Mössbauer isotopes. The x-rays generated in an undulator-based SR source can have especially high spectral density in particular nuclear resonance regions.

Before striking the nuclear target the SR light passes through a monochromator system to select a limited band of radiation, $\Delta\omega\hbar$ ($\Delta\omega\hbar$ lies usually in the eV- meV range). It is very narrow with respect to the primary spectrum of SR but extremely broad compared to the width of a nuclear resonance region. In practice one may assume that all the spectral components of the incident radiation have equal amplitudes within the selected band, i.e., $\varepsilon_\omega = \varepsilon_{\omega_0}$, ($\omega_0$ is the central frequency in the regarded resonance region). Thus, the spectrum of radiation of an individual electron is represented by a continuous set of coherent harmonics

$$E_i(\omega, t) = \varepsilon_{\omega_0} \exp(i\omega t) \qquad (1)$$

where the amplitude, ε_{ω_0}, is frequency independent. The amplitude is given by $\varepsilon_{\omega_0} = \sqrt{I_0/\Delta\omega}$ where I_0 is the intensity of SR within the frequency range selected by a monochromator system.

Electrons are grouped in bunches of finite size in a storage ring. Since the bunch size significantly exceeds the wavelength of x-radiation the wave packets from individual electrons cannot be regarded as coherent and the intensities from the individual electrons add up in the resulting wave field.

2.2 Energy and time domain presentations of resonance scattering. The overall process - the absorption of an x-ray photon by a nuclear target and the reradiation of a photon, - is a scattering process, due to the interaction with the target the incoming x-ray wave packet can be radically transformed.

In the theory of nuclear resonance scattering energy and time are new important variables in addition to the spatial ones familiar in x-ray scattering theory. Both the spectral composition and the time structure of a wave packet can be drastically transformed due to nuclear resonance scattering. The change of the spectral composition of the scattered radiation represents the steady-state scattering picture while the dynamics of scattering is revealed in the time domain. Both the *energy and time domain presentations* of resonance scattering are closely related. We shall use both presentations in the following to describe nuclear resonance scattering by an isolated nucleus and by a nuclear ensemble.

The SR sources provide an excellent opportunity to observe the time evolution of nuclear scattering. The excitation of a nuclear target by a SR pulse lasts for only about 0.2 ns. While the time response of nuclei lies in the microsecond range due to the sharpness of the resonance. Thus, the excitation and decay stages are well separated in time and the decay is so long on an absolute scale that it can be easily traced with modern electronics.

2.3 Elastic scattering by a bound nucleus. The nuclear resonance scattering amplitude is evaluated with the help of quantum mechanical perturbation theory. The x-ray can excite the nuclear transition current, which has the form of an oscillating multipole. This current is involved intrinsically in finding the scattering amplitude. The amplitude is derived as a quantum mechanical average over the nuclear ensemble. We limit our discussion to elastic forward scattering of x-rays by a bound nucleus having no hyperfine interaction with its surrounding. In this case we may simplify the definition of the scattering amplitude and consider it as a complex function, $f_0(\omega)$, which relates the amplitudes and phases of the relevant spectral components of the incident and scattered radiation

$$E_s(\omega) = f_0(\omega) \cdot E_i(\omega) \quad (2)$$

where the averaged scattering amplitude by a nucleus is

$$f_0(\omega) = -\frac{K}{4\pi} \cdot \sigma_0 \cdot \frac{\Gamma/2\hbar}{\omega - \omega_0 - i\Gamma/2\hbar} \cdot f_{LM}(\vec{K}) \cdot \chi \quad (3)$$

$\sigma_0 = \frac{2\pi}{K^2} \cdot \frac{1}{1+\alpha} \cdot \frac{2I_e+1}{2I_g+1}$ is the maximum resonance cross section, $\Gamma = \Gamma_\gamma + \Gamma_e$ is the natural width of nuclear excited level composed of the radiative and radiationless parts (the energy of the excited nucleus can also be lost by ejecting an atomic electron via internal conversion), $\alpha = \frac{\Gamma_e}{\Gamma_\gamma}$ is the internal conversion coefficient and I_g, I_e are nuclear spins of the ground and excited states (in case of the ^{57}Fe nucleus $\sigma_0 = 2.56 \times 10^{-18} \text{cm}^2$, $\alpha = 8.19$, $\Gamma = 5 \cdot 10^{-9}$eV, $I_g = 1/2$ $I_e = 3/2$); $K = \frac{\omega}{c}$ is the magnitude of the x-ray wave vector in free space,

$f_{LM}(\vec{K}) = \exp\{-\langle(\vec{K}\vec{u})^2\rangle\}$ is the Lamb-Mössbauer factor representing the recoilless fraction of the scattered radiation or the probability of the Mössbauer effect ($\vec{u}$ is deviation of the nucleus from its equilibrium position due to thermal vibrations) and χ is the isotopic enrichment of the target.

As a complex number the scattering amplitude may be written in the form $f_0 = |f_0(\omega)|\exp\{i\Phi(\omega)\}$, where $|f_0(\omega)|$ presents the attenuation of the magnitude of the amplitude of the scattered radiation component and $\Phi(\omega)$ is the phase shift of the component obtained in forward scattering. In Fig. 1 the variation of the scattering amplitude is displayed in the resonance region.

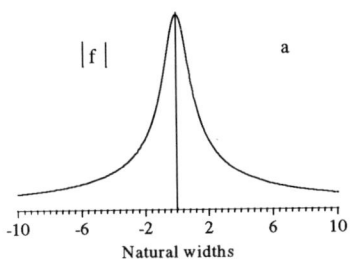
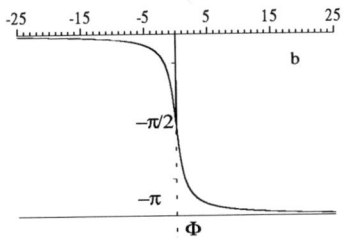
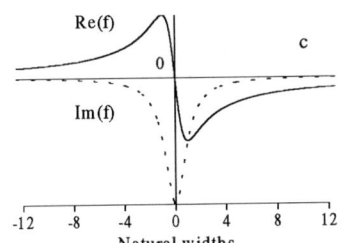

Fig 1. The amplitude of the elastic scattering by a bound nucleus in forward direction. The magnitude -(a), the phase - (b), the real and imaginary parts - (c) of the forward scattering amplitude in the vicinity of the resonance energy.

Dramatic changes of the scattering amplitude occur in the vicinity of a resonance, within a few natural level widths ($\approx 2\cdot10^{-8}$eV only in case of ^{57}Fe). The amplitude is sharply peaked at resonance, Fig 1a. and the phase drops from zero below the resonance down to $-\pi$ above the resonance, Fig 1b.

It is convenient to operate with the imaginary and real parts of the scattering amplitude: $f_0(\omega) = \text{Re}[f_0(\omega)] + i\,\text{Im}[f_0(\omega)]$ as shown in Fig. 1c. In the vicinity of the resonance *the amplitude is purely imaginary*, but in the wings $|\text{Re}(f_0)| \gg |\text{Im}(f_0)|$ as usual for x-ray scattering by electrons.

2.4 Scattered wave packet In Fig. 2 the scattering of a SR x-ray wave packet is displayed schematically. The upper panel shows the incoming wave packet, the bottom panel shows the outgoing wave. The left/right-hand columns present respectively the steady state and the dynamic pictures of scattering. The upcoming bundle of radiation approximated by $\delta(t)$ function can be decomposed into a continous set of oscillations of equal amplitudes over an infinite frequency range, Eq. (1) (see the upper left-hand corner). The components of SR are displayed as arrows (thick rods) in the complex plane determined by the axes Im(E), Re(E). The arrows are distributed along the ω axis.

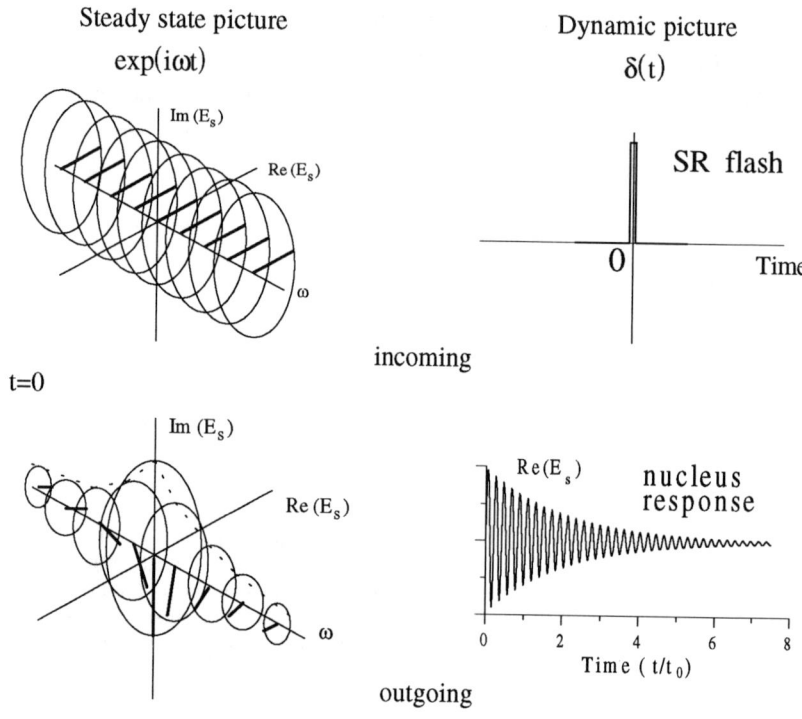

Fig. 2. Schematic view of nuclear resonance scattering of a SR x ray wave packet in the energy and time domains.

Each arrow is rotating with its angular frequency ω in the complex plane. At time zero they are all aligned in one direction producing a resultant sum with infinite amplitude, at all other times they are smeared homogeneously in phase over the whole period yielding zero amplitude.

The amplitudes and the phases of the elementary oscillations are changed due to the resonance scattering according to Eq. (2). In contrast to the incoming wave composition the oscillations are peaked now at the resonance frequency within a narrow band determined by the resonance width. Since all secondary components are coherent one can combine them back into a unit wave packet and thus obtain the nuclear response in the time domain. Geometrically this would mean performing the vector summation of all the arrows in Fig. 2 (bottom left-hand corner). The dephasing of the oscillations (due to rotation of the arrows at different angular frequencies) results in the decay of the wave packet amplitude. The sharper the resonance, the slower the dephasing and the response time of a nucleus is correspondingly longer.

Thus the decay of the scattering intensity can be interpreted as a fading of the interference signal which is due to the spectral components becoming more homogeneously distributed in phase with time.

Analytically this procedure is the standard integration of $E_s(\omega,t)$ in the frequency domain. Taking into account Eq. (1) and (2) one has

$$E_s(t) = \frac{\varepsilon_{\omega_0}}{2\pi} \int_{-\infty}^{+\infty} d\omega \cdot f_0(\omega) \cdot \exp(i\omega t) \qquad (4).$$

After substitution of Eq. 3 into Eq. 4 and performing the integration we obtain the wave packet of the scattered radiation

$$E_s(t) = \varepsilon_{\omega_0} f_0(t)$$

where $f_0(t) = -i\frac{1}{8\pi} \cdot \sigma_0 f_{LM} \chi \cdot \frac{1}{t_0} \exp(i\omega_0 t - t/2t_0)$ \qquad (5)

is the scattering amplitude as function of time, $t_0 = \hbar/\Gamma$ is the mean lifetime of the excited state.

The time response of a system to a δ-function excitation is usually called the response function; $f_0(t)$ thus presents the response function of an isolated nucleus. We can conclude now that due to nuclear scattering the initial δ-function wave packet is transformed into an exponentially decaying oscillation at the carrier frequency of the nuclear resonance.

The method of transition from time to frequency domain and back is widely used to find the time response of the scattering system. The relevant mathematical operations are Fourier transformations. However, when the response function is known it is possible to find the time response of the system to an arbitrary shaped wave packet with the aid of the response function technique given by the following operation

$$E_{sc}(t) = \int_{-\infty}^{t} dt' \cdot E_i(t') \cdot R(t-t') \qquad (6)$$

where $E_i(t')$ describes the wave packet of x-rays incident on a target, $R(t-t')$ is the response function of the target, t' and t are the excitation and deexcitation times respectively. Excitation and deexcitation are two subsequent events. The probability amplitude for the overall event is the product of the probability amplitudes for the constituent events. In particular, when $R(t)$ is a δ-function the scattered wave packet is identical to the incident one (if not taking into account its possible phase or frequency modulation in case of a mobile nucleus).

2.5 Coherent response of a nuclear ensemble.

Maxwell's wave equation. If there is a collection of identical nuclei the incident x-ray can interact resonantly with each of them. In the elastic scattering channel it is impossible to ascertain which nucleus in the system has been excited, therefore, all possible paths of scattering via individual nuclei have to be considered in the calculations. The nuclear excitation produced by an x-ray under these conditions is delocalized and the x-ray scattering process has a collective character.

The delocalized intermediate excited state created by a single x-ray quantum is understood as a spatially coherent superposition of the various excited states of all nuclei in the target. In each contributing term one nucleus is excited with a definite probability amplitude while all others are in the ground state. The spatial and temporal phasing of the nuclear excitations over a system is determined by the space and time coherence of the field associated with the incident x-ray. The superposition state of the nuclear excitations is called a ***nuclear exciton***.

When the excitation is distributed over the entire nuclear ensemble and when the phase correlation of the partial nuclear excitations is preserved during the lifetime of the excited state the interference of the wavelets reradiated by the nuclei can occur and coherent radiation fields can be built up in nuclear resonance scattering.

Maxwell's equations for a medium can be applied to find the steady state solutions for these fields. The existence of the distributed coherent nuclear excitation provides a physical basis for the use of *macroscopic polarization* in Maxwell's equations to treat the radiative effects of nuclei.

Maxwell's wave equation for a double space and time harmonic component of an x-ray field, a Fourier component, can be obtained in the following form

$$(k^2 - K^2)\vec{E} - \vec{k}(\vec{k}\vec{E}) = K^2 \frac{1}{\varepsilon_0} \vec{P} \qquad (8)$$

where $\vec{P}$ is the polarization of the medium induced by the electric field, $\vec{k}$ is a propagation vector inside the scattering medium and is in general complex to account for both refraction and absorption effects of the medium. The polarization represents the induced electric moment in a unit volume of the medium. It is derived as a quantum mechanical average of the induced nuclear current density.

Dynamical wave equations. The general scattering theory by Ewald takes into account multiple scattering of radiation by atoms. In the steady state, due to multiple scattering, self-consistency is established between the radiation field and the induced currents. The total field represents a coherent superposition of the waves allowed by a scattering system. The waves are dynamically coupled via currents feeding one another so that the total field must be considered as a single entity. The condition of the dynamical equilibrium between the field and currents means that each constituent wave generates the whole set of eigenwaves and vice versa each wave of the set contributes to the constituent wave.

In general a nuclear ensemble represents a strongly dispersive, optically active and diffracting medium. Therefore the eigenwaves can be in different polarization states and can have different propagation vectors, related by Bragg conditions. Maxwell's wave equation (8) is then split into a set of *dynamical wave equations*, connecting each constituent wave with all others

$$\left(\frac{k_d^2}{K^2} - 1\right) E_d^\xi = \sum_{d',\xi'} \eta_{dd'}^{\xi\xi'} E_{d'}^{\xi'} \qquad (9)$$

where η are the electric susceptibilities associating the polarization of the media and the eigen waves, the index d denotes the propagation direction and the index ξ labels the projection of the $\vec{E}_d$ -vector on the ξ-axis of the coordinate system, (for Cartesian coordinates ξ=x,y,z). In writing down Eq (9) we have assumed that the electromagnetic field inside a target is practically transverse, i.e. $\vec{E} \cdot \vec{k} = 0$.

Usually the basis of mutually orthogonal unit vectors $\vec{e}^\sigma$ and $\vec{e}^\pi$ is used to describe an arbitrary polarization state of radiation. By making use of this basis the set of dynamical equations can be reduced. The explicit form of the susceptibility, which is a matrix in general, should be derived for each particular scattering problem using the accepted polarization basis. The set of the dynamical equations together with the boundary conditions is sufficient for finding the propagation vectors, the polarization states and the scalar amplitudes of the constituent eigen waves.

In an irregular and optically isotropic medium only a single coherent wave is generated, that in the forward direction. The susceptibility is in this case a complex scalar - η. The main effect of the medium is a strong dispersion of waves in phase and in amplitude. In an irregular but optically anisotropic and active medium the waves of different polarization states propagating in the forward direction can be excited. The susceptibility is then a matrix of second rank, $\eta^{ss'}$, where s and s' take meanings σ, π. The optical activity of the nuclear system arises due to the hyperfine splitting of the nuclear transitions. A wealth of polarization phenomena in transmission of x-rays through the nuclear target can be observed, such as birefringence (dichroism, double refraction) and optical activity (Faraday effect).

In a regular nuclear array diffraction phenomena can occur. In the case of a single Bragg reflection two directions are permitted for coherent scattering. In absence of optical activity the susceptibility is again a (2×2) matrix, $\eta_{dd'}$, where the indexes d,d' denote the forward and the Bragg reflection directions. The Bloch waves of the radiation are formed consistent with the periodicity of the crystalline structure. They exhibit anomalous interaction of radiation with the scattering centers (Borrmann effect in case of interaction with electronic shells, Kagan-Afanas'ev effect in case of nuclear interaction). In the presence of optical activity the susceptibility turns out to be a (4×4) matrix, $\eta_{dd'}^{ss'}$.

In the next paragraphs we shall analyze the results of SR nuclear scattering for the particular case of forward scattering.

3. Transmission of a SR pulse through a nuclear target.

3.1 Response of the nuclear ensemble in the energy domain
The solution of the wave equation for a Fourier component is as follows (a single resonance is considered)

$$E_{tr}(\omega) = \varepsilon_{\omega_0} \exp[i(\omega t - KL)] \cdot \exp\left\{\frac{iT\Gamma/4\hbar}{\omega - \omega_0 - iq\Gamma/2\hbar}\right\} \cdot \exp(-T_e/2) \qquad (10)$$

where $T = \sigma_0 f_{LM} \chi nL = \mu_n L$ and $T_e = \sigma_{ph} nL = \mu_e L$ are the effective nuclear resonance and electronic thickness of the target, L is the physical thickness, n is the concentration of nuclei in the target, the factor q>1 accounts for a possible broadening of the nuclear resonance which however preserves the Lorentzian shape. In the following we shall omit the constant phase factor in the field amplitude.

The transmitted field Eq. (10) can be presented as the coherent superposition of the incident and forward scattered fields [26]. Therefore, the response of the target in the energy domain is given by the distribution of the scattered radiation components: $E_s(\omega) = E_{tr}(\omega) - E_i(\omega)$, where E_i presents the incident radiation given by Eq (1). From Eq. (10) it follows that the scattered radiation components are strongly dispersed due to resonance scattering by the target. The geometrical image of the distribution similar to that shown in Fig. 2 is strongly dependent on the target thickness.

In Fig. 3a the spectra of radiation scattered by targets of different thickness are displayed. Dramatic changes in the shape of the spectra occur with increasing thickness. In the limit of small thicknesses the curve is small and has a Lorentzian profile (curve 1). At larger thicknesses the single line is split and maxima are formed symmetrically at a distance from the resonance. The separation of the maxima increases with target thickness. The spectrum acquires **a *double hump or volcano-like profile***. This form of the spectra is a striking consequence of the enhancement effect due to multiple resonant scattering.

3.2 Response of a nuclear ensemble in the time domain.
The resultant wave packet is found by integrating all the time harmonics of the transmitted radiation. The result is as follows

$$E_{tr}(t) = \varepsilon_0 \frac{\Gamma}{\hbar} \exp(-T_e/2) \cdot [\delta(\tau) - \exp\{i\omega_0 t - q\tau/2\}(T/2)\alpha(T\tau)] \qquad (11)$$

where $\alpha(T\tau) = J_1(\sqrt{T\tau})/\sqrt{T\tau}$ with J_1 the Bessel function of first order ($\alpha(0) = 1/2$). Dimensionless thickness and time parameters are used in Eq (11): T was defined earlier and $\tau = t/t_0$ (the time response of a single unbroadened resonance target was obtained in [5]).

In the wave packet of transmitted radiation the incident and forward scattered fields are separated into prompt and the delayed parts (the first and the second term in the angular bracket of Eq. 11). The delayed part of the wave packet with the carrier frequency ω_0 is the coherent response of the nuclear target.

Inserting in Eq. 11 the explicit expression for ε_0 one obtains the expression for the time evolution of the forward scattering intensity

$$I_{fs}(T,\tau) = I_0 \frac{\Gamma}{\Delta E} \cdot \frac{1}{t_0} \exp(-T_e) \exp(-q\tau) \cdot \frac{T^2}{4} \sigma^2(T,\tau) \qquad (12)$$

where I_0 is the stationary intensity of the SR within the energy band ΔE determined by a monochromator system.

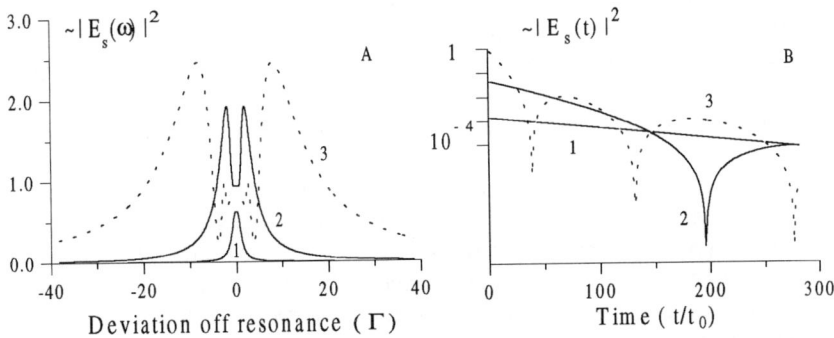

Fig. 3. The spectral composition - A, and the time dependence - B of the forward scattered radiation. (The curves 1, 2, and 3 correspond to the different target thicknesses: 0.2, 1.2, and 6μm SS enriched up to 95% by ^{57}Fe).

In Fig. 3b the time dependence is displayed for the same three target thicknesses. All the main features of the evolution of the forward scattering can be seen in this figure. First we consider the effects of the initial stage of decay. At this stage the time dependence can be approximated by $\approx T^2 \exp\{-\tau(q + T/4)\}$ (this is valid for time τ_1 found from the condition $T\tau_1 \approx 3$).

In the case of thin target, the curve 1, the decay is accelerated and characterized by the decay time $t'_0 = t_0/(1 + T/4) \cong 0.4 t_0$ (q=1.85, T=3.2). There are two different physical reasons for the faster decay of the forward scattered intensity. One reason is the speeding-up of the nuclear deexcitation which is related to the effect of enhancement of the radiative channel. The relevant speed-up is given by the exponential index T/4. Another reason for the drop in the coherent signal is the broadening of the resonance which leads to more rapid dephasing of the spectral components of the forward scattered radiation. The relevant acceleration is determined by the quantity (q-1). The physical difference of the two effects is directly seen from the fact that the coherent speed-up is accompanied by an increase in the time integral intensity while the acceleration of decay due to resonance broadening leads to a decrease. Both the differential and time integral intensities can be used to study different reasons for resonance broadening (like inhomogeneous distribution of hyperfine field, diffusion, relaxation).

As the thickness of the target increases the initial intensity and the decay rate grow (see curves 1, 2, and 3). Both effects correspond to the increase and broadening of the spectral compositions in Fig. 3a. The intensity is proportional to T^2, i.e. $\sim N^2$, where N is the number of nuclei involved in the process of collective coherent scattering (however, one should remember that the effective number of involved nuclei (per unit area) is limited by photoelectric absorption; the condition $\sigma_{ph} nd \approx 1$ defines $N_{max} \approx nd_{max} \approx 1/\sigma_{ph}$).

When the target thickness is large enough the intensity variations appear in the monotonic decay. According to our interference picture the variations can be interpreted as a beat between the two pronounced groups of time harmonics distributed symmetrically to the left and right of the resonance frequency (Fig 3a). The beat frequency is determined by the energy separation of the maxima. Both the distance between the maxima and the beat frequency grow as the target thickness increases.

The two effects of the initial stage of decay are described by the kinematical theory of scattering, i.e. in the approximation of a thin absorber. Actually, for a time short compared to the nuclear transition time, a nuclear ensemble of arbitrary thickness can be considered as a thin absorber. The effect of beating is typical for multiple nuclear scattering described by the dynamical scattering theory. The multiple scattering occurs in thick targets and becomes apparent only after some time of interaction of the γ-radiation with the nuclear ensemble. The multiple nuclear scattering in the resonance wings is responsible for the volcano profile of the spectrum in Fig 3a and respectively for the beat pattern in Fig 3b, which is termed ***dynamical beats*** (for a more detailed picture see [26]).

We note that space and time enters the σ-function on an equal footing: the argument of the function is a product of the relevant parameters. Therefore, the dynamical beat of the forward scattering intensity is a complicated space-time pattern: at a fixed instant one would observe zones of "darkness" and "brightness" at different depths in the volume of the target; these zones move along the propagation direction in time. This picture presents the dynamics of the decay of a single photon excitation.

Now we consider the effect of resonance splitting on the evolution of nuclear forward scattering of SR.

3.3 Resonance splitting, quantum beat. The hyperfine interaction splits nuclear transitions between the ground and the excited states into several components. The nuclear susceptibility under these conditions becomes more complicated, consisting of several terms related to different transition frequencies, and is polarization dependent. Let us take as an example the special but important case of magnetic hyperfine interaction in the ground and the first excited nuclear states in a ^{57}Fe hyperfine interaction in the ground and the first excited nuclear states in a ^{57}Fe target (these states have spins 1/2 and 3/2). In the case considered the magnetic dipole transition takes place between the states $1/2 \Leftrightarrow 3/2$. Due to the magnetic hyperfine interaction the nuclear levels are split into two sublevels in the ground state and four in the excited state. The new states are pure in spin projections with the quantization axis parallel to the internal magnetic field; the relevant magnetic quantum numbers are $m_g = \pm 1/2$ and $m_e = \pm 1/2, \pm 3/2$. By the selection rules six transitions between the sublevels of the ground and excited states with $\mu = m_e - m_g = 0, \pm 1$ are allowed. The nuclear susceptibility for forward scattering takes the form

$$\eta^{ss'}(\omega) = \text{const} \sum_{ge} \frac{\Gamma/2\hbar}{\omega - \omega_{eg} - i\Gamma/2\hbar} \cdot G^2(m_e, m_g, \mu) \cdot P^{ss'}(\mu) \tag{13}$$

where the frequencies of the allowed transitions are ω_{eg}, the probabilities of the transitions are given by the squares of the relevant Clebsch-Gordan coefficients, $G(m_e, m_g, \mu)$ and by the polarization factors $P^{ss'}(\mu)$, s and s' denote the polarization states of the incident and the forward scattered radiation respectively. The summation is performed over all allowed nuclear transitions (the explicit expressions are given in refs. [27,11,12]).

Polarized SR enables the selective excitation of different groups of nuclear transitions. In Fig. 4 the polarization of SR and the scattering geometry are shown.

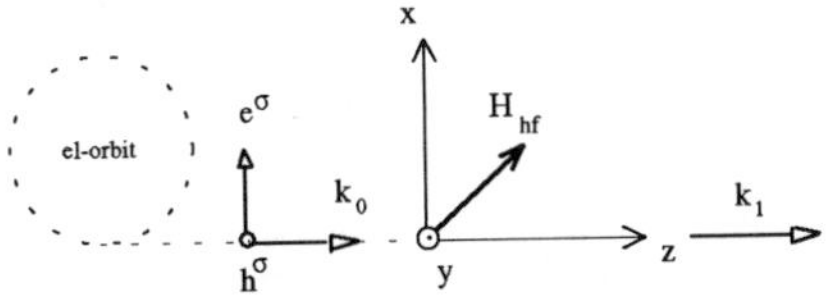

A simple practical rule for the selective excitation of the magnetic dipole transition is the following: *the magnetic polarization vector, $\vec{h}$, of the upcoming wave should have a component along the magnetic hyperfine field, $\vec{H}_{hf}$ in order to excite $\Delta m = 0$ transitions and it should have a component at right angle to $\vec{H}_{hf}$ to excite $\Delta m = \pm 1$ transitions.*

Fig 4. Polarization of synchrotron radiation, $\vec{e}^\sigma$ and $\vec{h}^\sigma$ are respectively electric and magnetic polarization vectors for a σ-polarized electromagnetic wave.

When $\vec{H}_{hf}$ lies in the xz-plane only the lines 1,3,4,6 in the Mössbauer spectrum of iron are excited and when $\vec{H}_{hf}$ is normal to the same plane only lines 2 and 5 are excited. In the first case the polarization of scattered radiation can vary from right/left circular, for $\vec{H}_{hf} \parallel z$, to σ-linear for $\vec{H}_{hf} \parallel x$; in the second case, $\vec{H}_{hf} \parallel y$, the scattered radiation is linearly σ-polarized.

The solution of the dynamical equations in these simple cases yields a forward scattered field consisting of separate groups of identically polarized oscillations with different carrier frequencies. The spectral compositions for the three mentioned sample magnetizations and the relative polarization of the oscillation groups are shown in Fig 5A. In the time domain the interference between the groups of identically polarized oscillations or between the partial wave packets of the scattered field gives rise to the so-called **quantum beats** of the forward scattered radiation [7] displayed in Fig. 5B.

The pure form of the quantum beat pattern can be seen in case of a thin target where multiple scattering is not yet important, the volcano profile is not formed and hence no dynamical beats are observed (see the curves 1 in Fig. 3A,B).

The most simple beat pattern, Fig. 5B bottom curve, is realized when $\Delta m=0$ nuclear transitions are excited. The beat occurs at a single frequency and with the highest contrast since only two subgroups of oscillators having equal strength (Fig. 5A bottom curve) interfere. The beat period is about 14 ns. By measuring the quantum beat period one immediately determines the energy separation of the excited nuclear transitions. As in Mössbauer spectroscopy the hyperfine splitting of both the ground and the excited states are

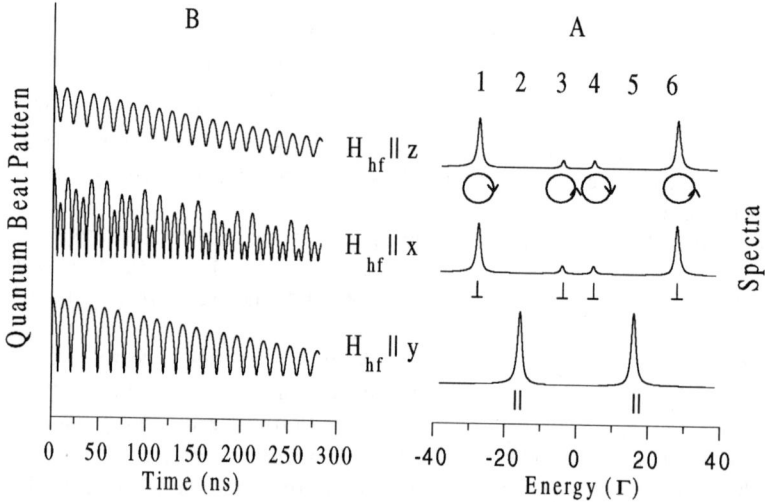

Fig. 5. The spectral compositions - A , and time dependences - B of nuclear forward scattering of SR by 0.2 μm thick ^{57}Fe foil for three different orientations of the foil magnetization (for the scattering geometry see Fig. 4).

involved . Therefore the quantum beat period contains information about the splitting of both communicating states.

Each wave packet related to the particular nuclear transition experiences the influence of the adjacent transition while propagating across the target. The presence of the neighboring transition increases or decreases the phase velocity due to resonance scattering depending on which side of the resonace the scattering occurs [28]. This effect is proportional to the target thickness and inversely proportional to the separation of the transitions (within the accepted assumption: $\Delta\omega\hbar \gg \Gamma$). In our example the beat shift derivative is about 0.14 ns/μm [29].

The quantum beat patterns are transformed when $\Delta m = \pm 1$ transitions are excited. If $\vec{H}_{hf} \parallel x$ the forward scattered wave packets corresponding to the 1,3,4,6 nuclear transitions have the same polarization and hence they interfere effectively. The corresponding spectral composition and beat pattern are shown in Fig. 5A,B (middle curves). In this case a more complicated beat (at three different frequencies) occurs. The most frequent beat is caused by the interference of the external 1 and 6 lines (the beat period is about 8.1 ns).

In the third example (Fig. 5A,B top curves), where $\vec{H}_{hf} \parallel z$, the same four nuclear transitions are excited. All wave packets related to these transitions are circularly polarized, however, the 1 and 4 transitions are polarized oppositely to the other two (transitions 3 and 6). Therefore, interference takes place within these pairs rather than between the pairs. In iron the separation of lines 1 and 4 is exactly the same as 3 and 6. The beat pattern has a simple form with a single beat period 14 ns. Note that the interference contrast in the top curve is less than that in the bottom one, because the amplitudes of the interfering waves are in the ratio 3:1.

It is of interest to note that the sum of the left and the right circularly polarized waves of equal amplitude, e.g. corresponding to the 1 and 6 transitions, is a linearly polarized wave with the carrier frequency $(\omega_1 + \omega_6)/2$ and the polarization vector precesses around the propagation direction with the frequency $\omega_6 - \omega_1$ (with period 8.1 ns). This time dependent Faraday rotation of the polarization plane was observed in [30]. It was proposed to use the optical activity of samples in combination with the polarization filtering technique to suppress the nonresonant component in NFS [30, 31].

We wish to emphasize that the spectral compositions corresponding to the last two cases (where 1,3,4 and 6 transitions are excited) are absolutely identical, while the time dependences of the nuclear forward scattering look quite different. This is because an interference technique reveals complex amplitudes of oscillations while a spectroscopic method only gives the magnitudes.

For a thick target the resonance lines in forward scattered spectra are broadened and acquire the volcano-like profile (Fig. 6A) due to multiple nuclear scattering. The interference between the two groups of oscillators belonging to one resonance yields the dynamical beats, as discussed above. The quantum and dynamical beats are well distinguished when the hyperfine splitting exceeds the width of the broadened resonance line. In this case the quantum beat is much more frequent than the dynamical beat (Fig 6B, for the experiments see [29,32]). However, in the example considered the interference between different resonances and between the two branches of one resonance is already of mixed character, the volcano profile becomes asymmetric (since the inner sides of resonances interfere destructively). With overlapping resonance lines the beat structure may have a complicated, strongly mixed character.

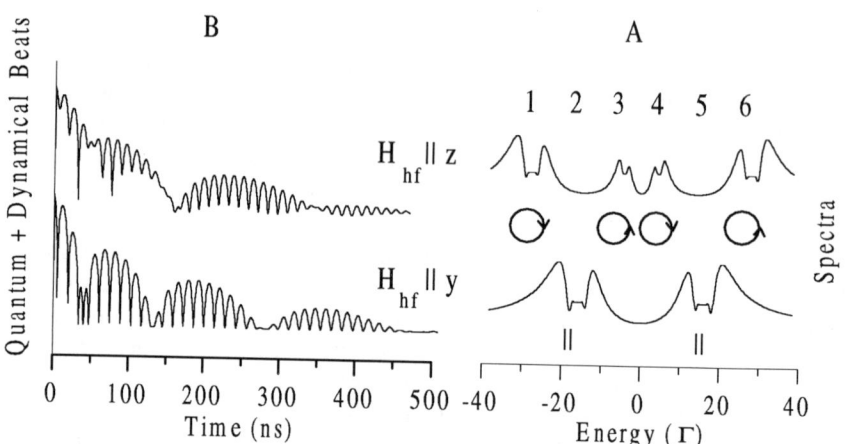

Fig. 6. Spectra - A, and time dependences - B of nuclear forward scattering of SR by 6 μm thick ^{57}Fe foil. Quantum beat are enveloped by the dynamical beats.

Concluding this section one can say that the time differential measurement is the main source of information about hyperfine interactions. More complicated cases of combined magnetic and electric quadrupole interaction, of combinations of several sets of hyperfine fields etc. can be investigated. The time dependences provide also information about the effective thickness parameters, in particular, about the Lamb-Mössbauer factor, and about the resonance broadening of both static and dynamic nature.

We can only mention here that the time integrated intensity of nuclear scattering measured as a function of such variables like incidence or scattering angle, sample temperature, magnetization, thickness, incident radiation energy, etc., can provide plenty of information characterizing the scattering system. Detailed information concerning the structure of the internal fields can be obtained by studying nuclear Bragg scattering.

There are many other channels of x-ray nuclear resonance scattering beside the coherent elastic one considered here. In the next section regarding applications of nuclear resonance scattering of SR we shall briefly touch on some of them.

4. Aspects of synchrotron x-ray nuclear scattering

Synchrotron x-ray sources provide various possibilities for probing a nuclear resonant sample. The variety is based on the unique combination of properties of synchrotron radiation: *well-defined time structure, pure polarization state, directionality, high beam intensity, small beam cross-section and tunable energy*. Here we give a brief description of the state-of-the-art synchrotron beamlines and mention some applications.

Recently third generation synchrotron radiation sources have come into operation: the European Synchrotron Radiation Facility (ESRF) in Grenoble, France [33], and the Advanced Photon Source (APS) in Argonne National Laboratory, U.S.A. Hard synchrotron radiation (in the x-ray range) is generated at these new sources by undulators providing especially enhanced spectral density at the resonance energies, in particular, for ^{57}Fe, ^{119}Sn, ^{151}Eu nuclei, where $\sim 10^4 - 10^5$ photons per second within the natural width of a nuclear level is reached. The monochromatization of the undulator x-ray radiation down to the 'eV' or 'meV' energy range is typically achieved by diffraction from Si single crystals. For further monochromatization nuclear resonant scatterers, i.e. energy selecting monochromators, are needed.

4.1 **Nuclear resonant diffraction.** The high directionality of SR immediately suggests interesting applications for studying diffraction from nuclear lattices in single crystals having very tight collimation requirements. Of special interest are the samples with complex structures of magnetic or electric fields in crystals. The sensitivity of nuclear scattering to the hyperfine environment can be effectively used for making structure determinations of the fields in crystals.

By exploiting the difference of structural factors for electronic and nuclear diffraction in such samples one is able to obtain systematically pure nuclear diffraction (observed first in the antiferromagnetic crystal, α–^{57}Fe$_2$O$_3$ [34]). Pure nuclear diffraction became a standard method for filtering the Mössbauer component of SR. The possibility of obtaining pure nuclear reflectivity from a crystal was used to successfully observe the nuclear scattering of SR for the first time [21]. For this purpose ^{57}Fe - enriched yttrium iron garnet crystals (ferrimagnetic) were employed. By proper choice of the reflection and magnetization of the crystal it became possible to observe pure nuclear diffraction of polarized synchrotron radiation from the selected sublattices of iron nuclei having particular hyperfine-field environments. The quantum beat pattern in the relevant time evolution yielded complete information about the hyperfine environment in the selected groups including the strengths and directions of the fields [35].

Pure nuclear diffraction of SR was observed also in hematite (a canted antiferromagnetic, with weak ferromagnetism) [36], in iron borate, ^{57}FeBO$_3$ (with a magnetic structure similar to that of hematite) [37], in ferric borate ^{57}Fe$_3$BO$_6$ (ferrimagnetic) [38].

Iron borate single crystals became standard pure nuclear scatterers because of their simple crystallographic and magnetic structure [14]. It is very easy to reach the magnetic phase transition in this crystal, the Néel-temperature is $T_N = 75.8^0 C$. Magnetization breakdown was observed by measuring the time distributions of the pure nuclear Bragg reflection (333) in the range from room temperature to T_N. The slowing down of the quantum beats was clearly seen when approaching T_N [39,40].

By making use of an iron borate single crystal set for pure nuclear diffraction in the immediate vicinity of T_N coherent Mössbauer radiation was generated with synchrotron x-rays [41]. Like a conventional Mössbauer source the new Synchrotron Mössbauer Source (SMS) emits single line radiation of about natural line width, but in addition the emitted radiation is fully recoilless, highly directed and of pure linear polarization. An extremely high suppression of the electronic scattering by a factor of about 10^{-10} is achieved. The SM source can be used for absorption and scattering measurements in the energy domain (in μeV range with neV resolution) with the help of the Doppler shift technique. Figure 7 shows the first Mössbauer spectra taken with the SMS which demonstrate the pure polarization state of the radiation.

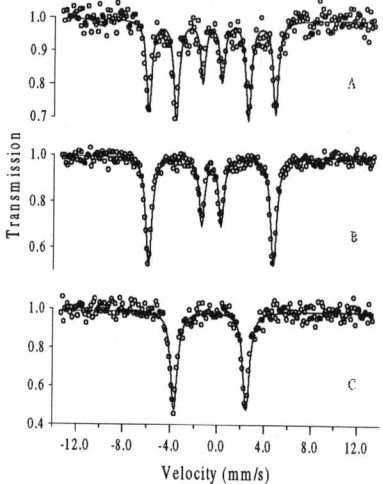

Fig. 7. Mössbauer spectra of 1 μm foil of ^{57}Fe taken with a Synchrotron Mössbauer source; A- no external field, B- $\vec{H}_{ex} \| x$, C- $\vec{H}_{ex} \| y$ [41].

The examples illustrate the capability of nuclear diffraction of SR for the investigation of hyperfine interactions and the structure of internal fields in single crystals. In certain cases the ability to employ polarization and crystallographic selection rules may be of great importance in these studies.

Special conditions for studying nuclear diffraction are offered by artificial layered structures. *Nuclear multilayers* are diffracting optical devices involving alternating layers where the resonant nuclei density is changed periodically in one direction. The structure and composition of nuclear multilayers may be varied within broad limits, and tailored to the experimental goals. Pure nuclear Bragg reflections of different nature were produced in synthesized multilayers by creating a ^{57}Fe isotope periodicity different from the electronic diffracting periodicity [42-44] and by creating antiferromagnetic coupling of the ^{57}Fe layers [45].

In [46] nuclear diffraction of SR was studied under conditions of extreme enhancement of the radiative channel provided by a nuclear multilayer. Tremendous speed up of the nuclear decay due to the enhancement of the radiative decay channel was observed. A lifetime of 4 ns (instead of 141 ns for an isolated nucleus) was measured which corresponds to a large coherent broadening of the resonance line.

4.2 Specular resonant scattering. Nuclear reflection of γ-rays at small glancing angles by an optically flat mirror presents an extreme case of coherent resonant scattering. The grazing incidence geometry increases the path length of radiation in a thin interface layer or in a sample-film and permits the use of total external reflection. It occurs in the energy range where the index of refraction involving both nuclear resonant and electronic parts is less than one.

A peaking of the nuclear reflection at the critical angle was observed while studying the specular resonant nuclear scattering of SR from a ^{57}Fe mirror [47]. This new phenomenon was explained by a rather complicated influence of the electronic scattering on the nuclear response of the system in the range of the critical angle and below it. The effect happens essentially because the extinction of the wave field due to electronic scattering reduces the illumination of the nuclei.

An example of the angle-resolved time evolution of the nuclear specular scattering was presented in [32]; changed speed-up and quantum beat patterns in accordance with the dynamical theory of scattering were observed.

Under total reflection conditions the coherent enhancement of the radiative channel causes considerable broadening of nuclear resonances and extends the energy range of nuclear reflectivity. On this basis a concept for a broadband nuclear monochromator using grazing incidence scattering was introduced [48]. An approach based on the optical theory was developed in [49] (for recent theoretical considerations see also [50]).

A new tunable x-ray source based on the specular nuclear reflection of SR from a ^{57}Fe-coated rotating mirror was recently designed [51]. The energy selectivity and magnetic optical activity of the nuclear reflector were effectively exploited. With the help of the polarization filtering technique [31] it became possible to select the resonantly scattered radiation with an energy band in the μeV range. The tunability in energy over the range of a few meV was achieved by fast rotation of the mirror, which acted as a Doppler shifter. The new technique makes μeV-resolved inelastic x-ray spectroscopy possible. The tunable x-ray source will enable the study of low energy excitations in solids such as soft phonons, magnons, rotational excitations and so on. The excitations can be probed in the way discussed in the section on inelastic scattering (see 4.5).

4.3 Nuclear resonant forward scattering. Several years passed after the first diffraction experiment before forward scattering was successfully observed. Cutting off a large part of nonresonant radiation with the aid of a high-resolution electronic monochromator allowed nuclear forward scattered radiation to be observed for the first time [52]. The monochromatization down to 6 meV was achieved with the help of dispersive high indexed reflections from Si(10,6,4) channelcuts. Later on an improved version of the high resolution monochromator, a so-called nested monochromator, combining symmetric high indexed and highly asymmetric reflections was introduced [53] which achieved high energy resolution with minimum sacrifice of angular acceptance. The performance of the forward scattering experiment was a break-through for applications of the SR nuclear scattering method. The geometry of the experiment is simple and allows the investigation of all kinds of different samples: single crystals, powders, glasses etc. Measurement of the time dependence of nuclear forward scattering (NFS) will probably become as frequently used to study hyperfine interactions as the standard

Mössbauer transmission experiment. The synchrotron radiation beam is so well collimated and has such a small size that it can be easily introduced into an oven, cryostat or diamond-anvil cell to study very small samples under extreme conditions.

The time evolution of NFS in magnetized ^{57}Fe enriched iron foils of different thicknesses demonstrated superimposed quantum and dynamical beats [29] (see also the further studies [32]). The foils were magnetized along the magnetic polarization vector of SR (X-axis in Fig. 4) to allow only $\Delta m = 0$ hyperfine transitions. At the same magnetization of an ^{57}Fe foil the dynamical and quantum beat patterns were investigated in the temperature range 8 -- 1048 K [54]. A dramatic development of the time dependence was observed due to magnetization breakdown on passing the Curie temperature and due to the decrease of the Lamb-Mössbauer factor.

First applications of NFS for high pressure experiments have been made [32,55]. The phase transition in iron metal from the ferromagnetic α-phase to the paramagnetic ε-phase under high hydrostatic pressure in a diamond cell had been studied [55]. A dramatic transformation of the time dependence of NFS from the pronounced quantum beat pattern in the presence of hyperfine splitting to a smooth exponential decay after the collapse of the splitting due to the phase transition was observed (Fig. 8).

In another experiment the temperature induced high spin ⇔ low spin transformation in a polycrystalline powder sample of [^{57}Fe(bpp)$_2$][BF$_4$]$_2$ was studied in the temperature range from 20 K to room temperature [56].

At the dedicated nuclear resonance beamlines at the ESRF and the APS the time needed for such measurements is greatly reduced and phase transitions can be followed on the time scale of minutes or even seconds.

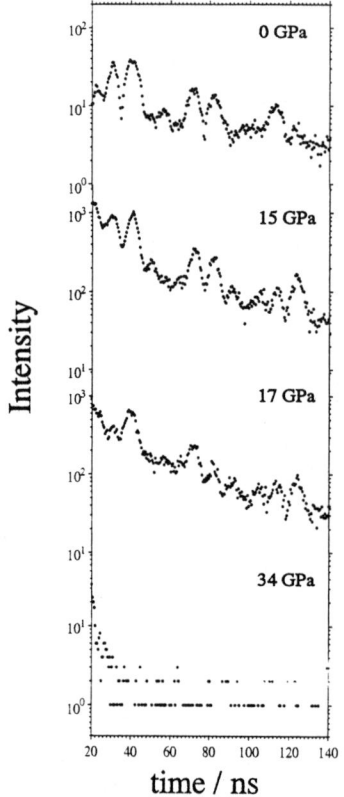

Fig. 8. Magnetic phase transition in iron under pressure as seen in the time dependence of nuclear forward scattering at 0.15-34 GPa [55].

4.4 Nuclear resonance small angle scattering. A plane wave of incident radiation is reproduced in forward scattering only if the lateral phase correlation length exceeds the size of the first Fresnel zone. Otherwise coherent scattering occurs with an angular dispersion of radiation. This small angle scattering (SAS) is characterized by the ratio of the radiation wave length to the phase correlation length, L, which is directly related to the sizes of the homogeneous regions in the scattering medium. X-ray and neutron small angle scattering are well established techniques for studying inhomogeneities in matter. A peculiarity of nuclear resonance scattering is that the parameter L is sensitive to the spatial homogeneity of the hyperfine environment. In particular, spatial variation of magnetization should cause scattering about the primary beam direction with an angular spread inversely proportional to the characteristic length of the magnetization variation. This effect was demonstrated in studies of nuclear resonance SAS of SR x-rays by iron foils [57]. The nuclear scattering was integrated over a delayed time window. The angular profiles of incident and the scattered radiation were analyzed by rocking a channel-cut Si(111) crystal. The results of the experiment are presented in Fig. 9. A considerable angular dispersion is observed in nuclear resonance scattering of x-rays by unmagnetized samples, in contrast to the case of scattering by magnetized ones. The observed effect is attributed to the magnetization inhomogeneity related to the domain walls. Additional information was obtained in measurements of the time differential dependences of nuclear resonance scattering. Quite different quantum beat patterns were observed at zero scattering angle and at that shifted by about 18 arcsec. This was direct proof that the signals arrived from the different subsystems of the sample. The last measurement illustrates quite new possibilities for nuclear resonance SAS compared to those involving x-ray and neutron atomic scattering processes.

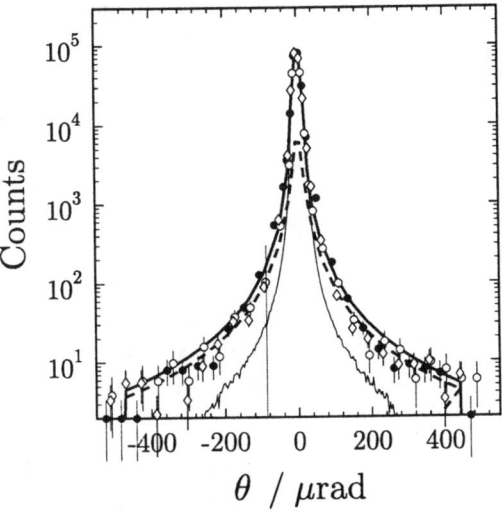

Fig. 9. Angular distributions of nuclear resonant scattering of SR from the 2.5 (○), 10 (●) and 36 (◊) μm unmagnetized ^{57}Fe foils. The thin solid line presents the profile of incident radiation. The dashed line shows the angular profile of nuclear resonance small-angle scattering [57].

4.5 Inelastic nuclear resonance scattering. It is well known, that the nuclear resonance can be considerably broadened and shifted in energy due to various excitations in condensed matter (thermal lattice vibrations, diffusion, spin waves, etc.). The energy transfer to the internal degrees of freedom can occur via nuclear recoil. By means of normal Mössbauer spectroscopy one can only cover a small range of shifted resonances.

A unique property of a SR source is the possibility to *excite nuclei with recoil* in a wide energy range. With the aid of a monochromator system a narrow energy band can be filtered from the broad spectrum of SR and this selected band can be easily used for scanning over a wide range of condensed matter excitations. The lack or excess of energy for the nuclear excitation is compensated by shifting the SR energy band. The relevant excitation spectra are measured by detecting the secondary radiation (γ-rays reradiated by nuclei, conversion electrons, or electronic shell fluorescence radiation) while the energy of the primary radiation is tuned. The excitation spectrum can be obtained with an energy resolution that is essentially determined by the monochromator system. Inelastic x-ray scattering can be studied in principle in both the spatially coherent, Bragg forward scattering and the incoherent, 4π-decay scattering channel.

Inelastic incoherent nuclear scattering of SR from solid, liquid and gaseous targets was observed in [58-62]. In [59] the phonon density of states was measured and compared for α-Fe (bcc local symmetry) and SS (fcc local symmetry) and for a number of samples of $SrFeO_x$. A dramatic difference in the phonon density of states for the oxide compound was observed while the system underwent structural changes. With the improved energy resolution [63] of about 4.4 meV the density of phonon states was measured in α-^{57}Fe over the temperature range 24-400 K [64] (Fig. 10). A considerable anharmonicity of the lattice vibrations was revealed. These first studies show the efficiency of the method for studying the low energy excitations in matter. An advantage of the new method is the

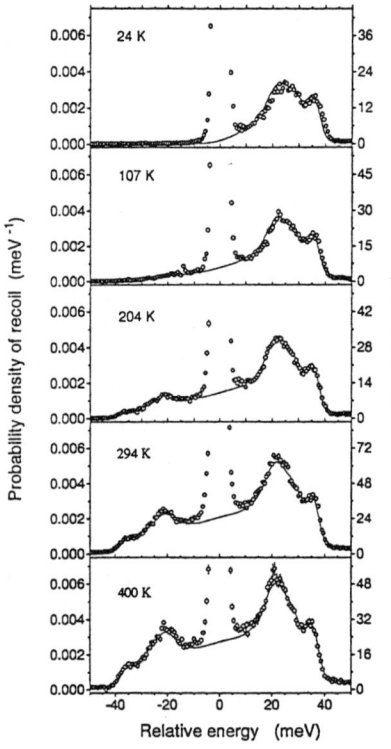

Fig. 10. Nuclear inelastic scattering of SR from α-^{57}Fe. The unshifted line is in the center. The side parts are due to recoil scattering with creation/ annihilation of phonons [64].

possibility to obtain the vibrational character of the local environment of the resonant nucleus.

In the method described above the nuclear resonance is used to obtain the scattering probability function from which the phonon spectrum can be extracted. Another technique suggested in [65] is based on using the nuclear resonance for an exit energy analysis of scattered radiation. In the new method the spectrum of the inelastically scattered x-ray radiation (by electrons in the target) is analyzed with the help of a nuclear resonance detector. The detector consists of a fast avalanche photodiode [66] with a 10 μm ^{57}Fe foil at the entrance as a resonance target. The energy bandpass of the detector was determined by the full width of the spectrum of the forward nuclear scattering in the iron foil (similar to those shown in Fig. 6a). The intensity of the nuclear scattered radiation within the detector bandpass was measured as a function of the difference between the energy of the incident beam and the nuclear transition energy. This measurement directly provided the spectrum of the x-ray radiation scattered inelastically by a sample. The spectrum is integrated over momentum transfer. The feasibility and energy resolution of the new technique are competitive with the energy analysis using crystal optics [65 and ref. therein]. Inelastic x-ray scattering spectra from water and gaseous Xe at room temperature were measured in [65]. The possibility to probe atomic dynamics in gases and the dynamics of density fluctuations in liquids was demonstrated.

Both techniques described above were successfully applied to study protein dynamics [67]. It was especially interesting to compare modes of motion in the whole sample (given by the exit analysis technique [65]) with those of the local environments of the resonant nuclei (given by the nuclear recoil scattering technique [58]).

It is clear that the methods described could be used to observe the effects of phase transitions (from liquid to gas or from solid to liquid) at the microscopic level as suggested in [62]. The possibility of performing neV scattering spectroscopy for studying inelastic or quasielastic scattering from non-resonant samples has been examined in [68].

4.6 Time-dependent perturbation of nuclear γ-resonance.

Dynamic perturbations in the nuclear γ-resonance can be either internal, inherent to the state of the system, or external artificially imposed perturbations. Thermal lattice dynamics, diffusion and different kinds of relaxation processes inherently cause changes in the strength, width and shape of the nuclear γ-resonance. External ultrasonic excitation of the target, or radiofrequency excitation of sample magnetization give rise to a sideband structure in the resonance. The dynamical effects take place either when the Mössbauer atom physically moves in space, or when the environment of the Mössbauer nucleus changes during the nuclear transition time. Therefore the dynamical effects can manifest themselves quite naturally in the time dependence of the coherent and incoherent nuclear scattering of SR.

The relevant transformations of the time dependence are predicted in the theory of nuclear resonant scattering of SR in the presence of diffusive motion of the nuclei [69]. Due to diffusive motion the phase correlation between the wavelets reradiated by the nuclei is lost with time, the higher diffusion rate the faster the decay of coherence. As a consequence the coherent signal fades more rapidly and the balance between the spatially coherent and incoherent channels is redistributed in favor of the incoherent one. Time-dependent SR nuclear scattering permits direct observation of diffusion effects. Recently a time-domain study of ^{57}Fe diffusion in the intermetallic alloy Fe_3Si was performed [70]. The transformation of the forward scattering time dependences from a simple exponential decay to a two-step exponential decay with an increase in the decay rate was clearly observed in the temperature range 876 K - 1013 K. The two different jumpy diffusion modes expected in the sample were revealed.

In contrast to the inherent excitations where the nuclear motion has a stochastic character, artificial dynamics has a regular character and can be manipulated.

In the forward scattering geometry the collective nuclear excitation extends over all targets through which the SR pulse was transmitted, even if they are spatially separated. This circumstance gives a unique possibility to perturb the collective excitation by manipulating one of the targets. The effect of ultrasonic vibrations on the nuclear scattering of SR was studied recently [71]. Two ^{57}Fe - enriched SS foils one of which was vibrating were excited by the SR pulse. When the vibration amplitude was comparable to or exceeded the wavelength of the radiation and the vibration phase was random with respect to time zero, the two components of the whole excitation decayed almost independently resulting in a decreased response. However, due to the periodic motion the initial phasing between the two foils was restored at times corresponding to multiples of the US period giving for a short time sharp emission spikes. In this way a 'nuclear exciton echo' was observed. In the kinematic approximation the echo spikes could also be interpreted as the effect of interference of the wave packets generated in the two nuclear targets. In the described scattering geometry the wave packets were not split spatially.

4.7 X-ray interferometry in nuclear scattering. The traditional interferometry technique where a photon passes two different optical paths and then recombines to give an interference pattern has been applied to x-ray nuclear resonance scattering in a series of works [72-75]. A conventional Laue-Laue-Laue (LLL) interferometer was used to demonstrate the interference between synchrotron x-ray beams affected by nuclear resonant forward scattering from two different nuclear targets inserted in each path of the interferometer in the gap between the beamsplitter and mirror. A Si phase shifter was put in both paths between the mirror and analyser [72]. The interference pattern between the beams reemitted with certain time delays and the effect of dispersion due to resonant nuclear scattering were observed [72-74]. The x-ray dispersion at resonance leads to both space and time variations in the scattered field amplitude (see Eq. (10,11)) [72]. This effect was demonstrated by putting nuclear targets of different thicknesses in the interferometer paths and analyzing the interference yield within different delayed-gate times [73]. The time distributions of the scattered photons (including both amplitude and phase variations) were measured with two nuclear targets having different hyperfine structures and with two identical targets one of which was moved to produce a Doppler shift [74]. The interference picture was in accord with the theory assuming the relevant wave packets to be superposed in the time domain.

A new wavefront-dividing interferometer was used to observe the interference effect in x-ray nuclear resonant scattering with an optical path difference of 2.4 mm [75]. It exceeds dramatically the coherence length provided by an electronically scattering monochromator, which is about 1μm, but is still much less than the coherence length delivered by a nuclear scattering monochromator, which may reach several meters.

The work described here is progressing rapidly and new applications of synchrotron radiation nuclear resonant scattering (**SRNRS**) are being found. The beam lines now available have the capability to do real material science. The following Mössbauer isotopes have been used up to now in synchrotron-based experiments: ^{57}Fe (14413 eV) [21], ^{169}Tm (8410 eV) [76], ^{119}Sn (23870 eV) [77], ^{83}Kr (9403 eV) [62,78], ^{181}Ta (6214 eV) [79], ^{151}Eu (21541.7 eV) [80,81], ^{161}Dy (25651.3 eV) [81].

REFERENCES

[1] G.T. Trammell, Proc. Int. Atomic Energy Agency Symp. on Chemical effects of Nuclear Transformations, Prague, 1960 (IAEA, Vienna, 1961) vol.1, p. 75.
[2] A.M. Afanas'ev and Yu. Kagan, Pis'ma Zh. Eks. Teor. Fiz. 2 (1965) 130 [JETP Lett. 2 (1965) 81].
[3] Yu. Kagan and A.M. Afanas'ev, Proc. Int. Atomic Energy Agency Symp. on Mössbauer Spectroscopy and its Application (IAEA, Vienna, 1972) p. 143.
[4] Yu. M. Kagan, A.M. Afanas'ev and V.G. Kohn, Phys. Lett. 68A (1978) 339.
[5] Yu. M. Kagan, A.M. Afanas'ev and V.G. Kohn, J. Phys. C 12 (1979) 615.
[6] J.P.Hannon and G.T. Trammell, Phys. Rev. 186 (1969) 306.
[7] G.T. Trammell and J.P. Hannon, Phys. Rev. B 18 (1978) 165; and B 19 (1979) 3835.
[8] J.P. Hannon and G.T. Trammell, Physica B 159 (1989) 161.
[9] J.P. Hannon and G.T. Trammell, in: Resonant Anomalous X-Ray Scattering, ed. G. Materlik, C.J. Sparks and K. Fischer (Elsevier, Amsterdam, 1994) p.565
[10] A.M. Afanas'ev and Yu. Kagan, Zh. Eksp. Teor. Fiz. 48 (1965) 327; [Sov. Phys. JETP 21 (1965) 215].
[11] Yu. Kagan, A.M. Afanas'ev and I.P. Perstnev, Zh. Eksp. Teor. Fiz. 54 (1968) 1530; [Sov. Phys. JETP 27 (1968) 819].
[12] Yu. Kagan and A.M. Afanas'ev, Z. Naturf. A28 (1973) 1351.
[13] J.P. Hannon and G.T. Trammell, Phys. Rev. 169 (1968) 315.
[14] G.V. Smirnov, V.V. Mostovoi, Yu.V. Shvyd'ko, V.N. Seleznev and V.V. Rudenko, Zh. Eksp. Teor. Fiz. 78 (1980) 1196 [Sov. Phys. JETP 51 (1980) 603].

[15] U. van Bürck, Hyp. Int. 27 (1986) 219.
[16] G.V. Smirnov, Hyp. Int. 27 (1986) 203.
[17] G.V. Smirnov, Hyp. Int. 72 (1992) 63.
[18] H. de Waard, Hyp. Int. 68 (1991) 143.
[19] G.V. Smirnov and A.I. Chumakov, in: Resonant Anomalous X-Ray Scattering, ed. G. Materlik, C.J. Sparks and K. Fischer (Elsevier, Amsterdam, 1994) p.609.
[20] S.L. Ruby, J. de Phys. 35 (1974) C6-209.
[21] E. Gerdau, R. Rüffer, H. Winkler, W. Tolksdorf, C.P. Klages and J.P. Hannon, Phys. Rev. Lett. 54 (1985) 835.
[22] E. Gerdau, R. Rüffer, H.D Rüter and J.P. Hannon Hyp. Int. 40 (1988) 49.
[23] J. Arthur, D.E. Brown, S.L. Ruby, G.S. Brown, and G.K. Shenoy, J. Appl. Phys. 67, (1990) 5704.
[24] R. Rüffer, Synchrotron Radiation News, 5 (1992) 25.
[25] E. Gerdau and U. van Bürck, in: Resonant Anomalous X-Ray Scattering, ed. G. Materlik, C.J. Sparks and K. Fischer (Elsevier, Amsterdam, 1994) p.589.
[26] G.V. Smirnov, Hyp. Int. 97/98 (1996) 551.
[27] G.T. Trammell, Phys. Rev. 126 (1962) 1045.
[28] A.I. Chumakov, G.V. Smirnov, M.V. Zelepukhin, U. van Bürck, E. Gerdau, R. Rüffer and H.D. Rüter, Europhys. Lett. 17 (1992) 269.
[29] U. van Bürck, D.P. Siddons, J.B. Hastings, U. Bergmann and R. Hollatz, Phys. Rev. B 46 (1992) 6207.
[30] D.P. Siddons, U. Bergmann and J.P. Hastings, Phys. Rev. Lett. 70 (1993) 359.
[31] U. Bergmann, D.P. Siddons and J.B. Hastings, in: Resonant Anomalous X-Ray Scattering, ed. G. Materlik, C.J. Sparks and K. Fischer (Elsevier, Amsterdam, 1994) p.619.
[32] S. Kikuta, in: Resonant Anomalous X-Ray Scattering, ed. G. Materlik, C.J. Sparks and K. Fischer (Elsevier, Amsterdam, 1994) p.635.
[33] R.Rüffer and A. Chumakov, Hyp. Int. 97/98 (1996) 589.
[34] G.V. Smirnov, V.V. Sklyarevskii, R.A. Voskanyan and A.N. Artem'ev, Pis'ma Zh.Eksp. Teor. Fiz. 9 (1969) 123 [Sov. Phys. JETP Lett. 9 (1970) 70] and Proc. Int. Conf. on Appl. Mössbauer Effect, Tihany, 1969 (Budapest, 1971) p. 73.
[35] H.D. Rüter, R. Rüffer, R. Hollatz, W. Sturhahn and E. Gerdau, Hyp. Int. 58 (1990) 2477.
[36] S. Kikuta, Y. Yoda, Y. Hasegawa, K. Izumi, T. Ishikawa, X.W. Zhang, S. Kishimoto, H. Sugiyama, T. Matsushita, M. Ando, C.K. Suzuki, M. Seto, H. Ohno and H. Takei, Hyp. Int. 71 (1992) 1491.
[37] U. van Bürck, R.L. Mössbauer, E. Gerdau, R. Rüffer, R. Hollatz, G.V. Smirnov and J.P. Hannon, Phys. Rev. Lett. 59 (1987) 355.
[38] A.I. Chumakov, G.V. Smirnov, M.V. Zelepukhin, U. van Bürck, E. Gerdau, R Rüffer and H.D. Rüter, Hyp. Int. 71 (1992) 1341.
[39] H.D. Rüter, R. Rüffer, E. Gerdau, R. Hollatz, A.I. Chumakov, M.V. Zelepukhin, G.V. Smirnov and U. van Bürck, Hyp. Int. 58 (1990) 2473.
[40] A.I. Chumakov, M.V. Zelepukhin, G.V. Smirnov, U. van Bürck, R. Rüffer, R. Hollatz, H.D. Rüter, E. Gerdau, Phys. Rev. B 41 (1990) 9545.
[41] G.V. Smirnov, U. van Bürck, A.I. Chumakov, A.Q.R. Baron and R. Rüffer, to be published.
[42] A.I. Chumakov, G.V. Smirnov, S.S. Andreev, N.N. Salashenko and S.I. Shinkarev, Pis'ma Zh. Eksp. Teor. Fiz. 55 (1992) 495; [Sov. Phys. JETP Lett. 55 (1992) 70].
[43] R. Röhlsberger, E. Witthoff, E. Lüken and E. Gerdau, J. Appl. Phys. 74 (1993) 1933.
[44] M.V. Gusev, A.I. Chumakov and G.V. Smirnov, Pis'ma Zh. Eksp. Teor. Fiz. 58 (1993) 251 [Sov. Phys. JETP Lett. 58 (1993) 257].
[45] T.S. Toellner, W. Sturhahn, R. Röhlsberger, E.E. Alp, C.H. Sowers, E.E. Fullerton, Phys. Rev. Lett. 74 (1995) 3475.
[46] A.I. Chumakov, G.V. Smirnov, A.Q.R. Baron, J. Arthur, D.E. Brown, S.L. Ruby, G.S. Brown and N.N. Salashenko, Phys. Rev. Lett. 71 (1993) 2489.

[47] A.Q.R. Baron, J. Arthur, S.L. Ruby, A.I. Chumakov, G.V. Smirnov and G.S. Brown, Phys. Rev. B 50 (1994) rapid comm. 1.
[48] J.P Hannon, G.T. Trammell, M. Mueller, E. Gerdau, H. Winkler and R. Rüffer, Phys. Rev. Lett. 43 (1979) 636.
[49] J.P. Hannon, G.T. Trammell, M. Mueller, E. Gerdau, R. Rüffer and H. Winkler, Phys. Rev. B 32 (1985) 6363, 6374.
[50] M.A. Andreeva, Hyp. Int. 97/98 (1996) 605.
[51] R. Röhlsberger, E. Gerdau, R. Rüffer, W. Sturhahn, T.S. Toellner, A.I. Chumakov, E.E. Alp, to be published.
[52] J.B. Hastings, D.P. Siddons, U. van Bürck, R. Hollatz and U. Bergmann, Phys. Rev. Lett. 66 (1991) 770.
[53] T. Ishikawa, Y. Yoda, K. Izumi, C.K. Suzuki, X.W. Zhang, M. Ando and S. Kikuta,Rev. Sci. Instrum. 63 (1992) 1015.
[54] U. Bergmann, S.D. Shastri, D.P. Siddons, B.W. Battermann and J.B. Hastings, Phys. Rev. B 50, (1994) 5957.
[55] H.F. Grünsteudel, H.J. Hesse, A.I. Chumakov, H. Grünsteudel, O. Leupold, J. Metge, R. Rüffer and G. Wortmann, Hyp. Int. (C) 1 (1996) 509.
[56] R. Rüffer, ESRF Newsletter 22 (1994) 12.
[57] Yu.V. Shvyd'ko, A.I. Chumakov, E. Gerdau, R. Rüffer, A.Q.R. Baron, A. Bernhard, J. Metge, Phys. Rev. B Rapid Comm. submitted (1996).
[58] M. Seto, Y. Yoda, S. Kikuta, X.W. Zhang and M. Ando, Phys. Rev. Lett. 74 (1995) 3828.
[59] W. Sturhahn, T.S. Toellner, E.E. Alp, X. Zhang, M. Ando, Y. Yoda, S. Kikuta, M. Seto, C.W. Kimball, and B. Dabrowskii, Phys. Rev. Lett. 74 (1995) 3832.
[60] X.W. Zhang, Y. Yoda, M. Seto, Yu. Maeda, M. Ando and S. Kikuta, Jpn. J. Appl.Phys. 34 (1995) L330.
[61] A.I. Chumakov, R. Rüffer, H. Grünsteudel, H.F. Grünsteudel, G. Grübel, J. Metge, O. Leupold, H.A. Goodwin, Europhys. Lett. 30 (1995) 427.
[62] A.Q.R. Baron, A.I. Chumakov, S.L. Ruby, J. Arthur, G.S. Brown, G.V. Smirnov and U. van Bürck, Phys. Rev. B 51 (1995) 16384.
[63] The best energy resolution today using wave selecting monochromators is achieved at the ESRF $\Delta E=1.65$meV and in APS $\Delta E=0.83$ meV, private communication.
[64] A.I. Chumakov, R. Rüffer, A.Q.R. Baron, H. Grünsteudel, H.F. Grünsteudel, Phys. Rev. B 54 (1996) 9596.
[65] A.I. Chumakov, A.Q.R. Baron, R.Rüffer, H. Grünsteudel, H.F. Grünsteudel and A. Meyer, Phys. Rev. Lett. 76 (1996) 4258.
[66] A.Q.R. Baron and S.L. Ruby, Nucl. Instrum. Methods A 343 (1994) 517.
[67] K. Achterhold, C. Keppler, U. van Bürck, W. Potzel, P. Schindelmann, E.-W. Knapp, B. Melchers, A.I. Chumakov, A.Q.R. Baron, R. Rüffer and F. Parak, European Biophys. J. Lett. 24 (1996).
[68] J.Z. Tischler, B.C. Larson, L.A. Boatner, E. Alp, T. Mooney and Q. Shen, in: Resonant Anomalous X-Ray Scattering, ed. G. Materlik, C.J. Sparks and K. Fischer (Elsevier, Amsterdam, 1994) p.647.
[69] G.V. Smirnov and V.G. Kohn, Phys. Rev. B 52 (1995) 3356.
[70] B. Sepiol, A. Meyer, G. Vogl, R. Rüffer, A.I. Chumakov, A.Q.R. Baron, Phys. Rev. Lett. 76 (1996) 3220.
[71] G.V. Smirnov, U. van Bürck, J. Arthur, S.L. Popov, A.Q.R. Baron, A.I. Chumakov, S.L. Ruby, W. Potzel and G.S. Brown, Phys. Rev. Lett. 77 (1996) 183.
[72] Y. Hasegawa, Y. Yoda, K. Izumi, T. Ishikawa, S. Kikuta, X.W. Zhang and M. Ando, Phys. Rev. B 50 (1994) 17748.
[73] Y. Hasegawa, Y.Yoda, K. Izumi, T. Ishikawa, S. Kikuta, X.W. Zhang, H. Sugiyama, and M. Ando, Jpn. J. Appl. Phys. 33 (1994) L772.
[74] K. Izumi, Y. Yoda, T. Ishikawa, X.W. Zhang, M. Ando and S. Kikuta, Jpn. J. Appl. Phys. 34 (1995) 4258.

[75] K. Izumi, T. Mitsui, M. Seto, Y. Yoda, T. Ishikawa, X.W. Zhang, M. Ando and S. Kikuta, Jpn. J. Appl. Phys. 34 (1995) 5862.
[76] W. Sturhahn, E. Gerdau, R. Hollatz, R. Rüffer, H.D. Rüter and W. Tolksdorf, Europhys. Lett. 14 (1991) 821.
[77] E.E. Alp, T.M. Mooney, T. Toellner, W. Sturhahn, E. Witthoff, R. Röhlsberger and E. Gerdau, Phys. Rev. Lett. 70 (1993) 3351.
[78] D.E. Johnson, D.P. Siddons, J.Z. Larese and J.B. Hastings, Phys. Rev. B 51 (1995) 7909.
[79] A.I. Chumakov, A.Q.R. Baron, J. Arthur, S.L. Ruby, G.S. Brown, G.V. Smirnov, U. van Bürck and G. Wortmann, Phys. Rev. Lett. 75 (1995) 549.
[80] O. Leupold, J. Pollmann, E. Gerdau, H.D. Rüter, G. Faigel, M. Tegze, G. Bortel, R. Rüffer, A.I. Chumakov and A.Q.R. Baron, Europhys. Lett. **35** (1996) 671.
[81] I. Koyama, Y.Yoda, X.W. Zhang, M. Ando and S. Kikuta, Jpn. J. Appl. Phys. to be published; see also S. Kikuta et al. in these proceedings.

Applications of X-Ray Nuclear Resonant Scattering with the Use of Synchrotron Radiation

S.Kikuta, Y.Yoda, I.Koyama, T.Shimizu, H.Igarashi, K.Izumi[1],
Y.Kunimune[2], M.Seto[a], T.Mitsui[b], T.Harami[b], X.Zhang[c] and M.Ando[c]

Department of Applied Physics, University of Tokyo, Bunkyo-ku, Tokyo 113, Japan
a Research Reactor Institute, Kyoto University, Kumatori-cho, Osaka 590-04, Japan
b Japan Atomic Energy Research Institute, Kamigoricho, Hyogo 678-12, Japan
c National Laboratory for High Energy Physics, Tsukuba, Ibaraki 305, Japan

Abstract. Over the past ten years theoretical and experimental studies on nuclear resonant scattering with synchrotron radiation have been carried out extensively and new fields of applications have been developed. We report on several studies related to nuclear resonant scattering that we have performed recently, as follows. 1) Nuclear excitation of ^{161}Dy and ^{151}Eu nuclei by using L X-rays emitted in the internal conversion process. 2) Nuclear forward scattering from a ^{57}FeBO$_3$ crystal platelet perturbed with the 180°- switching of magnetic field and the rf magnetic field. 3) Inelastic nuclear scattering for determining phonon energy spectra. 4) Three-beam nuclear Bragg scattering. 5) Two-photon correlations in the X-ray region.

Introduction

Studies on nuclear resonant scattering have been carried out using synchrotron radiation (SR) over the past ten years (1, 2). Outstanding features of SR such as the high brilliance, the pulsed time structure and the controllable polarization are very suitable for the study of nuclear resonance phenomena of X-rays. In particular the pulsed time structure of SR makes it possible to use the time differential measurement by which time-domain Mössbauer spectroscopy is realized in contrast to conventional γ-ray Mössbauer spectroscopy.

We report on several studies that we have performed recently.

There are many different Mössbauer isotopes that are suitable for nuclear resonant excitation with SR. We describe how nuclear excitation of ^{161}Dy and ^{151}Eu nuclei can be studied by using L X-rays emitted in the internal conversion process.

In both nuclear Bragg scattering and nuclear forward scattering the coherence-conserving decay caused by collective excitation with a SR pulse takes place. The study of a temporal behaviour of the decay process is an interesting problem if a certain perturbation is given to the nuclear scatterer. Nuclear forward scattering from a ^{57}FeBO$_3$ crystal platelet perturbed with the 180°-switching of magnetic field and the rf magnetic field is described.

---

[1] Present address : *Fundamental Research Laboratory, NEC, Tsukuba, Ibaraki 305, Japan*
[2] Present address : *ULSI Device Development Laboratory, NEC, Sagamihara 229, Japan*

On the other hand, nuclear resonant excitation with recoil became possible by using SR with the energy width of the order of meV. The phonon energy spectrum of a polycrystalline α-Fe foil has been observed by inelastic nuclear resonant scattering. The temperature-dependent changes of the phonon energy spectra of an α-Fe foil and fine particles of iron are reported here.

In addition, phenomena of multiple diffraction of resonant X-rays in a perfect crystal seem to be meaningful from a stand-point of X-ray optics. Here the time evolution of diffracted beams from a ^{57}FeBO$_3$ crystal are observed under the condition of three-beam nuclear Bragg scattering.

Finally, the study concerning X-ray coherence may be one of the most attractive fields. Several experiments on X-ray interferometry related to nuclear resonant scattering have been made. The successful observation of two-photon correlations in the X-ray region using a high resolution electronic monochromator is reported. As the next step, a nuclear monochromator will be used to make the effect of photon bunching more conspicuous.

The experiments were performed at the TRISTAN-Accumulation Ring (AR) (KEK, Tsukuba) except the experiment of two-photon correlations (3). The ring was operated at an energy of 6.5 GeV and a current of 30 mA in the single bunch mode. SR pulses were emitted on a repetition period of 1.26 μs with a very short width of about 100 ps. SR emitted from the in-vacuum type undulator was used. X-rays with an energy of 14.4 keV corresponding to the nuclear resonant energy of ^{57}Fe was obtained as the third harmonic.

1. NUCLEAR EXCITATION OF ^{161}Dy AND ^{151}Eu NUCLEI

The observation of nuclear excitation by SR has been reported in several isotopes such as ^{57}Fe (4), ^{119}Sn (5), ^{169}Tm (6), ^{83}Kr (7) and ^{181}Ta (8). In these isotopes resonant energies are moderately low. Here the observation of nuclear resonant excitation of ^{161}Dy and ^{151}Eu by using emission of time-delayed L X-rays is reported. Following nuclear resonant excitation, the de-excitation process may proceed via X-ray emission or conversion electron emission. The internal conversion process results in emission of an X-ray or an Auger electron. These emissions have a time-delayed characteristic of the exponential decay. We used the yield of L X-rays emitted in the L conversion process for the observation of nuclear excitation of ^{161}Dy and ^{151}Eu. This technique will be suitable for nuclei whose life-times are too short to use nuclear resonant scattering accompanied with the speed-up effect of the radiative decay.

A monochromator with an energy width of about 50 meV composed of two Si (660) channel-cut crystals in a dispersive setting was used. A tablet-shaped specimen of ^{161}Dy was made by pressing a mixture of ^{161}Dy$_2$O$_3$ powder and boron powder. A specimen of ^{151}Eu was also made by the same procedure using ^{151}Eu$_2$O$_3$ powder. Monochromatized X-rays were incident on the specimen with a glancing angle of about 6°. L X-rays emitted from the specimen in the internal conversion process were detected by an avalanche photodiode (APD) detector set nearly normal to the specimen at a distance of about 1mm. The APD detector had a sensitive area of 16 mm in diameter and a time resolution of better than 1ns. In Fig. 1 the time spectrum of ^{161}Dy L X-rays observed with a delay time later than 35 ns is shown, where the noise is negligibly small. The observed points in the decay curve of ^{161}Dy were fitted with an exponential curve by the least squares method. The decay constant τ, that is the half lifetime, of ^{161}Dy was obtained as

29.4 ± 1.9 ns. This value agrees well with the half lifetime of an isolated ^{161}Dy nucleus obtained by using the conventional Mössbauer γ-ray technique (9). Further the nuclear resonant energy of ^{161}Dy was determined as 25.65129 ± 0.00016 keV by using the X-ray diffraction technique. This value agrees fairly well with that reported in the literature (9).

The experiment of nuclear excitation of ^{151}Eu was performed by the same way as that of ^{161}Dy. The half lifetime was obtained as 9.3 ± 0.3 ns and the nuclear resonant energy as 21.54149 ± 0.00016 keV.

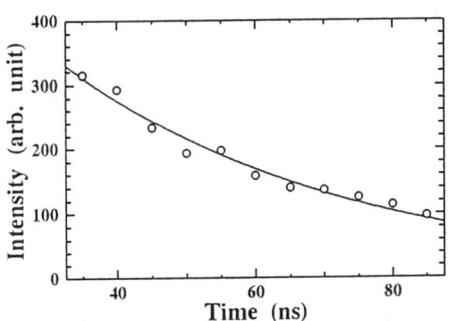

Figure 1. Time spectrum of ^{161}Dy L X-ray emission in the internal conversion process.

2. PERTURBED NUCLEAR FORWARD SCATTERING

The 180°- Switching of Magnetic Field

The studies of Mössbauer γ-ray spectroscopy in the time domain under external perturbations have attracted much interest. Recently the time evolution of nuclear Bragg scattering from ^{57}FeBO$_3$ crystals using SR under conditions of fast switching of the magnetic hyperfine field direction has been studied by Shvyd'ko et al. (10,11). At the 180°-switching the time reversal of the quantum beat pattern in nuclear resonant scattering was observed (12).

We report on the observation of time reversal of the quantum beat pattern in nuclear forward scattering from a ^{57}FeBO3 crystal platelet at the 180°-switching. Iron borate, FeBO3, is a weak ferromagnet (a canted antiferromagnet). A high-quality FeBO3 crystal can undergo a fast rotation of magnetization in the (111) easy plane in a few nanoseconds by means of the pulsed magnetic field of about 5 Oe. As shown in Fig. 2, σ-polarized X-rays were incident on a high resolution Si monochromator. It was composed of two channel-cut Si crystals nested with each other, in which the 422 asymmetric reflections and the 12 2 2 symmetric reflections were used to obtain highly monochromatic X-rays with an energy width of 6.4 meV. Next an enriched (95%) ^{57}FeBO$_3$ crystal platelet of 64 μm thickness was set perpendicular to the beam axis. The surface of the crystal was parallel to (111). A constant magnetic field Hc was applied parallel to the crystal surface in the direction of the magnetic polarization vector of X-rays. Under this condition nuclear forward scattering with the hyperfine transitions of Δm = 0 occurs. The pulsed magnetic field was produced by 10 mm diameter double-

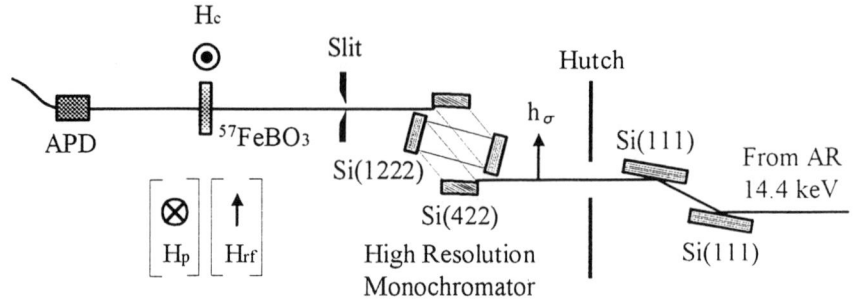

Figure 2. Schematic side view of the experimental setup for perturbed nuclear forward scattering.

wound Helmholtz coils connected to a current pulse generator. The system generating the pulsed magnetic field is based on the design of Ref. (13). The generator was operated at a radio frequency of 265 kHz which was obtained by dividing the radio frequency from the cavity of AR by three. The width of pulsed magnetic field was about 120 ns. Its rise time and fall time were 25.2 ns and 6.9 ns, respectively. The origin of the time scale was specified by the prompt signal of X-rays transmitted through the crystal platelet. The 180°-switching was produced by applying a pulsed magnetic field H_p = 27 Oe antiparallel to a static magnetic field H_c = 15 Oe. Here the trailing edge of the pulse was used. Nuclear forward scattered X-rays were observed by an APD detector with a time resolution of about 0.2 ns and a counting efficiency of about 3 %.

The time evolution of nuclear forward scattering from $^{57}FeBO_3$ was observed in the case that an abrupt magnetic field reversal was given in the decay process. Figure 3 shows the observed time spectra. For comparison, the time spectrum in the case that only a static magnetic field was applied is shown in Fig. 3(a). In the time spectrum the quantum beat is seen with a single period of about 15 ns, which arises from interference of two wave fields related to the hyperfine transitions of $\Delta m = 0$. The time, t_0, at which magnetization was reversed, was selected at three points 92 ns, 99 ns and 105 ns within one period of the quantum beat. The perturbed time spectra are shown in Fig. 3(b), (c) and (d). The period of the quantum beat after t_0 was about 15 ns except the region of one period around t_0. This is the same as the period of the quantum beat in the unperturbed time spectrum. However, the quantum beat patterns just after t_0 show drastic changes depending on t_0. In Fig. 3(b), just before t_0 the beat intensity is decreasing and at t_0 it begins to increase. The interval between two dips before and after t_0 widens to about 20 ns. On the contrary, in Fig. 3(c), the beat intensity which is increasing just before t_0 begins to decrease at t_0. The interval between two peaks before and after t_0 becomes 23 ns. In Fig. 3(d), t_0 is selected at the peak position of the beat pattern. The next peak arises after about 16 ns from t_0 and on the whole the beat pattern resembles in profile that in Fig. 3(a). These results show that the beat pattern is symmetric with respect to the switching time if attenuation due to the decay is neglected.

As Shvyd'ko et al. (11,12) reported, the time-reversal phenomena occur in

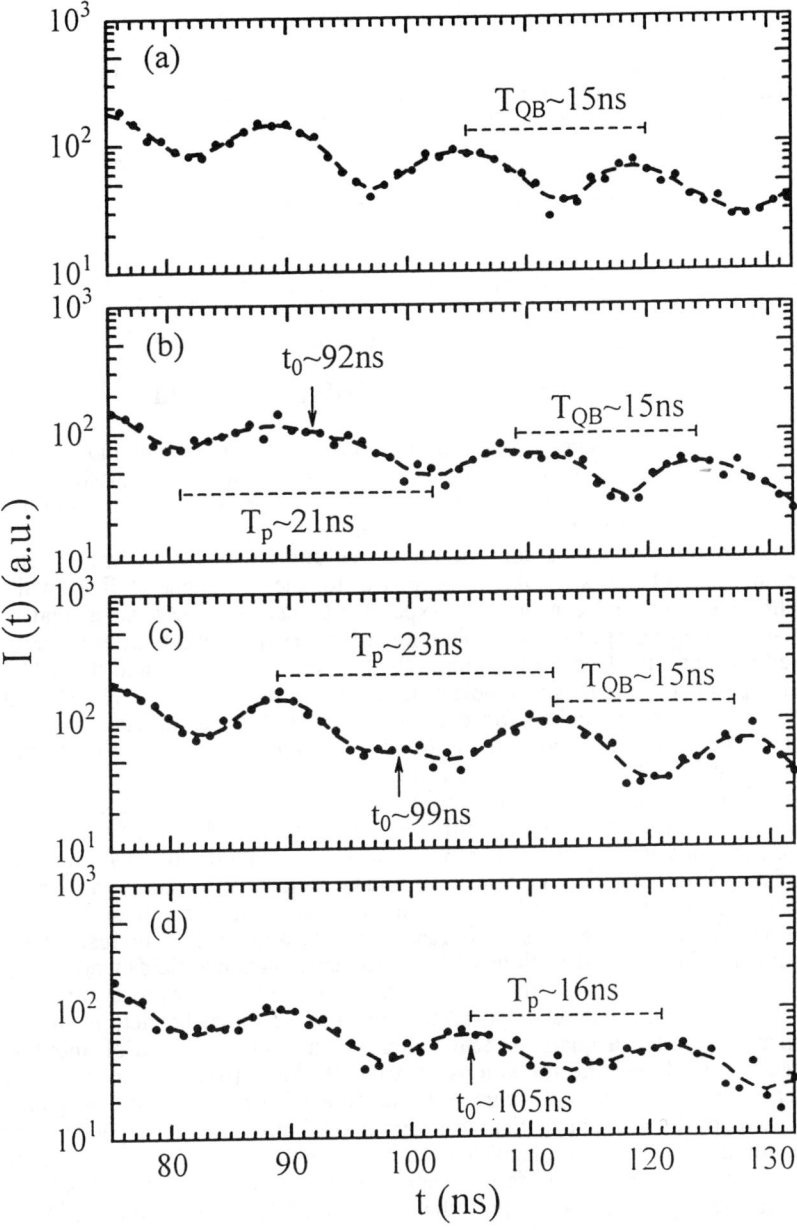

Figure 3. Time spectra of nuclear forward scattering from a $^{57}FeBO_3$ crystal platelet. (a): Unperturbed time spectrum. (b), (c) and (d): Perturbed time spectra for the cases that the 180°-switching was given at different times, t_0.

nuclear forward scattering as well as in nuclear Bragg scattering. The essential point is as follows. The hyperfine transitions $m_g \rightarrow m_e$ are denoted by $l = -3m_g - m_e$. At the moment of arrival of a SR pulse on the $^{57}FeBO_3$ crystal under the present experimental condition two hyperfine transitions $l = \pm 2$ corresponding to $\Delta m = 0$ can take place. At the moment of the 180°-switching the transitions l are transformed into the transitions $-l$. The relative transition frequencies arising from the hyperfine magnetic interaction change from Ω_l to Ω_{-l}. Since in this case $\Omega_{-l} = -\Omega_l$, no change in the period of the quantum beat will occur at the 180°-switching. The amplitude of the radiation reemitted by the excited nuclei has a time dependent factor which is related directly to the quantum beat, $\sum_l \exp[-i\Omega_l t]$. After the switching moment to it changes to $\sum_l \exp[-i\Omega_l (2t_0 - t)]$. It shows that the temporal behaviour of the quantum beat pattern is reversed at to.

Application of the rf Magnetic Field

The influence of magnetoelastic vibration on a $^{57}FeBO3$ crystal induced by the application of the rf magnetic field have been studied by Mössbauer γ-ray spectroscopy. It has also been analyzed by measuring the time evolution of nuclear Bragg scattering with the use of SR (14).

We report on the observation of the time evolution of nuclear forward scattering from an $^{57}FeBO3$ crystal platelet exposed to the rf magnetic field which is synchronized with a SR pulse. The experimental setup is the same as that in the 180°-switching except for the application of the rf magnetic field in place of the pulsed magnetic field, as shown in Fig. 2. The static magnetic field of 10 Oe was applied parallel to the magnetic polarization vector of the X-rays. The rf magnetic field was applied parallel to the crystal surface and perpendicular to the static magnetic field. It was produced by 20mm diameter Helmholtz coils, having the amplitude of 15 Oe and the frequency of 2.38 MHz which is three times as large as the radio frequency from the cavity of AR. The timing of applying the rf magnetic field was phase-locked with respect to the arrival time of the SR pulse. Three positions were chosen. The position 1 corresponds to zero amplitude of the rf magnetic field. The composite magnetic field of the static and rf magnetic fields oscillates between ±56° around the direction of static magnetic field. The positions 2 and 3 have phase differences of π/4 and −π/2 with respect to the position 1. The initial directions of the composite magnetic fields made angles of 35° and -56°. In Fig. 4(a), the time evolution of nuclear forward scattering from the $^{57}FeBO_3$ crystal platelet exposed to only a static magnetic field is shown for comparison. It has a quantum beat pattern with a single period of about 13 ns caused by two hyperfine transitions of $\Delta m = 0$. Fig. 4(b) is a time spectrum for the case of the position 1. Since at the time of arrival of the SR pulse the amplitude of the rf magnetic field was zero and the direction of magnetization was perpendicular to the static magnetic field, two transitions of $\Delta m = 0$ were excited. It is seen that there is no difference between the periods of the beat patterns in Fig. 4(a) and (b). It shows that the process of two transitions of $\Delta m = 0$ continues without change throughout the decay. On the other hand, a drastic change was observed in the intensity of the time spectrum in Fig. 4(b). The forward-scattered intensity is initially suppressed compared with the unperturbed case, but it is enhanced at about 210 ns after SR incidence. This time corresponds to the magnetization vector returning to its original direction. These phenomena may be considered as follows. At the moment of the SR pulse arrival all resonant nuclei scatter in phase in the forward direction. After that the positions of the resonant nuclei change periodically owing to magnetostriction caused by the rf magnetic field. In the crystal platelet resonant nuclei along the beam path will as a whole oscillate uniformly, since the wavelength of magnetoelastic vibration is large (~1mm) compared with the size relevant to scattering. However, it may be natural that there are relative displacements of the order of X-ray wavelength among individual nuclei. Then X-ray waves reemitted from individual nuclei become somewhat out of phase. Because of destructive interference among the reemitted X-ray waves the forward-scattered intensity will decrease. When the direction of magnetic field return to the initial direction, the positions of resonant nuclei are restored to their initial state. Accordingly, the constructive phase relation comes back and enhancement of the forward-scattered intensity takes place. In Fig. 4(c) and (d), since two transitions of $\Delta m = 0$ and four transitions of $\Delta m = \pm 1$ are related to nuclear forward scattering, the quantum beats show complex interference patterns. The intensity enhancements of nuclear forward scattering are seen at the times when the initial states of the scatterer are restored as in Fig. 4(b).

In the experiment of the 180°-switching, application of the pulsed magnetic field results in the abrupt switch of the crystal magnetizaion and the direction of the hyperfine magnetic field. This switching is realized under the condition that the duration of switching is much shorter than the mean life-time (141ns) and the period of Larmor precession (40ns). The switching of the hyperfine magnetic field causes a transformation of the nuclear transitions excited by SR. On the other hand, in the experiment of applying the rf magnetic field, the direction of the hyperfine magnetic field oscillates with a long period of the rf magnetic field (420ns).

3. INELASTIC NUCLEAR SCATTERING WTITH PHONON CREATION AND ANNIHILATION

The experiments of observing phonon energy spectra with radioisotope sources are extremely difficult because the spread of a phonon spectrum is much wider than even a very wide linewidth. However, recently we have observed the phonon energy spectrum by using the nuclear resonant scattering of SR for the first time (15). Furthermore, we have shown that this method is applicable not only to solid matter but also to liquid (16). Further, several measurements of inelastic scattering have been made (17, 18).

It is well known that measurements of the temperature dependence of phonon energy spectra are very effective for the study of condensed matter. The observations of the temperature dependent changes of the phonon energy spectra of an α-Fe foil (19) and fine particles of iron are described. The experimental setup shown in Fig. 5 is almost the same as previous measurements (15) except for the position of the X-ray detector. A cryostat or a furnace was used for changing temperature. A polycrystalline α-iron foil enriched in ^{57}Fe to 95.45% was used as a specimen. Another specimen was fine particles of ^{57}Fe, whose size was about 20 Å in diameter. The specimen was prepared by evaporating ^{57}Fe metal and depositing on the surface of oil containing a surfactant (20, 21). A high resolution monochromator produced a probe beam with a 6.4 meV energy width at the 14.413 keV nuclear resonance in ^{57}Fe; the peak energy of the beam was tuned with a relative accuracy of 0.6 meV. The nuclear resonant scattering

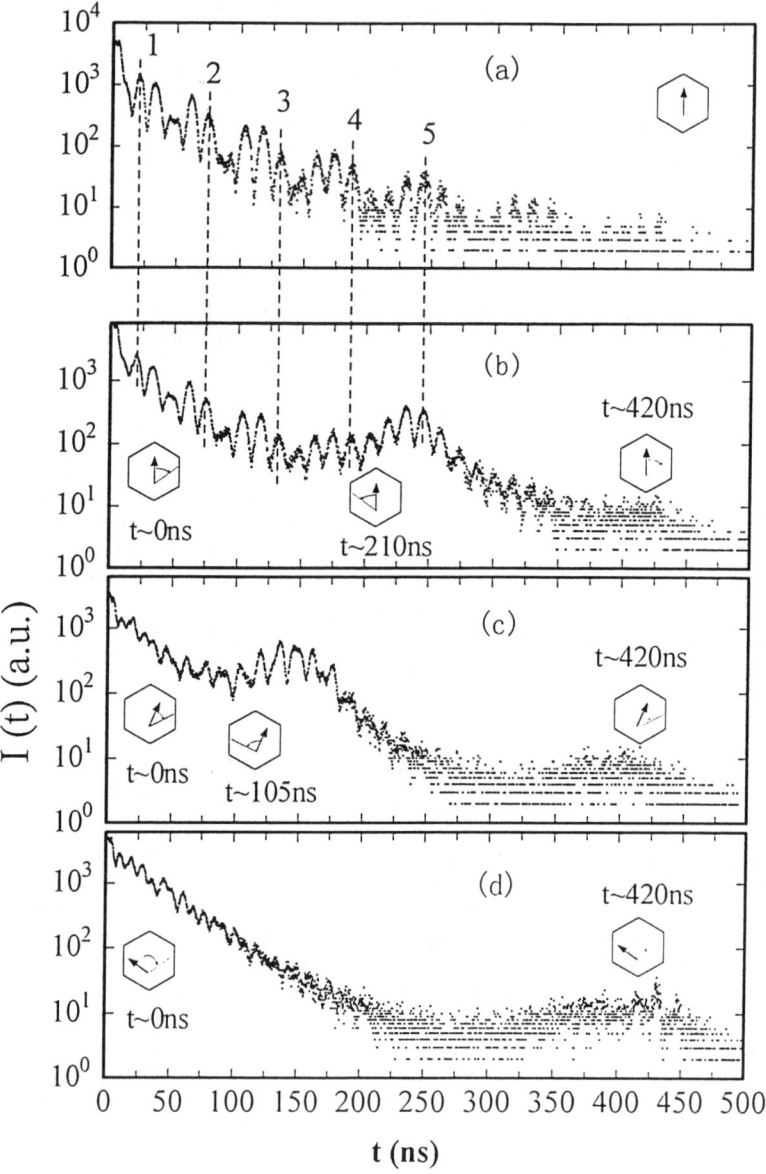

Figure 4. Time evolution of nuclear forward scattering from a $^{57}FeBO_3$ crystal platelet. (a): Only a static magnetic field was applied. (b), (c) and (d): rf magnetic fields were applied at the positions 1, 2 and 3, respectively, in addition to the static magnetic field.

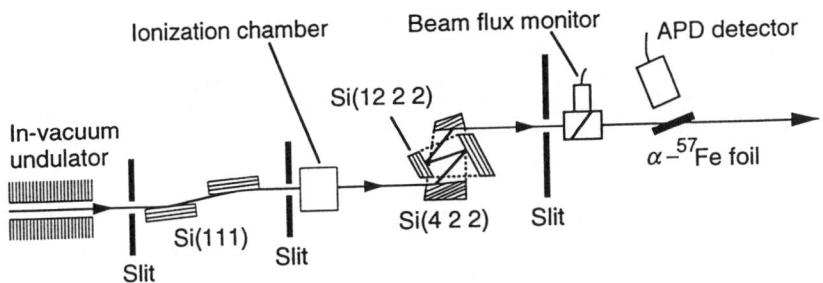

Figure 5. Schematic side view of the experimental setup for inelastic nuclear scattering.

from the specimen was observed with an APD detector. The energy spectra of the nuclear resonant excitation was observed as a function of the incident X-ray energy by counting the number of delayed photons. The incident photon energy was scanned by changing the diffraction angle of the high resolution monochromator.

The energy spectra of α-Fe observed at 300 K and 150 K are shown as filled circles in Fig. 6. In each spectrum, inelastic scattering due to phonon scattering is observed at both sides of the elastic peak. The measured energy spectra are well fitted with the spectra calculated from the phonon energy distribution function of

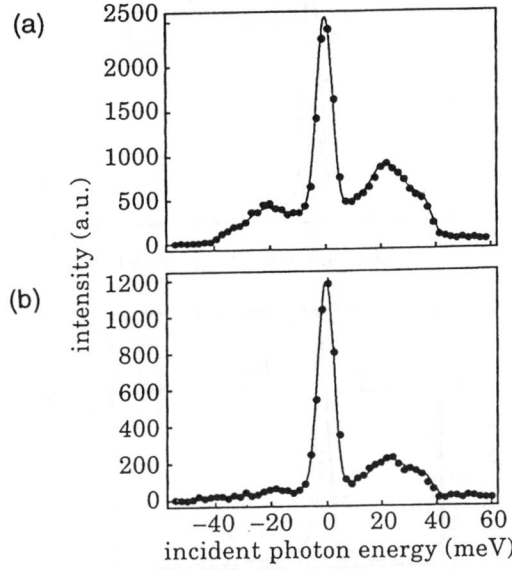

Figure 6. Energy spectra of nuclear resonant scattering from a polycrystalline α-^{57}Fe foil. The sample temperatures are (a) 300K and (b) 150K. The incident photon energy is relative to the energy of the first excited state of ^{57}Fe nucleus. The solid curves are the spectra calculated from the phonon energy distribution function of α-Fe.

α-Fe. The calculated spectra are shown in Fig. 6 as solid lines. The asymmetry of the intensity of the sidebands is observed in each spectrum. The observed asymmetry in the spectrum at 150 K is larger than that observed at 300 K. The area ratio of the high energy side to the low energy side is estimated to be about 2 at 300 K, and about 4 at 150 K. Inelastic scattering observed in the high energy side and in the low energy side corresponds to the nuclear resonant scattering with phonon creation and phonon annihilation, respectively. Phonon annihilation probability is proportional to the number of excited phonons. At lower temperatures there are fewer phonons in solids. Thus, the phonon annihilation probability decreases at low temperatures.

Furthermore, the observed ratio of the inelastic part to the elastic part at 150 K is smaller than that observed at 300 K. The temperature-dependent reduction of the ratio is directly correlated with the well-known fact that the recoilless fraction increases at low temperatures. In connection with the reduction of the ratio, decrease of multi-phonon scattering also occurs at low temperatures. Therefore, inelastic scattering at 150 K is almost due to one-phonon scattering.

Figure 7 shows the energy spectra of fine particles of Fe observed at 300 K and 373 K (solid curves). For comparison the energy spectra of nuclear forward scattering from an α-Fe foil are also shown (dotted curves). Comparing the two curves at 300 K, the peak positions nearly coincide. However, at 373 K the peak position of the solid curve shifts by 1.9 meV to the higher energy side of that of

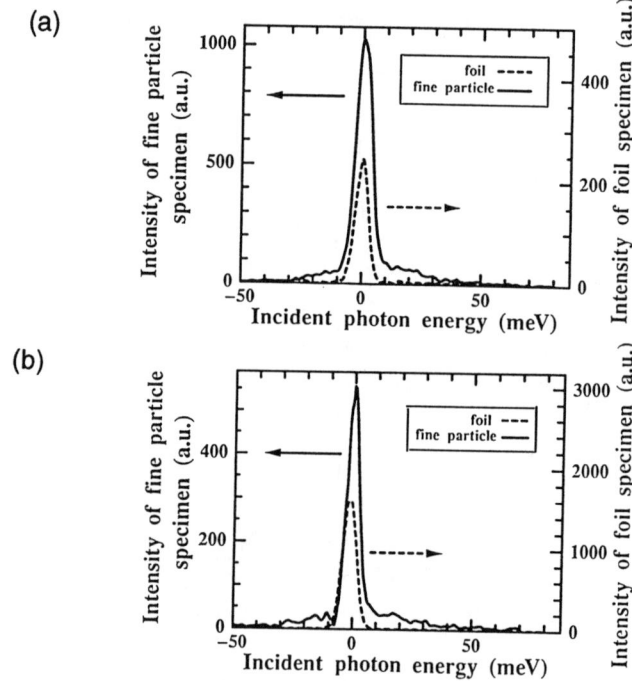

Figure 7. Observed energy spectra of inelastic nuclear scattering from fine particles of ^{57}Fe. The specimen temperatures are (a) 300 K and (b) 373 K. The solid curves are the spectra of fine particles of ^{57}Fe. The dotted curves are the spectra of nuclear forward scattering from an α-^{57}Fe foil.

the dotted curve. This energy shift is almost equal to the recoil energy of an isolated ^{57}Fe nucleus. It may be attributed to nuclear resonant scattering from ^{57}Fe nuclei situated at the surface layer of fine particles which are loosely coupled with each other. The profiles of inelastic scattering observed at both sides of the elastic peak are almost the same between two solid curves at 300 K and 373 K. On the other hand, comparing Fig. 7 with Fig. 6, a change can be seen in the profiles of inelastic scattering. In Fig. 7 the phonon peaks have disappeared. This fact suggests that the softening of lattice vibration takes place in fine particles.

It should be noted that this method gives the phonon energy spectra through resonant excitation and hence yields information about a specific element. If the incident photon energy is tuned to excite a particular nucleus, the vibrational character of that nucleus is obtained. Furthermore, it is possible to excite any nucleus of which the excitation energy is covered with SR even though the recoilless fraction is extremely small or zero. Therefore, this method will open a new field for studyng lattice dynamics.

4. THREE-BEAM NUCLEAR BRAGG SCATTERING

The dynamical theory of multi-beam X-ray diffraction in a perfect crystal has been developed as an extension of the usual dynamical theory of two-beam diffraction. The intensity changes in the diffracted beams caused by the interaction among them within a crystal has been studied theoretically and experimentally. In the case of nuclear Bragg scattering the phenomena of multi-beam diffraction show more characteristic behaviour compared with the case of electronic Bragg scattering, since it is related to the nuclear resonance condition. Theories for multiple diffraction of nuclear resonant X-rays have been given (22, 23).

We report on experimental studies of nuclear Bragg scattering in the case of three-beam. Variations of the time spectra of the diffracted beams with the diffraction conditions of three-beam case as well as two-beam case were observed. X-rays with an energy of 14.41 keV monochromatized by a Si (111) double-crystal monochromator were used as the incident beam. The symmetric 333 Bragg-case diffraction (θ_B = 15.5°) and the asymmetric 855 Bragg-case diffraction (θ_B = 38.6°) from a nearly perfect ^{57}FeBO$_3$ crystal (enriched to 95% in ^{57}Fe) took place on the crystal surface parallel to the (111) plane. The 855 reflected beam makes an angle of 53.4° with the crystal surface. The projections of the directions of 333 and 855 reflected beams onto the crystal surface makes an angle of 40.9°. The external magnetic field was applied parallel to the crystal surface and perpendicular to the scattering plane of the 333 reflection. In the 333 reflection pure nuclear scattering with two hyperfine transitions of $\Delta m = \pm 1$ occurs. In the 855 reflection nuclear scattering with six transitions occurs along with electronic scattering.

The time evolution of the 333 and 855 reflections was observed in the case of two-beam and three-beam, as shown in Fig. 8(a) and (b) respectively. A lifetime τ was estimated for each decay profile. In general, the speed-up effect appearing in nuclear resonant scattering corresponds to the broadening of energy relevant to scattering. It represents the collective character of the interaction of X-rays with resonant nuclei in a crystal. Comparing the 333 reflection with the 855 reflection in the two-beam case, we see that the lifetime for the 333 reflection is smaller than that for the 855 reflection. In the 333 reflection the time spectrum in the three-beam case is almost the same as that in the two-beam case. Because the fact that the scattering power of the 855 reflection is much smaller than that of the 333 reflection results in the weak influence of the 855 reflection on the 333 reflection. On the contrary, in the 855 reflection the influence of the 333 reflection is strong and the speed-up effect occurs markedly.

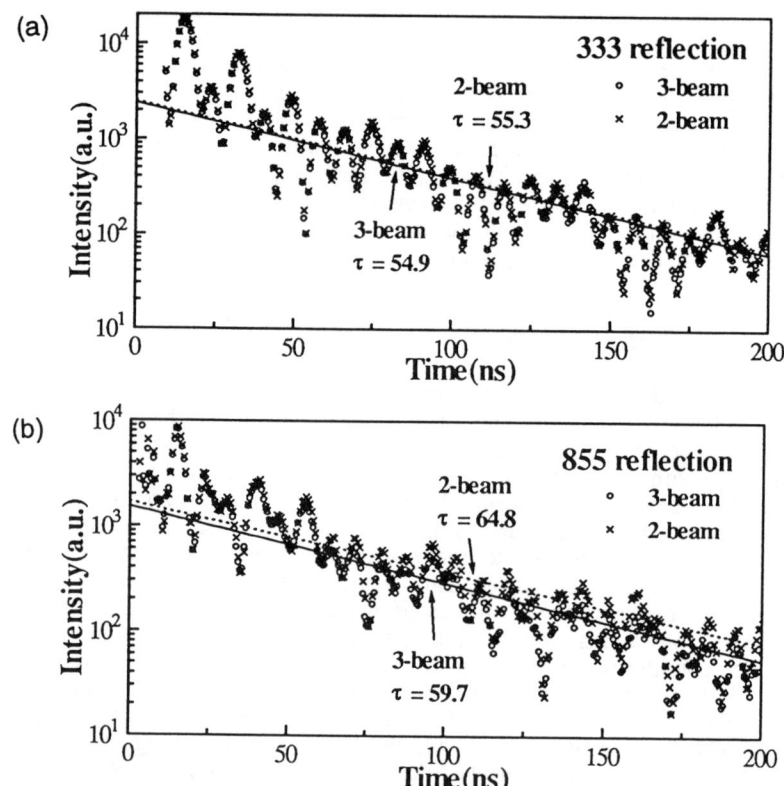

Figure 8. Time evolution of nuclear Bragg scattering. (a): The 333 reflection with and without the simultaneous 855 reflection. (b): The 855 reflection with and without the simultaneous 333 reflection.

5. TWO-PHOTON CORRELATIONS IN THE X-RAY REGION

The studies on X-ray interferometry related to nuclear resonant scattering have been developed (24-28). Here we report on the observation of two-photon correlations in 14.4 keV X-rays. As the first trial to raise the degree of temporal coherence a high resolution Si monochromator was used.

The intensity correlation experiment with a visible light was first demonstrated by Hanbury Brown and Twiss (29). They observed that photons in a monochromatic light beam from a thermal source did not arrive completely at random, but arrived in bunches. An extension of this experiment to the X-ray region using SR was proposed (30, 31). An essential condition for this experiment is to get X-rays with high temporal coherence and high spatial coherence. It was pointed out that the use of a nuclear monochromator will be effective for raising the degree of temporal coherence. X-rays with high spatial coherence will be available from third-generation SR rings because of the extremely small source size.

The usual interference experiments are related to the correlation of radiation field amplitudes. The first-order degree of coherence $\gamma_{12}^{(1)}(\tau)$ represents the correlation between the field at position r_1 and time t_1 and that at position r_2 and time t_2. It depends on time through $\tau = t_1 - t_2$. On the other hand, the intensity correlation caused by interference of two photons is expressed in terms of the second-order degree of coherence $\gamma_{12}^{(2)}(\tau)$. For thermal light $\gamma_{12}^{(2)}(\tau)$ is expressed as

$$\gamma_{12}^{(2)}(\tau) = 1 + \left|\gamma_{12}^{(1)}(\tau)\right|^2$$

When $\tau = 0$, the second-order degree of coherence becomes 2. It shows the bunching effect of photons. This effect is blurred owing to the incomplete spatial and temporal coherence. Taking account of the size of slit and the pulse width of SR which define spatial and temporal coherence, respectively, the integrated second-order degree of coherence $\gamma^{(2)}$ is obtained from $\gamma_{12}^{(2)}(\tau)$.

The experiment was carried out at the TRISTAN-Main Ring (MR) (KEK, Tsukuba). The MR was operated at an energy of 10 GeV and a current of 5~9 mA in the 8-bunch mode. The experimental setup is shown in Fig. 9. X-rays from an undulator was roughly monochromatized at 14.41 keV by a Si (400) double-crystal monochromator. Next a high resolution four-crystal Si monochromator was arranged, which is the same as that used in the section 2. The energy width of monochromatized X-rays 6.4 meV corresponds to the coherence time 0.66 ps. After the high resolution monochromator a precision slit was arranged, which determined spatial coherence. The horizontal and vertical nominal source sizes were 169 μm and 53 μm, respectively. The transverse coherence widths in the horizontal and vertical directions at a detector position located about 100 m from the source were calculated as 14 μm and 46 μm, respectively. The vertical slit width was fixed at 40 μm, while the horizontal slit width was changed from 20 μm to 1 mm. A Si crystal plate with 220 Laue-case diffraction served as a beam splitter. The diffracted and transmitted beams were incident on APD detectors with a time resolution of about 0.5 ns. Output pulses from two detectors were fed into the coincidence unit and the total coincidence count R was obtained. To normalize R the random coincidence count R_0 was needed. It was

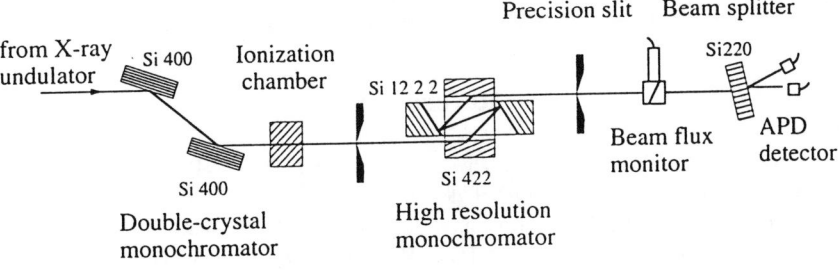

Figure 9. Schematic side view of the experimental setup for two-photon correlations.

obtained from the measurement of coincidence events between a photon emitted from a certain electron bunch and another photon emitted from the same electron bunch after making one revolution in its orbit. Actually to reduce an error of R_0 photons emitted from an electron bunch which makes two, three and four revolutions as well as one revolution were used. It is verified theoretically that the ratio R/R_0 is nearly equal to $\gamma^{(2)}$.

Figure 10 shows the variation of the integrated second-order degree of coherence $\gamma^{(2)}$ as a function of the horizontal slit width. The experimental results are plotted as points with their associated statistical errors. The solid line is the theoretical curve. The dotted straight line $\gamma^{(2)} = 1$ shows the random coincidence level. The region above this level is considered as contribution from the excess coincidence rate. At the slit size of 20 μm x 40 μm the excess part is about four times as large as the associated statistical error. As the slit size increases, the value of $\gamma^{(2)}$ decreases, approaching 1. At the slit size of 1 mm x 40 μm $\gamma^{(2)}$ is very close to 1. The data and the expected curve agree relatively well, giving clear evidence for the bunching effect.

This technique will be useful for characterization of the source size of the third-generation SR source just as Hanbury Brown and Twiss determined the angular diameter of visible stars. Further in the study of the X-ray laser action the beam diagnosis concerning coherence at the transition from an incoherent state to a coherent state may be made.

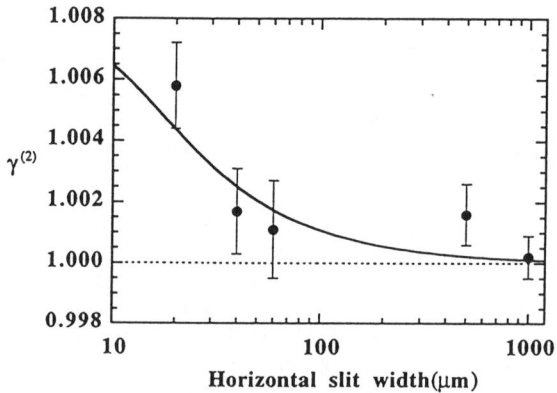

Figure 10. Variation of the integrated second-order degree of coherence with the horizontal slit width. The error bars show the statistical errors σ_γ.

References

1. van Burck, U. and Smirnov, G.V., *Hyperfine Interact.* 90, 313 (1994).
2. Smirnov, G.V., *Hyperfine Interact.* 97/98, 551 (1996).
3. Yamamoto, S., Zhang, X. W., Kitamura, H., Shioya, T., Mochizuki T., Sugiyama, H., Ando, M., Yoda Y., Kikuta, S. and Takei, H., *J. Appl. Phys.*, 74 ,500 (1993) .
4. Gerdau, E., Rüffer, R., Winkler, H., Tolksdorf, W., Klages, C.P., Hannon, J.P.,

Phys.Rev.Lett., 54, 835 (1985).
5. ALP, E.E., Mooney, T.M., Toellner, T., Sturhahn,W., Witthoff, E., Röhlsberger, R., Gerdau, E., *Phys.Rev.Lett.* 70, 3351 (1993).
6. Sturhahn, W., Gerdau, E., Hollatz,R., Ruffer, R., Ruter, H.D., Tolksdorf, W., *Europhys.Lett.* 14, 821 (1991).
7. Johnson, D.E., Siddons,D.P., Larese, J.Z., Hastings, J.B., *Phys. Rev.* B51, 7909 (1995).
8. Chumakov, A.I., Baron,A.Q.R., Arthur, J., Ruby, S.L., Brown, G.S., Smirnov, G.V., van Bruck, U., Wortmann, G., *Phys.Rev.Lett.* 75, 549 (1995).
9. Vetter, M., *Z.Phys.*225,336(1969).
10. Shvyd'ko, Yu.V., Chumakov, A.I., Smirnov, G.V., Hertrich, T., van Burck, U., Rüter, H.D., Leupold, O., Metge, J., Gerdau, E., *Europhys. Lett.* 26, 215 (1994).
11. Shvyd'ko,Yu.V., *Hyperfine Interact.* 90, 287 (1994).
12. Shvyd'ko, Yu.V., Hertrich, T., Metge, J., Leopold, O., Gerdau, E. and Ruter,H.D., *Phys. Rev.* B52, R711 (1995).
13. Kolotov, O.S., Pogozhev, V.A., Smirnov, G.V. and Shvyd'ko, Yu.V., *Instrum. Exp. Tech.* 26, 347 (1983).
14. Shvyd'ko, Yu.V., Chumakov, A.I., Smirnov, G.V., Kohn, V.G., Hertrich, T., van Burck, U., Gerdau, E., Ruter, H.D., Metge, J. and Leupold, O., *Europhys. Lett.*22, 305 (1993)
15. Seto, M., Yoda, Y., Kikuta, S., Zhang, X.W. and Ando, M., *Phys. Rev. Lett.*, 74, 3828 (1995).
16. Zhang,X. W., Yoda, Y., Seto, M., Maeda, Yu., Ando, M. and Kikuta,S., *Jpn. J. Appl. Phys.* 34, L330 (1995).
17. Sturhahn, W., Toellner, T.S., Alp, E.E., Zhang, X., Ando, M., Yoda, Y., Kikuta, S., Seto, M., Kimball, C.W. and Dabrowski, B., *Phys. Rev. Lett.* 74, 3832 (1995).
18. Chumakov, A.I., Rüffer, R., Grünsteudel, H., Gürnsteudel, H.F., Grübel, G., Metge, J., Leupold, O. and Goodwin,H.A., *Europhys. Lett.* 30, 427 (1995).
19. Seto, M., Yoda, Y., Kikuta, S., Zhang, X.W. and Ando, M., *IL Nuovo Cimento* 18D, 381 (1996).
20. Nakatani, I., Furubayashi, T., Takahashi, T. and Hanaoka, H., *J. Magn. Mater.* 65, 261 (1987).
21. Furubayashi, T. and Nakatani, I., *IEEE Trans. Magn.* 26, 1855 (1990)
22. Kohn, V.G., *JETP* 78, 357 (1994).
23. Hutton, J.T., Hannon, J.P. and Trammell, G.T., *Phys.Rev.* A37, 4269 (1988)
24. Hasegawa, Y., Yoda, Y., Izumi, K., Ishikawa, T., Kikuta, S., Zhang, X.W., Sugiyama, H. and Ando, M., *Jpn. J.Appl. Phys.* 33, L772 (1994).
25. Hasegawa, Y., Yoda, Y., Izumi, K., Ishikawa, T., Kikuta, S., Zhang, X.W. and Ando, M., *Phys. Rev.* B50, 17748 (1994).
26. Hasegawa, Y., Yoda, Y., Izumi, K., Ishikawa, T., Kikuta, S., Zhang, X.W. and Ando, M., *Phys.Rev.Lett* 75, 2216 (1995).
27. Izumi, K., Yoda, Y., Ishikawa, T., Zhang, X.W., Ando, M. and Kikuta, S., *Jpn. J. Appl. Phys.* 34, 4258 (1995).
28. Izumi, K., Mitsui, T., Seto, M., Yoda, Y., Ishikawa, T., Zhang, X.W., Ando, M. and Kikuta, S., *Jpn. J. Appl. Phys.* 34, 5862 (1995)
29. Hanbury Brown, R. and Twiss, R.Q., *Nature* 177, 27 (1956).
30. Ikonen, E. and Rüffer, R., *Hyperfine Interact.* 92, 1089 (1994).
31. Gluskin, E., McNulty, I., Vicarro, P.J. and Howells,M.R., *Nuclear Instrum. Meth.* A319, 213 (1992).

VI. ELECTRON MOMENTUM SPECTROSCOPY

3D Electron Momentum Densities of Solids.

F.F. Kurp[1], M. Vos[2], Th. Tschentscher[3], A.S. Kheifets[2],
H. Schulte-Schrepping[1], J.R. Schneider[1], E. Weigold[2] and F. Bell[4]

[1] *Hamburger Synchrotronstrahlungslabor (HASYLAB) at Deutsches Elektronen-Synchrotron (DESY), Notkestr. 85, 22603 Hamburg, Germany*
[2] *Electronic Structure of Materials Centre, Flinders University of South Australia, G.P.O. Box 2100, Adelaide, South Australia, 5001, Australia*
[3] *European Synchrotron Radiation Facility, B.P. 220, 38043 Grenoble, France*
[4] *Sektion Physik, Universität München, Am Coulombwall 1, 85748 Garching, Germany*

Abstract - The complete 3D-electron momentum density (EMD) of solids can be measured by the coincident detection of an inelastically scattered hard x-ray photon and its recoil electron (i.e. a so-called (γ,eγ) experiment). To avoid multiple scattering of the recoil electron the experiment must be performed with very thin self-supporting target foils. The experimental EMD is compared with a pseudopotential calculation. Experiments have been made both at the European Synchrotron Radiation Facility at Grenoble, France, and at the PETRA storage ring of DESY, Hamburg, Germany. In addition the EMD from the (γ,eγ) experiment is compared with that from a (e,2e) study.

I. INTRODUCTION

Traditionally, Compton scattering has been used to characterize the electron momentum density (EMD) of valence electrons in solids (1,2). The double differential cross section describing the energy and angular distribution of inelastically scattered x-rays is proportional to the so-called Compton profile which is defined as a twofold integration over the EMD. This integration results from the lack of information about the momentum distribution of the recoiling electrons. It can be avoided if the recoil electron is measured simultaneously with the scattered photon. The corresponding triple differential cross section is proportional to the EMD itself. Since integration averages over large volumes in momentum space, detailed information about solid-state effects like Fermi surfaces, localized states or electronic correlations might become difficult to obtain. It is therefore desirable to measure the EMD directly by a coincident detection of the scattered photon and the recoil electron. If the momenta of the primary and scattered photon and also that of the recoil electron are fixed experimentally, the momentum of the electron in its initial state can be reconstructed in a unique way. The experimental difficulty of such a (γ,eγ)-coincidence experiment depends primarily on the possibility to measure the recoil momentum undisturbed by multiple electron scattering within the target. Since the mean free path for elastic (nuclear) scattering for recoil electrons with energies of about 70 keV is 120 nm in a target like carbon (3), self-supporting targets are required which are as thin as possible. This has drastic implications for the coincident count rate: the mean free path for Compton scattering in carbon at a photon energy of 180 keV is 3 cm (4). Thus, in a 20 nm thin foil only one out of 10^6 photons will be scattered. To determine the photon and electron momenta, rather small solid angles of both photon and electron detector are required. Assuming as a guess an isotropic angular distribution of the scattered photons, a solid angle of the photon detector of 10^{-4} sr yields another factor 10^5. With a rather coarse angular resolution of the recoil momenta and a coincidence count rate of several Hz we end up with a desired monochromatized photon flux of 10^{12} photons/s at the target. This flux at photon energies of about 200 keV can only be delivered by modern synchrotron radiation sources with large lepton energies. Thus, experiments have been performed at the high energy x-ray scattering beam line ID15 of the ESRF (electron energy 6 GeV), which is equipped with an asymmetric wiggler with 7 periods and strong magnetic poles of 1.8 T and at the new undulator beamline at the PETRA storage ring (12 GeV) at DESY.

A method very similar to (γ,eγ) is the (e,2e) reaction where instead of a photon an energetic electron is used as projectile (5,6). Though both types of experiments yield the same information - the EMD of a solid - the experimental details are rather different contributing to advantages and disadvantages if the methods are compared. Again, multiple electron scattering within the target is the most severe problem. Comparing (γ,eγ) and (e,2e) experiments the situation is more relaxed in the former case since at least the photon will not be multiply scattered. The big advantage of the (e,2e) technique is its large cross section and its easily achieved monochromatic projectile flux. For the experimental situation to be described below we obtain from the Møller cross section 13 kb/sr/electron for the (e,2e) experiment (20 keV primary electron energy, 14° scattering angle) whereas the Klein-Nishina cross section yields 29 mb/sr/electron (185 keV photon energy, 140° scattering angle) only, thus a factor of $5 \cdot 10^5$ in favour of the (e,2e) experiment. In addition, a highly monochromatic electron flux of 10^{12} electrons/s (~ 100 nA) is quite easily achieved whereas comparable photon fluxes with considerably less monochromaticity are obtained only from synchrotron radiation facilities of the third generation. Altogether, this allows the introduction of electron spectrometers in (e,2e) experiments which in turn makes it possible to measure the EMD of solids as a function of the valence binding energy with a resolution of about 1.5 eV (7).

II. METHOD

Consider a projectile with energy E_0 and momentum $\mathbf{p_0}$ which is scattered at a target electron with initial momentum $\mathbf{q}$ resulting in a scattered projectile with energy E_S and momentum $\mathbf{p_S}$. The recoil electron will acquire momentum $\mathbf{p_R}$ and energy E_R. If the electron was bound with an energy $\varepsilon > 0$ relative to the vacuum level it holds

$$\varepsilon = E_0 - E_S - E_R$$
$$q = p_S + p_R - p_0 \qquad (1)$$

If all the momenta and energies are fixed experimentally, q and ε can be reconstructed in a unique way. Assuming the validity of the impulse approximation (6,8) the coincidence count rate will be proportional to the spectral electron momentum density (SEMD) $\rho_j(\varepsilon,q) = \rho_j(q)\delta(\varepsilon - \varepsilon_j(q))$, where $\rho_j(q)$ is the EMD of the jth band. In a crystalline solid the binding energy and momentum are correlated through the dispersion relation $\varepsilon = \varepsilon_j(q)$ where in the extended zone scheme $q = k + g$. k is the crystal momentum (i.e. restricted to the 1. Brillouin zone) and g a reciprocal lattice vector. Whereas in the (e,2e) experiment to be described the energy resolution is good enough to measure the SEMD for the valence electrons only, the (γ,eγ) experiment integrates over all binding energies, thus yielding the EMD $\rho(q)$ summed over all bands and binding energies.

III. EXPERIMENTS

Since most of the experimental results described in this paper have been obtained at the PETRA storage ring we will describe this experimental set-up in more detail (Fig.1). The undulator with 127 periods has its first harmonic at about 10 keV for a gap of 14 mm. At a photon energy of E_0 = 185 keV the essentially white beam

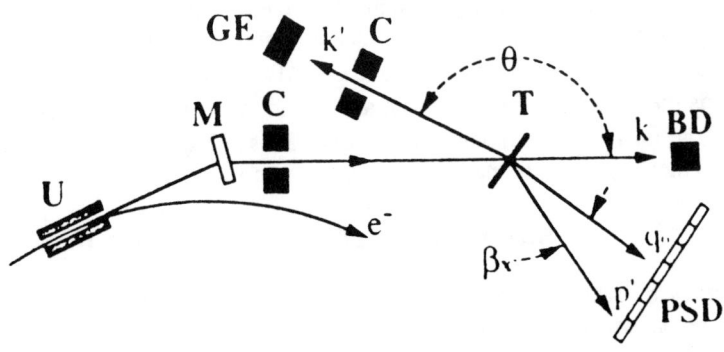

FIGURE 1. Experimental set-up: U: undulator, M: monochromator, C: collimator, T: target, BD: beam dump, GE: Ge-diode, PSD: position sensitive electron detector

was monochromatized by the (220) reflection of a plane "cooked" Si crystal in Laue geometry. The crystal had been carefully annealed at high temperatures to ensure small amounts of oxygen precipitation which combines a rather broad angular width of the rocking curve with high reflectivity (9). At a storage ring current of 20 mA a photon flux of about $3 \cdot 10^{12}$ photons/s with a monochromaticity of ΔE_0 = 660 eV (FWHM) in a beam spot of 3 (vertical) × 4 (horizontal) mm² was obtained. The monochromatized x-ray beam entered an evacuated target chamber with an externally mounted intrinsic Ge-diode (energy resolution ~ 600 eV FWHM, solid angle ~ 0.4 msr) at $\Theta = 140°$. For electrons initally at rest one obtaines for this scattering angle a scattered photon energy of E_{S0} = 112.8 keV, a recoil energy E_{R0} = 72.2 keV and a momentum transfer $K = |p_0 - p_S|$ = 75.3 a.u.. K makes an angle of 15° with the initial photon beam direction. Electrons with initial momenta different from zero will be ejected with a direction distribution more or less symmetric with respect to K. Therefore, a 2D position sensitive electron detector was placed with its center in the direction of K. The detector consisted of 33 individual PIN-diodes each equipped with its own electronic circuit (preamplifier, main amplifier, discriminator). Combined with the Ge-diode this allowed in a multi-parameter coincidence mode the simultaneous measurement of 33 coincident photon spectra. The poor energy resolution of the photodiodes (~ 5 keV FWHM) did not allow the measurement of the Doppler broadening due to the EMD in the electron channel. The target was a 22 nm thin self-supporting carbon foil made by laser plasma ablation. The ejected carbon atoms were deposited on a thin Betaine film ($C_5H_{11}NO_2 \cdot H_2O$) which had a fine crystalline-like structure that acted as a replica for the carbon film and guaranteed a high mechanical stability. Finally, the Betaine film was dissolved in water and the foil mounted on an aluminum frame with 12 mm diameter. Electron diffraction of these foils revealed graphite-like rings indicating isotropically distributed nanocrystals with no texture (10).

In the following the EMD will be given as a function of the initial electron momentum $q = (q_x, q_y, q_z)$ in a Cartesian coordinate system where the q_z-component is parallel K, q_x lies in the scattering plane made by the primary and scattered photon and q_y is perpendicular to that plane. In such a coordinate system q_z is essentially determined by the photon Doppler broadening $\Delta E_S = E_S - E_{S0}$ in the Ge-diode, where E_S is the scattered photon energy, and q_x and q_y by the angular deviation β_x and β_y of the final electron momentum p_R from K (see Fig. 1). One obtains from scattering kinematics (we use natural units, i.e. $\hbar = m = c = 1$)

$$q = (K\beta_x + E_0\sin\Theta \cdot \Delta E_S/K, K\beta_y, - E_0 \Delta E_S/(KE_{S0})) \qquad (2)$$

The accuracy by which these initial electron momentum components can be determined depends in a rather complex way on the angular and energy uncertainties of the experiment. They include the monochromaticity of the primary photon beam and the energy resolution of the photon detector on one hand, and the uncertainties of Θ, β_y and β_x on the other. In the determination of the latter the beam spot size has to be included. We estimate a resolution element of Δq_z = 0.60 a.u. (FWHM) and of $\Delta q_x \cong \Delta q_y$ = 0.85 a.u. (FWHM).

The data given in Fig. 2 were accumulated with an overall coincidence rate of about 3 Hz and represent a total of $2 \cdot 10^5$ coincidences. The 2D-plots give the experimental EMD in planes parallel to the scattering plane, i.e. for q_y as a parameter. Thin solid lines represent isodensity levels starting from 10 up to 90% of the maximum intensity at $\mathbf{q} = 0$, separated by 10% intervals. The dotted lines are at 5 and 95%, thick solid lines at 10, 50 and 90%. The finite extent of the experimental EMD, especially in q_x-direction, results from the limited size of the electron detector. The data are compared with a theoretical EMD of graphite obtained from a pseudopotential calculation (11). The theoretical results were spherically averaged

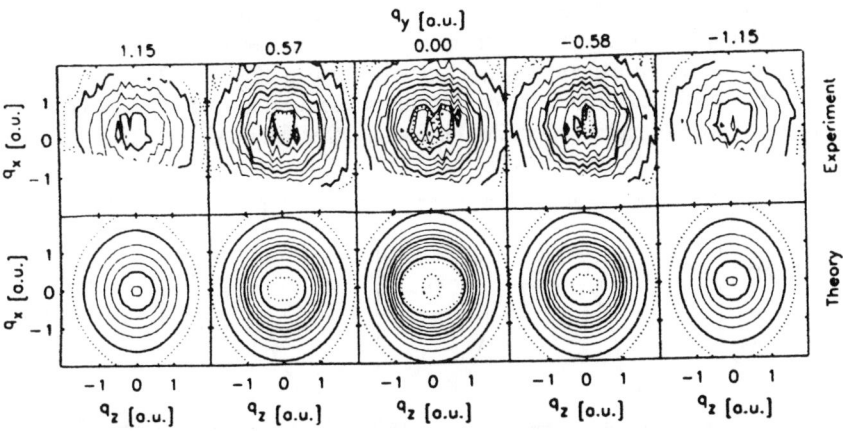

FIGURE 2. Contour plots of the experimental and theoretical EMD of graphite. The 2D-cuts through the 3D-EMD are (q_x, q_z)-planes parallel to the scattering plane.

to account for the isotropic nature of our foils and convoluted with the experimental resolution. Since the resolution in q_z-direction is different from that in the other two directions (see above) this convolution introduces again a slight anisotropy of the EMD which can be seen in Fig.2. In addition, theory has been corrected for the effect of electron multiple scattering. For the 22 nm thick carbon target and a mean free path of incoherent elastic scattering of 120 nm 91% of the electrons leave the foil unscattered, the scattering function of the remaining 9% has been obtained from a Monte Carlo calculation (12) which simulated the electron scattering. Finally, the theoretical EMD has been folded with this distribution according to a procedure described in more detail in ref. (12).

Since the theoretical EMD of ref. (11) holds for the valence electrons only, a $(1s)^2$ core contribution from (13) has been added. Whereas the contour plots of Fig. 2 demonstrate that in fact the complete 3D-EMD has been measured with a granularity indicated by the figure, 1D-cuts through the EMD give a better possibility to compare experiment with theory for preferential directions through the Γ-point. Figure 3 shows cuts along three orthogonal directions in q-space: Fig.3a holds for $(q_x,0,0)$, 3b for $(0,q_y,0)$ and 3c for $(0,0,q_z)$. While Fig. 3c is in essence a photon spectrum coincident with a specific photodiode of the electron detector (it is for the diode placed at $\beta_x = \beta_y = 0$; thus, according to eq.(2) the cut is more precise along $(-0.14\, q_z, 0, q_z)$) Figs. 3a and 3b represent events in a row (3a) or a column (3b) of the electron detector coincident with photons at $q_z = 0$. Due to the finite number of photodiodes the granularity in the q_x- and q_y-cuts is coarser than in the q_z-cut. Again, the experiment is compared with corresponding 1D-cuts through the theoretical EMD (11). We emphasize that theory has been fitted to the experiment by normalizing both to the same volume in momentum space. Therefore, in all 1D-cuts theory has been adapted to experiment by a single common scale factor. The experimental points in Fig. 3 have not been symmetrized. The most striking feature of the EMD is the dip at zero momentum predicted by theory and verified by the experiment as is clearly seen in Fig. 3c. Due to a coarser granularity of the electron detector this dip is less pronounced in Figs. 3a and 3b but even there the experimental EMD shows a rather flat maximum. This dip

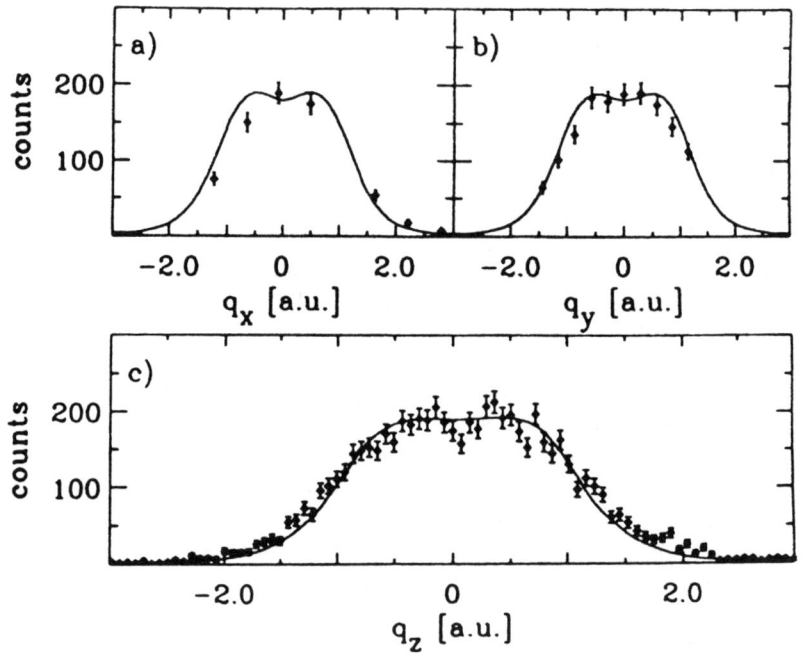

FIGURE 3. 1D-cuts of the EMD along specific direction of the electron momentum $q = (q_x, q_y, q_z)$. 3a: $(q_x, 0, 0)$; 3b: $(0, q_y, 0)$; 3c: $(-0.14\, q_z, 0, q_z)$. The solid line represents theory.

results from the contribution of π-electrons which are responsible for the weak van der Waals interlayer bonding in graphite. It is the p-character of these bonds from which a dip in the EMD for momenta parallel the c-axis results and which appear in all theoretical curves of Fig. 3 due to the spherical averaging. This contribution of π-electrons seems to be even more pronounced than predicted by theory, see Fig. 3c. A strong reduction of the EMD along the c-axis has also been observed in 2D positron annihilation data (14), though a quantitative comparison with our data seems to be problematic: an expected positron localization in the interlayer region may give a preferred weight for annihilation with π-states. Details as this dip, which are clearly seen in the coincident photon spectrum of Fig. 3c, are completely washed out in non-coincident Compton data of graphite and have neither been observed in our singles (i.e. non-coincident spectra) nor in earlier measurements (15-17). This statement holds also for what we call a "coincident Compton profile" $J_c(q_z)$: the experimental points of Fig.4 have been obtained by summing all coincidence events of the 2D electron detector. Due to the limited range of the detector (see Fig. 2) high momentum tails of the EMD are missing when compared with a true Compton profile. The solid line of Fig.4 was obtained by integrating the theoretical EMD over the corresponding finite momentum range. We would like to

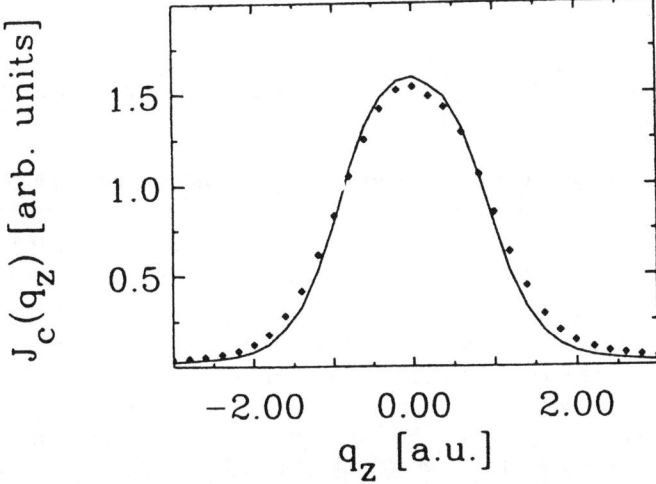

FIGURE 4. The "coincident Compton profile" $J_c(q_z)$ compared with theory (solid line). The experimental points have not been symmetrized.

stress that the normalization of experiment and theory in Figs. 2, 3 and 4 is always the same. The advantage of this coincident Compton profile compared to singles is that the trigger condition provides a spectrum which is free from the background of inelastically scattered photons originating primarily from the walls of the target chamber. The agreement of the experimental points - the statistical error is less than the diameter of the dots - with theory is good though not perfect.

In the (e,2e) experiment the two emerging electrons have been measured in the so-called asymmetric geometry (Fig.5): The fast electron left the backside of the foil with a scattering angle of 14° and the slow one at an angle of 76°. For a primary electron energy E_0 = 20 keV this means that the fast electron had an energy

$E_S = 18.8$ keV while $E_R = 1.2$ keV. The two spectrometers, a hemispherical analyser (fast electrons) and a toroidal analyser (slow electrons) were in an UHV-chamber ($2 \cdot 10^{-10}$ torr) (18). Electrons with pass energies of 100 eV (hemisphere) and 200 eV (toroid) were measured with position sensitive detectors over a range of azimuthal scattering angles: see the cones of polar angles and the range of azi-

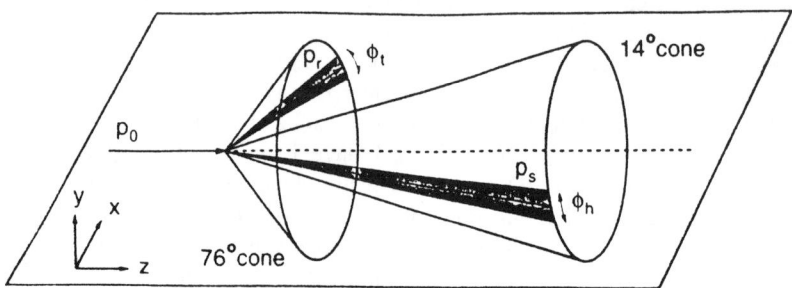

FIGURE 5. The (e,2e) set-up: the incoming beam with momentum p_0 scatters and outgoing slow and fast electrons with momenta p_s and p_r are detected over a range of azimuthal angles ϕ_t and ϕ_h.

muthal angles ϕ_t and ϕ_h in Fig.5. The overall energy and momentum resolutions of the (e,2e) experiment are estimated to be 1.6 eV and 0.2 a.u., respectively. From the range of azimuthal angles it is deduced that the component q_y of the initial electron momentum in a direction perpendicular to the incoming electrons can be determined over a range from -3.5 to $+3.5$ a.u. including the Γ point. The momentum transfer $K = 9.4$ a.u. is considerably smaller than in the (γ,eγ) experiment but large enough to ensure the validity of the binary encounter approximation (5,18). The strong enhancement of the cross section due to the asymmetric geometry compared to a symmetric geometry (5) - which maximizes the momentum transfer - allowed for beam currents less than 100 nA. The true overall coincidence count rate was a few Hz and the ratio of true to false coincidences about one. The total acquisation time was several days for each of the two experiments.

After the (γ,eγ) experiment the foil was transferred to Adelaide for the (e,2e) investigation. Before the experiment the foil was heated to 900 K to remove especially oxygen adsorbates from the surface. Fig. 6 shows (e,2e) spectra as a function of the binding energy ε where the value of q_y in a.u. is indicated in each panel. At zero momentum the peak at about 27 eV is attributed to the σ-electrons and shows the expected strong dispersion (7,19,20). The peak at smaller binding energies - at about 10 eV - disperses only moderately and can probably be identified as resulting from the π-electrons. This interpretation contradicts the experimental observation of the low energy peak at zero momentum, where π-electrons should have vanishing intensity (20,21). The π-electrons have maximum intensity along the ΓA

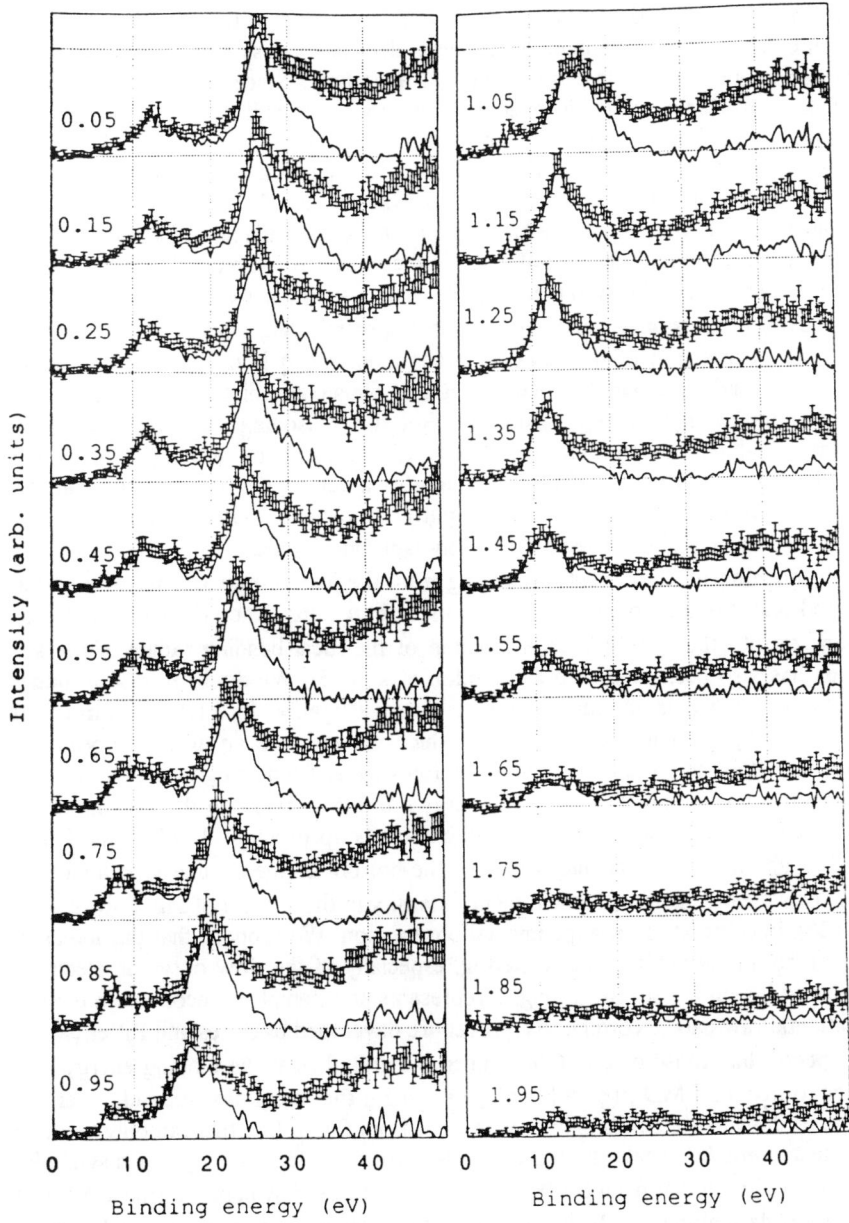

FIGURE 6. Experimental (e,2e) spectra for different electron momenta as indicated in each panel. The solid lines represent the data after correction for multiple energy losses.

direction of the first Brillouin zone (c-axis direction). It is the p-character of the bonding which causes the intensity of these electrons vanish for q = 0. Of course, this behaviour also survives the spherical averaging. Since a contribution from oxygen adsorbates can be excluded, a possible explanation may be the contamination of the $q_y = 0$ spectrum from spectra with larger q_y due to the diffraction of one of the outgoing electrons by a reciprocal lattice vector of a few atomic units. But we remark that Chao Gao et al. (22) have given another explanation for very similar experimental findings in diamond-like amorphous carbon foils. (Though they do not exclude the possibility of multiple elastic scattering (5)). They also observed in their (e,2e) data a strong peak at about 8 eV which was nearly despersionless and had the same spectral weight between $0 \leq q \leq 1.7$ a.u. The observation of this peak at larger momenta q is explained by a nearly 100% sp^2 bonding also in diamond-like carbon films contradicting the naive picture of sp^3 bonding in a diamond structure. But this reasoning was confirmed by a recent tight-binding molecular-dynamics study of amorphous carbon (23,24) which found a fraction of about 80% sp^2 bonding for a carbon density corresponding to diamond-like films produced by an ion sputtering technique. In contrast to our explanation, these authors interpret the non-vanishing intensity for $q \to 0$ as a s-p rehybridization of the π electrons arising from the loss of mirror symmetry of the twisted and wrinkled bonds in amorphous carbon films. The admixture of the s-character into the π-orbitals as a function of the local bending radius of the sp^2 bonds has also been discussed by Haddon et al. (25). We doubt if such a model would be applicable to our foil where an investigation by electron diffraction showed sharp graphite-like rings from which we estimate an average crystalline grain size of 10 nm. Thus, the state of the foil was far from being amorphous.

Since the zero point of the binding energy ε in Fig. 6 refers to the vacuum level, the range of the valence band in graphite extends up to about 26 eV if the Fermi level of about 5 eV is included (19,20) (the bottom of the σ-band is at about 21 eV). This means that there should be no intensity in the spectra of Fig. 6 for $\varepsilon > 26$ eV (26), in contrast to the experimental observation. We suppose that this intensity results from multiple inelastic scattering, especially of the slow outgoing electron. The solid line in the spectra of Fig. 6 represents an attempt to deconvolute the experimental raw data according to a procedure developed by Jones and Ritter (27).

Spectra like those of Fig. 6 have been integrated over the binding energy ε in order to get the EMD $\rho(\mathbf{q}) = \int \rho(\mathbf{q},\varepsilon) \, d\varepsilon$, where the integration extends over the range of the valence bands only. It is evident from Fig. 6 that this can only be done for the *deconvoluted* data. In Fig. 7 the EMD from the (e,2e) study (open symbols) is compared with that from the (γ,eγ) experiment (closed symbols) for an 1D cut of the (γ,eγ) data along ($-0.14\, q_z$, 0, q_z). In order to compare with the (e,2e) study the (γ,eγ) data have been corrected for the core contribution by subtracting a theoretical $(1s)^2$ EMD (13) from the experimental result. The solid line represents the theoretical EMD of Lou Yongming et al.(11) convoluted with the resolution of the (γ,eγ) experiment. The agreement of both sets of data is good though not

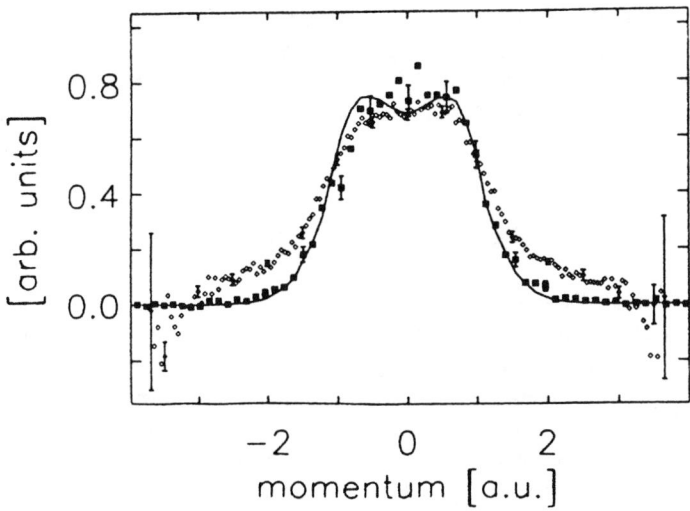

FIGURE 7. Comparison of the EMD from the (e,2e) experiment (open symbols) with that from the (γ,eγ) reaction (closed symbols).

perfect: The EMD of the (e,2e) experiment seems to be broader than that from the (γ,eγ) investigation. It is evident that this can not be explained by the poorer resolution of the latter since this would work in the opposite direction. Since the deconvolution procedure mentioned above (see also Fig. 6) only accounts for *inelastic* collisions, we suggest that the broadening of the (e,2e) data (Fig. 7) results from *elastic* multiple scattering. In view of the complicated nature of the multiple scattering problem - elastic and inelastic scattering, both coherent (i.e. Bragg scattering) and incoherent for the incoming and outgoing electrons - we admit that the deconvolution procedure might be only an intelligent guess (7).

V. SUMMARY AND PERSPECTIVES

The electron momentum densitiy of a polycrystalline graphite foil was measured either by (γ,eγ) or (e,2e) spectroscopy and the results were found to be in reasonable agreement. For the (γ,eγ) experiment the recoil electron energy was high enough to suppress almost quantitatively both elastic and inelastic multiple electron scattering within the target. The small scattering cross section (Klein-Nishina) and small photon fluxes resulted in rather poor energy and momentum resolution compared to the (e,2e) investigation. The asymmetric scattering geometry in the latter case guaranteed a large Møller cross section but caused also strong electron multiple scattering. Nevertheless, the comparison has demonstrated that solid state EMD's can be obtained by both methods with reasonable accuracy.

How could the experiments be improved? For the photon experiment we plan to use a multi-pixel 2D photon detector in place of the single pixel Ge diode. For the (e,2e) experiment the multiple scattering problem could be reduced greatly by going to higher primary electron energies. For $E_0 = 100$ keV and a symmetric geometry (i.e. a scattering angle of 43.7° for both outgoing electrons and $E_S = E_R = 50$ keV) the Møller cross section is reduced to 8 b/sr/electron, i.e. by a factor of $2 \cdot 10^3$. Nevertheless, it might be that such an experiment is feasible without losing too much resolution (5).

ACKNOWLEDGEMENTS

We thank Günther Dollinger an Peter Meier-Komor from the Target Laboratory of the Technical University of Munich for the graphite foil and Vejo Honkimäki from the University of Helsinki, Finnland, for his help during the beam time at the ESRF. The $(\gamma,e\gamma)$ experiment was supported by the Bundesministerium für Bildung, Wissenschaft, Forschung und Technologie under contracts 05 5WMAAI and 05 650WEA. The Electronic Structure of Materials Centre, Adelaide, is supported by a grant of the Australian Research Council.

REFERENCES

1. Williams, B. (ed): *Compton Scattering*, (McGraw-Hill, New York) 1977
2. Cooper, M.J., *Rep. Prog. Phys.*, **48**, 415 (1985)
3. Riley, M.E., MacCallum, C.J., and Biggs, F., *At. Data and Nucl. Data Tables*, **15**, 443 (1975)
4. Hubbel, J.H., Veigele, Wm.J., Briggs, E.A., Brown, R.T., Cromer, D.T., and Howerton, R.J., *J. Phys. Chem. Ref. Data*, **4**, 471 (1975)
5. Dennison, J.R., and Ritter, A.L., *J. Electr. Spec. Rel. Phenom.* **77**, 99 (1996)
6. Vos, M., and McCarthy, I.E, *Rev. Mod. Phys.* **67**, 713 (1995)
7. Vos, M., Storer, P., Canney, S.A., Kheifets, A.S., McCarthy, I.E., and Weigold, E., *Phys. Rev.* **B50**, 5635 (1994)
8. Eisenberger, P., and Platzman, P.M., *Phys. Rev.* **2**, 415 (1970)
9. Schneider, J.R., Gonçalves, O.D., Rollason, A.J., Bonse, U., Lauer, J., and Zulehner, W., *Nucl. Instr. Methods*, **B29**, 661 (1988)
10. Dollinger, G., Maier-Komor, P., and Mitwalsky, A., *Nucl. Instr. Methods*, **A303**,79 (1991)
11. Lou Yongming, Johansson, B., and Nieminen, R.M., *J.Phys.: Condens. Matter*, **3**, 1699 (1991)
12. Tschentscher, Th., Schneider, J.R., and Bell, F., *Phys. Rev.*, **B48**, 16965 (1993)
13. Clementi, E., and Roetti, C., *At. Data Nucl. Data Tables*, **14**, 177 (1974)
14. Lee, R.R., von Stetten, E.C., Hasegawa, M., and Berko, S., *Phys. Rev. Lett.*, **58**, 2863 (1987)
15. Cooper, M., and Leake, J.A., *Phil. Mag.*, **15**, 1201 (1967)
16. Issolah, A., Chomillier, J., Loupias, G., Levy, B., and Beswick A., *J.de Phys. Colloque*, **C9**, C9-847 (1987)

17. Tyk, R., Felsteiner, J., Gertner, I., and Moreh, R., *Phys. Rev.*, **B32**, 2625 (1985)
18. Storer, P., Clark, S.A.C., Caprari, R.C., Vos, M., and Weigold, E., *Rev. Sci. Instrum.* **65**, 2214 (1994)
19. Charlier, J.-C., Gonze, X., and Michenaud, J.-P., *Phys. Rev. B43*, 4579 (1991); Charlier, J.-C., Michenaud, J.-P., and Gonze, X., *ibid*, **46**, 4531, (1992)
20. Kheifets, A.S., and Vos, M., *J. Phys.: Condens. Matter* **7**, 3595 (1995)
21. Kheifets, A.S., Lower, J., Nygaard, K.J., Utteridge, S., Vos, M., and Weigold, E., *Phys. Rev.* **B49**, 2113 (1994)
22. Gao, C., Wang, Y.Y., Ritter, A.L., and Dennison, J.R., *Phys. Rev. Lett.* **62**, 945 (1989)
23. Wang, C.Z., Ho, K.M., and Chan, C.T., *Phys. Rev. Lett.* **70**, 611 (1993); *Phys. Rev.* **B47**, 14835 (1993)
24. Galli, G., Martin, R.M., Car, R., and Parrinello, M., *Phys. Rev. Lett.* **62**, 555 (1989); *Phys. Rev.* **B42**, 7470 (1990)
25. Haddon, R.C., Brus, L.E., and Raghavachari, K., *Chem. Phys. Lett.* **131**, 165 (1986)
26. Vos, M., Storer, P., Kheifets, A.S., McCarthy, I.E., and Weigold, E., *J. Electr. Spec. Rel. Phenom.* **76**, 103 (1995)
27. Jones, R., and Ritter, A., *J. Electr. Spec. Rel. Phenom.* **40**, 285 (1986)

Electron-momentum spectroscopy of solids by the (e,2e) reaction

A. S. Kheifets, M. Vos, S. A. Canney, X. Guo, I. E. McCarthy

Electronic Structure of Materials Centre, Faculty of Science and Engineering, The Flinders University of South Australia, Adelaide 5001, Australia

E. Weigold

Research School of Physical Sciences and Engineering, Institute of Advanced Studies, Australian National University, Canberra, ACT 0200, Australia.

Abstract. Recent developments in (e,2e)-momentum spectroscopy have resulted in the study of a diverse range of solid targets. These studies have revealed the electronic structure of solids in much more detail than was previously available using this technique. A summary of these results is presented here.

I INTRODUCTION

The (e,2e) reaction has been used for many years as a probe of target electronic structure of atoms and molecules [1]. The notation (e,2e) refers to a process in which an incident electron (energy E_o, momentum $\boldsymbol{p}_o$) knocks out a target electron, with subsequent detection of both outgoing electrons in time coincidence. After detection of the (e,2e) event, the energies (E_f and E_s) and momenta ($\boldsymbol{p}_f$ and $\boldsymbol{p}_s$) of both outgoing electrons are determined. Here indices f and s specify the faster and the slower of the two outgoing electrons, respectively. The binding energy ε and momentum $\boldsymbol{q}$ of the target electron *before* the collision are given by the following conservation laws:

$$\varepsilon = E_o - E_s - E_f, \qquad \boldsymbol{q} = \boldsymbol{p}_s + \boldsymbol{p}_f - \boldsymbol{p}_o \qquad (1)$$

Under conditions of a high incident energy and a large momentum transfer

$$E_o \gg \varepsilon, \qquad |\boldsymbol{p}_o - \boldsymbol{p}_f| > 1, \qquad (2)$$

the mechanism of an (e,2e) reaction is particularly simple: it is a free electron binary collision between the projectile and the target electron. We use atomic units of momentum 1 a.u. = a_o^{-1} in Eq. (2) and throughout the paper. Although the target electron is bound, it can be assigned the attributes of a free electron, i.e. the energy ε and momentum $\boldsymbol{q}$, with a certain probability $\rho(\varepsilon, \boldsymbol{q})$. Under this binary collision assumption the *(e,2e)* cross-section is given by a simple formula:

$$\frac{d\sigma}{d\boldsymbol{p}_s d\boldsymbol{p}_f} = (2\pi)^4 N \frac{p_s p_f}{p_o} f_{ee}\, \rho(\varepsilon, \boldsymbol{q}) \,, \qquad (3)$$

where the electron-electron factor f_{ee} is the Mott cross-section for the free electron collision, and N is the number of atoms in the target [2]. The spectral, or energy resolved, momentum density $\rho(\varepsilon, \boldsymbol{q})$ gives the probability of finding an electron with a given energy ε and momentum $\boldsymbol{q}$. It is obvious therefore that the *(e,2e)* technique directly determines the electron distribution in momentum space in the system of interest, and hence is commonly referred to as electron momentum spectroscopy (EMS). In solids it allows the determination of the dispersion relation between the electron energy and momentum $\varepsilon = E_j(\boldsymbol{q})$ in a particular energy band j.

Angle-resolved photoelectron spectroscopy (ARPES) is another experimental technique which is able to map the energy-momentum relation in solids, see for example Courths and Hüfner [3]. Although this technique has been very successful and is widely used, the theoretical understanding of the intensities of the observed peaks is far from complete. This is because the interaction processes are very complicated, involving a many-body problem in the initial and final states. Hence, it is not straightforward to measure the momentum densities of electrons using ARPES.

A technique that does enable the measurement of momentum densities is Compton scattering (see for example Cooper [4] for a review). This technique probes the target electrons with high energy photons. A measurement of the total momentum density of all target electrons integrated over a plane perpendicular to the scattering vector is obtained. Thus, only a projection of the momentum is obtained and information about the energy is not resolved.

The application of EMS to solids has been restricted primarily by the complications arising from their high atomic density. Both the incoming and outgoing electrons can undergo additional scattering other than the *(e,2e)* event itself. For solids one usually has to use a transmission mode in order to be able to access all possible values of the momentum $\boldsymbol{q}$. To obtain detailed information on the electronic structure of the target, high coincidence count rates and good energy and momentum resolution are required. The early *(e,2e)* experiments on solids [2,5,6] were unable to resolve any valence band structure because of poor energy resolution. It was not until the experiments of Ritter *et al.* [7], where the energy resolution had improved to 6 eV, that structure in the valence bands could be resolved. However, these experiments were still very restricted by coincidence rates and energy and momentum resolution.

Significant progress in *(e,2e)* experiments on solids has been achieved by the Flinders University group. The major difficulties involved in the measurements of solids were successfully overcome by utilising multi-parameter detection techniques and an improvement in the energy resolution [8]. A parallel development in theory based on the density-functional approach permitted calculations of energy-resolved electron momentum densities for a large number of crystalline and disordered materials [9,10]. This served as a basis for understanding and interpreting a vast amount of experimental data [11–14] (see also Dennison and Ritter [15] for a review).

Very recently a further improvement in energy resolution down to less than 1 eV was achieved by the Flinders group [16]. Target preparation techniques were also developed to a stage where monocrystalline targets became available for *(e,2e)* transmission experiments. On the theoretical side, a much better understanding of multiple scattering was achieved by Monte Carlo simulations of the experimental process which accounted for additional electron collisions other than the *(e,2e)* collision [17]. In addition, the finite life-time effects due to many-electron correlations in the ionized system were also included in the model. As a result of these developments in experiment and theory quantitative EMS spectroscopy of solids became a possibility for the first time. In a very recent EMS study on aluminum complete agreement has been achieved between the experimental results and theoretical calculations for the entire electronic structure (energy, momentum and intensity) [18].

The purpose of this paper is to give an overview of the latest achievements with the experiments and the theory of *(e,2e)* on solids. We concentrate primarily on the results of the Flinders group. In addition, we mention briefly recent work done elsewhere on reflection type *(e,2e)*.

II EXPERIMENTAL TECHNIQUE

The asymmetric non-coplanar *(e,2e)* reaction is utilized for EMS in transmission mode. During a typical *(e,2e)* experiment on valence electrons the incident energy is 20 keV, with the energy of the fast electron nominally 18.8 keV and the energy of the slow electron about 1.2 keV. Constant polar angles of 14° and 76° degrees are used for the detection of the fast and slow electron respectively; the geometry is shown in Fig. 1.

With this geometry the measured momentum q for a coplanar event, i.e. one with all three momentum vectors in the same plane, is zero. A range of azimuthal (out of plane) angles are measured simultaneously by both detectors. The fast electron is detected in a hemispherical analyser whilst a toroidal analyser detects the slow electron. Each electron analyser has a two-dimensional position-sensitive detector mounted at its exit, enabling a range of target electron binding energies ε and momenta q to be measured simultaneously. This is a major advantage as it greatly improves the *(e,2e)* coincidence rate.

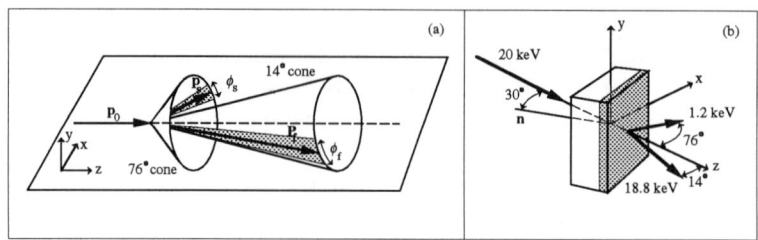

FIGURE 1. (a) Geometry of a transmission *(e,2e)* experiment. A range of outgoing azimuthal angles are highlighted to show how different target electron momenta can be measured. (b) An illustration of the surface that is studied in a transmission *(e,2e)* measurement on a thin film. The depth that is probed depends on the energy of the electrons and the angle of the target. This depth is dominated by the energy of the slowest electron that is detected.

Experiments have been proposed and attempted using diffracted beams in reflection from a solid surface. This experimental arrangement is advantageous for surface studies and does not require an elaborate and time-consuming preparation of a free-standing thin film. Recently Iacobucci *et al* have reported some *(e,2e)* measurements at 300eV incident beam energy in a grazing-angle reflection geometry [20]. A schematic of these experiments is shown in Fig. 2. These measurements from a graphite surface were the first successful grazing incidence measurements to be reported and they established the feasibility of the technique. The reaction in these measurements occurs between an incident beam electron that is specularly reflected from the graphite surface and the target electron.

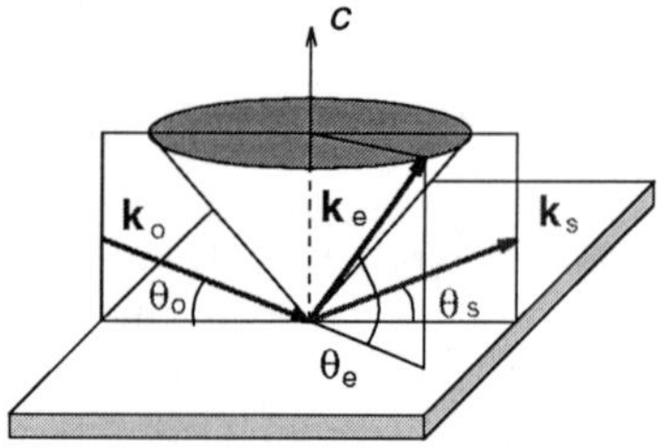

FIGURE 2. Geometry of a grazing-angle reflection *(e,2e)* experiment. Arrows indicate incident (o), scattered (s) and ejected (e) electrons. The shaded area shows the acceptance range of the ejected electron momenta

Analogous reflection type experiments were reported by Kirschner *et al.* in the backscattering geometry [21]. They used low-energy (14–25 eV) electrons and a time-of-flight measuring technique. Correlated electron pairs were detected from a W(100) single crystal and a LiF thin film. Compared to the higher energy *(e,2e)* experiments, interpretation of these low energy measurements is less straightforward and requires an elaborate model to describe the scattering.

III THEORETICAL MODEL

The spectral momentum density (SMD) which enters into Eq. (3) is defined according to [2] as

$$\rho(\varepsilon, \boldsymbol{q}) = \sum_\alpha |\langle \boldsymbol{q}\,(N-1,\alpha)|N,0\rangle|^2\, \delta(\varepsilon - E_{N-1,\alpha} + E_N)\,. \tag{4}$$

Here $|N,0\rangle$ and $|N-1,\alpha\rangle$ are the many-electron wave functions of the ground and ionized states of the crystal, respectively, and $|\boldsymbol{q}\rangle$ is the normalised plane wave. The summation is over all the final states α of the ionized system.

In the independent-particle approximation Eq. (4) simplifies to

$$\rho_j(\varepsilon, \boldsymbol{q}) = |\langle \boldsymbol{q}|\Psi_{j\mathbf{k}}\rangle|^2\, \delta(\varepsilon - E_j(\mathbf{k}))\,, \tag{5}$$

where we specify a one-electron state $\Psi_{j\mathbf{k}}$ by the band index j and crystal momentum $\mathbf{k}$. The spatial integration in Eq. (5) is performed over the crystal volume. Using Bloch's theorem it can be reduced to the unit cell Ω transforming Eq. (4) into

$$\rho_j(\varepsilon, \boldsymbol{q}) = (2\pi)^{-3} \sum_{\mathbf{G},\mathbf{k}\in 1^{\text{st}}\text{ BZ}} \left| \int_\Omega d^3r\, \psi_{j\mathbf{k}}(\boldsymbol{r}) e^{-i\mathbf{q}\cdot\boldsymbol{r}} \right|^2 \delta_{\mathbf{q},\mathbf{k}+\mathbf{G}}\, \delta(\epsilon - E_j(\boldsymbol{k}))\,. \tag{6}$$

Here the electron momentum $\boldsymbol{q}$ is translated to the first Brillouin zone by the reciprocal lattice vector $\boldsymbol{G}$.

To calculate the one-electron wavefunctions $\psi_{j\mathbf{k}}(\boldsymbol{r})$ and one-electron energies $E_j(\mathbf{k})$ one has to perform a band-structure calculation on the material of interest. The calculations reported in [9–14] relied on the linear muffin-tin orbital (LMTO) method as described by Skriver [22]. It is convenient to make an atomic-sphere approximation and substitute the unit cell with a number of atomic spheres each of which represents a non-equivalent atomic position. The total volume of the spheres is equated to the volume of the unit cell:

$$\sum_s 4\pi a_s^3/3 = \Omega\,, \tag{7}$$

where a_s is the muffin-tin (MT) radius of a sphere at site s.

The electron potential is spherically symmetric within the spheres. The tails of the LMTO orbitals outside the spheres are chosen to have zero kinetic energy. The Bloch sum of the tails is canceled within the spheres. So the one-electron wave function within any particular sphere centered at $\boldsymbol{R}_s$ can be written as

$$\psi_{j\mathbf{k}}(\mathbf{r} - \mathbf{R}_s) = \sum_{lm} b^{j\mathbf{k}}_{slm} i^l Y_{lm}(\hat{r}') \frac{1}{r'} P_{sl}(r') , \quad r' = |\mathbf{r} - \mathbf{R}_s| \leq a_s . \qquad (8)$$

Here $\mathbf{k}$ is the crystal wave vector, j is the band index, Y_{lm} is the spherical harmonic which depends on the orbital momentum l and its projection m. The expansion coefficients $b^{j\mathbf{k}}_{slm}$ for a given MT sphere s are found by solving the LMTO eigenvalue problem. The radial part of the wave function $P_{sl}(r)$ depends on the type of atom at site s and the orbital momentum l but does not depend on $\mathbf{k}$ and j which significantly increases the computational efficiency of the LMTO method.

By taking advantage of the central-field expansion (8) the SMD can be readily calculated as

$$\rho_j(\epsilon, \mathbf{q}) = \frac{2}{\pi} \sum_{\mathbf{Gk}} \left| \sum_s e^{-i\mathbf{q}\cdot \mathbf{R}_s} \sum_{lm} b^{j\mathbf{k}}_{slm} Y_{lm}(\hat{k}) \int_0^{a_s} dr\, j_l(qr) P_{sl}(r) \right|^2 \delta_{\mathbf{q},\mathbf{k}+\mathbf{G}}\, \delta(\epsilon - E_j(\mathbf{k})) ,$$
(9)

where $j_l(qr)$ is the spherical Bessel function.

The finite experimental energy resolution can be easily incorporated into the model by substituting the δ-function of energy entering Eqs. (6,9) by a normalised Gaussian function $\delta\epsilon^{-1}\Gamma\left[(\epsilon - E_j(\mathbf{k})/\delta\epsilon\right]$, where $\delta\varepsilon = (8\ln 2)^{-1/2}$· FWHM.

In practice it is of interest to calculate the SMD either in a particular direction ($\mathbf{q} = q\mathbf{e}$, $\mathbf{e}$ is the unit vector) or as a spherical average ($|\mathbf{q}| = q$) over the irreducible wedge of the Brillouin zone. These two types of calculations refer to either a crystalline oriented target, or a polycrystalline disordered target, respectively. In both cases the momentum dependence is reduced to a single scalar variable q and the momentum broadening can be included by a Gaussian convolution with the experimental momentum width.

Besides the instrumental effects, an additional energy broadening can be caused by the finite life-time of the hole created by the (e,2e) event. This effect can be described if instead of Eq. (5) we use a more general expression:

$$\rho_j(\varepsilon, \mathbf{q}) = |\langle \mathbf{q}|\Psi_{j\mathbf{k}}\rangle|^2 \, Im\, G_{j\mathbf{k}}(\varepsilon) , \qquad (10)$$

where the retarded, or one-hole, Green's function $G_{j\mathbf{k}}(\varepsilon)$ describes the final state of the ionized system after removing an electron with the crystal momentum $\mathbf{k}$ from the band j. The Green's function can be presented as

$$G_{j\mathbf{k}}(\varepsilon) = (\varepsilon - E_j(\mathbf{k}) - i\gamma(\varepsilon))^{-1} \qquad (11)$$

where $\gamma(\varepsilon)$ is inverse lifetime of the hole. To calculate γ from first principles one has to use a realistic model for the screening and relaxation caused by ionization. Alternatively, one can use Eq. (11) with an empirical dependence $\gamma(\varepsilon)$ obtained from independent experiments, i.e. from angle-resolved photoemission.

The results of the previously described LMTO calculations would describe the experimental data if all *(e,2e)* events were "clean" i.e. further elastic or inelastic scattering of either the incoming or both outgoing electrons does not occur. The Monte Carlo simulation gives an estimate of the rate at which certain momentum transfer (elastic scattering) and energy loss (inelastic scattering) combinations occur and how this in turn affects our measured spectra. For a detailed description of the Monte Carlo procedure as previously applied to *(e,2e)* spectroscopy of solids see the paper by Vos and Bottema [17].

The simulation involves the inclusion of both elastic and inelastic scattering events for any of the incident and two outgoing electrons. The depth t at which an *(e,2e)* event occurs is determined by a random number generator. One then needs to simulate each of the three trajectories for electrons with energy E_0, E_f and E_s and trajectory length t_0, t_f and t_s. For elastic scattering the cross-sections are calculated in the Born approximation from actual Hartree-Fock wavefunctions of the free atoms constituting the solid. For inelastic scattering only bulk plasmon excitations are considered. It is assumed that the plasmon distribution is Gaussian, centered at some mean plasmon energy. From these parameters the elastic and inelastic mean free paths for each of the three electrons involved are determined. In the simulation the probabilities of distances between subsequent elastic and inelastic scattering events vary as a Poisson distribution characterised by the mean free path. If the distance for elastic or inelastic scattering obtained from the distribution is smaller than the depth t, a contribution is made to the energy-resolved momentum density.

IV RESULTS

A Aluminum

Aluminum being a nearly free-electron metal, has a band structure that is well understood. Hence the agreement reached between the experiment and theory provides an excellent insight into the current understanding of *(e,2e)* in solids [18].

The energy-resolved momentum density for a polycrystalline aluminum film is shown in Fig. 3 (a). In this plot the dispersion curve of the valence band clearly stands out above the background. The densities are represented on a linear grey-scale with white corresponding to the highest intensity. The dispersion curve looks very similar to a free-electron parabola as is expected for metallic aluminum. The parabola extends from 4 eV binding energy at the top, down to 16 eV at the bottom. It should be noted that the experimental binding energies are referenced to the vacuum level. The Fermi level corresponds to 4

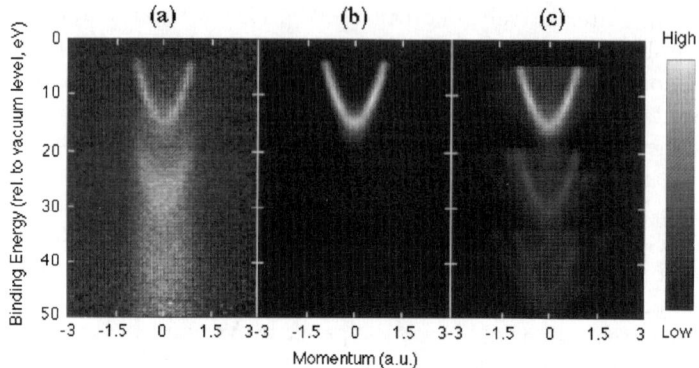

FIGURE 3. (a) Energy resolved momentum density from *(e,2e)* measurements on polycrystalline aluminum film, (b) as calculated using the LMTO method convoluted with Gaussians representing the finite experimental energy and momentum resolution and a Lorentzian taking into account energy broadening due to finite life-time effects, (c) simulated using the Monte Carlo procedure taking the LMTO momentum density as an input.

± 0.5 eV binding energy in the experiment. The measured bandwidth of 12 eV is close to that calculated for aluminum using the LMTO method (11.5 eV). The band is occupied in momentum space between approximately ± 1.0 a.u. which compares well with the known Fermi momentum of aluminum $p_F=$ 0.92 a.u. [19]

The energy-resolved electron momentum density of the LMTO calculation is presented in Fig. 3 (b). This calculation has been convoluted with an instrumental width of 1.0 eV to match the measured energy resolution of the *(e,2e)* experiment. The experiment also had a finite momentum resolution which was determined by the angular resolution of each of the three electron beams and was estimated to be 0.1 a.u. The additional energy broadening caused by the finite lifetime of the hole left behind by the ejected electron has also to be taken into account. Levinson *et al.* [23] measured the inverse lifetimes in their angle-resolved photoemission experiments on aluminum. These data have been used to obtain the function $\gamma(\varepsilon)$ which appears in Eq. (11) and produces a Lorentzian-type broadening.

The effects of elastic and inelastic scattering on the energy resolved momentum density were investigated using the Monte Carlo procedure. The Monte Carlo results are shown in Fig. 3 (c). The binding energy scale of the plots of Fig. 3 ranges from 0 to 50 eV so that a discussion on the multiple scattering events can also be made. Qualitative improvement of the agreement between theory and experiment is obtained after inclusion of the Monte Carlo simu-

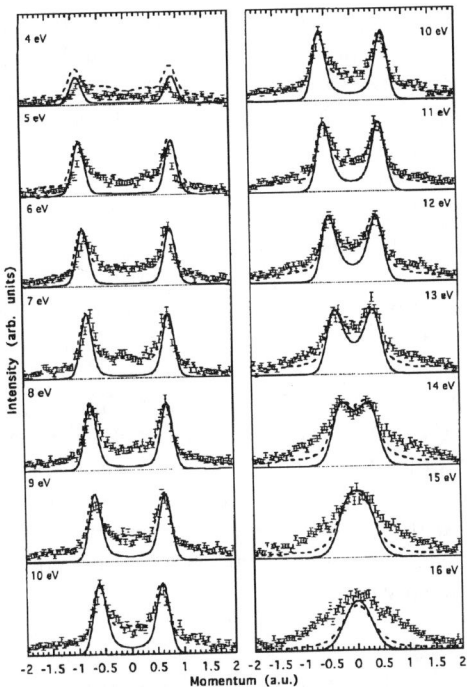

FIGURE 4. The momentum profiles of the measured aluminum sample for a series of energies are shown with error bars in each panel. Also shown is the LMTO calculation (solid lines) with a Gaussian convolution with 1 eV energy resolution, 0.1 a.u. momentum resolution, and a Lorentzian broadening for the finite lifetime. The dashed lines show the above LMTO calculation with the Monte Carlo simulation

lation. The valence band of the measurement is simulated very well. There is extra intensity inside and outside the valence band from the Monte Carlo simulation that is not seen in the LMTO calculation. In the (e,2e) measurement the plasmon dispersion of aluminum can clearly be seen. The plasmon dispersion is a replica of the aluminum valence band shifted to 15 eV lower binding energy, i.e. inelastic scattering resulting in energy loss. The Monte Carlo simulation also predicts this plasmon dispersion, however, the intensity of the plasmon distribution is not as high as the measurement. Also the plasmon distribution from the experiment appears to be "filled in" and this is not seen in the Monte Carlo simulation. The Monte Carlo simulation does predict a plasmon distribution of double excitation, i.e. 30 eV energy loss.

For a quantitative study, that is the determination of the energy–momentum position of the electrons and the relative intensities, a more thorough analysis is required. For this purpose a series of momentum profiles is constructed by taking slices through the valence band region (width of slices are 1 eV) with the centre of those slices being the binding energy indicated. The experimental momentum profiles are shown (with error bars) in Fig. 4. From these plots one can see two distinct peaks at the top of the band (4 eV) which are symmetric around zero momentum. As the energy increases these peaks disperse inwards, i.e. towards zero momentum, finally forming one peak centered at zero momentum at the bottom of the band (16 eV).

The momentum profiles for the LMTO calculation are shown as solid lines in Fig. 4. The agreement between the experiment and the LMTO calculation is excellent. The Monte Carlo simulation is shown in Fig. 4 by dashed lines. It is clear that the Monte Carlo simulation has significantly increased the intensity between the two peaks, i.e. where the LMTO predicted zero intensity. The mechanism responsible for this is elastic scattering. Up until 12 eV the final agreement is excellent. Above 12 eV the peak intensities themselves are in very good agreement although the measurement shows intensity at higher momentum that is not predicted in either of the theoretical models. This excess intensity is most likely a result of inelastic processes that are not accounted for in the Monte Carlo procedure. These processes include surface plasmon excitation and the effects of momentum transfer during plasmon excitation and single electron excitations.

B Natural graphite

The (e,2e) experiment on natural graphite [24] is the first reported EMS measurement on a single crystal target with a thoroughly controlled orientation and fairly well known electronic structure. Hence, it can be used to create a 3D momentum-density map complementary to the charge density distribution in real space.

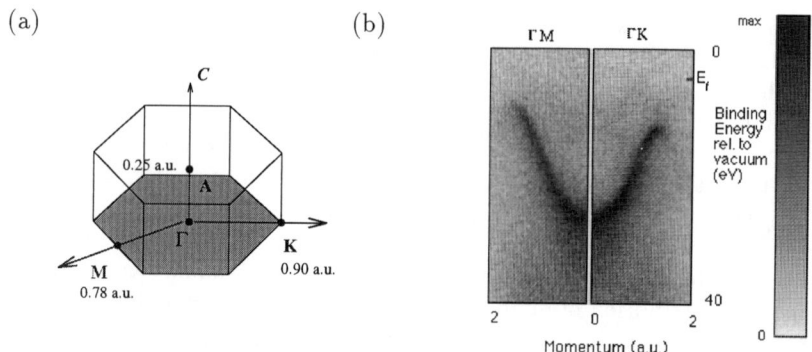

FIGURE 5. (a) Brillouin zone of graphite (upper-half). Distances from the zone centre Γ are indicated in atomic units. (b) Experimental spectral momentum density in ΓM and ΓK directions

Graphite is a layered compound with strong covalent bonding in plane and weak van der Waals bonding in between the planes. The large distance between the planes results in a quasi two-dimensional energy band structure with three of the carbon valence electrons in dispersing σ bands and one electron occupying a weakly dispersing π band. The orientation of the π bonding is such that its electrons have maximum momentum density along the c-axis and zero momentum density in the basal plane. The σ bands have non-zero densities in the basal plane.

The energy-resolved momentum density of graphite in the basal plane along the two high-symmetry directions ΓK and ΓM, indicated in Fig. 5 (a), is represented in Fig. 5 (b). As expected, only the σ-band contribution can be seen in these directions. Near the bottom of the valence band dispersion is parabolic. At larger momenta the anisotropy is clearly seen as predicted by the LMTO calculation of Ref. [10].

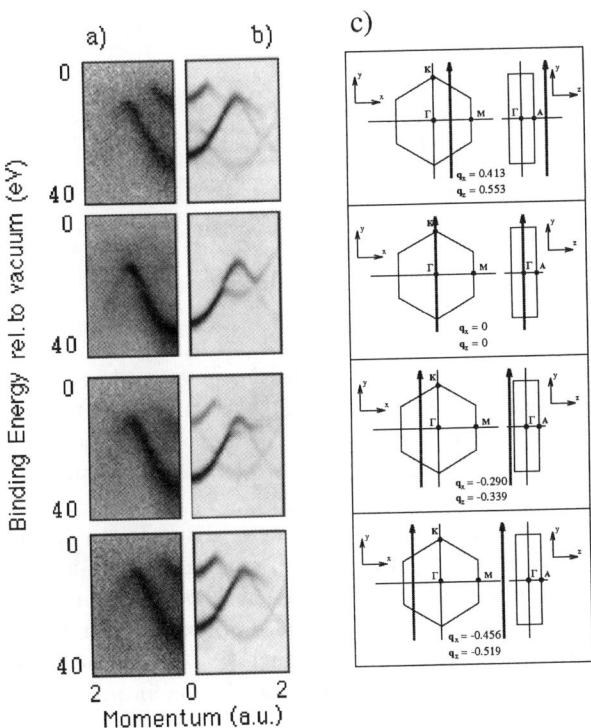

FIGURE 6. Spectral momentum density of natural graphite outside the basal plane. Experiment (a), LMTO calculation (b) and the Brillouin zone scheme (c). The measurement probes the electron momentum along the direction indicated as a solid arrow. The component q_y is varying in the experiment whereas q_x and q_z are fixed.

The out-of-plane SMD of graphite is shown in several panels of Fig. (6). The kinematic arrangement of the experiment is such that one component of the valence electron momentum q_y can be varied continuously whereas the others q_x and q_z are locked by the position of the slow electron detector. Several panels of Fig. (6) corresponds to various combinations of q_x, q_z indicated on the Brillouin zone scheme (c). When $q_z = 0$ (second top panel) the contribution from the π band is absent as expected. As $|q_z|$ increases the π band intensity increases too. This behaviour is in very good agreement with the LMTO calculation shown in Fig. (6b). Somewhat larger energy broadening in the experiment, especially near the bottom of the valence band, can be attributed to the finite lifetime effects which are not included in the calculation. An ad-

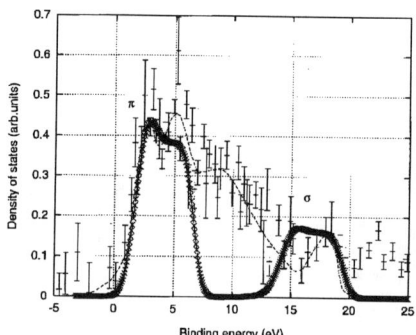

FIGURE 7. Partly integrated SMD from highly oriented pyrolytic graphite (HOPG). Reflection *(e,2e)* measurement from the basal plane (error bars), and the LMTO calculation (solid line).

ditional information about momentum densities in graphite can be extracted from the reflection *(e,2e)* measurements [20]. At the present stage the slow electron momentum is not fully resolved in this experiment, so the SMD integrated over the acceptance range of the slow electron momenta is measured. This partly integrated SMD is presented in Fig. (7) together with the LMTO calculation. Some distinctive features of the experiment (total width of the spectrum and relative intensity of the σ and π peaks) are in agreement with theory. However, some parts of the experimental spectrum are absent in the calculation. This can be attributed to more complex scattering dynamics not included in the model. A more definite conclusion can be made when fully resolved *(e,2e)* spectra are detected. This work is now in progress [25].

C Aluminum oxide

Aluminum oxide is the most complicated target studied so far by EMS [26]. In its trigonal form $\alpha-Al_2O_3$ has a unit cell with 10 non-equivalent atomic positions which results in an unusually complex pattern of the valence energy bands.

Figure 8 represents the measured spectral momentum density for the aluminum oxide polycrystalline sample (left panel) and the LMTO calculation of a spherically averaged $\alpha-Al_2O_3$ (right panel), respectively. Aluminum oxide has two distinct features in its valence band structure namely an upper valence band and a lower valence band. A previous (e, 2e) experiment [27] on an aluminum–aluminum oxide thin foil with an energy resolution of 4.5 eV reported dispersionless structures in both the upper and lower valence band. This is manifestly not the case in the present experiment. The present measured spectral momentum density plot shows a "dual parabola" dispersion pattern spanning 8 eV in the upper valence band (Fig. 5 left panel). The LMTO calculation shows a similar "dual parabola" dispersion pattern with nearly the same energy span. Near zero momentum both the EMS measurement and the LMTO calculation exhibit lower intensity throughout the upper valence band.

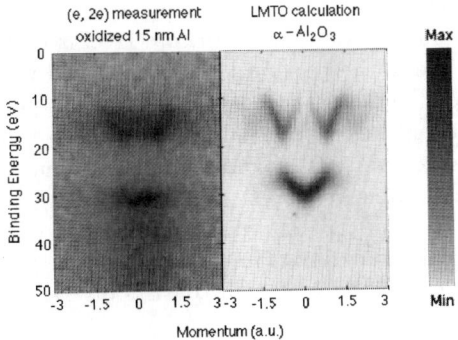

FIGURE 8. The spectral momentum density plots as measured for the aluminum oxide sample (left panel) and calculated using the LMTO method of spherically averaged $\alpha-Al_2O_3$ (right panel), respectively. The binding energy is relative to the vacuum level. The intensity scale is shown on the right hand side.

There is a gap between the upper valence band and lower valence band. The EMS measurement and the LMTO calculation give nearly the same gap. In the lower valence band, the LMTO calculation indicates a dispersion of about 5 eV. The EMS measurement shows a similar "bowl" shape in the spectral momentum density (the full width of momentum distribution becoming narrower with increasing binding energy in the lower valence band) but it shows less dispersion. Nevertheless, one can see that the measured major features in the valence band of aluminum oxide are qualitatively reproduced by the LMTO calculation.

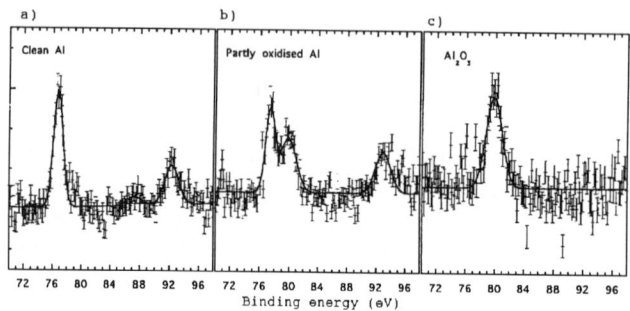

FIGURE 9. Core level spectra of the aluminum $2p$ state for (a) evaporated aluminum, (b) aluminum exposed to oxygen, and (c) oxidised aluminum

In addition to the valence band EMS important structural information can be obtained from the core atom-like states. These experiments are similar to the well-established technique of electron spectroscopy for chemical analysis (ESCA) except that an extra variable (electron momentum) becomes available.

An example of EMS of the $2p$ core-level region of aluminum is shown in Fig. 9. For the evaporated aluminum a $2p$ core level is evident at a binding energy of 77 eV relative to the vacuum level. Associated with this aluminum peak is a plasmon peak at 92 eV and a surface plasmon peak around 87 eV. Evidence of the surface plasmon peak indicates that the evaporated sample is not contaminated by oxygen. After exposing the aluminum film to oxygen a second peak appears, shifted by 2.6 eV to higher binding energy. This peak can be attributed to a chemical shift which arises when the surface becomes oxidized. When the aluminum film is fully oxidised there is no evidence of an aluminum $2p$ peak, instead the spectrum consists of a single core level peak of Al_2O_3.

V CONCLUSION AND FUTURE DIRECTIONS

We have shown that EMS of solids is past its demonstration phase and is capable of producing quantitative data on energy-resolved momentum densities of solids. A considerable amount of research has still to be done and it is likely that this technique will make significant and interesting contributions for many years to come. As an immediate task we consider the 3D electron momentum mapping in single crystals (graphite, silicon). We also plan experiments on polycrystalline films of $3d$ transition metals (Ni, Fe, Cu) and their oxides which show strong electron-correlation effects. Understanding and interpreting these experiments will require further development of theory beyond the independent-particle approximation.

REFERENCES

1. I.E. McCarthy and E. Weigold. *Rep. Prog. Phys.* 54, 789 (1991).
2. N. M. Persiantseva, N. A. Krasil'nikova, and V. G. Neudachin, *Sov. Phys. JETP* 49, 530 (1979).
3. R. Courths and S. Hüfner. *Phys. Rep.* 112, 53 (1984).
4. M.J. Cooper.*Rep. Prog. Phys.* 48, 415 (1985).
5. U. Amaldi Jr., A. Egidi, R. Marconero, and G. Pizzella. *Rev. Sci. Instrum.* 40, 1001 (1969).
6. R. Camilloni, A. Guardini-Guidoni, R. Tiribelli, and G. Stefani. *Phys. Rev. Lett.* 29, 618 (1972).
7. A.L. Ritter, J.R. Dennison, and R. Jones. *Phys. Rev. Lett.* 53, 2054 (1984).
8. P. Storer, R.S. Caprari, S.A.C. Clark, M. Vos, and E. Weigold. *Rev. Sci. Instrum.* 65, 2214 (1994).
9. A. S. Kheifets and Y. Q. Cai. *J. Phys.: Condens. Matter* 7, 1821 (1995).
10. A. S. Kheifets and M. Vos. *J. Phys.: Condens. Matter* 7, 3895 (1995).
11. M. Vos, P. Storer, S.A. Canney, A. S. Kheifets, I. E. McCarthy, and E. Weigold, *Phys. Rev. B* 50, 5635 (1994).
12. M. Vos, P. Storer, Y. Q. Cai, A. S. Kheifets, I. E. McCarthy, and E. Weigold, *J. Phys.: Condens. Matter* 7, 279 (1995).
13. Y. Q. Cai, M. Vos, P. Storer, A. S. Kheifets, I. E. McCarthy, and E. Weigold, *Phys. Rev. B* 51, 3449 (1995).
14. Y. Q. Cai, P. Storer, A. S. Kheifets, I. E. McCarthy, and E. Weigold, *Surface Science* 334, 276 (1995).
15. J.R. Dennison and A.L. Ritter, *J. Electron Spectr.* 77, 99 (1996).
16. S.A. Canney, M.J. Brunger, I.E. McCarthy, P.J. Storer, S. Utteridge, M. Vos, and E. Weigold, *J. Elect. Spect. and Rel. Phen.* (1996, in press).
17. M. Vos and M. Bottema, *Phys. Rev. B* (1996, to be published).
18. S. A. Canney, M. Vos, A. S. Kheifets, N. Clisby and I. E. McCarthy, *Phys. Rev. B* (1996, to be published).
19. C. Kittel, *Introduction to solid state physics*, Wiley, New York, 1986
20. S. Iacobucci, L. Marassi, R. Camilloni, S. Nannarone and G. Stefani, *Phys. Rev. B*, 51, 10252 (1995).
21. J. Kirschner, O.M. Artamonov, and S. N. Samarin, *Phys. Rev. Lett.*, 75, 2424 (1995).
22. H. L. Skriver. *The LMTO method.* Springer-Verlag, Berlin, 1984.
23. H.J. Levinson, F. Greuter, and E.W. Plummer. *Phys. Rev. B* 27, 727 (1983).
24. M. Vos, A. S. Kheifets, S. A. Canney, I.E. McCarthy and E. Weigold, *Phys. Rev. Lett.* (1996, to be published).
25. G. Stefani, private communication.
26. X. Guo, M. Vos, A. S. Kheifets, Z. Fang, S. Canny, S. Utteridge, I. E. McCarthy and E. Weigold, *Phys. Rev. B* (1996, to be published).
27. P. Hayes, M. A. Bennett, J. Flexman and J. F. Williams *Phys. Rev. B* 38, 13371 (1988).

Momentum-space Magnetism Studied by Magnetic Compton Scattering

Nobuhiko Sakai, Akihisa Koizumi, Naoki Miyamoto
and Yoshikazu Tanaka*

*Material Science Division, Himeji Institute of Technology,
1479-1 Harima Science Garden City, Ako-gun Hyogo 678-12, Japan
* The Institute of Physical and Chemical Research (RIKEN),
Wako, Saitama 351-01, Japan*

Abstract. The usefulness of magnetic Compton-profile spectroscopy to derive momentum-space information about the spin states of electrons is described with examples. The reconstructed three-dimensional momentum density of spin-polarized electrons in ferromagnetic iron reveals information on the spin states in momentum space in a straightforward fashion. The analysis of Mn-ion states in the intermetallic compound MnSb illustrates that atomic orbital wavefunctions in momentum-space representation can be used for the MCP analysis. The separate determination of the magnetic moments of $3d$ and $4f$ electrons together with wide band itinerant electrons in amorphous $Gd_xY_{1-x}Fe_{20}Al_{20}$ alloy are presented. The spin magnetic moments of Sm and Co atoms in $SmCo_5$ have been evaluated for the first time by utilizing the fact that the magnetic Compton-scattering cross section does not depend on atomic orbital angular momenta.

INTRODUCTION

The aim of this paper is to present the concept of momentum-space magnetism with examples from recent x-ray magnetic Compton profile (MCP) studies. Many MCP measurements have been made using synchrotron radiation (SR) following the X-90 conference where one of the authors (N.S.) reported on MCP[1]. One of the great advantages of SR sources over x-ray tubes or γ-ray sources is the high degree of polarization of the radiation. Linearly and circularly polarized x-rays having the desired energies are now available. Interesting information about electron spin states in magnetic materials has been obtained using MCP measurements in which the electronic states are reflected through their momentum-space wavefunctions. It should be emphasized that the fundamental quantity in MCP is not a Fourier component of the spin density of the electrons, but the momentum density of electron spins. Therefore the information obtained from MCP measurements, despite the present restriction to only ferro- and ferrimagnetic materials, is complementary to the information from magnetic Bragg diffraction. Another important difference is that the magnetic Compton-scattering cross section depends only on the individual electron *spin* vectors, and does not depend on the orbital angular momentum(2–5). A recent review article on MCP(6) is a useful reference which includes a theoretical derivation of the spin-dependent Compton-scattering cross section.

MOMENTUM-SPACE MAGNETISM

Hitherto electronic spin states have often been understood in relation with real space quantities or with energies. What kind of difference are there when electron states are observed, or described, in momentum space? Momentum–space spin density $n_{spin}(p)$ is related to wavefunctions in real space by the following equation,

$$n_{spin}(p) = n^{\uparrow}(p) - n^{\downarrow}(p) ,$$

$$n^{\uparrow/\downarrow}(p) = \sum_n \int |(2\pi)^{-3/2} \int \psi_{n,k}^{\uparrow/\downarrow}(r) \exp(i p \cdot r / \hbar) \, dr|^2 dk \qquad (1)$$

where $\psi_{n,k}(r)$ is a real-space electron wavefunction of wavevector **k** and band index n. The Fourier transform of $\psi_{n,k}(r)$ expresses the momentum–space wavefunction. The arrows ↑ and ↓ respectively represent the spin-up and the spin-down states. The sum over n and the integration with respect to **k** are made over occupied states. Unlike the wide spread of the itinerant electron spin density in real space, the itinerant spin density in momentum space is centered in the low momentum region, which enables the magnitude and the fine structures to be observed more readily. In real space, site positions of atoms are specified, while in momentum space, whole momentum distributions of electrons possess the same origin, and momentum distributions of atomic electrons overlap each other. When electrons belong to different atomic orbitals, 3*d* and 4*f* for example, they have characteristic momentum distributions corresponding to their orbitals. Now the MCP is defined as a double integral of $n_{spin}(p)$,

$$J_{mag}(p_z) = \iint n_{spin}(p) \, dp_x dp_y , \qquad (2)$$

where the z axis is taken to be parallel to the scattering vector of the x rays.
Thus the itinerant character of electrons and the characteristic momentum distribution of orbital electrons directly appear in MCP. When the scattering vector is taken to be parallel to a [*hkl*] direction of a single crystal sample, the MCP is said to be a directional MCP along the [*hkl*] direction. Since high momentum components of MCP reflect the core part of a real-space wavefunction, they do not change much when atoms are embedded in a different matrix. It is thus possible to

discriminate to some extent the 3d and 4f electron spin contributions to magnetic moments by a least-squares fitting procedure as shown below. It should be emphasized that MCP offers a new experimental method, which reveals electron spin states from the point of view of momentum space.

Angular correlation of annihilation γ rays of spin polarized positrons provides additional information on momentum-space magnetism. Ferromagnetic Gd(7), Fe(8,9) and Ni(10,11) have been measured. The spin-polarized positron experiment is superior to MCP experiments with regard to the experimental momentum resolution. However, the positron data provide momentum distributions of electron-positron pairs, and they are modified by the positron wavefunctions. Since positrons are repelled by the charge of the nucleus, information on 3d- or 4f-electron orbitals is severely limited.

Indispensable information can be obtained by means of the de Haas-van Alphen effect or angle-resolved photoelectron spectroscopy. The former gives important information on Fermi surfaces, and the latter reveals dispersion curves of electronic band structures. However, they are not as directly related to the momentum density of electrons as MCP.

It is well known that electron-impact ionization reactions provide unique information about the motion and correlation of valence electrons in atoms and molecules. The history, theory and experimental examples can be found in the review article given in reference (12).

TOPICS OF MCP

Circularly Polarized Synchrotron Radiation Sources

It is worth mentioning SR sources, because recent MCP measurements depend entirely on circularly polarized SR. The spin-dependent Compton scattering cross section is proportional to the degree of circular polarization and the energy of the incident x rays(13). Figure 1 shows the ratio between the spin-dependent scattering cross section Φ_{spin} and the charge-induced cross section Φ_0 as a function of the incident energy E_0. When the energy exceeds 400 keV, the spin-dependent and charge-induced cross sections become comparable.

Circularly polarized SR can be obtain in the directions slightly above or below the electron orbital plane of a storage ring. Since the intensity of circularly polarized high-energy SR from ordinary bending magnets is weak, many insertion devices have been designed and constructed to obtain intense circularly polarized x rays. A smaller orbital radius is better for higher energy SR, because the critical energy of SR emitted is inversely proportional to the radius of curvature. A wavelength shifter, in which the electron orbit is forced to bend with a smaller radius of curvature, emits more intense hard x-rays than a bending magnet. An asymmetric

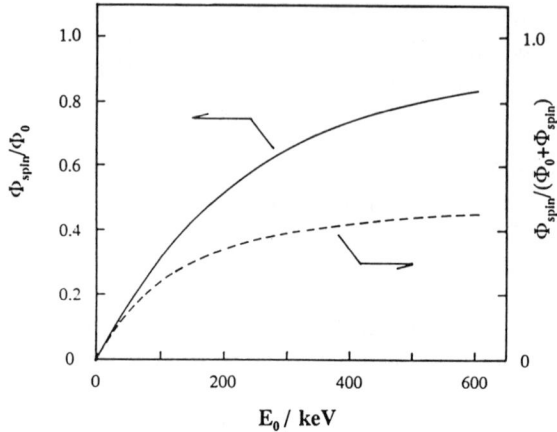

FIGURE 1. Energy dependence of the spin-dependent Compton-scattering cross section devided by the charge-induced scattering cross section. The right-hand ordinate is the one devided by the total intensity.

multipole wiggler (MPW), which in principle consists of a series of wavelength shifters, enhances the intensity by a factor equal to the number of shifters. All of them emit circularly polarized x-rays in off-plane directions. An elliptic MPW(14) has an electron orbit consisting of a series of tilted small half circles, and emits circularly polarized x-rays in the direction of the axis of the wiggler. An elliptic MPW is advantageous for intense high energy x rays with a smaller apparent source size than that of an asymmetric MPW. A helical undulator is superior to an EMPW regarding the heat load problem on the downstream optical elements. However, it only emits in the first harmonic, and at present, due to technical limitations on the magnetic field, the maximum energy is below 10 keV, which is not sufficient for MCP experiments.

In addition to insertion devices, x-ray phase plates can be used to produce circularly polarized x-rays from linearly polarized radiation(15,16). A trial has been made to produce hard x-rays using a Ge monolithic two-crystal phase plate, and 65 keV x-rays with a degree of circular polarization of 90 % have been produced(17).

Three-dimensional Momentum density of magnetic electrons in Fe

The most direct example of momentum-space magnetism will be a three dimensional momentum density of spins. This can be obtained from a set of directional Compton profiles along many crystal axes. If a scattering vector is set parallel to a crystal axis [hkl] of a single crystal sample, an observed $J_{mag}(p_{[hkl]})$ becomes a one-dimensional momentum distribution projected on the axis having the [hkl] direction,

$$J_{mag}(p_{[hkl]}) = \iint n_{spin}^{[hkl]}(\mathbf{p}) \, dp_x dp_y \quad . \tag{3}$$

Next, we make a quantity $B_{mag}(r_{[hkl]})$ as the Fourier transform of $J_{mag}(p_{[hkl]})$, where $r_{[hkl]}$ and $p_{[hkl]}$ denote that the vectors **r** and **p** are along a [hkl] direction.
Applying an interpolation procedure to the many directional experimental $B_{mag}(r_{[hkl]})$s, we obtain an empirical continuous three dimensional $B_{mag}(\mathbf{r})$ function. The experimental three-dimensional momentum density $n_{spin}(\mathbf{p})$ can thus be obtained by taking the inverse Fourier transform of this $B_{mag}(\mathbf{r})$.

From the beginning, ferromagnetic Fe has been an interesting material for MCP because of its narrow $3d$ bands and hybridization between $3d$ and s,p-like bands. Utilizing this reconstruction technique, a three-dimensional experimental momentum density of magnetic electrons in ferromagnetic Fe has been reconstructed from 11 directional MCPs(18).

Fig. 2(a) shows an experimental $n_{spin}(\mathbf{p})$ in the $p_z=0$ plane. Around the Γ point, a deeply negative density and a ring of positive density having 4-fold symmetry are found. The former is theoretically attributed to s,p-like itinerant electrons and the latter to $3d$-band electrons. It is clearly demonstrated that the negative spin polarization is almost within the first Brillouin zone.

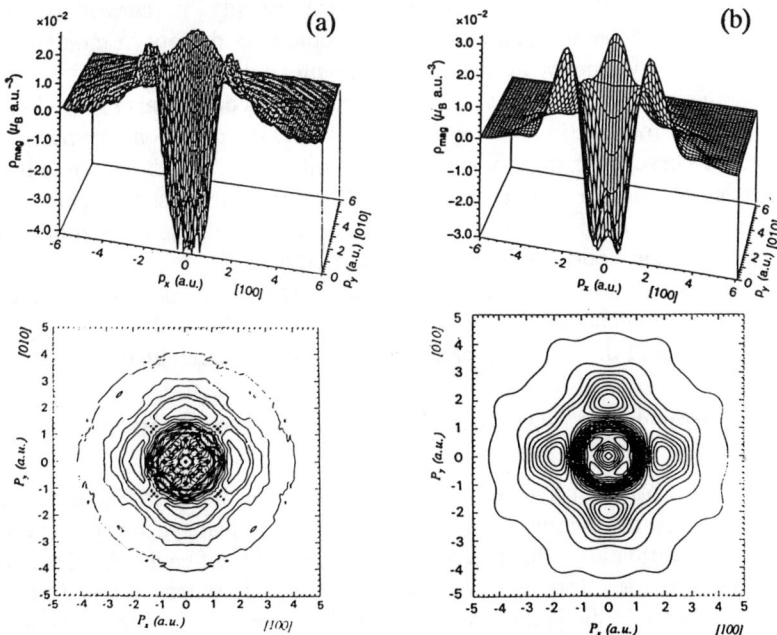

FIGURE 2. The momentum density of Fe+3wt% Si in the $p_z=0$ plane, where the origin is the Γ point. (a) is experiment and (b) is theory. (After Tanaka(18))

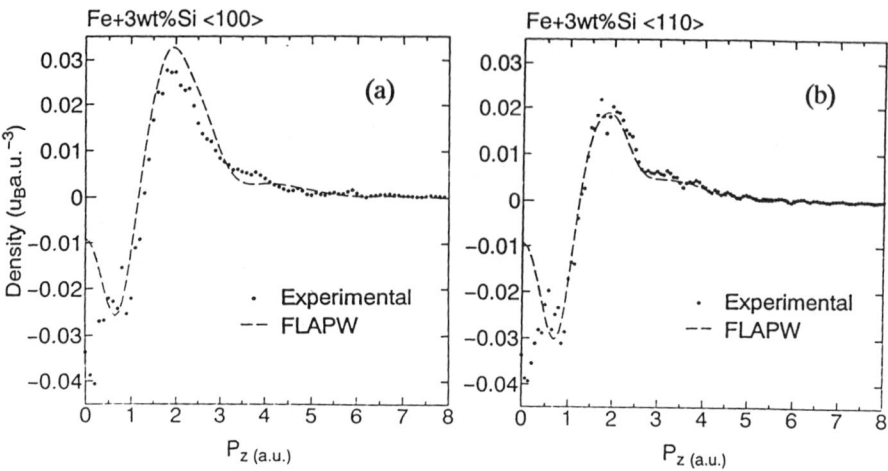

FIGURE 3. The cross sections of the momentum density: (a) along the [100] direction, (b) along the [110] direction.

A band theory calculation based on the fully linearized augmented plane wave approximation has successfully reproduced the details of the experimental results as shown in Fig. 2(b)(18,19). The theoretical density is convoluted with the experimental momentum resolution. Anisotropic spin density in momentum space reflecting the symmetry of the first Brillouin zones can be easily recognized. Quantitative comparisons between experiment and theory are also given in Figs.3(a) and 3(b) for cross sections along the [100] and [110] directions, respectively. The theoretical curve has a peak at $p_z=0$ atomic units (a.u.), which reflects a theoretical prediction of no spin polarization at the Γ point. The present experimental data could not confirm this prediction, because of severe systematic errors around the origin due to the Fourier transformation procedure.

The good overall agreement between the experiment and theory of the three dimensional momentum density of spins in iron confirms the reliability of MCP spectroscopy and the accuracy of the recent theoretical calculation.

Momentum–space wavefunctions and electron spin states in MnSb

Compton profiles of hydrogen–like $3d$ orbitals are shown in Table 1. It is worth mentioning that the symmetry of the wavefunction in real space is preserved by the Fourier transformation. The different Compton profiles of each orbital suggests that MCP may provide information on localized electronic states occupied by unpaired electrons. So far MnSb is known to be the only ferromagnetic compound among the intermetallic compounds which consist of $3d$ transition elements and Sb.

TABLE 1. Momentum-space wavefunctions $\phi(p)$ and their Compton profiles $J(p_z)$ of atomic 3d orbitals. The Compton profiles are normalized to unity. The parameter a is $Z_{eff}/3a_0$, where Z_{eff} is an effective charge of the atomic potential and a_0 is the Bohr radius.

$\psi(r)$	$\phi(p)$	$J(p_z)$
xye^{-ar}	$192\pi a p_x p_y/(a^2+p^2)^4$	$128a^9/[35\pi(a^2+p^2)^5]$
yze^{-ar}	$192\pi a p_y p_z/(a^2+p^2)^4$	$1280a^9 p_z^2/[35\pi(a^2+p^2)^6]$
zxe^{-ar}	$192\pi a p_z p_x/(a^2+p^2)^4$	$1280a^9 p_z^2/[35\pi(a^2+p^2)^6]$
$(x^2-y^2)e^{-ar}$	$192\pi a(p_x^2-p_y^2)/(a^2+p^2)^4$	$128a^9/[35\pi(a^2+p^2)^5]$
$(3z^2-r^2)e^{-ar}$	$192\pi a[3p_z^2-p^2]/(a^2+p^2)^4$	$256a^9/105\pi \cdot [60p_z^4/(a^2+p^2)^7 - 10p_z^2/(a^2+p^2)^6+1/(a^2+p^2)^5]$

A recent study on MnSb(20) reported that the observed MCP is well reproduced with the energy-level population of 3d electrons determined by a polarized neutron scattering experiment(21) in 1987. Moreover, as shown in Fig. 4, the observed anisotropy of the MCP between the a* and c* axes also confirmed the additional Sb-5p electron spin polarization in the a'_{1g} state as being -0.2 μ_B per Sb. This result demonstrates the feasibility of MCP spectroscopy for determining wavefunctions of localized states in magnetic compounds.

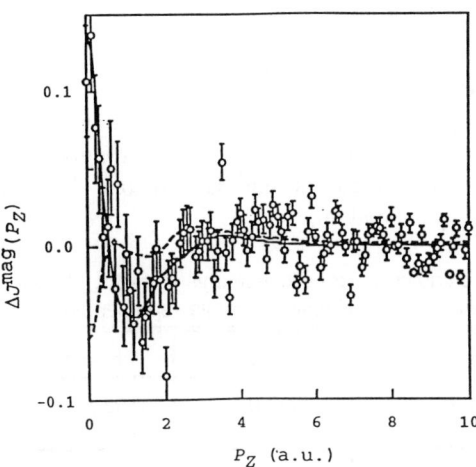

FIGURE 4. The anisotropy of MCP of $Mn_{1.1}Sb$ between the a* and the c* directions. The solid line is a calculated difference based on Sb-5p polarization of -0.2 μ_B in the a'_{1g}, and the dashed line in the e'_g state. (After Nakamura(20))

Evaluation of atomic spin moments in magnetic amorphous alloy

Although MCP does not contain local information in real space, it is possible to classify a profile into components such as 3d–spin and 4f–spin MCPs, because each component has a quite different momentum distribution. It is a recent progress that MCP is found to be useful for evaluating the amount of atomic spins in compounds or alloys, particularly in amorphous alloys which consist of 3d and 4f elements. In the case of amorphous structure, the experimental methods are limited to evaluate magnetic moments of the constituent atoms.

Amorphous $Gd_{6-x}Y_xFe_2Al_2$ (x=0 and 2) were prepared by hammer techique, and confirmed to be amorphous by X-ray diffraction. From a saturation magnetization measurement at 10 K under 5T, magnetic moments were found to be 18.5 μ_B per chemical formula for x=0 and 10.9 μ_B for x=2. Experimental MCPs shown in Fig. 5 were measured using 46.6 keV X-rays emitted from an elliptic MPW at the AR-NE1 beam line of KEK, Japan. Magnetization measurements on this amorphous alloys suggests no canted spin orientation. Using modified atomic Compton profiles of Fe–3d, Gd–4f(22) and itinerant electrons(Fe–4s,p and Gd–5d), the amounts of spin polarization of 3d, 4f and itinerant electrons were evaluated by means of a least squares fitting procedure, and their MCPs are shown in Fig. 5. Evaluated magnetic moments per atom for Fe and Gd and itinerant spin polarization per chemical formula are tabulated in Table 2. It can be easily recognized that, although the Gd-spin polarization per atom is not changed much by adding Y atoms, Fe-spin polarization per atom is considerably reduced by Y atoms. Assuming the following linear dependence of the spin polarization of Fe on x,

$$\mu_{Fe} = m_{Gd}(6-x) + m_Y x ,$$

we find m_{Gd} = -0.32 μ_B, and m_Y = 0.34 μ_B. This conclusion suggests that the occupation of neighboring sites of Fe by Gd or Y strongly influence the Fe spin polarization, and the replacement of one Gd by one Y reduces the Fe spin polarization by 0.33 μ_B per Fe atom.

TABLE 2. Spin moments and bulk magnetic moments of a-$Gd_{6-x}Y_xFe_2Al_2$. Symbols * denote that the assigned moments are per chemical formula represented in the left column.

sample	Gd-4f/atom	Fe-3d/atom	itinerant*	bulk*
$Gd_6Fe_2Al_2$ (x=0)	6.5 μ_B	-1.92 μ_B	0.92 μ_B	18.5 μ_B
$Gd_4Y_2Fe_2Al_2$ (x=2)	5.5 μ_B	-0.59 μ_B	0.53 μ_B	10.9 μ_B

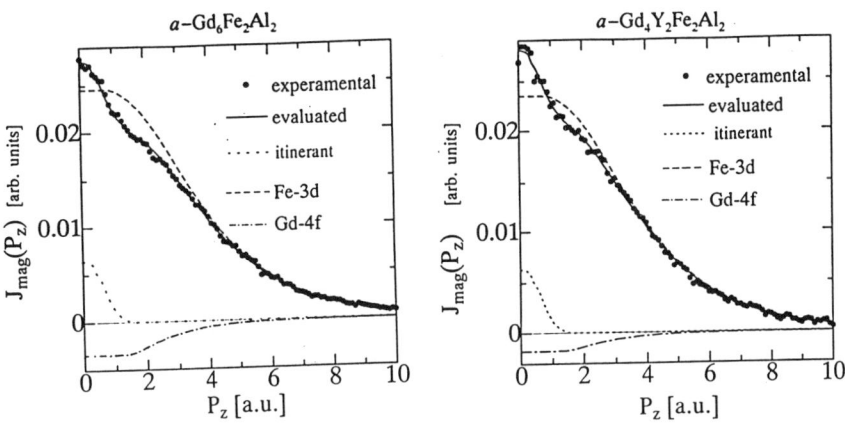

FIGURE 5. MCPs of a-$Gd_{6-x}Y_xFe_2Al_2$ (x=0, 2) at 10 K under 5 T. The lines are MCPs of Fe-3d, Gd-4f and itinerant electrons. The relative amounts of the components are determined by a least-squares fitting procedure.

Consequently, Fe atoms become nonmagnetic when x exceeds 3. This conclusion is consistent with the fact that in amorphous Y_xFe_{1-x} and Y_xNi_{1-x} alloys, the magnetic moment of Fe per atom is monotonously reduced by increasing x, and disappears above x=0.6 for Fe(23) and above x=0.2 for Ni(24). The reduction of Fe moment in a-$Gd_{6-x}Y_xFe_2Al_2$ at relatively low concentration of Y can be understood by the presence of Al which promotes the disappearance of Fe-spin polarization.

Spin moments of Sm and Co in SmCo$_5$

Samarium has a $(4f)^5$ electron configuration with a total spin S=5/2 and a total orbital angular moment L=5. Since S and L couple antiparallel, its magnetic moment $-\mu_B(L+2S)/\hbar$ is quite small. However, the intensity of MCP from Sm atoms is not weak, because the intensity of magnetic Compton scattering is related to *the spins* of the atomic electrons. The present result of MCP measurement on hcp SmCo$_5$ offers two pieces of new information. One is methodological and the other concerns the spin states of Sm and Co atoms.

SmCo$_5$ is known to be a hard magnet. Usually a MCP is obtained as a difference between two Compton profiles measured on the same sample magnetized in opposite directions. In order to alter the magnetization of the sample, a 1-ms pulsed magnetic field of 12T with an interval of 12 sec. was applied at room temperature along the c axis of SmCo$_5$. The Compton profiles were successively accumulated after each pulsed magnetic field. Figure 6 shows MCP of SmCo$_5$ along the c axis at room temperature. Two significant features can be observed. One is a dip around $p_z=0$ a.u., and the other is relatively weak intensity in the high momentum region above 5 a.u., where atomic 3d- and 4f-orbital momentum components should exist.

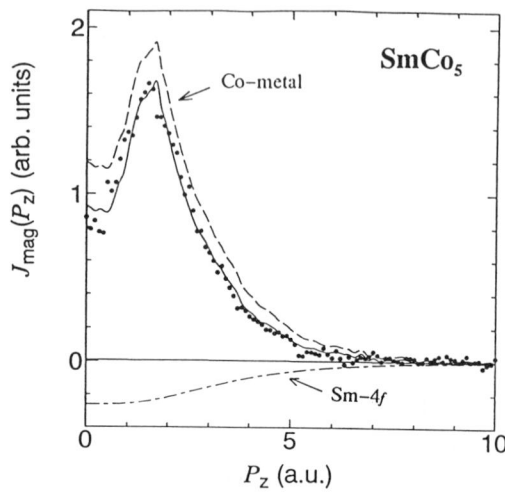

FIGURE 6. MCP of $SmCo_5$ at room temperature. The solid line is a derived profile from combining a directional experimental MCP of Co metal along the c axis(25) and a theoretical MCP of atomic Sm-4f electrons.

In order to evaluate the amount of Sm spin, the experimental MCP of $SmCo_5$ is analyzed using an experimental MCP of Co along the c axis(25), a theoretical 4f-orbit Compton profile(22) and an itinerant-electron MCP with a gaussian momentum distribution in the low momentum region.

According to the fit, the experimental MCP can be well reproduced without adding an itinerant-electron profile, that is, the experimental MCP of $SmCo_5$ can be represented by a Sm-4f part and a Co-metal part. The itinerant component seems to be included in the itinerant part of the experimental Co-metal MCP. This finding indicates that in $SmCo_5$ the amount of the spin polarization of itinerant electrons induced by the hybridization with 3d-electrons of Co is the same to that in Co metal, and is not affected much by the presence of Sm atoms. This is consistent with the previous fact that the dip of MCP of amorphous FeB around $p_z=0$ a.u. is almost the same to that of the MCP of polycrystalline Fe(26). The present conclusion does not exclude a small amount of spin polarization of itinerant electrons in ferromagnetic Sm(27), because of a relatively large statistical error around $P_z=0$ a.u. and neglect in the present analysis of a slightly better momentum resolution (0.66 a.u.) of the adopted Co-metal MCP than that (0.75 a.u.) of the present $SmCo_5$ MCP.

It should be noted that the least squares-fitting procedure clearly confirms the negative orientation of Sm spins with respect to Co spins. On the basis that each area of separated MCP is proportional to the total *spin* component of the corresponding atoms projected on the c axis, the evaluated relative amount of Sm spin with respect to Co spin is −0.19 to 1 per chemical formula. According to a neutron diffraction measurement, the amount of Co-3d spin in hcp Co is evaluated to be 1.86 μ_B. On the other hand reported magnetization measurement of hcp Co concluded the amount of Co-3d spin to be 1.56 μ_B, which was calculated by using a saturation magnetic moment of 1.715 μ_B and g factor of 2.20. As the deviation of 0.31 μ_B from the neutron value may be ascribed to the negative spin polarization of itinerant electrons, it is reasonable to assume that 1.56 μ_B should correspond to the total intensity of the MCP of Co. A spin moment of Sm -1.49 μ_B per atom at room temperature is thus deduced. This value is quite small in comparison with -5 μ_B for S=$-5/2$. Polarized neutron experiments(28) reported a temperature dependence of the Sm magnetic moment: 0.38 μ_B at 4.2 K and 0.04 μ_B at 300 K. The reduction of the magnetic moment at room temperature was explained by parametrization of crystal field mixing of excited multiplets (mainly J=5/2: L=5, S=$-5/2$; M_J=3/2) into the ground-state multiplet (mainly J=5/2: L=5, S=$-5/2$; M_J=5/2). The present Sm-spin moment is not consistent with the conclusion of the neutron experiment. Further investigations are required to settle the spin state of Sm atoms.

Similar 4f-spin determinations by MCP can be also found on Gd(29), HoFe$_2$(30) and CeFe$_2$(31).

Future prospects of MCP

As mentioned above, owing to circularly polarized SR, great progress has been made in the field of MCP spectroscopy. Today new big SR rings, ESRF and APS, have started operation, and the other ring, SPring-8, is under construction. It is clear that in the near future many valuable MCP measurements will be carried out by using higher energy x-rays. The new sources will permit improved momentum resolution. In conventional MCP experiments, solid-state detectors have been used for energy analysis of scattered x-rays, and momentum resolutions around 0.8 a.u. have been achieved. The amount of momentum resolution obtained by a solid-state detector is roughly inversely proportional to the square of the incident x-ray energy. Recently a significant improvement of momentum resolution up to 0.42 a.u. has been reported by using 167-keV SR(32). The best resolution using solid state detectors however will not exceed 0.3 a.u. When we recognize that the typical Fermi momentum of electrons in metals is about 1 a.u., a momentum resolution better than 0.1 a.u. will be advantageous to find delicate magnetic effects related to Fermi surfaces and electron-electron correlation effects. This challenge has been pursued since 1988 in Japan(33). A momentum resolution better than 0.2 a.u. has been reported in a 60-keV experiment(34), in which a Cauchois-type crystal analyzer and a position-sensitive integrating detector, the imaging plate (BaFBr:Eu^{2+} photostimulable phosphor screen)(35), were used. In order to reduce the statistical and systematic errors of MCP experiments, it will be indispensable to alter the integrating counter to a photon counting detector having an adequate counting efficiency for high energy x-rays.

ACKNOWLEDGMENTS

The authors would like to express their thanks to Prof. M.J. Cooper of University of Warwick, and Prof. Ino and Mr. Adachi of the University of Tokyo for providing their valuable data prior to publication.

REFERENCES

1. Sakai, N., AIP Conf. Proc. **215**, 749–759 (1990).
2. Timms, D.N., Zukowski, E., Cooper, M.J., Laundy, D., Collins, S.P., Itoh, F., Sakurai, H., Iwazumi, T., Kawata, H., Ito, M., Sakai, N. and Tanaka, Y., J. Pys. Soc. Jpn. **62**, 1716–1722 (1993).
3. Lovesey, S.W., Rep. Prog. Phys. **56**, 257–326 (1993).
4. Sakai, N., J. Phys. Soc. Jpn. **63**, 4655 (1994).
5. Carra, P., Fabrizio, M., Santoro, G., and Thole, B.T., Phys. Rev. B**53** R5994–R5997 (1996).
6. Sakai, N., J. Appl. Cryst. **29**, 81–99 (1996).
7. Hohenenser, C., Weigart, J.M. and Berko, S., Phys. Lett. **28A**, 41–42 (1968).
8. Mijnarends, P.E., Physica(Utrecht) **63**, 235–247, 248–262 (1973).
9. Genoud, P., Singh, A.K., Manuel, A.A., Jarlborg, T., Walker, E., Peter, M. and Weller, M., J. Phys. F: Met. Phys. **18**, 1933–1947 (1988).
10. Shiotani, N., Okada, T., Sekizawa, H., Mizoguchi, T. and Karasawa, T., J. Phys. Soc. Jpn. **35**, 456–460 (1973).
11. Gidley, D.W., Koymen, A.R. and Capehart, T.W., Phys. Rev. Lett. **49**, 1779–1783 (1982).
12. McCarthy, I.E. and Weigold, E., Rep. Prog. Phys. **54** 789–879 (1991).
13. Lipps, F. and Tolhoek, H.A., Physica(Utrecht) **20**, 85–98, 395–405 (1954).
14. Yamamoto, S., Kawata, H., Kitamura, H., Ando, M., Sakai, N. and Shiotani, N., Phys. Rev. Lett. **62**, 2672–2675 (1989).
15. Golovchenko, J.A., Kincaid, B.M., Lovesque, R.A., Meixner, A.E. and Kaplan, D.R., Phys. Rev. Lett. **57**, 202–205 (1986).
16. Mills, D.M., Nucl. Inst. and Methods in Phys. Research **A266**, 531–537 (1988).
17. Yahnke, C.J., Srajer, G., Haeffner, D.R., Mills, D.M. and Assoufid, L., Nucl. Instr. and Methods in Phys. Research **A347**, 128–133 (1994).
18. Tanaka, Y., Sakai, N., Kubo, Y. and Kawata, H., Phys. Rev. Lett. **70**, 1537–1540 (1993).
19. Kubo, Y. and Asano, S., Phys. Rev. B**42**, 4431–4446 (1990).
20. Nakamura, J., Takeda, T., Asai, K., Yamada, N., Tanaka, Y., Sakai, N., Ito, M. Koizumi, A. and Kawata, H., J. Phys. Soc. Jpn. **64**, 1385–1393 (1995).
21. Yamaguchi, Y. and Watanabe, H., J. Phys. Soc. Jpn., **45**, 846–854 (1987).
22. Biggs, F., Mendelsohn, L.B. and Mann, J.B., Nucl. Atom. Data Table **16**, 210–310 (1975).
23. Ishio, S., Aubertin, F., Limbach, T., Engelman, H., Dezsi, L., Gonser, U., Fries, S., Takahashi, M. and Fujikura, M., J. Phys. F: Met. Phys. **18**, 2253–2263 (1988).
24. Lienard, A. and Rebouillat, J.P., J. Appl. Phys. **49**, 1680–1682 (1978).
25. Lawson, P.K., Timms, D.N., Dixon, M.A., Cooper, M.J., McCarthy, J.E. and Tshentscher, T., to be published in J. X-ray Sci. Tech.
26. Sakai, N., Materials Sci. Forum **105–110**, 431–438 (1992).
27. Adachi, H., private communication.
28. Givord, D., Laforest, J., Schweizer, J. and Tasset, F., J. Appl. Phys. **50**, 2008–2010 (1979).
29. Sakai, N., Tanaka, Y., Itoh, F., Sakurai, H., Kawata, H. and Iwazumi, T., J. Phys. Soc. Jpn., **60**, 1201–1203 (1991).

30. Cooper, M.J., Zukowski, E., Timms, D.N., Armstrong, R., Itoh, F., Tanaka, Y., Ito, M., Kawata, H. and Bateson, R., Phys. Rev. Lett. **71**, 1095–1098 (1993).
31. Cooper, M.J., Lawson, P.K., Dixon, M.A.G., Zukowski, E., Timms, D.N., Itoh, F., Sakurai, H., Kawata, H., Tanaka, Y. and Ito, M., Phys. Rev. **B54**, 1–7 (1996).
32. Cooper, M.J., private communication.
33. Sakai, N., Shiotani, N., Ito, M., Itoh, F., Kawata, H., Amemiya, Y., Ando, M., Yamamoto, S. and Kitamura, H., Rev. Sci. Instr. **60**, 1666–1670 (1988).
34. Sakurai, Y., Tanaka, Y., Ohta, T., Watanabe, Y., Nanao, S., Ushigami, Y., Iwazumi, T., Kawata, H. and Shiotani, N., J. Phys.: Condens. Matter **6**, 9469–9475 (1994).
35. Amemiya, Y., Matsushita, T., Nakagawa, A., Satow, Y., Miyahara, J. and Chikawa, J., Nucl. Instr. and Methods in Phys. Research **A266**, 645–653 (1988).

VII. THEORETICAL ASPECTS

Inner-shell Ionization and Excitation

M.Ya. Amusia

Imperial College of Science, Technology and Medicine
London SW7 2BZ, UK
and
A.F. Ioffe Physical - Technical Institute
St Petersburg 194021, Russia

Abstract. The most important multi-electron correlation effects which are relevant for inner-shell ionization and excitation of atoms are presented, mechanisms of multiple ionization are considered, and the frequency dependence of the mean ion charge is discussed at length. Specific features of post-collision interaction near inner-shell thresholds are investigated. It is demonstrated that the state of the outer electrons considerably affects the oscillator strengths of inner-shell excitations and near threshold photoionization cross sections. Particular attention is given to single-charge ion production, which is determined by a complicated multielectron mechanism in the frequency region discussed. Attention is given to the simplest relativistic correction, namely to the non-dipole term and its cooperative macroscopic manifestation - the "drag" current. Some interesting phenomena connected with inner-hole decay processes, such as correlation effects and deviations from the exponential decay law are briefly considered.

INTRODUCTION

From the point of view of pure science the main interest in investigating inner atomic shells, their excitations and ionization is motivated by a desire to understand multielectron, or correlational, and relativistic effects. Correlation effects have been studied in depth for outer and intermediate electronic shells where the dominant feature is the polarization of these shells by the incoming photon. Theoretically such effects are reasonably well described by the random phase approximation with exchange - RPAE [1]. When going deeper into the atom the RPAE effects decrease, but other correlational corrections become more and more important, as do the relativistic corrections.

The initial state wave function for inner-shell ionization and excitation consists of an almost pure Coulombic wave function of the primary electron (which absorbs the incoming photon) and an admixture to it of virtual or real excitations of intermediate and outer shells. The latter complication is due either to configuration mixing or to inner-hole non-radiative decay. The structure of the final-state wave function depends upon its energy and if the energy is high enough a plane wave approximation is sufficient. For intermediate energies the primary photoelectron can be described by a Hartree-Fock wave function, while the complete final state includes also "direct knock-out" corrections, i.e., those configurations which are reached by inelastic scattering of the primary photoelectron by the outer shells. For low energies of the outgoing electrons, or excitation of discrete levels, the final state is a superposition of the Hartree-Fock photoelectron wave function, the "direct knock-outs" and many other correlational corrections arising from interaction among the outer electrons. In order to take into account relativistic corrections, it is necessary to include both dipole and quadrupole terms in the photon-electron interaction operator. It is important to bear in mind that the outer electrons are non-perturbatively affected by the creation of the inner hole and its subsequent decay.

Despite the previously mentioned complications, the problem of inner-shell ionization and excitation is considerably simplified compared to excitations in the outer shells. The inner electrons interact comparatively weakly and are well separated in space and energy from the outer and intermediate electrons. This is valid not only for atoms, but also for molecules and solids and permits the inner-shell processes and the corrections due to outer or intermediate shell effects to be factorized.

The creation of inner holes is accompanied and followed by electron emission and also emission of photons. The probability for radiative decay of inner holes increases very rapidly with nuclear charge Z. Radiative decay is therefore of considerable importance for the inner shells of medium and particularly of heavy atoms.

Inner-shell ionization leads to the formation of ions with different charges, electrons, and photons. The task of many-electron theory is to calculate the probability of forming ions with a given charge and the electron-photon distribution for the process. Although this aim is far from being achieved, in what follows we will describe some steps in the desired direction.

MANY-BODY EFFECTS NEAR INNER-SHELL THRESHOLDS

Multielectron effects are very important in inner-shell processes, but are generated by mechanisms other than in intermediate and outer shells. Following inner-hole formation, instead of the random phase approximation with exchange, different types of relaxational, or rearrangement effects, have to be considered [1]. When an inner hole is created, not only the photoelectron, but all other electrons are attracted to the hole. This increases the screening of the field acting upon the photoelectron thus decreasing the photoionization cross section, primarily near inner-shell thresholds. For photon energies well above the inner threshold this effect disappears. Another important effect is the non-radiative, or Auger decay, of an inner hole. As a result, the field acting upon the outgoing photoelectron changes asymptotically; instead of being (-1/r) it becomes (-2/r), or even stronger. The larger attraction due to the field results in an increase of the threshold cross section. The cross section near inner thresholds is thus determined by the competition of these two effects. It is difficult to account for these effects rigorously, but to estimate their influence we previously suggested the following two approximate approaches [2].

The increase of the screening due to the formation of an inner hole i may be taken into account by calculating the cross section $\sigma_i(\omega)$ using the following expression:

$$\sigma_{i\tilde{f}}(\omega) \sim |<i|d|\tilde{f}>|^2 \quad (1)$$

with $<i|$ and $|\tilde{f}>$ belonging to different sets of the wave functions. Here d denotes the photon-electron interaction operator, i is the Hartree-Fock (HF) wave function of an electron in the inner shell, while $\tilde{f}$ describes the photoelectron with energy ε_f moving in the self-consistent field of the residual ion with hole i instead of that of the initial atom.

The influence of inner hole decay can be estimated in the same way, leading to the cross section expression given by (1) with $\tilde{\tilde{f}}$ instead of $\tilde{f}$. The function $\tilde{\tilde{f}}$ describes the photoelectron in the HF self-consistent field of that ion state q which is created after Auger decay of the hole i:

$$\sigma_{i\tilde{\tilde{f}}}(\omega) \sim \sum_q \frac{\Gamma_{iq}}{\Gamma_i} |<i|d|\tilde{\tilde{f}}>|^2 \quad (2)$$

Here $\Gamma_i = \Sigma_{iq}\Gamma_{iq}$, where Γ_{iq} and Γ_i are the partial and total Auger widths of the hole i respectively.

Inner hole creation is accompanied by the formation of satellites and "shadows". One should distinguish between satellites which are monopole excitations of outer and intermediate shell electrons, and correlation satellites, which are by definition non-monopole. "Shadows" are quasi-forbidden (i.e. only energetically forbidden) near-resonance Auger decay processes. It was noted long ago that the cross section of the "shadow" level "i" ionization is closely connected to that of the "parent" level i [3]. It can be demonstrated that the following relation should hold:

$$\sigma_{"i"}(\omega) = \sigma_i^{RPAE}(\omega) \mathfrak{I}_{"i"} \sigma_i^{HF}(\omega - \Delta_{"i"}) / \sigma_i^{HF}(\omega) \quad (3)$$

$$\Delta_{"i"} = I_{"i"} - I_i$$

I_i, $I_{"i"}$ being the ionization potentials of the "parent" and "shadow" levels, respectively; $\mathfrak{I}_{"i"}$ is the so called spectroscopic factor of the state "i".

Recently, the "shadow" level cross sections were measured for "5s" in Xe [4], confirming the existence of a close correspondence between $\sigma_{"5s"}(\omega)$ and $\sigma_{5s}(\omega)$.

MECHANISMS OF MULTIPLE IONIZATION

In photoabsorption near and above inner-shell thresholds mainly multiply charged ions are created [5]. The simplest and most efficient way to do this is an Auger cascade, i.e., a sequence of Auger decays following the creation of the initial core hole by absorption of the incoming photon. In fact, usually a multitude of cascades is allowed. It is simple to estimate the highest $n_{max}^{(A)}(\omega)$ and the lowest $n_{min}^{(A)}(\omega)$ ion charge achievable after absorbing the photon of frequency ω. However, experiments demonstrate that the ion charge n lies outside of this range [6]. The role of mechanisms other than Auger decay is most clear for the creation of ions A^{+n} with $n > n_{max}^{(A)}(\omega)$ or $n < n_{min}^{(A)}(\omega)$. The increase of $\sigma^{(n)}(\omega)$ – the cross section of photoionization with creation of A^{+n} ions compared to $\sigma_A^{(n)}(\omega)$ – describing A^{+n} production via Auger cascades, is caused by a number of mechanisms: shake-off, direct knock-out, ground-state correlation, virtual, and double Auger decay.

Shake off is the result of an instant variation of the self-consistent field acting upon the outer electrons due to hole creation. In shake-off mainly slow electrons are emitted with an almost isotropic angular distribution. For $\omega \gg I_i$ shake-off adds to the cross section $\sigma^{(n)}(\omega)$ the value $\Delta\sigma_{so}^{(n)}(\omega)$ which has almost the same ω dependence as $\sigma_A^{(n-1)}(\omega)$.

Direct knock-out is a process of inelastic scattering of the primary photoelectron by atomic electrons, which leads to their emission from the atom. The contribution to $\sigma^{(n)}(\omega)$ due to a direct knock-out - $\Delta\sigma_{ko}^{(n)}(\omega)$, decreases as a function of ω if ω is well above the ionization potential of the target atom inner shells. In this case the energies of the electrons emitted are of the order of the ionization potential of the knocked out electrons and their angular distribution is close to being isotropic.

Ground state correlation is a mechanism, which accounts for the deviation of the initial-state wave function from a pure HF wave function due to the interactions of the electrons in the initial state of the target atom. All other above mentioned mechanisms represent the effect of final state correlations. The correction $\Delta\sigma_{gs}^{(n)}(\omega)$ as a function of ω is similiar to $\Delta\sigma_{so}^{(n)}(\omega)$, as are the energy and angular distributions of the extra emitted electrons.

Virtual Auger - decay is a process in which an additional electron is emitted via virtual creation of an inner hole, i.e. below the threshold necessary for its real formation. This mechanism leads to a correction $\Delta\sigma_{va}^{(n)}(\omega)$ which increases when the threshold of the primarily ionizing shell is approached from below.

In **double Auger-decay** two electrons are emitted simultaneously instead of one as in an ordinary Auger process. Predominantly, one of the emitted electrons is fast, while the other is slow. The correction $\Delta\sigma_{da}^{(n)}(\omega)$ has almost the same ω dependence as $\sigma_A^{(n-1)}(\omega)$, but is much smaller of course.

In a real photoionization process all the mechanisms described above interfere, leading to the observed cross section $\sigma^{(n)}(\omega)$.

For primary holes in an innermost shell the expressions for the corrections to $\sigma^{(n)}(\omega)$ can be factorized. It can be shown that for the shake-off contribution one has[1]:

$$\Delta\sigma_{so}^{(n)}(\omega) = \sigma_{A,i}^{(n-1)}(\omega)\left\{\sum_{p\leq F}\int_{\varepsilon>0}\left|<p|r^{-1}|\tilde{\varepsilon}>/(I_{p(i)}+\varepsilon)\right|^2\right\} \quad (4)$$

Here the summation is performed over those occupied states $(\leq F)$ with the ionization potential in the presence of hole i being $I_{p(i)}$, for which $I_{p(i)} \leq \omega$. The state $|\tilde{\varepsilon}>$ is calculated in the field, which has an extra $(-1/r)$-potential compared to that acting upon $<p|$.

The contribution $\Delta\sigma_{ko}^{(n)}(\omega)$ for a primary hole created deep inside the atom is given by:

$$\Delta\sigma_{ko}^{(n)}(\omega) = \sigma^{(n-1)}(\omega)\left\{1-\sum_{l'=l\pm 1}\exp\left[-2\,\widetilde{\text{Im}}\,\delta_{l'}(\varepsilon)\right]\right\} \quad (5)$$

where δ_l is the elastic scattering phase shift of the photoelectron and its energy is equal to $\varepsilon = \omega - I_i$. The symbol $\sim$ above Im in (5) denotes that only that fraction of the imaginary part of the phase shift $\text{Im}\,\delta_l(\varepsilon)$ is taken into account, which corresponds to ionization of any of the outer atomic electrons by the photoelectron.

The processes, discussed in this section affect also the angular distributions and spin orientation of the outgoing electrons.

The cross sections $\sigma^{(n)}(\omega)$ with $n < n_{min}^{(A)}(\omega)$ presents the contribution of outer shell ionization affected by virtual ionization or excitation of the inner electrons. The cross section $\sigma^{(n)}(\omega)$ with $n > n_{max}^{(A)}(\omega)$ is determined by hole multiplication mechanisms, such as shake off, direct knock-out, ground state correlation and double Auger decay.

MEAN ION CHARGE YIELD

Different mechanisms of multiple ionization are subsumed by the mean ion charge $N(\omega)$, which is determined by the following relation:

$$N(\omega) = \sum_{n\geq 1} n\sigma^{(n)}(\omega)\bigg/\sum_{n\geq 1}\sigma^{(n)}(\omega). \quad (6)$$

---

[1] The atomic system of units is used in this paper; $e = \hbar = m_e = 1$.

In the vicinity of the inner shell i threshold a comparatively simple expression for $N(\omega)$ holds:

$$N(\omega) = \frac{\bar{n}_{i+1}\sigma_{i+1}(\omega) + \bar{n}_i \sigma_i(\omega)}{\sigma_{i+1}(\omega) + \sigma_i(\omega)}, \qquad (7)$$

with the average ion charge $\bar{n}_i$ being determined by the expression

$$\bar{n}_i = \sum n_i^{(q)} \Gamma_i^{(q)} / \Gamma_i \qquad (8)$$

where $\Gamma_i^{(q)}$ is the partial width of the hole i Auger decay into the channel q, with $\Gamma_i = \sum_q \Gamma_i^{(q)}$ if the radiative widths are negligible.

In the one-electron approximation $N(\omega)$ "jumps" at the threshold $\omega = I_i$, the magnitude of the jump being equal to $\Delta N(I_i) = (\bar{n}_i - \bar{n}_{i+1}) \sigma_i(I_i) / (\sigma_i(I_i) + \sigma_{i+1}(I_i))$. In a real multi-electron atom $N(\omega)$ even at $\omega \approx I_i$ is a smooth function which starts to increase monotonically well below $\omega \approx I_i$ [6]. This was explained in [7] by accounting for the contribution $\sigma_{\cdot_i\cdot}(\omega)$ of virtual Auger-decays of hole i which is nonzero even at $\omega < I_i$. In this frequency region $\sigma_i(\omega)$ in (7) must be substituted by $\sigma_{\cdot_i\cdot}(\omega)$ given by the following expression [7]:

$$\sigma_{\cdot_i\cdot}(\omega) \cong \sigma_i(I_i)\left[\frac{1}{2} + \frac{1}{\pi}\arctan\left(\frac{\omega - I_i}{\frac{1}{2}\Gamma_i}\right)\right] \qquad (9)$$

As a result, one obtains for $\Delta N_i(\omega)$:

$$\Delta N_i(\omega) = (\bar{n}_i - \bar{n}_{i+1})\frac{\beta}{\varepsilon + \beta}, \qquad (10)$$

where $\varepsilon = (I_i - \omega)/\Gamma_i$ and $\beta = \sigma_i(I_i)/2\pi\sigma_{i+1}(I_i)$.
This formula proved to be in agreement with the results of measurements in the vicinity of the 1s threshold in Ar[6], leading to a comparatively slow increase in $\Delta N_i(\omega) \sim \varepsilon^{-1}$ as ε decreases.

INNER-SHELL POST COLLISION INTERACTION

Near intermediate and inner-shell thresholds the well-known and extensively studied post collision interaction (PCI) [8] becomes considerably modified due to the cascade of sequential Auger decays instead of a single decay. This cascade is the reason why the photoelectron will finally find itself moving in the Coulomb field of a residual ion with the charge $n^{(A)}$, i.e. in the field $(-n^{(A)}/r)$ instead of $(-1/r)$ which it would be without the Auger decay. Several publications on this problem appeared recently [9,10]. We will present here a rather simple approach developed some time ago [11].

Qualitatively, multistep PCI can be understood in the following way. Each time an extra Auger decay takes place in the Auger cascade, the primary photoelectron feels an instantaneous increase in the field in which it is moving. As a result, the photoelectron loses energy step by step. To calculate the total photoelectron energy shift, let us consider this process classically. If the Auger cascade consists of N events, one can determine the distance which the primary photoelectron moves before the 1st, 2nd... Nth decay using:

$$\begin{aligned} r_1 &= v_1 T_1 \\ r_2 &= v_2 T_2 + r_1 \\ r_t &= v_t T_t + r_{t-1} \\ r_N &= v_N T_N + r_{N-1} = \sum_{t=1}^{N} v_t T_t, \\ T_t &= \Gamma_t^{-1} \end{aligned} \quad (11)$$

It is assumed that the alteration of the field acting upon the primary electron proceeds by steps

$$\Delta U_t = -1/r_t \quad (12)$$

and therefore $n_A = N$. At each step the photoelectron's energy and speed decrease in accordance with energy conservation.

$$v_t^2/2 + \Delta U_t = v_{t+1}^2/2. \quad (13)$$

The experimentally observed speed of the photoelectron is $v_t = v_N$ and the total photoelectron energy shift is given by:

$$\Delta\varepsilon \equiv \Delta U \equiv \sum_{t=1}^{N}(-1/r_t) \quad (14)$$

For two-step decay, equations (13) and (14) lead to the following expression:

$$\Delta U = -\frac{\Gamma_1}{v_1}\left[1 + \frac{\Gamma_2}{\Gamma_2 + \frac{v_2}{v_1}\Gamma_1}\right], \quad (15)$$

$$\frac{v_2}{v_1} = (1 - \frac{2\Gamma_1}{v_1^3})^{1/2}$$

If the second step transition is very fast i.e. for $\Gamma_2 \to \infty$, one has $\Delta U = -2\Gamma_1/v_1$ instead of the usual $-\Gamma_1/v_1$. If the widths at every step are equal to each other $\Gamma_t = \Gamma$, the following formula gives $\Delta\varepsilon$,

$$\Delta\varepsilon = -\Gamma \sum_{t=1}^{N}(1/\sum_{i=1}^{t} v_i). \quad (16)$$

The expressions presented above can be improved by accounting for the finite Auger-electron speed. The simplest and most crude way to do this is to substitute ΔU_t from (12) by the expression $\Delta \tilde{U}_t = -r_t^{-1}(1 - v_t/v_{At})$, where v_{At} is the t th step Auger electron speed. The expression for corrected energy shift $\Delta\tilde{\varepsilon}$ is given by the sum of $\Delta\tilde{U}_t$ and instead of (13) the equation $v_t^2 + 2\Delta\tilde{U}_t = v_{t+1}^2$ must be used.

For a comparatively slow Auger-electron, the three-body interaction effect [8] must be included to account for the relative velocity direction of the Auger and the photoelectron. This can be achieved semi-classically, by substituting ΔU_t from (12) by the following expression:

$$\Delta\tilde{\tilde{U}}_t = -r_t^{-1} + \left|\mathbf{r} - \mathbf{v}_{At} \mathbf{T}_t\right|^{-1} \qquad (17)$$

which leads to a modification of the energy shift $\Delta\tilde{\tilde{\varepsilon}} = \sum_{t=1}^{N} \Delta\tilde{\tilde{U}}_t$ and the velocities at each step $v_t^2 + 2\Delta\tilde{\tilde{U}}_t = v_{t+1}^2$.

In general, the alteration of the field acting upon the photoelectron at the t-th step can differ from $(-1/r)$, because of double Auger-decay, or shake-off processes accompanying the Auger decay. If at the t-th step there are two competing decay channels, 1 and 2, which produce residual ion charges q_1 and q_2 the energy shift at this step is given by the expression

$$\Delta\varepsilon_t \approx -\frac{1}{r_t}\frac{q_1\Gamma_t^{(1)} + q_2\Gamma_t^{(2)}}{\Gamma_t^{(1)} + \Gamma_t^{(2)}} \qquad (18)$$

To distinguish these channels in order to obtain energy shifts either $(-q_1/r_t)$ or $(-q_2/r_t)$, the photoelectron and residual ion with the charge q_1 or q_2 must be observed in coincidence. This was done recently when investigating PCI effects near the 2p-threshold in Ar [13]. The photoelectron's energy shift in coincidence with Ar^{+3} proved to be larger than that for Ar^{+2}, however, it was not equal to $-2\Gamma_{2p}/v$ but instead it was $(-1.73\Gamma_{2p}/v)$. If one assumes, that Ar^{+3} is produced via the double Auger-effect, the energy distribution of both decay electrons must be taken into account. It is known [12], that predominantly one of them is fast and the other slow. So, by using the following expression for the energy shift

$$\Delta\varepsilon(v_{A_1}, v_{A_2}) = -\frac{2\Gamma}{v}[1 - \frac{v}{2}(\frac{1}{v_{A_1}} + \frac{1}{v_{A_2}})] \qquad (19)$$

(with $v_{A_1}^2 + v_{A_2}^2 = 2\varepsilon_A$, ε_A being the total energy of the Auger-decay) and averaging $\Delta\varepsilon(v_{A_1}, v_{A_2})$ over the Auger-electron energy distribution the $\Delta\varepsilon$ value can be calculated.

It appears, however, that accounting for the real Auger electrons energy distribution is not sufficient to explain the difference mentioned above. This fact may signal the possible role of shake-off in Ar^{+3} ion production. If shake-off takes place before Auger-decay, it will not affect the photoelectron energy shift, but a shake-off after Auger-decay will increase the energy shift. It is hard to see how these two kind of shake-off could be distinguished purely experimentally. On the other hand, the fact that the photoelectron's PCI energy shift increases with the number of Auger electrons emitted has been confirmed already by several experiments [14, 15].

INFLUENCE OF OUTER ELECTRONS

The inner-shell discrete excitations are strongly affected by the outer shell electronic structure. This becomes evident by comparing the inner-shell oscillator strengths for inert gas atoms and their alkali neighbors. The ratio of oscillator strengths with the same quantum numbers is extremely sensitive to the effective charge of the field acting upon the excited electron [16].

Let us start with a hydrogenic model in which the inner and excited electrons are moving in different fields with effective charges Z_i and Z_f, respectively. In this model the inner-shell discrete excitation oscillator strengths are given by the expression:

$$F_{if} = A Z_i^{-5} Z_f^{5} \tag{20}$$

To estimate the difference in F's for 1s-np transitions, let us use Slater's effective charges for Ar, K and Ca: $Z_{4p}^{Ar} = 2.2$, $Z_{4p}^{K} = 2.85$, $Z_{4p}^{Ca} = 3.5$ and $Z_{5p}^{Ar} = 1$, $Z_{5p}^{K} = 1.15$, $Z_{5p}^{Ca} = 1.3$. There is no screening of a 4p electron from a 4s electron and therefore Z_{4p}^{K} is considerably larger than Z_{4p}^{Ar}. As a result of the rather fast 1s-hole decay at least one unit must be added to these values, leading to the following ratios:

$F_{1s-4p}^{K} / F_{1s-4p}^{Ar} = 1.9$, $F_{1s-5p}^{K} / F_{1s-5p}^{Ar} = 1.09$ and $F_{1s-4p}^{Ca} / F_{1s-4p}^{Ar} = 3.22$, $F_{1s-5p}^{Ca} / F_{1s-5p}^{Ar} = 1.18$.

The corresponding ratios for n=6 are close to one. With subsequent growth of n they approach the limits $(Z_{1s}^{K} / Z_{1s}^{Ar})^{-5} \approx 0.8$ and $(Z_{1s}^{Ca} / Z_{1s}^{Ar})^{-5} \approx 0.6$. Accurate calculations for 1s - 4p, 5p, 6p and 7p transitions in Ar and Kr [16] confirm the qualitative results obtained in the hydrogenic approximation.

Excitation of outer electrons considerably affects the inner-electron oscillator strengths. For instance, the effective 4p charge is almost the same for K and excited Ar* (3p^{-1}4s), and therefore $F_{1s-4p}^{K} / F_{1s-4p}^{Ar^*} \approx 1$.

By moving the outer electron from 4s to higher n* levels, inner-electron excitations with n<n* become exposed to a stronger effective charge leading to $F_{1s-np}^{K^*} / F_{1s-np}^{K} > 1$ for n<n*, while for n>n* this ratio is close to one. If the excited electron is located close to the outer electron, even a small variation in the state of the latter, for example due to a 4s-4p transition in K when 2p-3d or 3p-3d levels are excited, can considerably change the inner transition oscillator strengths; in our example, F_{2p-3d}^{K} and F_{3p-3d}^{K}.

The inner-shell excitations are of autoionizing character, their shape being represented by the Fano-profile $\sigma(\varepsilon) \approx \sigma_0 (q+\varepsilon)^2 / (1+\varepsilon^2)$ with $\varepsilon = (\omega - \omega_R)/\Gamma_A$. The position of the resonance ω_R, i.e. the inner-shell excitation energy, is weakly affected by the effective charge acting upon the outer electron, while Γ_A is proportional to F_{if}.

The strong dependence of F_{if} upon Z_f demonstrated by (20) opens the possibility to affect inner-shell excitations simply by moving the outer electron from one level to another, or modifying the outer electron structure by going from isolated atoms to multiatomic aggregates, such as solids or clusters.

The alteration of screening or the effective charge can affect the oscillator strength indirectly when the electron correlations which lead to interaction between different excitations from a given inner or intermediate shell are strong enough. The outer-shell structure determines to a large extent the inner-shell ionization cross section near threshold. To clarify this, calculations were performed for two sequences - K^-, K and K^+ near the 1s threshold and Ba, Ba** and Ba^{++} near the 4d threshold [17]. In the first case Hartree-Fock and RPAE results were obtained which show that differences in the outer shell modify the cross section over quite a broad frequency region (up to 15 Ry above threshold). The Ba case was also treated in the generalised RPAE frame, i.e. taking into account the outer shell rearrangement due to inner hole creation. It appeared that modification of the outer shell, not only the removal of two 6s electrons but just their excitation into $8s^2$ states affects the $4d^{10}$ and $5p^6$ photoionization cross section dramatically by increasing the 4d-4f oscillator strength from 0.0007 (Ba) to 0.608 (Ba**) and 0.705(Ba^{++}) and the threshold value of the 4d- cross section from 1.28Mb(Ba) to 40.6Mb (Ba**) and 44.6(Ba^{++}). On the other hand, the 5p-cross section at threshold decreases rapidly, from 83Mb(Ba) to 24.4Mb(Ba**) and 19.8Mb(Ba^{++}). It was found that from the intermediate electrons point of view the excitation $6s^2$ - $8s^2$ is already similiar to ionization.

SINGLE-CHARGED ION PRODUCTION

Single-charged ions are produced only if an outer electron is emitted. This is a process with a comparatively simple final state, which includes a fast electron and an outer shell hole for ω near an inner-shell threshold.

The cross section for direct outer electron emission in the frequency range considered is small and a smooth function of ω. The contribution of indirect ionization, i.e. that which proceeds via virtual excitation of inner electrons, can be considerable and has a strong ω dependence. This effect can be taken into account by using an effective length operator $\tilde{r}$, describing the photon-electron interaction

$$\tilde{r} = r[1 + \frac{\alpha^{(in)}(\omega)}{r^3}] \qquad (21)$$

Here $\alpha^{(in)}(\omega)$ is the inner-shell dipole dynamic polarizability. The inclusion of the second term in (21) accounts for the peculiarities in the vicinity of the inner-shell threshold, for instance a maximum which was observed in $\sigma^+(\omega)$ near the 1s-threshold in Ar [6] and reproduced, at least qualitatively, by recent calculations [18]. However, in the same atom near the 2p-threshold the first and second term in (21) interfere destructively, leading to a minimum in $\sigma^+(\omega)$ [19] instead of the experimentally observed maximum. In this connection another mechanism of $\sigma^+(\omega)$ formation was suggested [19], namely the recapture of a slow photoelectron into a discrete excited state due to PCI. One should note, however, that most probable is the recapture to highly excited levels, which can then decay via autoionization by emitting slow electrons from the outer shell. Therefore, the recapture cross section is larger than that of A^+ formation. The difference is particularly large for the innermost shells.

The qualitative picture of the recapture with subsequent autoionizational decay was recently confirmed in studies of Ar^+ and Ar^{++} yield near the 2p- threshold [20]. It was demonstrated straightforwardly, that 67% of electrons are recaptured to those states, which then decay via autoionization, thus increasing the Ar^{++} instead of the Ar^+ yield. It was demonstrated also, that maxima in the Ar^+ yield correspond to the maxima observed in the fluorescence yield, thus confirming directly the physical picture of the recapture process.

RELATIVISTIC CORRECTIONS AND THE "DRAG" CURRENT

To the lowest order in v / c (v and c being the speeds of the electrons and light, respectively) corrections to the photoionization process may be taken into account by substituting the dipole photon-electron interaction operator by one which includes quadruple corrections:

$$\frac{1}{c}\varepsilon\nabla \to \frac{1}{c}\varepsilon\nabla(1+i\mathbf{k}\mathbf{r}), \tag{22}$$

where k= ω /c, and ε is the light polarization vector.

The term proportional to k in (22) leads to interference of dipole and quadruple photon components, qualitatively modifying the angular distributions, which acquire the following form [21]:

$$\frac{d\sigma(\omega)}{d\Omega} = \frac{\sigma_\gamma(\omega)}{4\pi}\left\{1-\frac{\beta(\varepsilon)}{2}P_2(\cos\theta)+\frac{\omega}{c}[\gamma(\varepsilon)P_1(\cos\theta)+\delta(\varepsilon)P_3(\cos\theta)]\right\} \tag{23}$$

where $\sigma_\gamma(\omega)$ is the total photoabsorption cross sction, $P_s(\cos\theta)$, s=1,2,3 are Legendre polynomials, ε is the photoelectron energy. Coefficients $\gamma(\varepsilon)$ and $\delta(\varepsilon)$ for the case when a shell with angular momentum ℓ is ionised are expressed via products of dipole $\ell \to \ell \pm 1$ and quadrupole $\ell \to \ell \pm 2,0$ transition matrix elements and the corresponding phase differences [2]. On the contrary, $\beta(\varepsilon)$ is expressed via products of dipole $\ell \to \ell \pm 1$ matrix elements and $(\ell+1),(\ell-1)$ phase differences only. It appears, that in some cases the quadrupole terms are considerably larger than the dipole ones overcoming at least partly the smallness of the parameter $\omega r_s/c$, r_s being the ionizing shell radius. In the pure dipole approximation, the terms $\sim \omega/c$ are neglected. Note that electron correlations affect all of the transition matrix elements, but differently, depending of course on the multipole nature of the transitions. Systematic measurements of $\delta(\varepsilon)$ and $\gamma(\varepsilon)$ coefficients for Ar above the 1s and Kr above the 2p ionization thresholds were performed recently [22].

The presence in (23) of the term with $P_1(\cos\theta)$ leads to a "drag" current, which is a macroscopic cooperative motion of photoelectrons along or opposite to the photon flux direction. Indeed, after averaging the projected photoelectron velocity along the photon beam direction, $v \cos\theta$, over the distribution (23), an "electromotive force" is found, which after dividing by the "resistance" gives an expression for a current $j(\omega)$:

$$j(\omega) = -\frac{\omega}{3c} W \frac{\gamma(\varepsilon)\sigma_\gamma(\omega)}{\sigma_{el}(\varepsilon)} \qquad (24)$$

Here W is the photon flux and $\sigma_{el}(\varepsilon)$ is the "electron-target atom" elastic scattering cross section. The "drag" current $j(\omega)$ is enhanced at Ramsauer minima in the cross section $\sigma_{el}(\varepsilon)$. The factor $\gamma(\varepsilon)$ describes how the total momentum ω/c of the photon flux is distributed among the photoelectrons and the recoil ions. This distribution must be kept in mind, because the linear momentum conservation law determines the total momentum of the outgoing electrons and the residual ions, while the current is determined by the electron momentum only.

In deriving (24) it was assumed that as soon as a photoelectron collides with another target atom it is scattered with equal probability in any direction, thus leaving the current. While $\gamma(\varepsilon)$ is determined by a quite complicated expression, for a rough estimate of the current, it can be substituted by $\gamma(\varepsilon) \sim r_s$. One should keep in mind that the relativistic corrections to the total cross section $\sigma_\gamma(\omega)$ start to be important at higher frequencies, because the ω/c term vanishes in $\sigma_\gamma(\omega)$.

Recently the "drag" current created by photons of near 1s-threshold energies in Ar and Kr was estimated [23]. For $\varepsilon = 0.16$ Ry above the 1s-threshold contributions of all Ar subshells were found. It appeared that for $j(\omega)/W$ from (24) one has in 10^{-23} Coulomb units 2; 1.26; 0.67; 0.20; 0.08 for photoelectrons removed from 1s, 2s, 2p, 3s and 3p subshells, respectively. The 1s contribution is significantly increased, namely by a factor of twenty in Ar and fifty in Kr for a photoelectron energy at the Ramsauer elastic scattering minima. For a flux $W=10^{13}$ photons/sec-cm^2 the maximum values of the current are $0.6 \cdot 10^{-8}$ amp/cm^2 and $3 \cdot 10^{-8}$ amp/cm^2 for Ar and Kr, respectively, which may be observable.

DECAY OF DEEP HOLES

In the light and medium heavy atoms the inner holes decay via multi-step Auger-processes while in heavy atoms radiative processes dominate, which are followed by predominantly Auger decays in the subsequent steps. Of interest is the investigation of the spectrum of secondary emitted particles, i.e. electrons and photons, and their energy and angular distributions. Quite informative, but difficult to perform are coincidence experiments, which would observe for instance an emitted electron and an ion of a given charge (see [24]).

The double-hole states created in inner-hole decay can decay in turn via many channels; of special interest are decays where either a single electron or a single photon are emitted, since they exhibit very specific spectral lines. The electron and photon lines, corresponding to the decay of two 1s - holes in Ar, were observed many years ago in heavy ion-atom collisions, where it was almost impossible to calculate the probability of double-hole state production. Moreover, in the ion collision experiment the two inner-shell holes were accompanied by an almost unknown number of emitted outer electrons, which made the comparison of the results of measurements and calculations difficult. On the other hand, if the two-hole state is created in a photoabsorption process, the calculations may be performed with good accuracy and the measurements can be directly linked to a given ion state. One must bear in mind, that these lines are comparatively weak, particularly for inner shells where they are of about 10^{-6} of the corresponding one-hole decay line.

In connection with inner holes it is of interest to study the time-dependence of the decay process. The main feature of the decay process is described by the exponential law

$$N(t) = N_e \exp(-\Gamma t), \qquad (25)$$

where $N(t)$ is the number of excited atoms in a sample, which initially at $t=0$ had N_e atoms ready to decay. It has been known for a long time, that this law can have considerable and observable corrections [25]. It has been demonstrated that, if at $t=0$ a particle is confined within a three-dimentional potential barrier, as is assumed for the α-decay of nuclei, then the time dependence of the barrier penetration will not be described precisely by (25) but must be corrected by including a non-exponential power term, so that instead of (25) one has:

$$N(t) = N_e (e^{-\Gamma t} + \frac{b}{tE_o}). \qquad (26)$$

Here E_0 is the energy of the decaying resonance, Γ is its width and b is a constant ($b \sim 1$). The contribution of the second term in (26) is larger than that of the first term for $t < b/E_0$ and $t > \Gamma^{-1} \ell n(E_0/\Gamma)$.

The nonexponential correction in (26) is of the order of Γ/E_0. For inner holes this can be important, provided that the picture of barrier penetration is applicable to Auger, radiative, or autoionizational decay. The magnitude and perhaps the very presence of the correction to the exponential law (25) described by the second term in (26) is determined by the preparation of the initial decaying state. In other words, (25) is incorrect if a moment $t=0$ exists, when the existed atom (ion) together with subsequently emitted electrons and/or photons are confined in a finite sized volume.

Experimental investigations of (25) and its possible corrections presented by (26) would be of interest.

CONCLUDING REMARKS

Present day storage-ring light sources produce intense beams of high frequency electromagnetic radiation and enable experimental studies on photoionization and photoexcitation of innermost shells of even the heaviest atoms.

Inner holes decay predominantly by emitting many electrons, thus producing a kind of an avalanche. Different mechanisms contribute to this multiple ionization and it is therefore of interest to investigate the energy and angular distribution of the emitted electrons. The investigation of the mean ion charge as a function of photon frequency has proven to be quite informative. Special attention should be given to inner-shell PCI, which is strongly affected by multistep Auger decay of an inner hole. The state of the outer electrons and of all the electronic surrounding of an atom in a multiatomic aggregate strongly affects inner-shell excitation and ionization. This opens a way to manipulating the processes by exciting outer electrons.

Detailed information on the ionization and hole decay processes is desirable, which could be obtained from different coincidence experiments, in which, for example, a photoelectron and an ion, a photoelectron and an Auger-electron, or decay photon, are observed simultaneously.

Of interest is the process of continuous spectrum photon emission, or internal Bremsstrahlung, as well as the hole decays with simultaneous emission of an electron and a photon. An example is the so-called radiative Auger decay. These processes can proceed directly, for instance as the usual Bremsstrahlung of an electron in the static field of an atom, or via virtual excitation of some other atomic electrons. In the latter case strong collective effects can manifest themselves, if several atomic electrons are excited virtually and coherently.

The search for relativistic effects in photoelectron angular distributions has been started and the investigation of corresponding cooperative effects like e.g. "drag" currents will start, hopefully, soon. Of fundamental interest is the study of the time dependence of the decay of holes.

In total, there is still a lot to be done in the field of inner-shell photoionization and excitation of atoms. Step by step this field is becoming an important area of research in the domain of photo-process studies.

ACKNOWLEDGMENT

I am grateful to the Alexander-von-Humboldt Foundation for financial support which made my participation in the X-96 conference possible.

REFERENCES

1. Amusia, M. Ya., *Atomic Photoeffect*, New York-London, Plenum Press, 1990.
2. Amusia, M. Ya, Ivanov, V. K, and Kupchenko, V.A., *J. Phys. B: At. Mol. Phys.*, **14**, L667 (1981).
3. Amusia, M. Ya., *Advances in Atomic and Molecular Physics*, **17**, 1 (1981).
4. Whitfield, S.B., Langer, B., Viefhaus J., Wehlitz, R., Berrah, N., Mahler, W. and Becker U., J. Phys. B: *At. Mol. Opt. Phys,* **27**, L359 (1994).
5. Tawara, H., Hayaishi, T., Koizumi, T., Matsuo, T., Shima, K., and Yagishita, A., *J. Phys. B. At. Mol.- Opt. Phys.* **25**, 1476 (1992).
6. Doppelfield, J., Anders N., Esser B., von Busch F., Scherer H., and Zinz S., *J. Phys. B: At. Mol. Opt. Phys.* **26**, 445 (1993).
7. Amusia, M. Ya., *Phys. Lett,* **A183**, 201-204 (1993).
8. Kuchiev, M. Yu., and Sheinerman, S.A., *Sov. Phys: Uspekhi* **32**, 569-587 (1989), in Russian.
9. Koike, F., *Phys. Letters* **193**A, 173 (1994).
10. Sheinerman, S. A., *J. Phys. B:, At. Mol. Opt. Phys* **27**, L571 (1994).
11. Amusia, M. Ya., Lectures on photoionization at the University of Reno, unpublished, 1992.
12. Amusia, M. Ya., Lee, I. S., and Kilin, V. A., Phys. Rev **A45**, 4576 (1992).
13. Kjeldsen, H., Thomas, T. D., Lablanquie P, Lavollee, M., Penent, F., Hochlaf, M., and Hall, R. I., *J. Phys. B. At. Mol. Opt. Phys,* **29**, 1689 (1996).
14. Hayaishi, T, Murakami, E., Morioka, Y., Shigemasa, E., Yagishita, A., and Koike, F., *J.Phys.B: At. Mol. Opt. Phys.* **27**, L115 (1994).
15. Hayaishi T, Murakami E., Lu, Y, Shigemasa, E., Yagishita, A., Koike, F., and Morioka, Y., *J.Phys.B: At. Mol. Opt. Phys.(1996)* to be published.
16. Amusia, M. Ya., Baltenkov, A .S., and Zhuraleva, G. I., *J. Phys.B: At. Mol.Opt. Phys.* **29**, L153 (1996).
17. Amusia, M. Ya., and Cherrysheva, L. V., prepared for publication, 1996.
18. Amusia, M. Ya., and Baltenkov, A. S., unpublished, 1995.
19. Amusia, M. Ya., Kuchiev, M. Yu., Sheinerman, S. A., and Sheftel, S.I., *J.Phys.B:* At. Mol. Phys. **10**, L535, (1977).
20. Samson, J. A. R., Stolte, W. C., He, Z. X., Cutler, J. N., and Hansen, D., *Phys Rev. A*, (1996), in press.
21. Amusia, M. Ya., and Cherepkov, N. A., *Case Studies in Atomic Physics,* **5** 47 (1975),
22. Krassig, B., Jung, M., Gemmel, D. S., Kanter, E. P., Le Brun, T., Southworth, S. H., and Young, L., *Phys. Rev. Lett.* **75**, 4736 (1995).
23. Amusia, M. Ya., and Baltenkov, A. S., X-96 Book of Abstracts, Hamburg, Germany, Tu Po56 (1996).
24. Arp, U., Le Brun, T., Southworth, S. H., Cooper, J. W., MacDonald, M. A., and Jung, M., X-96 Book of Abstracts, Hamburg, Germany, Tu Po 31 (1996).
25. Khalfin, L. A., *Zh Exp. Teo. Fiz.* **33**, 1371 (1956), in Russian.

X-Ray Natural Widths, Level Widths and Coster-Kronig Transition Probabilities

T. Papp[1], J. L. Campbell[2] and D. Varga[1]

[1] Institute of Nuclear Research of the Hungarian Academy of Sciences,
Debrecen, Bem tér 18/c, P.f. 51, H-4001 Hungary
[2] Guelph-Waterloo Program for Graduate Work in Physics, University of Guelph, Guelph, Ontario, N1G2W1, Canada

Abstract. A critical review is given for the K-N7 atomic level widths. The experimental level widths were collected from x-ray photoelectron spectroscopy (XPS), x-ray emission spectroscopy (XES), x-ray spectra fluoresced by synchrotron radiation, and photoelectrons from x-ray absorption (PAX). There are only limited atomic number ranges for a few atomic levels where data are available from more than one source. Generally the experimental level widths have large scatter compared to the reported error bars. The experimental data are compared with the recent tabulation of Perkins et al. and of Ohno at al. Ohno et al. performed a many body approach calculation for limited atomic number ranges and have obtained reasonable agreement with the experimental data. Perkins et al. presented a tabulation covering the K-Q1 shells of all atoms, based on extensions of the Scofield calculations for radiative rates and extensions of the Chen calculations for non-radiative rates. The experimental data are in disagreement with this tabulation, in excess of a factor of two in some cases.

A short introduction to the experimental Coster-Kronig transition probabilities is presented. It is our opinion that the different experimental approaches result in systematically different experimental data.

1. INTRODUCTION

An x-ray transition is characterized by the transition probability, the polarization and the energy distribution of the photons. Absolute photon intensity measurements are limited in accuracy at the present level of experimental technique, and even the results of relative measurements, e.g., x-ray branching ratios show large scatter for the Kα/Kβ ratio (1) or for the L shell x-rays (2). Polarization measurements frequently result in large values (3) an observation which is not yet explained by theoretical calculations. The energy distribution is usually described as a Lorentzian distribution (Breit-Wigner formula (4)), which is a very good approximation. Deviation from the Lorentzian shape is expected when the so called dynamical relaxation is important (5). The Lorentzian distribution is characterized by two parameters, namely the energy centroid (E_0) and the Lorentzian width (Γ, which is the sum of the two relevant atomic level widths involved in the X-ray transition). Since a wide range of analytical techniques is based on characteristic x-rays it is widely assumed that the x-ray energies are well known. The two commonly used binding energy tables of Bearden and Burr (6) and Sevier (7) give energy values differing by more than 10 eV and in some cases they deviate by more than 25 eV. This can introduce difficulties in x-ray analytical work and in atomic physics measurements (8) as well. Accurate x-ray energy values are also important in nuclear and particle physics when x-rays are used for energy calibration (9). For the determination and application of x-ray energies (10) the x-ray lineshape is necessary information. The last compilation on x-ray line widths and level widths was published around twenty years ago (11). Meanwhile technological advance has yielded new data and at the same time has created a demand for an accurate data base. Not only the high resolution spectroscopic measurements but also everyday analytical work demands this. To support our assertion, a PIXE (proton induced x-ray emission) L shell spectrum of thorium

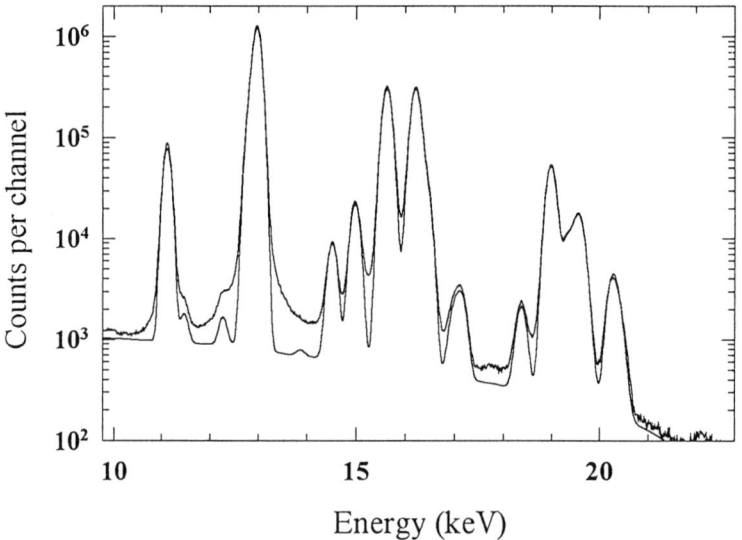

FIGURE 1. L x-ray spectrum of thorium target bombarded by 1MeV protons and measured with a Si(Li) detector. The continuous line represents a fit, where the Lorentzian broadening was left out from the best fit. It demonstrates the importance of the natural lineshape even in the case of moderate resolution detectors like Si(Li) detectors.

metal measured by a Si(Li) detector is shown in figure 1. along with a model fit where the Lorentzian broadening of the x-rays was neglected (8). The wing of the Lorentzian distribution is visible and the potential for confusion between this phenomenon and the continuous background is well demonstrated. Accurate level energy and widths data may be important in fields other than x-ray spectroscopy, such as the description of electron capture processes or even for neutrino mass studies.

To have a general view on the availability and accuracy of level width data we have assembled the experimental data from a variety of sources and examined its consistency. The data are taken mainly from work on elemental solids; it should be noted that the theoretical work is strictly applicable only to free atoms, and solid state effects are often important. In addition, level widths can be altered from the elemental values if the electrons are influenced by chemical bonding. The present database is also of value in identifying where improvements to our presently incomplete knowledge of natural widths are most critical.

2. THEORETICAL CALCULATIONS OF LIFETIME BROADENING

The width of a level is the sum of its radiative and non-radiative widths. Recently Perkins et al (12), using the results of single-particle-model (SPM) predictions, presented a tabulation covering K-Q1 shells of all atoms, based upon extensions of the Scofield calculations (13) for radiative rates and extensions of the Chen calculations (14) for non-radiative rates. For the radiative width the results of Dirac-Fock type calculations are widely accepted. It has been reviewed by several authors (15), and found to be quite accurate for medium and heavy elements. Except for the K-shell of heavy elements, the Auger process is the dominant decay mode of inner shell vacancy state. There are two approaches to calculate the Auger widths. Chen and coworkers have performed extensive systematic Dirac-Hartree-Slater calculations, and Ohno and Wendin have carried out a series of nonrelativistic many-body calculations.

3. SURVEY OF ATOMIC LEVEL WIDTHS

3.1. Measurements

Campbell and Papp (16) have recently presented a very detailed examination of the internal consistency of an extensive database of experimentally measured atomic level widths. This in turn permitted an assessment of the accuracy of presently available single-particle-model (SPM) and many-body (MB) theories.

In this review we will not reproduce the detailed arguments of Campbell and Papp; instead we will select broad conclusions and major issues raised by them.

The data were assembled from a variety of spectroscopic techniques employing both electron and photon detection. The majority of the data came from X-ray photoelectron spectroscopy (XPS), a technique of inherently very high resolution; subtraction of the natural width of the exciting radiation (typically Al Kα) from the measured XPS width provides directly the width of the level excited. Because the width of the exciting radiation must be small compared to the width of the level under study, the use of XPS is confined to levels of low binding energy. A very few data were provided by the PAX method (17), where the energy distribution within an X-ray line is determined by using the X-rays to eject photoelectrons from a thin foil, and then performing electron spectroscopy to determine their energy distribution. The width of the ionized level in the foil atoms must be subtracted from the measured linewidth to provide the width of the X-ray transition. A very few data were from Auger electron spectroscopy (AES); in this case the width of the measured line reflects the widths of three atomic levels, and so the uncertainty tends to be greater than in the case of XPS.

In terms of volume of data, X-ray emission spectroscopy (XES) was the second ranking contributor to the database. Many of the results came from the early days of X-ray spectroscopy (1930s and 1940s), and their reliance on photographic techniques for recording the diffracted spectra gives cause for concern regarding their accuracy. Turning to modern XES results, based on counting techniques for recording the spectra, there are some very accurate systematic studies of particular levels (e.g., L2 and L3 in the middle of the periodic table).

Finally, a few results have been obtained indirectly by bombarding thin foils or gas columns with synchrotron radiation (SYNC), the photon energy being scanned over absorption edges of the foil or gas atoms. Synchrotron experiments have provided two different types of result. In the first case the absorption edge profile is measured, from which the level widths can be determined directly; in the second, ratios of Coster-Kronig transition rates to fluorescent decay rates are obtained, thereby allowing the experimenter to deduce the total level width from the theoretical fluorescence width.

To trace the source of possible systematic errors, the data obtained by the different techniques are represented by different symbols; curly ones correspond to data obtained by XES, while symbols with sharp edges represent XPS data. The aim of the paper was to present the existing data rather than to make a critical selection, although the internal consistency of the data was checked and are discussed.

3.2. Assessment of results

Figures 2-10 present a comparison of measured, derived and predicted

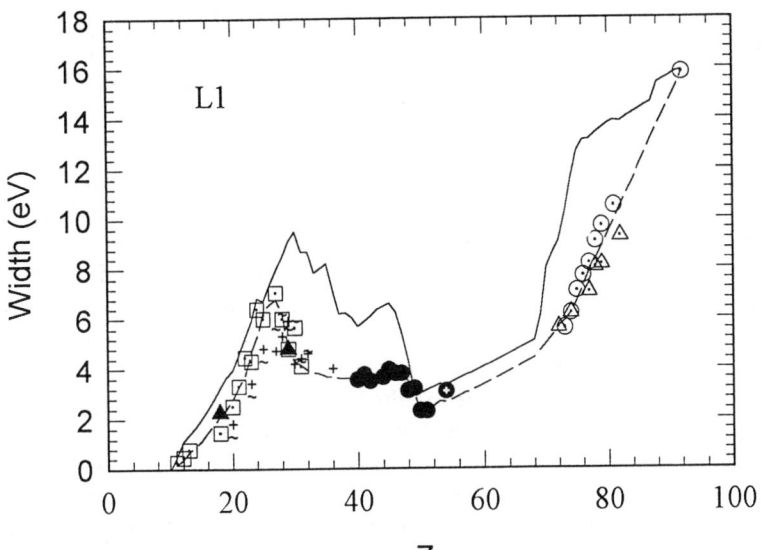

FIGURE 2. L1 widths. The continuous line represents the SPM results and the dashed line is the empirical fit. Manybody predictions (35) are shown as plus signs (atomic calculation) and tilde signs (solid state calculation). Notations for experimental data are: open square, ref (20); filled triangle, ref (23,36); open circle, ref (41); filled circles, ref (25,38); triangle up, ref (44).

widths for the L1-N6 subshells. In each case, the full curve is the prediction of the SPM (12), employing Scofield's approach for radiative transition rates and the Chen approach for Auger and Coster-Kronig rates. Because of their complexity, many-body calculations are limited to a few interesting subshells and regions of atomic number; they are shown by the symbols of tilde, plus sign, and horizontal bars in several of the figures, and by dotted lines in the M2 and M3 cases where they are more numerous. The dashed lines represent an empirical fit to the overall trend of the measured data. Values of the K, L2 and L3 level widths may be obtained directly from XPS measurements on L2 and L3 lines and indirectly from XES measurements on $K\alpha_1$ and $K\alpha_2$ X-rays. Salem and Lee, and Pessa (18) reviewed a large volume of experimental $K\alpha_1$ and $K\alpha_2$ X-ray linewidths. Krause and Oliver (11) compared these two data sets with semi-empirical K X-ray widths derived from Scofield's Dirac-Fock radiative level widths and a set of recommended K-shell fluorescence yields (11). There is a more recent set of accurate K X-ray widths at medium to high Z measured by Kessler et al. (19).

There is a fairly good agreement between experimental and SPM values at medium (above Z=30) and high Z simply indicates that the K level widths there, which in turn reflect mainly Scofield's radiative widths, (13) are accurate at a 1-3% level. The data do not provide a stringent test of the L subshell level widths. Nor does the comparison provide a stringent test of K width at low Z; this is because the X-ray lines of the 3d elements are affected by various non-lifetime broadening effects such as multiplet splitting of the L2 and L3 levels. These effects, are discussed in some detail by Krause and Oliver (11), and are important when the level concerned is other than a deep core level; thus they can have a role for the K level when Z<13 and for the L level at somewhat higher Z values. There are recent measurements (10) on oxygen and copper K shell and it is concluded that level widths data cannot be derived from experimental data alone but requires theoretical input on the multiplet splitting and the rates of all possible transitions.

The XPS widths in the figures are essentially measured directly. However, various of the other widths presented in the figures are deduced through combinations of different experimental data. For example, the open circles with dots at Z>70 in the L1 case of fig. 2 are derived by combining XES and XPS results. The XES results are from pre-1950 wavelength-dispersive spectroscopy measurements of the width of the L1N23 X-ray line. From these, Campbell and Papp subtracted the best-fit N2,3 level widths of fig. 9; these are based upon XPS measurements. (The resulting L1 widths are in modestly good agreement with indirect results deduced from SYNC experiments). In similar fashion, M1 widths at high Z are obtained by subtracting the SPM L3 width from the measured XES width of the L3M1 X-ray line, this being justified by the good agreement between the measured and SPM L3 widths. It must be remembered that errors compound significantly in this approach. The only L3 subshell results that deviate significantly from theory are those of the absorption edge profile measurements at Z=56,62 for the L3 subshell.

Before any more detailed examination of specifics is undertaken, it is immediately obvious that the SPM is successful only for the highest subshells within any given shell, i.e., L3, M5, N6, N7. In these cases, Coster-Kronig (C-K) or super Coster-Kronig (sCK) transitions are energetically not possible. The L2L3M4,5 C-K transition is not allowed in atomic states in the 20<Z<30 range. However, the experiments provide evidence that it is open in the solid state. The L2L3M4,5 Coster-Kronig transition does not become energetically possible again until Z~90. In the region 50<Z<90 there is a well-established discrepancy of about 10% between measured and theoretical L2-L3 Coster-Kronig probabilities (27). It is our opinion that the different experimental approaches result in systematically different experimental data. It is also of importance to take into account the many body satellites in the determination of Coster-Kronig transition probabilities, even for the case of moderate resolution Si(Li) detectors (27). Here, the Coster-Kronig probability accounts for some 10-20% of the L2 width.

When the C-K or sCK transitions are allowed the widths given by the evaluated atomic data base of Perkins et al. can be too large, since the C-K yields were modified to yield agreement for the fluorescence yield with the atomic num-

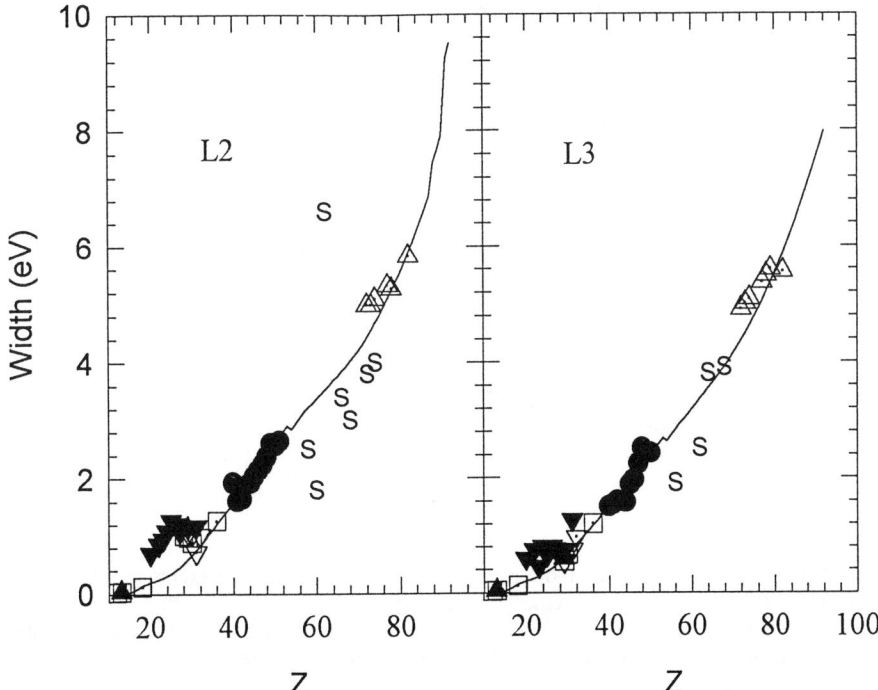

FIGURE 3. L2 and L3 widths. Notations: continuous line, SPM prediction; open squares with a dot, ref. (20) own measurement, filled triangle down, ref. (20) pre 1981 measurement; open triangle down AES of ref (24) filled circles, ref (25, 26); open triangle up, SYNC of ref (44); S, SYNC of ref (45).

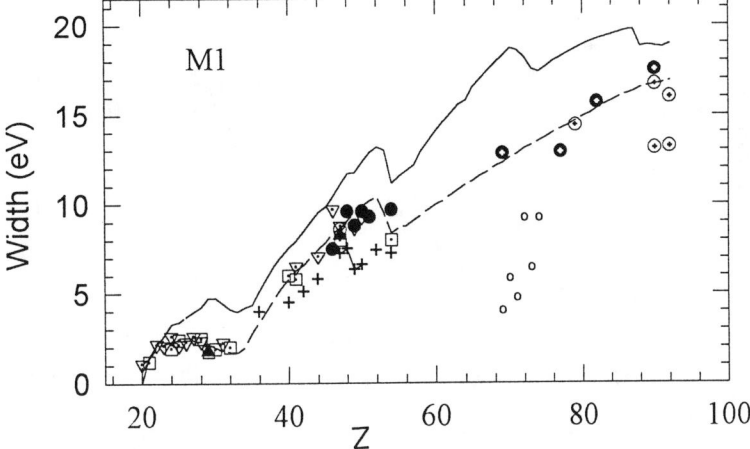

FIGURE 4. M1 widths. Plus signs represent many body calculations (46). Open squares and triangle up are from ref (20); filled square up, ref (23); filled circles, ref (47); filled circles with white crosses, ref (8) open circles with crosses, ref (42, 43, 48, 51, 57); small open circles, ref (50).

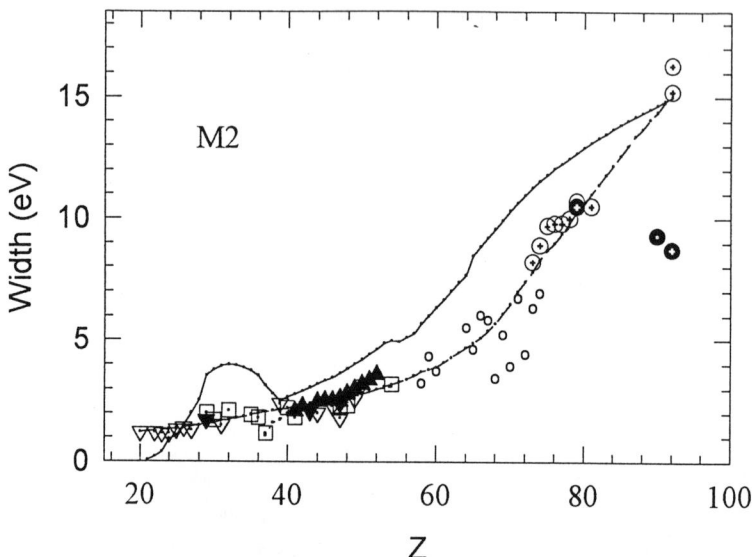

FIGURE 5. M2 widths. The dotted curve in the interval 37<Z<54 represents the manybody prediction discussed in the text. Open squares and triangle down, ref (20); full triangle up, ref (54); full triangle down, ref (23,52); open circles with crosses, ref (39, 40, 42); filled circles with white cross, ref (43,49,51,57,); small open circles, ref (50).

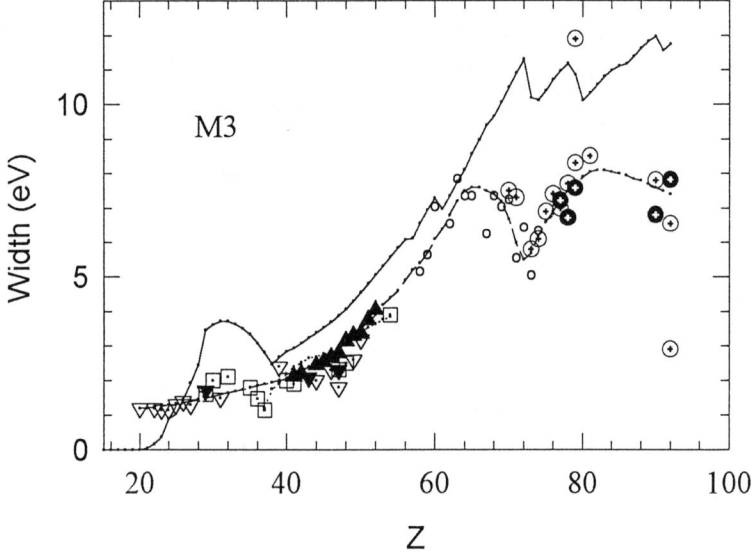

FIGURE 6. M3 widths. Notations are the same as in fig. 6. with the exception of additional data from ref. (48), open circles with crosses.

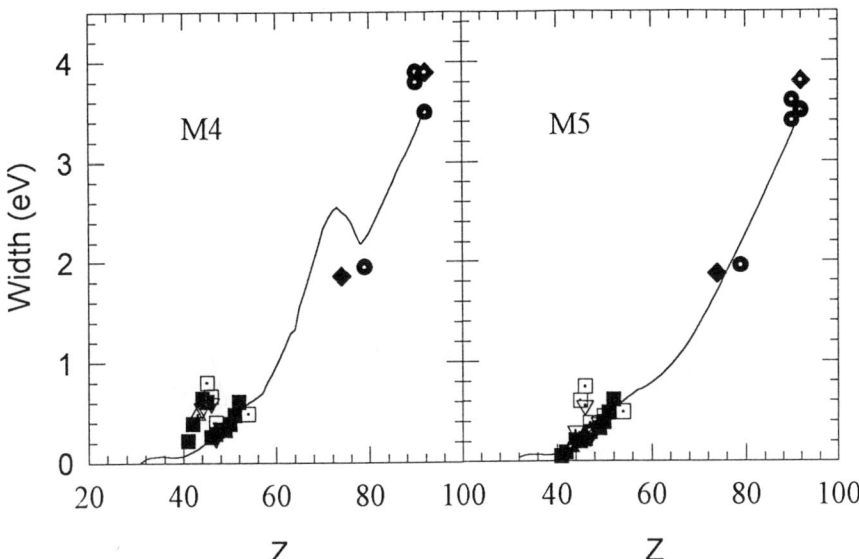

FIGURE 7. M4 and M5 widths. The continuous line is the SPM prediction. Open square, ref. (20,56); filled square, ref (54); triangles, ref (22,52); dots, ref (51,58,59,60); diamond with white dot, ref (17); full diamond from the present measurement.

ber systematics. For example the Perkins table predicts an opening of the M4M5N C-K channel in the Z=70-80 range. On the other hand the calculation of Chen et al. give very similar values for the M4 and M5 widths, and less than four per cent value for the f_{45} C-K probability. We have performed a PAX measurement on tungsten Mα and Mβ lines and it gave the same widths for the $Mα_1$ and $Mβ_1$ transitions (filled diamonds in fig. 7), supporting the result of the Chen et al. calculation and indicating that the M4,5 C-K channel has much smaller strength at least in the solid state. Another example is the N1 case in fig. 8; the SPM predicts very large transition rates for the sCK

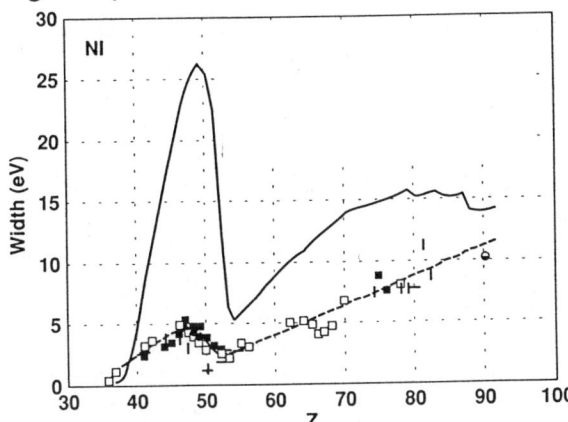

FIGURE 8. N1 widths. Horizontal bar represent many-body calculations (64). Notations: vertical bar and open square, ref (20); filled square, ref (54); filled triangle, ref (23); filled circle, ref (47); half filled circle, ref (51).

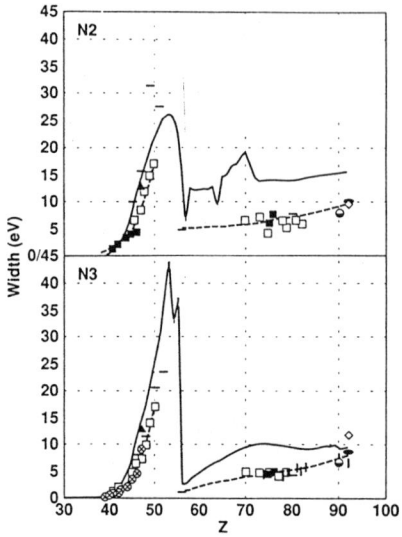

FIGURE 9. N2 and N3 widths. Horizontal bar represent manybody calculations (64). Vertical bar and open square, ref (20); filled square, ref (54,62); filled triangle, ref (23); circle with cross, ref (65); half filled circle, ref (51); filled ellipses, ref (42), open diamond, ref (61).

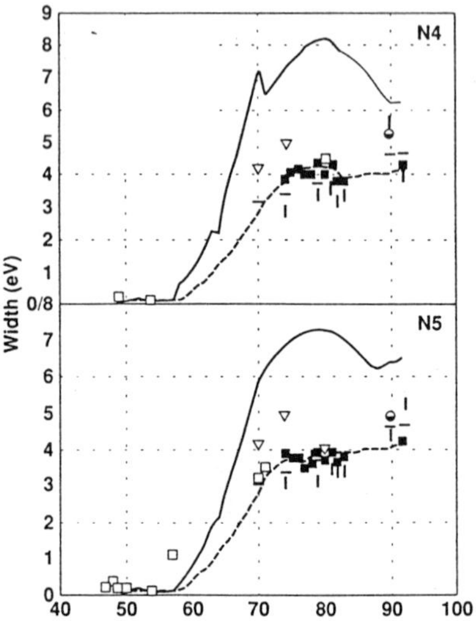

FIGURE 10. N4 and N5 widths. Horizontal bar represent manybody calculations (64). Vertical bar and open square, ref (20); filled square, ref (63); open triangle down, ref (64); circle with cross, ref (65); half filled circle, ref (51).

transition N1N2,3N4,5 transitions; but the data show that the actual rates are some five times smaller. Perkins et al. provide a clear warning of the possibility of overestimate of sCK rates, and our survey here enables this to be quantified quite accurately. Again using the N1 example, even at high atomic numbers where ordinary CK processes dominate the decay, an overestimate persists, demonstrating the difficulty in describing these non-radiative processes within the SPM when the energies involved are very low.

Where available, MB predictions are quite successful, the M2, M3, and N2-N5 cases being good examples. But in the M1 case, the MB predictions tend to fall below the measurements by about as much as the SPM exceeds them.

The reliability of the various experimental techniques is an important factor in planning future measurements to fill gaps in the database or to resolve some of the various evident inconsistencies. Consistency among XPS data from different sources is generally quite good. The M2 case provides an example of this, with good agreement among the XPS results depicted by open and filled squares, and open and filled triangles in the atomic number region $20<Z<55$. But it is well worth noting in the M3 case that a few of the open triangle down symbols lie low enough that they are clearly erroneous by up to 1 eV. In the N4 and N5 cases, one set of XPS data (vertical bars) runs systematically 0.5-1 eV below another set (filled squares). These uncertainties notwithstanding, the XPS accuracy is good enough for the M4 and M5 cases that one can discern in the 40<Z<50 region an excess of about 0.5 eV over the SPM width; this must be attributed to the M4M5N4,5 CK transition becoming energetically possible at a lower atomic number than the SPM predicts.

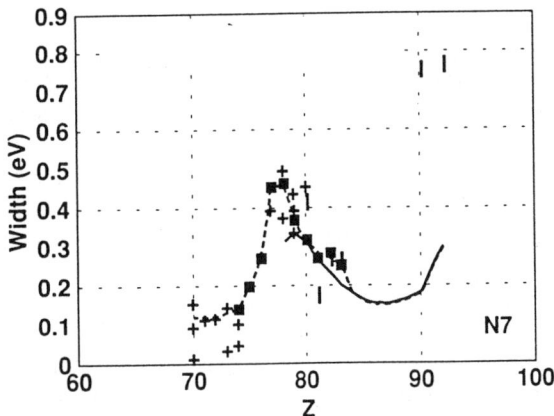

FIGURE 11. N7 widths. Vertical bar, ref (20); plus sign and full square; ref (63).

3.3 Comparison of experimental techniques

In regions where XPS data overlap with modern XES results, although such regions are very limited, the agreement between these methods is good. An example is found in fig. 4 for the M1 level in the region 40<Z<54. Here the filled circle, XES data fit well with the open triangle, XPS results.

There are inconsistencies within the XES data. Again the M1 results of fig. 4 provide an example. The small open circle data, which are from one particular experiment using wavelength-dispersive spectroscopy (50) of the weak L3M1 X-ray line, fall well below recent energy-dispersive data (8) in the region around Z=70; yet, a small number of other wavelength-dispersive results in the region above Z=80 tends to support the latter. This is clearly a case that merits further measurement. The most serious inconsistencies occur at the highest values of atomic number. The M2 and M3 cases are especially troubling. For example, at Z=92, a very sophisticated 1994 measurement agrees with the old measurement, both of them depicted by open circles with crosses. Yet, in the M3 case, the analogous values differ by 5 eV. Two other older measurements differ 6 eV from the 1994 one for M2, while for M3 they are in good agreement. Despite these problems at the highest Z values, it is interesting to note how the plethora of data in the M3 case allow one to deduce the existence of a local minimum in the widths at Z=75, as predicted by SPM theory; a second minimum predicted at Z=81 is not verified.

3.4 PAX spectroscopy

To understand the reason of the discrepancies in x-ray linewidth, x-ray energies, and polarization data, measurements with different detection technique can be useful. An alternative approach to the application of crystal diffraction spectrometers is the PAX spectroscopy. We have recently extended the PAX method pioneered by Krause and coworkers (17), to higher X-ray energies (up to 10 keV). PAX, which has inherently different spectrometer transmission function than crystal diffraction spectrometers, provides a *complementary means* for determining the intrinsic x-ray line shape, the Auger widths and the XPS linewidths on the same spectrometer. The PAX spectroscopy was applied on several x-ray transitions and the lineshape, as well as the natural widths of the Kα and Lα,β transitions of copper, Lα transition of silver and Mα,β transition of tungsten were determined. Various materials such as carbon, silicon, copper and silver were used as converters. The level widths and x-ray line widths were determined relative to the 3d level width of silver. The background originating from inelastically

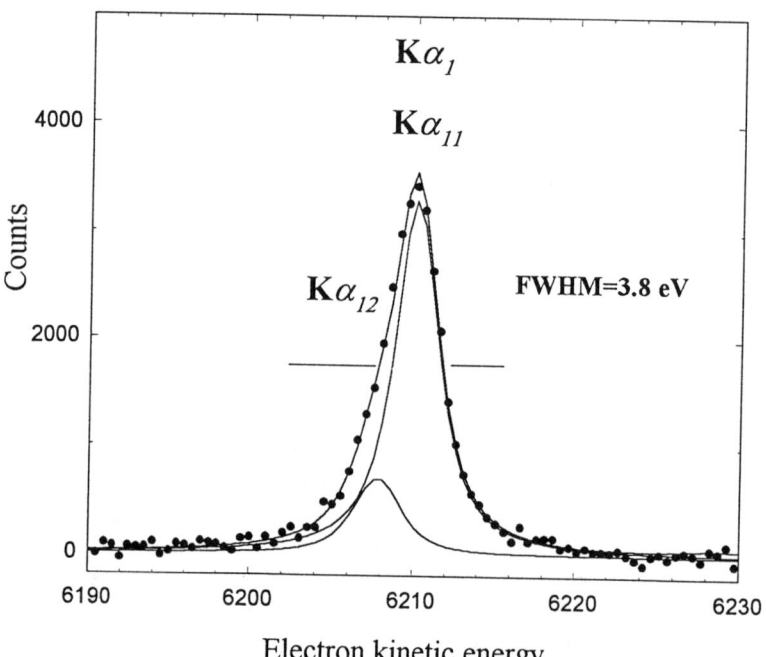

FIGURE 12. Lineshape of the copper Kα1 transition measured by PAX. Dots are the measured data after background subtraction. The continuous lines are the best fit with the sum of two Lorentzians, and the two Lorentzian components.

scattered electrons can be reduced if nanolayers are used as converters. The copper $K\alpha_1$ spectrum is presented in fig 12. The $K\alpha_1$ transition has pronounced asymmetry. It was observed that in this energy range the resolution and luminosity of the PAX spectroscopy can compete with the diffraction spectrometers.

4. CONCLUSIONS

The exercise of Campbell and Papp (16) provides level widths over certain regions of the periodic table and for certain subshells, with much greater accuracy than does the SPM theory. It also shows that the few existing MB calculations meet with reasonable although not complete success. The database reveals significant discrepancies among X-ray linewidths measured by diffraction spectroscopy, and suggests the need for further effort in this area, especially for the heaviest elements.

ACKNOWLEDGEMENT

Support from the Hungarian Research Foundation under contract T-7274/1993, T-016636 and from the Natural Sciences and Engineering Research Council of Canada.

REFERENCES

1. E. Schönfeld and H Janßen, *Physicalish-Technische Bundesanstalt*, PTB-Bericht Ra-37 (1995)
2. T. Papp, J. L. Campbell and S. Raman, *J. Phys. B: At. Mol. Opt. Phys.* **26**, 4007 (1993)
3. J. Hrdý, A. Hennins and J. A. Bearden, *Phys. Rev.* **A 21**, 708 (1970) ;
   K. S. Kahlon, N. Singh, R. Mittal, K. L. Allawadhi and B. S. Sood, *Phys. Rev.* **A 44**, 4379 (1991);
   V. P. Petukhov, Török, P. Závodszky, J. Pálinkás L. Sarkadi and S. M. Blokhin, *Nucl. Instrum. Meth.* **B 75**, 17 (1993)
4. V. Weisskopf and E. Wigner, *Z. Phys.* **63**, 54 (1930) and **65**, 18 (1930);
   E. Arnous and W. Heitler, *Proc. Roy. Soc.* (London) **A220** , 290 (1953)
5. M. Ohno and R. E. LaVilla, *Phys. Rev.* **A 45**,4713 (1992) and ref. therein
6. J. A. Bearden and A. F. Burr, *Rev. Mod. Phys.* **39**, 125 (1967)
7. K. D. Sevier, *At. Data Nucl. Data Tables* **24**, 323 (1979)
8. T. Papp and J. L. Campbell, *Nucl. Instrum. Meth.* **B 114**, 225 (1996);
   T. Papp, J. Campbell, J.A. Maxwell, J.-X. Wang and W.J. Teesdale, *Phys. Rev.* **A45**, 1711 (1992).
9. J. J. Simpson, *Phys. Rev.* **D 23**, 649 (1981)
10. A. Menzel, S. Benzaid, M. O. Krause, C. D. Caldwell, U. Hergenhahn and M. Bissen, *Phys. Rev.* A (1996) in press;
    M. Deutsch, G. Hölzer, J. Härtwig, J. Wolf, M. Fritsch and E. Förster, *Phys. Rev.* **A 51**,

283 (1995);
M. Deutsch, O Gang, G. Hölzer, J. Härtwig, J. Wolf, M. Fritsch and E. Förster, *Phys. Rev.* **A 52**, 3661 (1995)

11. M. O. Krause, *J. Phys. Chem. Ref. Data* **8**, 307 (1979);
    M.O. Krause and J.H. Oliver, *J. Phys. Chem. Ref. Data* **8**, 329 (1979).
12. S.T. Perkins, D.E. Cullen, M.H. Chen, J.H. Hubbell, J. Rathkopf and J.H. Scofield, Tables and graphs of atomic subshell relaxation data derived from the LLNL evaluated atomic data library Z = 1-100. Lawrence Livermore National Laboratory report UCRL50400 Vol. 30, 1991.
13. J.H. Scofield, *At. Data Nucl. Data Tables* **14**, 121 (1974); J.H. Scofield, *Phys. Rev.* **A9**, 1041 (1974); *Phys. Rev.* **A10**, 1507 (1974).
14. M.H. Chen, B. Crasemann and H. Mark, *Phys. Rev.* **A21**, 436 (1980); *Phys. Rev.* **A24**, 177 (1981); *Phys. Rev.* **A21**, 449 (1980); *Phys. Rev.* **A27**, 2989 (1983).
15. B. Crasemann, Proc. 2nd workshop on High-Energy Ion-Atom Collisions (Aug. 27-28, Debrecen) Akadémiai Kiadó Budapest 1985, ed. D. Berényi and G. Hock, p. 199;
    M. H. Chen, Fifteenth International Conf. on X-Ray and Inner Shell Processes (July 9-13, 1990) Knoxville, Tennessee, AIP Conference Proceedings 215, ed. T. A. Carlson, M. O. Krause and S. T. Manson, p. 391;
16. J. L. Campbell and T. Papp, *X-Ray Spectrometry* **24**, 307 (1995)
17. M.O. Krause, *Advances in X-Ray Analysis*, Vol. **16**, pp. 74-89 Plenum Press, New York (1973)
    M.O. Krause, *Physica Fennica*, Vol. **9**, Supplement S1, 281 (1974)
18. S.I. Salem and P.L. Lee, *At. Data Nucl. Data Tables* **18**, 234 (1976); V.M. Pessa, *X-Ray Spectrom.* **2**, 169 (1973).
19. E.G. Kessler, Jnr., R.D. Deslattes, D. Girard, W. Schwitz, L. Jacobs and O. Renner, *Phys. Rev.* **A26**, 2696 (1982).
20. J.C. Fuggle and S.F. Alvarado, *Phys. Rev.* **A22**, 1615 (1980).
21. W. Theis and K. Horn, *Phys. Rev.* **B47**, 16060 (1993).
22. P.H. Citrin, G.K. Wertheim and Y. Baer, *Phys. Rev.* **B27**, 3160 (1983).
23. M. Ohno, J.-M. Mariot and C.F. Hague, *J. Elect. Spect. Rel. Phen.* **36**, 17 (1985).
24. E. Antonides, E.C. Janse and G.A. Sawatzky, *Sol. State Comm.* **17**, 417 (1975).
25. P. Putila-Mantyla, M. Ohno and G. Graeffe, *J. Phys. B: At. Mol. Phys.* **17**, 1735 (1984).
26. H. Juslen, M. Pessa and G. Graeffe, *Phys. Rev.* **A19**, 196 (1979).
27. T. Papp, J.L. Campbell and S. Raman, *Phys. Rev.* **A49**, 729 (1994);
    W. Jitschin, R. Stötzel, T. Papp, M. Sarkar and C. D. Doolen, *Phys. Rev.* **A 52**, 977 (1995)
28. J.A. Leiro, M. Heinonen, F. Werfel, E.G. Nordstrom and K.H. Karlsson, *Phil. Mag. Lett.* **68**, 153 (1993).
29. P.H. Citrin, P.M. Eisenberger, W.C. Marra, T. Aberg, J. Utriainen and E. Kallne, *Phys. Rev.* **B10**, 1762 (1974).
30. L.I. Yin, I. Adler, M.H. Chen and B. Crasemann, *Phys. Rev.* **A7**, 897 (1973).
31. J. Vayrynen and S. Aksela, *J. Elec. Spect. Rel. Phen.* **23**, 119 (1981).
32. R. Nyholm, N. Martensson, A. Lebugle and U. Axelsson, *J. Phys. F: Met. Phys.* **11**, 727 (1981).
33. S.L. Sorensen, S.J. Schaphorst, S.B. Whitfield, B. Crasemann and R. Carr, *Phys. Rev.* **A44**, 350 (1991).
34. M. Ohno, *J. Phys.* **C17**, 1437 (1984).
35. M. Ohno, *J. Phys. B: At. Mol. Phys.* **17**, 195 (1984).
36. P. Glans, R.E. LaVilla, M. Ohno, S. Svensson, G. Bray, N. Wassdahl and J. Nordgren, *Phys. Rev.* **A47**, 1539 (1993).
37. S. Fritzsche, B. Fricke and W.-D. Sepp, *Phys. Rev.* **A45**, 1465 (1992).

38. M. Ohno and R.E. Lavilla, *Phys. Rev.* **A45**, 4713 (1992).
39. J.N. Cooper, *Phys. Rev.* **61**, 234 (1942).
40. J.H. Williams, *Phys. Rev.* **37**, 1431 (1931).
41. J.N. Cooper, *Phys. Rev.* **65**, 155 (1944).
42. J. Hoszowska, J.-Cl. Dousse and Ch. Rheme, *Phys. Rev.* **A50**, 123 (1994).
43. F.K. Richtmeyer, S.W. Barnes and E. Ramberg, *Phys. Rev.* **46**, 843 (1934).
44. U. Werner and W. Jitschin, *Phys. Rev.* **A38**, 4009 (1988).
45. U. Arp, G. Materlik, M. Richter and B. Sonntag, *J. Phys. B. At. Mol. Opt Phys.* **23**, L811 (1990);
    O. Keski-Rahkonen, G. Materlik, B. Sonntag and J. Tulkki, *J. Phys.* **B17**, L121 (1984).
46. M. Ohno, *Phys. Rev.* **B29**, 3127 (1984).
47. M. Ohno, P. Putila-Mantyla and G. Graeffe, *J. Phys. B: At. Mol. Phys.* **17**, 1747 (1984).
48. P. Amorim, L. Salgueiro, F. Parente and J.G. Ferreira, *J. Phys. B: At. Mol. Phys.* **21**, 3851 (1988).
49. J.J. Merrill and J.W.M. Dumond, *Annals. Phys.* **14**, 166 (1961).
50. S.I. Salem and P.L.Lee, *Phys. Rev.* **A10**, 2033 (1974).
51. W. Meierkord, T. Blumke, M. Brussermann, J. Hofste, H.-U. Menzebach, J.F. Pennings, Z. Stachura, W. Vollmer, J. Wigger and B. Cleff, *Z. Phys.* **D18**, 75 (1991).
52. L. Kövér, I. Cserny, V. Brabec, M. Fiser, O. Dragoun and J. Novak, *Phys. Rev.* **B42**, 643 (1990);
    L. Kövér and I. Cserny, *J. Elec. Spect. Rel. Phen.* **63**, 31 (1993).
53. G.C. Nelson and B.G. Saunders, *Jour. de Physique, Colloque C4* (**10**), 97 (1971).
54. N. Martensson and R. Nyholm, *Phys. Rev.* **B24**, 7121 (1981).
55. L.I. Yin, I. Adler, T. Tsang, M.H. Chen, D.A. Ringers and B. Crasemann, *Phys. Rev.* **A9**, 1070 (1974).
56. S. Hufner and G.K. Wertheim, *Phys. Rev.* **B11**, 678 (1976).
57. L.G. Parratt, *Phys. Rev.* **54**, 99 (1938).
58. M. Ohno, A. Laakonen, A. Vuoristo and G. Graeffe, *Phys. Scripta* **34**, 146 (1986).
59. A. Laakonen, A. Vuoristo and G. Graeffe, *Proc. 19th Ann. Conf. Finnish Physical Society* **3:4** (1985).
60. A. Laakonen and G. Graeffe, *Jour. de Physique 48*, **C9-405** (1987).
61. O. Keski-Rahkonen and M.O. Krause, *Phys. Rev.* **A15**, 959 (1977).
62. A. Berndtsson, R. Nyholm, N. Martensson, R. Nilsson and J. Hedme, *Phys. Stat. Sol. (b)* **93**, K103 (1979).
63. R. Nyholm and N. Martensson, *Phys. Rev.* **B36**, 20 (1987).
64. M. Ohno and G. Wendin, *Phys. Rev.* **A31**, 2318 (1985).
65. M.O. Krause, *Chem. Phys. Lett.* **10**, 65 (1971).
66. S. Svensson, N. Martensson, E. Basilier, P.A. Malmqvist, U. Gelius and K. Siegbahn, *Phys. Scripta* **14**, 141 (1976).
67. R.E. Lavilla, *Phys. Rev.* **A17**, 1018 (1978).
68. M. Ohno, *Phys. Scripta* **21**, 589 (1979).
69. P. Citrin, G.K. Wertheim and Y. Baer, *Phys. Rev. Lett.* **41**, 1425 (1978).
70. Handbook of X-ray Photoelectron Spectroscopy, ed. J. Chastain, Perkin-Elmer Corp. 1992

Giant Resonances in Photon Emission Spectra of Atoms, Clusters, and Solids

L.G.Gerchikov[1], A.V.Korol[2], A.G.Lyalin[3] and A.V.Solov'yov[4]

[1] *St Petersburg Technical University, Politechnicheskaya 29, St Petersburg 195251, Russia.*
[2] *Department of Physics, Maritime Technical University, Leninskii prospect 101, St Petersburg 198262, Russia.*
[3] *Institute of Physics, St Petersburg State University, Ulianovskaya 1, St Petersburg, Petrodvoretz 198904, Russia.*
[4] *A.F.Ioffe Physical-Technical Institute, RAS, St Petersburg 194021, Russia.*

Abstract. We consider a photon emission in collisions of electrons with atoms and clusters under the conditions when the radiation of atomic electrons gives significant or even dominating contribution to the total spectrum. This occurs, for example, when the frequency of the emitted photon is comparable with the energy of giant collective or plasmon resonance in an atom or cluster. We consider the manifestation of the effect for atomic and solid targets demonstrating the significant role of the photon self-absorption and electron energy loss in a solid target. We present results of recent calculations of the cross sections of the process for a number of targets. Theoretical results are compared with the available experimental data.

INTRODUCTION

In this paper we review the recent developments achieved during the last years in the theory of polarizational bremsstrahlung (BrS) of atoms and clusters.

Consider two mechanisms of the photon radiation during the collision. The first mechanism, called in the present paper as ordinary BrS, is the photon emission by a charged projectile accelerated in a static field of a target. This process is a well known quantum mechanical process (see, for example, [1]). A number of review papers have been devoted to various aspects of the ordinary BrS in collisions of charged particles with atoms and nuclei (see, for example, [2] and references therein).

The second mechanism, known as the polarizational BrS, accounts for the photon emission due to the virtual excitations (polarization) of target electrons during the collision. This kind of radiation was recognized relatively recently [3–7]. Results obtained in this field before 1990 have been reviewed in [8–10].

In the present paper we are mainly concentrated on the role of the dynamic electron structure of a target on the spectrum of photon emission in the region of the giant dipole resonances. This range of frequencies is of particular interest since the polarizability of a target in this region is very large and, consequently, almost all the radiation is formed via the polarizational mechanism. We consider the polarization radiation effect for isolated atoms, solid targets and clusters. In the case of a solid target we demonstrated that the influence of the self-absorption of the emitted photons and the projectile electron energy loss on the profile of the giant resonance is very important.

Atomic system of units is used.

BREMSSTRAHLUNG OF ELECTRONS ON ISOLATED ATOMS AND IN DENSE MEDIA

Consider the BrS of a non-relativistic electron on an isolated many-electron atom. The differential cross section, which characterizes the spectral distribution of radiation, is given by

$$\omega \frac{d\sigma}{d\omega} = \int d\Omega_\gamma \frac{\omega d^2\sigma}{d\omega d\Omega_\gamma} = \frac{1}{(2\pi)^4} \frac{\omega^4}{c^3} \frac{p_2}{p_1} \int d\Omega_{\mathbf{p}_2} \int d\Omega_\gamma \sum_\lambda |f^{\text{tot}}|^2 \quad (1)$$

Here $c \approx 137$ is the velocity of light. The integral is carried out over the solid angle of the vector of final momentum $\mathbf{p}_2$, $d\Omega_{\mathbf{p}_2}$, and over the directions of photon emission, $d\Omega_\gamma$. The sum is taken over the photon polarizations λ. The quantity $\omega \, d^2\sigma/d\omega d\Omega_\gamma$ is the two-fold differential BrS cross section, which characterizes the spectral and the angular distribution of the radiation. The quantity f^{tot} is the total amplitude of the BrS process equal to the sum of the ordinary, f^{ord}, and the polarizational, f^{pol}, BrS amplitudes $f^{\text{tot}} = f^{\text{ord}} + f^{\text{pol}}$.

An adequate description of the BrS emission formed in the intermediate energy electron-atom collision one obtains using the first order distorted partial-wave approximation (DPWA) [11]. Within the frame of the DPWA we consider the BrS process in the lowest order of the non-relativistic perturbation theory in the electron—dipole-photon interaction and in the Coulomb interaction, $\hat{V} = |\mathbf{r} - \mathbf{r}_a|^{-1}$, between the projectile and the atomic electrons which leads to a virtual atomic excitation.

The amplitudes f^{ord} and f^{pol} are written as follows:

$$f^{\text{ord}} = <\mathbf{p}_2|\mathbf{er}|\mathbf{p}_1> \quad (2)$$

$$f^{\text{pol}} = \sum_n \left\{ \frac{<0|\mathbf{eD}|n><\mathbf{p}_2\,n|\hat{V}|\mathbf{p}_1\,0>}{\omega - \omega_{n0} + i0} - \frac{<\mathbf{p}_2\,0|\hat{V}|\mathbf{p}_1\,n><n|\mathbf{eD}|0>}{\omega + \omega_{n0}} \right\} \quad (3)$$

Here $|p_1>$ and $|p_2>$ are the wave functions of the incident and the scattered electrons with asymptotic momenta $\mathbf{p}_1$ and $\mathbf{p}_2$. Vector $\mathbf{D}$ in (3) is the operator of the dipole momentum of the atomic electrons, $\omega_{n0} = E_n - E_0$ is the energy of the atom's transition from the ground state 0 to the excited state n (including excitations into continuum), ω and $\mathbf{e}$ are the photon's energy and the polarization vector.

Figure 1 (left) presents results of calculations of the BrS spectra for 250 eV electron on Ba, La and Eu in the vicinity of ionization potentials of the 4d-subshells (marked with vertical lines in the figure). The calculations have been done according to the algorithm described earlier in [11].

The choice of the targets and the regions of photon frequencies were determined by the fact [4] that a wide maximum in the emission spectrum can appear in the region of ω close to the ionization potentials of many-electron atomic subshells. It was demonstrated [8] that the position of the BrS maximum is closely related to the maximum in the photoionization spectrum, $\sigma_\gamma(\omega)$.

It is known from studying the photoionization of Ba, La and Eu both experimentally [12] and theoretically [13], that in vicinities of the 4d-thresholds there are powerful maxima in $\sigma_\gamma(\omega)$.

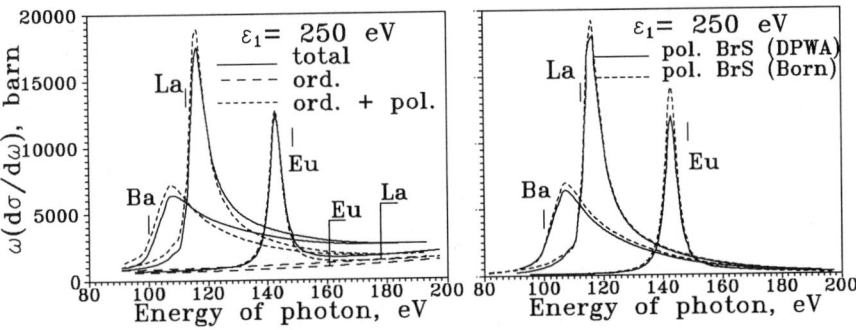

FIGURE 1. Left picture: results of calculation of the BrS spectra for 250 eV electron on Ba, La and Eu atoms in the vicinity of ionization potentials of the 4d-subshells (marked with vertical lines). Interference between ordinary BrS (long-dashed curves) and polarizational BrS represented by a sum of ordinary and polarizational contributions (short-dashed curves) alters shape of the resonance in the total cross section (solid curves). Right picture: dependencies of $(\omega d\sigma/d\omega)^{pol}$ calculated within the frame of DPWA (solid curves) and the plane-wave Born approximation (dashed curves) for Ba, La and Eu atoms.

These maxima manifest themselves as well in the spectrum of the total BrS (solid curves in figure 1 (left)). In the maximal points the magnitudes of $(\omega d\sigma/d\omega)^{tot}$ greatly exceed the magnitudes of the ordinary BrS, - long-dashed curves in figure 1 (left) (we did not plot the dependence $(\omega d\sigma/d\omega)^{ord}$ for Ba since it almost coincides with that for La).

The solid curves in figure 1 (left) were calculated using various methods: (i) for Ba and La we used the RPAE with relaxation scheme (GRPAE) [14], (ii) for Eu the spin-polarized RPAE (SP RPAE) method [15] was applied.

As it was noted in the cited papers these approximations take into account the major part of the correlation effects and, being applied to the description of the photoionization process, provide a satisfactory agreement with the experimental data. However, the calculation of the BrS spectrum undertaken within the frame of the DPWA and with the simultaneous accounting for the correlation effects in the RPAE (or/and its variations - GRPAE, SP RPAE) are rather laborious and computer time consuming.

It is thus desirable to construct an approximate simplified method which will allow to compute the BrS spectrum more efficiently but without significant loss of accuracy.

To start with we note, firstly, that in the vicinity of the maximum almost all the BrS radiation is formed through the polarizational mechanism. Secondly, the polarizational part of the BrS spectrum is more sensitive to the model of the atomic electron-photon interaction rather than to the choice of the projectile's wave function (see e.g. [8,11]) . To illustrate this in figure 1 (right) the dependencies $(\omega d\sigma/d\omega)^{pol}$, calculated within the frame of the DPWA (solid curves) and the plane-wave Born approximation (dashed lines), are compared.

It is seen that for the incident energies as low as 250 eV the Born approximation gives almost the same result as the DPWA. The reason for this coincidence lies in the fact that, in contra to the process of ordinary BrS, the polarizational radiation is formed mainly at large distances, $r \sim p_1/\omega$, between a projectile and an atom [16,17], where the distorting influence of the atomic potential on the projectile's movement is comparatively small. Hence, to calculate the polarizational component of the BrS spectrum one may use the Born approximation which results in the formula [4]:

$$\left(\omega \frac{d\sigma}{d\omega}\right)^{pol}_B = \frac{16}{3} \frac{\omega^4}{c^3 p_1^2} \int_{p_1-p_2}^{p_1+p_2} \frac{dq}{q} |\alpha(\omega,q)|^2 \qquad (4)$$

The only characteristics in (4), dependent on the internal dynamics of the target, is the generalized polarizability $\alpha(\omega, q)$. Recently [18] we introduced a simple approximate method for the calculation of $\alpha(\omega, q)$ and, consequently, of $(\omega d\sigma/d\omega)^{pol}_B$. This method allows to avoid rather complicated direct numerical computations of the many-electron correlation effects.

To obtain the magnitude of the generalized polarizability with accounting for the many-electron correlation effects, we first write $\alpha(\omega, q)$ as:

$$\alpha(\omega, q) = \alpha(\omega) \cdot G(\omega, q) \qquad (5)$$

This equality is just a definition of a new function $G(\omega, q)$ equal to the ratio $\alpha(\omega,q)/\alpha(\omega)$ of the *exact* generalized and dipole polarizabilities.

Now let us assume that all the information about many-electron correlation effects is contained in the dipole polarizability $\alpha(\omega)$, while the factor $G(\omega, q)$ is not that sensitive to the correlation effects and can be calculated in the *Hartree-Fock* approximation. Hence, instead of (5) we obtain the following approximate formula

$$\alpha(\omega, q) \approx \alpha(\omega) \, \frac{\alpha^{\mathrm{HF}}(\omega, q)}{\alpha^{\mathrm{HF}}(\omega)} \equiv \alpha(\omega) \cdot G^{\mathrm{HF}}(\omega, q) \qquad (6)$$

This relation is the essential point of the method. Provided the approximate equality in (6) is true, it allows to avoid the complicated direct calculations of the generalized polarizability by means of the many-body theory and reduces the problem to a simpler one: the calculation of the factor $G^{\mathrm{HF}}(\omega, q)$ in the Hartree-Fock approximation.

Substituting (6) into (4) one gets:

$$\left(\omega \frac{d\sigma}{d\omega} \right)^{pol}_{B} = \frac{16}{3} \frac{\omega^4}{c^3 p_1^2} |\alpha(\omega)|^2 \int_{p_1 - p_2}^{p_1 + p_2} \frac{dq}{q} \, |G^{\mathrm{HF}}(\omega, q)|^2 \qquad (7)$$

To check the validity of the proposed method we calculated the polarizational part of the spectrum for $0.25 - 10$ keV electron on Ba, La and Eu using the exact Born formula (4) and the approximate one (7). In all considered cases the results are close. Figure 2 illustrates this fact for the collision of 250 eV electron with Ba. The discrepancy between the solid curve, corresponding to (4), and the short-dashed one, representing (7), is almost negligible. For the sake of comparison we present also the polarizational cross section calculated via (4) with $\alpha(\omega, q)$ obtained in the independent-particle model with no correlations included (a long-dashed curve).

In figure 2 the short-dashed curve was obtained by using the relation (6) where the factor $\alpha(\omega, q)$ was calculated directly with the GRPAE corrections taken into account.

Since the total amplitude of the process contains two terms then the total BrS cross section is given by a sum of the ordinary and the polarizational parts and the interference term as well. The plane wave first Born approximation, being suitable for the accurate description of the polarizational part of the BrS cross section (as demonstrated above), provides poor results for the ordinary BrS and the interference parts of the total BrS cross section of an intermediate energy electron.

It looks attractive to modify the above formalism in a way that the approximate method (6) of the calculation of the generalized polarizability could be applied to the computation of the total BrS spectrum rather than its polarizational part only. One may do it as follows.

Supposing the incident energy ε_1 is large enough compared with the ionization potential of those atomic subshells which, for a given photon frequency,

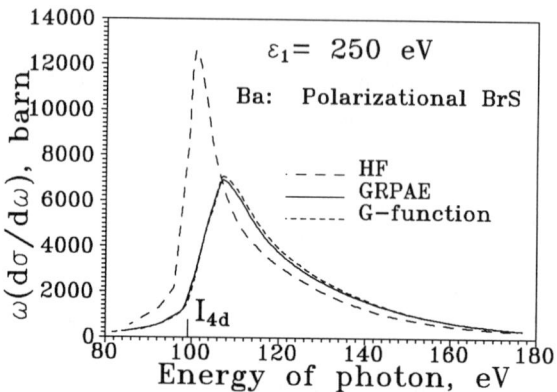

FIGURE 2. Polarizational BrS cross section in collision of 250 eV electron with Ba calculated using the exact Born formula (5) (solid curve) and the approximate one (8) (short-dashed curve). For the sake of comparison we present also the polarizational cross section calculated with $\alpha(\omega, q)$ obtained in the independent-particle model with no correlations included (long dashed curve).

give the main contribution to the sum over the excited states in (3), it is possible to neglect the exchange between the projectile and the atomic electrons. Then, one may express the amplitude of the polarizational BrS through a generalized atomic polarizability [18] as follows:

$$f^{pol} = -\frac{i}{2\pi^2} \int d\mathbf{Q} \, \frac{(e\mathbf{Q})}{Q^2} <\mathbf{p}_2|e^{-i\mathbf{Q}\mathbf{r}}|\mathbf{p}_1> \cdot \alpha(\omega, Q) \qquad (8)$$

The advantage this formula, derived in the DPWA approximation, is that the only characteristics in (8), dependent on the internal dynamics of the target, is the generalized polarizability. It allows to apply the approximate method (6) to the calculation of $f^{tot} = f^{ord} + f^{pol}$ and, consequently, to the calculation of the total BrS spectrum.

Results of the DPWA calculations of the BrS spectrum formed in the electron–Eu collision are presented in figure 3. The curves show the dependencies of the total BrS cross section on the incident electron energy ε_1 calculated for three photon energies, which are (i) $\omega_1 = 142.9$ eV, - the maximum point of the spectrum (see figures 1), (ii) $\omega_2 = 145$ eV, - the ω-point on the right slope of the peak, (iii) $\omega_3 = 163$ eV, - i.e. the ω-point far beyond the maximum.

All the curves were calculated using eqs. (1), (2) and (8). The generalized polarizability of Eu was calculated (i) directly within the SP RPAE (solid lines), and (ii) using the approximation (6), - dashed lines.

Comparing the sets of the solid and the dashed curves we may state that the approximation (6) combined with the DPWA formula (8) could be effectively used for the calculations of the total BrS spectrum in the whole ω-interval.

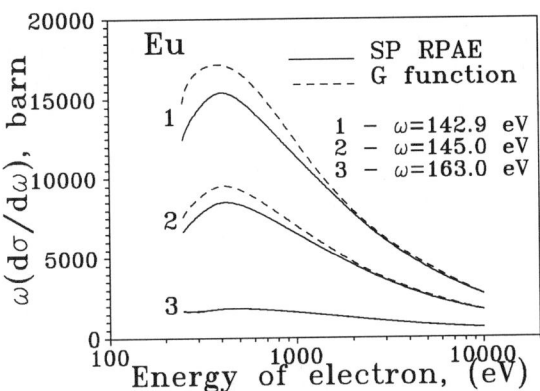

FIGURE 3. BrS spectra. Total BrS cross section in electron-Eu collision versus the incident electron energy ε_1 calculated for three photon energies as indicated. The generalized polarizability of Eu was obtained in the SP RPAE (solid lines) and using the approximation (7), - dashed lines.

When using the approximate formula (6) to obtain the generalized polarizability $\alpha(\omega, q)$ it is possible to avoid completely the direct calculation of the factor $\alpha(\omega)$ within the framework of some many-body scheme. To do this one may use the relationship between $\operatorname{Im}\alpha(\omega)$ and the photoabsorption cross section, $\sigma_\gamma(\omega)$:

$$\operatorname{Im}\alpha(\omega) = \frac{c}{4\pi\omega}\sigma_\gamma(\omega) \qquad (9)$$

Provided the dependence $\sigma_\gamma(\omega)$ is known in a wide ω-region one can restore the real part of $\alpha(\omega)$ via the dispersion relation. Such a way of obtaining $\alpha(\omega)$ and, consequently, of $\alpha(\omega, q)$, is especially useful when the direct calculations can hardly be performed, as for example, when the BrS process is investigated in a dense media (see below) rather that in a pure "one electron — one atom" collision.

In figure 4 (left) we present results of calculation of the two-fold differential cross section $d^2\sigma/d\omega d\Omega_\gamma$ for 0.5 and 4 keV electrons on *isolated* La atom in the vicinity of the 4d-subshell ionization potential. The calculations have been done within the DPWA framework using the algorithm described in [18]. We have used the experimental data for the photo-ionization spectrum [19,20] to calculate the dynamic atomic polarizability $\alpha(\omega)$.

It is clearly seen from the figure that for all energies of the projectile the magnitude of the total BrS exceeds greatly that of the ordinary BrS in the vicinity of maximum.

The BrS cross section of an electron scattered in an infinitely *thick target* is related to the BrS cross section (1) on an isolated atom as follows:

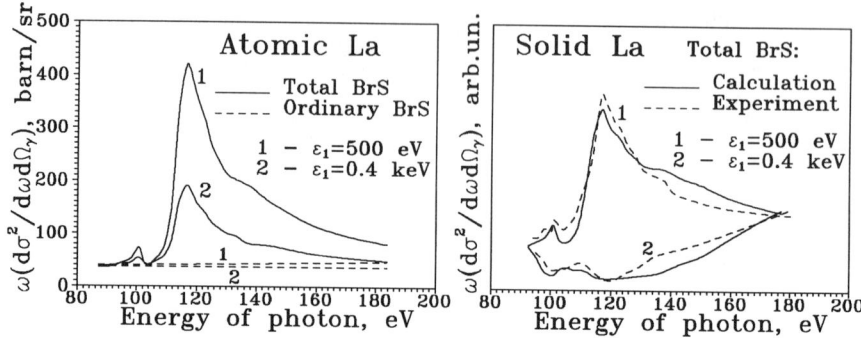

FIGURE 4. Left picture: two-fold differential cross section for 0.5 keV and 4 keV electrons on *isolated La* atom calculated in the vicinity of ionization potentials of the 4d-subshells (marked with vertical lines in the figure). Angle of the photon emission $\Theta = 45°$. The calculations have been done using "G-function" approach (eq. (9)) using the tabulated data on Im $\alpha(\omega)$ from [20]. Right picture: same as for left figure but for *metallic La* target. Dashed lines present the experimentally measured spectra [22]. All curves are normalized in the point $\omega = 176$ eV.

$$\frac{\omega\, d^2\tilde{\sigma}}{d\omega d\Omega_\gamma} = N_{at} \int_0^\infty \int_\omega^{\varepsilon_1} \frac{\omega\, d^2\sigma}{d\omega d\Omega_\gamma} \exp\left(-\frac{N_{at}\sigma_\gamma(\omega)x}{\cos\Theta}\right) f(x,\varepsilon)\, dx d\varepsilon \qquad (10)$$

Here $f(x,\varepsilon)$ is the distribution function of electrons scattered in a solid target. It depends on the electron energy and the penetration distance x. To describe the energy distribution of electrons scattered in a thick target we used a model function $f(x,\varepsilon)$, introduced in [21]. $\sigma_\gamma(\omega)$ - is the atomic photo-absorption cross section, N_{at} - is the density of atoms in the target; Θ - is the photon emission angle with respect to the surface normal. Thus the factor $\exp\left(-\mu(\omega)x/\cos\Theta\right)$ takes into account the photon attenuation.

In (10) it is assumed that the cross section $d^2\tilde{\sigma}/d\omega d\Omega_\gamma$ represents by itself the sum of the BrS cross sections $d^2\sigma/d\omega d\Omega_\gamma$ on individual atoms. Such an approximation is valid provided the size of the spatial region where the radiation is formed (a coherence length) is less than the mean interatomic distance, and thus the process of the photon emission during the projectile's collision with any individual atom occurs independently. We also assume that the target is an amorphous solid.

Results of calculations of the BrS cross section (10) for metallic La are presented in figure 4 (right) (solid curves).

Comparing theoretical curves in figures 4 we note that for low incident electron energy, $\varepsilon_1 = 0.5$ keV, the BrS spectra on the isolated and on the metallic La targets are much alike in their shapes, exhibiting intensive maxima in the vicinity of the 4d-threshold. With the increase of ε_1 up to 4 keV the shape of the solid La BrS curve changes drastically: the maximum transfers into the minimum in the same ω-region. Such an effect is due to the process of radiation absorption in solid target.

The dashed lines in figure 4 (right) correspond to the experimentally measured [22] emission spectra of metallic La. In the cited papers the experimental data are given in arbitrary units, and the 4 keV spectrum was plotted level with the 0.5 keV curve in some ω-point on the right wing of the spectra. Therefore, to compare our calculated cross sections with the experiment we matched the 0.5 and 4 keV dependencies $d^2\sigma^{SB}/d\omega d\Omega_\gamma$ in the point $\omega = 176$ eV.

With the accuracy of 10% there is an agreement between the calculated and the measured electron BrS spectra on metallic La target. Some discrepancy may be attributed to an error of the absolute magnitudes of the photoabsorption cross sections [19,20], as well as to the mistake introduced in the calculations by the use of the model function of scattered electron energy distribution.

BREMSSTRAHLUNG OF ELECTRONS ON METALS, CLUSTERS AND FULLERENES

Fullerenes and metallic clusters are the other types of highly polarizable systems. This is born out by the extremely large madnitude of the giant resonance photoionization cross section for clusters as compared with atoms [23,24]. Electrons of a suitable energy can thus excite a giant dipole oscillation, which will radiate, and this radiation, due to the dynamic polarizability of the target, is superimposed on the usual BrS background. The specific feature of this process for clusters is that the dominance of the polarizational BrS in the giant resonance region is much stronger than that for atoms.

There are two possible kinds of giant resonances, which may be excited in clusters (see e.g. [23,25]). Firstly, one may have giant resonances of the electrons localized on individual atoms. Secondly, there are collective oscillations of delocalized electrons belonging to the whole metal cluster and these oscillations give rise to giant resonances of a new kind (as compared with an atom case) known also as plasmon resonances [24,26].

The energy of the giant resonance of the first type in a cluster is comparable with the corresponding giant resonance energy of an isolated atom, which is about $100 eV$ (see the above section), while the energy of the plasmon resonance is much lower, typically being in the range 2–5 eV for metal clusters [32] and of about 19 eV for fullerenes [24]. The width of the second type resonances is usually much smaller than that of the first type (especially for metal clusters) and hence these resonances are less damped out than the giant resonances in atoms.

Further, we discuss only the manifestations of the collective motion of the delocalised electrons in the spectra of photons emitted in collisions involving clusters.

The significance of the dynamic polarization in the radiative processes with metallic clusters has been recently stressed in [27]. The application of this idea to the radiative electron capture and the BrS has been considered in these papers. Photon emission spectrum generated in electron-fullerene collision was described in the main logarithmic approach in [28] and more accurately in [29]. Recently, experiments on the collisions of electrons with metal clusters and fullerenes were performed [30], but giant resonance photon emission was apparently not searched for, although the cross section for this process in is much larger then for atoms as demonstrated in the present paper.

The description of metal clusters within the framework of the jellium model [31] is quite similar to the description of atoms or nuclei. Indeed, the equations of the jellium model are analogous to the Hartree-Fock equations for atoms in which the Coulomb potential of the nucleus is replaced by the mean field potential of the ionic background. This means that one can apply with minor modifications the approximations and methods known for atom to the description of the polarizational BrS process in collisions of electrons with clusters [27].

The correct treatment of the polarizability, $\alpha(\omega; q)$ is rather important for the correct description of the polarizational BrS cross sections. As has been demonstrated in [29] calculation of the polarizabilities for metal clusters and fullerenes in the vicinity of plasmon resonances differs from that for atoms. The specific feature arises due to the small parameter, $a/R \ll 1$, where R is the radius of the cluster and a is the atomic size. Parameter a characterizes the depth of the surface layer in a cluster or the thickness of the fullerene shell, where the plasmon excitation occurs.

When calculating generalized polarizability, $\alpha(\omega, q)$, in the vicinity of plasmon resonance, the main contribution to the matrix elements arises from a thin layer of the width a near the surface, where the oscillation of the charge density takes place . This is rather natural assumption for a fullerene due to its spherical like surface shell structure. However, the statement is correct for metal clusters too [29], [32]. Using this fact, one can express the generalized dynamic polarizability via the dipole dynamic polarizability as [29]

$$\alpha(q,\omega) = \frac{3j_1(qR)}{qR}\alpha(\omega) \qquad (11)$$

This equation is valid if the transferred momentum is smaller compared to the characteristic atomic one, $q \ll 1/a$. Here $j_1(x)$ is the spherical Bessel function. Equation (11) allows to describe analytically the momentum dependence of the generalized dipole polarizability and calculate the cross section of the process.

The cluster dipole dynamic polarizability can be written in the resonance plasmon pole approximation [32]

$$\alpha(\omega) = \frac{R^3 \omega_p^2}{\omega_p^2 - \omega^2 - i\omega\Gamma} \tag{12}$$

Here R is the cluster radius, ω_p is the plasmon resonance frequency. For metal clusters, $R = r_s N^{1/3}$, where r_s is the core radius adjusted to the bulk equilibrium condition ($r_s = 3.96$ for sodium), N is the number of atoms in a cluster. The resonance width, Γ, in the denominator of Eq.(12) is connected with plasmon dumping. For example, $\Gamma \approx 0.4 eV$ for the sodium Na_{40} cluster. In metal clusters according to the classical Mie theory the dipole plasmon resonance frequency $\omega_p = \sqrt{N_e/R^3}$, where N_e is the number of delocalised electrons. For a sodium cluster, this equation gives $\omega_p = 3.5 eV$ which is quite close to the experimental value $\omega_p = 3.22 eV$ [32].

In the fullerene case $\omega_p = \sqrt{2N_e/3R^3}$ [33]. For C_{60} ($N_e = 240$ and $R = 6.8$), it gives $\omega_p = 18.5\ eV$ which is quite close to the experimental value $\omega_p = 19$ [34]. The imaginary part of the polarizability is proportional to the photoabsorption cross section. The experimental data on the photoabsorption by fullerenes [34] can be reproduced, if one inputs in (12) the frequency $\omega_p = 19$ eV and the plasmon resonance width $\Gamma = 11\ eV$. The radius of the fullerene and the width of the plasmon resonance can, in principle, be determined pure theoretically (see e.g. [32]). However, we shall not further discuss this issue in our paper, considering Γ and R for fullerenes as the given constants.

As demonstrated for atoms in the previous section [18], the cross section for polarizational BrS near giant resonances depends more strongly on many-electron correlations in the target rather than on the choice of the wavefunction for the projectile electron. Therefore the correct treatment of the dynamic response of a cluster on an electric field of the projectile performed in the previous subsection in a certain sense is more essential for the cross section calculation then choice of the projectile wave function.

The projectile electron effectively excites a plasmon, when the time of the collision with the cluster is close to the plasmon oscillation period. This occurs at the collision velocity $v \sim \omega_p R$, being much larger then the characteristic velocities of electrons in a cluster. Therefore describing photon emission in the vicinity of the giant plasmon resonance, we can use the Born approximation. The cross section of the polarizational BrS process calculated in the Born approximation is given by [29]

$$\frac{d\sigma}{d\omega} = \frac{16 e^2 \omega^3}{3c^3 v^2} |\alpha(\omega)|^2 \left(S(q_{min} R) - S(q_{max} R) \right), \tag{13}$$

$$S(x) = \frac{1}{8x^6} \left(6 + 9x^2 - (12x + 2x^3 - 4x^5) \sin(2x) - \right.$$

$$(6 - 3x^2 + 2x^4)\cos(2x) - 8x^6 Ci(2x))$$

Here $q_{min} = v(1 - \sqrt{1 - 2\omega/v^2})$, $q_{max} = v(1 + \sqrt{1 - 2\omega/v^2})$ are the lower and upper limits of the transferred momentum. $Ci(x)$ is the cosine integral function. For actual projectile electron velocities $v \sim \omega_p R$ and $\omega \sim \omega_p$, the energy of the projectile electron at is $\omega_p R^2/2$ times larger than ω and therefore $q_{min} \approx \omega/v$ and $q_{max} \approx 2v$. Note that equation (13) is applicable, if $v \gg \omega a$. In the region of small $x = \omega R/v \ll 1$, one derives $S(x) \approx ln(C_1/x)$, where $C_1 = e^{7/4-\gamma}/2 \approx 1.61549$, $\gamma = 0.577216$ is the Euler constant.

In this case large impact parameters are kinematically allowed in the collision and therefore the process mainly takes place at large distances. Therefore the cross section (13) becomes similar to the cross section of the polarizational BrS of an electron on an atom obtained in the main logarithmic approximation [8]. In the opposite case $x \gg 1$, the asymptotic of $S(x)$ contains oscillating terms: $S(x) \approx (9/8x^4)(1 - (2/x)\sin(2x))$. Origin of these terms in the cross section is physically clear. Indeed, large $x \gg 1$ correspond to the situation, when the wave length, π/q_{min}^{-1}, becomes less than the size of the cluster. Since the cluster is a system possessing an edge, diffraction of an electron at an edge occurs. This diffraction phenomenon described first by [27] manifests itself in the oscillatory behaviour of the cross section (13) as a function of parameter $x = \omega R/v$. Thus the oscillatory behaviour takes place in a wide range of frequencies, $v/R \ll \omega \ll v/a$. It is easy to estimate that the magnitude of the cross section (13) at these frequencies exceeds the magnitude of the background radiation cross section.

Let us now compare the contributions of the ordinary and polarizational mechanisms to the spectrum of photon emission. Estimation of the ordinary BrS cross section is given by $d\sigma/d\omega \sim 16e^4 N_e/3c^3 v^2 \omega$. This estimate states a simple fact that the ordinary BrS background is formed as incoherent sum of the ordinary BrS contributions generated in collisions of the projectile with the atoms in the target cluster. Such an estimate of the ordinary BrS background including radiative processes accompanying ionization of the cluster is applicable both for metal clusters [27] and fullerenes [28,29].

Substituting (11) and (12) to (13) and comparing it with the estimate of the ordinary BrS background at the resonance frequency, $\omega = \omega_p$, we derive the ratio, R, of the polarizational to the ordinary BrS for metal clusters and fullerenes [29],

$$A(\omega_p) \sim N_e \left(\frac{\omega_p}{\Gamma}\right)^2 S(\frac{\omega_p R}{v}).$$

The ratio $\omega_p/\Gamma \sim 10$ is rather large for metal clusters and it practically does not depend on the cluster size providing the constant width Γ. For fullerene C_{60}, $\omega_p/\Gamma \sim 2$ and it tends to decrease for larger fullerene sizes, since the resonance frequency in this case is not constant, $\omega_p^2 \sim 1/N_e^{1/2}$. These estimates show that the polarizational BrS dominates by orders of magnitude over the ordinary BrS background in the vicinity of the plasmon resonance in both metal cluster and fullerene cases. We made this conclusion for the neutral clusters, however it is valid also for cluster ions as well [27]. The dominating role of the polarizational BrS was established for fullerenes in [28] and for metal clusters in [27].

Origination of all factors in $A(\omega_p)$ is quite clear. Factor N_e characterizes the coherence of radiation of the cluster electrons in the polarizational BrS process. Factor $(\omega_p/\Gamma)^2$ arises due to the resonance nature of the polarizational BrS in the vicinity of the plasmon excitation. Function $S(\omega_p R/v)$ describes the kinematics of the collision. It depends on the scaling parameter $\omega_p R/v$. For fullerenes, additional coefficient 4/9 arises in $A(\omega_p)$ as a result of the topological difference between metal clusters and fullerene molecules.

Now let us illustrate the behaviour of the cross sections written above for the metal clusters Na_{40} and the fullerene C_{60}.

The polarizational BrS spectra for electron-C_{60} collision calculated using (13) are plotted in figure 5 (left) [29]. Dashed, solid and dotted curves correspond to the projectile electron velocities 2, 3.5 and 6. The polarizational mechanism manifests itself in the resonance behaviour of the spectrum at $\omega \approx \omega_p$. The shape of the distribution is determined by two factors: first, by the dynamic polarizability of the fullerene which has the resonance character near $\omega \approx \omega_p$, and also by the parameter $\omega R/v$, characterising the distance at which the dynamic polarization of fullerene occurs. This dependence is described by the factor S in (13). The factor S logarithmically diverges if characteristic collision distance is much larger than the fullerene's radius $v/\omega R \gg 1$ and very rapidly decreases, if $v/\omega R \lesssim 1$. It means that this factor cuts off the photon spectrum at $\omega \sim v/R$. Obviously, intensity of the BrS in the resonance region $\omega \approx \omega_p$ is highest, if $v \approx \omega_p R$ (solid curve in figure 5 (left)). At smaller collision velocities, $v < \omega_p R$, the position of the maximum in the photon emission spectrum shifts towards the low value $\omega \sim v/R$ (dashed curve in figure 5 (left)). At higher velocities $v > \omega_p R$ the cutoff of the photon spectrum lies at higher frequencies than the polarizability maximum, which rises the right shoulder of resonance maximum (dotted curve in figure 5 (left)). Note that dependence of the position of the maximum of the giant resonance in the photon emission spectrum on the collision velocity has been reported experimentally for the Xe atomic target [35].

Results of similar calculations for the cluster Na_{40} are presented in figures 5 (right) [29]. The emission spectrum is shown in figure 5 (right) for the projectile electron velocities $v = 0.7$ (dashed curve), 1.5 (solid curve) and 3 (dotted curve). It is clearly visible that the emission spectrum has a resonance peak narrower than in the case of the fullerene target. This feature of the BrS spectrum is connected with a smallness of the plasmon resonance width in a metal cluster, $\Gamma/\omega_p \approx 0.13$ for Na_{40}, whereas this ratio is about 0.6 for the fullerene C_{60}. The plasmon resonance frequency in clusters is lower than the ionization

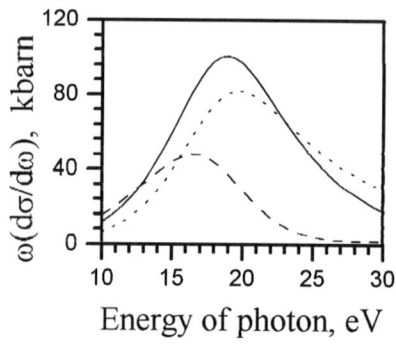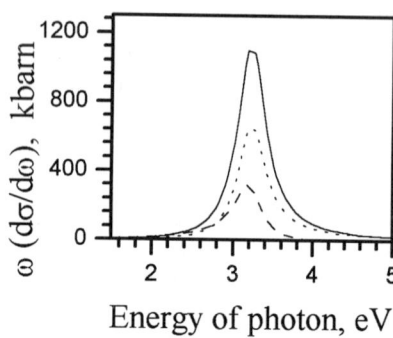

FIGURE 5. Left picture: Polarizational BrS cross section for electron-fullerene C_{60} collision versus the photon energy and calculated for various projectile electron velocities: $v = 3.5$ (solid line), $v = 2$ (dashed line), $v = 6$ (dotted line. Right picture: same as for the left figure but for the target cluster Na_{40} and the projectile electron velocities $v = 1.5$ (solid line), $v = 0.7$ (dashed line), $v = 3$ (dotted line).

potential and the resonance width is determined only by the damping processes. For fullerene, ω_p exceeds the ionization potential resulting in the large value of the plasmon resonance width due to the photoionisation process. As a consequence, the shape of the emission spectrum in electron-cluster collision is governed mainly by the frequency dependence of the dipole polarizability. Alteration of the altitude of the resonance maximum with variation of the collision energy is much stronger in this case than shift of the position of the resonance.

The plasmon giant resonances appear in the emission spectra of cluster ions too [27]. The shapes of these resonances is similar to the profiles arising in the corresponding photoabsorption cross sections. The two peaks structures of the resonances are associated with the deformation and the non-spherical shape of the cluster targets.

The absolute value of the polarizational BrS cross sections for clusters is rather large compared to that for atoms. There are several reasons for the large value of the cross sections. First, plasmon resonances in clusters, particularly metal clusters, are much more narrower than giant resonances in atoms. Second, the number of valence electrons involved in the collective motion in a cluster -N_e can be much larger then in an atom. These electrons radiate coherently in the polarizational BrS process [36] and thus the cross section of the process gains a large enhancement factor- N_e^2. The third reason of the large cross sections consists in the possibility for the process to take place at relatively small collision velocity.

CONCLUSION

Our consideration shows, that the polarizational mechanism dominates in radiative spectra resulting from collisions of electrons with atoms, metal clusters and fullerenes in the vicinity of giant resonances, which can have various nature: single particle or collective excitation in the case of an atom or collective plasmon excitation in the case of metal cluster or fullerene. The polarizational BrS effect is much stronger in collisions of electrons with metal clusters and fullerenes rather than with atoms. A characteristic spectral signature of the emitted light should be searched for to indicate whether the polarizational radiation process is occurring.

The polarizational BrS problem is rather broad, because this kind of radiation can be emitted in any collision involving structured particles- nuclei, atoms, molecules, clusters. The number of various colliding pairs, different interaction forces between particles, kinematical conditions and the frequency ranges make this problem quite spacious and interesting.

ACKNOWLEDGMENTS

The financial support of (i) the Joint Project with Imperial College London funded by the Royal Society of London, (ii) the Joint Project with Institute for Theoretical Physics Frankfurt am Main University funded by the Volkswagen Foundation and NATO (iii) the ISTC Grant No 076, (iv) the RFBR Grant No 96-02-17922-a is greatly acknowledged.

REFERENCES

1. Akhiezer, A.I. and Berestetskii, V. B., *Quantum electrodynamics*, Nauka: Moscow, 1981.
2. Pratt, R.H., in *Fundamental Processes in Energetic Atomic Collisions* ed. Lutz, H.O., Briggs, J.S., and Kleinpoppen, H., New York: Plenum, 1984; Pratt, R.H., and Feng, I.J., in *Applied atomic collision physics* Academic press, 1984, vol.2; in *Atomic Inner-Shell Physics*, ed. Craseman, B., New York: Plenum, 1985; Nakel, W., *Phys.Rep.* **243**, 317 (1994).
3. Buimistrov, V.M., and Trakhtenberg, L.I., *Sov. Phys. – JETP* **42**, 55 (1975).
4. Amusia, M.Ya., Baltenkov, A.S., and Paiziev, A.A., *Sov. Phys. – JETP Lett.* **24**, 332 (1976); Amusia, M.Ya., Baltenkov, A.S., and Gilerson, V.B., *Sov.Tech.Phys.Lett.* **3**, 455 (1977).
5. Pindzola, M.S., and Kelly, H.P., *Phys.Rev.* **A14**, 204 (1976).
6. Zon, B.A., *Sov.Phys. – JETP* **46**, 65 (1977).
7. Wendin, G., and Nuroh, K., *Phys.Rev.Lett.* **39**, 48 (1977).
8. Amusia, M.Ya., *Comments At.Mol.Phys.* **11**, 123 (1982); *Phys.Rep.* **142**, 269 (1988).

9. Kogan, V.I., Kukushkin, A.B., and Lisitsa, V.S., *Phys.Rep.* **213**, 1 (1992).
10. Amusia, M.Ya., Buimistrov, V.M., Tsytovich, V.N., Zon, B.A., *et al*, ed. Tsytovich, V.N., and Ojringel, I.M., *Polarizational bremsstrahlung*, New York: Plenum, 1993.
11. Amusia, M.Ya., Chernysheva, L.V., and Korol, A.V., *J.Phys.B: At.Mol.Opt.Phys.* **23**, 2899 (1990); Amusia, M.Ya., and Korol, A.V., *ibid.* **24**, 3251 (1991) ; Chernysheva, L.V., Avdonina, N.B,, Amusia, M.Ya., and Korol, A.V., unpublished (1989).
12. Fomichev, A.V., Zimkina, T.M., Gribovskii, S.A., and Zhukova, I.I., *Sov.Phys.Solid State* **9**, 1163 (1967); Becker, U., Kerkhoff, H.G., Lindle, D.M., Kobrin, I.H., Ferret, T.A., Heimann, P.A., Truesdale, C.M., and Shirley, D.A., *Phys.Rev.A* **34**, 2585 (1986); Richter, M., Meyer, M., Pahler, M., Prescher, T., Raven, S.V., Sonntag, B., and Wetzel, H.E., *ibid.* **40**, 7007 (1989); Kutzner, M., Zikri Altun, Kelly, H.P., *ibid.* **41**, 3612 (1990).
13. Kelly, H.P., *Physica Scripta* **17**, 109 (1987); Amusia, M.Ya., Ivanov, V.K., and Kupchenko, V.A., *Z.Phys.D: At.Mol.Cl.* **14**, 219 (1989).
14. Amusia, M.Ya., ed. Burke, P.G., and Kleinpoppen, H., *Atomic photoeffect*, New York: Plenum, 1990.
15. Amusia, M.Ya., Dolmatov V.K., and Ivanov, V.K., *Sov. Phys. - JETP* **58**, 67 (1983).
16. Zon, B.A., *Z.Exp.Teor.Phys.* **77**, 44 (1979).
17. Amusia, M.Ya., Zimkina, T.M., and Kuchiev, M.Yu., *Sov. Phys.-Techn. Phys.* **27**, 866 (1982).
18. Korol, A.V., Lyalin, A.G., Shulakov, A.S., and Solov'yov, A.V., *J.Phys.B: At.Mol.Opt.Phys.* **28**, L155 (1995); *Sov. Phys. - JETP* **82**, 631 (1996); Korol, A.V., Lyalin, A.G., and Solov'yov, A.V., *J.Phys.B: At.Mol.Opt.Phys.* **28**, 4947 (1995); *Phys.Rev.A* **53**, 2230 (1996); in *Proceedings of Farago Symposim*, ed. Kleinpoppen, H., New York: Plenum, 1996.
19. Zimkina, T.M., and Gribovskii, S.A., *J. de Physique*, Coll C-4, Sup.10 **32** C4-282 (1971); Richter, M., Meyer M., Pahler, M., Prescher, T., Raven, S.V., Sonntag, B., and Wetzel, H.E., *Phys.Rev.A* **39**, 5666 (1989).
20. Henke, B.L., Gullikson, E.M., and Davis, J.C., *At.Data Nucl.Data Tables* **54**, 2 (1993).
21. Borivskii, I.B., and Rydnik, V.I., *Izv. USSR Acad. Scie.*, **31**, 1009 (1967).
22. Zimkina, T.M., Shulakov, A.S., Brajko, A.P., Stepanov, A.P., and Fomichev V.A., *Fiz.Tverd.Tela* **26**, 1981 (1984); Zimkina, T.M., Shulakov, A.S., and Brajko, A.P., *Izvest. AN SSSR*, ser. Fiz., **48**, 1263 (1984).
23. Bréchignac, C., and Connerade., J.P., *J.Phys.B:At.Mol.Opt.Phys.* **27**, 3795 (1994).
24. Bertsch, G.F., Bulgac, A., Tomanek, D., and Wang, Y., *Phys.Rev.Lett.* **67**, 1991 (1992).
25. Brack, M., *Rev.Mod.Phys.* **65**, 677 (1993); der Heer, W.A., *ibid.* **65**, 611 (1993).
26. Bréchignac, C., Cahuzac, Ph., Carlier, F., and Leygnier, J., *Chem.Phys.Lett.* **164**, 433 (1989); Selby, K., Kresin, V., Masui, J., Vollmer, M., de Heer, W.A., Scheidemann, A., and Knight, W.D., *Phys.Rev.B* **43**, 4565 (1991).

27. Connerade, J.P. and Solov'yov, A.V., 1995, XIX International Conference on the Physics of Electronic and Atomic Collisions, Abstracts, Whistler, Canada; V European Conference on Atomic and Molecular Physics, Contributed Papers, Part II, ed. R.C.Tompson, Edinburgh, United Kingdom; XI International Conference on the Physics of Vacuum Ultraviolet, Abstracts, Tokyo; *J.Phys.B:At.Mol.Opt.Phys* **29** 365 (1996); *ibid.* **29**, in press (1996)
28. Amusia, M.Ya., and Korol, A.V., *Phys.Lett. A* **186**, 230 (1994).
29. Gerchikov, L.G., and Solov'yov, A.V., 1996, 17 Int. Conf. X-Rays and Inner Shell Processes, Hamburg, Germany (1996); submitted to *Z.Phys.D* (1996).
30. Kresin, V., Scheidemann, A., Knight, W.D., *Proc.Int. Symp. Electron Collis. with Mol., Clust. and Surfaces*, London, England (1993); Keller, J.W., and Caplan, M.A. *Chem.Phys.Lett.* **193**, 89 (1992); Huang et al, *J.Phys.Chem.* **99**, 1719 (1995).
31. Ekardt, W., *Phys.Rev.Lett.* **52**, 1925 (1984); *Phys.Rev.B* **31**, 1558 (1985); *ibid.* **32**, 1961 (1985); Guet, C., and Johnson, W.R., *ibid.* **45**, 283 (1992); Ivanov, V.K., Ipatov, A.N., Kharchenko, V.A., and Zhizhin, M.L., *Pis'ma JETP* **58**, 649 (1993); *Phys.Rev. A* **50**, 1459 (1994).
32. Kreibig, U., and Vollmer, U., *Optical Properties of Metal Clusters*, Springer-Verlag:Berlin, 1995.
33. Rotkin, V.V., and Suris, R.A., *Molecular Materials* **4**, 87 (1994); *Sov.Solid State Phys.* **36**, 3569 (1994).
34. Hertel, I.V., Steger, H., de Vries, J., Weisser, B., Mezel, C., Kamke, B., and Kamke, W., *Phys.Rev.Lett.* **68**, 784 (1992).
35. Verkhovtseva, E.T., Gnatchenko, E.V., Zon, B.A., Nekipelov, A.A., and Tkachenko, A.A., *Zh.Eksp.Teor.Fiz.* **98**, 797 (1990).
36. Amusia, M.Ya., Kuchiev, M.Yu., and Solov'yov, A.V., *Sov.Tech.Phys.Lett.* **11**, 577 (1985).

Double ionization of a helium–like atom or ion by a high energy photon

T. Surić*, K. Pisk* and R. H. Pratt[†]

*Rugjer Bošković Institute, P. O. Box 1016, Zagreb, Croatia
[†]Department of Physics and Astronomy, University of Pittsburgh, Pittsburgh, PA 15260, USA

Abstract. We review recent progress in calculations of the ratio of double to single photoionization of helium–like atoms and ions at high energies. The importance of relativistic effects is discussed, with particular emphasis on recent relativistic calculations of double ionization of helium–like atoms and ions by Compton scattering. These calculations, based on the sudden approximation or the impulse approximation, show that, for low Z elements, relativistic effects do not alter the high energy limit of the Compton double to single ratio R_C, even though they do affect single and double ionization Compton scattering separately.

INTRODUCTION

Double ionization of two electron atoms and ions by one photon is attracting significant attention both experimentally and theoretically. The motivation is the investigation of electron–electron correlation effects. The interest in the use of photons in ionizing two electron systems (particularly helium) is based on the observation that the interaction between electrons and the radiation field is described by a one–body operator. This means that, in the lowest order for a given ionization process, a photon interacts with only one electron. Hence the simultaneous ejection of two electrons is due to the electron–electron interaction between the two electrons, and therefore double ionization by a photon provides a clear probe of electron–electron correlation. In a two electron system no other correlations are present. For this simplest many–body system we may be able to develop more exact theoretical approaches and more detailed experimental methods. We wish to understand eventually all possible correlation effects in this relatively simple system, with the purpose of better understanding of these effects in more complex systems.

The photoionization processes are photoeffect (incident photon absorbed) and Compton scattering (incident photon scattered). Photoeffect dominates the total cross section at lower photon energies and its relative contribution in photon atom interaction decreases as photon energy increases. For a He–like system, at energies for which the incoming photon momentum k is comparable to the average momentum of the bound electrons p_{av} (of a two electron atom or ion in its ground state), photoeffect and Compton scattering are comparable. For higher energies Compton scattering dominates. In the case of helium Compton scattering dominates above, roughly, 6.5 keV.

The simplest observable, for both experimental and theoretical treatment of the correlation effects, has been the ratio between probabilities for double and single photoionization. This ratio is nonzero because of electron–electron correlation; its value depends on incident photon energy. Also, the value depends on the process under consideration. Therefore we talk about the Compton double to single ionization ratio R_C and the photoeffect ionization ratio R_P. The energy dependence of these observables, and particularly their high energy limits, are subjects of extensive studies. In the last several years significant progress has been achieved [1]. However, the relatively large disagreements in theoretical predictions as well as in experimental data indicate that we still do not understand all relevant aspects of the processes which are involved in determining the ratio for the double to single ionization for both Compton scattering and photoeffect.

As we will see, much understanding of these processes can be obtained within the shake–off approximation, *i.e.* a sudden approximation. By shake–off approximation we mean the assumption that the final state electron–electron interaction can be neglected, and that one electron scatters the high energy photon and accepts almost all momentum transfer [2].

PHOTOEFFECT RATIO R_P

Until recently it was believed that the energy dependence of R_P for helium was well understood over the whole nonrelativistic energy region. The theoretical predictions were consistent with measurements showing that R_P increases with energy, reaching a maximum (R_P=0.05) at photon energies of $\omega = 150 - 250$ eV, and then decreases to the constant shake–off value $R_P = 0.017$ for energies above roughly 3 keV.

New measurements [3], performed using cold target recoil spectroscopy, have found the peak of R_P to be lower than previously assumed, by about 25 %. These new data are consistent with calculations [4] relating the photoeffect ratio to that for fast charged particle impact.

The shake–off ratio is given in nonrelativistic calculations [5–8] as

$$R_P = 1 - \frac{\sum_B |\int \Phi_B^*(\mathbf{r}_1)\Psi_i(\mathbf{r}_1,0)d^3r_1|^2}{\int |\Psi_i(\mathbf{r}_1,0)|^2 d^3r_1}. \tag{1}$$

Here $\Psi_i(\mathbf{r}_1,\mathbf{r}_2)$ represents the initial state wave function, $\Phi_B(\mathbf{r}_1)$ is a bound state hydrogen–like electron wave function (in the potential of charge Z), and the summation is over all bound states. The shake off formula Eq. (1) has usually been derived in the nonrelativistic dipole approximation without retardation. It reflects the fact that, in the dipole approximation, the photon can be absorbed only if nucleus takes additional momentum (note that one of the electrons in Eq. (1) is removed from near the nucleus, $r_2 = 0$). Namely, in the dipole approximation, a two–electron system, in the absence of nuclear field, cannot absorb a photon. Most theoretical analysis uses the dipole approximation. However, the shake–off result Eq. (1) is not changed if retardation effects (i.e. higher multipole effects) are included, given that for low Z two–electron systems retardation has little effect on ground state wave function Ψ_i and hydrogenic bound states Φ_B. In explicit calculations, using wave functions consistent with shake–off assumptions, Kornberg and Miraglia [9] have shown that, even when retardation alters the cross sections for single and double photoionization by significant factors, the same factor enters both processes, and so the ratio R_P is unchanged.

We may likewise expect that relativistic effects do not alter the photoeffect ratio R_P, based on the shake–off argument. The shake–off process, when applied to double photoionization, can be viewed as a two–step process. In the first step one electron is suddenly removed, corresponding to the single ionization process. Here we may allow very large, even relativistic, energies and therefore relativistic effects can be important. The second step is shake–off of the other electron which is (at least for small Z) most often a slow, nonrelativistic electron. Calculating the ratio between double and single photoionization cross sections, under the assumption of the validity of the sudden approximation, the contributions of the first step (the single ionization cross section), including relativistic as well as retardation effects, cancel out completely, leaving R_P as in Eq. (1).

However, at high energies, where the shake–off mechanism was assumed to be valid, Drukarev [10,11] has called attention to an additional mechanism [12–14] for double photoeffect, which would first manifest itself as a linear rise of the ratio with energy, causing a roughly 10% correction to the shake–off limit for R_P at about 12 keV. This prediction awaits experimental verification. The point is that, while a photon cannot be absorbed by one free electron in the absence of the atomic nucleus (and momentum must be transferred to the atomic nucleus), a photon can be absorbed by two free electrons (though not in the dipole approximation). This leads to a mechanism for double ionization which is not available for single ionization. The analysis was performed within nonrelativistic quantum mechanics without assuming dipole approximation, therefore allowing two electrons to absorb the photon even in the absence of the atomic nucleus [15].

A pair of photoelectrons produced through this alternative mechanism share the photon energy nearly equally (and we may call this mechanism an "equal energy sharing mechanism"). The residual ion is left with small momentum. This is different from the case when photoelectrons are produced through the shake-off mechanism, in which one electron takes almost all the photon energy and the residual ion takes almost all transfered momentum. Note that in the experimental situation where photoeffect events are distinguished from Compton events by measuring the recoil of the residual ion [16,17], the events produced through the equal energy sharing mechanism are not counted in photoeffect events, but in Compton events. These experiments count as photoeffect events only those produced in the shake-off regime (one electron takes almost all available energy while the other shakes-off), and therefore such experiments would not observe the term in R_P caused by the equal energy sharing mechanism.

Even though the shake-off ratio may therefore not be the high energy limit of R_P, it is a physically well-defined quantity which we may measure. This observable, which we may call the shake-off ratio $R_P^{s.o.}$, gives us information about the initial state electron–electron correlation when one of the electrons is near the nucleus [18,19].

Likewise, if the equal energy sharing mechanism is of the estimated magnitude, it may be within the scope of today's experimental methods. Its measurement should provide information on electron–electron correlation when the electrons are close to each other. This mechanism may also be important in double–electron capture processes in heavy (highly–charged) ion–atom collisions when a single high energy photon is emitted. For low Z ions the theoretical predictions are reported to be in good agreement with experimental data [20]. However, for highly charged ions a surprisingly enhanced cross section for double electron capture is found [21].

COMPTON RATIO R_C

There are now several calculations of the Compton scattering ionization ratio R_C [22–29]. For finite energies (up to 20 keV) these calculations, performed for He, disagree significantly. However in the high energy limit these theoretical results appear to be in agreement, as discussed by Surić, Pisk and Pratt [18].

Andersson and Burgdörfer [23] performed Compton calculations for helium using highly correlated initial state wave functions and approximately correlated or uncorrelated final state wave functions. Their various calculations all predicted the ratio decreasing toward the same constant ratio of $R_C = 0.8\%$. Hino, Bergstrom and Macek performed lowest order MBPT calculations for He with a basis set constructed using the V^{N-1} potential. In their first paper [24] they asserted that their results reached an asymptotic limit of around $R_C = 1.7\%$ somewhat above 10 keV. However in their later work [25], performed with energies up to 20 keV, they found that their results for the ratio were decreasing and had not reached an asymptotic limit. They concluded their results were consistent with the high energy limits of Andersson and Burgdörfer [23] and Surić et al. [26]. Surić et al. [26,2], using generalized shake–off model or impulse approximation (IA) have derived a high energy limit formula for the double to total ionization ratio for Compton scattering

$$R_C = 1 - \sum_B \int d^3r \left| \int \Phi_B^*(\mathbf{r}_1) \Psi_i(\mathbf{r},\mathbf{r}_1) d^3r_1 \right|^2. \tag{2}$$

Here $\Psi_i(\mathbf{r},\mathbf{r}_1)$ represents the initial state wave function, $\Phi_B(\mathbf{r}_1)$ is a bound state hydrogen like electron wave function (in the potential of charge Z), and the summation is over all bound states. Using highly correlated initial state wave function they obtain $R_C = 0.008$ (0.8%) for He. At finite energies their results with IA approach the limiting shake–off value from below. The shake–off ratio for Compton scattering, given by Eq.(2), differs from the photoeffect shake–off ratio, given by Eq.(1), because different regions of the initial state contribute to these two processes. Namely, when the incoming photon energy goes to infinity the photoeffect shake–off ratio will be determined exclusively by the region where one of the electrons is at the nucleus. By contrast, the Compton process can occur on free electrons, and all regions contribute in proportion to the probability amplitude that an electron can be found there.

Using the lowest order MBPT based on hydrogenic wave functions and introducing an effective nuclear charge, Amusia and Mikhailov [27,28] obtained $R_C = 0.0168$ for helium. Subsequently, Surić, Pisk and Pratt [18] showed that this introduction of effective charge in the calculation was unjustified. If bare charge is used in the calculation, as is more appropriate, the result would be consistent with other high energy predictions for R_C.

Kornberg and Miraglia [29] studied the energy dependence of R_C using L and V form of the A^2 interaction operator, correlated initial state and uncorrelated final electron states. Their results are form dependent for finite energies, but, for both forms, approach the limit predicted by Eq. (2) ($R_C = 0.008$ for helium). The L-form results support the results of Andersson and Burgdörfer (who used the same form of the A^2 operator, but more correlated initial state and final states) that the asymptotic limit is approached from above. However, the V-form results approach the asymptotic limit from below, as do the IA results of Surić et al..

In the energy region between 6 keV and 20 keV, experimental data are not yet definitely conclusive. At energies between 6 keV and 11 keV the theoretical predictions of Andersson and Burgdörfer seem to be in better agreement with measurements, performed by Levin et al. [30] and by Spielberger et al. [17], than are the other predictions. At higher energies, 15-20 keV, the measurements of Levin et al. [30] seem to be in quite good agreement with the predictions of Bergstrom, Hino and Macek [25], which are higher then the results of Andersson and Burgdörfer as much as 50%. Recent measurement by Spielberger et al. [31] with 58 keV photons found $R_C = (0.84^{+.08}_{-.11})\%$ in good agreement with the theoretical predictions for the shake–off limit. Another measurement by Wehlitz et al. [32] performed with 57 keV photons is less conclusive. They found $R_C = (1.25 \pm 0.3)\%$.

More extensive study, in a wide energy range, is clearly required in order to understand the energy dependence of R_C. There does seem to be agreement among the theoretical predictions that the high energy limit of R_C is given by Eq. (2), and this is supported by the measurement of Spielberger et al. [31]. However, more measurements in this and even higher energy regions are required. As stated by Spielberger et al [31], the experiments with photons above 100 keV up to '....150 keV are clearly within reach'. This is based on the facts that Compton scattering cross section slowly varies with energy, in this energy region, and on the availability of new–generation high energy synchrotron radiation facilities.

As one considers this higher energy region, it is clear relativistic effects must be examined. (All existing predictions are nonrelativistic). Recently, we have extended our theoretical analysis of double ionization by Compton scattering, based on sudden approximation [2], to relativistic regions. This approach should be appropriate for the relativistic treatment of double ionization Compton scattering from He-like systems. We have demonstrated that for low Z elements relativistic effects do not alter the high energy limit of the Compton double to single ratio R_C, although these effects do affect single and double ionization Compton scattering separately. (The shake–off argument that relativistic effects do not alter the ratio R_C is the same as for the shake–off for photoeffect).

Justification of the sudden approximation for double ionization by Compton scattering at high energies comes from the observation that, in single ionization Compton scattering at high incident photon energies, most outgoing electrons have large energy. Figure (1) illustrates that, in high energy Compton scattering, most electrons do have enough energy to apply the sudden approximation. We show the singly differential cross section (with respect to outgoing electron energy) in single ionization Compton scattering from helium, calculated using using an exact numerical approach or IA (which can not be distinguished on this figure for these energies). Results for incident photon energies of 25 keV and 150 keV are displayed. We see that most electrons have energies below the edge determined by the Compton energy for scattering from free electrons at rest. Between zero energy and this edge, the differential cross section is a slowly varying function of energy, and all these energies are, approximately, equally likely. If we assume that 3 keV outgoing electron energy is required to apply the sudden approximation (the shake–off ratio R_P in photoeffect, for helium, is approached at this energy) then we see that at 150 keV this approximation is valid for most of the Compton events.

Assuming the validity of sudden approximation we have concluded that relativistic effects are the same in the double and single ionization cross sections. Therefore, in treating double ionization cross sections by Compton scattering at high, relativistic energies it is appropriate to apply the shake–off factor to the single ionization cross sections. At these relativistic energies, and for low Z elements, the single ionization cross sections can be treated, within a high de-

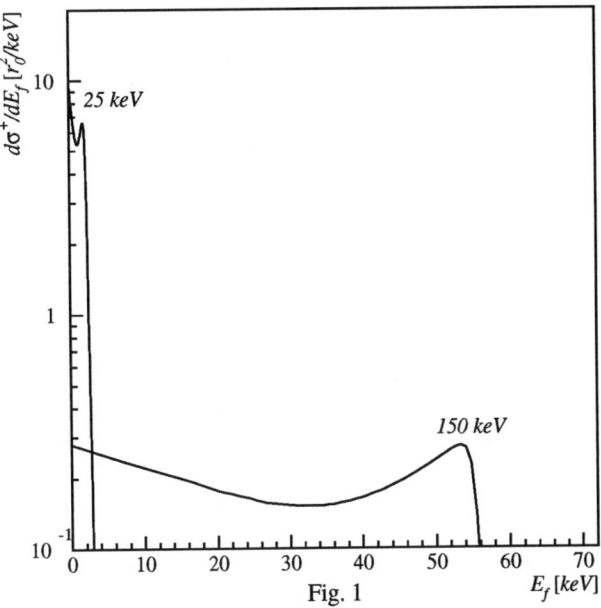

FIGURE 1. Singly differential cross section (with respect to outgoing electron energy) in single ionization Compton scattering from helium. Results for incident photon energies of 25 keV and 150 keV are displayed. The maximum outgoing electron energy (free Compton scattering edge), in Compton scattering from a free electron at rest, is 2.2 keV (for 25 keV incident photons) and 55.5 keV (for 150 keV incident photons). Above these energies the singly differential cross sections decrease rapidly with increasing energy.

gree of accuracy, using impulse approximation. This is well established [33–36] for the treatment of cross sections up to the level of detail of doubly differential cross section (with respect to outgoing photon angles and energy) [37]. Therefore, in evaluating such differential cross sections, in this energy region we may use IA. Conversely, Surić et al. have verified that if one calculates double and single ionization singly differential cross sections the ratio is 0.008 (for He) for all, except very small angles, where both sudden approximation and IA are not valid.

An interesting application of shake–off related ideas in nonrelativistic Compton scattering has been presented by Wang et al. [38]. These authors present calculations of the ratio between double and single ionization singly differential cross sections (with respect to energy transfer) by Compton scattering. These calculations show that, except for small energy transfers (where sudden approximation is not supposed to be good) this ratio is constant (the shake–off constant R_C) up to the free Compton scattering edge. This is consistent with with the usual Compton shake–off picture. However, for larger energy transfers (which contribute negligibly to the total cross section) the ratio changes and rises toward the shake–off ratio for photoeffect R_P, reflecting the fact that large energy transfers, as for photoeffect, are only possible when large momenta are transfered to the atomic nucleus.

CONCLUSION

We have reviewed recent progress in calculations of the ratio of double to single photoionization of helium–like atoms at high energies. These ratios, both for ionization through photoeffect and Compton scattering, are largely to be understood through the shake–off mechanism, which assumes that one electron is suddenly removed from a two–electron system allowing the other only to shake off, as the residual system relaxes. Retardation does not change the photoeffect shake–off ratio (retardation is already included in the Compton process). Relativistic effects do not significantly change the ratio for either process. Both retardation and relativity affect the separate cross sections for double ionization and single ionization processes significantly. In the case of photoeffect we have also called attention to the recently proposed additional double ionization mechanism, involving equal energy sharing, which modifies the shake–off prediction for R_P at high energy.

REFERENCES

1. See the recent extensive review of J. H. McGuire, N. Berrah, R. J. Bartlett, J. A. R. Samson, J. A. Tanis, C. L. Cocke and A. S. Schlachter, J. Phys. B **28**, 913 (1995).
2. T. Surić, K. Pisk, B. A. Logan and R. H. Pratt, to be published.
3. R. Dörner, T. Vogt, V. Mergel, H. Khemliche, S. Kravis, C. L. Cocke, J. Ullrich, M. Unverzagt, L. Spielberger, M. Damrau, O. Jakutzki, I. Ali, B. Weaver, K. Ullmann, C. C. Hsu, M. Jung, E. P. Kanter, B. Sonntag, M. H. Prior, E. Rotenberg, J. Denlinger, T. Warwick, S. T. Manson, and H. Schmidt–Böcking, Phys. Rev. Lett. **76**, 2654 (1996).
4. S. T. Manson and J. H. McGuire, Phys. Rev. A **51**, 400 (1995).
5. A. Dalgarno and A. L. Stewart, Proc. Phys. Soc. London **76**, 49 (1960).
6. F. W. Byron, Jr., and C. J. Joachain, Phys. Rev. **164**, 1 (1967).
7. T. Åberg Phys. Rev. A **2**, 2 (1970).
8. T. Åberg, in *Photoionization and Other Probes of Many Electron Interactions*, edited by F. J. Wuilleumier (Plenum, New York, 1976) p. 49.
9. M. A. Kornberg and J. E. Miraglia, Phys. Rev. A **52**, 2915 (1995).
10. E. G. Drukarev, Phys. Rev. A **51**, R2684 (1995).
11. E. G. Drukarev, Phys. Rev. A **52**, 3910 (1995).

12. The contribution of this mechanism to the spectrum of outgoing electrons in double ionization was estimated by M. Ya Amusia, E. G. Drukarev, V. G. Gorshkov and M. P. Kazachkov, J. Phys. B **8**, 1248 (1975). Its contribution to the double ionization total cross section and its effect on the photoeffect double to single ionization ratio was not discussed in detail. Subsequently, E. G. Drukarev and F. F. Karpeshin, J. Phys. B **8**, 399 (1976), discussed the contribution of this mechanism to the double ionization total cross section and calculated its effect on the photoeffect double to single ratio. In this work they extended the discussion to the relativistic region.
13. For further discussion see M. Ya Amusia, Comments At. Mol. Phys. **10**, 155 (1981) and *"Atomic Photoeffect"*, M. Ya. Amusia, Plenum, New York, 1990 (Chapter 8).
14. We are aware of an attempt by K. Pisk (*"Double photoeffect"*, Masters thesis, University of Zagreb, Zagreb 1969, in Croatian, unpublished) and by K. Pisk and K. Ilakovac (*Proceedings of the Fifth Congress of the Mathematicians, Physicists and Astronomers of Yugoslavia*, Book II (Physics), page 97–100, Skopje 1972) to consider double ionization of two K-shell electrons. The analysis is numerical, for photons of 662 keV on Al, Zn and Ag. They present results for differential cross sections with respect to energies and angles of outgoing electrons. It is interesting to note that they consider the equal sharing mechanism as the main mechanism (dominant mechanism) for double ionization. The shake–off mechanism is not considered nor are the results presented in that kinematical region. The total cross sections were not considered.
15. It should be noted that in his paper [10] Drukarev, discussing this mechanism, after extracting the C/p^4 behavior (for large relative electron momenta p), used factorized (for example Hartree-Fock) type two electron wave functions in evaluating the constant C. These wave functions in fact do not have the $1/p^4$ behavior required for the mechanism to contribute significantly. Instead, these functions have $1/p^8$ behavior (M. A. Kornberg, J. E. Miraglia, T. Surić, R. H. Pratt, *"Comment on the asymptotic high–energy limit of the ratio of double to single photoionization of helium"*, unpublished). However, as clarified later (E. G. Drukarev, private communications) these factorized wave functions are used simply as approximations to the exact wave functions to evaluate integrals which have been derived from expressions which do exhibit the proper $1/p^4$ behavior. Further aspects of this behavior for the helium ground state wave function and proper approximate wave functions are discussed in E. G. Drukarev, R. H. Pratt, T. Surić, M. Kornberg, J. Miraglia (to be published).
16. J. A. R. Samson, Z. X. He, R. J. Bartlett, and M. Sagurton, Phys. Rev. Lett. **72**, 3329 (1994).
17. L. Spielberger, O. Jakutzki, R. Dörner, J. Ullrich, U. Meyer, V. Mergel, M. Unverzagt, M. Damrau, T. Vogt, I. Ali, Kh. Khayyat, D. Bahr, H. G. Schmidt, R. Frahm, and H. Schmidt–Böcking, Phys. Rev. Lett. **74**, 4615 (1995).
18. T. Surić, K. Pisk and R. H. Pratt, Phys. Lett. A **211**, 289 (1996).
19. T. Surić, in: *Proceedings of Indo-US Workshop on Radiation Physics, 1996, Darjeeling, India*, to be published.

20. V. L. Yakhontov, private communications; V. L. Yakhontov and M. Ya. Amusia, Phys. Lett. A, accepted for publication.
21. A. Warczak, M. Kucharski, Z. Stachura, H. Geissel, H. Irnich, T. Kandler, C. Kozhuharov, P. H. Mokler, G. Münzenberg, F. Nickel, C. Scheidenberger, Th. Stöhlker, T. Suzuki, P. Rymuza, Nucl. Instrum. Methods, **98**, 303 (1995).
22. L. R. Andersson and J. Burgdörfer, Phys. Rev. Lett. **71**, 50 (1993);
23. L. R. Andersson and J. Burgdörfer, Phys. Rev. A **50**, R2810 (1994).
24. K. Hino, P. M. Bergstrom, Jr., and J. H. Macek, Phys. Rev. Lett. **72**, 1620 (1994).
25. P. M. Bergstrom, Jr., K. Hino and J. H. Macek, Phys. Rev. A **51**, 3044 (1995).
26. T. Surić, K. Pisk, B. A. Logan and R. H. Pratt, Phys. Rev. Lett. **73**, 790 (1994).
27. M. Ya. Amusia and A. I. Mikhailov, Phys. Lett. A **199**, 209 (1995).
28. M. Ya. Amusia and A. I. Mikhailov, J. Phys. B **28**, 1723 (1995).
29. M. A. Kornberg and J. E. Miraglia, Phys. Rev. A **53**, R3709 (1996).
30. J. C. Levin, G. Bradley Armen and I. A. Sellin, Phys. Rev. Lett. **76**, 1220 (1996).
31. L. Spielberger, O. Jakutzki, B. Krässig, U. Meyer, Kh. Khayyat, V. Mergel, Th. Tschentscher, Th. Buslaps, H. Bräuning, R. Dörner, T. Vogt, M. Achler, J. Ullrich, D. S. Gemmell, and H. Schmidt-Böcking, Phys. Rev. Lett. **76**, 4685 (1996).
32. R. Wehlitz, R. Hentges, G. Prümper, A. Farhat, T. Buslaps, N. Berrah, J. C. Levin, I. A. Sellin, and U. Becker, Phys. Rev. A **53**, R3720 (1996).
33. P. Eisenberger and W. A. Reed, Phys. Rev. B **9**, 3237 (1974).
34. R. Ribberfors, Phys. Rev. B **12**, 2067 (1975).
35. T. Surić, P. M. Bergstrom, Jr., K. Pisk, R. H. Pratt, Phys. Rev. Lett. **67**, 189 (1991).
36. P. M. Bergstrom, Jr., T. Surić, K. Pisk, R. H. Pratt, Phys. Rev. A **48**, 1134 (1993) and references therein.
37. However, it has been found (Z. Kaliman, T. Surić, K. Pisk, and R. H. Pratt, to be published) that for the triply differential cross section (when the outgoing electron is also observed) the IA approach is less justified at these energies.
38. J. Wang, J. H. McGuire, J. Burgdörfer and Y. Qiu, Phys. Rev. Lett **77**, 1723 (1996).

Double Ionization of Helium by Photons and Charged Particles

J. Burgdörfer and Y. Qiu

*Department of Physics, University of Tennessee
Knoxville TN 37996-1200, USA*
and
*Physics Department, Oak Ridge National Laboratory
Oak Ridge, TN 37831-6377, USA*

J. Wang and J.H. McGuire

*Department of Physics, Tulane University
New Orleans, LA 70118*

Abstract. We discuss double ionization of helium by charged particles, photoionization, and Compton scattering. With the help of an analysis of the cross section as a function angular momentum transfer (partial waves) or momentum transfer, distant Coulomb collisions can be related to photoionization while hard Coulomb collisions can be related to Compton scattering. Within this framework, accurate tests for recent photoionzation and charged-particle data are possible. We also study the behavior of the differential Compton and charged particle cross sections at high energies and show that beyond the binary encounter limit all processes approach the photoionization limit.

1. Introduction

Interactions of both photons and charged particles with matter are common in nature. Despite some different physical characteristics, both photons and charged particles interact with matter through electromagnetic fields. This similarity has been recognized and exploited in physical description of these interactions. Most importantly, the Weizsäcker and Williams method [1,2] of virtual photons has found applications in various areas of physics. In this method, the field of the moving charged particle is considered to consist of virtual photons. Interactions can therefore be described in terms of emission and absorption of virtual photons. An advantage of the method is that it can help circumvent difficulties encountered when dealing directly with interacting charged particles.

There, however, are important differences between photons and charged particles. In the photoelectric effect (photonionization), photons are annihilated. This usually gives rise to selection rules such as the optically allowed (dipole) atomic transitions. On the contrary, charged particles are not annihilated in atomic interactions. In additions to dipole allowed transitions, charged particles can cause non-dipole transitions as well. These transitions require large energy and momentum transfers, often termed "hard" collisions. A well known example of hard collisions is the production of binary encounter electrons generated by head-on collisions between a massive particle and an electron. Typically, many partial waves (or multipoles) are needed to describe these collisions. Only recently, a relation between inelastic photon scattering, i.e. Compton scattering, which is mostly non-dipole and charged particle scattering was explicitly established within the first-order Born approximation [3].

The renewed interest in such relations was stimulated by recent experimental studies of double ionization of helium by charged particles, photoionization, and Compton scattering. These two-electron processes probe electron-electron correlations in the initial and final state of this fundamental three-body Coulomb problem which are not yet well understood. As these different ionization processes access different final states of the three-body Coulomb continuum, they probe different aspects of the correlation problem. Interrelating data for different processes provide therefore new insights into two-electron processes.

We briefly review in this contribution progress along two different lines: We discuss a new relationship which connects data for near-threshold photoionization data with high-energy charged particle data and which can be considered as a generalization of the Weizsäcker and Williams theory. We also discuss the behavior of double ionization by photoionization, by Compton scattering and charge particle scattering at high energies.

2. Generalized Weizsäcker-Williams Method

The idea of relating excitation and ionization by charged particles to corresponding photon processes goes back to Weizsäcker and Williams (WW) [1,2]. Their method consists of representing the interaction of charged particles with charge Z with an atom by an electromagnetic field pulse generated by the projectile passing by at impact parameter b with velocity v,

$$F_\perp(t) = Z \frac{\gamma b}{(b^2 + \gamma^2 v^2 t^2)^{3/2}}$$
$$F_\parallel(t) = -Z \frac{\gamma v t}{(b^2 + \gamma^2 v^2 t^2)^{3/2}}. \tag{1}$$

The spectral distribution is determined by the Fourier transform of Eq.(1),

$$\tilde{F}_\perp(\omega) = Q \frac{2\omega}{\gamma v^2} K_1\left(\frac{b\omega}{\gamma v}\right)$$
$$\tilde{F}_\parallel(\omega) = -iQ \frac{2\omega}{\gamma^2 v^2} K_0\left(\frac{b\omega}{\gamma v}\right) \qquad (2)$$

In the first Born approximation for charged particle impact [4],

$$A^{B1}(i \to f, b) = -i \int_{-\infty}^{+\infty} dt \langle f|V_p|i\rangle \exp[i(\epsilon_f - \epsilon_i)t]$$
$$= -i\langle f| \int_{-\infty}^{+\infty} dt V_p \exp[i(\epsilon_f - \epsilon_i)t]|i\rangle. \qquad (3)$$

For large impact parameters, a dipole approximation can be used for the perturbation by the projectile,

$$V_p(r,t) = -\frac{Z}{|r - R(t)|} \simeq -\frac{Z}{R(t)} - \vec{r} \cdot \vec{F}(t), \qquad (4)$$

where $\vec{F}(t) = Z_p \vec{R}(t)/R^3(t)$ is the force experienced by the electron which resembles a time-dependent electric field. Using $\vec{R}(t) = b\hat{y} + vt\hat{z}$ and $\omega = (\epsilon_f - \epsilon_i)$

$$A^{B1}(i \to f, \vec{b}) \simeq i\langle \vec{r}\rangle_{fi} \cdot \tilde{F}(\omega) \qquad (5)$$

with

$$\langle \vec{r}\rangle_{fi} = \langle f|\vec{r}|i\rangle \qquad (6)$$

and $\tilde{F}(\omega)$ given by Eq.(2). In the method of virtual quanta, the field in Eq.(5) is replaced by that arising from an equivalent flux photon with energy $E = \hbar\omega$ and, therefore, the collisional ionization process becomes equivalent to the photoionization process.

As evident from Eqs. (3) and (5), the Weizsäcker-Williams method contains two essential limitations: the validity of the first-order Born approximation and the restriction to "distant", soft collisions. Progress has been made in overcoming the second limitation. We show in the following that hard collisions can also be related to photon processes, however photon scattering rather than photoabsorption. The method discussed in the following can be considered a generalization of the Weizsäcker-Williams method.

Our starting point is the separation of the dipole from the non-dipole contributions in a multipole expansion of the cross section for ionization differential in the energy transfer, ϵ, namely,

$$\frac{d\sigma(\epsilon)}{d\epsilon} = \sum_{\ell=0}^{\infty} \frac{d\sigma_\ell(\epsilon)}{d\epsilon}$$
$$= \frac{d\sigma_1(\epsilon)}{d\epsilon} + \sum_{\ell \neq 1} \frac{d\sigma_\ell(\epsilon)}{d\epsilon} = \frac{d\sigma_d(\epsilon)}{d\epsilon} + \frac{d\sigma_{nd}(\epsilon)}{d\epsilon} \qquad (7)$$

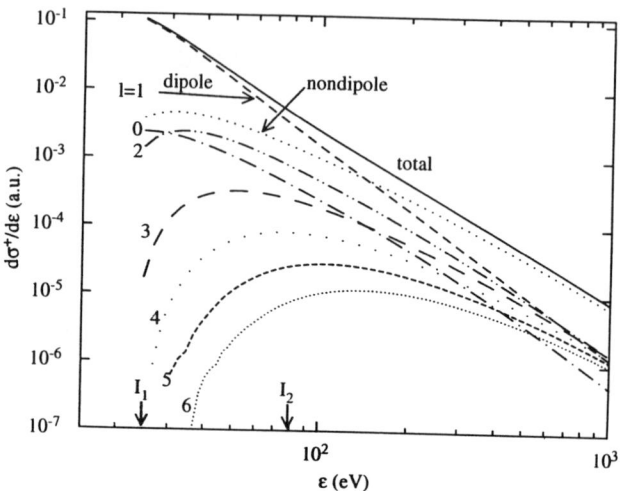

FIGURE 1. Single ionization cross section of helium as a function of energy transfer ϵ by 20 MeV proton collisions with He using a Roothan-Hartree-Fock wavefunction (see text): shown are the dipole term ($l = 1$); non-dipole $l = 0, 2, \ldots 6$; and total cross sections (Σ_l); nondipole term ($\Sigma_{l \neq 1}$).

where 'd' and 'nd' represent the dipole ($\ell = 1$) and non-dipole ($\ell \neq 1$) contributions to the cross sections. Fig. 1 displays the partial-wave decomposition of the energy differential cross section for single ionization of He by 20 MeV protons calculated in the first Born approximation [4,5]. A five-term Roothan-Hartree-Fock wavefunction [6] is used for the initial state. We also performed calculations using a fully correlated CI-initial state [7]. The results are within 5% of each other, showing that the single ionization cross section (unlike double ionization) is insensitive to electron correlation in the initial state at high energies.

In order to relate the dipole component (Eq.(7)) to the WW expression (Eq.(2)) and all non-dipole terms neglected in the WW to photon scattering, we show that dipole contributions originate, indeed, from soft collisions while non-dipole terms originate from hard collisions. In Fig. 2, the cumulative single ionization cross section of the dipole and non-dipole terms as a function of the momentum transfer is shown. The quantity σ_Q/σ graphed in Fig. 2 is defined as

$$\sigma_Q/\sigma = \int_Q^{Q_{max}} \frac{d\sigma}{dQ'} dQ' \Big/ \int_{Q_{min}}^{Q_{max}} \frac{d\sigma}{dQ'} dQ' \tag{8}$$

where σ is the single ionization cross section at the threshold and Q_{min} and Q_{max} are the minimum and maximum momentum transfers. The value of σ_Q/σ at a given Q represents the percentage contribution to the cross section from all momentum transfers $Q' \geq Q$. Figure 2 shows that most of the dipole contribution comes from small momentum transfers and most of non-dipole

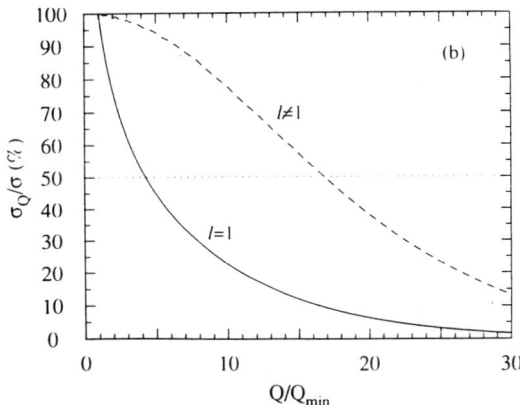

FIGURE 2. The cummulative single ionization cross section for 10 MeV protons on He as a function of the momentum transfer Q at the single ionization threshold. For a given Q, σ_Q/σ denotes the percentage of the contribution from momentum transfers $\geq Q$ for the dipole term ($l = 1$, solid curves); the non-dipole terms ($l \neq 1$, dash curve), respectively.

contribution from large momentum transfers. Half (50%) of the dipole term is from momentum transfers $Q/Q_{min} \leq 4.2$ compared to only about 5% of the non-dipole terms. Due to the dominance of small momentum transfers in the dipole term, the Bethe term which is equivalent to the WW method (Eq.(5)) accounts for over 95% of the total dipole contribution.

It is useful to define an energy distribution for singly ionized states, $\rho_\ell(\epsilon) \equiv \frac{1}{\sigma^1} \frac{d\sigma_\ell^1(\epsilon)}{d\epsilon}$, where σ^1 is the total cross section for single ionization. Note that $\int \rho_\ell(\epsilon) d\epsilon \leq 1$, but $\int \rho(\epsilon) d\epsilon = \int \Sigma_\ell \rho_\ell d\epsilon = 1$. The fraction of this energy distribution for each partial wave is defined by $f_\ell(\epsilon) = \rho_\ell(\epsilon)/\rho(\epsilon)$. Clearly, the non-dipole fraction is related to the dipole fraction by $f_{nd}(\epsilon) = 1 - f_d(\epsilon)$.

The most frequently studied quantity in double ionization of helium, a paradigm for studying the role of electron correlation [8], is the ratio of double to single ionization R. The ratio, R_Z, for protons and electrons has been probed experimentally well into the regime of relativistic energies [9]. The ratio, R_γ, for photons is being measured at ever increasing energies, the highest energy being 58 keV at present [10]. In the following, we use R_Z and R_γ to denote ratios for charged particles and photons, respectively. In particular, R_γ is used in the generic sense that it represents ratios for ionization by photons. Specific processes are denoted explicitly by R_{PE} for photoelectric effect (photoionization) and R_C for Compton scattering.

The ratio of double to single cross sections, differential for ϵ for each partial wave is defined as,

$$R_\ell(\epsilon) \equiv \left[\frac{d\sigma_\ell^2(\epsilon)}{d\epsilon} / \frac{d\sigma_\ell^1(\epsilon)}{d\epsilon}\right]. \qquad (9)$$

Accordingly,

$$R(\epsilon) = \Sigma_{\ell=0}^{\infty} f_\ell(\epsilon) R_\ell(\epsilon) = f_d(\epsilon) R_d(\epsilon) + f_{nd}(\epsilon) R_{nd}(\epsilon). \quad (10)$$

Relation (10) holds for any process, including photoionization, Compton scattering and scattering by charged particles. Applied to charged particles, Eq. 10 reads

$$R_Z(\epsilon) = f_{Z,d}(\epsilon) R_{Z,d}(\epsilon) + (1 - f_{Z,d}(\epsilon)) R_{Z,nd}(\epsilon). \quad (11)$$

which is still exact. The dipole term in (11) may be related for fast charged particles ($Z/v \ll 1$, where v is the velocity of the particle) to the ratio of cross sections for photoionization in line with the WW method,

$$R_{Z,d} = R_{PE} = \sigma_{PE}^2 / \sigma_{PE}^1 \quad (12)$$

with $\epsilon = E - I$, E is the energy of the incident photon, and I is the binding energy of the target electron(s). Note that the ratio R_{PE} is a ratio of total cross sections.

Generalizing the WW method, the non-dipole term of (11) may be related to cross sections for inelastic photon scattering (Compton scattering). In first-order perturbation theory one finds [3],

$$\left(\frac{d^2\sigma}{d\epsilon dQ^2} \right)_C = \frac{r_0^2}{8} \frac{v^2}{k^2} \frac{Q^4}{Z^2} \left[1 + \left(1 - \frac{Q^2}{2k^2} \right)^2 \right] \left(\frac{d^2\sigma}{d\epsilon dQ^2} \right)_Z \quad (13)$$

where Q^2 is the square of the momentum transfer of the projectile and k is the initial photon momentum. This relation is valid for arbitrary transitions, including both single and double ionization. We note that Compton scattering tends to dominate over photoionization when $k\, r_{target} > 1$ where the multipole expansion breaks down and many partial waves contribute. Integrating over the momentum transfer and using a peaking approximation valid at high energies [7], one obtains

$$R_Z(\epsilon) = R_C(\epsilon). \quad (14)$$

This holds for each partial wave, so that

$$R_{Z,nd}(\epsilon) = R_{C,nd}(\epsilon). \quad (15)$$

In Compton scattering the maximum transfer at energy $\epsilon = 2E_\gamma^2/c^2$, corresponding to the maximum value of $\Delta \lambda = \lambda_C(1 - \cos\theta)$ represents of binary encounter (BE) like backscattering ($\theta = 180°$) of the photon.

Using (12) and (15) in (11), one has

$$R_Z(\epsilon) = f_{Z,d}(\epsilon) R_{PE} + (1 - f_{Z,d}(\epsilon)) R_{C,nd}(\epsilon). \quad (16)$$

This is the final result for the ratio for charged particles, expressed in terms of the corresponding ratio for photoionization and Compton scattering differential in energy transfer (or virtual photon energy). Fig. 3 schematically summarizes the connection between charged-particle collisions, photoionization, and Compton scattering in terms of their contributions as a function of energy and momentum transfer.

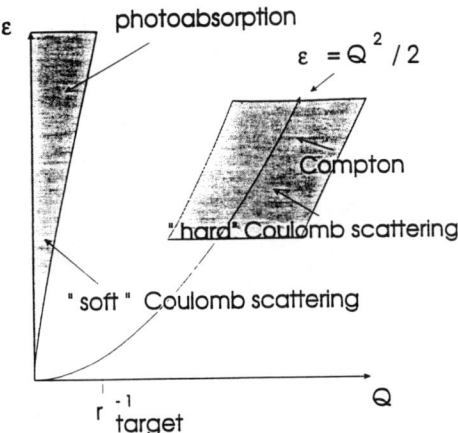

FIGURE 3. Dominant regions in the energy transfer (ϵ)-momentum transfer (Q) plane for Compton scattering, photoabsorption and charged particle scattering, schematically.

Intergrated over all energy transfers, the corresponding expression for the total cross section ratio becomes

$$R_Z = \sigma_Z^{++}/\sigma_Z^{+} = \int_{I_2}^{\infty} \frac{d\sigma_Z^{++}}{d\epsilon} d\epsilon \Big/ \int_{I_1}^{\infty} \frac{d\sigma_Z^{+}}{d\epsilon} d\epsilon$$

$$\simeq \int_{I_2}^{\infty} R_{PE} \rho_{Z,d}^{+}(\epsilon) d\epsilon + \int_{I_2}^{\infty} R_C \rho_{Z,nd}^{+}(\epsilon) d\epsilon \qquad (17)$$

where $\rho_{Z,d}^{+}(\epsilon)$ and $\rho_{Z,nd}^{+}(\epsilon)$ are respectively the dipole and non-dipole spectral density of states for single ionization by charged particles

$$\rho_{Z,d}^{+}(\epsilon) = \frac{d\sigma_{Z,d}^{+}}{d\epsilon} \Big/ \int_{I_1}^{\infty} \frac{d\sigma_Z^{+}}{d\epsilon} d\epsilon \qquad (18)$$

$$\rho_{Z,nd}^{+}(\epsilon) = \frac{d\sigma_{Z,nd}^{+}}{d\epsilon} \Big/ \int_{I_1}^{\infty} \frac{d\sigma_Z^{+}}{d\epsilon} d\epsilon. \qquad (19)$$

Application of Eq.(17) to *high-energy* charged-particle data permits very accurate tests of recent data for photoionization *near threshold*.

Shown in Fig. 4 are results for the ratio of double to single ionization of He by protons from the convolution of the energy distributions of single ion-

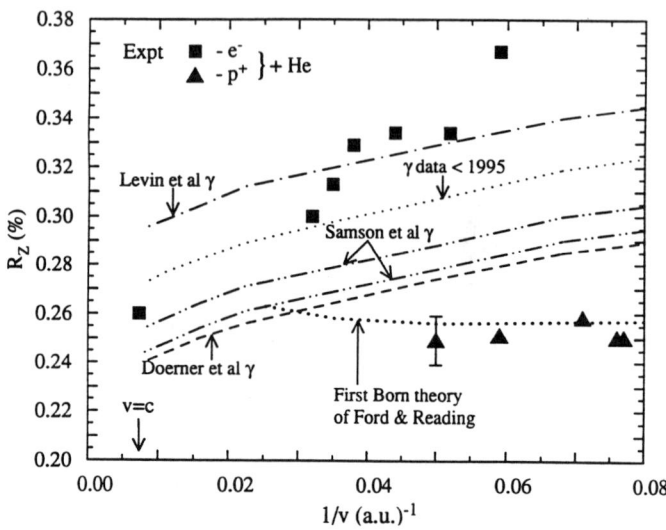

FIGURE 4. Ratio, R_Z, of double to single ionization of He by protons as a function of the inverse impact speed $1/v$. Starting from the top, the theory curves represent results using: photoionization (γ) data by Levin et al. [13] (dash-dot- line); γ data before 1995 [12] (thin dot line); γ data sets of Samson et al. [15] (dash-dot-dot line) taken at different facilities, ALS (upper one) and BNLS (lower one); γ data by Dörner et al. [14]; first Born calculation by Ford and Reading including partial waves s, p and d [39] (thick dot line). Symbols: Experimental data for electrons (squares) and protons (triangles) from [9,23]. The arrow marks the speed of light c.

ization with ratios by photoionization and Compton scattering. In addition to the photon data available prior to 1995 [12], we use four sets of new photoionization data R_{PE}: data by Levin et al. [13], by Dörner et al [14], and two sets by Samson et al. [15]. For the Compton ratio R_c we use our previous calculation [16,17]. The latter has been found to be in good agreement with the direct measurement of the ratio at high energies [10,18] and with measurements of Samson et al. for single ionization at low energies [19], shown in Fig. 5. We note that at low energies ($E \lesssim$ 5keV) minor differences to other calculations for ionization by Compton scattering [20,21] are not significant and do not influence the results of Fig. 4 since the Compton contribution to R_Z is less than $\approx 25\%$.

Use of R_{PE} data by Levin et al., increases the ratio R_Z when compared to the convolution based on photoionization data prior to 1995. The increase is entirely due to the larger values of R_{PE} very close to the threshold ($\lesssim$ 100 eV) compared to previous data. On the other hand, Dörner et al.'s data [14] are in accord with the original set near threshold but lower near the maxima and beyond. They result in an overall reduction of R_Z, but less than the reduction of R_{PE} near the maximum would suggest, for the reason just discussed. The theoretical data of Tang and Shimamura [22] give values

R_Z (not shown in Fig. 4) very close to the experimental data of Dörner et al, and are almost indistinguishable from the latter on the scale of Fig. 4. The two sets of data of Samson et al., are taken at different facilities and times (ALS and BNLS) and give results in between the other two experimental data sets.

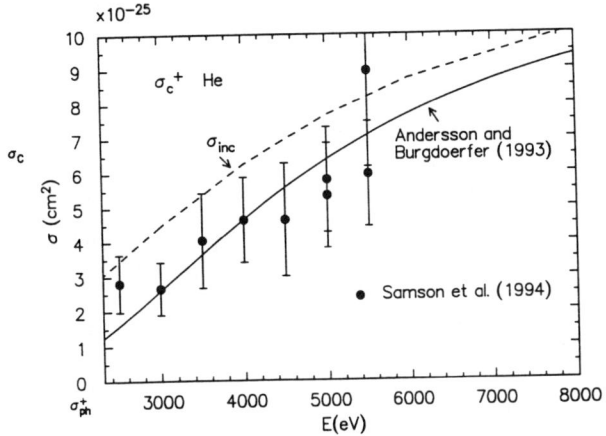

FIGURE 5. Comparison between normalized data for single ionization of helium by Compton scattering [19] and the theoretical prediction [16] using a highly correlated initial state and an uncorrelated Coulomb final state.

The high-velocity ratio of R_Z at 400 MeV is found to be 0.295%, 0.240%, 0.254%, and 0.244%, respectively, using the four sets of new experimental data. They are to be compared with the value of 0.26% [9,23] for charged particle data. Assuming this accepted value to be accurate within less than 10%, it would indicate that the R_{PE} data of Levin et al. are too high in the immediate region above threshold. It would also indicate Dörner et al.'s data may represent the lower bound of the accepted value. The data of Samson et al., give results in between the other two data sets and are closer to the asymptotic value of R_Z. Recent theoretical photoionization data have also been used as input to calculating R_Z at 400 MeV. The data by Tang and Shimamura [22] yield 0.240%, and the data by Meyer and Greene [24] in the acceleration and velocity gauges give 0.223% and 0.254%, respectively. The energy range covered by the calculation of Proulx and Shakeshaft [25] is insufficient for input to the convolution.

We emphasize the important characteristics of the ratio R_Z include not only the asymptotic value of R_Z, but also the way in which it is approached. Within the validity of the first Born approximation, or equivalently, within the framework of the generalized Weizsäcker-Williams theory, the R_Z values reconstructed from photoionization and Compton scattering data are correct only to the order Z^2. They should lie in between R_Z's for protons and electrons, thereby averaging out the sign dependence of higher order Z^3 contributions ("Barkas effect") to charged particle impact. Both the recent data by Dörner et al. [14] and Samson et al. [15] pass this test. Summarizing our discussion, we note that the application of the generalized WW method permits the reconstuction of the high-energy charged-particle rate of R_Z by photoionization at a level of accuracy of $\approx 10\%$ which can serve as a sensitive test for photoionization data and calculations near threshold.

3. The High-Energy Limit for Photoionization

Major experimental progress has been made toward the experimental determination of the high-energy limit of the ratio $R_{ph}(\omega \to \infty)$. Several measurements have been performed above 1 keV photon energies using synchrotron radiation [8]. As pointed out by Samson et al. [26], one difficulty with the measurements is the increasingly important admixture of ionization by Compton scattering which dominates photoionization above 6 keV. The difficulty was overcome recently by the COLTRIMS technique pioneered by the Frankfurt group [14] which permits the kinematic separation of photoionization and Compton scattering in the recoil ion momentum spectrum. Fig. 6 presents a comparison of the data with several calculations. The calculations involve a highly correlated initial state wavefunction, a 20-parameter

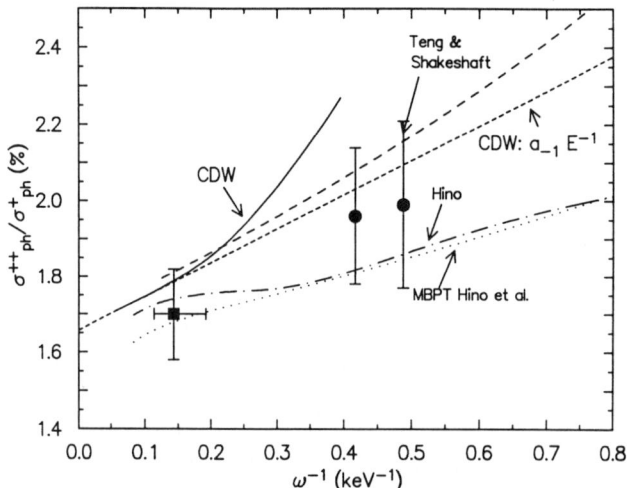

FIGURE 6. Comparison of the ratio $R_{ph} = \sigma_{ph}^{++}/\sigma_{ph}^{+}$ with recent experimental data [10,18,13] dash-dotted curve, independent-electron final state (ie); solid curve, final state including electron-electron distortion (e-e); dashed curve, linear extrapolation of the CDW e-e distortion calculation curve correct to order E^{-1}. Dash-double-dotted curve, MBPT calculations by Hino et al. [41] and Hino [42]. Arrows represent asymptotic calculations using Eq.[22].

Hylleraas-type wavefunction which satisfies the cusp condition at the origin to a good degree of approximation in order to accurately represent ground-state correlation. For the final state, both uncorrelated states of two Coulomb waves as well as approximate forms of correlated wavefunctions have been used.

At high energies, the direct treatment of the three-body Coulomb continuum is avoided by considering instead excitation-ionization and employing sum rules [27,28]. The final state with one continuum electron of momentum $\vec{k}$ ($k = \sqrt{2\epsilon}$) and one bound electron in a He$^+$(nlm) state is here represented by the wavefunction

$$\Psi_f^{(-)}(\vec{r}_1, \vec{r}_2) = \qquad (20)$$
$$(1\sqrt{2})[\phi_{nlm}(\vec{r}_1)\phi_{\vec{k}}^{(-)}(\vec{r}_2) \times D^{(-)}(\vec{k}_{12}, \vec{r}_{12}) + \vec{r}_1 \leftrightarrow \vec{r}_2],$$

where ϕ_{nlm} and $\phi_{\vec{k}}^{(-)}$ are bound and continuum states of He$^+$, respectively, and

$$D^{(-)}(\vec{k}_{12}, \vec{r}_{12}) = \qquad (21)$$
$$e^{-\pi\alpha/2}\Gamma(1 - i\alpha) \times_1 F_1[i\alpha, 1, -i(k_{12}r_{12} + \vec{k}_{12} \cdot \vec{r}_{12})]$$

is a Coulomb distortion factor describing the electron-electron interaction. $\vec{r}_{12}$ denotes the interelectronic position vector, $\vec{k}_{12} = \vec{k}/2$ is the relative momentum of the electron-electron system, and $\alpha = 1/k$. This final-state wavefunction is the bound-free analog to the two-electron continuum wavefunction used by Brauner, Briggs, and Klar [29] in their treatment of ($e, 2e$) processes, and by Maulbetsch and Briggs [30] in a study of the asymmetry parameter for double photoionization of He close to threshold. The uncorrelated final state is recovered from (21) by setting $D^- = 1$. The calculations with both correlated and uncorrelated final state wavefunctions converge to the same, universal, high photon energy value of $R(E = \infty) = 1.66\%$ in perfect agreement with the early calculations of Byron and Joachain [27] and Åberg [28], using the accurate 39-parameter Hylleraas-type initial state of Kinoshita [31], as well as the analysis by Dalgarno and Sadeghpour [32]. This confirms the conclusion that the exact value of $R(\infty)$ is independent of correlation in the final state and can be obtained by using an accurate representation of the ground-state wavefunction in the shake-off (or sudden-approximation) expression for photoionization:

$$R_{ph} = \frac{\Sigma_c |\int d^3\vec{r} \phi_c(\vec{r})\Psi_i(0,\vec{r})|^2}{\Sigma_b |\int d^3\vec{r} \phi_b(\vec{r})\Psi_i(0,\vec{r})|^2}. \qquad (22)$$

Here, the sum in the numerator includes all final states of the ionized second electron in the continuum (labeled c, with wavefunction ϕ_c) and the sum in the denominator extends over all bound states (labeled b, with wavefunction ϕ_b) of the spectator electron in the case of single ionization. The shake-off probability is given by the probability of finding the spectator electron at a position, $\vec{r}$, while the photoabsorbing electron is in the vicinity of the nucleus ($\vec{r} \simeq 0$). Because of the required momentum exchange, photoabsorption can take place only when the electron is localized in the vicinity of the nucleus (see Fig. 3).

Furthermore, we find that the leading-order correction as $E \to \infty$ is proportional to E^{-1} down to about 5 keV. Unlike $R(\infty)$, we find the asymptotic coefficient a_{-1} in the expression $R(E) = R(\infty) + a_{-1}E^{-1} + ...$ to be very sensitive to final-state correlations. For uncorrelated final states a_{-1} is very small (if E is measured in keV, a_{-1} =0.03 keV). For correlated final states we find $a_{-1} = 0.90$ keV in good agreement with the experimental high-energy data [18].

It is important to note that the analysis of the high-energy behavior was performed in the non-relativistic dipole approximation neglecting retardation effects. Numerical investigations of relativistic wavefunctions [33] have shown that R_{ph} is unaffected by retardation effects. A simple explanation for the independence can be given with reference to Eq. (22): The detailed form of the transition matrix element does not enter the high-energy limit at all. The only relevant input to Eq.(22) is that the primary photoabsorption process takes place near the nucleus because the recoil of the nucleus must balance the momentum transfer associated with the energy transfer to the atom by a high-energy photon. Since the photon momentum $k = E/c$, explicitly taken into account by retardation, remains small compared to the mechanical momentum of the electron $p_e \simeq \sqrt{2E}$,

$$k \ll p_e \tag{23}$$

for all $E \ll 2c^2$, retardation effects should be negligible for R and Eq. (22) should remain valid for photon energies small compared to the pair production threshold (1.022 MeV).

4. Double Ionization by Compton Scattering

Compton scattering differs from photoabsorption in several important aspects: (i) the relevant interaction operator is to first order A^2 rather than $\vec{p} \cdot \vec{A}$ for high but nonrelativistic photon energies $\omega \gg \epsilon_i$ (ϵ_i: binding energy of helium), (ii) Compton scattering intrinsically is a two-body rather than a three-body process (see Fig. 3). The role of the He nucleus is primarily that of a spectator rather than that of a participant balancing energy and momentum. One therefore expects R_c to be different from the corresponding ratio for photoabsorption. The doubly differential Compton scattering cross section as a function of the energy transfer ϵ and momentum transfer Q plane for inelastic scattering from an arbitrary initial state $|i\rangle$ to an arbitrary final state $|f\rangle$ may be expressed [1] to first order in A^2 as

$$\left(\frac{d^2\sigma}{d\epsilon dQ^2}\right)_c = \frac{\pi r_0^2}{2k^2}\left[1 + \left(1 - \frac{Q^2}{2k^2}\right)^2\right] F_I(\epsilon, Q^2), \tag{24}$$

with

$$F_I(\epsilon, Q^2) = \int d\Omega_e \sqrt{2(\epsilon + \epsilon_i)} \left|\langle f | \Sigma_{j=1}^N e^{i\vec{Q}\cdot\vec{r}_j} |i\rangle\right|^2 \tag{25}$$

the inelastic transition form factor (proportional to the generalized oscillator strength [4]) for the N-electron atom integrated over all emission angles of the emitted electrons and weighted with the density of continuum final states.

The dominant contribution to Compton scattering in Eq. (24) stems from the "Bethe ridge," i.e., from the region near the free- electron dispersion curve $\epsilon \sim Q^2/2$ [3] (see Fig. 3.). This contrasts with photoabsorption where the optically allowed region corresponds to $Q \approx 0$ or $\epsilon \gg Q^2/2$. Translated into coordinate space this implies that the entire probability distribution of the electron contributes to Compton scattering while only the near-nuclear region contributes to high-energy photoabsorption. Furthermore, a large number of partial waves contribute to the cross section. These differences manifests themselves directly in the behavior of two-electron transitions.

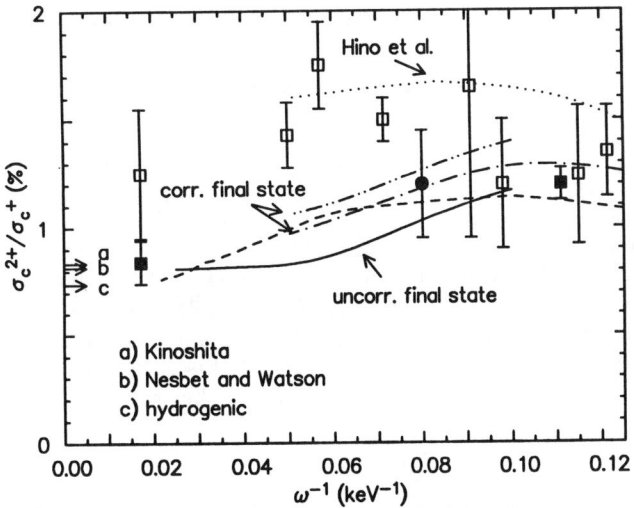

FIGURE 7. Ratios of $R_c = \sigma_c^{2+}/\sigma_c^+$ of double to single ionization by Compton scattering as a function of the ionized photon energy from Ref. 17: ———, uncorrelated final state; — · —, correlated final state [Eq.(21)]; — · · — correlated final state (see Ref. 17); · · ·, Hino et al. [20]; - - -, Anderson and Burgdörfer [16]. Asymptotic ratios (Eq.(26)) with initial state, (a) Kinoshita [31], (b) Nesbet and Watson [34], and (c) hydrogenic, □: Spielberger et al. [10]; ■: Wehlitz et al. [11], ●: Sagurton et al. [43].

The ratio of double to single ionization by Compton scattering R_c, as given by several calculations, together with most recent experimental data is shown in Fig. 7. The variation of R_c for different choices of final states gives an indication for the error introduced by the approximation of the final state at photon energies where the Compton electron is still comparably slow ($\lesssim 1$ keV). All our calculations tend monotonically to a value $R_c \simeq 0.8$ as $1/E \to 0$, or equivalently $E \to \infty$, in agreement with the shake limit for Compton scattering, as discussed below. This limit was independently found by Surić et al. [21] employing an impulse approximation. We also show in Fig. 3 the recent calculation by Hino et al. [20] for double ionization by Compton scattering employing a many-body perturbation theory approximation. Their values lie within the energy range shown systematically higher. Initially they were thought to converge to the asymptotic value $R_c(E \to \infty)$=1.67% for photoabsorption. Later it was realized that the high-energy limit had not been reached. The most recent data by Spielberger et al. [10] agree with our calculation while the data of Wehlitz et al. [11] appear to lie in between these two calculations. Because of the large error bar, definite conclusions are difficult to draw.

In analogy to Eq.(22), the shake limit for Compton scattering can be found by removing the constraint $r_1 = 0$ due to the fact that Compton scattering is a two-body scattering process that does not rely on the close proximity or recoil of the nucleus. Consequently, the entire radial distribution of the Compton scattered ("fast") electron contributes and one finds

$$R_c = \frac{\Sigma_c |\int d^3\vec{r}_2 \phi_c(\vec{r}_2)\Psi(\vec{r}_1,\vec{r}_2)|^2}{\Sigma_b |\int d^3\vec{r}_2 \phi_b(\vec{r}_2)\Psi(\vec{r}_1,\vec{r}_2)|^2} . \qquad (26)$$

Numerical evaluation of Eq.(26) yields values R_c=0.815% using the CI initial state wavefunction by Nesbet and Watson [34] and R_c =0.835% for the more accurate wavefunction of Kinoshita [31], also shown in Fig. 7. Using the hydrogenic wavefunction with Z=1.688 leads to the value $R_c = 0.73$% [16]. The shake-off value R_{ph} is approximately twice the value of R_c. Surić et al. [35] have recently shown that $R_{ph}/R_c \simeq 2$ holds along the isoelectronic sequence from H$^-$ to helium-like neon. Amusia and Mikhailov [36], on the other hand, found a Z dependence R_{ph}/R_c of this ratio which is near zero at $Z = 1$ (H$^-$), reaches 1 for helium, and approaches the asymptotic value 2 for $Z \to \infty$.

5. Differential Double Ionization of He by Compton Scattering and Charged Particles at Large Energy Transfers

The connection between double ionization by Compton scattering and charged particles can be explored in more detail when studied differentially as a function of energy transfer. While measurements for charge particles have very recently become available, investigations for Compton scattering should become possible with the availability of new synchrotron radiation sources.

The ratio of double to single ionization as a function of energy transfer ϵ is displayed in Fig. 8. The ratios of $R_c(\epsilon)$ for different incident photon energies, ω, are almost identical to each other below the binary encounter energy. They rise quickly from threshold and level off around 0.86% for $I_2 \ll \epsilon \leq \epsilon_{BE}$. $R_Z(\epsilon)$ shows a maximum near 200 eV and approaches the same value as Compton scattering in agreement with Eq. (14). In comparison with recent experimental data [37,38], also shown in Fig. 8, our results are in good agreement with the experiment for $\epsilon \geq 3$ keV. Our Compton ratio approaches the value 0.8% of the ϵ-integrated ratio discussed above which is dominated by large ϵ [41]. Considerable differences exist between theory and experiment toward lower ϵ. Experimental data show a decrease from about 2% at 1 keV down to 1% at 3 keV and are about twice as high as our theory in this energy range. Two sources of possible discrepancies are present in our theory: the lack of final state correlation and higher order terms of the Born series. The second order Born term, in particular the Z^3 effect, is known to be important for the total cross section in the energy range studied here [39,40]. However, the total cross section is dominated by small energy transfers. We believe that the final state correlations are most likely the cause of discrepancy between theory and experiment for ϵ between 0.5 and 3 keV, consistent with experimental observations.

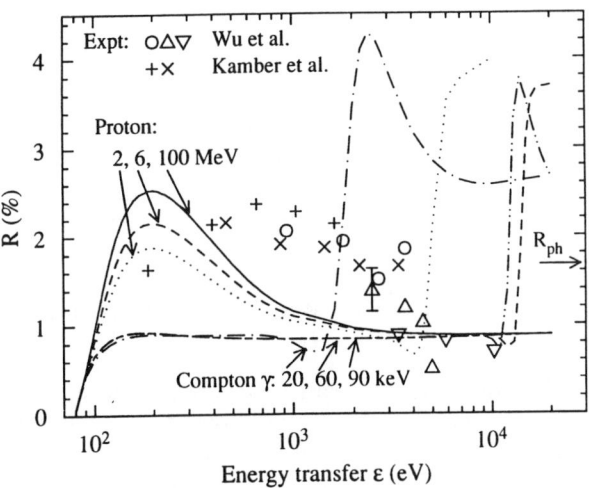

FIGURE 8. The ratio, R, of double to single ionization of He as a function of energy transfer ϵ by Compton scattering (20 keV, ----; 60 keV, --·--·--; 90 keV, — - — - —) and protons (2 MeV, ——; 6 MeV, - - - ; 100 MeV, ·····). Experimental data are from Wu et al. [37] (2 MeV, o; 3MeV, $\triangle$; 6 MeV, ∇) and from Kamber et al. [38] (3 MeV, +; 6 MeV, ×). R_{ph} denotes the high-energy photoionization limit 1.66%.

Fig. 8 displays another remarkable and surprising feature: a small dip followed by a sharp rise of both $R_c(\epsilon)$ and $R_Z(\epsilon)$ at $\epsilon \gtrsim \epsilon_{BE}$. The sharp rise is partially due to two-electron kinematics in double ionization. The rise in $R_c(\epsilon)$ could be experimentally tested provided sufficient photon flux is available since the cross section is small. At $\epsilon > \epsilon_{BE}$ the behavior of both $R_c(\epsilon)$ and $R_Z(\epsilon)$ is similar to the corresponding ratio for photoionization at much lower energies. These similarities have a simple explanation: for $\epsilon < \epsilon_{BE}$, the Compton photon and the charged particle can deliver sufficient momentum Q to balance ϵ such that the process is localized on the Bethe ridge $\epsilon = Q^2/2$. For energies beyond the binary encounter limit $\epsilon > \epsilon_{BE}$, the maximum momentum transfer Q^c_{BE} for Compton scattering is insufficient while the minimum momentum transfer $Q^Z_{min} = \epsilon/v$ for charged particles exceeds the Q value required for the Bethe ridge. In other words, for $\epsilon > \epsilon_{BE}$ both processes "fall off" the Bethe ridge, however, in opposite directions.

However, as Compton and charged particle scattering move away from the Bethe ridge, they cease to exist as two-body processes. Instead, they emerge as three-body processes with the recoil of the He nucleus required to bring the scattering processes back on the energy shell, just as for photoionization.

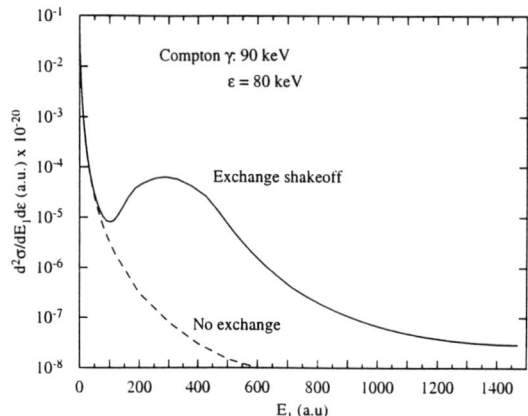

FIGURE 9. Electron-energy distribution for Compton scattering ($\hbar\omega = 90$ keV) within energy transfer $\epsilon = 80$ keV with and without exchange shake-off term (Eq.27).

The direct link to photoionization can be easily shown analytically in the limit of very large but non-relativistic energy transfer $\epsilon_{BE} \ll \epsilon \ll c^2$. In this limit, the cross section for emitting a fast electron with momentum, $\vec{k}_1$ and a slow electron with $\vec{k}_2$ by Compton scattering is proportional to

$$\sigma^{++} \propto |\int d^3r_2 \phi^*(\vec{k}_2, \vec{r}_2) \int d^3r_1 e^{i(Q^c_{BE} - \vec{k}_1)\cdot \vec{r}_1} \Psi_i(r_1, r_2, r_{12})$$
$$+ \int d^3r_2 \phi^*(\vec{k}_2, \vec{r}_2) e^{iQ^c_{BE}\cdot \vec{r}_2} \int dr_1^3 e^{-i\vec{k}_1\cdot \vec{r}_1} \Psi_i(r_1, r_2, r_{12})|^2 \tag{27}$$

where we have used the Hylleraas coordinates (r_1, r_2, r_{12}) to represent the initial state Ψ_i. Since for $\epsilon \gg \epsilon_{BE}$ we have $1 \ll Q^c_{BE} \ll k_1$, the first term in Eq.(27) is equivalent to the shake-off limit of photoionization (Eq. (22)). Note, however, that in addition to this direct shake-off term, there is also an "exchange shake-off" term. For the case of 20 keV photons shown in Fig. 8, $Q^c_{BE} \approx 8$. For this momentum transfer, the exchange shake-off term is important and causes the ratio to lie above the high energy limit of photoionization $\approx 1.66\%$. The exchange shake-off term peaks at $\vec{Q}^c_{BE} \simeq \vec{k}_2$, when the transferred momentum matches the momentum of the "slow" electron. This term causes a dramatic deviation for the U-shaped energy distribution of the emitted electrons (Fig.9). This effect should not be confused with the recently proposed contribution from equal energy sharing in photoionization at high energies [44]. However, an exchange shake-off contribution as discussed above is expected in photoionization, if retardation is taken into account. Its contribution to the transition amplitude relative to the direct term is, however, suppressed by the factor (k_2/k_1). In the limit $Q_{BE}^c \to \infty$, the exchange shake-off term becomes negligible. The present analysis uses only the A^2 interaction term. For $\epsilon > \epsilon_{BE}$ beyond the binary ridge the cross section due to the A^2 term rapidly decreases. It is therefore not obvious that it still dominates over the second-order contribution from the $\vec{p}\vec{A}$ term. A simple estimate involving the closure approximation indicates however that the $(\vec{p}\vec{A})$ amplitude should be smaller by a factor (ϵ/E) relative to the A^2 term. Therefore, the present approximation should be valid beyond the binary encounter energy $(\epsilon > \epsilon_{BE})$ as long as $\epsilon \ll \hbar\omega$. We emphasize that the contribution of the region $\epsilon > \epsilon_{BE}$ to the total cross section is very small and does not affect the integrated ratios [10,16,17]. This region is, however, of strong conceptual interest as it provides a unified description for double ionization by all three processes.

6. Summary

We have studied double ionization of He by photoionization, Compton scattering, and charged particle impact. In a generalization of the Weizsäcker-Williams method we relate the charged-particle cross section to photon processes, where soft collisions are related to photoionization while hard collisions can be related to Compton scattering. High-energy data for charged-particle collisions can serve as sensitive tests for photon data near threshold. We also discuss characteristic differences between the high-energy behavior for double ionization by photoionization and Compton scattering which result in different values for the ratio of double to single ionization. We also have presented results for differential ratios as a function of energy transfers. The ratios of double to single ionization for Compton scattering and for charged particles are shown to approach the same limit of 0.86% for large ϵ up to the binary encounter limit. For ϵ exceeding the energy transfer achievable in a binary encounter we find the ratio rises sharply. At higher ϵ the exchange shake-off mechanism is found to be important. The differential ratios for Compton and charged particle scattering are shown to approach the photoionization limit when the exchange shake-off term becomes negligible at very high ϵ. The emerging interconnection bewteen photons and charged particles impact can serve as guidance to further differential studies of multiple ionization processes.

Acknowledgements

This work was supported by the NSF and by the U.S. Dept. of Energy, Off. of BES, Div. of Chem. Sci., under Contract No. DE-AC05- 96OR22464 with LME Res. Corp.

References

1. W. Heitler, *The Quantum Theory of Radiation*, (Dover, New York, 1984) p. 414.

2. J.D. Jackson, *Classical Electrodynamics*, (Wiley, New York, 1975), Ch. 15.

3. J. Burgdörfer, L.R. Andersson, J.H. McGuire, and T. Ishihara, *Phys. Rev. A*50, 349 (1994).

4. M.R.C. McDowell and J.P. Coleman, *Introduction to the Theory of Ion-Atom Collisions*, (North-Holland, Amsterdam, 1970), Chpt. 7.

5. J. Wang, J.H. McGuire, and J. Burgdörfer, *Phys. Rev. A*54, 613 (1996).

6. E. Clementi and C. Roetti, *At. Data Nucl. Tables*, 14, 177 (1974).

7. J. Wang, J. McGuire, and J. Burgdörfer, *Phys. Rev. Lett.*, in press (1996).

8. For a review, see J.H. McGuire, N. Berrah, R.J. Bartlett, J.A.R. Samson, J.A. Tanis, C.L. Cocke, and A.S. Schlachter, *J. Phys. B*28, 913 (1995).

9. A. Müller, W. Groh, U. Kneissl, R. Heil, H. Ströher, and E. Salzborn, *J. Phys. B*16, 2039 (1983).

10. L. Spielberger et al., *Phys. Rev. Lett.*76, 4685 (1996).

11. R. Wehlitz et al., *Phys. Rev. A*53, R3720 (1996).

12. N. Berrah, F. Heiser, R. Wehlitz, J. Levin, S.B. Whitfield, J. Viefhaus, I.A. Sellin, and U. Becker, *Phys. Rev. A*48, R1733 (1993).

13. J.C. Levin, G.B. Armen, and I.A. Sellin, *Phys. Rev. Lett.*76, 1220 (1996).

14. R. Dörner, T. Vogt, V. Mergel, H. Khemliche, S. Kravis, C.L. Cocke, J. Ullrich, M. Unverzagt, L. Spielberger, M. Damrau, O. Jagutzki, I. Ali, B. Weaver, K. Ullmann, C.C. Hsu, M. Jung, E.P. Kanther, B. Sonntag, M.H. Prior, E. Rotenberg, J. Denlinger, T. Warwick, S.T. Manson, and H. Schmidt-Böcking, *Phys. Rev. Lett.*76, 2654 (1996).

15. J. Samson et al. *private communication*, (1996).

16. L.R. Andersson and J. Burgdörfer, *Phys. Rev. Lett.*71, 50 (1993).

17. L.R. Andersson and J. Burgdörfer, *Phys. Rev. A*50, R2810 (1994).

18. L. Spielberger et al., *Phys. Rev. Lett.*74, 4615 (1995).

19. J. Samson, Z. He, R. Bartlett, and M. Sagurton, *Phys. Rev. Lett.* **72**, 3329 (1994).
20. K. Hino, P. Bergstrom, and J.H. Macek, *Phys. Rev. Lett.* **72**, 1620 (1994).
21. T. Surić, K. Pisk, B.A. Logan, and R.H. Pratt, *Phys. Rev. Lett.* **73**, 790 (1994).
22. J.Z. Tang and I. Shimamura, *Phys. Rev.* A**52**, R3413 (1995).
23. L.H. Andersen, P. Hvelplund, H. Knudsen, S.P. Moller, A.H. Sorensen, K. Elsener, K.G. Rensfelt, and E. Uggerhoj, *Phys. Rev.* A**36**, 3612 (1987).
24. K.W. Meyer and C.H. Greene, *Phys. Rev.* A**50**, R3573 (1994).
25. D. Proulx and R. Shakeshaft, *Phys. Rev.* A**48**, R875 (1993).
26. J.A.R. Samson, C.H. Greene, and R.J. Bartlett, *Phys. Rev. Lett.* **71**, 201 (1993).
27. F.W. Byron Jr. and C.J. Joachain. *Phys. Rev.* **164**, 1 (1967).
28. T. Åberg, *Phys. Rev.* A**2**, 1726 (1970).
29. M. Brauner, J.S. Briggs, and H. Klar, *J. Phys.* B**22**, 2265 (1989).
30. F. Maulbetsch and J.S. Briggs, *Phys. Rev. Lett.* **68**, 2004 (1992).
31. T. Kinoshita, *Phys. Rev.* **105**, 1490 (1957).
32. A. Dalgarno and H.R. Sadeghpour, *Phys. Rev.* A**46**, R3591 (1992).
33. M. Kornberg and J. Miraglia, *Phys. Rev.* A**52**, 2915 (1995).
34. R. Nesbet and R. Watson, *Phys. Rev.* **110**, 1073 (1958).
35. T. Surić et al., *Phys. Lett.* **211**, 289 (1996).
36. M. Amusia and A. Mikhailov, *Phys. Lett.* A**199**, 309 (1995).
37. W. Wu, S. Datz, N.L. Jones, H.F. Krause, B. Rosner, K.D. Sorge, and C.R. Vane, *Phys. Rev. Lett.* **76**, 4324 (1996).
38. E.Y. Kamber, C.L. Cocke, S. Cheng, and S.L. Varghese, *Phys. Rev. Lett.* **60**, 2026 (1988).
39. J.F. Reading and A.L. Ford, *Phys. Rev. Lett.* **58**, 543 (1987).
40. J. Ullrich, R. Moshammer, H. Berg, R. Mann, H. Tawara, and R. Dörner, J. Euler, H. Schmidt-Böcking, S. Hagmann, C.L. Cocke, M. Unverzagt, S. Lencinas, and V. Mergel, *Phys. Rev. Lett.* **71**, 1697 (1993).
41. K. Hino et al., *Phys. Rev.* A**48**, 1271 (1993).
42. K. Hino, *Phys. Rev.* A**47**, 4845 (1993).
43. M. Sagurton et al., *private communication* (1996), and M. Sagurton et al., *Phys. Rev.* A**52**, 2829 (1995).
44. E. Drukarev, *Phys. Rev.* A**51**, R2684 (1995).

VIII. X-RAY ABSORPTION SPECTROSCOPY

X-ray Absorption and Dichroism of Transition Metal Compounds

Frank de Groot

Solid State Physics, University of Groningen,
Nijenborgh 4, 9747 AG, Groningen, Netherlands

Abstract. The analysis of x-ray absorption spectra to determine the electronic and magnetic structure of transition metal compounds is discussed. The models to describe the ground state of transition metal compounds (single-particle, impurity, crystal field) are introduced. Some basic aspects of the interaction of x-rays with matter are recapitulated and the description of x-ray absorption is separated into single-particle models for the 1s edges and multiplet models for the 2p edges. Magnetic circular dichroism is introduced and the six Thole sum rules are discussed. The complications and experimental problems of the sum rules are outlined. The last section briefly mentions some aspects of resonance studies, for which a detailed knowledge of x-ray absorption is crucial.

INTRODUCTION

Before the x-ray absorption process is discussed, first the models to describe the ground state of transition metal compounds are introduced. In the series of increased correlation transition metals are intermediate between the strongly correlated rare earths and the only weakly correlated sp-metals. The degree of correlation is roughly given by the ratio of the radial extension of the partly filled wave function and the distance to the nearest neighbours [1]. Within compounds, clusters, surfaces, adsorbates and molecules, the nearest neighbour distances are different, and more importantly the neighbour itself is a different element. For example, in case of a transition metal oxide the oxygen atoms dilute the system as far as the $3d$ electrons are concerned, with the result that the atomic correlated nature of the $3d$ electrons is increased.

To shed light on the ground state of transition metal compounds, one can start from the single-particle picture, which has been strongly developed within the concept of the local-spin-density approximation (LSDA) to density-functional theory. The ab-initio LSDA calculations are successful for the determination of the ground state crystal structure. Also a large series of experimental data can be nicely explained using LSDA as a starting point.

Experimental data more sensitive to the correlated nature of the electrons however, are not explained. This is true for example of the magnetic moments, and related properties, and in particular for the description of x-ray and electron spectroscopic data. The errors in the description of x-ray spectroscopies is caused by the incorrect treatment of final state effects, but also deficiencies in the ground state description can play a role. In order to describe this experimental data it is crucial to explicitly include atomic correlation effects. An important experiment in this respect is the photoemission - inverse photoemission (XPS-BIS) experiment on NiO, which revealed a band gap of about 3.5 eV, in contrast to a gap of only a fraction of an eV in normal LSDA calculations [2]. The basic model to account for this discrepancy is the (Anderson) impurity model which will be described below.

The Impurity Model

A transition metal compound, for example an oxide, is characterized by a strongly bonding/anti-bonding combination of the oxygen $2p$ states and the metal $4sp$ states. This generates an occupied valence band and some 5-10 eV higher, the empty conduction band states. In between, the strongly correlated metal $3d$ states are found, which are anti-bonding due to interaction with the oxygen $2p$-states.

To describe this system the impurity model singles out the $3d$ states and includes explicitly their atomic correlation effects. Because of this, the occupation of the $3d$ states is important. Using second quantization, the occupation number operator n_d is given by annihilating a $3d$ electron (a_d) and then re-creating it $(a_d^\dagger)$. The correlation energy is given by U_{dd}. The first two terms of the impurity Hamiltonian are the average energy of the $3d$ states, $\varepsilon_d n_d$, and the $3d$ correlation energy $U_{dd} n_d n_d$. Note that this second term is not included correctly within single-particle models such as LSDA. This strongly correlated, i.e. narrow, $3d$ band is positioned in between two broader bands, which are the valence band which has mostly ligand p character and the empty conduction band dominated by metal $4sp$ character. One can expect two types of charge excitations involving the narrow band: (a) from the $3d$-band to the empty conduction band, (b) from the filled valence band to the $3d$-band. Some older papers assumed excitation (a) to be dominant, but it has been shown several times that excitation (b) can explain most experimental data of the transition metal compounds. To include this charge excitation two extra terms must be added to the Hamiltonian: the energy of the valence band, summed over the wave vector k and the hopping term (t_{pd}) coupling the $3d$-band and the valence band. The total Hamiltonian is:

$$\mathcal{H}_I = \varepsilon_d n_d + \sum_k \varepsilon_{pk} n_{pk} + t_{pd} \sum_k (a_d^\dagger a_{pk} + a_{pk}^\dagger a_d) + U_{dd} n_d n_d$$

This impurity model was originally developed by Anderson to describe a magnetic transition metal impurity in a non-magnetic host [3]. It later turned out that the Impurity model could also be used in concentrated systems, in which case the model neglects the direct $3d3d$ overlap.

The excitation of the valence band to the localized $3d$ state costs an energy $\varepsilon_d - \varepsilon_p = \Delta$. Another excitation is a $3d3d$ charge fluctuation ($3d^N 3d^N \rightarrow 3d^{N-1} 3d^{N+1}$) which costs the energy U_{dd}. Depending on which of these two basic energies is lowest, a material can be classified as a Hubbard-system, if $U < \Delta$, or as a charge-transfer-system, if $\Delta < U$. This classification is known as the Zaanen-Sawatzky-Allen model [4].

The Impurity model turned out to be very fruitful for its application to core excitations [5,6]. The basic idea behind its use is that the screening of the core hole is most effective for localized states, which are the $3d$-states in transition metal compounds. The higher order terms of the two-electron interactions can also be included within the Impurity model. These atomic multiplet interactions form the basis of atomic multiplet and crystal-field multiplet models and will be discussed below.

The Crystal-Field Model

A further simplification of the model used to describe transition metal compounds can often be made. If the main interest of research focuses on the symmetry properties of the transition metal ion, and related properties such as the local magnetic moments, it is appropriate to use a model which includes only the metal $3d$ states. The focus is on the intra-atomic interactions, that is the two-electron integrals coupling the $3d$ electrons. Within second quantization these two-electron integrals are given as:

$$\mathcal{H}_M = \sum_{d_1,d_2} \sum_{d_3,d_4} G_{dd}\, a^\dagger_{d_1} a^\dagger_{d_2} a_{d_3} a_{d_4}$$

These two electron integrals give rise to three so-called Slater integrals F^0, F^2 and F^4, of which direct interaction F^0 controls the value of the Hubbard U. F^2 and F^4 control the energy levels of the $3d^N$ configuration and they give rise to Hund's rules. Hund's rules state that the configuration with lowest energy has the maximum spin and if more than one configuration exists with this maximum spin, the state with the maximum orbital quantum number is the ground state. $3d$ spin-orbit coupling gives rise to the third Hund's rule which determines the J value of the ground state according to coupling of S and L, being parallel for more than half-filled shells. The theory of a transition metal atom in a crystalline field has been treated in various textbooks [7,8]. Often the dominating crystalline field is the octahedral or 8-fold cubic surroundings. These fields belonging to the O_h subgroup of the spherical symmetry SO_3 have been studied in detail by Tanabe and Sugano [7]. The Tanabe-Sugano diagrams show the energies of the states as a function of the cubic crystal field. Crystal fields have some important consequences for the ground state properties of transition metal compounds. The field decreases the energy the t_{2g} states, which can give rise to low-spin ground states instead of the Hund's rule high-spin ground states. Lower symmetries, such as tetragonal fields, can cause the same effect, for example for divalent nickel ($3d^8$). Another consequence of the crystalline field is the quenching of the $3d$ spin-orbit coupling. If, in octahedral symmetry, the t_{2g} shell is partly filled, $3d$ spin-orbit coupling is still important, but if the e_g shell is partly filled the $3d$ spin-orbit coupling does not split the ground state energy in first order perturbation theory [7]. This has also some bearing on the Jahn-Teller effect, according to which a degenerate ground state is always split by symmetry distortions. It implies that the degenerate e_g states are sensitive to the Jahn-Teller effect and often elongate in the z-direction and split the energy levels of the $3d_{x^2-y^2}$ and $3d_{z^2}$ orbitals.

Within the crystal field multiplet model a transition metal ion in a solid is characterized by the number of $3d$ electrons and the symmetry of the atom in the solid. The ground state configuration is determined by the multiplet effects ($\mathcal{H}_M$) and the cubic crystal field. Both multiplet effects and crystal-field effects can be introduced in the impurity model [9,10].

The detailed symmetry of this configuration is determined by three interactions which are often of the order of ambient temperatures: crystal field distortions, the $3d$ spin-orbit coupling and (super)exchange interactions. The (super)exchange interactions are the inter-site magnetic couplings of the transition metal ions moderated by the oxygens ions. The magnetic interactions themselves can lower the symmetry from cubic but the structural effects will be minimal in that case. Another situation arises when the electronic ground state has a symmetry lower than cubic. In that case necessarily the magnetic structure will also have a symmetry lower than cubic, in other words there will be an anisotropy in the moments. As will be discussed below, x-ray absorption and x-ray dichroism, are very important tools to determine the moments of transition metal ions, and their anisotropy.

Comparison of Ground State Models

Three ground state models have been discussed: single-particle approaches, in particular LSD, the impurity model and the crystal-field multiplet model. In table 1 they are compared as far as their ground state and final state description are concerned. LSD can be performed with a complete basis-set (alternatively there are LSD methods not using basis-sets and it might be better to denote them as full-potential), while the basis-set for the impurity model and the crystal-field model are limited, in the crystal-field model only the $3d$ electrons are considered.

	Local-Spin-Density	Impurity	Crystal-field
GROUND STATE Basis Set	METAL *spdf* LIGAND *spdf*	METAL *d* LIGAND *p*	METAL *d*
dd-correlation	'J' spin	F^0, F^2, F^4 'U'+spin+orbit	F^2, F^4 spin+orbit
extensions:	$(LSD+U;+orbit)$		
FINAL STATE *pd*-correlation	F^0 core potential	F^0, F^2, G^1, G^3 core potential + multiplets	F^2, G^1, G^3 multiplets

TABLE 1. Comparison of local-spin-density, the impurity model, and the crystal-field model

X-RAY ABSORPTION

Independent of the model of the electronic structure used for the ground state, the x-ray absorption process and, in general, the x-ray excitation process can be largely described in the same way.

It has been described in many textbooks that the interaction of x-rays with matter ($\mathcal{H}_{int}$) can be treated as a small perturbation to the Hamiltonian of the system [11]. The transition probability W between a system in its initial state Φ_i and final state Φ_f is given by the Fermi golden rule:

$$W_{fi} \sim |\langle \Phi_f | \mathcal{T} | \Phi_i \rangle|^2 \, \delta_{E_f - E_i - \omega}$$

The δ-function takes care of the energy conservation and a transition takes place if the energy of the final state equals the energy of the initial state plus the x-ray energy. The squared matrix element gives the transition rate. The transition operator $\mathcal{T}$ is related to the interaction Hamiltonian $\mathcal{H}_{int}$ by the Lippmann-Schwinger equation:

$$\mathcal{T} = \mathcal{H}_{int} + \mathcal{H}_{int} \frac{1}{E_i - \mathcal{H} + i\Gamma/2} \mathcal{T}$$

Γ is the lifetime broadening of an excited state and $\mathcal{H}$ is the Hamiltonian of the unperturbed system. The Lippmann-Schwinger equation is solved iteratively and in first order $\mathcal{T}_1$ describes one-photon transitions, such as x-ray absorption, x-ray emission and x-ray photoemission. The electromagnetic field is described with both annihilation and creation operators a_{kq} and $a_{kq}^\dagger$ times a series of plane waves e^{ikr} with a polarization $\hat{e}_{kq}$. In case of x-ray excitations the creation operator can be safely neglected. The electromagnetic field acts on the momentum operator **p** of the electron.

$$\mathcal{T}_1 = \frac{e}{mc} \sum_{k,q} a_{k,q}(\mathbf{p} \cdot \hat{\mathbf{e}}_{\mathbf{kq}}) e^{ikr}$$

The golden rule can be approximated further if the wave-vector $\mathbf{k}$ is larger than the atomic distances $\mathbf{r}$. In case of soft x-rays $\mathbf{k} \cdot \mathbf{r} \leq e^2 \ll 1.0$ and all terms in the Taylor expansion of e^{ikr} vanish, i.e. e^{ikr} can be approximated to 1. If, for the moment, we integrate over the polarization degrees of freedom, this leaves for the dipole transition only $\mathcal{T}_1(E1) = e/mc \cdot \mathbf{p}$. Thus, in the expression of the golden rule only the momentum operator $\mathbf{p}$ appears. Because the momentum operator commutates with the position operator $\mathbf{r}$, also the position operator can be used in the golden rule.

Selection Rules in X-ray Absorption

The selection rules in x-ray absorption are caused by the matrix element contained in the Golden rule: $|\langle \Phi_f | \mathbf{r} | \Phi_i \rangle|^2$. This matrix element can be separated into a radial part and an angular part. The angular part is usually written in terms of a 3J-symbol. For a ground state with quantum numbers J and M_J it is:

$$\begin{pmatrix} J & 1 & J' \\ -M & q & M \end{pmatrix}^2$$

The polarization q couples the various M_J states. The triangular relations for the 3J-symbol give the selection rules: The J quantum number can be changed with maximally one ($\Delta J = +1, 0, -1$) and the polarization determines the change in M ($\Delta M = q$). Within the single-particle picture the ΔJ selection rule can be divided into two selection rules: the conservation of spin ($\Delta S = 0$) and a change of the orbital moment by exactly one ($\Delta L = \pm 1$).

Single-particle versus Multiplets

Knowing this basic expression for x-ray excitations, the problems occur in the determination of the ground state and final state wave functions and their energies. The previous section discussed the approximations for the ground state. The final state poses additional complications if similar accuracy is desired. The basic problem is that the final state contains a core hole. While all occupied deep core levels can be safely taken out of the ground state calculation, the core level with the hole must be included in the final state calculation. A problem arises from the fact that the core hole is a local phenomenon, which is a major complication to electronic structure calculations working in k-space. It can be solved, in principle, by increasing the unit-cell to a size for which the core holes do not interact with each other [12]. The extra potential of the core hole is only part of the problem and a far more complex problem is to deal with the two-electron integrals coupling the core level to the $3d$ states. Within the impurity model the core-hole induced terms of the Hamiltonian are:

$$\mathcal{H}_C = \varepsilon_c n_c + \sum_{p_1,p_2} \sum_{d_1,d_2} G_{pd}\, a^\dagger_{p_1} a^\dagger_{d_1} a_{p_2} a_{d_2} + \sum_{p_1,p_2} \xi_p a^\dagger_{p_1} a_{p_2}$$

ε_c is the core energy, G_{pd} contains the core hole potential and also the core hole multiplet effects (F^0_{pd}, F^2_{pd}, G^1_{pd} and G^3_{pd}). ξ_p is the $2p$ spin-orbit coupling splitting the $2p$ edge into the L_3 and L_2 edges. It turns out that in the case of a $2p$ core hole the multiplet terms of G_{pd} have energy effects of the order of 5 to 10 eV. Note that this is a surprisingly large value for a core state at 500 to 800 eV.

This multiplet interaction has until now never been included in ab-initio electronic structure methods. Within single-particle models it is completely neglected. The solution of single-particle methods is to calculate the final state with the core hole but without its multiplet interactions. Even methods which try to include correlation effects (LSD+U) and/or orbital-polarization effects completely neglect the multiplet effects of the core hole.

The importance and success of the crystal-field model and the Impurity model for core spectroscopy is that the core hole can be included in the calculation at the same level of description as the valence states. It is included as an additional localized state and all its two-electron integrals can be included in the same manner as the $3d3d$ integrals. Thus although the crystal-field model gives a rather poor description of the ground state its importance is that it correctly takes into account the core hole. Because the multiplet effects turn out to be so large in case of a $2p$ core hole, it makes in fact little sense to use a single-particle approach in order to determine the spectral shape. In case of a $1s$ core hole the multiplet interactions with the valence band is weak. There is a $1s3d$ exchange interaction but this is of the order of meV's, and can be safely neglected for the description of the spectral shape. Thus in case of K edge x-ray absorption it is possible to use single-particle models to describe both the ground state and, with inclusion of only the core hole potential, the final state. Note that the limitations of LSDA itself remain important. If correlation effects and/or orbital polarization are important for the ground state they must also be included to describe the x-ray absorption spectral shape.

Thus as a rule one can state that the single-particle approach is better for K edges, while $2p$ core states make it necessary to use a multiplet approach. In the next two section the problems within these two models are discussed.

The Single-Particle Approach

Within the single-particle approach the Golden rule becomes equal to a matrix element coupling the core wave function to the valence state, multiplied by the density-of-states.

$$W(\omega) \sim |\langle \phi_v(\omega)|\mathbf{x}|\phi_c\rangle|^2 \cdot \rho(\omega)$$

There are many methods to determine the density-of-states ($\rho(\omega)$). A recent overview of band structure methods applied to x-ray absorption can be found in [13]. For the application to x-ray absorption the unoccupied states are of special interest. Because most LSDA calculations are performed with the goal of determining the ground state emphasis is given to the correct description of the occupied states. For the calculation of the unoccupied states it will be necessary to give special attention to the calculation. In particular if basis-set methods, for example LMTO, are used it is important to increase the basis-set [12]. As already mentioned another important inclusion is the core-hole potential [12]. Concerning the extensions of LSDA with the inclusion of correlations, orbital polarization, or concerning the closely related self-interaction-correction [14], one can remark that to a first approximation these additions affect the occupied states. Their effect on the unoccupied states is only indirect due to the modifications in the occupied states [10].

Multiple Scattering versus Band Structure

The most important technique for calculating the x-ray absorption spectral shape is multiple scattering. Multiple scattering can be viewed as a LSDA method, and if performed within k-space it is a band structure technique [13]. It has been shown that full multiple scattering leads in fact to exactly the same result as band structure techniques like LAPW [15]. However usually multiple scattering is performed in real space. Real space calculations have the advantage for x-ray absorption that the core hole can be easily included. Other advantages are that one can calculate up to higher energies without problems and one uses the same method for calculations of the extended-fine-structure (EXAFS). It is also the most obvious point of view to describe the x-ray absorption process as a scattering process of an electron with a certain energy as given by the x-ray excitation. Also one can try to isolate the importance of particular scattering paths, etc. Viewed as electron scattering the x-ray absorption process has connections to other electron scattering experiments, for example LEED, and phenomena such as finite probing depths can be included. Recent reviews and papers on the progress of multiple scattering can be found in [15,16]. The applications of multiple scattering to circular dichroism can be found in [17].

The Multiplet Approach

As already mentioned the large two-electron integrals necessitate the multiplet approach. The impurity model for the final state includes the valence state interactions ($\mathcal{H}_I$), the 3d3d two-electron integrals ($\mathcal{H}_M$), the crystal field effects and the effects of the core hole ($\mathcal{H}_C$). The crystal-field multiplet model does not include the interactions with the delocalized bands and only contains the two-electron integrals of the valence states and between core and valence states, the core-hole spin-orbit coupling and the crystal-field.

Additional interactions, which can be added to both the crystal-field model and the impurity model, are the 3d spin-orbit coupling, lower symmetry crystal field effects and exchange effects. These interactions are smaller and they determine the precise nature of the ground state. The interplay of 3d spin-orbit coupling, local symmetry and exchange interactions determine all local properties of the transition metal ion, for example its spin and orbital moments and their asymmetries.

X-ray Absorption versus X-ray Photoemission

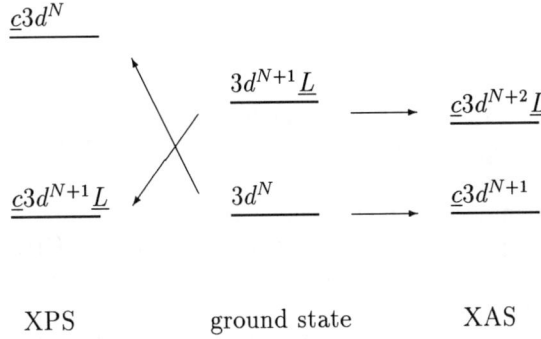

In XPS one detects the energy of the emitted electron, in other words one measures the core ionized state. In XAS one excites the core electron to an empty state and the final state of XAS can be assumed neutral. This difference between ionized state versus neutral state comes back if both XPS and XAS are described with the impurity-multiplet model, as visualized above for a typical charge-transfer-insulator.

In this scheme each case (ground state, XAS and XPS) is described with two configurations. For all three cases the actual states are formed by hybridizing the two states to bonding and antibonding combinations. In XAS the excited state levels with the core hole ($\underline{c}$) are reached by the dipole transition from the ground state levels as indicated by the vectors. The energy difference of the XAS final states equals $\Delta_{XAS} = \Delta - U + Q$. Because Δ_{XAS} is only a little smaller than Δ, assisted also by the fact that the final state hopping (t_{pd}) tends to be smaller than the ground state, the bonding combinations for both the ground state and the XAS state form a similar combination of the two states. This has the consequence that almost all transition strength goes to this bonding combination, while the antibonding combination gains only little intensity, in other words there are only weak charge-transfer satellites in x-ray absorption [9,10].

This is completely different in XPS where the importance of screening effects can be seen from the crossing transition vectors. The so-called well screened state gains no intensity in an ionic system for which all intensity goes to the poorly screened state. The stronger the hybridization, the stronger the well screened peak. There is a question concerning the lowest state of an XPS spectrum compared with an XAS spectrum. In a metal this lowest core hole state (Φ_C) can be reached both by XAS and XPS, in other words the XAS and XPS edges must be at the same energy. Often the situation is complex because the spectral shapes of XAS and XPS show significant structure, viz. satellites. For example, it is possible that the lowest state in XAS has little or zero intensity due to selection rules, or that the lowest state in XPS has little intensity due to poor screening.

MAGNETIC CIRCULAR DICHROISM

X-ray absorption experiments are carried out at synchrotron radiation sources producing x-rays which, at a bending magnet, are linearly polarized in the plane of the ring, with increasing asymmetry out-of-plane. With crossed undulators or asymmetric wigglers the intensity and degree of circular polarization can be significantly increased. This makes possible a series of new experiments making use of magnetic circular dichroism (MCD), the difference between left ($q = -1$ or $x - iy$) and right ($x + iy$) circularly polarized x-rays. Circular dichroism is sensitive to the local magnetic moment. Linear dichroism is the polarization dependence of normal and grazing incident x-rays, i.e. the difference between $q = \pm 0$ and $q = 0$. Linear dichroism is sensitive to layered structures and in general to all structures with a symmetry lower than cubic, either induced by crystallographic or magnetic interactions.

The Atomic Single-Electron Model

The basic model to discuss MCD effects is the atomic, single electron model as worked out by Erskine and Stern [18,19]. It describes the transitions of the $2p$ core state to the $3d$ valence state, neglecting the $3d$ spin-orbit coupling. The relative intensities of the reduced transition probabilities for left and right circular polarized x-rays are given in Table 2. The X-MCD signal, defined as the normalized difference in absorption between right and left polarized x-rays, can be calculated with the Table.

For the L_3 edge the MCD is $\frac{25-15}{25+15} = \frac{1}{4}$, while for the L_2 edge the MCD is $-\frac{1}{2}$. Together with the L_3:L_2 ratio of 4 : 2 this gives an X-MCD signal of $+1$ for the L_3 edge and -1 for the L_2 edge. Applying the orbital sum rule as will

	$\lvert\tfrac{3}{2},\tfrac{3}{2}\rangle$	$\lvert\tfrac{3}{2},\tfrac{1}{2}\rangle$	$\lvert\tfrac{3}{2},-\tfrac{1}{2}\rangle$	$\lvert\tfrac{3}{2},-\tfrac{3}{2}\rangle$
$Y_{22}^{\downarrow}$	-	**6**	-	-
$Y_{21}^{\downarrow}$	-	-	6	-
$Y_{20}^{\downarrow}$	-	1	-	**3**
$Y_{2-1}^{\downarrow}$	-	-	6	-
$Y_{2-2}^{\downarrow}$	-	-	-	18
	$Y_{11}^{\uparrow}$	$\sqrt{\tfrac{2}{3}}\cdot Y_{10}^{\uparrow}$	$\sqrt{\tfrac{1}{3}}\cdot Y_{1-1}^{\uparrow}$	
		$\sqrt{\tfrac{1}{3}}\cdot Y_{11}^{\downarrow}$	$\sqrt{\tfrac{2}{3}}\cdot Y_{10}^{\downarrow}$	$Y_{1-1}^{\downarrow}$

TABLE 2. *Relative matrix elements (squared) from the 3d ground states $Y_{lm}^{\downarrow}$ to the 2p states $\lvert j, m_j\rangle$ for the L_3 edge. Left circular polarization is indicated with boldface. The m_j states are written into their Y_{lm} functions at the bottom.*

be discussed below yields an orbital moment of zero in accord with the neglect of spin-orbit coupling.

Thole Sum Rules

An important step for MCD was made when Theo Thole and coworkers discovered that the integrated intensity of the MCD signal could be related to the *ground state* expectation value of the orbital moment [20]. In doing so, one encounters the problem that the absolute cross sections are normally not measurable. This problem was solved by normalizing the MCD signal by the x-ray absorption cross section, which in turn is related by a sum rule to the number of unoccupied states. This sum rule is also known as the initial state rule, relating the intensities of XAS spectra to the initial state, in contrast to the energy-positions (spectral shapes) which are given with the final state rule.

In memory of the late Theo Thole the sum rules used in the field of core spectroscopies will be named *Thole Sum Rules*. Thole, Carra and van der Laan have developed an important unification in the sum rules, by relating the measured quantities to a series of tensors. It turns out that from the combination of x-ray absorption, x-ray linear and circular dichroism, and spin-polarized photoemission, it is possible to determine all of the simple tensors [21,22].

The sum rules will be explained with the example of the $3d$ valence band of transition metals and the $L_{2,3}$ x-ray absorption spectrum. The x-ray absorption cross section (σ) is divided into three components: left (L, i.e. $x - iy$), right (R, i.e. $x + iy$) and z-polarized. The integrated sum over the complete $L_{2,3}$ edge can be indicated with the $\mathcal{T}_{00}$ Thole sum rule [22]. The first number refers to the orbital moments (dipole=1, quadrupole=2) and the second number refers to the spin moment (spin-averaged=0, spin-polarized=1). The dipole, or orbital, moment can be deduced from circular polarized x-rays ($\mathcal{T}_{10}$), the quadrupole moment from linear polarized x-rays ($\mathcal{T}_{20}$) and the spin moment from spin-polarized detection ($\mathcal{T}_{01}$). These three notions can be combined to form the six basic Thole sum rules [21,22].

The $\mathcal{T}_{00}$ sum rule applies to unpolarized x-ray excitations and spin-averaged detection. It determines the number of holes $\langle n_h \rangle$ in the bands to which the core electron is excited. This is the initial state rule:

$$\langle n_h \rangle = \int_{edge} (\sigma_L + \sigma_R + \sigma_Z)$$

Apart from excitation to the $3d$ states, a $2p$ electron can also be excited to $4s$ states and to higher excited s and d states. Thus in order to determine the number of $3d$-holes it is necessary to separate the spectrum into a $3d$-part and a part related to all other states. Fortunately the $2p \rightarrow 3d$ matrix elements are much larger than the $2p \rightarrow 4s$ matrix elements. Additionally the $4s$-states are assumed to be band-like and structureless. The usual approach is to cut the spectrum at some energy above the edge and to subtract the edge jumps of the L_3 and the L_2 edge. The assumption then is that the $4s$ and all higher states are subtracted as being part of the background. An implicit assumption is that the x-ray absorption cross section is related to the number of empty states, i.e. that the matrix element coupling the core state to the valence state is constant over the energy range under consideration and is equal for $2p_{3/2}$ and $2p_{1/2}$ core states.

The $\mathcal{T}_{10}$ sum rule for circular dichroism determines the orbital moment $\langle L_z \rangle$:

$$\langle L_z \rangle = 2 \int_{edge} (\sigma_L - \sigma_R)$$

The orbital moment is given by two times the integrated difference between left and right circular polarized x-rays. Note that this sum rule assumes the determination of the absolute x-ray absorption cross sections, which then can be directly related to the orbital moment.

The $\mathcal{T}_{20}$ sum rule for linear dichroism determines the quadrupole moment $\langle Q_{zz} \rangle$:

$$\langle Q_{zz} \rangle = 3 \int_{edge} (2\sigma_Z - \sigma_L - \sigma_R)$$

The quadrupole moment, or in other words the quadrupole charge distribution, is given by three times the integrated difference between grazing and normal incident x-rays. Again this sum rule is exact for absolute cross sections. The problem of the unknown absolute x-ray absorption cross sections can be overcome by dividing the T_{10} orbital moment and T_{20} quadrupole moment sum rules by the overall absorption cross section. This normalized sum rule was developed in the original paper [20]. It gives the orbital moment per hole.

Sum Rules in Spin-Polarized Photoemission

The most direct way to determine the spin-expectation value of the $3d$-electrons is by spin-polarized photoemission. The integrated difference between spin-up and spin-down electrons emitted from the valence band is a direct measure of the spin-moment $\langle S_z \rangle$, given by the T_{01} sum rule:

$$\langle S_z \rangle = \int_{edge} (\sigma_{L+} + \sigma_{R+} + \sigma_{Z+}) - (\sigma_{L-} + \sigma_{R-} + \sigma_{Z-})$$

One can combine the spin sum rule with the circular dichroism sum rule. This implies circular dichroic excitation and spin-polarized detection of the valence ($3d$) electrons. According to the T_{11} spin-orbit sum rule:

$$\langle \sum l_z(i) s_z(i) \rangle = 2 \int_{edge} (\sigma_{L+} - \sigma_{R+}) - (\sigma_{L-} - \sigma_{R-})$$

$\langle \sum l_z(i) s_z(i) \rangle$ is the sum of the spin-orbit interactions of the individual electrons. This sum rule can be viewed as a combination of the sum rules T_{10}, which is a measure of $\langle L_z \rangle = \sum l_z(i)$, and T_{01} which determines $\langle S_z \rangle = \sum s_z(i)$. Note that the T_{11} sum rule assumes photoemission from the $3d$-band. Recently experiments have been carried out which make use of circular dichroic x-rays and the detection of the spin-polarization of the electrons. These spin-polarized circular-dichroic (SP-CD) resonant photoemission experiments have been performed on copper oxides [23]. The difference between MCD and SP-CD is very important: MCD is given with $L^\uparrow + L^\downarrow - R^\uparrow - R^\downarrow$ divided by their sum, which equals $\frac{1}{4}$, using the numbers of table 2. Note that SP-CD signal can also be calculated from table 2 by multiplying the squared matrix elements with the percentage of spin-up respectively spin-down character of the $2p$ states. This SP-CD, or T_{11} signal, is $L^\uparrow - L^\downarrow - R^\uparrow + R^\downarrow$ divided by their sum, which equals $\frac{5}{12}$ [23,24]. An important difference between MCD and SP-CD is that the latter is internally referenced, hence sensitive to the local magnetic moment, irrespective of long-range order. It shares this internal referencing with techniques like local-spin-selective x-ray absorption (see below) [25] and spin-polarized photoelectron diffraction [26].

The combination of linear dichroism and spin-polarized detection gives the spin-quadrupole sum rule T_{21}:

$$\left\langle \sum q_{zz}(i,j) s_z(i) \right\rangle = 3 \int_{edge} (2\sigma_{Z+} - \sigma_{L+} - \sigma_{R+}) - (2\sigma_{Z-} - \sigma_{L-} - \sigma_{R-})$$

The spin-quadrupole operator, also called the magnetic dipole operator, is usually indicated by the symbol $\langle T_z \rangle$, and is important in the study of magneto-crystalline anisotropy [27]. It is mostly discussed in relation to the effective spin ($\langle S_e \rangle$) sum rule which is discussed below.

Effective Sum Rules

Apart from the six basic sum rules, additional sum rules have been developed. These sum rules use the fact that the $2p$ edges of the transition metal compounds are split into the L_3 and L_2 edge, separated by the $2p$ spin-orbit splitting. It has been shown by Carra and coworkers that two additional sum rules can be found from a linear combination of the L_3 and L_2 edges. These sum rules do not measure directly a single moment, but they measure a combination of two moments. The first of these effective sum rules determines a linear combination of the spin-moment and the spin-quadrupole interaction. This is because the spin-moment and spin-quadrupole moment both couple to an effective spin-operator $\langle S_e \rangle$, which equals $\sum \langle s_z(i) \rangle + \frac{7}{2} \sum \langle t_z(i) \rangle$. The effective-spin sum rule $\mathcal{T}_{01-21}$ is:

$$\langle S_e \rangle = \frac{3}{2} \int_{L_3} (\sigma_L - \sigma_R) - 3 \int_{L_2} (\sigma_L - \sigma_R)$$

This sum rule implies some additional approximations. The sum rule is only correct if the L_3 and L_2 edge have a ratio of two to one. Final state multiplet effects are important for the L_3:L_2-ratio if they are large compared with the core hole spin-orbit coupling. This is the case for the early $3d$ transition metals and the error in the effective-spin sum rule is more than 50% for the early $3d$ transition metals. The error decreases to values less than 10% for iron, cobalt and nickel. $4d$ and $5d$ transition metals have a larger $2p$ spin-orbit coupling, hence multiplet effects do not affect the L_3:L_2-ratio [28] and the sum rule is expected to hold as far as multiplet effects are concerned. A different complication arises from the ground state effect of the d spin-orbit coupling. Because the $4d$ and $5d$ spin-orbit coupling is larger, the spin-orbit effects on the L_3:L_2 intensity ratio are also larger. This creates potentially large values of T_z. The effective sum rule remains correct but the difference between the spin and the effective spin can become large due to the large T_z value. In $3d$ systems which are highly asymmetric with respect to the crystal symmetry, and hence potentially have large quadrupole moments, the spin-quadrupole coupling has been shown to be very weak, essentially because of the small $3d$ spin-orbit coupling [27]. This implies that the effective spin moment is a good measure of the spin moment. The averaged spin moment $\langle S \rangle$ in asymmetric systems can then be determined by angle averaging [27].

A second effective sum rule exists for linear dichroism with an effective operator $\langle P_e \rangle$, which can be related to a linear combination of the $\mathcal{T}_{11}$ spin-orbit operator and T_{31} operator, as has been discussed by Carra and coworkers [21].

Sum Rules in Experiments

To apply the T_{10} orbital moment sum rule to an experimental spectrum a number of problems arise. As discussed above a first approximation for the correctness of the Thole sum rules are fixed radial matrix elements. If radial matrix elements change with energy the correlation of x-ray absorption with the density-of-states breaks down, in other words this implies a break-down of the initial state rule(s). Another implicit assumption is that the core state from which the electron is excited is pure. Apart from shallow core levels this is correct to high accuracy. As discussed above absolute cross sections cannot be measured, necessitating the normalization of the T_{10} with the T_{00} sum rule to the familiar sum rule of the orbital moment per hole:

$$\frac{\langle L_z \rangle}{\langle n_h \rangle}(3d) = 2\frac{\int_{edge}(\sigma_L - \sigma_R)}{\int_{edge}(\sigma_L + \sigma_R + \sigma_Z)}$$

In experiment one determines σ_L and σ_R. σ_Z is approximated as the average of σ_L and σ_R, which is exact within single-particle models, but is only approximate within the multiplet model. In principle the only experimental parameter one has to know is the degree of circular polarization, defined as α. Then one can determine the $3d$ orbital moment/hole of the system under consideration.

An experimental complication is the requirement to determine which part of the spectrum to assign to the edge and which part to the background. The usual approach is to divide the spectrum into edge, background and EXAFS portions. The L_3 and L_2 parts both contain an edge-jump assumed to have the statistical ratio 2:1. By normalizing far from the edge and subtracting the background one retains the edge spectrum. A further decision one has to make is to decide where the $3d$ part of the spectrum stops and the EXAFS starts, in other words at which energy to stop the integration. For correlated systems with significant charge-transfer (satellites) this choice is not obvious.

A further set of complications arises with the detection technique used. The conceptually, though not technically, easiest measurement is by transmission. Then the measured signal should reflect directly the MCD. If one reverses polarization there is always the problem of the degree of polarization-detection, which includes potential differences in efficiencies α' for detection of σ_L and σ_R. The problem of detection efficiency is more important for electron-yield and fluorescence-yield techniques. In principle one can correct for this efficiency by normalization. A problem which occurs regularly is that the overall experimental system, consisting of synchrotron, monochromator, sample and detector is not stable with respect to the relative efficiencies for σ_L and σ_R.

An improvement to this instability has been found by the use of the so-called "mangle" [29].

Another detection problem is the occurrence of saturation effects. In fluorescence-yield this is a well-known problem and for the magnetically interesting materials (rare earths, Fe, Co and Ni, their alloys and multilayers) the $2p$ edges are always strongly affected by saturation. More subtle saturation effects exist, also in electron-yield detection. It turns out that the penetration depth of soft x-rays is not that much longer than the escape depth of the electrons. Furthermore there can be intrinsic errors in yield detection due to the energy-dependence of the Auger decay channels [30].

Another complication in spin-polarized detection is spin-filtering, the phenomena that in a magnetic medium the electron scattering properties are different for spin-up and spin-down. In MCD one does not detect the spin of the electron, but still there will be a correlation between the use of σ_L or σ_R and the creation of spin-up or spin-down electrons, in which case spin-filtering will affect the detector-efficiency for electron-yield.

A problem of a different type occurs if one wishes to determine the total orbital moment of the $3d$ states, that is one wishes to remove the number of $3d$-holes from the formula. In principle ab-initio band structure as well as quantum-chemistry calculations can determine the number of holes per site and per state. The number of holes at a particular site is, however, not uniquely defined. In particular one can ask the question which region of space to assign to a particular atom and where to stop the integration.

In principle there is the possibility to avoid all problems related to the normalization as well as the number of holes, by dividing the orbital sum rule by the effective spin sum rule [21]. In doing so however one introduces all the complications of the effective spin sum rule (being a linear combination of $\langle S_z \rangle$ and $\langle T_z \rangle$) and its intrinsic errors of up to 100% due to the incomplete separation of the L_3 and L_2 edge, as discussed above.

Once all sources of error have been accounted for correctly one has a last point of concern, which is the precise nature of the sample. Assuming electron-yield detection for the moment, there are a series of experimental boundary conditions, such as a UHV conditions and minimal stray-fields inside the chamber. Electron-yield has a probing depth of a few nanometers so one measures necessarily a surface effect. This implies the necessity of a well-defined surface system, both chemically and magnetically.

Given this series of complications, ambiguous procedures and potential experimental sources of error, combined with the small signals, it is no surprise that there is a significant scatter in the values of the orbital moment as determined from experiment. This is even more so for the effective-spin sum rule due to the additional intrinsic deviations.

An important parameter is the angular variation of the moments, in other words the difference between, for example, $\langle S_z \rangle$ and $\langle S_x \rangle$ in case of a low-symmetry system. These variations are smaller than the moments themselves, demanding an even better accuracy.

MCD Spectral Shapes

The six Thole sum rules are general results which can, in principle, be straightforwardly applied to experiment. As discussed above a series of complications arise for MCD. The additional complications for linear dichroism and particularly for spin-detection will not be discussed in this paper.

It is often not realized that the spectral shape itself can provide important additional information. A complication is that the spectral shape is dominated by final state effects, which as discussed are dominated by multiplet effects. Hence an interpretation of the spectral shape assumes a correct inclusion of these multiplet effects. Because of this complication, only few studies have attempted a detailed description of the MCD spectral shape of $3d$ transition metal systems [29,31].

RESONANCES

Recently we have seen a strongly increased interest in resonance studies. In resonance studies the x-ray absorption step is used to populate intermediate states which are probed with respect to their decay products. Both radiative and non-radiative channels are investigated. A radiative channel implies resonant fluorescence or in other words resonant x-ray emission. Because the overall process is a photon-in photon-out experiment this can also be denoted as resonant (inelastic) x-ray scattering. Another name used for experiments with excitation energies below the edge is resonant Raman scattering. Non-radiative decay involves Auger matrix elements, giving rise to resonant photoemission processes. If the final state cannot be reached via a direct photoemission channel, for example $2p3p3p$ decay, this can also be denoted as resonant Auger.

It is clear that a detailed understanding of resonance processes has as a prerequisite a detailed understanding of the x-ray absorption step. In the following some of the topics and unsolved questions in resonance studies will be mentioned.

Resonant Photoemission and Resonant Auger

Detecting the valence band at excitation energies in the region of the $2p$ x-ray absorption edge, multiplets are important for the description of the intermediate states with the $2p$ hole. The resonant photoemission spectrum changes shape according to the symmetry of the intermediate states. This can be combined with the use of MCD for the excitation and/or the detection of spin for the decay [23,32]. If the $3p$ photoemission or the $2p3p3p$ Auger spectra are measured at resonance, multiplet effects have to be included at both the intermediate and final states. A theoretical simulation including multiplet and charge transfer effects has been given for $3s$, $3p$ and valence band resonant photoemission for a series of transition metal oxides [34].

In case of resonant photoemission/Auger, charge transfer is important as the Auger decay ionizes the system and the configurations change their energy ordering significantly. This has been analyzed in detail for the case of $2p3p3p$ resonant Auger of CaF_2 [35]. An important result of resonance studies is that due to the increased number of spectral features the semi-empirical parameters used in the model Hamiltonians such as the Anderson impurity model can be determined with higher accuracy. An important phenomenon which can be studied is the ratio of resonant Auger to normal Auger, which signals the importance of Coster-Kronig and possibly other core-relaxation channels of the $2p$ core hole. These core-relaxation channels occur at energies lower than those predicted by the present models (see below).

Resonant X-ray Scattering

Resonant x-ray scattering is described with the generalized Kramers-Heisenberg formula [33]:

$$I(\Delta\omega) = \sum_{3d^9} \left| \sum_{2p,q,q'} \frac{<3d^9|r_{q'}|2p^5><2p^5|r_q|3d_0^9>}{E_{3d_0^9}+\omega-E_{2p}-i\Gamma_{2p}} \right|^2 \cdot L_f$$

The formula as given describes the resonant inelastic scattering of the $3d^9$ ground state to a series $3d^9$ states at the $2p$ edge. q denotes the polarization of the incoming x-ray and q' the emitted polarizations. There is interference within the intermediate states if (1) they reach the same final state and (2) they are separated less than their lifetime broadening (Γ_x). In the following some examples of resonant scattering are given:

2p3s Inelastic Scattering of CaF_2

The $2p3s$ resonant scattering of CaF_2 is described with the combined x-ray absorption and x-ray emission process $3d^0 \rightarrow 2p^53d^1 \rightarrow 3s^13d^1$. The dipole matrix elements have been calculated with the crystal field multiplet program. The ligand field multiplet calculations agree well with the experimental data, though some questions remain open concerning the possibility of differences in crystal field strength of singlet and triplet states [37].

The constant final state spectrum $I(\omega)|_{\omega-\omega'}$ maps out a specific final state while scanning through the absorption spectrum. In other words one can map out special symmetries of the intermediate states, that is the x-ray absorption final states. In case of CaF_2 the $3s^13d^1$ final states make it possible to map, for example, the E_g character or the *triplet* character of the $2p^53d^1$ intermediate states. The detection of the excitation spectrum obtained at a fixed decay energy $I(\omega)|_{\omega'}$ is in general a complex feature which depends on both the intermediate and the final state multiplets. It has been shown that interference effects in the intermediate states are not negligible. Apart from the resonant process in the experimental spectra there is clear evidence of the non-resonant $2p3s$ x-ray emission process [36]. As for the $2p3p3p$ Auger processes also in the case of $2p3s$ resonant x-ray emission experiments the non-resonant processes occur at energies lower than those predicted by the present models [37].

Consequences for Fluorescence Yield Detection

A similar description can be given for the resonant x-ray emission of the $2p3d$ ($L\alpha$) decay following the $2p3d$ excitation. An important difference between $2p3s$ x-ray emission and $2p3d$ x-ray emission is that the latter involves the partially empty valence band. The consequence is that, whereas the integrated $2p3s$ x-ray emission is constant for all $2p^53d^{N+1}$ intermediate states, the $2p3d$ x-ray emission can be strongly dependent on the symmetry of the intermediate states. With a fluorescence detector one measures essentially the $2p3d$ decay, thus a fluorescence yield experiment of $2p$ x-ray absorption contains the state dependent decay. Auger decay dominates over fluorescence for soft x-rays and also the Auger decay will be dependent on the symmetry of the intermediate states if it involves the $3d$-band. That is both $2p3p3d$ and $2p3d3d$ Auger decay channels are state-dependent. In the case of divalent nickel, it has been shown that the energy dependence of overall Auger decay is small. The $2p3d$ x-ray emission, being approximately two orders of magnitude weaker than the overall Auger decay is more than three times as large at the high energy side of the multiplet states as compared with the first states. This large variation gives rise to a distortion of the fluorescence yield detected spectrum from the x-ray absorption spectral shape [30]. The peaks at higher energy in the L_3 edge appear stronger than they are, and this is repeated for the L_2 edge. This phenomenon occurs even in the infinitely dilute limit, i.e. without saturation and self-absorption effects.

Resonant 1s3p X-ray Emission and Local Spin-Selective X-ray Absorption

The final states of $1s2p$ ($K\alpha$) and $1s3p$ ($K\beta$) x-ray emission spectra are dominated by strong multiplet effects, in contrast to their intermediate states with the $1s$ hole, which can be described with a single particle description. The final states are in principle the same multiplet states as for the $2p$ and

$3p$ x-ray absorption final states. However, the configurations reached will be different due to the screening of the $1s$ core hole in the intermediate state. Also the $1s2p$ transition matrix elements are different from the $2p3d$ matrix elements in x-ray absorption process.

In the case of divalent manganese the spectral shape can be described easily. The $3d^5$ ground state has all five $3d$ electrons parallel, hence a symmetry of $^6A_1$. Exciting the $1s$ electron to a state above the Fermi level does not affect the alignment of the five $3d$ electrons and though there will be screening of the $1s$ core hole (which can be described with the charge transfer model) the symmetry of the intermediate states will be essentially given by a $1s$ electron coupled to five parallel $3d$ electrons, i.e. $^5A_1$ and $^7A_1$. The $^5A_1$ and $^7A_1$ can be related to the spin-character of the emitted electron. The $^5A_1$ state relates to 100% spin-up, while 6/7 of the $^7A_1$ state relates to spin-down. The final state after the $1s3p$ ($K\beta$) x-ray emission process can be described with a $3p^53d^5$ configuration, split by a 15 eV $3p3d$ exchange splitting. Because of the relation with an escaping spin-up or spin-down electron, these decay channels can be used as a spin-detector. This makes it possible to determine the spin-up and spin-down selected intermediate states, within the single particle model equal to the spin-polarized density-of-states.

This technique is named local spin selective x-ray absorption spectra, because one detects the spin of the escaping electron referenced to the local situation of the spins of the $3d$-electrons. A signal is found in all cases where there is a local moment, irrespective of any ordering of these moments, in other words local-spin-selective x-ray absorption is internally referenced. For a ferromagnetic material, by comparison of this local spin selective x-ray absorption technique with MCD one can obtain information about the so-called Fano-factor P, the ratio between MCD signal and spin-polarization (for each energy ω) is equal to $\frac{\sigma_L - \sigma_R}{\sigma_\uparrow - \sigma_\downarrow}$. It has been shown that the Fano factor is strongly dependent on energy [25]. Instead of the $1s3p$ also the $1s2p$ ($K\alpha$) decay channels can be used. $K\alpha$ makes possible a separation between $j=3/2$ and $j=1/2$, which in turn can be related to spin-up and spin-down.

2p3d and 2p4d X-ray Emission of Gd Excited with Circularly Polarized X-rays

One can describe the $2p4d$ off-resonant x-ray emission of Gd in a similar way to the $1s3p$ x-ray emission of Mn. The calculation assumes a Gd ground state $4f^7[^8S]$ and a $2p$-photoemission process leaving the $4f$ electron in its $[^8S]$ symmetry. (Calculations show that this is 99% correct due to the small $2p4f$ multiplet interactions). The total symmetry of the $2p^54f^7$ intermediate states can be related to left and right circular excitation, much like the relation to spin-up and spin-down as described above. This makes it possible to relate a linear combination of $2p^54f^7$ intermediate states to left and right circularly polarized x-rays respectively. Calculation of the $2p^54f^7 \to 3d^94f^7$ transitions then gives the $2p3d$ spectral shapes and their MCD. Exactly the same procedure can be followed for the $2p4d$ spectral shapes. The $3d^94f^7$ final states are dominated by a large $3d$ spin-orbit splitting of about 32 eV. This separates the spectra into $3d_{5/2}$ and $3d_{3/2}$. The $4d^94f^7$ final states are dominated by the large $4d4f$ exchange splitting, separating the spectrum into spin-up and spin-down, much like the case of Mn described above. The atomic multiplet calculation describe the experimental spectra in great detail, including the MCD effects [38].

The local spin-flip spectral distribution obtained by resonant x-ray scattering

A last consequence of resonant scattering is described in this section: the possibility to detect spin-flip dd transitions by making use of the spin-orbit coupling at the $2p$ or the $3p$ resonance [39]. It can be shown that resonant inelastic x-ray scattering can be used to study the local spin flip excitation spectral distribution in magnetically ordered $3d$ transition metal compounds. In magnetically ordered materials the interatomic exchange interactions are quite large leading to average spin-flip energies, for example in high T_c compounds, of as much as 0.25 ev. Depending on the scattering geometry (i.e excitation/decay via left, right or z polarized xrays), one can determine the selection rules for the resonant scattering processes. It has been shown that in cubic symmetry a $3d^9$ system does show spin-flip transitions for $e_g(3d_{z^2})$ states, for example:

$$3d_{z^2}^\uparrow \xrightarrow{left} 2p_{3/2} \xrightarrow{z} 3d_{z^2}^\downarrow$$

In case of a tetragonal distortion, as in the high T_c superconductors, the spin-flip scattering is absent because of the symmetry properties of the $3d_{x^2-y^2}$ ground state, but the spin-flip scattering remains visible for the crystal field excitations [39].

ACKNOWLEDGMENTS

This research has been made possible by a fellowship of the Royal Netherlands Academy of Arts and Sciences.

REFERENCES

1. D. van der Marel and G.A. Sawatzky *Phys. Rev. B.* **37** 10674 (1988).
2. G.A. Sawatzky and J.W. Allen, *Phys. Rev. Lett.* **53**, 2339 (1984).
3. P.W. Anderson, *Phys. Rev.* **124**, 41 (1961); P.W. Anderson in: *Transition Metal Compounds*, Ed. E.R. Schatz, (Proc. Buhl Int. Conf. Pittsburgh, 1963).

4. J. Zaanen, G.A. Sawatzky and J.W. Allen, *Phys. Rev. Lett.* **55**, 418 (1985); J. Zaanen, G.A. Sawatzky and J.W. Allen, *J. Magn. Magn. Mat.* **54**, 607 (1986); J. Zaanen, PhD. thesis: *The electronic structure of transition metal compounds in the impurity model* (University of Groningen, 1986).
5. A. Kotani and Toyozawa *J. Phys. Soc. Japan* **35**, 1073 (1973); A. Kotani and Toyozawa *J. Phys. Soc. Japan* **37**, 912 (1974).
6. O. Gunnarsson and K. Schönhammer, *Phys. Rev. B.* **28**, 4315 (1983); O. Gunnarsson and K. Schönhammer, *Phys. Rev. B.* **31**, 4815 (1985); O. Gunnarsson, O.K. Andersen, O. Jepsen and J. Zaanen, in: *Core-Level Spectroscopy in Condensed Systems* Ed. by J. Kanamori and A. Kotani, (Springer Verlag, Berlin), 82 (1988).
7. S. Sugano, Y. Tanabe and H. Kitamura, *Multiplets of Transition Metal Ions* (Academic Press, New York, 1970).
8. J.S. Griffith, *The Theory of Transition Metal Ions* (Univ. Press, Cambridge, 1964); chapter 9.
9. K. Okada, A. Kotani and B.T. Thole, *J. Elec. Spec.* **58**, 325 (1992); K. Okada and A. Kotani, *J. Phys. Soc. Japan* **61**, 449 (1992).
10. F.M.F. de Groot, *J. Elec. Spec.*, **67**, 529 (1994).
11. M. Weissbluth, *Atoms and Molecules*, (Academic Press, New York, 1978).
12. M.T. Czyżyk and R.A. de Groot, in: 2nd *European Conference on Progress in X-ray Synchrotron Radiation Research, Rome, November 1989* Eds. A. Balerna, E. Bernieri and S. Mobilio, (SIF, Bologna, 1990), page 47; M.T. Czyżyk, R. Potze and G.A. Sawatzky *Phys. Rev. B.* **46**, 3729 (1992).
13. R. Zeller in: *Unoccupied Electronic States*, Ed. J.C. Fuggle and J.E. Inglesfield (Springer, Berlin, 1992), page 25.
14. Z. Szotek, W.M. Temmermann and H. Winter, *Phys. Rev. B.* **47**, 4029 (1993).
15. J.J. Rehr, *Proc. X-ray Absorption Fine Structure VII, Kobe, Jpn. J. Appl. Phys.* **32**, 8 (1993).
16. D.D. Vvedensky, in: *Unoccupied Electronic States*, Ed. J.C. Fuggle and J.E. Inglesfield (Springer, Berlin, 1992), page 139.
17. C. Brouder, M. Alouani and K.H. Benneman *Phys. Rev. B.* **54**, 1-sept (1996).
18. J.L. Erskine and E.A. Stern, *Phys. Rev. B.* **12**, 5016 (1975).
19. J. Stöhr and Y. Wu, *X-ray Magnetic Circular Dichroism: Basic concepts and theory for 3d transition metal atoms*, eds. A.S. Schlachter and F.J. Wuilleumier, NATO-ASI series E, volume 254 (Kluwer, Dordrecht, 1994), page 221.
20. B.T. Thole, P. Carra, F. Sette and G. van der Laan, *Phys. Rev. Lett.* **68**, 1943 (1992).
21. P. Carra, B.T. Thole, M. Altarelli and X. Wang, *Phys. Rev. Lett.* **70**, 694 (1993); P. Carra, H. König, B.T. Thole and M. Altarelli, *Physica B* **192**, 182 (1993).
22. G. van der Laan and B.T. Thole, *Phys. Rev. B.* **48**, 210 (1993); G. van der Laan, "Dichroic photoemission for pedestrians" *Core level spectroscopies for magnetic phenomena*, 153 (NATO-ASI vol. 345, 1995).
23. L.H. Tjeng, B. Sinkovic, N.B. Brookes, J.B. Goedkoop, R. Hesper, E. Pellegrin, F.M.F. de Groot, S. Altieri, S.L. Hulbert and G.A. Sawatzky (unpublished).

24. The relative intensities are respectively $L^\uparrow = 67$, $L^\downarrow = 8$, $R^\uparrow = 27$ and $R^\downarrow = 18$. This gives for the normalized MCD effect: $\frac{67+8-27-18}{67+8+27+18} = \frac{1}{4}$. The normalized SP-CD effect is: $\frac{67-8-27+18}{67+8+27+18} = \frac{5}{12}$.
25. F.M.F. de Groot, S. Pizzini, A. Fontaine, K. Hämäläinen, C.C. Kao and J. Hastings, *Phys. Rev. B.*, 51, 1045 (1995).
26. B. Hermsmeier, C.S. Fadley, M.O. Krause, J. Jimenez-Mier, P. Gerard and S.T. Manson, *Phys. Rev. Lett.* **61**, 2592 (1988).
27. J. Stöhr and H. König, *Phys. Rev. Lett.* **75**, 3748 (1995).
28. F.M.F. de Groot, Z.W. Hu, M.F. Lopez, G. Kaindl, F. Guillot and M. Tronc, *J. Chem. Phys.*, **101**, 6570 (1994).
29. E. Pellegrin, L.H. Tjeng, F.M.F. de Groot, R. Hesper, G.A. Sawatzky, C.F.J. Flipse, J.D. O'Mahony, Y. Moritomo, Y. Tokura and C.T. Chen (unpublished).
30. F.M.F. de Groot, M.A. Arrio, Ph. Sainctavit, Ch. Cartier and C.T. Chen, *Solid State Comm.*, **92**, 991 (1994).
31. G. Schütz, P. Fischer, K. Attenkofer, M. Knülle, D. Ahlers, S. Stähler, C. Detlefs, H. Ebert and F.M.F. de Groot, *J. Appl. Phys.*, **76**, 6453 (1994).
32. Articles in: *Core Level Spectroscopies for Magnetic Phenomena*, eds. P.S. Bagus, G. Pacchioni and F. Parmigiani. NATO-ASI series B vol. 345, (Plenum, New-York, 1995).
33. J. Tulkki and T. Åberg, *J. Phys. B.* **13**, 3341 (1980); J. Tulkki and T. Åberg, *J. Phys. B.* **15**, L435 (1982); T. Åberg and B. Crasemann, in: *X-ray Anomalous (Resonance) Scattering: Theory and Experiment*, eds. K. Fischer, G. Materlik and C. Sparks (Elsevier, Amsterdam, 1994).
34. A. Tanaka and T. Jo, *J. Phys. Soc. Japan* **63**, 2788 (1992).
35. F.M.F. de Groot, R. Ruus and M. Elango *Phys. Rev. B.* **51**, 14062 (1995).
36. J.E. Rubensson, S. Eisebitt, M. Nicodemus, T. Böske and W. Eberhardt, *Phys. Rev. B.*, **49**, 1507 (1994); J.E. Rubensson, S. Eisebitt, M. Nicodemus, T. Böske and W. Eberhardt, *Phys. Rev. B.*, **50**, 9035 (1994).
37. F.M.F. de Groot, *Phys. Rev. B.* **53**, 7099 (1994).
38. F.M.F. de Groot, M.H. Krisch and F. Sette (unpublished).
39. F.M.F. de Groot and G.A. Sawatzky (unpublished).

New Applications of X-Ray Magnetic Circular Dichroism

G. Schütz, P. Fischer, K. Attenkofer, D. Ahlers

*University of Augsburg, Institute for Physics, Experimental Physics II
Memmingerstr. 6 D-86135 Augsburg, Germany*

Abstract. X-ray magnetic circular dichroism (X-MCD) in core-level absorption is intimately related to the local spin and orbital polarization distribution in the final states, as it is based on angular-momentum conservation and the interplay of exchange and spin-orbit interaction. It provides symmetry and element-selective information about magnetic aspects of electronic structure. In favourable cases spin and orbital contributions to the local magnetic moments can be deduced directly by applying the so-called "sum-rules".

Recently studies of the dichroic contributions in the EXAFS energy range have attracted considerable attention. From systematic studies in various systems a direct proportionality between the strengths of the magnetic signal and the spin moment has been found. This can be easily explained by an exchange contribution to the Coulomb backscattering amplitude and gives important new insights into the exchange interactions of low-energy electrons in solids.

With the advent of new intense x-ray sources a variety of other x-ray methods, which involve core-level absorption, can be used to study magnetism. Recent examples are magnetic anomalous small-angle x-ray scattering and imaging magnetic domains with a zone-plate x-ray microscope, which both provide a high-resolution quantitative details of the spatial magnetization contributions.

INTRODUCTION

The effect of magnetic circular dichroism (MCD) is well known in the visible spectral region as the magneto-optical Kerr effect and in the γ-ray energy range as spin-dependent Compton scattering. In the x-ray energy range, however, where core-level photoabsorption is the dominant interaction process of photons with matter, magnetic circular dichroism (X-MCD) was first observed ten years ago at the K-edge of metallic iron [1].

The dichroic effects are generally defined either as the difference of the absorption coefficients for photon helicity antiparallel (μ^+) and parallel (μ^-) to the magnetization direction $\mu_c = (\mu^+ - \mu^-)/2$ or as the normalized contribution $\mu_c/\mu_0 = (\mu^+ - \mu^-)/(\mu^+ + \mu^-)$. While at K-edges μ_c/μ_0 is smaller than

10^{-2}, the dichroic effects at L-edges appear to be significantly stronger as it has been observed in Gd and Tb metal [2]. *Giant* dichroic signals of more than 20% were reported for the first time at the $L_{2,3}$-edges of 5d-impurities [3] in iron. As expected from these pioneering studies which were performed in the hard x-ray region with energies above 5 keV the existence of strong X-MCD effects in the soft x-ray range have been observed at $L_{2,3}$-edges of Fe, Co and Ni and at the rare earth $M_{4,5}$-edges [4] which was predicted earlier by atomic multiplet calculations [5].

In the meantime a variety of theoretical models have been developed to calculate the near-edge X-MCD as described in the contribution to this volume by de Groot and in [6]. Nowadays, one of the most interesting aspects is the possibility of extracting local spin and orbital moments. To do this the magnetic L_2- and L_3 or M_4 and M_5 dichroic spectra are interpreted by applying *sum rules* which have been derived either on the basis of an atomic approach [7,8], or in the vector-coupling model [9].

The occurrence of a dichroic contribution to the EXAFS (spin-polarized EXAFS = SPEXAFS) has also proven to be a universal phenomenon [6,10,11]. As demonstrated in this report, the systematics observed in several studies show that, based on a two-step model, interesting correlations of the SPEXAFS to the local magnetic short-range order, the magnetic moment and the spin density of the neighbouring atom, have been found. This demonstrates the potential of the method to provide novel insights into the exchange phenomena of spin-polarized electrons with ferromagnetic materials.

SIMPLE THEORETICAL MODEL

In this section we briefly outline in a simple two-step vector-coupling model the basic concepts of X-MCD, which also provides a satisfactory interpretation of the dichroic effects in the EXAFS energy range. In this naive picture similar to the approach by Erskine and Stern [12] the sensitivity of X-MCD on the polarization of the final states stems from the *Fano-effect* [13]. After absorption of a circularly polarized photon the outgoing photoelectron acquires in the free atom limit an expectation value of its spin $\langle \sigma_z \rangle$ in photon beam direction $\hat{z}$. Thus for an initial p-state the value of $\langle \sigma_z \rangle$ amounts to $-1/2$ and $+1/4$ after L_2- and L_3-absorption, respectively, resulting from angular momentum conservation and spin-orbit splitting of the initial $p_{1/2}$ and $p_{3/2}$ states. Consequently in this simple model the spin-polarized photoelectron which is ejected after the absorption of a circularly polarized photon acts as a probe, which monitors the spin direction of the final majority (ρ^+) and minority (ρ^-) states in accordance with Pauli's exclusion principle.

If the energy dependence of the radial matrix element $M(E)$ and its sensitivity to the final-state spin or orbital quantum numbers can be neglected, as it is assumed by the formulation of the sum rules, one can use the ansatz

$\mu^{\pm} = \mu_0 \pm \mu_c \sim |M(E)|^2(\rho \pm \langle\sigma_z\rangle \cdot \Delta\rho_s)$, where ρ and $\Delta\rho_s$ denote the total density of states $\rho = \rho^+ + \rho^-$ and $\Delta\rho_s = \rho^+ - \rho^-$. Thus the normalized dichroic signal can be directly correlated to the normalized spin-density of the absorbing atom by:

$$[\mu_c/\mu_0](E) = \langle\sigma_z\rangle \cdot [\Delta\rho_s/\rho](E). \qquad (1)$$

Consequently in this approach the ratio $[\mu_c/\mu_0]_{L_2} / [\mu_c/\mu_0]_{L_3}$ between the X-MCD at the L_2- and L_3-edges should equal exactly -2 (or $\mu_c(L_2)/\mu_c(L_3) = -1$ since the dipole transition strength is twice as large at the L_3 edge than at the L_2 edge).

In magnetic systems, which can be described as pure *spin-band* ferromagnets, the ratio of -2 is indeed observed. A deviation from this value can be understood by the significant influence of an orbital polarization present in the final states, since the ejected photoelectron has also an expectation value of the orbital moment of $\langle l_z \rangle = +3/4$ for an initial *p*-state due to the constraint $\Delta m_l = 1$ independent of its spin-orbit configuration. Thus the photoelectron probes both the spin of the final states and their orbital polarization.

For initial *s*-states, e.g. K and L_1 edges, the interpretation of X-MCD is still an open problem. An estimate of the Fano parameter deduced from the systematics observed experimentally gives an only very small value of $\langle\sigma_z\rangle \sim 10^{-2}$ [1], which on the other hand is correlated to weak MCD effects of less than 1 % correlated with the final *p*-state spin polarization. In general for these more delocalized states the orbital moment is nearly quenched. However, the sensitivity of the dichroic signal to an orbital polarization is approximately one order of magnitude larger since at K-edges $\langle l_z \rangle$ is 100% introducing a complex relation of the K-edge MCD to the spin and orbital moments.

The *Fano effect* can be also used to explain the occurrence of a magnetic contribution to the EXAFS. The EXAFS oscillations occuring in the absorption profiles about 30 eV beyond an edge result from an interference of the outgoing electron wave with the incoming waves being backscattered at the next neighbors. Since as outlined above for circular polarized x rays the outgoing photoelectron wave is spin-polarized, in the case of an alignment of the magnetic moment of the backscattering atom an exchange contribution in addition to the Coulomb scattering potential is involved in the scattering process. These exchange effects, which are the origin of the magnetic EXAFS can be well described by an additional exchange scattering contribution to the overall scattering amplitude as shown recently [14,15].

EXPERIMENTAL RESULTS AND APPLICATIONS: NEAR-EDGE X-MCD

Fe-metal layers

To demonstrate, that especially for the heavier 3d transition metals the sum rules work rather well, they are applied to $L_{2,3}$-MCD spectra of Fe-metal layers. The exact thickness of the 9.25 Å and 24 Å Fe-layers (deposited on 300 Å Au on a glass substrate and protected by a 30 Å Al coverlayer) and magnetic moment per atom of $2.07(3)\mu_B$ for the thin and $2.14(3)\mu_B$ for the thicker sample were determined via XFA and SQUID measurements.

Figure 1 shows a strong deviation from the ratio $\mu_c(L_2)/\mu_c(L_3) = -1$ which would be expected for pure spin moments. This indicates the existence of an orbital moment coupling parallel to the spin. Applying the sum rules

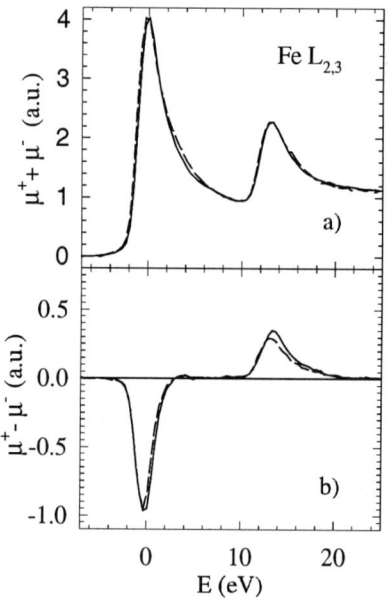

FIGURE 1. Fe $L_{2,3}$ absorption profile $(\mu^+ + \mu^-)/2$(upper part) and corresponding dichroic signal $(\mu^+ - \mu^-)/2$(lower part) of an Fe layer on Au with Fe thicknesses 24 Å(solid line) and 9.25 Å(dashed line)

described in [9] a spin and orbital moment of $m_S \sim 2.02\mu_B$ $(2.08\mu_B)$ and $m_L \sim 0.20\mu_B$ $(0.15\mu_B)$ can be deduced with an uncertainty of about 5%. In particular for the spin sum rule the complex additive term T_z, which is not well known, can be set to zero, as it is expected for Fe, Co and Ni. These values are in excellent agreement with the results of the macroscopic magnetic measurements and confirm the expected increase of m_L with decreasing layer thickness. However, the absolute values of m_L seems to be systematically larger than expected from neutron and g-factor measurements indicating the restricted reliability of the sum rules with respect to absolute numbers.

Magnetic imaging of domains via Fe $L_{2,3}$ X-MCD

In principle the dichroic effect can provide magnetic contrast in all spectroscopic x-ray imaging techniques involving element-specific absorption close to an absorption edge. Recently we have developed a novel domain imaging technique using the transmission x-ray microscope (TXM) at BESSY I in combination with a contrast enhancement based on x-ray magnetic circular dichroism [15,16]. The target was an amorphous GdFe layer with an easy

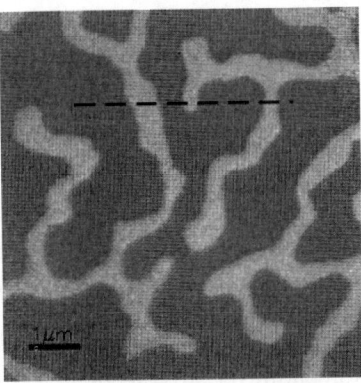

FIGURE 2. Image of magnetic domains at a magnetic field $12mT$ below $-H_c$.

magnetization direction perpendicular to the surface. As shown in Figs. 2 and 3 this method provides a high-resolution study of the hysteresis properties of the domain pattern and their long-term variation in real-time transmission mode with a spatial resolution of less than $50 nm$.

While at photon energies below the Fe L_3 absorption edge the spatial intensity distribution is homogenous, significant structures appeared at $E = 706 eV$ (see Fig. 2). As the photons are left circularly polarized the magnetization of the darker/lighter shaded domains point antiparallel/parallel to the photon beam direction.

In Fig. 3 the absorption profile $(\mu - \mu_{sat})x$ along the dashed line marked in Fig. 2 is shown where μ_{sat} corresponds to an image with maximum absorption. An analysis of the wall profiles proves that their widths are smaller than the experimental resolution of $50 nm$, which is in good agreement with the esti-

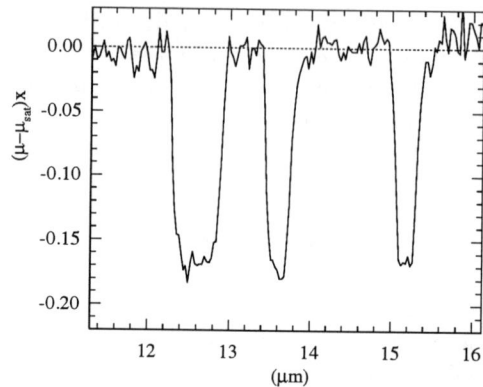

FIGURE 3. Absorption scan across the region indicated by the dashed line in Fig. 2. The displayed absorption is normalized to the intensity for saturation magnetization.

mate of the wall width $\delta_{dw} = \pi\sqrt{A/K_u} \sim 30\,nm$ depending on the exchange and the anisotropy constant A and K_u.

The difference of the absorption for reversed domain magnetization yields values of $\sim 17\%$. This proves the complete alignment of the Fe magnetic moment in the photon beam direction, i.e. perpendicular to the plane of the target.

These results demonstrate that the combination of X-MCD with TXM allows imaging magnetic domains with a spatial resolution comparable to that of magnetic force microscopy, scanning electron microscopy with polarization analysis (SEMPA) and Lorentz microscopy measurements. The major advantages of our new technique, however, are the element specificity and quantitative nature and the fact that this method is undisturbed by varying magnetic fields applied to the sample.

CrO_2

Going to the lighter $3d$ elements the experimental separation of the corresponding L_2- and L_3- contributions is not possible due to the decrease of the $2p$ spin-orbit splitting below 10eV, which on the other hand is indispensable for the application of the spin sum rules. In contrast to Fe,Co and Ni a theoretical description of the X-MCD in these elements based on an atomic model is more appropriate. As demonstrated in Fig. 4 (left panel) a comparison with the calculation of the dichroic effects for transition metal ions in the ligand-field multiplet approach is indeed able to reproduce the global characteristics of the experimental findings [17]. Setting the spin-orbit splitting to zero in these calculations gives the best agreement between theory and experiment. The vanishing influence of an orbital momentum and correlated spin-orbit effects can also be verified by the application of the orbital sum rule, which corresponds to an integration of the complete L_3 and L_2-spectra. One finds an upper limit for m_L to be less than $10^{-2}\mu_B$.

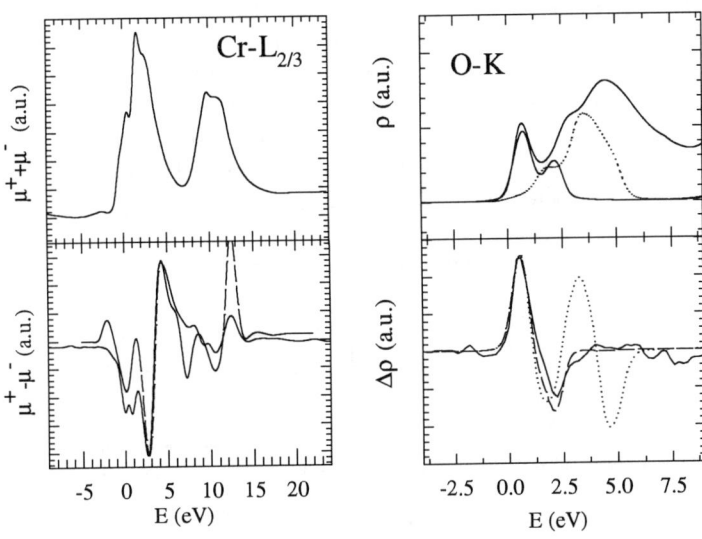

FIGURE 4. Left panel: Cr $L_{2,3}$ absorption profile (upper part) and corresponding dichroic signal (lower part) in comparison with atomic ligand-field multiplet calculations (dashed). Right panel: Oxygen K-spectrum (upper part) and corresponding dichroic signal (lower part) in comparison with the p-projected unoccupied DOS (dashed line in upper part) and the corresponding spin-density difference (dashed line in lower part).

The rather small theoretically calculated magnetic p-moment of O of $0.04\mu_B$ is induced via super-exchange and the correlated spin polarization of the unoccupied p-band is only weak. Nevertheless, it was possible to detect the corresponding rather small dichroic signal at the O K-edge as shown in Fig. 4 (right panel). The experimental data are compared with results of spin polarized density-of-states (DOS) calculations, separated into contributions from the c– and a–axis [18]. Obviously the contributions from the c-axis alone are reflected in the dichroic signal following the simple two-step model for spin-only systems. Furthermore, as it has been found for the description of the Cr L-edge X-MCD, the orbital polarization of the oxygen p-states is negligible and relation (1) can thus be applied. An estimate of the local p-moment of $(1/\langle\sigma_z\rangle) \cdot \mu_B \int (\mu_c/\mu_0)(E)\rho(E)dE \approx 0.06\mu_B$ is in good agreement with the theoretical predictions [18]. The spin density projected onto the a–axis is only weak. This reflects possibly the small alignment of this magnetic component under the experimental conditions of the finite external field and the elevated temperature.

Gd $L_{2,3}$ X-MCD

The $L_{2,3}$ X-MCD in rare earths represents the limits of the applicability of the sum rules. The ratios of the normalized Gd $L_{2,3}$ MCD profiles as seen in Fig. 5 are close to -2, as it is expected for pure spin polarization of the final states. Applying the sum rules with $T_z = 0$ yields a spin and orbital moment of $m_S = -0.24\mu_B$ and $m_L = -0.004\mu_B$ in significant disagreement to the theoretical calculations, which predict $\mu_S = +0.47\mu_B$ for the spin and $\mu_L = -0.04\mu_B$ for the orbital d-moment [19].

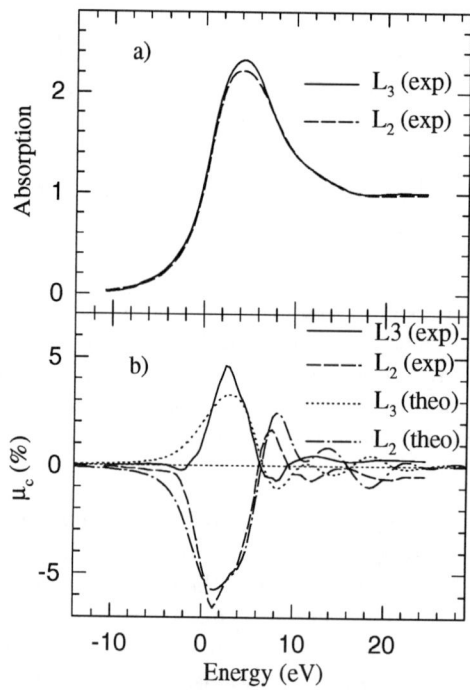

FIGURE 5. Upper part: Gd L_3(solid line) and L_2-absorption (dashed line). Lower part: Corresponding dichroic contribution in comparison with the fully-relativistic band-structure calculations for the X-MCD at the L_3 (dotted line) and the L_2-edge (dash-dotted line).

As discussed in [11] the physical origin of this failure can be attributed to a strong energy and spin dependence of the matrix element, since close to E_F the overlap of initial and final spin-up wavefunctions is much larger for the corresponding states of minority character in the sum rules. It is also possible that the term T_z, which has been neglected up to now, plays an important role. Nevertheless the MCD calculations shown in Fig. 5 using single particle band-structure approaches [20] are able to reproduce the experimental findings.

Magnetic small-angle x-ray scattering

One common way to study size distributions and correlations of particles in the nm range is small angle x-ray scattering (SAXS). Using tunable x rays available at synchrotron sources the technique of contrast enhancement (anomalous SAXS = ASAXS) can be applied, which exploits the variation of the element-specific anomalous scattering amplitude $f'(E)$. Since this varies with energy by about 20% in the vicinity of absorption edges, more detailed structural information can be obtained. The basic idea of magnetic anoma-

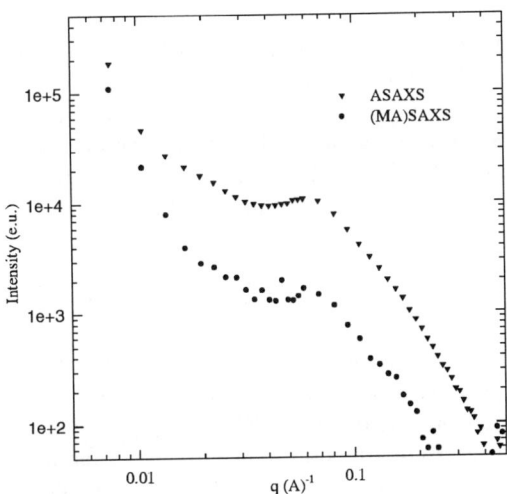

FIGURE 6. ASAXS at the Fe K-edge in $Gd_{25}Fe_{75}$ in comparison with the results of MASAXS at the Gd L_3 edge.

lous small angle x-ray scattering (MASAXS) is now to achieve the contrast enhancement via the effect of X-MCD in the vicinity of an absorption edge [15]. As the dichroic signal depends on the magnetization of the absorbing atoms the magnetic scattering curves reflect the size distribution and correlation lengths of the magnetic scattering precipitates. Since the magnetic absorption coefficient for parallel/antiparallel alignment of magnetic electrons relative to the photon propagation direction given by $\mu^{\pm}(E) = \mu_0(E) \pm \mu_c(E)$ is known, the corresponding scattering amplitudes $f'^{\pm}(E)$ and $f''^{\pm}(E)$ both acquire additive magnetic contributions $f'_c(E)$ and $f''_c(E)$. Similarly, as $f''(E)$ and $f'(E)$ can be related to the absorption coefficient via the optical theorem and a Kramers-Kronig relation, this can be done similarly for their magnetic counterparts.

The magnetic contrast scattering curve presented in Fig. 6 was obtained by recording two scattering profiles with parallel/antiparallel orientation of the magnetization relative to the photon propagation direction at an incident photon energy 3 eV above the Gd L_3 inflection point (see Fig. 5) where the maximum dichroic effect for Gd is expected.

Figure 6 shows the magnetic scattering curve intensity obtained in that way plotted against the momentum transfer q measured at the Gd L_3 edge ($E = 7243$ eV). It is compared with the ASAXS profile obtained by taking scattering profiles between 6843 and 7109 eV, i.e. below the corresponding Fe K-edge at 7111 eV The similarity of both profiles is obvious. The pronounced maximum at $q \sim 0.06$ Å^{-1} can be interpreted as a correlation maximum with $l_c \sim 100$ Å of the Gd precipitates. Although the magnetic profile in Fig. 6 follows roughly the non-magnetic scattering curve, slight deviations can be observed for small q-values, i.e. for large particle sizes. This can be attributed to an enhanced magnetization in the formation of magnetic domains with a corresponding thickness of less than ~ 100 nm, which was observed by magnetic x-ray microscopy as shown in Fig. 2 [16].

MAGNETIC EXAFS

The validity of the model for the SPEXAFS sketched in Section is nicely demonstrated by the experimental data taken at the L-edges of pure hcp-Gd metal (see Fig. 7). Here the magnetic EXAFS $\chi_c(k)$ given as a function of the photoelectron k-vector correspond directly to the difference $\chi_c(k) = (\mu^+(k) - \mu^-(k))/2$. As expected one observes that the frequencies of the SPEXAFS oscillations are very similar to those of the spin-averaged EXAFS since in pure metals the magnetic and electronic neighbors are identical. This results in a nearly identical position of the prominent peak in the Fourier transform (FT) indicating the distance of the next neighboring shell. A comparison of the relative amplitudes yields a magnetic contribution of 8% for the L_2, 4% for the L_3- and 2.4% for the L_1-edge. The relative SPEXAFS strengths χ_c/χ_0 at the three L-edges corresponds to the ratio of the photoelectron spin polarization $\langle \sigma_z \rangle$ of $L_3 : L_2 : L_1 \approx +0.25 : -0.5 : -0.15$.

These results and those of magnetic $L_{2,3}$-absorption studies in various other systems as RE–alloys and 5d-impurities in iron give strong hints that the relative strength of the SPEXAFS normalized to the photoelectron spin-polarization scales with the magnetic *spin* moment μ_S of the neighboring

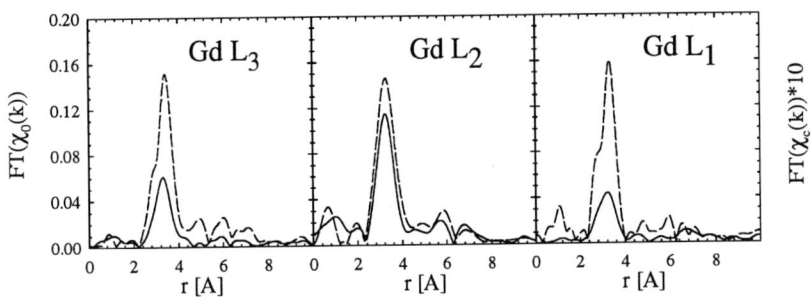

FIGURE 7. Fourier transform of the magnetic EXAFS FT $\chi_c(k)$ at the Gd metal $L_{3,2,1}$ - edges (solid line, right scale) compared to the spin-averaged EXAFS Fourier transform (dashed line, left scale)

atoms: $\frac{1}{\langle\sigma_z\rangle}\frac{FT(\chi_c)}{FT(\chi_0)} = +2.4(2)\% \cdot \mu_S[\mu_B]$. The direct proportionality, which does not depend on the charge Z of the backscattering atom, represents the low-energy limit of the scattering theory, where the mean backscattering amplitude is nearly independent of Z in the range of the FT. Simply spoken the outer atomic electrons, which carry the local magnetic moment, contribute predominantly to the backscattering potential. Thus the linear dependence of χ_c/χ_0 on the spin moment can be well understood.

The SPEXAFS at K-edges are very weak as expected from the small value of $\langle\sigma_z\rangle \sim 10^{-2}$. Nevertheless, from our highly accurate data [14,15] we could extract the k-distribution of the magnetic elastic backscattering amplitude F_c. While the Coulomb backscattering amplitude in case of Co and Ni exhibits a very broad k-distribution as shown in Fig. 8 its magnetic counterpart decreases much faster from about 5 Å^{-1}.

The result of the calculation on the basis of a Born-Ockhur approximation for spin-dependent elastic scattering cross section for forward scattering are marked by dots [21]. Since it is estimated that the corresponding effects for backscattering are somewhat smaller, our result gives somewhat higher values than expected. Nevertheless, an excellent agreement concerning the energy, or k-dependence is found. Forming the ratio between the maxima of the magnetic scattering amplitudes one gets 2.3. Comparing this to the ratio of the atomic spin-moments ($\mu_{sp}(Co) = 1.57$ and $\mu_{sp}(Ni) = 0.61$) of 2.5 the agreement is satisfactory. A magnetic contribution to the electron mean free path was not found for Co and Ni within the statistical uncertainties. This is in agreement with theoretical approximations which predict a ratio of 10^{-2} between the inelastic and elastic scattering amplitudes [21].

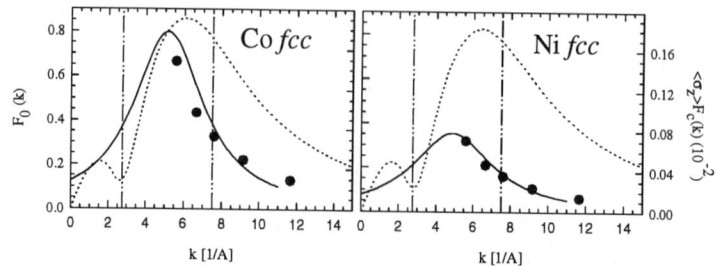

FIGURE 8. Magnetic backscattering amplitude F_c (solid line) deduced from the experimental magnetic EXAFS at the fcc Co metal (left panel) and fcc Ni metal (right panel) K-edge in comparison with the theoretical backscattering amplitude F_0. The dots mark the results of the calculation on the basis of a Born-Ockhur approximation [21].

ACKNOWLEDGMENTS

We like to thank all colleagues and coworkers involved in this work, in particular our long-term co-worker H. Ebert (LMU Munich) for the successful theoretical support and G. Bayreuther (Univ. Regensburg) for providing the Fe layers and their magnetic characterization. Many thanks to D. Raasch and G. Much (Philips Res. Lab., Aachen) who prepared the GdFe samples and H. Jakusch (BASF) for the CrO_2 powder. We thank K. Schwarz, P. Blaha for the CrO_2-bandstructure and F.M.F. de Groot for the LFM calculations. The successful collaboration with C.T. Chen at the Dragon beamline (NSLS), with G. Schmahl and P. Guttmann for the magnetic imaging work and with G. Goerigk and H.-G. Haubold for the MASAXS studies is highly appreciated. We thank the technical staffs at the TU Munich, the University of Augsburg, at HASYLAB, BESSY and NSLS for their help. This work is supported by the BMBF under contract 05 621 WAA and the DFG.

REFERENCES

1. G. Schütz, W. Wagner, W. Wilhelm, P. Kienle, R. Zeller, R. Frahm, and G. Materlik, *Phys. Rev. Lett.* **58**, 73 (1987); G. Schütz, *HASYLAB, annual report* (1985).
2. G. Schütz, M. Knülle, R. Wienke, W. Wilhelm, W. Wagner, P. Kienle and R. Frahm, *Z. Phys.* **B 73**, 67 (1988); G. Schütz and R. Wienke, *Hyperfine Interactions* **50**, 457 (1989).
3. G. Schütz, R. Wienke, W. Wilhelm, W. Wagner, P. Kienle, R. Zeller and R. Frahm, *Zeitschrift für Physik B - Cond. Matt.* **75**, 495 (1989).

4. C.T. Chen, F. Sette, Y. Ma and S. Modesti, *Phys. Rev.* **B42**, 7262 (1990); F. Sette, C. T. Chen, Y. Ma, S. Modesti and N. V. Smith, Conf. Proc. *6th Int. conf. on X-ray Absorption Fine Structure*, ed. by S.S. Hasnain (Ellis Horwood Ltd., 1991) p.96.
5. G. van der Laan, B. T. Thole, G. A. Sawatzky, J. B. Goedkoop, J. C. Fuggle, J.-M. Esteva, R. Karnatak, J. P. Remeika, H. A. Dabkowska, *Phys. Rev.* **B34** 6529 (1986).
6. G. Schütz, D. Ahlers, in: H. Ebert, G. Schütz (Eds.) *Spin-Orbit Influenced Spectroscopies of Magnetic Solid*, Springer Verlag, Berlin (1996).
7. B.T. Thole, P. Carra, F. Sette and G. van der Laan, *Phys. Rev. Lett.* **68(12)** 1943 (1992).
8. Carra, P., Thole, B. T., Altarelli, M., and Wang, X., *Phys. Rev. Lett.* **70** 694 (1993).
9. R. Wienke, G. Schütz and H. Ebert, *J. Appl. Phys.* **69**(8), 6147 (1991); G. Schütz, P. Fischer, S. Stähler, M. Knülle, K. Attenkofer Proc. 7th Int. Conf. X-ray Absorption Fine Structure, Kobe 1992, *Jpn. J. Appl. Phys.* **32** suppl. 32-2, 869 (1993); G. Schütz, M. Knülle and H. Ebert, *Physica Scripta* **T49**, 302 (1993).
10. G. Schütz, R. Frahm, P. Mautner, R. Wienke, W. Wagner, W. Wilhelm and P. Kienle, *Phys. Rev. Lett.* **62(22)**, 2620 (1989).
11. G. Schütz, P. Fischer, K. Attenkofer, M. Knülle, D. Ahlers, S. Stähler, C. Detlefs, H. Ebert, F.M.F. de Groot, *J. Appl. Phys.* **76**, 6453 (1994).
12. J. L. Erskine and E. A. Stern, *Phys. Rev.* **B12**, 5016 (1975).
13. U. Fano, *Phys. Rev.* **178**, 131 (1969).
14. M. Knülle, D. Ahlers and G. Schütz, *Sol. St. Comm.* **94**, 267 (1995).
15. P. Fischer, R. Zeller, G. Schütz, G. Goerigk, H.-G. Haubold Proceedings of XAFS IX (Grenoble 1996), *Journal de Physique IV* (Colloques), in press.
16. P. Fischer, G. Schütz, G. Schmahl, P. Guttmann, D. Raasch, *Z.f. Physik* (1996) in press.
17. F. M. F. de Groot, J. C. Fuggle, B. T. Thole, G. A. Sawatzky, *Phys. Rev. B* **42**, 5459 (1990).
18. K. Schwarz, *J. Phys. F- Metal Physics* **16** L221 (1986).
19. J. Sticht, J. Kübler, *Sol. Stat. Comm.* **53**, 529 (1985).
20. H. Ebert, G. Schütz and W.M. Temmerman, *Sol. State Comm.* **76**, 475 (1990).
21. J.A.D. Matthew, *Phys. Rev.* **B25**, 3326 (1992).

Achievements and Prospects in X-ray Absorption Spectroscopy

Yizhak Yacoby

Racah Institute of Physics, Hebrew University Jerusalem Israel 91904

Edward A. Stern

Department of Physics, University of Washington Seattle Wa. 98125

In this paper we discuss the developments in X-ray absorption fine structure (XAFS) data analysis. We show that these developments have made Extended XAFS into a powerful technique for the investigation of local structural properties of condensed matter systems. The state of data analysis in the X-ray absorption near edge structure (XANES) region and in Diffraction Anomalous Fine Structure (DAFS) are briefly discussed. As an example, the contribution of EXAFS experiments to our fundamental understanding of structural phase transitions is presented showing the power of EXAFS experiments when combined with detailed data analysis.

I. INTRODUCTION.

In the past twenty years following the pioneering work of Stern et al. (1,2) X-ray absorption fine structure spectroscopy (XAFS) has become a wide and diverse field with important contributions in most areas of condensed matter science. International conferences on this subject are held every two years with many new and interesting contributions. This field has also its pitfalls and in a number of cases unreliable conclusions have been published. However, a great deal of progress has been made in recent years in developing XAFS into a highly reliable and highly useful tool. A number of extensive reviews and many overviews have been written on this subject. A partial list appears in references (3–16)

An important part of the progress made in this field is due to the development of the theory of XAFS to the extent that experimental measurements can be compared to theory either quantitatively or semiquantitatively. This development has led to very reliable and important results in condensed matter physics, chemistry, biology and geology. It should be emphasized that many interesting results and conclusions have also been obtained from careful systematic studies of the variation of the absorption spectra with different parameters. However, the full power of this experimental technique can be realized only when used in conjunction with a reliable theory of the absorption process.

This paper is neither a general review of the subject nor a review of all the developments and important contributions made in this field. The methods of measuring XAFS have been fairly advanced for a number of years now and are used by many researchers (17). So in this paper we shall discuss mainly the contributions made in developing our ability to analyze experimental measurements on firm theoretical grounds and provide some examples in which such measurements and analyses led to interesting results in condensed matter physics.

XAFS Spectroscopy is usually divided into two spectral regions: X-ray absorption near edge structure (XANES) and extended x-ray absorption fine structure (EXAFS). XANES covers the range from about 10 eV below the main edge up to 50 eV above it, whereas EXAFS covers the range from about 30 eV to about 1500 eV above the edge. Both EXAFS and XANES are strongly affected by the structure. However XANES is also strongly affected by the chemistry of the material and the valence of the probe atom. Unfortunately the full chemical and structural information in XANES is rather difficult to obtain because the spectrum requires calculating a stronger interaction between the excited photoelectron and the surroundings than does EXAFS. Significant progress has been made in calculating XANES spectra from first principles. These calculations allow *semi-quantitative* comparisons between theory and experiment. However, at present, there is no general theoretical formulation that will allow quantitative comparison with experiment for any material. On the other hand, EXAFS spectra which depend only weakly on the chemistry are easier to calculate in a way accurate enough to compare quantitatively with experiments. These calculations can now be routinely done for practically any material. The combination of EXAFS experiments and their quantitative analysis has recently developed into a very powerful tool which, when properly used, provides reliable local structural information on condensed matter systems.

The analysis of both XANES and EXAFS measurements is particularly difficult when the probe atom is present in more than one inequivalent site. This problem exists in many homogeneous materials and in cases where the phase of interest is embedded in another material which contains the same probe atom. To overcome this difficulty a number of researchers have been studying the anomalous fine structure which accompanies the diffraction peaks. This method known as the Diffraction Anomalous Fine Structure (DAFS) method is quite difficult to implement both experimentally and theoretically. So far, it has been used in a rather limited way. However, if the present difficulties in measurement and analysis are overcome, it may become a powerful complement to XAFS.

Very interesting work has been done in recent years in X-ray magnetic circular dichroism (XMCD) and spin polarized EXAFS (SPEXAFS). These measurements provide information on the magnetic properties of the absorbing atom and its neighbors. This work has been pioneered by G. Schütz et al.

(18,19) and has been followed by the discovery of a sum rule which relates the integral of the circular dichroism signal to the expectation value of the ground state orbital angular momentum (20). This work will not be further discussed in this paper; more details can be found in the articles of F. deGroot and G. Schütz in this volume.

Finally, to illustrate the power of the combined use of EXAFS experiments and their detailed analysis we shall discuss a new area of research which has developed over the last few years, namely, the study of local disordered structural distortions in systems with long range order. EXAFS experiments have shown that such distortions are present in many types of crystals, in particular in crystals undergoing structural phase transitions. These distortions seem to play a very important role in driving structural and possibly other phase transitions.

II. THEORETICAL CONSIDERATIONS AND DATA ANALYSIS.

X-ray absorption fine structure spectroscopy measures the probability for absorption of an X-ray photon by exciting the system from its ground state to an excited state. In the one-electron picture this can be described as the excitation from an inner core level to an excited state of the system. Naively speaking one could consider two approaches to the theory of XAFS within the single-electron approximation. The first can be described in the band structure model in which the photoelectron is excited from the initial inner core state to an unoccupied state in the conduction band (21). The second would be to consider the excitation of the photoelectron from the initial state to the final state, where the final state is described by an unoccupied state of the absorbing atom modified by the scattering of the photoelectron from surrounding atoms in the system (2). In principle, the two approaches should lead to the same result, but in practice the latter permits a simpler method to take into account the important physics. Since the excitation energies above the Fermi level are large, up to 1500 eV, the excited electron interacts very efficiently with other excitations, such as plasmons, and lattice vibrations. In addition it interacts with the core hole left behind in the absorbing atom. As a result, the excited electron tends to be localized in the vicinity of the absorbing atom favoring the second approach.

Initially the scattering of the photoelectron was calculated by the small atom approximation assuming only single scattering events from the neighbors. These approximations led to the basic EXAFS formula (1,27):

$$\chi(k) = \sum_\Gamma \frac{S_0^2}{kR_\Gamma^2} \mid f_{eff}^\Gamma(k) \mid \sin[2kR_\Gamma + \phi^\Gamma(k) + 2\delta_c(k)]$$
$$\times e^{-2\sigma_\Gamma^2 k^2} e^{-2R_\Gamma/\lambda(k)}. \tag{1}$$

where, $\chi(k)$ is the actual absorption minus the absorption of the embedded

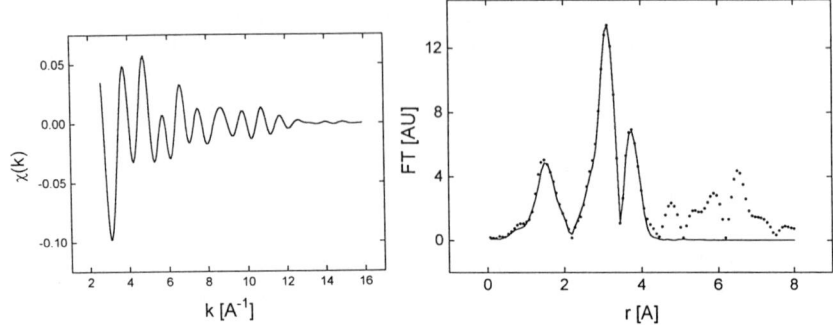

FIG. 1. (left) Normalized experimental Cu XAFS spectrum of $YBa_2Cu_3O_7$. X-rays polarized along the c-axis. (right) Absolute value of the Fourier transform of the $YBa_2Cu_3O_7$ spectrum on the left multiplied by the wave number k. (Experiment - dots; Theory - solid line).

probe atom normalized to the smooth atomic absorption. The summation is over all scattering paths Γ. S_0^2 is an amplitude correction factor which accounts for the change in the passive electrons overlap. k is the photoelectron wave number

$$\hbar k = \sqrt{2m(E - E_0)} \qquad (2)$$

Here E and E_0 are the X-ray photon energy and the absorption threshold energy, respectively. R is half the scattering path length; $|f_{eff}|$ and ϕ are the effective scattering amplitude and phase shift; δ_c is the probe atom phase shift; σ^2 is the mean squared relative displacement (MSRD) and λ is the coherence length of the photoelectron. Notice that χ is essentially a sum of sine functions. Thus the Fourier transform of $\chi(k)$ is loosely related to the pair distribution function. An example of $\chi(k)$ and the absolute value of its Fourier transform are shown in Fig. 1. The first oxygen shell is fairly well isolated but all other shells are mixed with each other and with multiple scattering contributions preventing their direct identification.

The small atom and single-scattering approximations were quickly found to be inadequate. For a number of years the use of experimental standards provided a way to analyze data, especially the part contributed by the first nearest neighbors, in a fairly reliable way (22). However, it was soon realized that a more fundamental way based on deeper theoretical knowledge was necessary in order to analyze data that would give reliable information out to 4th, 5th and 6th nearest neighbors. Based on the pioneering work of Lee and Pendry (23) and theoretical contributions of several authors (24–28) a theory has evolved which takes into account the finite size of the scattering atoms and more accurately the curved nature of the photoelectron wave function and includes multiple scattering contributions. Luckily, the EXAFS spectrum can still be expressed in the form of Eq. (1) (27). The difference is that the summation is carried over all scattering configurations single and multiple and the more sophisticated account of the photoelectron wave function is lumped into the scattering amplitudes and phase functions.

A number of software packages, EXCURVE (25), FEFF (27), UWXAFS (29) (which uses FEFF), GNXAS (30) and EDA (31) are now available. They allow experimentalists to calculate EXAFS spectra from first principles in terms of a structural model of the system, MSRDs and possibly higher cumulants (32). The packages differ from each other in various aspects. We have had excellent experience with FEFF. In particular, the fact that the various scattering contributions are handled separately facilitates a linear expansion of the slowly varying amplitude and phase functions in terms of the bond lengths and bond angle of each scattering configuration while the fast varying part which is due to $2kR$ can be treated exactly. This feature is important when one tries to determine a local structure which is unknown both qualitatively and quantitatively. However, other groups have successfully used other packages.

The actual data analysis is done by comparing the experimental results to the theoretically calculated EXAFS spectrum. The fit is done either in k space or in real space. By varying the structural parameters of the model and the parameters that describe the thermal and static disorder in the model, one attempts to obtain the best fit between theory and experiment. From it one obtains the structural and dynamical parameters. An example of the fit between theory and experiment is shown in Fig 1. This quality of fit is typically achieved when the experimental results are of high quality.

It should be emphasized that in many real cases the phase shifts and amplitudes are not accurately provided by theory. It is often necessary to introduce some corrections. The most important correction is due to the fact that the zero energy levels of the various atoms in the system are not the same. These energies may differ from each other by a few eV. As a result, the phase shifts calculated from theory may be off by a value that is inversely proportional to the photoelectron wave number (33,34).

One of the problems encountered in the analysis of XAFS spectra is the background associated with the absorption of the embedded atom excluding the scattering effects. In the past, this background was manually subtracted using cubic splines. More recently automatic ad-hoc methods were developed to subtract the background on the basis of the assumption that any relatively smooth signal which cannot be accounted for by the theory is background (35,36). In the last couple of years real attempts to actually calculate the absorption of the embedded absorbing atoms have been made (30,37). A comparison between an experimental atomic XAFS spectrum and the calculated spectrum are shown in Fig. 2 (37). The comparison shows that while the main features are theoretically reproduced, a quantitative comparison is still not at hand.

Another feature of XAFS which can be important in certain cases is multi-electron absorption processes. An example of this effect is shown in Fig. 3

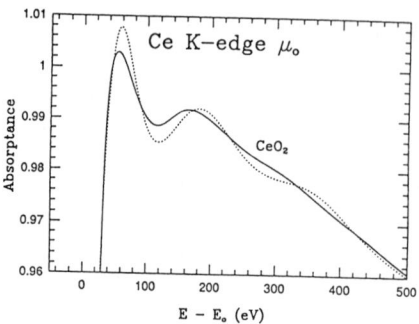 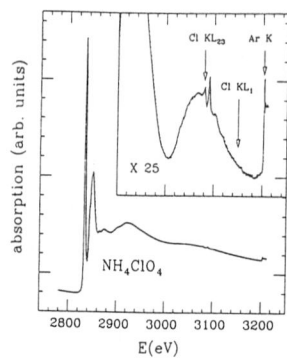

FIG. 2. Atomic XAFS in CeO_2 (Experiment - dots; Theory - solid line) (37)
FIG. 3. Cl K-edge X-ray absorption spectrum of NH_4ClO_4. The autoionizing KL-edge resonances in the 3100 eV region are clearly visible in the inset. (38)
Reprinted from *Physica B*, 208 & 109 29, A. Filipponi, 1995 with kind permission from Elsevier Science - NL, Sara Burgerhartstraat 25, 1055 KV Amsterdam, The Netherlands

(38).

In conclusion, using modern EXAFS data analysis it is possible to obtain very detailed and reliable information on the structure surrounding the absorbing atoms of a great variety of condensed matter systems. Further progress is needed. In particular, better calculations of the phase shifts, the mean free path, the mean squared relative displacements and the embedded probe absorption are necessary. In addition, multi electron effects may be playing a role in certain cases (38). Adding these processes to the calculated EXAFS may also improve the reliability of EXAFS data analysis.

When the excited photoelectron energy is small, as in the XANES region, its interaction with the surroundings is strong and the electronic properties of the system become important and strongly influence the XANES spectra. Thus, the XANES spectra depend both on the electronic as well as the structural properties of the system. Since the photoelectron energy is comparable to the Fermi energy the absorption depends on the detailed distribution of unoccupied excited states. Furthermore, the scattering of the electron by other excitations is less efficient so the electron travels longer distances before it loses its phase coherence. This means that multiple scattering must be taken to higher orders. The scattering is very sensitive to errors in the scattering potentials and in the electron self-energy. Finally, in this regime the decay of the core-hole plays a more important role and even the one-electron approximation and the muffin-tin model for the calculation of the potentials are questionable. The result is that present theoretical calculations of the XANES spectra cannot be carried out to a level of accuracy that allow quantitative comparison with theory. However, important progress in this field is being made.

Curvedwave multiple scattering calculations do provide qualitative agreement with experiment. An example is shown in Fig. 4 (39) calculated using

FEFF6 package. As can be seen the calculations reproduce the main features of the spectra. Even when quantitative comparison with experiment cannot be made it is possible to draw important conclusions by clever use of the experimental results and qualitatively valid theoretical calculations.

Attempts to break away from the one-electron approximation have been recently made. One worth particular mention is the multichannel theory of Natoli et al (40). When one considers relaxation effects as an integral part of the absorption process, the system after absorbing a photon can be in a number of different final states. The transition to each one of these states is known as a channel. This approach has been used in the past to discuss scattering in atoms and has been adapted by Natoli et al to extended systems. This approach can in principle account for multi-electron effects, the decay of the core-hole and screening effects. The simplest situation is the sudden approximation limit where the channels do not interfere with each other. Of course the more interesting situation is when the channels do interfere. At present it is not clear whether these theoretical considerations will lead to a more accurate understanding of experimental results in practice.

III. DIFFRACTION ANOMALOUS FINE STRUCTURE (DAFS).

In many systems the probe atom may occupy more than one inequivalent position in the unit cell. The EXAFS and XANES spectra in this case are a combination of contributions of the absorbing atoms in all different sites and are therefore affected by many more structural dynamical and electronic parameters. One way to increase the amount of information has been to use powder samples with the c-axes of all particles lined up and the measurements done with polarized X-ray radiation parallel and perpendicular to the 'c' axis. This was done for example in the study of YBCO samples (41). However, this method cannot always be used and is limited to doubling the information content in EXAFS and XANES spectra.

A more general approach is DAFS. The intensity of a diffracted beam is proportional to the Fourier transform of the autocorrelation function of the spatial dielectric function of the system. Usually, when the X-ray photon frequency is far from any of the atomic resonances, the dielectric function is approximately proportional to the electron density. However, if one of the atoms is at or close to resonance the dielectric constant will depend on the photon energy in such a way that its imaginary part is in fact the usual XAFS spectrum. Clearly, as in all dielectric constants, the real and imaginary parts are related to each other through the Kramers-Kronig relation. Thus, the intensity of the diffracted beam measured as a function of photon energy at a Bragg point will have an oscillatory component similar to that of EXAFS and XANES. However, there is also a fundamental difference between the two. The total observed signal at any given Bragg direction is a result of the coherent interference of all the signals generated by both resonant and non-resonant

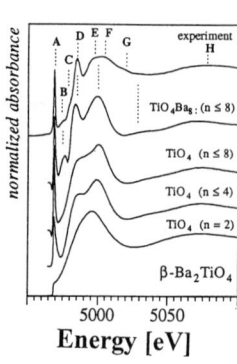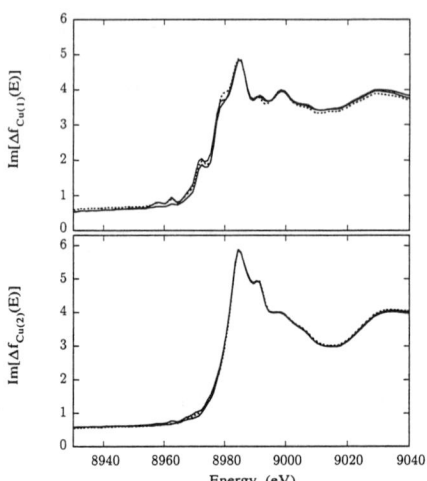

FIG. 4. Comparison of the experimental Ti K-XANES spectrum of $\beta - Ba_2^{[4]}TiO_4$ with theoretical FEFF6 multiple scattering calculations on TiO_4 and TiO_4Ba_8 clusters. (n is the scattering order). (39)

FIG. 5. Site separated $YBa_2Cu_3O_{6.8}$ $f''(E)$ in the near threshold region for (a) Cu(1) and (b) Cu(2). (44)

atoms with a phase relationship which depends on the particular diffracted beam chosen. Thus different diffracted beams will have different coherent admixtures of the signals contributed by the different resonant atoms at the different inequivalent sites. This enables the contributions of the different inequivalent probes to be separated.

Unfortunately, in spite of the fact that anomalous diffraction has been known for about 40 years (42,43), very few successful attempts have been made to implement it practically. The reason is that both the experimental measurements and their detailed analysis are very difficult. The measurement is best done on single crystals and as the energy varies the incident beam to crystal angle must also vary so as to maintain the Bragg condition. In making DAFS measurements one must correct for the usual instrument errors but in addition it is necessary to correct for fluorescence background and for the varying absorption lengths of the incident and diffracted beams (44). However, these difficulties are not insurmountable as seen from the work of Cross and Sorenson (45,44).

The data analysis, while more complicated than that of XAFS, is not intractable. First using the Kramers-Kronig relation it is possible to obtain the real and imaginary parts of the resonant dielectric function $f'(E)$ and $f''(E)$ without the need to resort to a model. Second, in some cases it is even possible to separate the contribution of different sites using the spectra measured at different Bragg points. Fig. 5 shows the separated contributions of Cu(1) and Cu(2) XAFS of $YBa_2Cu_3O_7$. Please notice that the two are indeed very different from each other. Finally, a generalized fit program that fits the theoretical spectra of a number of Bragg points to the experimental spectra is feasible and underway.

In conclusion, DAFS spectroscopy has significant promise of providing very useful information on complicated systems. However, a great deal of work is still needed to make this method practical for wide use.

IV. LOCAL STRUCTURAL DISTORTIONS IN CRYSTALS, AN EXAMPLE OF A EXAFS STUDY.

One area in which EXAFS has been particularly useful is the study of local structural distortions in systems undergoing structural phase transitions. The information provided by EXAFS in this area led to a deeper understanding of the mechanism driving various structural phase transitions.

A. Order-disorder and displacive structural phase transitions.

Classically structural phase transitions have been divided into order-disorder and displacive type transitions (46). Far below T_c, a system undergoing an order-disorder transition is fully ordered, namely, the local and average structure are the same. In contrast above T_c the average and local structures are different. The local structure would have the low symmetry of the low temperature phase whereas the average structure will have the higher symmetry of the high temperature phase. In a displacive phase transition the system is ordered both below and above T_c so that except for lattice vibrations and critical fluctuations near T_c, the local and average structures are essentially the same. In spite of the fact that this division of structural phase transitions into order-disorder and displacive is not a result of any basic symmetry requirement, it is helpful in understanding the mechanism driving the phase transition. In particular, it is important to know whether the local distortion from the high symmetry structure is a result of the phase transition or is actually involved in driving it.

Obviously, EXAFS is particularly suited to provide information on the local structure because, in contrast to diffraction methods, it provides information on the local pair distribution function in spite of the fact that the sense of the local distortion may be different at different unit cells. In addition, as seen below, it may also provide information on bond angles (namely, three body distribution function).

B. Structural phase transitions in perovskite crystals.

Most of the work discussed below has been done on oxygen perovskite type crystals with a chemical formula of the form ABO_3, but work done on other crystal families indicate that the results have a more general validity. Ideal perovskites have a simple cubic structure with the B atoms at the cube center, the oxygens at the face centers forming an octahedron around the B atoms and the A atoms at the cube corners. These crystals undergo various structural phase transitions. The most important types of distortions are the ferroelectric, in which the cations move relative to the oxygens rendering a net dipole moment per unit cell; the antiferroelectric, in which the dipoles of neighboring cells are inverted and the antiferrodistortive distortion, which involves the rotation of the oxygen octahedra. The last two transitions involve a multiplication of the unit cell.

C. Qualitative evidence of local disorder in the high symmetry phase.

The first experiments to detect inconsistencies with the simple displacive quasi-harmonic picture were diffuse x-ray scattering (47). These experiments measure the spatial Fourier transform of the electron density density correlation function. In crystals with a structure that can be described by harmonic or quasi-harmonic models, the Fourier transform has delta function peaks at the inverse lattice points and a weak broad structure around them. The Fourier transform is approximately zero throughout the rest of the reciprocal space. The results in a number of pure oxygen perovskite ferroelectrics showed that in the cubic, tetragonal and orthorhombic phases the Fourier transform is not zero also on planes perpendicular to [001] type axes (48). One interpretation of the results suggested that the metal atoms are displaced along [111] type directions both in the cubic phase and in the tetragonal and orthorhombic phases. The displacements are disordered but have relatively large correlation lengths along [001] type axes (49). However, other interpretations to explain the results were consistent with the displacive model.

Raman scattering experiments provide additional information from a different point of view. Assuming the quasi-harmonic picture, each lattice point of a perovskite crystal in the cubic phase has a center of inversion symmetry. Thus, first order Raman scattering is forbidden by symmetry. Raman (50) and differential Raman (51) experiments on pure and mixed crystals showed that this selection rule is violated. For example in $KTaO_3$:6%Nb first order Raman lines were measured up to 120°K above T_c (52) and disappeared at higher temperatures. Theoretical and experimental considerations have shown that the local symmetry is violated below 120K above T_c on a time scale larger than $\sim 10^{-10}$sec (52).

Infra-red reflectivity measurements (53) and Raman (54) and neutron (55) scattering measurements have been used to study the temperature dependence of the soft mode frequency in both pure and mixed ferroelectrics. All three techniques show that the soft mode frequency does not vanish as T approaches T_c but saturates at a finite value. In addition Raman (56) and neutron (57) scattering experiments showed that a quasi-elastic peak grows as T_c is approached. The interpretation of these results showed that the system has a degree of disorder near T_c and its long wavelength slow component grows critically as T approaches T_c (57). However, as mentioned above, it has been argued that these results of the dynamics in the materials can still be interpreted as having a displacive driving mechanism.

The experiments discussed in this section suggest that ferroelectric crystals, both pure and mixed, which have clear displacive characteristics are disordered above T_c. These experiments however, do not provide unambiguous quantitative information about the temperature dependence of the disordered distortions. In addition, these results raise the question whether this behavior is limited just to ferroelectrics or whether crystals with other types of phase transitions also display structural disorder.

D. EXAFS of locally distorted systems.

1. EXAFS measurements and data analysis.

The details of the experimental measurements and data analysis cannot be provided here but some general remarks should be made. Most of the measurements were done in transmission. The samples were usually powdered, then mixed with graphite powder and pressed into pellets. Special attention was paid to use small enough particles so as to avoid the thickness effect but not too small so as to avoid surface effects on the phase transition properties.

In analyzing the data we used the FEFF5 package of Rehr et al (27). In most cases phase correction parameters of the type C/k were necessary. However in almost all cases the number of parameters was less than or equal to two thirds of the number of relevant experimental points N_{exp}, where $N_{exp} = \frac{2}{\pi}\Delta R \Delta k$. ΔR and Δk are the ranges in real and wave number spaces where the experimental and theoretical spectra are being fitted. This reduces the chance that a good fit is obtained with the wrong model. The handling of structural distortions deserves special attention. Notice that even one distortion parameter, for example the displacement of the probe relative to all other atoms, may affect many or all single and multiple scattering paths of the photoelectron. The programs we developed (58) calculate the theoretical EXAFS spectra for any structural model expressed in terms of an initial model modified by a number of structural distortion parameters, provided the distortions are not too large (of the order of 0.5Å in length and 15° in angle). This calculation is carried out by linearly (and in certain cases quadratically) expanding the scattering amplitude and phase functions of each scattering path in terms of its bond lengths and bond angles. The values of the MSRDs of the single scattering paths are taken as fit parameters and the values of the multiple scattering MSRDs are expressed in an approximate form in terms of these parameters.

Special attention has been paid to estimate the error brackets of the different parameters calculated in this way (59,29). In estimating the errors we have taken into account the number of fit parameters relative to the number of independent relevant experimental points, the mismatch between theory and experiment and the interdependence of various parameters. Different models have been tried and only if the fit quality was clearly superior was the corresponding model pronounced as a better model.

2. $PbTiO_3$ - a pure ferroelectric crystal.

$PbTiO_3$ undergoes a ferroelectric phase transition at $\sim 760K$. Both above and below this temperature it displays underdamped transverse optical phonons, the frequencies of which decrease as T_c is approached. For this and other reasons this crystal was believed to undergo a displacive type phase transition, namely, it was expected that the structure of this crystal above T_c is cubic and undistorted both on the average and on the local scale.

EXAFS measurements were carried out on this crystal in order to determine the temperature dependence of its local structure. Both Pb and Ti were used as probes (60) . Examples of the EXAFS spectra after removing the background are shown in Fig. 6. At 12K, the range of useful data extends all the way to $k = 16 Å^{-1}$. It decreases at higher temperatures down to $k = 9.5 Å^{-1}$ at $T = 848K$. The Fourier transforms of the data and the theoretical fits are shown in Fig. 7. At 12K the first 4 shell configurations were fit to the data. However, since at high temperatures only the first 2 shells gave significant contributions the data was analyzed for the first 2 shells at all temperatures. The main results obtained from these fits are as follows: A comparison between the average unit cell dimensions measured by X-ray diffraction and the local unit cell dimensions measured from EXAFS are shown in Fig. 8 (60), (61). Notice in particular that above T_c the average cell is cubic whereas locally it remains tetragonal. Furthermore, even below T_c the average and local dimensions are different showing that the system has local disorder. The local and average Pb displacements from the center of the unit cell are shown in Fig. 9 (60), (62), (63) . Notice that at low temperatures the average and local distortions are equal to within the experimental error. But as the temperature rises the difference between the two increases and above T_c, while the average distortions become zero, the local distortions are at least 50% of the distortions at very low temperatures. Finally, from the values of the average Pb displacement relative to the nearest oxygen, its MSRD obtained from the Debye-Waller factor and a third cumulant, one can calculate the Pb - nearest oxygen probability distribution function shown in Fig. 10. Notice that at low temperatures, the distribution function is a Gaussian centered at 2.5Å. As the temperature rises the distribution function moves slightly towards the center at 2.85Å, but even at 848K, almost 90K above T_c the peak of the distribution is clearly off center. Another interesting feature is the fact that the low edge of the distribution at about 2.2Å is

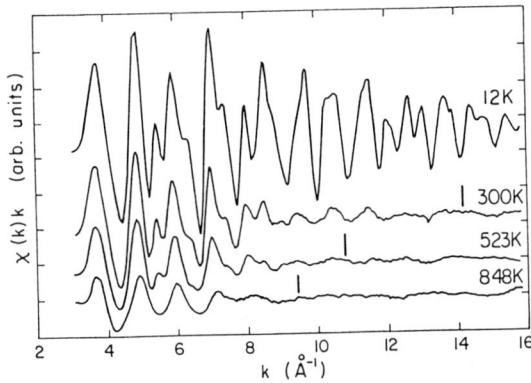

FIG. 6. The normalized Pb-edge XAFS of PbTiO$_3$ multiplied by the photoelectron wavenumber k as a function of k for several representative temperatures. The vertical marks indicate the limits of useful XAFS data.

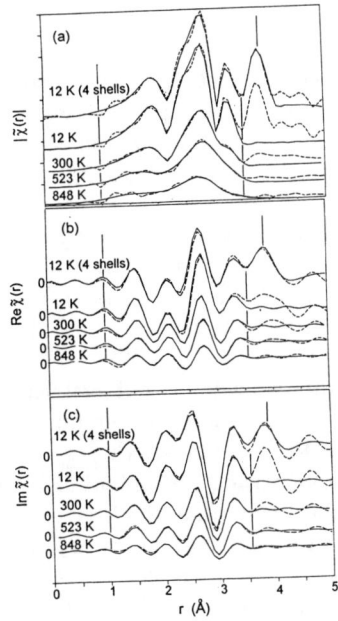

FIG. 7. The Fourier transform of the Pb XAFS of PbTiO$_3$. (Experiment - dashed; Theory - solid line).

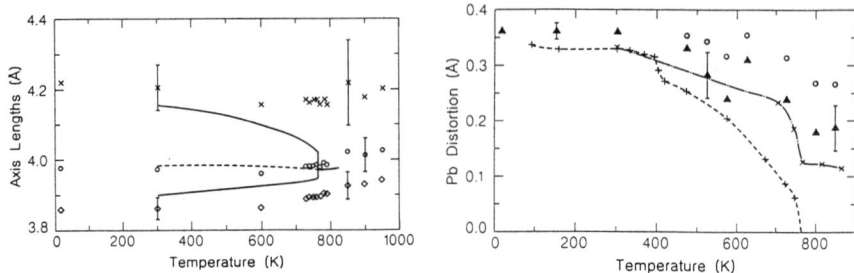

FIG. 8. Temperature dependence of the a and c unit cell dimensions. Solid lines represent the average dimensions obtained from X-ray diffraction. The dashed line represents the geometric average of the lattice constants. The x's and the diamonds represent the c and a axes obtained from the Ti XAFS. The circles are the average lattice constants obtained from XAFS.

FIG. 9. The temperature dependence of the Pb distortions obtained from the Pb EXAFS. The open circles represent the distortions calculated from the peaks of the distortion distribution functions. The solid triangles are the average local distortions obtained from EXAFS. The +'s are from X-ray diffraction and the x's are from neutron diffraction.

essentially temperature independent. This is consistent with the fact that at very short distances the interatomic repulsion rises very sharply.

In conclusion, these results show that even in a crystal considered to be the classical example of displacive ferroelectrics, the crystal is locally distorted even far above T_c and the distortions are large, more than 50% of the distortions at 0K. These distortions are disordered above T_c and are partially disordered below T_c indicating that the transition is not purely displacive. More specifically the local distortion, or at least a large fraction of it, is not due to the transition and in fact it may be involved in driving the phase transition.

Indications are that this behavior is not restricted to this crystal alone. $KTaO_3$ is a paraelectric perovskite which remains cubic at all temperatures. Replacing some or all the Ta by Nb makes this crystal ferroelectric at temperatures that increase with increasing Nb content. EXAFS measurements using both Nb and Ta as probes show that the Ta atoms are within experimental accuracy on-center both below and above T_c. In contrast the Nb atoms are displaced to off-center positions both below and far above T_c. The Nb - O probability distribution function is similar to that of the Pb - O but in this case the probability to find the Nb on center even at 300K, namely at

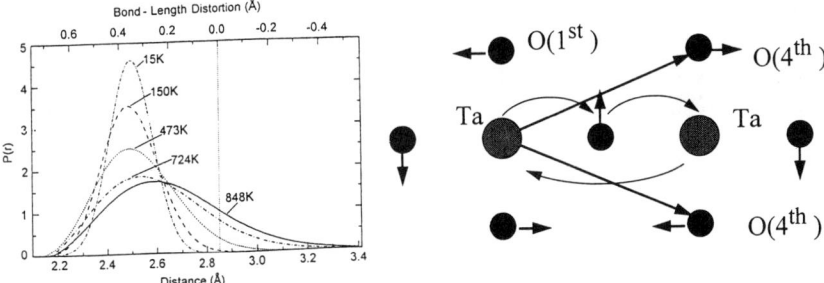

FIG. 10. The Pb to nearest oxygen distance probability distribution function in PbTiO$_3$ calculated from the Pb EXAFS at various temperatures. The vertical line indicates half the diagonal distance between two oxygens.

FIG. 11. The effect of oxygen octahedra rotations on the double and triple scattering from the Ta 3rd neighbor and the splitting of the 4th oxygen shell.

$T \simeq 3.5 T_c$, is negligibly small. In addition, the Nb displacements are found to be independent of the Nb concentrations. These results show that the Nb displacements are not a result of the ferroelectric transition but may be a cause of it.

Finally, it turns out that this behavior is not a property of perovskite ferroelectrics alone. Pb$_x$Ge$_{1-x}$Te and PbTe$_{1-x}$S$_x$ have very interesting anomalous properties. Their basic structure is rock salt. PbTe is paraelectric but GeTe becomes ferroelectric below $\sim$700 K. The mixed crystal is also ferroelectric where T_c is a non-linear function of the concentration. Both end materials of PbTe$_{1-x}$S$_x$ are not ferroelectric but the mixed crystals do become ferroelectric. EXAFS measurements in these materials using both Pb and Ge as probes in Pb$_x$Ge$_{1-x}$Te (64) and Pb as a probe in PbTe$_{1-x}$S$_x$ (65) show that Ge and S, respectively, are displaced to off-center positions both below and above T_c. Wang and Bunker (65) suggest that the off-center displacements above T_c are due to the large mismatch in ionic radii between Pb and Te on one side and Ge and S respectively on the other. The authors further suggest that the off-center displacements are an essential element in rendering the mixed crystals ferroelectric and the transition is a mixture of order-disorder and displacive.

3. Disordered local distortions in crystals undergoing antiferrodistortive phase transitions.

NaTaO$_3$ is an oxygen perovskite crystal that undergoes a series of structural phase transitions (66). Above 900K it is simple cubic. As the temperature decreases it undergoes transitions to a tetragonal and two different rhombohedral phases at 900, 820 and 750K. The most important distortion in these transitions is the rotation of the oxygen octahedra around the [100], [110] and [111] type axes, respectively. These distortions are called antiferrodistortive. Mixed $K_{1-x}Na_xTaO_3$ crystals also undergo antiferrodistortive transitions with T_c decreasing as the concentration of Na decreases. The question whether systems with antiferrodistortive transitions are also structurally disordered above T_c is particularly interesting. Oxygen octahedra rotations in different cells are coupled because every two neighboring octahedra share an oxygen atom. Thus, the rotation of one necessarily involves the rotation of the other. Simple statistical mechanics considerations show that if the octahedra are assumed to be approximately rigid, as many crystallographers do, the system must be ordered in all its phases. These systems were therefore assumed to undergo displacive-type phase transitions.

EXAFS measurements were used to determine the local structure of $K_{1-x}Na_xTaO_3$ crystals as a function of temperature (67). The way the oxygen octahedra rotations were measured is shown in Fig. 11. The probe used was Ta. Single scattering paths from the nearest neighbor (nn) oxygens are hardly affected by the rotations because the Ta-O distance hardly changes. On the other hand, the colinear double and triple paths to the next Ta are strongly affected by the rotations because collinear paths make large contributions to the EXAFS spectrum and are very sensitive to deviations from colinearity. In addition, the single scattering paths to the 4th oxygen neighbors are also strongly affected by the rotation. In order to increase the reliability of the results we used two independent distortion parameters, one to describe the nn oxygen, the other the 4th neighbor oxygen rotations.

The fits between theory and experiment in $K_{1-x}Na_xTaO_3$ are even better than in $PbTiO_3$. The oxygen octahedra rotation angles as a function of temperature and Na concentration are shown in Fig. 12. The values of the rotation angle parameter of the 1st and 4th shell oxygens are given explicitly for 82% Na. As seen the two parameters agree well with each other. The rotation angles are for rotations around [111] type axes. We probed the possibility that at $750 < T < 900K$ the rotations are about [110] or [100] type axes. The first was consistent with the data but the local rotations are definitely not around [100] type axes. At low temperatures the average and local rotation angles in pure $NaTaO_3$ agree very well with each other. However, in all samples the rotation angles do not vanish above the transition temperature to the cubic phase. The angles at the highest temperatures measured are more than 60% of their values at very low temperatures. This means that antiferrodistortive materials are also distorted in a disordered fashion hundreds of degrees above T_c. These results lead to two conclusions: First, the oxygen octahedra cannot be considered as solid. Second, the rotations are not a result of the structural phase transition and may be involved in causing it.

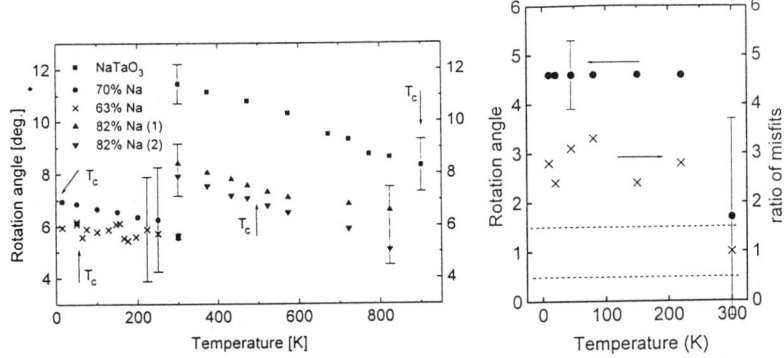

FIG. 12. Rotation angles in four crystals as a function of temperature. For the 0.82% Na crystal, the two independent measurements of the rotation angle are shown. The phase transition temperatures to the cubic phase are indicated for each crystal.

FIG. 13. Temperature dependence of the rotation angle about the [111] type axes in $Ba_{0.6}K_{0.4}BiO_4$.

4. Disordered structural distortions in superconductors.

EXAFS measurements on $Ba_{0.6}K_{0.4}BiO_3$ (68,69) have revealed that disordered structural distortions are present in this system in spite of the fact that neutron diffraction results (70) shows that this system is cubic at all temperatures. $BaBiO_3$ has two main structural distortions: First the oxygen octahedra surrounding neighboring Bi atoms have different sizes. This distortion is known as the breathing mode. The second involves oxygen octahedra rotations around [110] type axes. In contrast $Ba_{0.6}K_{0.4}BiO_3$ is cubic down to the lowest temperatures measured and undergoes a superconducting transition at 32K. The EXAFS results show that the local and average breathing mode distortions in $BaBiO_3$ are in agreement. On the other hand the oxygen octahedra rotations are different. The local rotations are about the [111] type axes and not like the average rotations which are about [110] type axes. This means that the component of the rotations about the [001] axis is disordered. In $Ba_{0.6}K_{0.4}BiO_3$ the average rotation is zero at all temperatures whereas the local rotations shown in Fig. 13 are large and persist up to at least 220K. It is premature at this time to say if these distortions are involved in rendering the rather high superconducting transition temperature. However, models by Plakida et al (71) and Hardy and Flocken (72) suggest that dynamical disordered local distortions may contribute to the carrier pair binding.

V. SUMMARY AND CONCLUSIONS.

X-ray absorption fine structure spectroscopy has come a long way since it was first used to obtain structural and electronic information. The developments in the experimental technique made it possible to obtain highly reliable experimental data on samples with large as well as dilute probe atom concentrations. Furthermore, it is possible to obtain now excellent data on magnetic dichroism providing information on the magnetic properties and various surface sensitive techniques have been highly developed and provide detailed local structural information on materials surfaces. However, there are also very interesting challenges ahead: The development of DAFS into a generally useful technique is one of them.

The developments in our theoretical understanding of XAFS and in particular the ability to calculate EXAFS spectra in terms of local structural parameters has developed EXAFS into a highly reliable tool for the study of the local structure of materials. Additional developments are still necessary, in particular, the elimination of the need to use phase and amplitude correction parameters. The elimination of these parameters will reduce the number of parameters needed in order to fit theory and experiment thus increasing the accuracy of the structural and dynamical parameters obtained from the fit. Much more effort will be needed to bring XANES to a level similar to that of EXAFS. Here the problem is much more complex, especially because indications are that the one-electron and the muffin-tin approximations may not be good enough.

One area in which EXAFS has made a special contribution has been briefly reviewed here. EXAFS measurements have shown that local structural distortions are present in various materials. So far local disordered distortions have been found and quantitatively investigated in materials undergoing ferroelectric, antiferrodistortive and antiferroelectric structural phase transitions but they may be present in other systems as well as seen from the results on $Ba_{0.6}K_{0.4}BiO_3$. The study of the local structural distortions in the ferroelectrics has already given us a new insight into the mechanism driving these transitions. A model that assumes the existence of local disordered distortions which are a result of local interatomic interactions has been proposed recently (73). In this model the local distortions are dynamic and interact with the soft mode phonon. This model leads to ferroelectric phase transitions at a temperature higher than the temperature where the soft mode frequency vanishes. Furthermore this model was shown to quantitatively explain the Raman, hyper-Raman, infrared and dielectric measurements, providing, in addition, an explanation of the central peak phenomenon observed in these systems.

The availability of third generation synchrotrons opens up exciting new possibilities. The high brilliance X-ray beams may be used to study the structural and electronic properties with high spatial and temporal resolutions with implications both on our fundamental understanding as well as practical applications.

REFERENCES

1. Sayers D.E., Stern E.A. and Lytle F., Phys. Rev. Lett. **27**, 1204 (1979).
2. Stern E.A., Phys. Rev. **B10**, 3027 (1974).
3. Koningsberger D.C. and Prins R., *X-Ray Absorption*, New York, John Wiley & sons (1987).
4. Oyanagi Y., Researches of the Electrotechnical Laboratory, No. 966 (1994).
5. Goulon J., Physica **B158**, 5 (1989).
6. Kizler P., Physics Letters, **A172**, 66 (1992).
7. Petiau J. and Calas G., J. de Physique Colloque, **48**, 1085 (1987).
8. Fonda L., J. of Physics (condensed matter), **4**, 8269 (1992).
9. Crozier E.D., Physica **B158**, 14 (1989).
10. Garner C.D., Physica **B208& 209**, 1 (1995).
11. Hasnain S.S., Progress in Biophysics & Molecular Biology, **50**, 47 (1987).
12. Citrin P.H., Surface Science, **299**, 199 (1994).
13. Watson P.R., J. of Physical and Chemical Data, **21**, 123 (1992).
14. Kizler P., J. of Noncrystalline Solids, **150**, 342 (1992).
15. Sayers D.E. and Paesler M.A., J. de Physique Colloque, **47**, 349 (1986).
16. Krill G., J. de Physique Colloque, **47**, 907 (1986).
17. Stern E.A. and Heald S.M., *Handbook of synchrotron radiation*, edited by Koch E.E., (North Holland, Amsterdam, 1983) vol 1, ch. 10.
18. Schütz G., Wagner W., Wilhelm W. and Kienle P., Phys. Rev. Lett. **58**, 737 (1987).
19. Schütz G., Frahm R., Mautner P., Wienke R., Wagner W., Wilhelm W. and Kienle P., Phys. Rev. Lett. **62**, 2620 (1989).
20. Thole B.T., Carra P., Sette F. and van der Laan G., Phys. Rev. Lett. **68**, 1943 (1992).
21. Kronig R. de L., Z. Phys., **70**, 317 (1931).
22. Sayers D.E., Bunker B.A., edited by Koningsberger D.C. and Prins R., *X-Ray Absorption*, New York, John Wiley & sons (1987) ch 6.
23. Lee P.A., Pendry J.B., Phys. Rev. **B27**, 95 (1975).
24. Müller J.E., Schaich W.L., Phys. Rev. **B27**, 6489 (1983).
25. Gurman S.J., Binsted S. and Ross I., J. Phys. **C17**, 143 (1984); ibid, **C19**, 1845 (1986).
26. Mustre J. Zabinsky S.I. and Albers R.C., J. Am Chem. Soc., **113**, 5135 (1991).
27. Rehr J.J., Albers R.C. and Zabinsky S.I., Phys. Rev. Lett. **69**, 3397 (1992); Rehr J.J., Zabinsky S.I., and Albers R.C., Phys. Rev. **B 41**, 8139 (1990).
28. Natoli C.R., Physica **208&209**, 5 (1995).
29. Stern E.A., Newville M., Ravel B., Yacoby Y. and Haskel D., Physica **B 208& 209**, 117 (1995).
30. Filipponi A., Di Cicco A., Tyson T.A. and Natoli C.R., Solid State Comm., **78**, 265 (1991).
31. Kuzmin A., Physica **B 208&209**, 175 (1995).
32. Bunker G., Nuclear Inst. and Meth., **207**, 437 (1983).
33. Hanske-Petitpierre O., Yacoby Y., Mustre de Leon J., Stern E.A. and Rehr J.J., Phys. Rev. **B44**, 6700 (1991).

34. Newville M., Ravel B., Rehr J.J., Stern E.A. and Yacoby Y., Physica **B 208**& **209**, 154 (1995).
35. Newville M., Livins P., Yacoby Y., Rehr J.J. and Stern E.A., Phys. Rev. **B47**, 126 (1993).
36. Brigges F., Both C.H. and Li G.G., Physica **B 208**& **209**, 121 (1995).
37. Zabinsky S.I., Rehr J.J., Ankudinov A., Albers R.C. and Eller M.J., Phys. Rev. **B52**, 2995 (1995).
38. Filipponi A., Physica **B 208**& **209**, 29 (1995).
39. Farges F., Brown G.E. and Rehr J.J., unpublished.
40. Natoli C.R., Benfatto M., Brouder C., Ruiz Lopez M.F. and Foulis D.L., Phys. Rev. **B42**, 1944 (1990).
41. Heald S.M., Tranquada J.M., Moodenbaugh A.R. and Xu Y., Phys. Rev. **B 38**, 761 (1988).
42. Cauchios Y. and Bonnelle C., Compte Rendus de l'Academie Des Sciences (Paris) **242**, 1596 (1956).
43. Heno Y., Compte Rendus de l'Academie Des Sciences (Paris) **242**, 1596 (1956).
44. Stragier H., Cross J.O., Rehr J.J., Sorensen L.B., Bouldin C.E. and Woicik J.C., Phys. Rev. Lett. **69**, 3064 (1992); Sorensen L.B., Cross J.O., Newville M., Ravel B., Rehr J.J., Stragier H., Bouldin C.E. and Woicik C.E., in *Resonant anomalous X-ray Scattering: Theory and Applications*, Materlik G., Sparks C.J. and Fischer K. editors, Elsevier Science (1994) pp. 389.
45. Cross J.O., PhD. thesis submitted to the University of Washington,
46. Lines M.E. and Glass A.M.,*Principles and Applications of Ferroelectrics and Related Materials*, Clarendon Press, Oxford, (1977).
47. Comes R., Lambert M. and Guinier A., Acta Crystallographica, **A26**, 244 (1970).
48. Comes R., Currat R., Denoyer F., Lambert M. and Quittet A.M., Ferroelectrics, **12**, 3 (1976).
49. Comes R. and Shirane G., Phys. Rev. **B5**, 1886 (1972).
50. Quittet A.M. and Lambert M., Solid State Commun., **12**, 1053 (1973).
51. Yacoby Y. and Just S., Solid State Commun., **15**, 715 (1974).
52. Yacoby Y., Z. Physik B Condensed Matter, **31**, 275 (1978).
53. Gervais F., Ferroelectrics, **53**, 91 (1984).
54. Scott J.F., Rev. of Mod. Phys., **46**, 83 (1974)
55. Axe J.D., Harada J. and Shirane G., Phys. Rev., **B1**, 1227 (1970).
56. Fontana M.D., Idrissi H., Kugel G.E. and Wojcik K., J. Phys. Condensed Matter, **3**, 8695 (1991).
57. Shapiro S.M., Axe J.D. and Shirane G., Phys. Rev. **B6**, 4332 (1972).
58. Muster J., Yacoby Y., Stern E.A. and Rehr J.J., Phys. Rev. **B42**, 10843 (1990).
59. N. Sicron and Y. Yacoby, unpublished.
60. Sicron N., Ravel B., Yacoby Y., Stern E.A., Dogan F. and Rehr J.J., Phys. Rev. **B 50**, 13168 (1994).
61. Glazer A.M. Mabud S.A., J. Appl. Cryst. **12**, 49 (1979).
62. Kuprianov M.F. Saitsev S.M. Gagarina E.S. Pesenko E.G., Phase Trans. **4**, 55 (1983).
63. Nelmes R.J. Politz R.O., Tun Z. Kuhs W.S. and R. Retori, Ferroelectrics, **108**, 91 (1990).
64. Islam Q.T. and Bunker B.A., Phys. Rev. Lett. **59**, 2701 (1987)
65. Wang Z. and Bunker B.A., Phys. Rev. **B 46**, 11277 (1992).

66. Ismailzade I.G., Kristallografiya, **7**, 718 (1963) [soviet Phys. Crystallogr. **7**, 584 (1963).]
67. Rechav B., Yacoby Y., Stern E.A., Rehr J.J. and Newville M., Phys. Rev. Lett. **72**, 1352 (1994).
68. Heald S.M., Dimarzio D, Croft M., Hedge M.S., Liand S. and Greenblatt M, Phys. Rev. **B 40**, 8828 (1989).
69. Yacoby Y., Stern E.A. and Heald S.M., XAFS IX international conference, Grenoble (1996).
70. Pei S., Jorgensen J.D., Dabrowski B., Hinks D.G., Richards D.R., Mitchell A.W., Newsam J.M., Sinha S.K., Vaknin D. and Jacobson A.J., Phys. Rev. **B 41**, 4126 (1990).
71. Plakida N.M., Aksenov V.L.A. and Drechsler S.L., Euro-Physics Letters, **4**, 1309 (1987).
72. Hardy J.R. and Flocken J.W., Phys. Rev. Lett. **60**, 2191 (1988).
73. Y. Girshberg and Y. Yacoby, unpublished.

Linear Polarization Effects in X-Ray Emission and Absorption Spectra

G. Dräger

Fachbereich Physik der Martin-Luther-Universität, D-06108 Halle, Germany

Abstract. A selective survey is given on polarization and direction effects in X-ray spectra which can be observed by working with single-crystal samples and taking advantage of linearly polarized radiation. The first section deals with the polarization effects in the X-ray absorption spectra (XAS) arising from linear X-ray dichroism and from the angular dependence of the 1s→3d quadrupole transition. Recently obtained results on NiO, FeO, $FeCO_3$, $Bi_2Sr_2CaCu_2O_8$ and α-Fe_2O_3 are reported.
In the second part the polarized X-ray emission spectra (XES) are considered. Recently obtained results at $YBa_2Cu_3O_{7-\delta}$ are presented and linear magnetic dichroism (LMXD) in XES has been found.
Finally, the polarization dependence of resonantly excited X-ray emission spectra (RXES) is considered. First results of a complete polarization analysis of resonantly excited Cu Lα-XES of La_2CuO_4 are reported.

INTRODUCTION

X-ray emission spectra (XES), absorption spectra (XAS) and resonantly excited X-ray emission spectra (RXES) can show an angular dependence if they are measured from single-crystal samples. This means the recorded spectral distribution will change if the polarization and direction of the incident radiation (in the case of XAS and RXES), or of the emitted radiation (for XES and RXES), is varied relative to the sample axis.

The theoretical interpretation of the measured angular dependence shows that it is a polarization effect determined by the polarization $\vec{e}$ of the incident (exciting) and emitted radiation. Additionally, if quadrupole transitions are involved, the angular dependence is determined by the beam direction $\vec{k}$ of the incoming or outgoing radiation. Such effects can be observed and investigated if the incident radiation is highly polarized and if the emitted radiation can be analyzed with respect to its polarization state. Therefore, investigations of polarization and beam direction effects are performed preferably by taking advantage of synchrotron radiation as the incident or exciting radiation and by using suitable equipment to analyze the outgoing or emitted radiation.

POLARIZATION EFFECTS IN XAS

Polarization and beam direction effects can be studied in a quantitative way using a transmission geometry and thin single-crystal absorbers (1,2). If thick crystals or only surface layers are to be investigated other signals like fluorescence, electron yield or sample current can be recorded instead of transmitted intensities. However, quantitative interpretation becomes more complicated.

The incident (exciting) radiation should be linearly polarized, which is best attained using synchrotron radiation from the plane of orbit. The first polarization effects in XAS were investigated in the sixties using continuous radiation from an X-ray tube after Bragg reflection close to 45° (see Ref. 1 and the references in it).

The angular dependent XAS can be advantageously interpreted using the detailed calculations of Brouder (3) for anisotropic XAS. Looking at a general expression for the cross sections in X-ray absorption (for example equation 3.16 in Ref. 3) the angular dependence of XAS is determined by the matrix elements

$$\langle i | \vec{e} \cdot \vec{r} | f \rangle \text{ and } \langle i | \vec{e} \cdot \vec{r}\vec{r} \cdot \vec{k} | f \rangle.$$

These are the electric dipole and quadrupole transition amplitudes for the transitions between the initial and final states i and f, respectively. The $\vec{e}$ and $\vec{k}$ dependence of the transition matrix elements means that for all experimental arrangements the angular dependence of the XAS is determined only by the orientation of the single-crystal sample axes with respect to the polarization vector $\vec{e}$ and the X-ray wave vector $\vec{k}$.

Dipole and quadrupole transitions can be distinguished unambiguously from their angular dependence implicit in the matrix elements. A good example is the pre-edge structure of 3d metal absorption spectra (2).

In 1993 and 1994 Poumellec et al. published investigations of the polarized XANES at the V K edge of V_2O_5 and $VOPO_4$ (4) and of VO_2 (5). They obtained valuable information on the asymmetry of local environments and on the phase transition of VO_2 by the analysis of the V K pre-edge absorption with respect to the appearance of dipole and quadrupole transitions.

In the group of the author, an angular dependence of the pre-edge structure was recently found in the Ni K absorption spectra of single-crystal NiO (Fig. 1). The Ni 1s→4p dipole transitions are isotropic because of the cubic structure of NiO. Therefore, the angular dependence of the metal pre-edge absorption can result only from Ni 1s→3d quadrupole transitions. The interpretation of the $\vec{e}$- and $\vec{k}$-dependence of the spectra on the basis of the quadrupole matrix elements reveals the predominantly e_g ($d_{x^2-y^2,z^2}$)-like symmetry character of the empty states above the NiO band gap. The analogous experiment with FeO showed the existence of

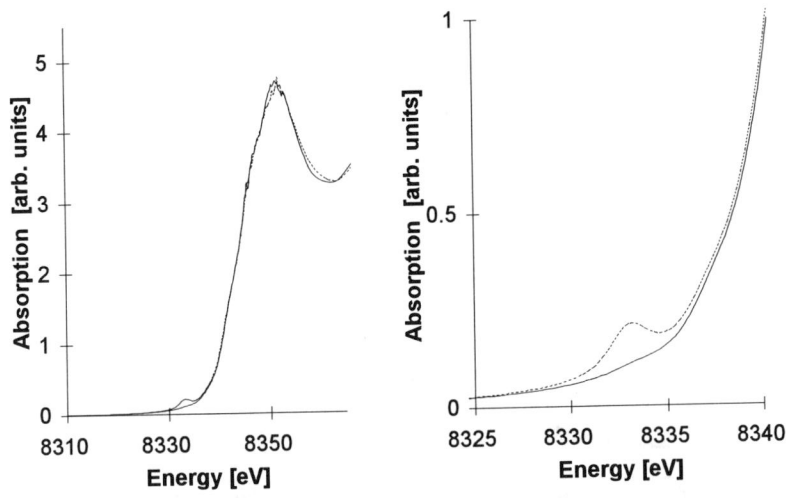

FIGURE 1. Angular dependent Ni K absorption spectra of single-crystal NiO. Two spectra are shown taken for the wave vector $\vec{k} \parallel [100]$ and $[110]$ and for the polarization vector $\vec{e} \parallel [010]$ and $[1\bar{1}0]$ (the full and the dashed lines, respectively). The spectra show orientation dependent changes resulting from 1s → 3d quadrupole transitions only in the pre-edge absorption area at about 8330-8335 eV (left). The large pre-edge absorption for $\vec{k} \parallel [110]$, $\vec{e} \parallel [1\bar{1}0]$ and the small absorption for $\vec{k} \parallel [100]$, $\vec{e} \parallel [010]$ (right: the enlarged spectra from the left side) show predominantly e_g ($d_{x^2-y^2, z^2}$) -like symmetry and less t_{2g} ($d_{xz, yz, xy}$)-like symmetry, respectively, of the empty states near the Fermi level.

empty t_{2g} ($d_{xz,yz,xy}$) and e_g states. Both results are in good agreement with electronic band structure calculations of Terakura et al. (6).

In these experiments, for the first time, an angular dependence of the XAS was found in highly symmetric cubic crystals. In 1995 Heumann et al. (7) published the linear polarized Cu K and Fe K pre-edge absorption spectra of single-crystal $Bi_2Sr_2CaCu_2O_8$ and $FeCO_3$ measured in the transmission geometry. The spectra displayed a strong angular dependence from which some qualitative conclusions could be drawn with respect to the distribution of suborbital m-resolved 3d-like states. Now some quantitative results can be reported (Figs. 2 and 3). In the top of Fig. 2 the orientation-dependent Fe K pre-edge absorption of $FeCO_3$ is shown superimposed on the beginning slope of the Fe K main absorption edge. The contributions of the different 3d suborbital states to the Fe 1s→3d quadrupole transitions are also shown. These contributions with different suborbital m-symmetry (m = xy, xz, yz, x^2-y^2, z^2) were evaluated by calculating the angular part of the quadrupole matrix element for several sample orientations with respect to the polarization $\vec{e}$ and the beam direction $\vec{k}$ of the synchrotron radiation. In the lower part of Fig. 2 the m-resolved densities of 3d-like states are shown, revealing

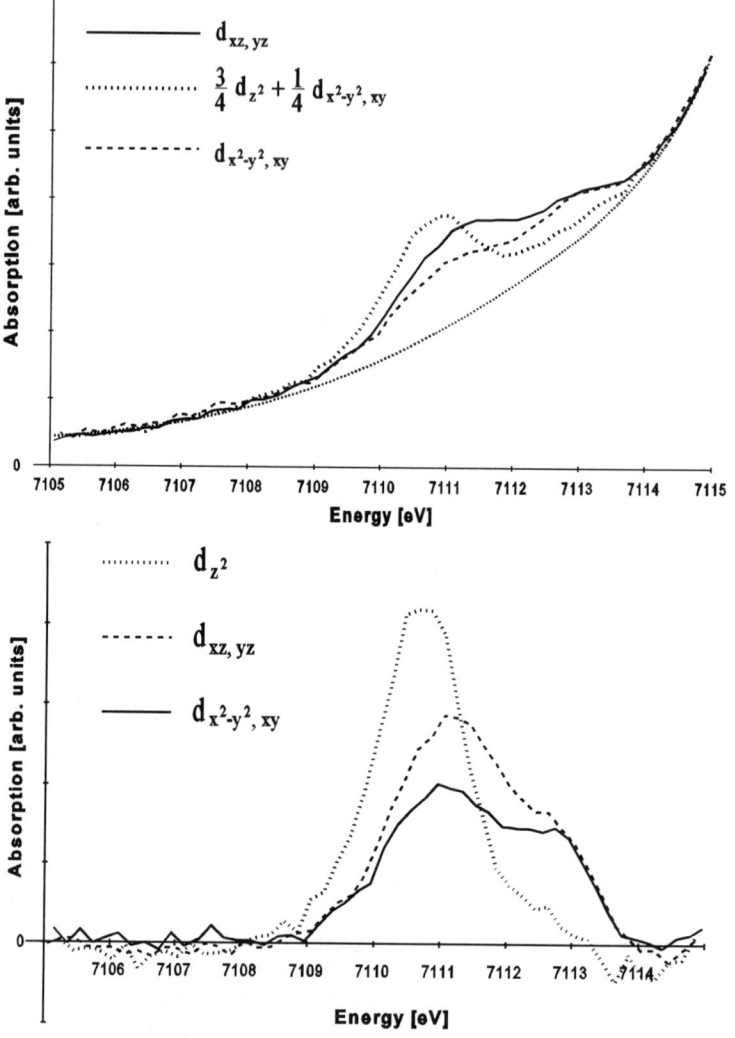

FIGURE 2. The polarization and direction dependent Fe K pre-edge absorption of $FeCO_3$ (at the top). The separation of the Fe K pre-edge absorption into partial components with different orbital symmetries d_{z^2}, $d_{xz,yz}$ and $d_{x^2-y^2,xy}$ (at the bottom). The separation is proceeded after subtraction of the dipole transition component (dotted weak line at the top) so that only the possible quadrupole 1s→3d transitions are considered.

great differences depending on the partial suborbital symmetry. A complete symmetry resolution of d-like states can only be achieved by evaluating and interpreting quadrupole transitions. Dipole transitions as measured in the L absorption spectra are not able to completely resolve the symmetry.

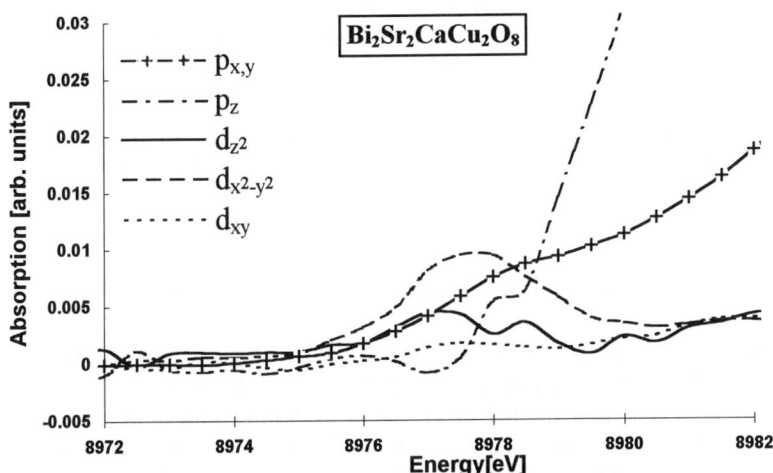

FIGURE 3. The separation of the Cu K pre-edge absorption of $Bi_2Sr_2CaCu_2O_8$ into the contributions of the different m-resolved p- and d-like DOS as calculated from the experimental data. The existence of empty $d_{x^2-y^2}$-like states is clearly shown as well as the absence of d_{xy}-like hole states. The controversial existence of d_{z^2}-like hole state is also proved by the exact symmetry resolution of the Cu K pre-edge absorption. The contributions of $p_{x,y}$- and p_z-like DOS at the beginning slope of the Cu K edge are strongly different.

A detailed quantitative analysis of the Cu K XANES of $Bi_2Sr_2CaCu_2O_8$ for five selected sample orientations has provided the partial DOS components $p_{x,y}$, p_z, d_{z^2}, $d_{x^2-y^2}$, and d_{xy} for Cu as shown in Fig. 3. Some results of the symmetry analysis are briefly discussed in the caption of Fig. 3. Furthermore, it must be emphasized
- the high discrimination of the method with respect to the symmetry of states and
- the strong bulk sensitivity of the transmission mode which can be applied to K absorption as opposed to the surface sensitive yield modes which are usually used when measuring L absorption spectra.

In 1985 B. T. Thole et al. (8) predicted a linear magnetic X-ray dichroism (LMXD) which is the difference in cross section for light polarized perpendicular and parallel to a magnetic moment $\langle M \rangle$ and depends on $\langle M^2 \rangle$ of the ions. LMXD can therefore be observed in any system with collinear magnetic ordering. The first experimental proof was obtained by G. van der Laan et al. (9) at the $M_{4,5}$ edges of rare earth materials. Calculations by van der Laan and Thole (10) predicted strong LMXD at $L_{2,3}$ edges of 3d transition metal ions which will appear besides the crystal field induced linear dichroism considered above.

P. Kuiper et al. (11) in 1993 reported strong linear magnetic dichroism at the Fe $L_{2,3}$ edge of α-Fe_2O_3 (hematite). Utilizing a magnetic spin flip transition first discovered by Morin they could differentiate between the linear dichroism induced by the crystal field and the magnetic moments. In hematite magnetic dominate crystal field effects.

POLARIZATION EFFECTS IN XES

Polarization effects in XES can be advantageously studied in single-crystal samples having tetragonal, hexagonal, trigonal, and lower symmetry.

The emitted radiation excited by electrons or X-rays should be measured as a function of the angle between its polarization $\vec{e}$ and the crystal axes of the sample. This can be achieved by using a suitable arrangement of the emission sample relative to the crystal monochromator or diffraction grating of the spectrometer (12). The best polarization analyzer is a spectrometer crystal which reflects and monochromatizes the emitted radiation at a Bragg angle as near as possible to 45°. Using such a monochromator the principal intensities having a polarization parallel to the axes of the emitting crystal can be measured directly. Partially polarized XES can further be evaluated for well defined geometrical configurations of the sample and monochromator and considering the angular dependence of dipole transitions (12). Since the transition matrix elements for X-ray absorption and emission are in principle the same, the XAS results of C. Brouder (3) can be used for interpreting polarized XES.

Recently, the method of polarized X-ray emission spectroscopy has been used to investigate the electronic structure of complex materials. For example, the polarized emission spectra of the high-T_c superconductor $YBa_2Cu_3O_{7-\delta}$ (YBCO) was investigated by F. Werfel et al. (13), W. Czolbe et al. (14) and more recently by A. Kottmann et al. (15). In Ref. 14 the polarized Cu $K\beta_{2,5}$ emission of YBCO is compared with partial Cu p_m - like DOS (m = x,y,z) calculated by Ching (16). As can be seen in Fig. 4 the experimental spectra are well reproduced by the broadened Cu 4p partial DOS but without the shoulder structure 3 (σ) for which the corresponding Cu $4p_{x,y}$- like DOS is obviously missing.

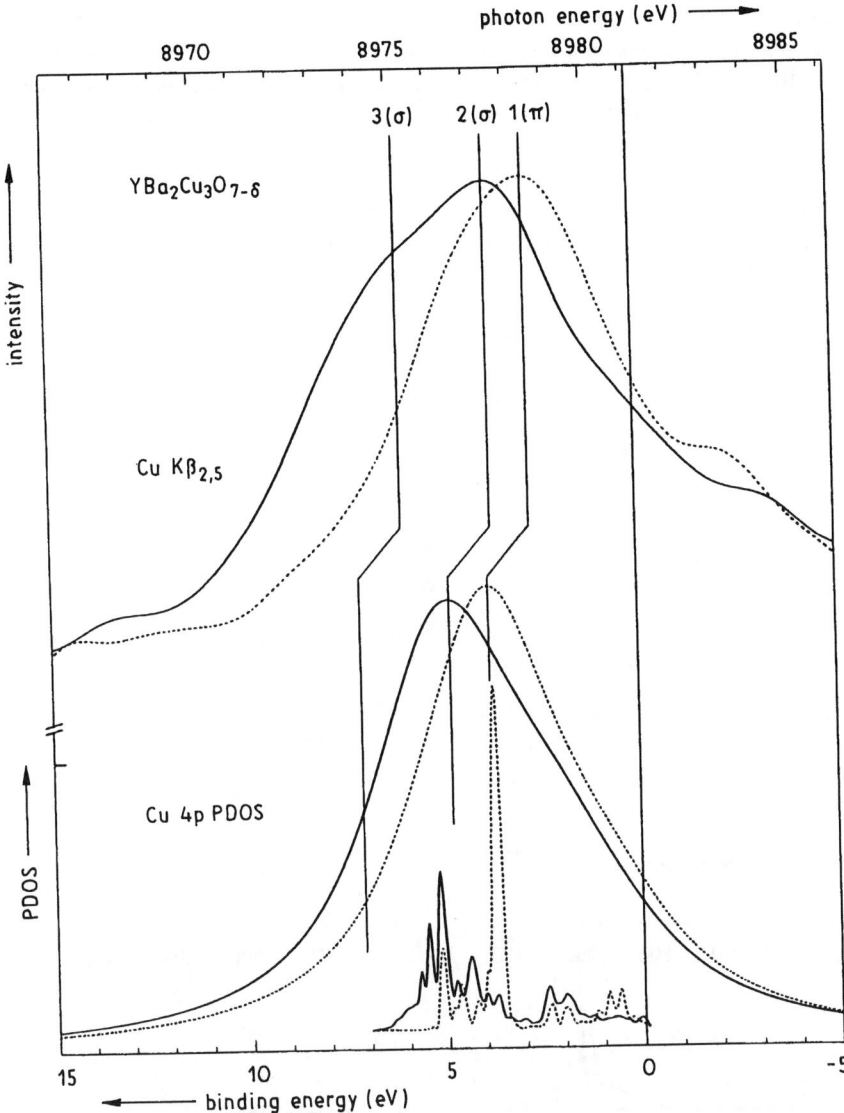

FIGURE. 4. Polarized Cu K$\beta_{2,5}$ emission spectra of YBa$_2$Cu$_3$O$_{7-\delta}$ together with a calculated local l,m-resolved Cu p DOS. Two energy scales are drawn, the photon energy scale of the Kβ emission at the top and the binding energy scale used for the alignment of both the Kβ emission spectra and the theoretical DOS at the bottom of the figure. Solid curves denote the σ and $p_{x,y}$ components and dotted ones the π and p_z components (14).

In Ref. 15 the polarized O Kα and Cu Lα X-ray emission spectra of YBCO are measured and compared with electronic structure calculations. The comparison of the band-like O 2p states with the results of LDA - calculations works well. For the interpretation of the Cu Lα emission, however, improved theoretical methods in the frame of cluster calculations and using the "Anderson impurity" model should be done because of strong correlation effects (15).

By correlating several polarized spectra of one compound in the common binding energy scale valuable information can be obtained on anisotropic bonding and hybridization.

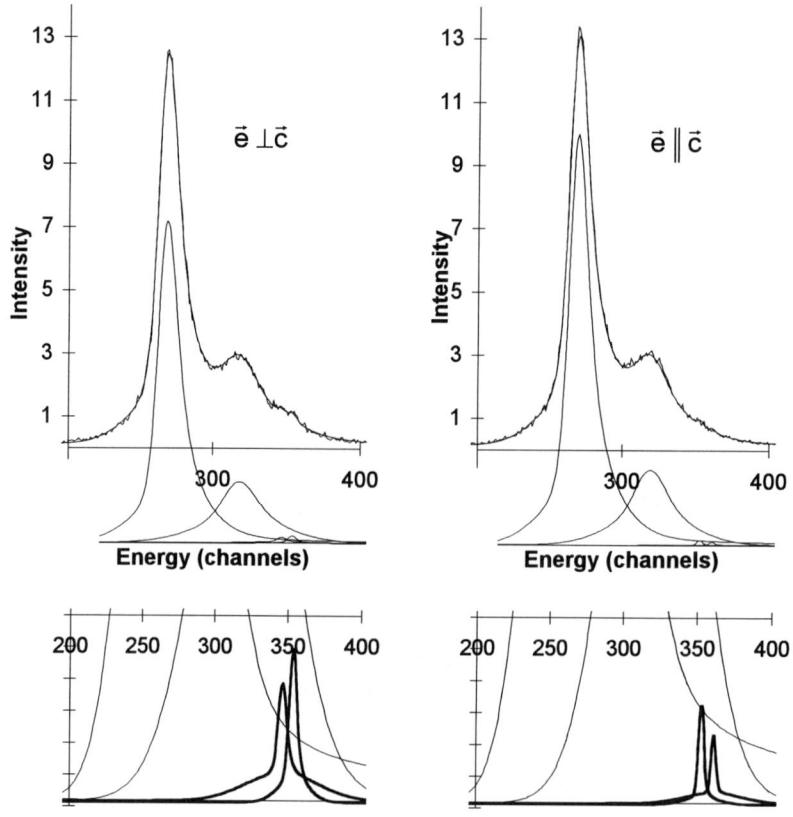

FIGURE 5. The linearly polarized Fe $K\beta_{1,3}, \beta'$ emission of α-Fe_2O_3 for two different geometrical configurations of the emission polarization $\vec{e}$ and the trigonal crystal axis $\vec{c}$. At the top: the measured spectra together with integral fit function. Middle: the composition of partial components forming the integral fit function. At the bottom: the middle part of the figure, but enlarged by the factor of 20, and displaying the polarization dependent components (heavy lines).

Very recently, in the group of the author the linearly polarized Fe K$\beta_{1,3}, \beta'$ of α-Fe$_2$O$_3$ (hematite) was investigated. At the low energy side of Kβ' small changes of the intensity were found for the polarization $\vec{e}$ of the emitted radiation parallel and perpendicular to the trigonal axis of Fe$_2$O$_3$ (Fig. 5). The observed polarization dependent changes are shown in an enlarged scale at the bottom of Fig. 5. Since the linear dichroism appears at 3p core level states having a strong exchange splitting and multiplet coupling but only small crystal field interaction, it seems to be a LMXD effect measured here in an emission experiment.

POLARIZATION EFFECTS IN RXES

For incident photon energies near the core binding energy, inelastic X-ray scattering theory predicts a coherent absorption-emission process which provides resonantly excited X-ray emission spectra (RXES). The transition probability and the polarization of the exciting (incident) and emitted (outgoing) radiation is determined by the expression

$$\frac{\langle f | \vec{e}_{out} \cdot \vec{r} | m \rangle \langle m | \vec{e}_{in} \cdot \vec{r} | i \rangle}{E_m - E_i - h\nu_{in} - i\Gamma/2}$$

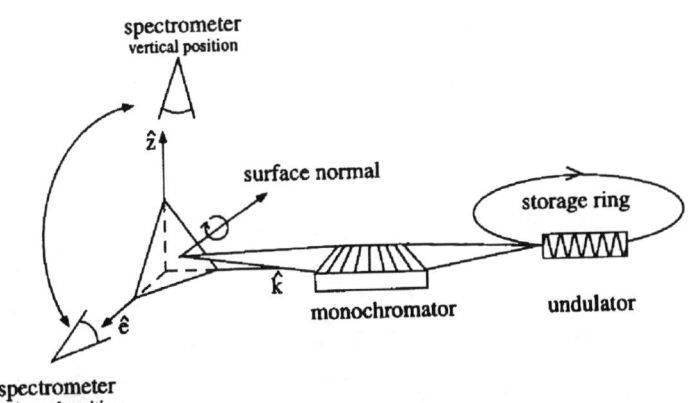

FIGURE 6. Schematic of the experimental arrangement. Linearly polarized X-rays from an undulator insertion device are monochromatized and focused onto the sample. The coordinate system is defined by the polarization vector $\vec{e}$, the incident beam direction $\vec{k}$, and a third orthogonal vector $\vec{z}$. The sample axes coincide with the axes of this Cartesian coordinate system and may be permutated by rotating through 120° around the surface normal. The spectrometer is positioned with its optical axis either horizontally or vertically. Incidence and take-off angles are equal in all configurations.

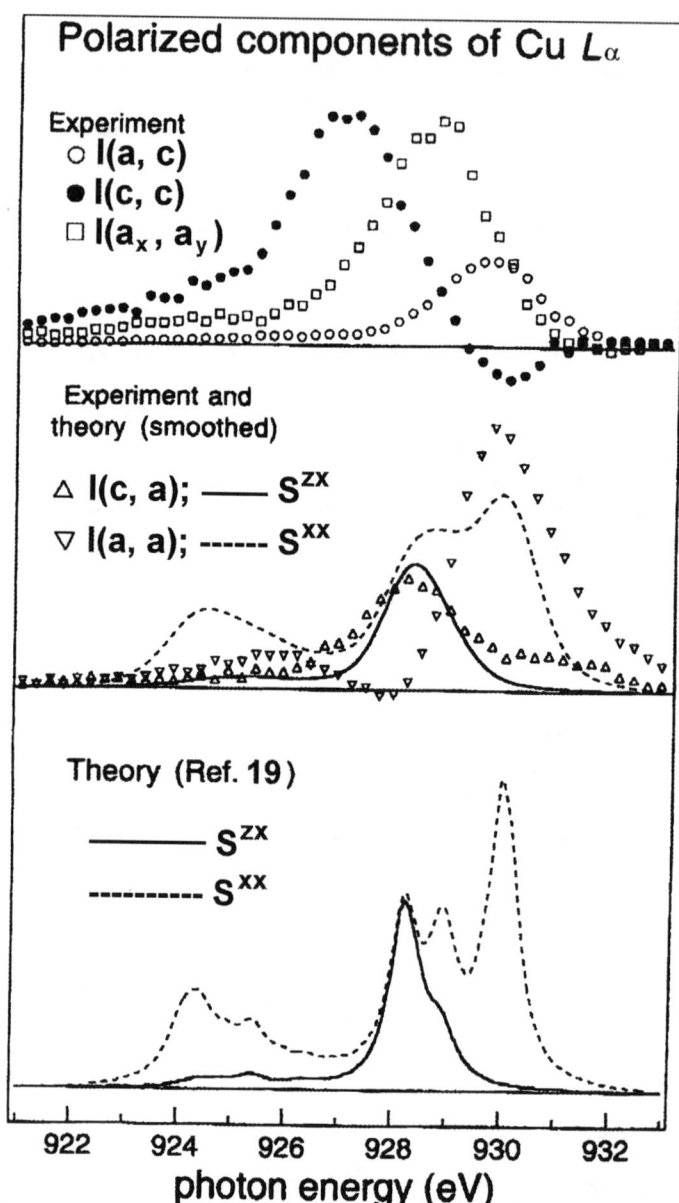

FIGURE 7. Polarized basic components (upper and middle panel) extracted from the measured Cu $L\alpha$-spectra of La_2CuO_4. In the middle panel the results are compared with the corresponding theoretical spectra (after broadening by a gaussian). The lower panel shows the theoretical spectra from Tanaka and Kotani (19). For further explanations and details see Ref. 20.

Here the initial state i is the ground state, whereas the final state f has an electron in the conduction band and a hole in the valence band. The intermediate state m has a core hole and an electron in the conduction band. The resonance term in the denominator contains the energy E_m of the intermediate state, E_i of the initial or ground state, the photon energy $h\nu_{in}$ of the incident (exciting) radiation and the core level width Γ.

The product of the two dipole transition matrix elements for the coherent absorption-emission process causes a complicated polarization and angular dependence of the incident and emitted radiation if one works with single-crystal samples. On the other hand, the evaluation and interpretation of resonant X-ray emission spectra excited with definite polarization and their polarization analysis provides a lot of information on the electronic structure. First experiments to study the resonantly excited inelastic scattering process and for receiving more information on the electronic band structure of graphite were reported by Skytt et al. (17) and Carlisle et al. (18).

Very recently, the group of J. Nordgren /Uppsala and the group of the author performed a common RXES experiment at the beamline BW3 at HASYLAB/Hamburg. In this experiment the polarized subbands in resonant Cu Lα emission of single-crystal La$_2$CuO$_4$ following the excitation with linearly polarized synchrotron radiation were investigated. The experimental arrangement and some explanations are given in Fig. 6. By evaluating the spectra taken for selected geometrical configurations of the sample axes, the polarization of synchrotron radiation and the position of the X-ray spectrometer a set of principal emission intensities $I(\vec{e}_{out}, \vec{e}_{in})$ could be extracted. Here the polarization of the emitted (outgoing) and exciting (incident) radiation are defined in the coordinate system of the sample crystal. It means for example, the emission intensity $I(a, c)$ is polarized parallel to the crystal axis $\vec{a}$ and has been excited by synchrotron radiation polarized parallel to the crystal axis $\vec{c}$.

In Fig. 7 the five principal intensities are shown and two of them, $I(c, a)$ and $I(a, a)$, are compared with the theoretical spectra S^{zx} and S^{xx} calculated by Tanaka and Kotani (19). For further details and explanations see Ref. 20.

CONCLUSIONS

Linear polarization effects in X-ray emission and absorption spectra are highly interesting in developing and improving new methods and techniques of X-ray spectroscopy. They open a number of new possibilities of investigating the electronic and magnetic structure of crystalline solids and of verifying the theory.

REFERENCES

1. Brümmer, O. and Dräger G., p*hys. stat. sol.* 14, K175 (1966)
2. Dräger, G., Frahm, R., Materlik, G., and Brümmer, O., *phys. stat. sol.* (b) 146, 287 (1988)
3. Brouder, C., *J. Phys.* C2, 701 (1990)
4. Poumellec, B., Cortes, R., Sanchez, C., Berthon, J., and Fretigny, C., *J.Phys. Chem. Solids* 54, 751 (1993)
5. Poumellec, B., Cortes, R., Loisy, E., and Berthon, J., *phys. stat. sol.* (b) 183, 335 (1994)
6. Terakura, K., Oguchi, T., Williams, A. R., and Kübler, J., *Phys.Rev* B 30, 4734 (1984)
7. Heumann, D., Hofmann, D., Dräger, G., *Physica* B 208&299, 305 (1995)
8. Thole, B. T., van der Laan, G., and Sawatzky, G. A., *Phys.Rev Lett.* 55, 2086 (1985)
9. van der Laan, G., Thole, B. T., Sawatzky, G. A., Goedkoop, J. B., Fuggle, J. C., Esteva, J.-M., Karnatak, R., Remeika, J. P., and Dabkowska, H. A., *Phys.Rev Lett.* 34, 6529 (1986)
10. Van der Laan, G., Thole, B.T., *Phys.Rev* B 43, 13401 (1991)
11. Kuiper, P., Searle, B. G., Rudolf, P., Tjeng, L. H., and Chen, C. T., *Phys. Rev Lett.* 70, 1549 (1993)
12. Dräger G., and Brümmer, O., *phys. stat. sol.* (b) 124, 11 (1984) and Proceedings of the X-84 Conference, Leipzig 1984, p. 337
13. Werfel, F., Dräger, G., Leiro, J. A., and Fischer, K., *Phys. Rev.* B45, 4957 (1992)
14. Czolbe, W., Dick, U., Dräger, G. and Fischer, K., *phys. stat. sol.* (b) 174, 91 (1992)
15. Kottmann, A., Siebinger, R., Lamparter, P., and Steeb, S., *Z. Naturforsch.* 50a, 935 (1995)
16. Ching, W. Y., *J. Amer. Ceram. Soc.* 73, 3135 (1990)
17. Skytt, P., Glans, P., Mancini, D. C., Guo, J.-H., Wassdahl, N., Nordgren, J. and Ma, Y., *Phys. Rev.* B50, 10457 (1994)
18. Carlisle, J. A., Shirley, E. L., Hudson, E. A., Torminello, L. J., Calcott, T. A., Jia, J. J., Ederer, D. L., Perera, R. C. C., and Himpsel, F. J., *Phys.Rev Lett.* 74, 1234 (1995)
19. Tanaka, S.and Kotani, A., *J. Phys. Soc. Jpn.* 62, 464 (1993)
20. Duda, L.-C., Dräger, G., Guo, J.-H., Heumann, D., Bocharov, S., Hergert, W., Wassdahl, N., Nordgren, J., Tanaka, S., and Kotani, A., in preparation

From the S_2 Molecule to its Condensed Molecular and Polymerized Phases: An X-ray Absorption Study at the S K Edge

R.C. Karnatak

Laboratoire de Spectroscopie Atomique et Ionique
Université Paris-Sud, 91405 ORSAY Cedex, FRANCE

Abstract. X-ray K absorption spectra of sulfur have been studied in its vapour and molecular and polymerized solid phases. The spectrum of the S_2 molecule is discussed in the light of the p orbital energies and the s-p orbital energy differences of some second and third period elements. The cyclic and chain forms of sulfur show some similarity in σ^* line structure and a marked difference in their near edge structure due to different number of S atoms surrounding a central atom in these allotropic forms. The results are discussed in the light of different vapour to condensed phase deposition processes. The role of planar *cis* and *trans* isomers of an intermediate tetrasulfur cluster, crucial for the formation of a chain or cyclic variety, is stressed.

INTRODUCTION

Sulfur is an element which is known to exist in a large number of allotropic forms. This is due to the fact that in solid, liquid, or vapour phases of sulfur one observes a great amount of flexibility in the S-S bond lengths and bond angles leading to various stable structures. From the structural point of view the observed structures in solid sulfur can broadly be classified into two categories: cyclic oligomers (cyclic S_n) and chain polymers (1-5). The study of the chain variety of sulfur is of great interest in the rubber industry due to the important role of sulfur chain links in the process of rubber vulcanization.

The S_2 molecule, the first member of the cyclic S_n and polymer series attracted much scientific interest in the past. Because of its high stability, optical spectra of the S_2 molecule were widely studied (6). In these studies the S_2 vapour is obtained by heating either FeS at high temperature or native sulfur (α-sulfur) in a furnace at low temperature and low pressure conditions. In each case a yellow deposit is observed at the exit of the heating element. This product is identified as α- sulfur.

The polymerized variety of sulfur, usually called as ω-sulfur, is also obtained from the sulfur vapour, but at high temperatures. Instead of slow cooling as in the case of α-sulfur, the high temperature sulfur vapour is quenched in CS_2 vapour (1,3). The cyclic phases are dissolved away in CS_2. The undissolved product contains ω-sulfur, which transforms very slowly to α-sulfur at room temperature.

The previous mass spectrometric studies indicate that a substantial quantity of molecular S_n (n=1-12) clusters are present in sulfur vapour. These studies were used to determine the ionization potentials of the clusters. From the structural point of view, S_3 (thiozone) and S_4 (tetrasulfur) molecules attracted much experimental and theoretical interest. In the context of the stable isomers, matrix trapped small clusters of sulfur had also been studied (7) by infrared (ir) and Raman spectroscopy in the past. An important and detailed theoretical study (8) of the isomers of the tetrasulfur molecule tried to resolve the long standing controversial question on its open or closed structure. It is now believed that ordinarily the stable isomers of sulfur clusters with n ≥ 5 are all cyclic. Then an interesting question about the formation of the polymerized (chain) variety of sulfur arises. In response to this question one might think that there exist two seed isomers for a particular S_n molecule which grow up with the addition of other atoms or small molecules and may decide to follow two distinct paths to form cyclic or chain varieties of sulfur.

The clusters are considered to be intermediates between the atoms or molecules and the corresponding solid. Generally the properties of the atoms and molecules are well understood. For sulfur, the S_2 molecules can be separated in pure state. The condensed varieties of sulfur are also well known. The intermediate clusters and their isomers are usually identified by their ir spectra. A particular method of preparation of clusters may be of great help for their identification. For example one observes in matrix-trapped clusters the presence of S_3 (7) only when the clusters are prepared in a discharge tube. By simple thermal heating these clusters are not observed due to the extremely low concentration of S atoms which could combine with the relatively abundant S_2 species to yield S_3 molecules. Theoretical models developed for the isomers and the calculated ir spectra are of great help for the identification of the isomers.

In this paper we present a detailed study of the S_2 molecule and its condensed molecular and polymerized phases by X-ray absorption spectroscopy. After a brief description of the experimental methods (spectroscopic instrumentation and sample preparation) we will present our recently observed K absorption spectrum of the S_2 molecule. The observed K absorption spectroscopic feature will be interpreted and discussed in the light of the systematics of 1s—π^* and 1s—σ^* resonances observed in the K absorption spectra of simple molecules of first period elements. We will discuss the condensation behaviour of S_2 in comparison to that of the O_2 molecule. In this context we will discuss the formation of condensed molecular and polymerized condensed phases of sulfur obtained by condensation of sulfur vapour. Our X-ray spectroscopic results on α- and ω-sulfur will be interpreted in the light of existing theoretical results on the structures and stability of the intermediate molecular clusters which, in fact, are the building blocks for the formation of the solid.

EXPERIMENTAL

The X-ray absorption measurements were performed at the SA 32 beam line of the SUPER ACO storage ring, LURE, Orsay. A focussing toroidal mirror is used on this beam line to eliminate the higher harmonics and to focus the beam on the sample. A double crystal monochromator (9) employing Si (111) (2d=6.271 Å) and Ge (111) (2d=6.532 Å) crystal pairs were used to monochromatize the incident beam. We used the Si (111) pairs in the case of K absorption spectra of molecular and polymerized sulfur and Ge (111) pairs were employed for the study of sulfur vapour. In these spectra the broadening of the spectral features is due to two factors: (i) monochromator and (ii) inherent core hole lifetime width. The theoretical broadening due to the monochromator crystals pairs (10) is estimated to be 0.3 and 0.8 eV respectively for Si(111) and Ge(111) in the S K edge region (2500 eV). The theoretical 1s level width of S given in ref (11) is 0.59 eV. If we suppose that the instrumental broadening is simulated by a Gaussian function and take the 1s core hole lifetime as Lorentzian, we obtain an estimate of total broadening of about 0.66 and 0.99 eV for Si(111) and Ge (111) crystal pairs respectively.

The K absorption spectra of S in the 2450-2520 eV range were obtained in the photoelectron yield mode for α and ω-sulfur samples by measuring the photocurrent from the isolated metallic sample holder with a picoammeter. This mode is particularly useful for samples like sulfur with a high partial pressure which might contaminate a detector like a channeltron. It also maintains a stable background due to a slight negative charging of the sample and thus avoids largely the detection of stray electrons. For the vapour phase, the transmitted intensity through the vapour column was detected by a Si diode. In both cases an incident beam monitor placed between the monochromator and the sample was used to account for beam intensity variations during the experiment. The relative energy of the spectra was calibrated against the intense white line from α-sulfur (2472.0 eV). The accuracy of the absolute energy scale is estimated to be better than 0.2 eV.

The samples of α and ω sulfur for spectroscopic measurements were obtained by spreading the powders on adhesive tape fixed to an Al plate. The ω-sulfur was prepared by quenching the orthorhombic sulfur vapour in CS_2. The quenching is followed by an annealing of the condensate for a period during which the polymer acquires its interesting properties: the structure and thermal stability. The condensed product was further purified by dissolving the residual α-sulfur in CS_2 solution (1). Finally the insoluble product was dried and analyzed by X-ray diffraction. This resulting product is known as ω-sulfur. Here we will not go into the details of the ω-sulfur crystal chemistry and simply refer to some articles (1,3,5) on this subject.

A 1 m long stainless steel tubular furnace was used to confine the vapour column. Details of this furnace will be described elsewhere (12). This custom-made furnace is being used to obtain the spectra of volatile or nonvolatile substances from the optical to the X-ray range between room temperature and 1200° C. About 10 g of commercially available flower of sulphur is placed inside a cylindrical sample holder tube containing an opening at each end to accommodate a 80 cm long stainless tube with several small holes at the middle of the tube. The sample holder is slid over up to the centre of the stainless steel effusion tube in such a way that the holes are on the upper side and remain enclosed inside the sample holder. This assembly is then put inside a 1 m long stainless steel heating tube. A number of ceramic rings were used to isolate electrically the tube and sample holder from the heating tube which is clamped to two electrical feed throughs. The two ends of the heating tube are water cooled. These are connected to vacuum pumps and also carry a buffer gas (He) inlet system. The heating tube is surrounded by water cooled thermal screens which are in turn housed inside a 10 cm diameter outer tube which is pumped separately. The furnace is isolated from the monochromator vacuum system by a thin plastic window. On the other end of the 1 m heating tube the detector assembly carrying a Si diode is fixed.

In the actual experiment the furnace containing the sulfur sample is pumped down to 10^{-4} - 10^{-5} torr with a mild heating to about 60° C for more than two hours to eliminate the water vapour and other volatile impurities. At this stage the buffer gas He is introduced inside the heating tube and maintained at a pressure of about 1 torr. The spectra are obtained by continuous repeated scans while slowly raising the temperature. One notices an evolution in some spectral features as a function of the temperature. Near the melting point (95° C) of sulfur the spectral features did not show any evolution and the vapour density was found to be satisfactory so that several good quality spectra were obtained.

RESULTS AND DISCUSSION

The K absorption spectrum of sulfur vapour is shown in figure 1. In this spectrum the main line at 2472.0 eV is followed by two weak and broad structures at 7.4 and 9.2 eV above it. The K edge spectra from α and ω-sulfur normalized for the absorption jump at 2483.0 eV are given in figure 2. These two spectra also show an intense main line at 2472.0 eV. The main line appears quite narrow in the vapour spectrum (figure 1). It becomes broader and asymmetric in α-sulfur and in ω-sulfur this line shows a distinguishable high energy shoulder (figure 2). The α-sulfur shows a broad feature at 2479.4 eV and in ω-sulfur this feature becomes still broader and its centre of gravity is found to be at 2478.9 eV.

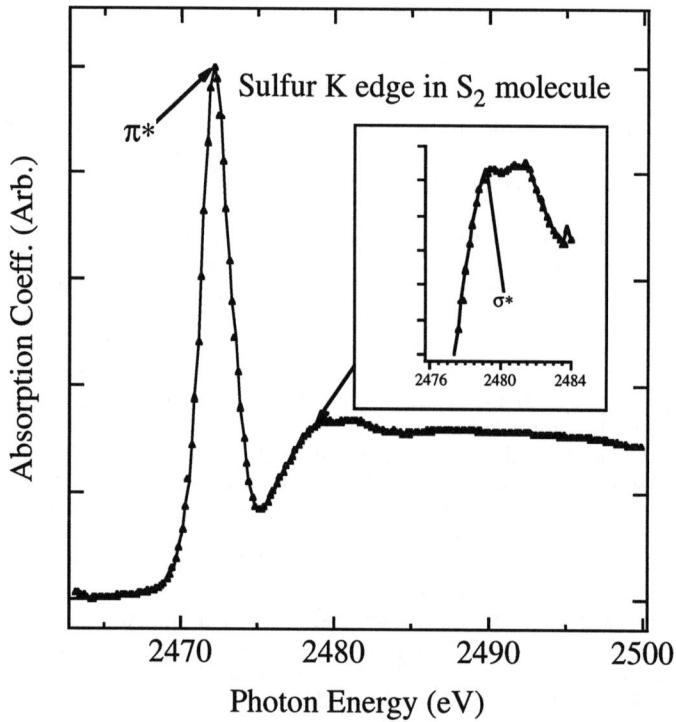

Figure 1. K edge spectrum of S_2 molecule. In the insert the details of the σ* line are shown.

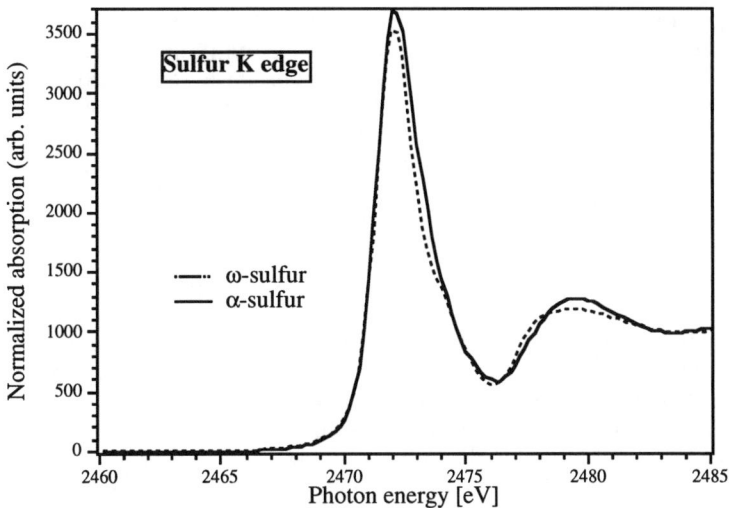

Figure 2. Sulfur K edge of α- and ω- sulfur.

To our knowledge, the K absorption spectrum of the S_2 molecule has not been reported so far. The K edge spectrum of α-sulfur has already been measured in the past (13). An interesting study of sulfur cross-link analysis in rubber by X-ray near edge structure was recently reported in a series of papers (14,15). In connection with the environment of S in Ge-S and GeS_2-Ag_2S glasses α and ω-sulfur K edge structures were also studied (16). In the present publication, we will discuss only the near edge features of sulfur in vapour and in molecular and polymerized solid phases. The EXAFS features of α and ω-sulfur will be described elsewhere (17). In order to understand the vapour to solid transition in sulfur and to interpret the spectral features observed in the spectra we first discuss the absorption spectrum of the S_2 molecule.

Sulfur vapour

Various thermodynamic and mass spectroscopic studies (1,3) show that sulfur vapour consists of a number of sulfur clusters denoted by S_n (n=2-12). The relative concentration of a particular species of S_n critically depends upon temperature and ambient pressure. It is well known that at pressures less than 1 torr in the presence of excess He, S_2 is predominant. In the past this later condition has been widely used to study the optical spectra of the S_2 molecule (6). Recently, the Rydberg states of S_2 molecules were studied (6) by using the same furnace as used in the present investigation. All the lines of the Rydberg series observed originated from the $^3\Sigma_g^-$ ground state of the S_2 molecule. We used similar pressure and furnace heating conditions for our X-ray absorption study: The partial pressure of sulfur vapour much less than 0.1 torr. Under these conditions the vapour should predominantly consist of S_2 molecules. Therefore we should expect that the absorption spectrum from our sulfur vapour should be similar to that observed for O_2, since the outer valence configuration of S_2 is similar to that of the homologous O_2 molecule. In fact, we do observe a K spectrum of sulfur vapour which is similar to the K absorption spectrum of O_2 (18-21) in the 535-545 eV range.

The molecular ground state of the S_2 molecule (22) is:

$$KKLL(\sigma_g 3s)^2 (\sigma_u 3s)^2 (\sigma_g 3p)^2 (\pi_u 3p)^4 (\pi_g 3p)^2 \ {}^3\Sigma_g^-$$

The description of the ground state is substantiated by the observed optical spectra of S_2 and other isoelectronic molecules. As in O_2, the outer two electrons in two degenerate $\pi_g 3p$ orbitals of the S_2 molecule have parallel spins and account for the magnetic interaction. If we assume that the spins of these two electrons are up then the transition of a spin down electron from a 1s level to a vacant $\pi_g 3p$ level is allowed. So by analogy with the O_2 molecule, we interpret the first peak in the S vapour spectrum as transition of a spin down 1s electron to the π^* level and the following two structures at about 8 eV above this peak as σ^* resonances corresponding to allowed spin up and down transitions from the 1s level of sulfur.

It is interesting to compare the systematics of the positions of the π^* and σ^* resonances in low Z molecules with our new results on the S_2 molecule and to see whether the resonance positions can be extrapolated to this molecule. In this connection it is worth mentioning that Hitchcock and Brion (23) constructed a diagram in which they plotted for some π bonded molecules, the positions of π^* and σ^* resonances relative to the ionization potential (IP) as a function of Z, the sum of the atomic numbers of the bonded atomic pairs, and their bond lengths. They found a much higher slope for σ^* resonance positions than for π^* positions. This behaviour has further been discussed by Stöhr (24) for a number of molecules. He observed a large scatter of data points for σ^* relative to π^* resonances. In fact, the π and π^* orbitals originate from p_x, p_y orbitals and are independent of the sp_z hybridization. Their positions depend on the energy of the atomic p orbital. The different behaviour of σ^* compared to π^* resonances is explained by invoking different sp_z hybridization along the bonded atomic pairs of the second period elements. If we simply extrapolate these results to Z=32, would we expect an inversion of the σ^* and π^* resonance positions for the S_2 molecule? A straightforward answer to this question comes from the s and p valence orbital energies and their difference along the third period elements and their trends compared to those of the second period elements. We have plotted in Figure 3(a) and (c) the relevant data for np (n=2,3) orbital energies and ns-np orbital energy differences for some elements along the second and third period elements. The

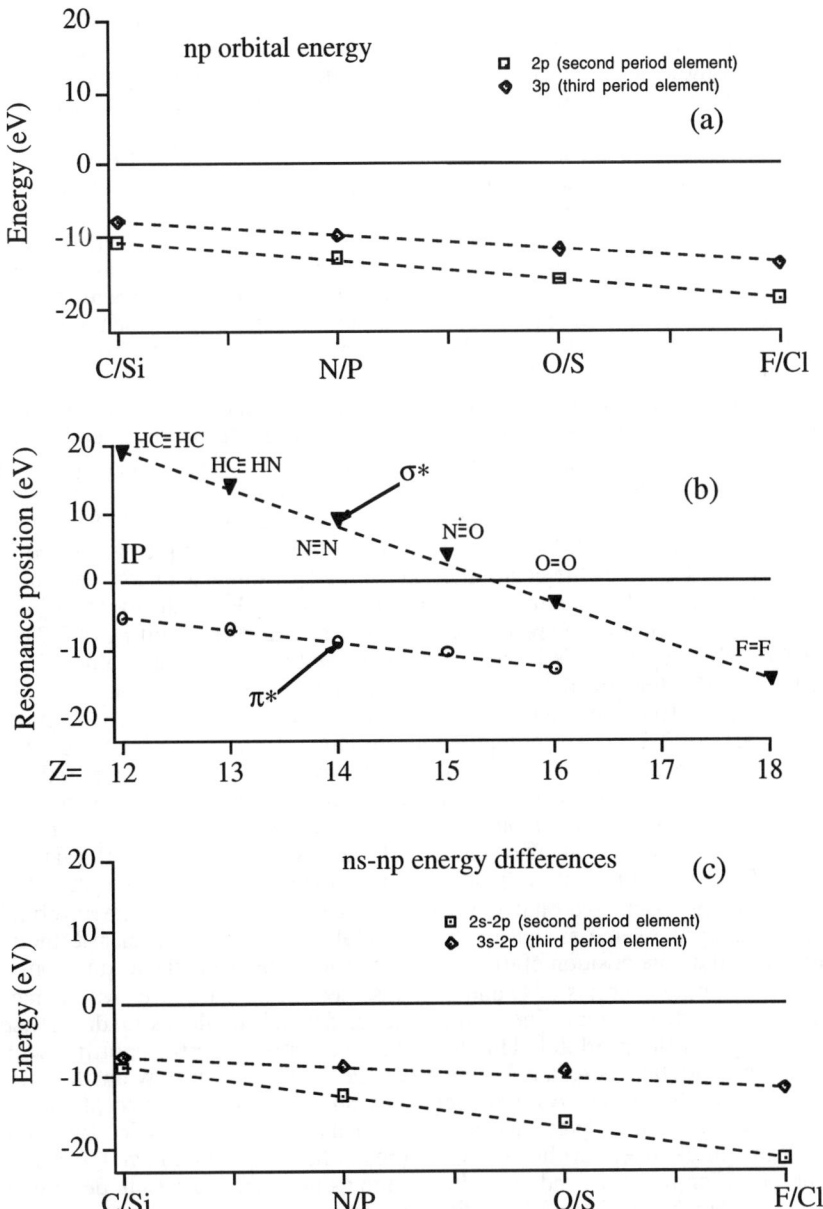

Figure 3. (a) np orbital energies for 2nd and 3rd period elements (ref. 25.); (b) σ* and π* resonances of some selected molecules (ref. 24) relative to the ionization potential, plotted against Z, the sum of the atomic number of the bonded pairs; (c) ns-np energy difference for 2nd and 3rd period elements.

theoretical ns and np orbital energies are obtained from reference (25). For comparison we have plotted in Figure 3(b) the observed π^* and σ^* resonances of some selected molecules relative to their 1s IP. One can immediately see that 2p orbital energies and 2s-2p differences in second period elements are systematically below the corresponding 3p energies and 3s-3p difference in third period elements. So the sp orbitals are closer in third period elements and lead to more sp_z hybridization between the bonded pairs than for those in the second period. As seen in Figures 3 (a) and (b) one expects that the σ^* resonances of third period elements will be systematically above those of second period elements and will have a different slope. Similarly, the p orbital energies which govern the positions of the π^* resonances also move up towards the IP for the third period bonded pairs relative to those of second period elements. The most important conclusion which we draw from this discussion is that the π^* and σ^* resonances will both move upward and will not cross, if we go from the O_2 to the S_2 molecule.

In S_2 the separation between the π^* peak and the first σ^* peak is observed to be 7.4 eV whereas it is 9.8 eV in the case of the O_2 molecule. So from the known positions of π^* and σ^* resonances relative to the IP for O_2, we can obtain a qualitative estimate of these quantities for the S_2 molecule. From the measured values of IP-π^* = 13 eV (18), we estimate (IP-π^*) for S_2 to be 13x(11.7/15.9)=9.56 eV, where the quantities inside the brackets are the p orbital energies for S and O respectively. This gives us the position of the IP for S_2. Similarly the σ^* resonance of S_2 is estimated to be situated at 3.2x(9.2/16.5)=1.8 eV, by using sp orbital energy differences for S and O respectively. This yields a σ^* - π^* splitting of 7.7 eV for the S_2 molecule, which is in good agreement with the observed value of 7.4 ±0.2 eV. The K edge position in the S_2 spectrum (Figure 1) which we were guessing at about 10 eV above the sharp π^* line is also in fair agreement with the estimated value of 9.6 eV. The complex line structure of the σ^* resonance near the threshold makes an accurate measurement of the IP of the 1s level of S impossible.

Our qualitative estimation of π^* and σ^* resonance positions of S_2 relative to the IP needs some more discussion regarding the different behaviour of π and π^* orbitals as compared to that of σ and σ^* orbitals. As the π and π^* orbitals are independent of sp_z hybridization their splitting is expected to remains symmetric with respect to the atomic p orbital. Consequently the Z dependence of the π^* orbital is approximately the same as that of the π orbital. So a simple linear arithmetic operation to obtain the position of the π^* resonance in S_2 by using the π^* position of O_2 and the p orbital energies of O and S, is justified. In the case of σ^* resonances the situation is quite different. The σ^*-σ difference depends on the ns-np difference and the energy of the p orbital. This leads to an asymmetric σ^*-σ splitting with respect to the p orbital. As seen in Figure 3 (b), the 3s-3p curve is now much flatter than that of the 2s-2p one. As a consequence smaller Z dependence of the σ^* position is expected. In the previous paragraph we have obtained a crude estimation of σ^* - IP by simple linear arithmetic operation. In fact we should have taken into account the different slopes and intercepts of the ns-np energy curves to determine σ^*-IP for S_2 from that of O_2 and theoretical energies.

α and ω-Sulfur

The ring and chain forms of these allotropic varieties of sulfur have similar basic building blocks. The sequence of the covalently bound sulfur atoms in these allotropes is not linear. For any three successive atoms in the chain or cycle, α defines the bond angle between two consecutive sulfur bonds. In a five-atom sequence, if the two terminal S atoms are on the same side of the plane defined by three central atoms, the linking leads to a *cis*-conformation yielding a cyclic microstructure. If the two terminal atoms are situated on opposite sides of the plane, the resulting structure is in *trans*-conformation. In this case the sulfur atoms are linked into a helical chain. The S-S-S bond angle in α-sulfur as given in the literature (5) is 107.3°-109.0°. For the polymerized variety 105°-107° values are reported. In a separate EXAFS study (17) we determined the S-S bond lengths in these two varieties of sulfur. We found the S-S distance in α-sulfur to be 2.035±0.02 Å. In ω-sulfur it was 2.066±0.02 Å.

α-sulfur is the best known variety of sulfur. It is described as the packing of the crown shaped S_8 rings to form a face centred structure. The K absorption and emission spectra of α-sulfur studied earlier (13) were compared with the calculation of the molecular orbitals of S_8 molecules (26). The absorption spectrum was reported to be in good agreement with the molecular orbital calculation (26). In this study the absorption spectrum (13) contained a main peak which shows a broad doublet structure which probably was due to saturation of the peak intensity due to a sample thickness effect. This is in contrast to our α-sulfur spectrum, obtained by employing the photoyield method, which probes only a few hundred angströms inside the sample. As a result, in our experiments, the broadening and saturation effects due to sample thickness are minimised. The main line of our α-sulfur spectrum is quite narrow and asymmetric.

It is interesting to emphasize again on the fact that the K spectrum from the double bonded S atoms in the S_2 is characterized by a π^* main line followed by a σ^* doublet. In α and ω-sulfur the S atoms are single bonded and the π levels are filled up by the electrons from the adjacent S atoms. The first accessible levels are of σ symmetry. The K absorption spectrum of ω-sulfur is characterized by a main σ^* line with a distinguishable high energy shoulder as compared to the asymmetric α-sulfur main σ^* line. In ring and polymer allotropes the S-S bond lengths and S-S-S bond angles are nearly the same and the interactions among rings and among chains are very weak. The immediate environment of a given S atom is identical in both cases. One may easily think that the asymmetry and the shoulder on the 1s-σ^* lines respectively of α and ω-sulfur seem to have the same origin. Our deconvolution (not shown in this paper) of these spectra yields a σ^* line splitting of about 1.1 eV. This may be compared with the theoretical separation (1.9 eV) (26) obtained between the two lowermost vacant molecular orbitals. Taking into account uncertainties as large as ±0.5 eV in the calculation of orbital energies, we find a fairly good agreement between the observed σ^* splitting and the theoretical result. In α and ω-sulfur the main line may be interpreted as transitions to relatively localized σ^* states due to 3s-3p mixing. This interpretation is supported by the results of the deconvolution of the 1s-σ^* main lines of α and ω-sulfur, respectively, into two components. This deconvolution results in a narrow intense line and a broad high energy line 1.1 eV apart in each case. In agreement with the theoretical calculation, the broader line may be due to transitions to hybridized states resulting from the mixing between the electrons from two σ pairs and two lone pairs. The present interpretation is at variance with the attribution (16) of the asymmetry and the shoulder to transitions to 4p atomic type states.

One may also think that this splitting may be due to σ–σ bond interactions in these two allotropes. It is worth mentioning here that such splittings had been observed in hydrocarbon chain paraffins (24). The observed splitting diminishes with increasing bond length and increasing Z (sum of atomic number of two elements involved in bonding). In cyclic compounds, owing to bond strains, the split structures are diffuse and weak. The bond strains in cyclic α-sulfur may smear the structure and contribute to the asymmetry of the main line. In ω-sulfur, the chain configuration has freedom to relax along the chain and, as a result, the shoulder on the 1s-σ* line may appear.

The molecular orbital calculation (26) also predicted six other vacant orbitals above the first two empty orbitals already mentioned in the previous paragraph. These orbitals are spread over 2 eV and are approximately 4 eV above the first empty orbital. They are formed by complex combinations of s and p type orbitals. Individual transitions to these orbitals are probably very weak and are not observed. It is tempting to attribute the high energy near edge structure at about 7 eV above the main line in both allotropes to the transitions to 4p type states. It seems to be quite sensitive to the environment of the central S atom. In ω-sulfur it is much broader (5.3 eV width) than the 3.8 eV width observed for α-sulfur. This may be due to more significant bond length disorder in ω-sulfur than in α-sulfur. The 2.066±0.02 Å average S-S bond length of the polymerized ω-sulfur chain determined by our EXAFS measurements (17) is slightly longer than that observed (2.035±0.02 Å) for crown shaped α-sulfur rings. The centre of gravity of the broad structure in the ω-sulfur spectrum near 2479 eV is about 0.5 eV below the corresponding structure of α-sulfur. The difference in the absorption spectra, or the shapes and positions of the structures should reflect the difference in the surroundings of the S atoms in these two allotropic forms. In the S_8 ring a central S atom has 2 equivalent symmetrical groups of 4 neighbours at different distances. In the helical chain polymer, there are many S atoms symmetrically situated around a given S atom of the helix, directly in the line of sight. The helical chain configuration may be considered as a sequence of atoms in trans- conformation at increasing distances from the central absorbing atom as opposed to the cis- conformation in the rings where the atoms are situated at different fixed distances depending on the number of atoms in the ring.

We wish to point out that the 1s→ π* transition in the S_2 molecule is coincident with that of the main line in α and ω-sulfur observed at 2472 eV. The reason for this coincidence most probably arises from the fact that S atoms in solid ring or chain configurations undergo sp^3 mixing and thus form the two directional bonds and creating two active pairs $3s^2$ and $3p^2$ (two directions of a tetrahedron). The double bonded S atoms which are 1.889 Å apart in the S_2 molecule, are now separated by 2.035 Å and 2.066 Å for single bonded S atoms in α and ω-sulfur respectively. Thus, in the solid, the two σ bonds are less stabilized in energy and approach the energy of π orbitals in the S_2 molecule. This may explain why the 1s→ π* transition in S_2 is close in energy to 1s→ σ* in the solid. It is interesting to compare the spectral shift of the σ* resonances from the doubly bonded atoms in S_2 molecule and the singly bonded atoms in cyclic or chain variety of sulfur to the σ* resonances from the gas phase O_2 and H_2O_2. In H_2O_2 the O-O bond are the single bonds. The K absorption spectra of these two molecules(27,24) show that the σ* resonance from the singly bonded oxygen atoms in H_2O_2 approaches the position of the π* resonance of O_2. The discrepency is probably due to the fact that in condensed sulfur one S atom is surrounded by two other S atoms whereas in H_2O_2 the singly bonded oxygen atoms have two terminal hydrogen atoms.

From the S_2 molecule to its condensed molecular or polymerized phases

It is interesting to mention that the observed K absorption spectrum of the S_2 molecule is similar to that of O_2 whereas their condensed phases have different spectra. Physisorbed O_2 layers condensed on Pt (111) crystal surface yield a spectrum (28) which is similar to that of the molecule in the gas phase. By contrast the yellow product which was deposited on the two cold ends of our furnace showed an α-sulfur like spectrum. This means that in condensed O_2 layers, the interactions between the O_2 molecules are of van der Waals type whereas, in the case of condensed sulfur, the less electronegative S atom starts sharing its two external electrons with the two adjacent neighbour S atoms to complete its octet and form covalent bonds. In this connection it should be remembered that except in ozone, the alkali-metal ozonates and a number of peroxides and trioxides, the covalent linking of O to O does not proceed beyond O=O or -O-O- (29). Sulfur, on the other hand, easily forms chains of S atoms in its elemental form. Thus an interesting question arises: How does an S_2 molecule, which has two unpaired p electrons, form a chain or ring structure with other molecules on condensation? The present results suggest that the S_2 units are directly or indirectly building a portion of the ring or chain blocks to form different varieties of sulfur.

Basically the formation of α-sulfur from S_2 vapour and the chain sulfur obtained by quenching high temperature sulfur vapour involve similar mechanisms. In the former case sulfur vapour has enough time to cool and crystallize as orthorhombic sulfur. In this case the route for the formation of stable S_8 rings is straightforward since intermediate S_4 and S_6 clusters may be involved. The presence of S atoms produced by fragmentation was detected by the observation of atomic lines in the optical absorption spectrum of S_2 vapour in our stainless steel tube furnace. So the formation of S_8 rings from S_3, S_5 and S_7 clusters with S atoms and S_2 molecules may also be expected.

In the industrial process for the preparation of ω-sulfur, the sulfur vapour at about 800 K is quenched in saturated vapour of the solvent CS_2 kept at its boiling point. It is well known that sulfur vapour at high temperature and normal pressure contains mainly small S_n (n=2,3,4) molecules. For the formation of rings or a chains the essential condition is that there should exist a cluster that has two isomers which become two different seeds to grow to either ring or chain form sulfur. Another condition which one can envisage is that they should have slightly different stability. On going through the available literature on the stability of S_n clusters we find that tetrasulfur had a long standing problem about its structure both from theoretical and experimental point of view. The controversy hinged over whether the S_4 molecule is an open or closed structure. We will not go into details of the tremendous effort to understand the structure of this molecule. We will simply refer to two important papers on this subject. For the experimental work we will refer to infrared and structure studies of matrix trapped isotopically enriched S_3 and S_4 clusters by Barbson et al (7). On the basis of these studies two infrared bands were assigned to two different open-chain S_4 isomers. On the theoretical side Quelch et al. (8) studied the potential energy surface of S_4 using *ab initio* electronic theory. For the global minimum of S_4, they predicted singlet *cis* planar structure. They found a singlet *trans* isomer (≈ 10 kcal mol⁻ above the *cis* isomer) and several other isomers close to it. They had been able to assign most of the vibrational spectra previously attributed to S4. They also found a surprisingly small energy barrier to internal rotation between the cis and trans isomers.

The theoretical study on sulfur vapours in ref.(3) deals with the stability of cyclic and chain clusters. The SCF calculations were performed in the context of the stability of chains in ω-sulfur. In this calculation the sulfur chain is constructed in a sequential way by adding a sulfur atom at each step. The molecular parameters used for the helical chain are d = 2.07 Å; α = 106°; β = 84°, where d, α and β are bond length, S-S-S bond angle and S-S-S-S dihedral angle respectively. Starting from the S_2 molecule considered as first chain fragment, one additional sulfur atom is added to it and the energy per atom of the new fragment is calculated. This preliminary calculation was limited to S_8 only. In Figure 4 we have plotted the energy per atom for chain fragments obtained from ref.(3) against the number of atoms in the fragment. We have also included in this figure the corresponding data obtained from ref.(3) for most stable isomers of S_n clusters. They correspond to the optimized geometries of isomers obtained from all electron calculations (30,31)

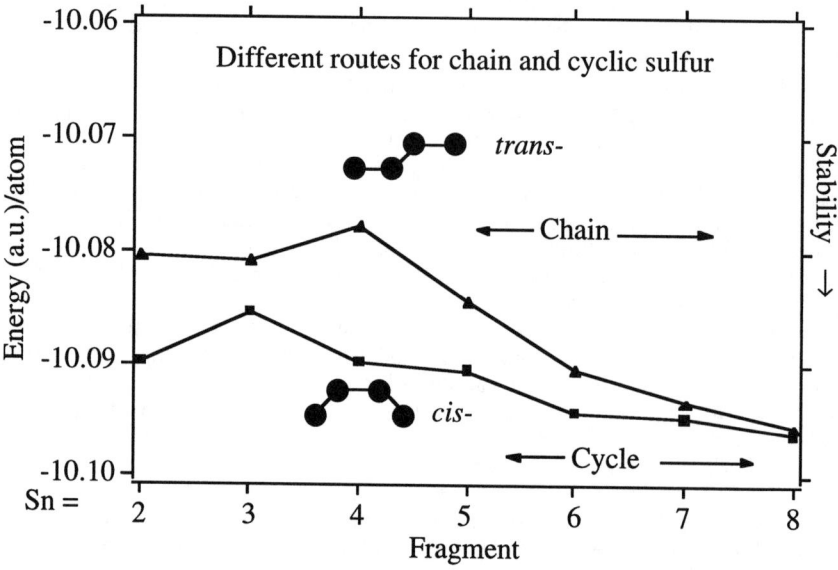

Figure 4. The calculated energy (a.u)/atom plotted against the number n of S atoms in a cluster fragment (data obtained from ref. 3). In the upper curve the parameters such as bond length d, bond angle a and the dihedral angle b are those of an infinite chain in ω-sulfur. In the lower curve the points represent the energy/atom for most stable isomers (ref. 3).

Now we can analyse the different condensation behaviour of α-sulfur as compared to that of ω-sulfur which is obtained by quenching. We see immediately that the upper curve calculated for chain configurations using fixed parameters shows a maximum at S_4. This is the most unstable chain configuration. As the chain grows it becomes more and more stable. This is the same for the lower curve. The energy per atom in the upper curve is expected to converge towards an asymptotic value approaching that of the S_8 cyclic cluster. It is well known that ω-sulfur slowly transforms to the cyclic variety when left at room temperature for a long time. It may be noted that the separation between the two curves is maximum at S_4. The undissolved product in CS_2 solution maintains its chain configuration. This means that there exists always an internal energy barrier to rotation between helix or *trans* to *cis* isomers. The magnitude of the barrier seems to be crucial for further growth of cluster into chain or cyclic species. It is worth mentioning here that the theoretically predicted lowest energy *cis* or *trans* isomers of S_4 which are planar and have dihedral angles 0° and 180° become cyclic and a helicoidal chain on adding one sulfur atom. For the cyclic species the dihedral angles change to different values depending upon the number of S atoms in the cluster.

In relation with the present study we also investigated chain and cyclic forms of selenium (32) by X-ray absorption spectroscopy. While the chain form of selenium is quite stable the cyclic form is difficult to obtain in pure form. A similar high temperature vapour quenching method was used to prepare cyclic selenium. As the sulfur cyclic species the cyclic form of selenium is soluble in CS_2. Selenium is separated by slow evaporation of CS_2 solution. Both varieties give spectra similar to those of the corresponding sulfur species. We will discuss the Se vapour condensation behaviour and the finer details of these spectra in a future paper.

In conclusion, the most stable gas phase S_2 molecule, cyclic S_8 and the polymer chain S_n variety of sulfur were investigated by X-ray absorption spectroscopy at the K edge of sulfur. We observed a great similarity between the K spectra of O_2 and S_2 molecules indicating that they have similar unoccupied orbitals available for transitions from the K level. The cyclic and chain forms of sulfur show some similarity in their σ* line structure and a marked difference in their near edge structure due to different number of S atoms surrounding the central atom in these allotropes. At present it is rather difficult to obtain the X-ray absorption spectra from other cyclic clusters in the gas phase. This will require a study of the precise conditions of the vapour pressure and temperature required to maintain a specific species for a time long enough for measurements, and suggests that, in the future detailed mass spectroscopy experiments will be required.

ACKNOWLEDGMENT

The research has been performed under an European Research Network Programme (GDRE) "Chalcogenides" initiated by the Centre National de la Recherche Scientifique (France). It is great pleasure for me to thank my coworkers J.M. Durand, J. Olivier-Fourcade, J.C. Jumas, M. Womes, C.M. Teodorescu, A. Elafif and J.M. Esteva for their essential contributions to the results presented here. I am grateful to Alain Pellegatti, Centre de Thermodynamique et de Microcalorimétrique, CNRS, UPR 7461, Marseille, for illuminating discussion and helpful suggestions on the theoretical aspects of the stability of the cyclic and chain varieties of sulfur

REFERENCES

1. Michaud F., Ph.D. Thesis, "Synthesis by quenching and physico-chemical characterisation of three allotropic forms of polymerized sulfur" (in French), Chimie Physique-Chimie Théorique, Université Scientifique et Technique du Languedoc, Montpellier .(unpublished), 1989
2. Michaud F., Fourcade J., Philippot E., Maurin M, Discours M. Proceedings "Symposium International du soufre Elementaire en Agriculture", Acropolis, Nice, **2**, 789-796 (1987)
3. Ezzine M., Chastel R., Bergman C., J. Alloys and Compounds **220**, 206-211(1995) and references therein
   Ezzine M., Ph.D. Thesis, "Experimental and theoretical approaches to the stability of molecular and polymerized varities of sulfur" (in French), Université de Provence (Aix-Marseilles), (unpublished), 1992
   Ezzine M., Pellegatti A., Dandey J-P., Proceedings "Première Conférence Magrébine de Génie des Procédés (COMAGEP 1), Marakech, Morocco, 1994
4. Coppens P., Yang V.W., Blessing R.H., Cooper W.F., Larsen F.K. J. Am. Chem. Soc. **99**, 760-766 (1977)
5. Tuinstra F., Thesis: Structural aspects of the allotropy of sulphur and the other divalent elements. Eds., Delft (The Netherlands), V. Waltmann, 1967
   Tuinstra F. Physica **34**, 113 (1967)
6. Bredohl H., Breton J., Dubois I., Esteva J.M., Macau-Hercot D., Remy F., Somé E., J. Molec. Spectr. **173**, 49-54 (1995) and references therein
7. Barbson G.D., Mielke Z., Andrews L., J. Phys. Chem. **95**,79-86 (1991)
8. Quelch G.E., Schaefer H.F. III, Marsden C.J., J. Am. Chem. Soc. **112**, 8719-8733 (1990)
9. Lemonnier M., Collet O., Depautex C., Esteva J.M., Raoux D.,Nuclear Instruments and Methods **195**, 133-139 (1982)
10. Cowan P.L., Brennan S., Deslatte R.D., Henins A., Jach T., Kessler E.G., Nuclear Instruments and Methods in Physics Research **A246**, 154-158 (1986)
11. Krause M., Oliver J.H.,J. Phys. Chem. Ref. Data **8-2**, 2689 (1979)
12. Esteva J.M. et al. unpublished
13. Sugiura C., J. Phys. Soc. Japan **30**, 1766 (1971)
14. Chauvistré R., Hormes J., Brück D., Sommer K., Engels H.W., Kautschuk+Gummi **45**, 808-813 (1992)
15. Chauvistré R., Hormes J., Brück D., Sommer K., Kautschuk+Gummi **47** 481-484 (1994) and references therein.
16. Armand P., Ibanez A., Philippot E., J. Solid State Chemistry **104**, 308-318 (1993)
17. Durand J.M. et al. (unpublished)
18. Hitchcock A.P., Brion C.E., J. Electron Spectrosc. **18** 1-21 (1980)
19. Ma Y., Chen C.T., Meigs G., Randall K., Sette F., Phys. Rev. **A44** 1848-1858 (1991)
20. Kosugi N., Shigemasa E., Yagishita A., Chem Phys. Lett. **190**, 481-487 (1992)
21. Yagishita A. , Shigemasa E., Kosugi N., Phys. Rev. Lett. **72** 3961-3964 (1994)
22. Herzberg G., *Molecular Spectra and Molecular Structure, 1. Spectra of diatomic molecules,* D. Van Norstrand Co. Inc, New York, 2nd Edition, 1950, p.343
23. Hithcock A.P., Brion C.E., J. Phys. B. At. Mol. Phys. **14**, 4399-4413 (1981)
24. Stöhr J., *NEXAFS Spectroscopy*, Springer-Verlag, New York, 1992
25. Slater JC, *Quantum Theory of Atomic Structure*, Vol. 1, McGraw-Hill, New York, 1960, p. 206
26. Chen I., Phys. Rev. **B2**, 1053-1060 (1970)
27. Rühl E., Hitchcock A.P., Chem. Phys. **154**, 323 (1991)

28. Wurth W., Stöhr J., Feulner P., Pan X., Bauchspiess K.R., Baba Y., Hudel E., Rocker G., Menzel D., Phys. Rev. Lett. **65,** 2426-2429 (1990)
29. Wells A.F.,*Structural Inorganic Chemistry,* Oxford Science Publications, Oxford, 5 th Edition, 1990, p. 495
30. Dixon D.A., Wasserman E., J. Phys. Chem. **15,** 5772 (1990)
31. Raghavachari K., Mcmichael Rohlfing C., Brinkley J.S., J. Chem. Phys. **93,** 5862 (1990)
32. Olivier-Fourcade J., Jumas J.C., Karnatak R.C., unpublished

EXAFS and Diffraction Studies on High T_c Superconductors: Mesoscopic Stripes in the CuO_2 Plane

N. L. Saini, A. Lanzara and A. Bianconi

Dipartimento di Fisica, Università di Roma "La Sapienza", 00185 Roma, Italy

Abstract. The modulated structure of the CuO_2 plane in $Bi_2Sr_2CaCu_2O_{8+y}$ (Bi2212) superconductor has been solved by joint polarized Cu K-edge extended x-ray absorption fine structure (EXAFS) and Cu K-edge anomalous x-ray diffraction. The statistical distribution of the Cu-O pairs has been measured by EXAFS and the spatial distribution of the lattice distortions of the CuO_2 plane by anomalous diffraction. The incommensurate superstructure involves a displacement of Cu ions along the c-axis with a large anharmonic character giving a split distribution of the Cu-O(apical) bonds in agreement with the statistical distribution obtained by EXAFS. The EXAFS results show an anomalously large Cu-O(planar) distance, 1.96 Å, assigned to distorted CuO_2 stripes of width W inbetween undistorted stripes of width L. From the measurement of L=15±0.5 Å we have calculated the energies E_n of the bottom of the one-dimensional subbands of the superlattice and found that the Fermi level E_F is tuned to a "shape resonance" $E_F - E_n < \hbar \omega_D$, where ω_D is the Debye frequency, providing a mechanism for T_c amplification.

INTRODUCTION

High T_c superconductors are layered materials with CuO_2 planes as a key feature for the superconducting properties which are separated by insulating rocksalt oxide layers. The importance of the CuO_2 planes has created major interest to study the electronic and structural behaviour of these structural elements since the discovery of the high T_c superconductors.

The structure of the CuO_2 planes determined by diffraction techniques has shown a homogeneous distribution of the Cu-O structural units. However, the average structure might differ from the atomic arrangement on the local scale. The inhomogeneous nature of the CuO_2 plane is a natural aspect due to the existing 1) steric effects introduced by the dopants in the rocksalt layers; 2) polaronic lattice distortions associated with the doped holes in the CuO_2 plane; 3) lattice mismatch between the Cu perovskite layers and rocksalt layers with a tensile stress within the rocksalt layers producing a compressive stress in the CuO_2 lattice. The inhomogeneity of the CuO_2 plane at the mesoscopic scale length (10-100 Å) has direct implications for the understanding of superconductivity with small coherence length ξ_0 ($\xi_{a,b} \sim 25$ Å, $\xi_c \sim 3$ Å) in the high T_c materials, and hence is one of the points of research activities during the recent years [1-5].

Experimental methods sensitive to the local atomic environment have clearly shown that the local structure of the CuO_2 plane is different from the average in most of the superconducting cuprate perovskites [4, 5]. The electronic instability of the inhomogeneous CuO_2 plane has been indicated by lattice anomalies observed with doping and temperature. Local probes like X-ray absorption [6-14], neutron diffraction PDF [15] have been able to show temperature dependent structural anomalies in most of the superconductors. The anomalous behavior of the high T_c superconductors has also been indicated by thermal expansion coefficient [16-17], elastic coefficients [18], ion channeling measurements [19-21], and inelastic neutron scattering [22,23].

In this paper we show that joint analysis of EXAFS and diffraction can provide important information on the microscopic structural modulations in the high T_c superconductors. We discuss how EXAFS and diffraction can be exploited to derive the mesoscopic structure of the CuO_2 plane. The EXAFS technique is briefly described followed by an introduction to anomalous diffraction. The $Bi_2Sr_2CaCu_2O_{8+y}$ (Bi2212) system has been taken as a representative example.

POLARIZED EXAFS : BASIC ASPECTS

The x-ray absorption coefficient $\mu(E)$ is generally given by the product of the matrix element times the joint density of states for the electronic transitions from the initial to final states. It can be solved in real space for electronic transitions from an initial localized core level to a final state described by an outgoing spherical wave which interferes with the waves backscattered from the neighbouring atoms [24-28]:

$$\mu(E) = \mu_0 [1 + \sum_{n \geq 2} \chi_n(E)] \quad (1)$$

where $\mu_0(E)$ is the so-called atomic absorption coefficient for the selected atomic core level and $\chi_n(E)$ represents the contribution arising from all multiple scattering pathways beginning and ending at the central absorbing atom and involving (n-1) neighboring atoms.

The modulation function χ can be extracted from the experimentally recorded absorption coefficient and given by:

$$\chi(E) = \sum_{n \geq 2} \chi_n(E) = \frac{\mu(E) - \mu_0(E)}{\mu_0(E)} \quad (2)$$

The shorter scattering pathway is the one involving the first shell and only the single scattering term χ_2. This term can be isolated by Fourier filtering because multiple scattering pathways for the first shell, that contribute to χ_n with $n \geq 3$ as well as all contributions from further shells, are longer.

The single scattering EXAFS signal for the first shell can be written as:

$$\chi_2(k) = \sum_i \frac{S_0^2 e^{-2R_i/\lambda}}{kR_i^2} A_i(k) \cdot e^{-2k^2\sigma_i^2} \sin[2kR_i + \delta_i(k)] \quad (3)$$

where R_i is the radial distance and $\delta_i(k)$, the phase function determined by both the absorber and backscatterer. S_0^2 is an amplitude correction factor due to photoelectron correlation and is also called the passive electron reduction factor. $A_i(k)$ is the scattering power, λ is the photoelectron mean-free path and σ_i^2 is the correlated Debye-Waller factor of the photoabsorber-backscatterer pairs. The scattering power is given by $A_i(k) = N_i^* F_i(k)$, where N_i^* is the average coordination number and $F_i(k)$ is the backscattering amplitude of the neighbouring atoms. The photoelectron wave vector (k in $Å^{-1}$) is given by

$$k = \frac{p}{h} = \frac{\sqrt{2m(E-E_0)}}{h} \quad (4)$$

E (eV) is the incident photon energy while E_0 (eV) is zero of the ejected photoelectron energy.

For oriented single crystal samples the EXAFS signal has a dependence on the angle between the preferred sample direction and the x-ray polarization vector. The EXAFS equation for polarized K-edge EXAFS can be written as:

$$\chi(k) = \sum_i 3N_i \cos^2(\theta_i) \frac{S_0^2 e^{-2R_i/\lambda}}{kR_i^2} F_i(k) e^{-2k^2\sigma_i^2} \sin[2kR_i + \delta_i(k)] \quad (5)$$

where N_i is the equivalent number of neighbouring atoms at a distance R_i sitting at the angle (θ_i) with respect to the direction of the electric field of the polarized x-ray beam [28].

To extract the structural parameters from an EXAFS signal, the shift of the photoelectron energy origin E_0 and phase shifts should be known. These parameters can be either fixed or allowed to vary when analyzing EXAFS data.

In the high T_c materials the Cu K-edge EXAFS probes a region about 6 Å around the central Cu atom. Therefore, the Cu-K-edge EXAFS is an ideal probe to investigate microscopic changes in the local lattice structure of the CuO_2 planes in high T_c compounds [6-14]

ANOMALOUS DIFFRACTION : BRIEF OVERVIEW

Inelastic atomic scattering processes increase drastically when the incident photon energy reaches an atomic absorption edge. The change in the atomic scattering factor at the absorption edges is called anomalous scattering [29]. The atomic scattering factor (f) can be written in the following form:

$$f(S,\lambda) = f_0 + f'(S,\lambda) + if''(S,\lambda) \quad (6)$$

The anomalous contribution to the scattering factor is given by the last two terms. The two anomalous components of the scattering factor are interrelated through the Kramers-Kronig transform.

$$f'(\omega_0) = \frac{2}{\pi}\int_0^\infty \frac{\omega}{\omega^2 - \omega_0^2} f''(\omega)d\omega \qquad (7)$$

For element-selective anomalous diffraction the measurements are made at two different energies at which the atomic scattering factor of the selected atom is significantly different [29]. The strong variation of the Cu scattering factor in the vicinity of the Cu K-edge can be exploited to obtain local information on the Cu lattice.

The local information is obtained by taking the difference between two sets of intensities (structure factors), i. e.

$$\Delta F_{hkl}(E_1, E_2) = \Delta F_{hkl}(E_1) - \Delta F_{hkl}(E_2) \qquad (8)$$

where $\Delta F(E_1)$ and $\Delta F(E_2)$ are structure factors at the two energies. The structure factor is given by summation of the atomic scattering factors over all n atoms in the structural unit cell [29].

$$F_{hkl}(E) = \sum_n f_{n,E} \exp(2\pi i H \cdot r_n) T_n \qquad (9)$$

where $H = ha^* + kb^* + lc^*$ and T_n is the temperature parameter of the nth atom. The difference between the structure factors at two energies can be written using (6), (8) and (9).

$$\Delta F_{hkl}(E_1, E_2) = \Delta F_r - i\Delta F_i \qquad (10)$$

ΔF_r and ΔF_i are the real and imaginary parts of the anomalous scattering contribution. Equation (10) represents the difference between two vectors in the complex plane. The experimentally observable quantity is the difference δ_F, the difference in magnitude of two vectors. In centrosymmetric, or almost-centrosymmetric structures, and small values of f'', δ_F is almost equal to the real part of the difference, ΔF_r. Thus the anomalous part of the structure factor can be extracted directly by measuring two different diffraction patterns at different energies.

MODULATION OF THE CuO₂ PLANE IN Bi2212 BY POLARIZED EXAFS AND ANOMALOUS DIFFRACTION

An incommensurate modulation of the type $q_s = \beta\ b^* + (1/\gamma)\ c^*$ (in the orthorhombic notation where $b^* = 1/(\sqrt{2}d)$ and d is the average Cu-Cu distance) appears to be a common feature of optimally doped cuprate superconductors [8]. However, the satellite reflections are often weak and diffuse, indicating ordering over small domains, and are better resolved at temperatures lower than 200 K-100 K. The best characterized situation is that of the $Bi_2Sr_2CaCu_2O_{8+y}$ (Bi2212) superconductor in which the modulation gives rise to sharp and intense satellite peaks up to high temperatures. Bi2212 has also been considered a prototype material to investigate several properties of high T_c superconductors such as the transport and magnetic properties [30], the normal band structure, the superconducting gap and its anisotropy [31]. The modulated structure of the

Bi2212 system has been studied by several groups using electron, x-ray and neutron diffraction techniques [32-42]. The modulation of the BiO layer is established in the literature [32, 33, 38-41]. Although various diffraction experiments report a modulation of the Cu sub-lattice [32, 33, 38-41], the superstructure is assigned to the BiO layers and the CuO_2 lattice is considered to be rigid. We have undertaken the present work to study this problem by joint Cu K-edge polarized EXAFS and Cu K-edge anomalous x-ray diffraction allowing us to overcome the limitations of the x-ray diffraction to solve the anharmonic incommensurate superstructure with a large unit cell of Bi2212, due to the large number of parameters required for the structural refinement.

The sample used for the study was a single crystal grown by travelling floating zone method [43] with $T_c = 84$ K. The EXAFS measurements were performed on the beam-line BL-4C at Photon Factory at Tsukuba. The synchrotron radiation emitted by the 2.5 GeV storage ring at a typical current of 350mA was monochromatized by a fixed-exit double crystal Si(111) monochromator and sagittally focused on the crystal. The spectra were recorded by detecting the fluorescence yield (FY) using 9 NaI(Tl) x-ray detectors. The crystal temperature was monitored with an accuracy of ±0.5K. The EXAFS signal $\chi=(\alpha-\alpha_0)/\alpha_0$, where α is the absorption coefficient and α_0 is the so-called atomic absorption, was extracted from the absorption spectrum using standard procedures and corrected for fluorescence self-absorption [44].

The signal due to the Cu-O(planar) [and Cu-O(apical)] is well isolated in the $E//a$ [and $E//c$] spectra as it can be seen in the Fourier transform (FT) signals shown in Fig. 1. The FT has been performed between $k_{min}=3$ Å^{-1} to $k_{max}=17$ Å^{-1} using a Gaussian window and corrected with theoretically calculated phase shifts.

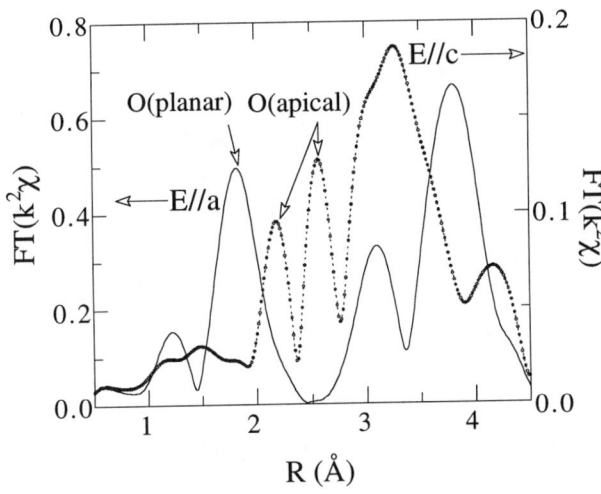

FIGURE 1. Fourier transform of the experimental $E//a$ [$E//c$] Cu K-edge EXAFS.

Figure 2 shows the Cu-O(planar) [and Cu-O(apical)] EXAFS signal obtained by standard Fourier filtering from the E//a [and E//c] Cu K-edge EXAFS. Multiple scattering signals are not present because any multiple scattering contribution will have a longer effective photoelectron pathway. The first shell EXAFS data were analyzed by non-linear least squares fitting using the curved wave EXAFS theory [45] in the range 3-13 Å^{-1} where the EXAFS amplitude due to scattering by oxygen atoms is large. The number of independent parameters that can be extracted is $N_{ind} \sim (2\Delta k \Delta R)/\pi \sim 5$, where $\Delta k = 10$ Å^{-1} and $\Delta R = 0.8$ Å, are the ranges in k and R space over which the first shell data are fitted. The best fits have been obtained by a two shells fit with two distances R_{long} and R_{short} and effective coordination numbers N_{long} and $N_{short} = N_{tot} - N_{long}$ (where N_{tot} is fixed), the Debye-Waller factors were taken to be the same and it was found that they can be fixed to the

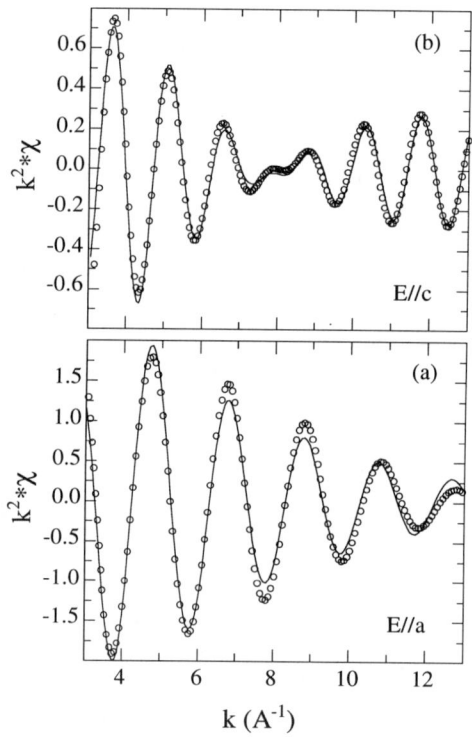

FIGURE 2. (a) Fourier filtered EXAFS signal (multiplied by k^2) due to the Cu-O(planar) pairs (circles) and its fit with two distances (solid line) and; (b) due to the Cu-O(apical) pairs (squares) and its fit with two distances (solid line).

expected values for the correlated Debye model. Thus the two distances fitting is essentially a 3 parameters (R_{long}, R_{short} and N_{long}) fit. All the other parameters were fixed to the values determined by using standard model compounds. A typical fit to the experimental E//a and E//c EXAFS is shown by a solid line in Fig. 2. We can see directly the presence of two main Cu-O(apical) distances separated by $\Delta R \sim 0.19 \pm 0.02$ Å from the beat in the E//c EXAFS oscillation at $k \sim 8.5 \pm 0.5$ Å^{-1}.

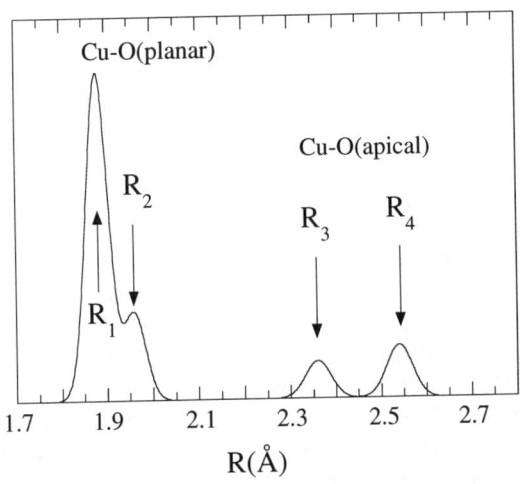

FIGURE 3. Pair distribution function (PDF) of the Cu-O lattice. The PDF of the Cu-O(planar) pairs obtained by **E//a** Cu K-edge EXAFS, is normalized to $N(R_1)+N(R_2)=4$ while the PDF of the Cu-O(apical) pairs obtained by **E//c** Cu K-edge EXAFS is normalized to $N(R_3)+N(R_4)=1$.

The pair distribution function (PDF) for the Cu-O pairs extracted from the EXAFS data at 30K is shown in Fig. 3. There are two well-separated Cu-O(planar) distances ($\Delta R \sim 0.08$Å) for the CuO$_4$ square planes. These two distances are within the range of the average Cu-O(planar) distances in the crystallographic structures of all synthesized cuprate superconductors [46]. The short Cu-O(planar) bonds at $R_1=1.88$ Å are the expected distances for the average crystallographic structure. The long anomalous Cu-O(planar) bonds, $R_2=1.96$ Å in Fig. 3 are associated with tilting of the CuO$_4$ square planes in the (110) direction, of the CuO$_5$ pyramids, where two oxygen atoms per CuO$_4$ square plane get displaced along the c-axis giving a rhombic distortion with two long distances R_2 and two short distances R_1 as in the low temperature tetragonal (LTT) like structure (Fig. 4). We obtain a tilt angle $\theta=16°$ (or $\theta=14°$) of the CuO$_4$ square plane using $\cos\theta=R_1/R_2$ (or $\cos\theta=<R>/R_2$, where $<R>=a/\sqrt{2}$). The presence of two Cu-O(apical) distances, $R_3=2.36$ Å and $R_4=2.54$ Å, separated by $\Delta R \sim 0.18 \pm 0.02$ Å is revealed by the damping of the E//c EXAFS oscillations at $k \sim 8.5 \pm 0.5$ Å^{-1}. The anomalously short Cu-O(apical) bonds R_3 are associated with the distorted CuO$_5$ pyramids. The short

Cu-O(apical) bond (0.2 Å shorter) may be due to modulation of the Cu z-coordinate.

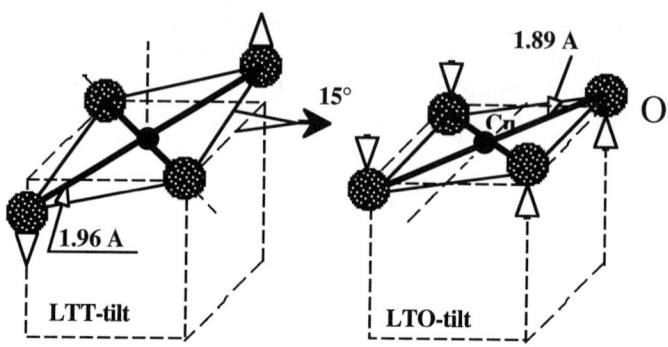

FIGURE 4. Pictorial view of the low temperature tetragonal (LTT) and low temperature orthorhombic (LTO) kind of distortions.

Anomalous diffraction measurements on the same crystal were performed at the European Synchrotron Radiation Facility (ESRF), Grenoble on the wiggler beamline ID11. The sample was given a 24° oscillation around the b-axis and the diffraction images were recorded on a 35x43 cm (A3 size) image plate detector. This geometry permitted a precise evaluation of the amplitude and anharmonic content of the modulation of the z component of Cu atoms. For the anomalous diffraction two patterns were recorded sequentially at two wavelengths (λ_1=1.3788Å, at the rising edge of the Cu K threshold and λ_2=1.4086 Å, about 200 eV below the edge chosen to have a large variation in the real part of the Cu anomalous scattering factor ($\Delta f'$=6.2 electrons) with no variation in the imaginary part f''.

The peak indexing was made using the four dimensional approach for incommensurate 1D modulations [47], considering that for each Bragg reflection the diffraction vector **H** can be written as $\mathbf{H} = h\,\mathbf{a}^* + k\,\mathbf{b}^* + l\,\mathbf{c}^* + m\,\mathbf{q}_s$, ($\mathbf{q}_s = \beta\,\mathbf{b}^* + (1/\gamma)\,\mathbf{c}^*$, γ=1). The superstructure period (given by $\lambda_p = 1/\beta$) has been found to be temperature independent (λ_p=4.73±0.01). The data were corrected for the Lorentz factor, for polarization, and for absorption. The structure factor was calculated in the four-dimensional approach using the $N_{1,\bar{1},1}^{Bbmb}$ superspace group and an isotropic thermal factor. In this situation, having an almost centrosymmetric structure, the measurable quantity $\delta_F = |F(\lambda_2)| - |F(\lambda_1)|$ may be assumed equal to the variation of the Cu contribution (real part) only [29]. It is given, for permitted reflections, by the formula

$$\delta_F = 8 \Delta f'_{Cu} \, e^{-B_{Cu}\left(\frac{\sin\theta}{\lambda}\right)^2} \int_0^1 \cos\left(2\pi h x_{Cu}(t) + \pi\frac{k}{2}\right) \cdot$$

$$\cos(2\pi(k+m\beta)y_{Cu}(t))\cos\left(2\pi(l+m)z_{Cu}(t) - \pi\frac{k}{2}\right)\cos(2\pi m t)\, dt$$

The particular symmetry of the metal sites requires z(t) = z(-t) [32]. We have refined the z component values of Cu atoms independently from the structural parameters of all other atoms, fitting δ_F values with $z_{Cu}(t) = c_0 + c_1 \cos(2\pi t) + c_2 \cos(4\pi t)$ (where t= βy). The value of $\Delta f'_{Cu}$, of an overall scale factor, and a correction factor which account for small differences in the relative scale of the λ_1 and λ_2 data sets, were refined as part of the refinement. In general, the Fourier coefficients of $x_{Cu}(t)$ and $y_{Cu}(t)$ as well as the B_{Cu} thermal parameter should also be refined. However, the intensity of the Bragg reflections collected during our experiments are poorly sensitive to displacements along either the a- or b-axes. In addition, due to the limited reciprocal space range, B_{Cu} is strongly correlated with the scale factor. Therefore, all these parameters were fixed to the values obtained by Yamamoto et al [32]. As expected, attempts to refine these parameters did not improve the fit. As a self consistency check, the agreement factor between F_{calc} and $F_{obs}(\lambda_2)$ (the observed structure factor away from the edge) was minimized with respect to the z-coordinates and the modulation of the heavy atoms.

Figure 5 shows the modulation of the Cu z component determined by the anomalous diffraction at low temperature in the superconducting phase (T<T_c).

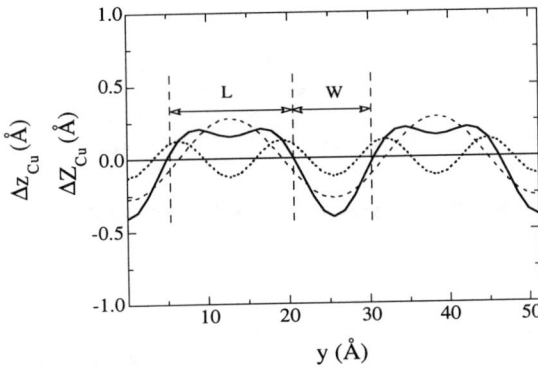

FIGURE 5. Modulation of the z coordinate of Cu in the CuO_2 layers along the b axis (solid line) of the Bi2212 structure. The dashed and dotted lines represent first and second harmonic contributions.

The amplitude of the first harmonic component, $c_1=-0.009$ ($c_1=-0.276$ Å), is comparable with previous results. On the other hand the amplitude of the second harmonic component, $c_2=-0.0042$ ($c_2=-0.128$ Å), is quite large. The large amplitude of the second harmonic gives rise to an anharmonic modulation with the Cu lattice having almost a constant z coordinate separated by stripes with a distorted lattice where the z coordinate of Cu differs by about 0.6 Å.

It should be mentioned that in spite of the intrinsic differences of the two techniques both EXAFS and anomalous diffraction show a distribution of the Cu-O(apical) distance from 2.3 to 2.6 Å. Thus the presence of the anomalously short Cu-O(apical) bond (0.2 Å shorter) observed in EXAFS is clearly due to the modulation of the Cu z-coordinate and hence the PDF of the Cu-O(apical) can be assigned to the anharmonic displacement of Cu.

Figure 6 shows the spatial distribution of the flat and distorted Cu lattice where the modulation of copper is plotted along with modulations of other heavy atoms. It can be clearly seen that the modulation amplitude of the Cu atoms in the c-axis direction is larger than that of the other atomic species. It decreases from Cu to Bi

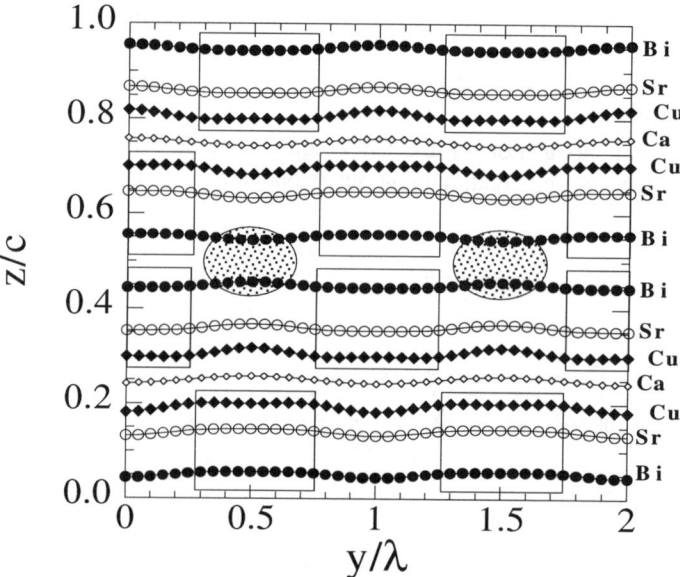

FIGURE 6. Modulation of z/c (c= 30.653 ± 0.002 Å) for copper and the heavy atoms over two periods of the superstructure. The squares indicate sections of wires of undistorted CuO_2 stripes and BiO rocksalt blocks (U-blocks). The modulations of two neighbour parallel CuO_2 planes are out of phase so that flat Cu stripes are surrounded by bent stripes.

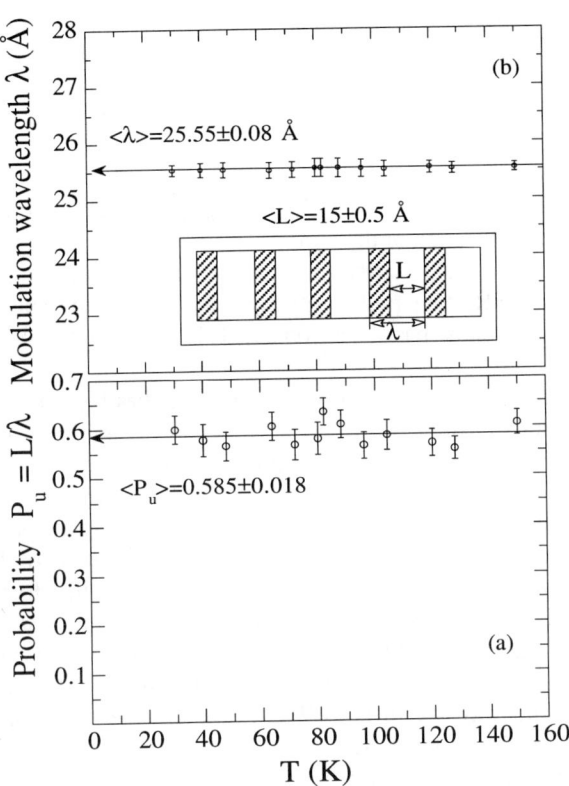

FIGURE 7. Temperature dependence of the probability of the undistorted CuO_5 lattice (P_u) given by $N(R_4)/(N(R_3)+N(R_4))$ measured by $E//c$ EXAFS (panel a) and the superstructure period λ measured by anomalous diffraction (panel b). The pictorial view of the stripes in the CuO_2 plane is shown as insert.

and it cannot be assigned simply to a steric effect due to the Bi layer. The large c-axis displacement of the Cu planes is correlated with the strong b-axis modulation of the BiO layers. The projection of the Bi2212 crystal in the y-z plane is shown over an area of 51 Å (along the b-axis) times 30.65 Å (along the c-axis). The rectangles indicate the section of wires of undistorted BiO rocksalt lattice and flat CuO_2 stripes (U stripes). They are separated by distorted blocks of stretched BiO lattice and bent CuO_2 plane (D stripes). This complex structure of Bi2212 can be described as wires of distorted lattice, where dopants (substituted and/or interstitial atoms) are located giving negative localized charges (acceptors) as indicated by shadowed circles, and wires of undistorted lattice indicated by squares in Fig. 6. In each Cu bilayer the U stripes in a first layer are close to the D stripes in the second layer.

The probability (P_d) of the minority distorted CuO_5 pyramids characterized by one short Cu-O(apical)=R_3 and two Cu-O(planar)=R_1 and two long Cu-O(planar)=R_2 has been found to be 0.415. The probability P_d is given by $N(R_3)/N_{tot}$ (from E//c data) = $2N(R_2)/N_{tot}$ (from E//a data). The measured probability P_u as a function of the temperature is shown in Fig. 7a. Therefore, the majority undistorted CuO_5 pyramids characterized by one Cu-O(apical)=R_4 and four Cu-O(planar)=R_1 have probability P_u=1-P_d =0.585. The spatial distribution of the distorted Cu sites has been obtained by joint EXAFS and diffraction experiments.

The one dimensional modulation of the Cu lattice has been determined by X-ray anomalous diffraction at the Cu K-edge. We report in Fig. 7b measurements of the wavelength λ of the 1D anharmonic modulation in the direction of the b axis in the orthorhombic direction, at 45° from the Cu-O-Cu in the plane as a function of temperature. The large anharmonic character is indicated by the presence of intense second order diffraction peaks. The anharmonic modulation of the CuO_2 plane gives stripes of distorted lattice (D-stripes) of width W that form linear domain walls inbetween the stripes of undistorted lattice (U-stripes) of width L (L≠W) running along the a-axis direction, as shown in the insert in Fig. 7b. The probability of undistorted Cu sites in the U-stripes is given by P_u=L/(L+W) and the superstructure wavelength is λ=W+L: We find the width of the undistorted lattice to be L =λ P_u = 15±0.5 Å.

In summary, the modulation of the Cu plane in the Bi-2212 system has been determined by EXAFS and anomalous diffraction. The measured Cu-O distances show a large distribution with an elongation of the Cu-O(planar) bonds. The large contribution of the second harmonic merely indicates that the CuO_2 plane is made up of flat and bent stripes. The distortions in the D-stripes are large enough to give a local electronic structure different from the U-stripes. In this structure the distorted D stripes (bent CuO_2 plane) would act as potential barrier between the U stripes formed by the undistorted lattice (flat CuO_2 plane). In fact the elongation of the Cu-O(planar) bonds, by about 0.08 Å, in the D-stripes directly changes the hopping integral in the plane [48] and modifies the band structure mainly at the M point of the Brillouin zone (BZ) giving a potential barrier between the stripes. Therefore the electronic structure of the modulated CuO_2 plane can be described as superlattice of quantum wires [49] with the formation of superlattice subbands [50].

We have analyzed the electronic structure of this system with a simplified model assuming an infinite potential barrier between the stripes, and so an electron wave vector in the y direction (perpendicular to the stripes) quantized k_y (n) =$n\pi$/L. For each integer value n, a one-dimensional (1D) subband is defined with energy minima E_n that depends on the actual electron effective mass i. e. on the band dispersion. Recently a phenomenological tight-binding fit of the experimental band dispersion has provided the hopping coefficients for the CuO_2 band of Bi2212 [31], that we have used for evaluation of subbands and the relative density of states. Each 1D subband (n) is obtained by cutting the 2D dispersion with fixed k_y (n) = $n\pi/L$, where L=15±0.5 Å is given by the present experiment. The resulting density of states (DOS) for the first 3 subbands is shown in Fig. 8. The Fermi level is near to the divergence of the DOS in the second subband, and it is separated from its bottom by (E_F-E_2) ~ 70 meV, moreover it is close the bottom of the third subband (E_F-E_3) ~40 meV, therefore the shape resonant condition is verified with $E_F-E_n < \hbar\omega_\Delta$, where ω_D ~ 50 meV is the Debye frequency. From this result it follows that the electron-electron coupling is strongly enhanced and the critical temperature of the homogeneous 2D plane is amplified by a factor of the order of 15.

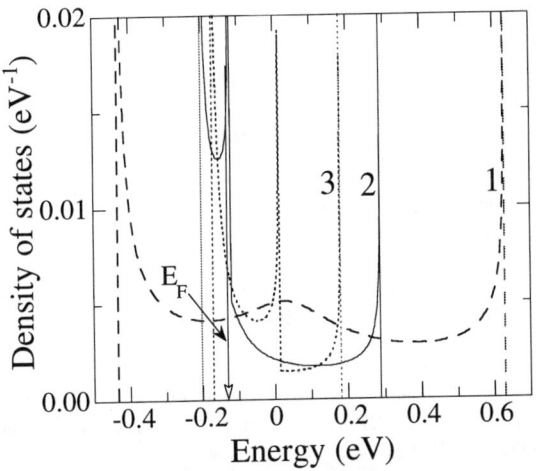

FIGURE 8. The calculated density of states of the n=1, 2 and 3 subbands for an ideal stripe of width L=15 Å and infinite potential barrier, running in the (π,π) direction following the tetragonal notation for a square lattice. The position of the Fermi level E_F in Bi-2212 is indicated by an arrow.

ACKNOWLEDGEMENTS

This work was partially supported by Istituto Nazionale di Fisica Nucleare (INFN), Istituto Nazionale di Fisica della Materia (INFM) and Consiglio Nazionale delle Ricerche (CNR). We thank M. Lusignoli, T. Rossetti, P. Bordet, Å. Kvick, and P. G. Radaelli, M. Missori, A. Perali and H. Oyanagi for help. One of us (AB) would like to thank the director of the CNRS Laboratoire de Crystallographie in Grenoble for warm hospitality.

REFERENCES

[1] *Lattice Effects in High-T_C Superconductors,* edited by Y. Bar-Yam, T. Egami, J. Mustre de Leon and A. R. Bishop. (World Scientific Publ. , Singapore, 1992)

[2] *Phase Separation in Cuprate Superconductors,* edited by K. A. Müller & G. Benedek (World Scientific Publ. , Singapore, 1993)

[3] *Phase Separation in Cuprate Superconductors* , edited by E. Sigmund and A. K. Müller (Springer Verlag, Berlin-Heidelberg (1994)

[4] T. Egami, and S. J. L. Billinge *Progress in Material Science* **38**, 359 (1994)

[5] T. Egami, and S. J. L. Billinge in *Physical Properties of High Temperature Superconductors* Vol. **V**, edited by D. M. Ginsberg (World Scientific, 1996) and references therein.

[6] H. Oyanagi, K. Oka, H. Unoki, Y. Nishihara, K. Murata, H. Yamaguchi, T. Matsushita, M. Tokumoto, and Y. Kimura, *J. Physical Soc. Japan* **58**, 2896 (1989).

[7] J. B. Boyce, F. Bridges, T. Claeson, T. H. Geballe, C. W. Chu, and J. M. Tarascon *Phys. Rev. B* **35** 7203 (1987); *Physica Scripta* **37**, 912 (1988); J. M. Tranquada, S. M. Heald, A. R. Moodenbaugh, and M. Suenaga, *Phys. Rev. B* **35** 7187 (1987).

[8] A. Bianconi, N. L. Saini, A. Lanzara, M. Missori, T. Rossetti, H. Oyanagi, H. Yamaguchi, K. Oka and T. Ito, *Phys. Rev. Lett.* **76**, 3412 (1996) and references therein; N. L. Saini, A. Lanzara, A. Bianconi, H. Oyanagi, H. Yamaguchi, K. Oka and T. Ito, *Physica C.* (in press, 1996).

[9] D. Haskel, E. A. Stern, D. G. Hinks, A. W. Michell, J. D. Jorgensen, J. I. Budnick *Phys. Rev. Lett.* **76**, 439 (1996)

[10] A. Lanzara, N. L. Saini, T. Rossetti, A. Bianconi, H. Oyanagi, H. Yamaguchi, Y. Maeno, *Sol. State Commun.* **97**, 93 (1996)

[11] N. L. Saini, T. Rossetti, A. Lanzara,M. Missori, A. Perali, H. Oyanagi, and A. Bianconi, *J. Supercon.* . **9** 343 (1996); A. Bianconi, N. L. Saini, T. Rossetti, A. Lanzara, A. Perali, M. Missori, H. Oyanagi, H. Yamaguchi, Y. Nishihara and D. H. Ha, *Phys. Rev. B* (in press, 1996).

[12] J. Mustre de Leon, S. D. Conradson, I. Batistic, and A. R. Bishop, *Phys. Rev. Lett.* **65** 1675 (1990); S. D. Conradson, I. D. Raistrick, and A. R. Bishop *Science* **248** 1394 (1990); J. Mustre de Leon, I. Batistic, A. R. Bishop S. D. Conradson, S. A. Trugman *Phys. Rev. Lett.* **68**, 3236 (1992).

[13] J. Röhler, A. Larisch and R. Schafer, *Physica C* **191** (1991) 57; J. Röhler in *Materials and Crystallographic Aspects of HTc-Superconductivity* edited by E. Kaldis (Kluwer Academic Publ. , Dordrecht, 1994) 353.

[14] H. Oyanagi, K. Oka, H. Unoki, Y. Nishihara, K. Murata, H. Yamaguchi, T. Matsushita, M. Tokumoto, and Y. Kimura, *J. Physical Soc. Japan* **58**, 2896 (1989).

[15] S. J. L. Billinge, G. H. Kwei, and H. Takagi *Phys. Rev. Lett.* **72**, 2282 (1994); S. J. L. Billinge, T. Egami, D. R. Richards, D. G. Hinks, D. Dabrowski, J. D. Jorgensen, and K. Violin *Physica C* **179** 279 (1991); S. J. L. Billinge, and T. Egami *Phys. Rev. B* **47**, 14 386 (1993)

[16] M. Lang, R. Kürsch, A. Grauel, C. Geibel, F. Steglich, H. Rietschel, T. Wolf, Y. Hidaka, K. Kumagai, Y. Maeno, and T. Fujita *Phys. Rev. Lett.* **69**, 482 (1992).

[17] M. Braden, O. Hoffels, W. Schnelle, B. Büchner, G. Heger, B. Hennion, I. Tanaka, and H. Kojima *Phys. Rev. B* **47**, (1993).

[18] M. Nohara, T. Suzuki, Y. Maeno, T. Fijita, I. Tanaka, and H. Kojima *Phys. Rev. Lett.* **70**, 3447 (1993); *Phys. Rev. B* **52**, 570 (1995).

[19] R. P. Sharma, L. E. Rehn, P. M. Baldo, J. Z. Liu, Phys. Rev. Lett. **62**, 2869 (1989); Phys. Rev. B **40** 11 396 (1989).

[20] T. Haga, K. Yamaya, Y. Abe, T. Yajima and Y. Hidaka, Phys. Rev. B **41** 826 (1990).
[21] J. Remmel, O. Meyer, J. Geerk, J. reiner, G. Linker, A. Erb and G. Müller-Vogt, Phys. Rev. B **48** 16168 (1993).
[22] M. Arai, K. Yamada, Y. Hidaka, S. Itoh, Z. A. Bowden, A. D. Taylor, and Y. Endoh *Phys. Rev. Lett.* **69**, 359 (1992)
[23] M. Arai, K. Yamada, Y. Hidaka, A. D. Taylor,and Y. Endoh *Physica C* **181**, 45 (1991); M. Arai, K. Yamada, S. Hosoya, A. C. Hannon, Y. Hidaka, A. D. Taylor,and Y. Endoh *Bullettin of the Electrotechnical Laboratory* **58**, n° 6, pag. 22 (1994)
[24] A. Bianconi, J. Garcia, A. Marcelli, M. Benfatto, C. R. Natoli and I. Davoli *Journal de Physique (Paris)* **46**, Colloque C9, 101 (1985).
[25] M. Benfatto, C. R. Natoli, A. Bianconi, J. Garcia, A. Marcelli, M. Fanfoni, I. Davoli *Phys. Rev. B* **34**, 5774 (1986)
[26] A. Bianconi in *X Ray Absorption: Principle, Applications Techniques of EXAFS, SEXAFS and XANES* edited by R. Prinz and D. Koningsberger, J. Wiley and Sons, New York 1988.
[27] A. Bianconi, J. Garcia, M. Benfatto in *Topics in Current Chemistry* **145** Springer Verlag Berlin edited by E. Mandelkow pag. 29 (1988)
[28] *X Ray Absorption: Principle, Applications Techniques of EXAFS, SEXAFS and XANES* edited by R. Prinz and D. Koningsberger, J. Wiley and Sons, New York 1988.
[29] P. Coppens, D. Cox, E. Vlieg and I. K. Robinson *Synchrotron Radiation Crystallography*, Academic Press, London (1992) pag. 118.
[30] A. Maeda, M. Hase, I. Tsukada, K. Noda, S. Takebayashi, and K. Uchinokura, *Phys. Rev. B* **41**, 6418 (1990)
[31] M. R. Norman, M. Randeira, H. Ding, and J. C. Campuzano *Phys. Rev.* B **52** 615 (1995) and references cited therein.
[32] A. Yamamoto, M. Onoda, E. Takayama-Muromachi, F. Izumi, T. Ishigaki, and H. Asano, *Phys. Rev. B* **42**, 4228 (1990).
[33] A. I. Beskrovnyi, M. Dlouhà, Z. Jiràk, S. Vratislav and E. Pollert, *Physica C* **166**, 79 (1990); A. I. Beskrovnyi, M. Dlouhà, Z. Jiràk and S. Vratislav, *Physica C* **171**, 19 (1990).
[34] H. Maeda, Y. Tanaka, M. Fukutomi and T. Asano *Jpn. J. Appl. Phys.* **27**, L209 (1988); ibid. *Jpn. J. Appl. Phys.* **27**, L727 (1988); Y. Matsui, H. Maeda, Y. Tanaka, and S. Horiuchi, *Jpn. J. Appl. Phys.* **27**, L372 (1988); Y. Matsui and S. Horiuchi, *Jpn. J. Appl. Phys.* **27**, L2306 (1988); H. Maeda and H. Koizumi, *Jpn. J. Appl. Phys.* **27**, L807 (1988); ibidem ; 27, L1172 (1988); A. Maeda, M. Hase, I. Tsukada, K. Noda, S. Takebayashi, and K. Uchinokura, *Phys. Rev. B* **41**, 6418 (1990).
[35] Y. Hirotsu, O. Tomioka, T. Ohkubo, N. Yamamoto, Y. Nakamura, S. Nagakura, T. Komatsu, and K. Matsushita, *Jpn. J. Appl. Phys.* **27**, L1869 (1988).
[36] E. A. Hewat, M. Dupuy, P. Bordet, J. J. Capponi, C. Chaillout, J. L. Hodeau, and M. Marezio *Nature* **333**, 53 (1988).
[37] H. -W. Zandbergen, W. A. Groen, F. C. Mijlhoff, G. van Tendeloo, S. Amelinckx *Physica C* **156**, 325 (1988); G. van Tendeloo, H. W. Zandbergen, S. Amelinckx, *Solid State Commun.* **66**, 927 (1988).
[38] Y. Gao, P. Lee, P. Coppens, M. A. Subramanian, and A. W. Sleight *Science* **241** 954 (1988).
[39] P. Lee, Y. Gao, H. S. Sheu, V. Petricek, R. Restori, P. Coppens, A. Darovskikh, J. C. Phillips, A. W. Sleight and M. A. Subramanian *Science* **244**, 62 (1989).
[40] V. Petricek, Y. Gao, P. Lee, and P. Coppens *Phys, Rev. B* **42**, 387 (1990).
[41] G. Calestani, C. Rizzoli, M. G. Francesconi and G. D. Andreetti *Physica C* **161**, 598 (1989).
[42] X. B. Kan and S. C. Moss *Acta Cryst.* **B48**, 122 (1992).
[43] D. H. Ha, K. Oka, F. Iga, H. Unoki, and Y. Nishihara in *Advances in Superconductivity V,* Ed. by Y. Bando and H. Yamauchi, Springer-Verlag, Tokio, 1993, pag. 323

[44] L. Tröger, D. Arvanitis, K. Baberschke, H. Michaelis, U. Grimm and E. Zschech, *Phys. Rev. B* **46**, 3283 (1992); J. Goulon, C. Goulon-Ginet, R. Cortes and J. M. Dubois, *J. Phys. I (France)* **43**, 539 (1982).

[45] S. J. Gurman, N. Binsted and I. Ross, *J. Phys. C* **17**, 143 (1984); S. J. Gurman, N. Binsted and I. Ross, *J. Phys. C* **19**, 1845 (1986).

[46] H. Ihara, *Bulletin of Electrotechnical Lab. Vol.* **58**, 64 (1994); also see e. g. F. Izumi and E. Takayama-Muromachi in *High Temperature Superconducting Materials and Engineering,* edited by. D. Shi (Pergamon, 1995) pag. 81.

[47] P. M. De Wolff *Acta Cryst.* **B32**, 521 (1974); P. M. De Wolff, T. Janssen, A. Janner, *Acta Cryst.* **A37**, 625 (1981).

[48] Y. Seino *J. Phys. Soc. Japan* **59** (1990)815.

[49] A. Bianconi and M. Missori, *Sol. State Commun.* **91**, 287 (1994); A. Bianconi *Sol. State Commun.* **91**, 1 (1994); A. Bianconi, M. Missori, H. Oyanagi, H. Yamaguchi, D. H. Ha, Y. Nishiara and S. Della Longa *Europhys. Lett.* **31**, 411-415 (1995); A. Bianconi and M. Missori *J. Phys. I (France)* **4**, 361 (1994); A. Bianconi *Sol. State Commun.* **89**, 933 (1994); A. Bianconi *Physica C* **235-240**, 269 (1994); A. Bianconi, M. Missori, N. L. Saini, H. Oyanagi, H. Yamaguchi, D. H. Ha, Y. Nishihara *J. Superconductivity* **8** 545 (1995); A. Perali, A. Bianconi, A. Lanzara, N.L. Saini *Sol. State Commun.* **100**, 181 (1996)

[50] G. Bastard, *Wave Mechanics Applied to Semiconductor Heterostructures* (Les edition de Physique, Les Ulis, France 1988).

IX. ELECTRON AND X-RAY EMISSION SPECTROSCOPY

Auger Electron and X-ray Spectroscopy of Hollow Atoms

R. Morgenstern

KVI Atomic Physics, Rijksuniversiteit Groningen
Zernikelaan 25, 9747 AA Groningen, Netherlands

Hollow atoms as formed during collisions of multiply charged ions on metallic, semiconducting and insulating surfaces have in recent years successfully been investigated by various spectroscopic methods: low- and high-resolution X-ray spectroscopy as well as high resolution Auger electron spectroscopy have been applied to study the electronic structure and the dynamic processes involved in formation and decay of hollow atoms.

Spectroscopy is often associated with extremely high precision measurements. However, such a precision can only be obtained if a system is observed in a well defined stationary state which exists for an appreciable time. In that respect hollow atoms are difficult objects, since they are extremely unstable and exist only for periods in the order of fs (1). Nevertheless, X-ray and Auger spectroscopy are powerful tools to investigate these atoms and especially to obtain information on the dynamic processes involved in the formation and the decay of hollow atoms.

Hollow atoms, i.e. atoms with most of their electrons in outer orbitals and empty or sparsely filled inner shells, can be formed in different ways. The very state selective method of multi-electron photon excitation has e.g. been applied to excite and characterize Li$(2s^2 2p)^1$P states (2). In this contribution we will discuss hollow atoms that are formed when multiply charged ions come into contact with neutral matter, and especially with metals with their practically inexhaustible reservoir of weakly bound electrons. Such hollow atoms might have intriguing properties. Imagine an Ar atom with all its 18 electrons in the orbital with principal quantum number $n=18$. All electrons could e.g. have their spins parallel, resulting in an Ar atom with huge spin S. Also, since these $n=18$ orbitals are nearly degenerate, a strong configuration interaction would result and this could give rise to a nearly classical collective electron motion. Would it be possible to observe plasmons in hollow atoms like those excited in the conduction band of metals? Last but not least such an atom carries a lot of excitation energy, which could possibly be released during the interaction with matter or radiation. There is in fact a lot of speculation that hollow atoms might be an appropriate medium for X-ray lasers, for atomic lithography and for other fancy applications (3-5). Sources for highly charged ions like the electron cyclotron resonance ion source (ECRIS) (6), electron beam ion source (EBIS) or ion trap (EBIT) presently allow (7) to routinely produce multiply charged ions up to U^{82+}.

TABLE 1. Critical distance r_c, energy gain DE due to the image charge acceleration and the maximum time T_{max} available for hollow atom decay in front of the surface.

	N^{6+}	Ar^{17+}	Th^{71+}
r_c (a.u.)	18.8	31.7	64.8
T_{max} (fs)	64	83	140
ΔE (eV)	17.3	82.6	705

HOLLOW ATOM FORMATION AT SURFACES

The most basic features of hollow atom formation at metallic surfaces are well described by the classical overbarrier model (8-10). The Coulomb barrier between projectile ion and surface drops until at a critical distance $z = R_c = \sqrt{2q}/W$ electron transfer becomes possible. With further decreasing projectile-surface distance additional electrons are transferred, resulting in a further decrease of the effective projectile charge q_{eff}. There will in fact be a dynamic equilibrium between decreasing distance and decreasing charge such that the Coulomb barrier remains just at the level of the workfunction W. For distances $z < r_c$ this results in a distance dependent effective charge given by

$$q_{eff} = \tfrac{1}{2}(Wz)^2$$

The most direct proof for this picture comes from measurements of the image charge acceleration (11): since the ions are slow as compared to the Fermi velocity of the electrons, there is sufficient time to form a negative image charge $-q_{eff}(z)$, and the attraction between charge and image charge leads to an energy gain given by

$$\Delta E = \int_{\infty}^{0} \frac{(q_{eff})^2}{(2z)^2} dz = \frac{\sqrt{2}}{6} W q^{\frac{3}{2}}$$

One can show that 75% of the energy gain occurs before the first electron capture, whereas the last 25% of the kinetic energy increase is gained during the neutralisation process. To get a feeling for the various quantities and their dependence on the primary projectile charge q some typical values are summarised in table 1 for the critical distance r_c, the energy gain ΔE and an upper limit T_{max} for the time between first electron transfer and touch-down on the surface. As one can see from this table typically a time interval of 100 fs is available between the first electron transfer and touch down. During this period a lot of electron dynamics evolves, leading in particular to the emission of a vast amount of low energy electrons. An impression of the number of electrons involved in the whole formation and deexcitation process can be obtained from measurements of electron emission statistics which have been performed in the group of Aumayr and Winter (12-15). They used an electron detector in which the number of emitted electrons per incident ion is converted into a pulse with a specific height. The pulseheight distribution therefore allows a direct determination of not only the average electron yield γ, but also the probability $p(n)$ for the emission of n electrons. An example for such a measurement is shown in fig.1. On the average there are 11.2 electrons emitted per incoming Ar^{9+} ion. The distribution is well described by a binomial distribution and this allows the additional conclusion that on the average 32 electrons are involved in the hollow atom formation and decay process.

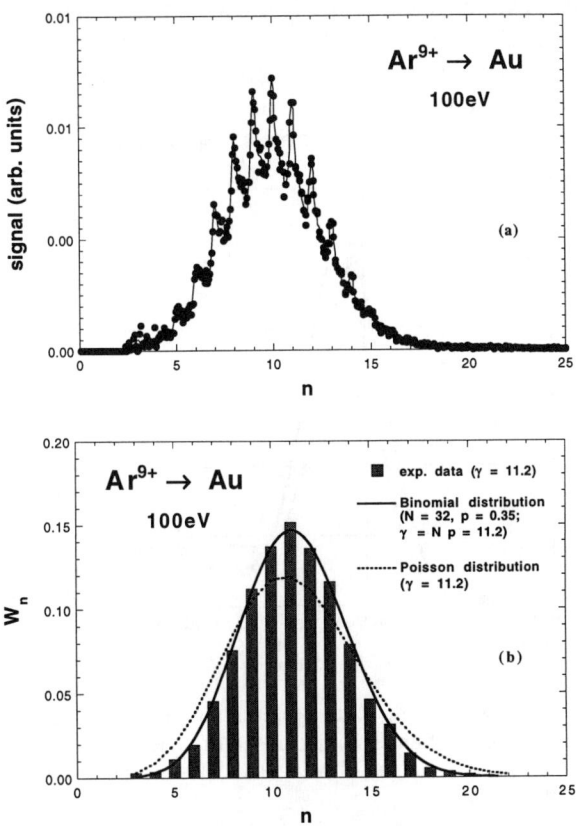

FIGURE 1. Spectrum of the number statistics for slow electrons emitted during 100 eV collision of Ar^{9+} ion on a gold target (from (16)). The bar diagram (fig. from Aumayr, private communication) represents a fit with a binomial distribution .

LOW-RESOLUTION X-RAY SPECTROSCOPY

One way to observe the decay of hollow atoms - or at least the final steps of this decay - is the detection of X-rays. For low-Z atoms fluorescence yields are rather low, but since radiative transition rates scale proportional to Z^4, an appreciable photon yield of several percent can e.g. be expected for radiative transitions into the K-shell of hollow Ar atoms. For this reason hydrogenlike Ar^{17+} and Kr^{35+} ions or bare Ar^{18+} and Kr^{36+} ions have in the past often been used as projectiles to study formation and decay of hollow atoms by photon analysis. Fig.2 shows spectra of K_α and K_β X-rays, resulting from Ar^{17+} impact on an insulating SiO_2 surface at two different angles of incidence. The spectral resolution of a Si(Li) detector with which these spectra were taken is insufficient to resolve structures within the K_α- and K_β-lines. However one can clearly see a shift of the lines towards higher energies when the angle of incidence of the projectiles is decreased. The energy of the K_α-line is mainly influenced by the number of L-shell electrons and therefore one can conclude that the radiative transition for the 15° spectrum takes place in the presence of fewer L-electrons than for the 45° spectrum.

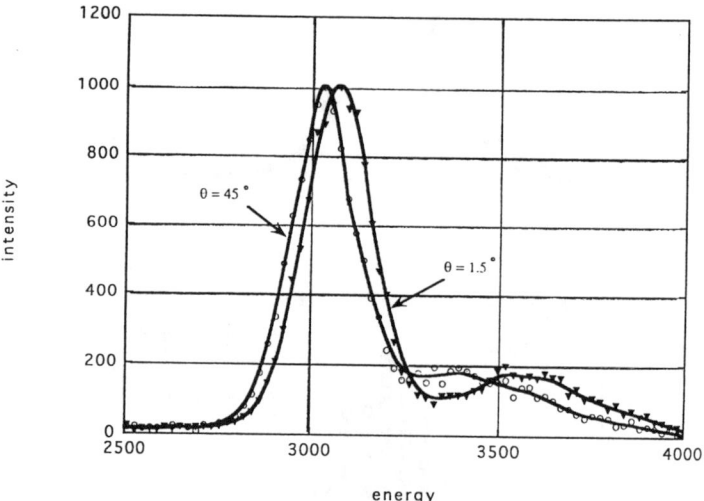

FIGURE 2. Spectrum of K_α and K_β X-rays resulting from impact of 340 keV Ar^{17+} ions on a SiO_2 target for two different angles of incidence (from Briand et al (17)). The shift to higher photon energies at small angles of incidence indicates decay from a projectile with an incompletely filled L-shell.

This shift can be exploited to gain information on the L-shell filling at the moment of the radiative transition. Winecki et al (18) have done this in a systematic way for collisions of Ar^{17+} on carbon surfaces. Fig.3 shows the energy positions of K_α- and K_β-lines as a function of the angle of incidence. One can clearly see the incomplete L-shell filling at the moment of decay, especially for those cases where the angle of incidence is below the critical angle where the projectile does no longer penetrate into the solid but is reflected. Charge state measurements on the other hand clearly show, that practically all reflected projectiles are completely neutralised. Winecki et al have used the experimental data to model the neutralisation process with calculations in which the M-shell filling rate is the main parameter. The binding energy of the target electrons is in fact such that transfer to the projectile proceeds mainly via the Ar M-shell, from where the electrons relax to lower orbitals via Auger and radiative transitions. The rates of Auger and radiative transitions involved in the further decay are known to a sufficient degree and therefore Winecki et al could obtain a rather detailed picture of the time dependent M- and L-shell filling of the Ar-projectiles. A 'sidefeeding' as already invoked by Folkerts and Morgenstern (19) - in the present case into the projectile M-shell - seems to be the most important channel for projectile neutralisation. The X-ray shifts resulting from the model calculations are also shown in fig.3.

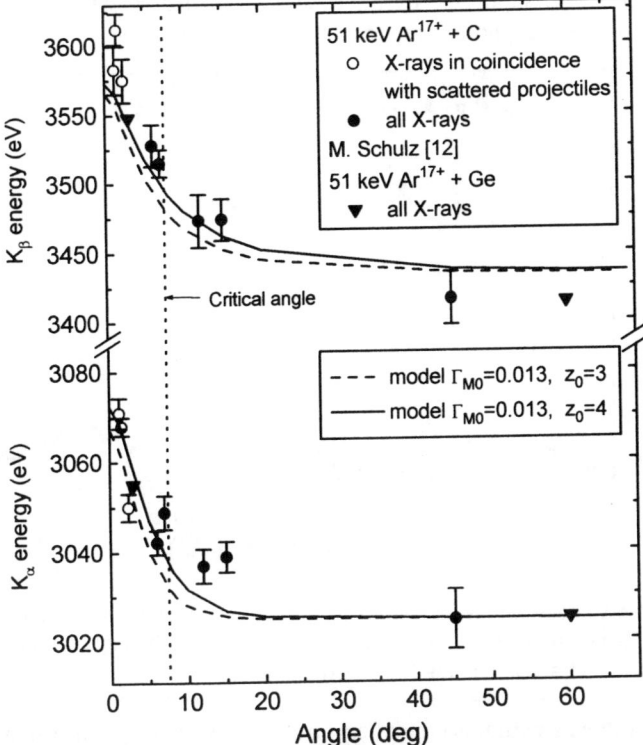

FIGURE 3. Energy positions of K_α- and K_β-lines arising from Ar^{17+} collisions on a carbon surface at different angles of incidence. The significant increase of X-ray energies for angles below the critical one where projectile reflection takes place is due to an incomplete L-shell filling outside the surface (from Winecki et al (18))

Even more detailed information about the characteristic L-shell filling times can be obtained by using the "clock-property" of bare ions. In such ions two radiative electron transitions to the K-shell take place sequentially, and the time interval between these two transitions is reasonable well known. If one is able to extract the number of L-electrons present at the moment of each of these transitions one can determine the filling rate of the L-shell. An example of a spectrum resulting from bare Kr^{36+} ions on a metallic target is shown in fig.4

From the spectrum shown in fig.4 Briand et al (17) deduced the average number of electrons present in the L-shell during the first and the second transition to be 2 and 3.2, respectively. Taking into account the transition of one electron from L to K during the first process this implies that during the average lifetime of the second K-vacancy, which is only 0.15 fs, on the average, 2.2 more electrons have arrived in the L-shell. This corresponds to an average filling rate of 1.5×10^{16} s^{-1}. This is in the order of magnitude of the LMM Auger transition rate and therefore also in this case the L-shell filling very likely proceeds via Auger cascades from the M-shell. This conclusion is also supported by the successful measurement of coincidences between K^0LM and K^1LL photons reported by Briand et al (17).

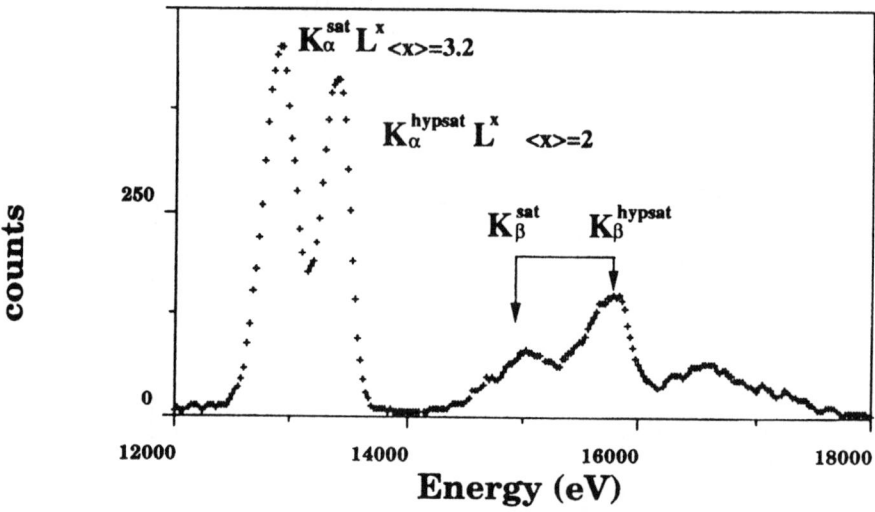

FIGURE 4. K X-ray spectrum resulting from 7 keV/q collisions of bare Kr^{36+} ions on a metallic target. Two K_α and two K_β lines are observed, those at the higher energies corresponding to the first transition to the still empty K-shell); (from (17))

One of the problems with this type of spectroscopy is the fact that formation and decay of hollow atoms takes place nearly simultaneously and that there is a competition between radiative decay of a partly filled L-shell and a further L-shell filling via additional electron capture from the target. Yamazaki et al (20) have recently tried to separate these two processes and to observe "hollow atoms" after their formation at a surface. To realise multiple electron transfer and at the same time avoid violent collisions they prepared a thin (10 μm) foil of Al_2O_3 with straight microcapillaries of about 100 nm diameter in a honeycomb pattern. This foil was bombarded with Ne^{9+} ions and the resulting hollow atoms or ions were observed via their X-ray decay on the entrance- and exit-side of the capillaries. In view of the limited length of these capillaries it was hoped to observe hollow atoms behind the foil that survived the interaction with the foil.

Fig.5 shows two X-ray spectra, measured on the entrance and the exit side of the foil respectively. The fact that the latter one is significantly shifted to higher photon energies already indicates that the radiative transitions observed here take place in a highly ionised projectile. In fact Yamazaki et al could identify the ions $Ne^{7+}(1s2s2p)^4P$ and $N^{8+}(1s2p)^3P$ as the most probable candidates for the observed emission. This identification is supported by the observed intensity decrease downstream from the foil which corresponds to an average lifetime of 0.8 ns. Although it can not completely be excluded that some "spectator electrons" still reside in high Rydberg states, the most probable scenario is that the hollow atoms have already decayed to a large extent, and that besides groundstate particles mostly Li-like ions in a long living high spin state with one K-shell vacancy leave the capillaries.

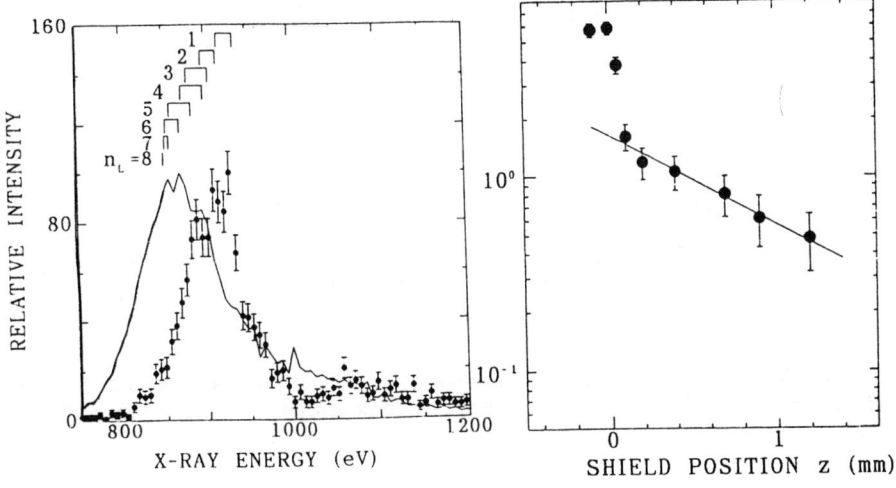

FIGURE 5. X-ray spectra resulting from collisions of Ne^{9+} on a sieve-like foil with 100 nm capillaries. Solid line: spectrum emitted on the entrance side; points: spectrum emitted on the exit side. (b) X-ray intensity as a function of distance downstream the foil (from Yamazaki et al, (20))

HIGH-RESOLUTION X-RAY SPECTROSCOPY

More details on hollow atom formation and decay can be obtained when high resolution crystal spectrometers are used, allowing for a separation of the various components of the K_α-line. Fig.6 shows a representative spectrum obtained from 340 keV collisions of Ar^{17+} ions on a Ag surface. In this case the most intensive line is the one on the low energy side labelled KL^8, corresponding to a completely filled L-shell. This implies that the L-shell filling rate is significantly higher than the K-vacancy decay rate. However, in view of the high projectile energy it is clear that practically all decay processes take place inside the solid and that binary collisions between the projectile ions and target atoms might play a significant role in filling the L-shell.

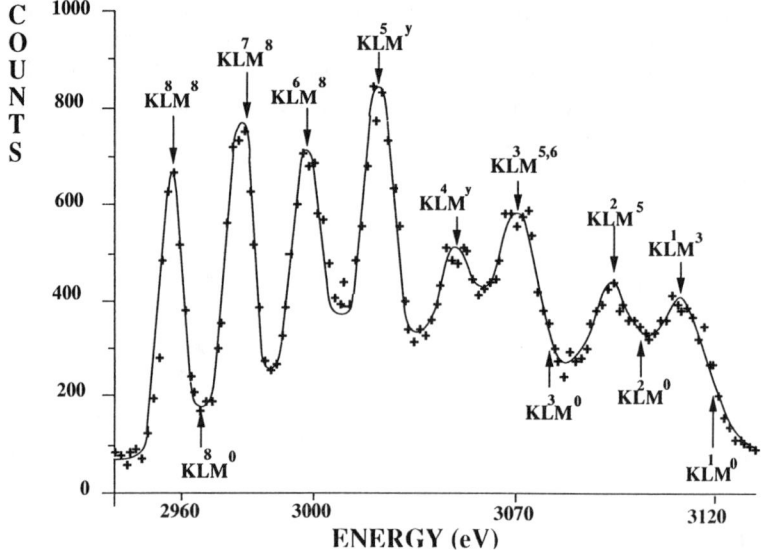

FIGURE 6. X-ray spectrum obtained from 340 keV Ar^{17+} ions on a Ag surface at an angle of incidence of 90°. K_α-lines corresponding to different degrees of L-shell filling are clearly resolved. (from (21))

It is tempting to view a high resolution spectrum as shown in fig.6 as kind of a snapshot with the line intensities representing state populations at a certain moment. A schematic view of this idealised picture is shown in fig.7, where hypothetical populations of L-shell configurations with different numbers of electrons are shown as function of the time. A measurement at the moment t_1 would yield the a spectrum as shown in the inset of the drawing. However one has to keep in mind that in reality measurements extend over a longer period, given by the lifetime of the K-vacancy.

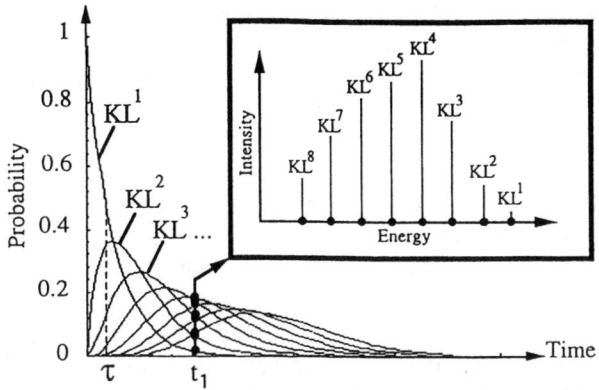

FIGURE 7. Schematics of subsequent L-shell filling and structure of the measured X-ray spectrum as a function of time (from (17))

In order to form hollow atoms in a more gentle way Briand et al have modified the collision processes in two different ways. First of all they have lowered the collision energy as far as possible, and secondly they have chosen insulating targets. Two spectra resulting from collisions on an insulating SiH target with energies of 1 and 200 eV/q respectively are shown in fig.8 (the 1 eV/q energy has to be regarded with caution in view of a possible influence of image charge acceleration effects). The dominance of the KL^1 line in the 1 eV/q spectrum clearly indicates that now the L-shell filling rate is significantly lower. An accurate comparison with the 200 eV/q spectrum moreover reveals some subtle details: at the higher collision energy the KL^1 line is broadened and shifted to a lower photon energy by about 10 eV. This broadening and shift can be ascribed to additional M-electrons. At the low collision energy of 1 eV/q therefore it is probably not a hollow atom which is formed but rather an excited ion in a still relatively high charge state with only very few electrons in excited orbitals.

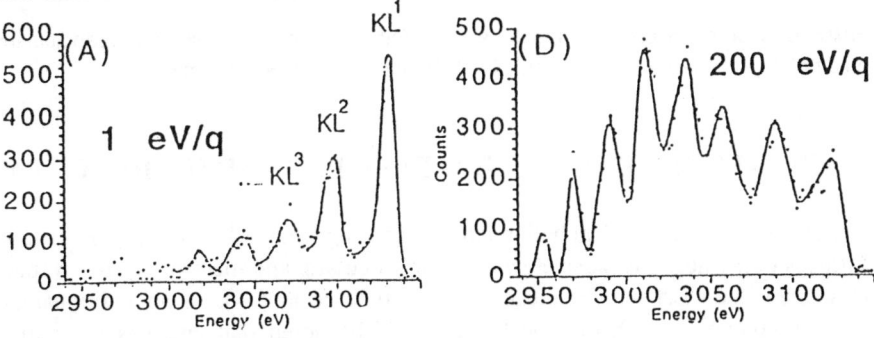

FIGURE 8. K X-ray spectra from Ar^{17+} ions impinging perpendicularly on a SiH surface at two different collision energies, 1eV/q and 200 eV/q respectively (from (22))

Also for a metallic target Briand et al have lowered the collision energy as far as possible, and fig.9 shows a comparison between two spectra obtained from 1-2eV/q collisions of Ar^{17+} on a Au and a SiH target respectively. Although there are clear differences in the spectra they have in common that again the KL^1 line is rather narrow and at an increased energy of 3133 eV, indicating again that the decay does not take place in a multiply excited "hollow atom" but rather in an excited ion. As is pointed out by Briand et al the projectiles are probably reflected from the target in case of the SiH target.

The pronounced variation of the spectra when different projectile energies and target materials are used indicate that the picture implied by the classical overbarrier model has to be applied with some care. If there were always a fast neutralisation of the projectiles in high orbitals, the characteristic Auger cascade times would govern the filling of the M- and L-shell. The fact that the spectra exhibit dramatic variations is a direct indication for other filling processes, which are especially important for high collision energies and for insulating targets.

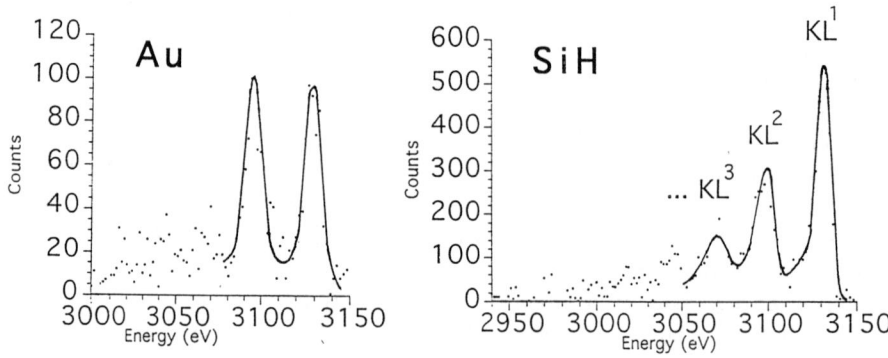

FIGURE 9. Comparison of K_α spectra arising from 1-2eV/q Ar^{17+} colliding perpendicularly on a polycrystalline Au and an insulating SiH target respectively (from (23)).

HIGH-RESOLUTION AUGER ELECTRON SPECTROSCOPY

For low-Z projectiles the hollow atoms resulting from collisions with solid state surfaces decay nearly exclusively via Auger processes, and therefore electron spectroscopy is the most suitable means to study their decay (24). As an additional advantage compared to X-ray spectra it should be noted that energies of emitted electrons are strongly influenced by the charge state of the electron emitting particle, and therefore this charge state can clearly be identified from the spectra. A disadvantage of this method however is caused by the fact that Auger processes are two-electron processes. This implies that the energy of the ejected electrons depends on the initial binding energy of *two* electrons, resulting in energy spectra which are considerably more complicated than X-ray spectra.

Peak-Identifications

Fig.10 shows an energy distribution of electrons resulting from 2 keV N^{6+} collisions on an Al(110) surface. One can clearly distinguish LMM and KLL electrons. The KLL structure around 350 eV exhibits several sharp peaks, and one can try to assign these to various configurations or states of the hollow atoms. First of all it can be stated that all electrons in this structure are emitted from a completely neutralised projectile. More detailed Hartree-Fock atomic-structure calculations (25) using the Cowan Code (26) allow to identify the lowest and the highest peak in this structure as being due to configurations N($1s2s^23l^4$) and N($1s2s^22p^4$) respectively. The structures in between can not uniquely be ascribed to a certain configuration because for all cases where $2s$ as well as $2p$ electrons are available several peaks - corresponding to L_1 or $L_{2,3}$ transitions - result from each initial state.

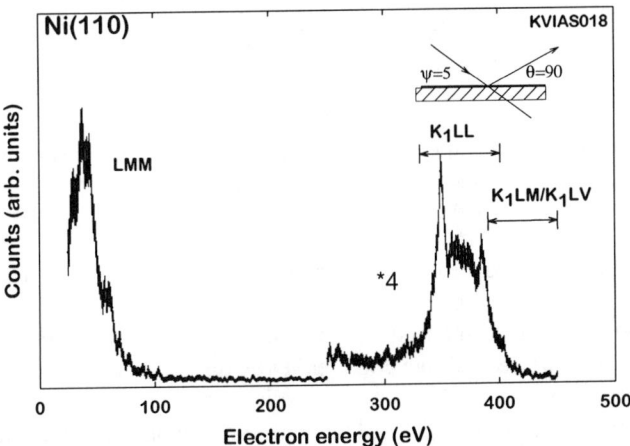

FIGURE 10. Energy spectrum of Auger electrons resulting from N^{6+} colliding on a Ni(110) surface at 250 eV. One can distinguish electrons from LMM and KLL Auger processes (from Limburg et al (27)). Reprinted from *Surface Science*, 313, J. Limburg et al., The Interaction of Hydrogenic Ions, 355- 364, 1994 with kind permission from Elsevier Science - NL, Sara Burgerhartstraat 25, 1055 KV Amsterdam, The Netherlands

Hollow atom formation seems to depend on the initial projectile charge in a characteristic way. This can be concluded from spectra arising from different projectiles. The KLL-part of spectra arising from collisions of hydrogenlike C^{5+}, O^{6+}, O^{7+}, F^{8+}, and Ne^{9+} on a Si(100) surface are shown in fig 11.

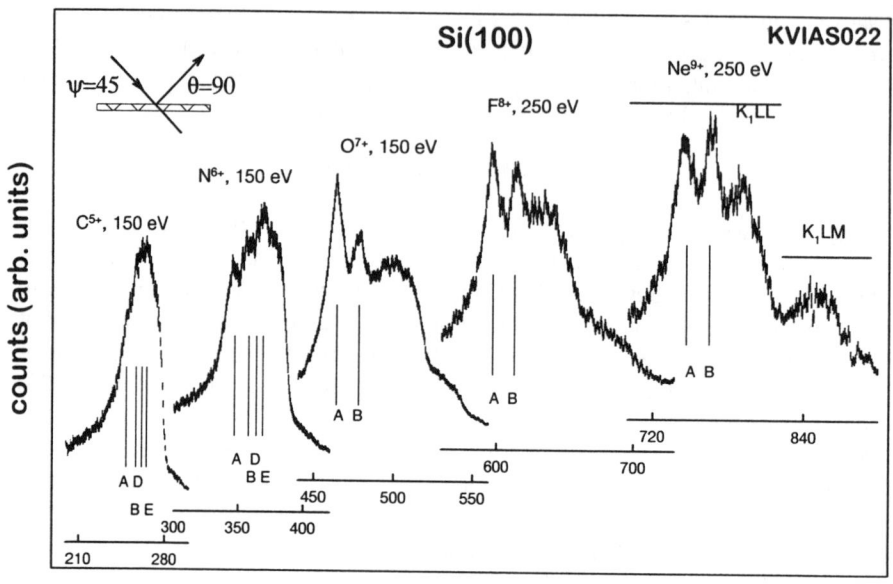

FIGURE 11. KLL-part of electron energy spectra arising from collisions of various H-like projectiles on a Si(100) target surface. There is a pronounced difference between the spectra arising from C- and N-ions as compared with those from O-, F- and Ne-ions.

There is a significant difference between spectra for C- and N-ions on the one hand and O-, F- and Ne-ions on the other hand. In the latter spectra there are two pronounced low-energy peaks, the first one of which corresponds to the $(1s2s^2)$ L-shell configuration. The second one could either be due to an L-shell configuration with *three* electrons or to a $(2s2p)$ or $(2p^2)$ configuration.

To distinguish between these possibilities one can compare spectra arising from H-like ions with those from bare ions with the same charge. In the latter case the $2s^2$ and the $2s2p$-states are nearly degenerate because there is no core electron. In fig. 12 such a comparison of KLL spectra is shown for hydrogen-like O^{7+} and bare N^{7+} ions. The K_0LL part of the N^{7+} spectrum which should comprise the same capture and decay processes as the O^{7+} spectrum consists of essentially a single peak. From this one can conclude that the two peaks in the O^{7+}-spectrum both correspond to 2-electron configurations. Moreover atomic-structure calculations allow a definite identification of the two peaks as $(2s^2)^1S$ and $(2s2p)^3P$ respectively.

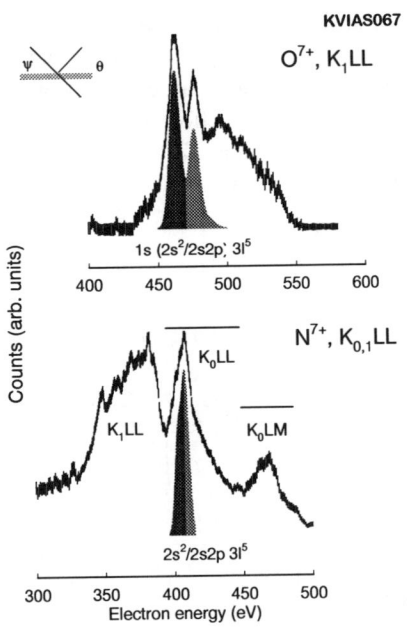

FIGURE 12. Comparison of KLL Auger spectra from 250 eV collisions of hydrogen-like O^{7+} and bare N^{7+} ions on Si(100) at an angle of incidence of 15°.

Relative Peak Intensities and their Velocity Dependence

In the first place one might wonder why the $(2s^2)^1S$ peak has such a high intensity. For a statistical population of two-electron states one would expect higher intensities for the $2s2p$ and the $2p^2$ L-shell configurations. But a closer look to the spectra reveals that especially in the initial phase of L-shell filling the population is by no means statistically. Limburg et al (28) and Schippers et al (25) have investigated this in more detail, whereby they not only used H-like projectile ions, but also metastable He($1s2s$)-like ones which already have a $2s$-electron to begin with. A comparison of two spectra arising from H-like O^{7+} and He-like O^{6+} collisions is shown in fig.13a. In both spectra the ratio of the two low-energy peaks, corresponding to a $(2s^2)^1S$ and a $(2s2p)^3P$ configuration respectively, are about the same: roughly 3:1, whereas a ratio of 1:9 would be expected on purely statistical arguments. The dominance of the $2s^2$-configuration can be ascribed to Coster-Kronig transitions in the projectile, during which the L-shell relaxes to the lowest configuration, whereby loosely bound M-electrons are emitted. In fig.13b a decay scheme of Coster-Kronig transitions is shown. Atomic structure calculations indicate that for oxygen projectiles the energy separation between the various core configurations is sufficiently high to allow an emission of M-electrons which are bound by roughly 5 eV, i.e. the workfunction of the target material. The same is true for F- and Ne-projectiles. For C- and N-ions however the energy separations are not sufficient for Coster-Kronig transitions.

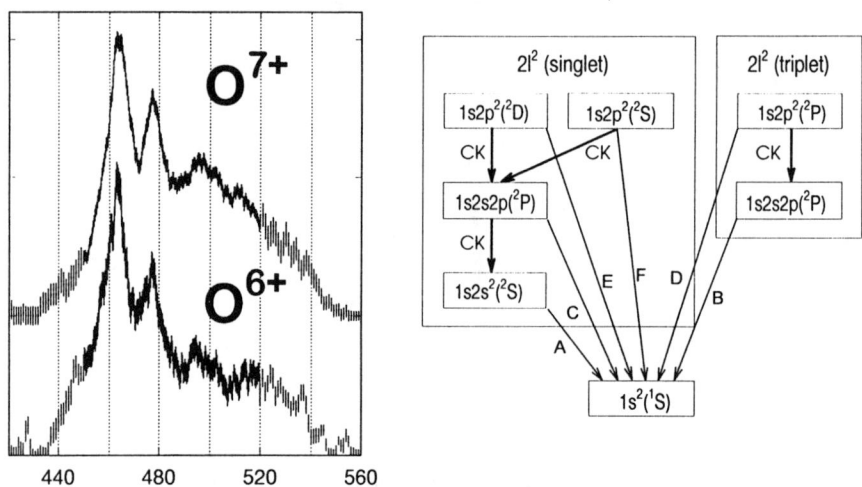

FIGURE 13. a) Comparison of KLL Auger electron spectra arising from 250 eV collisions of H-like O^{7+} (upper spectrum) and metastable He-like O^{6+} projectiles on a Si(100) surface. b) schematic diagram of Coster-Kronig relaxation processes in a hollow atom with a Li-like $1s2l^2$ core and Z-3 M-electrons. Since the M-electrons are regarded as spectator electrons the states are labelled corresponding to the LS-coupling scheme of the core. Bold solid arrows denote Coster-Kronig transitions (from (25)).

For the actually observed electron intensities not only the population of the various states is important, but also their decay rate. States with a low decay rate will appear with low intensity in those cases where only a short observation time is available. There are two effects which can limit the effective observation time: (i) modification or destruction of the initially formed states upon 'touch down' on the surface in those cases where the hollow atoms are formed in front of the surface, and (ii) further filling of the L-shell with additional electrons. In both cases the decay is interrupted, resulting in a relatively lower intensity of long-lived states. Schippers et al described the population and decay of the L-shell configurations in fig.13b with a set of rate equations from which the relative intensities of the two peaks of fig.13a could be deduced for different total observation times. From a comparison with the experimental spectra an effective observation time of 200 fs was found. Since under the experimental conditions relevant for the spectra of fig.13a the time between first electron capture and 'touch down' is less than 100 fs this implies that the electrons are partly ejected from inside the solid.

In the past there has been a lot of discussion whether the electrons in the low energy peaks are emitted above or below the surface (29-33).

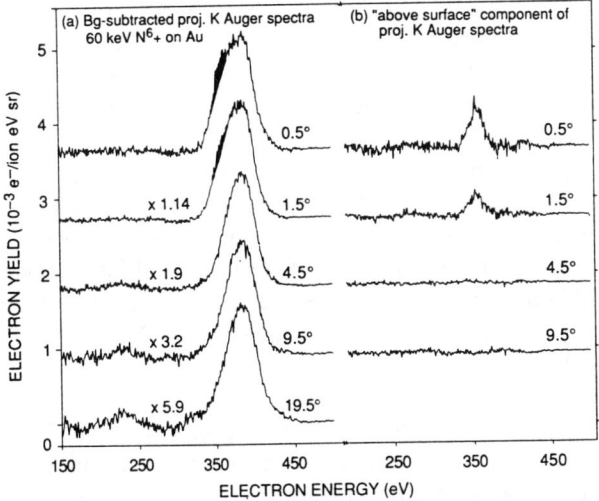

Figure.14. (a) Background subtracted KLL Auger spectra measured for 60 keV N^{6+} ions colliding on a Au(110) surface; (b) above surface components (from (29)).

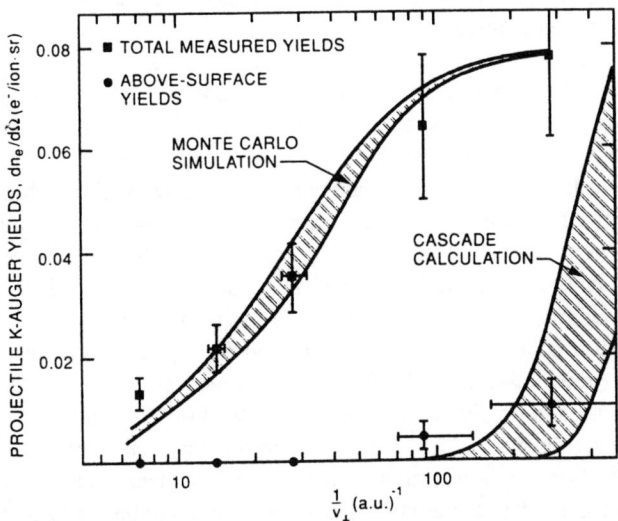

Figure.15. Comparison of measured and calculated electron yields as a function of the inverse perpendicular projectile velocity. The hatched areas labelled "Monte Carlo" and "cascade" are numerical simulations of the subsurface and above surface components of the KLL electron spectra, respectively (from (29))

The argument in favour for an above surface process was the fact that these peaks become prominent - or are even exclusively appearing - at low collision energies, and that their relative intensity saturates at decreasing vertical velocity - a fact which is connected with the image charge acceleration. In fig.14 spectra obtained by Meyer et al (29) are shown in which the above surface structure has been identified. In fig.15. the measured intensities asre compared with results from Monte Carlo and cascade calculations which show the various intensities as a function of the inverse perpendicular projectile velocity. These calculations strongly support the picture of above surface emission.

For collisions with trajectories penetrating into the solid however recent measurements of angular electron distributions however have shed some doubt on the hypothesis of above surface emission. Grether et al (33) have measured such distributions for collisions of Ne^{9+} on an Al(111) surface and found deviations from an isotropic distribution, which they ascribed to angle dependent electron attenuation within the solid. We will come back to the question of above- or below-surface decay during the discussion of different target materials.

So far electron spectra from rather slow collisions have been discussed, for which the classical overbarrier model can be expected to be valid - at least for metallic targets. For higher collision energies several effects might lead to a modification of the hollow atom formation and decay:
(i) insufficient time in front of the surface for hollow atom formation, (ii) insufficient time for the observation of hollow atom decay due to projectile penetration deep into the solid, and (iii) modification of innershell populations due to hard, binary projectile-target collisions.

In order to study these effects Limburg et al (34) measured KLL Auger spectra at different collision energies. In fig.16 spectra from N^{6+} collisions on Al at collision energies of 150 eV, 8 keV and 60 keV respectively are shown. Obviously the low energy peak corresponding to the $(2s^2)$ configuration, which is very prominent at the lowest collision energy, has practically disappeared at the high collision energy. Instead the high energy peak corresponding to a maximum filling of the L-shell becomes very prominent in 60 keV spectrum. At high projectile velocities apparently there is a fast L-shell filling mechanism active such that there is too little time for the configurations with a doubly filled L-shell to decay.

Limburg et al used a simple L-shell filling model to simulate the measured distributions. In this model two L-shell filling mechanisms were taken into account: one via an Auger cascade involving valence band (LVV) or M-shell (LMM) electrons, and one via quasi resonant electron transfer from inner shell target orbitals to projectile L-shell orbitals. It was assumed that the former mechanism is velocity independent, whereas the latter one is proportional to the number of L-shell vacancies and to the collision frequency, i.e. proportional to v/d with d the relevant lattice constant. At a velocity of 60 keV nitrogen ions the latter mechanism results in an L-shell filling rate of roughly 3 electrons per fs.

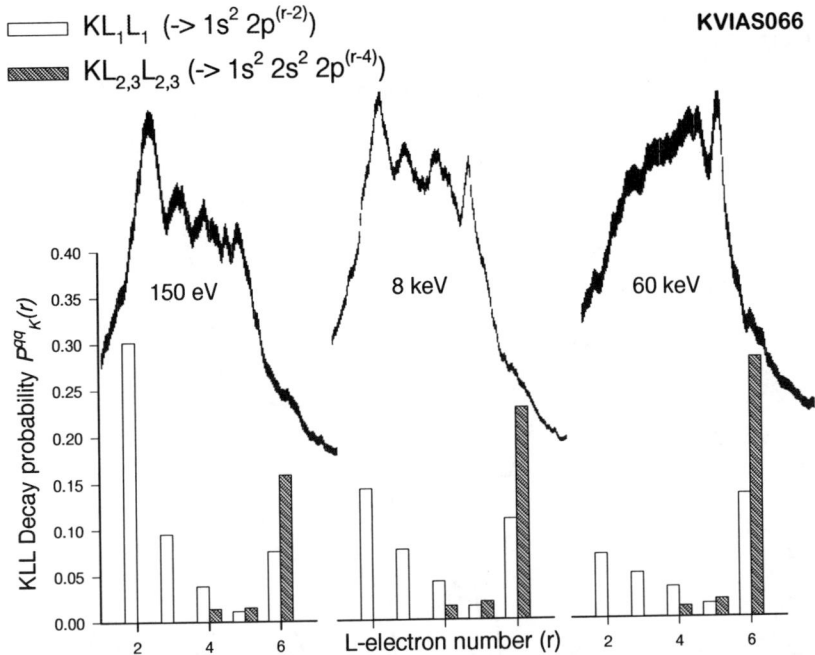

FIGURE 16. Velocity dependence of KLL spectra resulting from N^{6+} on Al(110). The measured spectra are compared with simulated relative intensities obtained from a model description for L-shell filling and decay (data from Limburg et al (34))

The bars in fig.16 indicate the relative contributions to the various parts of the KLL spectra, arising from Auger transitions with differently filled L-shells, involving s- and p-electrons, respectively. The qualitative agreement with the measured spectra indicates that indeed the side-feeding process, i.e. the direct electron transfer into the projectile L-shell, is mainly responsible for the L-shell filling at higher collision velocities. For high collision velocities the low-energy electron peak disappears because the corresponding configuration with a doubly filled L-shell does not exist sufficiently long to decay, i.e. the L-shell filling rate becomes higher than the Auger decay rate. Therefore electron emission from an L-shell with maximum filling dominates the spectrum. Two more quantitative descriptions of L-shell filling have recently be given by Burgdörfer et al (35) and by Stolterfoht et al. (36) respectively. Burgdörfer et al treat the direct quasiresonant capture from a Au surface into the L-shell of O^{q+}-projectiles within the classical overbarrier model, whereby the neutralization is significantly enhanced by the upward shift of the energy levels at the surface due to dynamical screening. Stolterfoht et al treat the capture in terms of binary collisions, whereby the relevant transfer cross sections are calculated within a molecular orbital treatment. In both cases it was found that the direct L-shell filling is much faster than the one via Auger cascades.

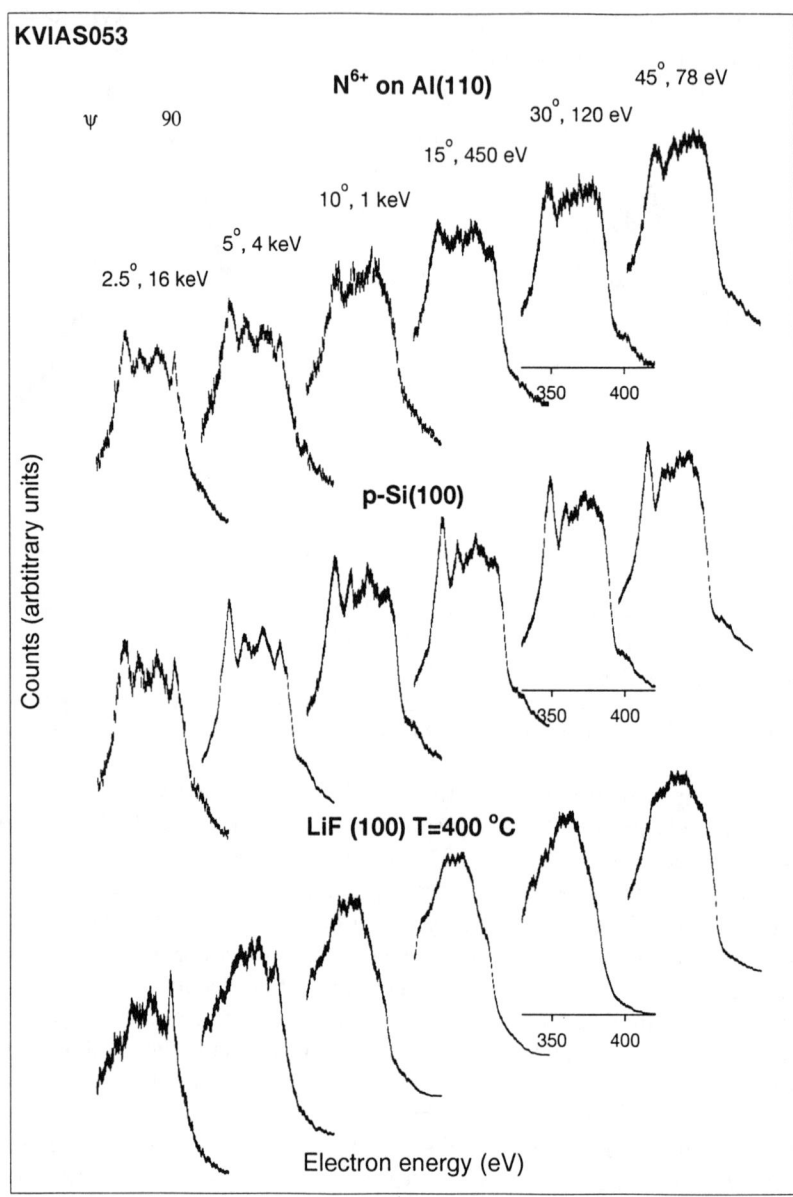

FIGURE 17. KLL Auger electron spectra from N^{6+} collisions on Al(110), Si(100) and LiF(100) targets. Collision energy and angle of incidence are chosen such that the surface is initially approached with the same velocity for all spectra.

Hollow Atom Formation at Different Target Surfaces

In the preceding sections electron spectra from different projectiles and a variety of target species have been presented, but the spectral features were exclusively discussed in connection with projectile properties such as charge state and electronic configuration. We will now discuss the influence of the target material. For this purpose three series of spectra, which were taken under the same experimental conditions are shown in fig.17 for N^{6+} ions colliding on a metallic Al(110), a semiconducting Si(100) and an insulating LiF(100) target respectively (37, 38). Within each series the collision energy and the angle of incidence were varied simultaneously in such a way that the initial vertical velocity is the same for all cases. Whereas the spectra for the Al(110) and the Si(100) are rather similar, those for the insulating LiF(100) surface are significantly different. The most obvious difference is the absence of the low-energy peak corresponding to the $(2s^2)$ L-shell configuration which occurs in all the other spectra. As discussed before this peak will no longer be observed when at high collision velocities the L-shell filling rate becomes much higher than the KLL Auger rate. The fact that for the Al and Si targets this peak remains significant even at high collision energies is a clear indication that the corresponding atomic configuration -a completely neutralised projectile with a $(2s^2)$ core - is formed before the close collision region is entered. The low-energy peak can therefor be taken as a signature for above surface formation and decay of hollow atoms, which obviously takes place in front of metallic and semiconducting surfaces as long as the vertical velocity is sufficiently low. The absence of these peaks in case of the LiF target therefore implies that hollow atom formation in front of the LiF surface does not take place. This was discussed in more detail by Limburg et al (37). Another - more subtle - difference of the LiF spectra is the fact that they exhibit a low energy tail not present in the other spectra. This can be ascribed to KLL processes in an incompletely neutralised projectile.

Finally there are also slight differences between the spectra from Al and Si targets: for Al the various spectral structures are much more 'washed out' than in Si. This is probably due to the significantly higher free electron density in Al as compared to Si, resulting in an increased scattering of electrons emitted inside the solid. This conclusion is supported by the observation that for small angles of incidence - where the probability of projectile penetration into the solid decreases - also for the Al target the spectral structures become more pronounced.

CONCLUSIONS

Hollow atom formation at conducting surfaces is well understood. The anslysis of K_a and K_b X-rays as well as KLL Auger electrons emitted from projectiles neutralized in front of the surface allow to study the time dependent L-shell population and the decay of these atoms. At insulator surfaces electron capture is reduced and excited ions instead of hollow atoms are formed.

Unfortunately hollow atom formation and decay are not processes clearly separated in time: electron capture and emission occurs nearly simultaneously. In order to produce "free hollow atoms" one could try either to decrease the formation time and/or to increase the decay time. To realise the first option one can try to use metallic clusters or C_{60} bucky balls as targets instead of solid state surfaces. Since electron transfer occurs at large impact parameters hollow atoms might be formed 'en passent' without the complications arising from direct filling of innershell vacancies. First experiments in this direction have been performed by Briand et al (39). Also one could try to use highly magnetised targets such that high spin states are formed with a higher probability which might have significantly longer lifetimes.

Regarding applications of "hollow atoms" one has to realise, that so far most experiments have been done with the aim to *observe and understand* their formation and decay. It is still a long way to *handle* them in a controlled way.

ACKNOWLEDGEMENTS

It is a pleasure to thank R.Hoekstra, J.Limburg and S.Schippers for their essential ideas and contributions to the "hollow atom" work in Groningen, and for the lively discussions we have had. The work performed in Groningen is part of the research program of the 'Stichting voor Fondamenteel Onderzoek der Materie' (FOM) which is financially supported by the 'Nederlandse Organisatie voor Wetenschappelijk Onderzoek' (NWO). Cooperation with various European groups was supported by the HCM network "Interaction of slow highly charged ions with solid surfaces" of the European Union (ERBCHRXCT930103)

REFERENCES

1. Vaeck, N., Hansen, J., J.Phys. B: At.Mol.Opt.Physics **28**, 3523-3543 (1995)
2. Kiernan, L., Kennedy, E.T., Mosnier, J.P., Costello, J.T., Sonntag, B., Phys.Rev.Lett. **72**, 2359-2362 (1994)
3. Morgenstern, R., Das, J., Europhysics News **25**, 3-6 (1994)
4. Hughes, I., Physics World **8**, 43-47 (1995)
5. Schmieder, R.W., Bastasz, R.J., in AIP Conf.Proc.274: VIth Int.Conf. on the Phys. of Highly Charged Ions, eds. P.Richard, M.Stöckli, C.L.Cocke, C.D.Lin, 675-681 (1993)
6. Geller, R., Jacquot, B., Nucl.Instr. and Meth. **202** 399-401 (1982)
7. Schneider, D.H.G., Briere, M.A., Physica Scripta **53**, 228-242 (1996)
8. Burgdörfer, J., Lerner, P., Meyer, F.W., Phys.Rev. **A44** 5674-5685 (1991)
9. Burgdörfer, J., Meyer, F.W., Phys.Rev. **A47**, R20-R22 (1993)
10. Burgdörfer, J., in 'Fundamental Processes and applications of atoms and ions', C.D.Lin, ed., World Scientific (1993)
11. Winter,H., Auth, C., Schuch, R., Beebe, E., Phys.Rev.Lett.**71**, 1939-1942 (1993)
12. Aumayr, F., Kurz. H., Schneider, D., Briere, M.A., McDonald, J.W., Cunningham, C.E., Winter, HP., Phys.Rev.Lett.**72,** 1943-1946 (1993)

13. Aumayr, F., Winter, HP., Comments At.Mol.Phys.**29**, 275-303 (1994)
14. Kurz. H., Aumayr, F., Schneider, D., Briere, M.A., McDonald, J.W., Cunningham, C.E., Winter, HP., Phys.Rev. A **49**, 4693-4702 (1994)
15. Aumayr, F., in "The Physics of Atomic and Electronic Collisions" (L.J.Dube, J.B.A.Mitchell, J.W.McConkey, C.Brion eds. AIP Conf.Proceedings 360), 631-645 (1995)
16. Kurz, H., Aumayr, F., Lemell, C., Töglhofer K., Winter, HP., Phys.Rev. A **48**, 2182- (1993)
17. Briand, J.-P., d'Etat-Ban, B., Schneider, D., Briere, M.A., Decaux, V., McDonald, J.W., Bardin, S., Phys.Rev. A**53** 2194-2199 (1996)
18. Winecki, S., Cocke, C.L., Fry, D., Stöckli, M.P., Phys.Rev. A**53**, 4228-4237 (1996)
19. Folkerts, L., Morgenstern, R., Europhys.Lett. **13**, 377-382 (1990)
20. Yamazaki, Y., Ninomiya, S., Koike, F., Masuda, H., Azuma, T., Komaki, K., Kuroki,K., Sekiguchi, M., J.Phys.Soc.Japan **65**, 1199-1202 (1996)
21. d'Etat, B., Briand, J.P., Ban, G., de Billy, L., Desclaux, J.P., Briand, P., Phys.Rev. A**48**, 1098-1106 (1993)
22. J.-P.Briand, G.Giardino, G.Borsoni, M.Froment, M.Eddrief, C.Sébenne, S.Bardin, D.Schneider, J.Jin, H.Khemliche, Z.Xie, M.Prior, Phys.Rev. A **54** (1996) in press
23. Briand, J.-P., Thuriez, S., Giardino, G., Borsoni, G., Froment, M., Eddrief, M., Sébenne, C., Phys.Rev.Lett.**77**, 1452-1455 (1996)
24. Das, J., Morgenstern, R., Comments At.Mol. Phys.**29**, 205-227 (1993)
25. Schippers, S., Limburg, J., Das, J., Hoekstra, R., Morgenstern, R., Phys.Rev. A**50** 540-552 (1994) and Phys.Rev. A**50** 4429-4430 (1994)
26. Cowan, R.D., The Theory of Atomic Structure and Spectra (California University Press, Berkeley, 1981)
27. Limburg, J., Das, J., Schippers, S., Hoekstra, R, .Morgenstern, R., Surf.Sci.**313**, 355-364 (1994)
28. Limburg, J., Das, J., Schippers, S., Hoekstra, R., Morgenstern, R., Phys.Rev.Lett.**73** 786-789 (1994)
29. Meyer, F.W., Overbury, S.H., Havener, C.C., Zeijlmans van Emmichoven, P.A., Zehner, D.M., Phys.Rev.Lett. **67**, 723-726 (1991)
30. Andrä, H.J., Simionovici, A., Lamy, T., Brenac, A., Lamboley, G., Bonnet, J.J., Fleury, A., Bonnefoy, M., Chassevent, M., Andriamonje, S., Pesnelle, A., Z.Phys. D - Atoms, Molecules and Clusters **21**, S135-S142 (1991)
31. Andrä, H.J., Simionovici, A., Lamy, T., Brenac, A., Pesnelle, A., Europhys.Lett. **23**, 361-366 (1993)
32. Das, J., Morgenstern, R., Phys.Rev.A **47**, R755-R758 (1993)
33. Grether, M., Spieler, A., Köhrbrück, R., Stolterfoht, N., Phys.Rev. A**52**, 426-432 (1995)
34. Limburg, J.,Schippers, S., Hughes, I., Hoekstra, R., Morgenstern, R., Hustedt, S., Hatke, N., Heiland, W., Phys.Rev.A**51**, 3873-3882 (1995)
35. Burgdörfer, J., Reinhold, C., Meyer, F.W., Nucl.Instr.and Meth.in Phys.Research B **95**, 415-419 (1995)
36. Stolterfoht, N., Arnau, A., Grether, M., Köhrbrück, R.Spieler, A., Page, R., Saal, A., Thomaschewski, J., Bleck-Neuhaus, J., Phys..Rev.A **52**, 445-456 (1995)
37. Limburg, , Schippers, S., Hoekstra, R., Morgenstern, R., Kurz, H., Aumayr, F., Winter, HP., Phys.Rev.Lett. **75**, 217 -220 (1995)
38. Limburg, J., Schippers, S., Hoekstra, R., Morgenstern, R., Kurz, H., Vana, M., Aumayr, F., Winter, HP., Nucl.Instr.and Meth.in Phys.Research B **115** 237-240 (1996)
39. J.-P.Briand, L.de Billy, J.Jin, H.Khemliche, M.Prior, Z.Xie, M.Nectoux, D.H.Schneider, Phys.Rev. A **53**, R2925-R2928 (1996)

Photoexcitation and Decay of Hollow Lithium States

F. J. Wuilleumier, S. Diehl, D. Cubaynes, and J.-M. Bizau

Laboratoire de Spectroscopie Atomique et Ionique, Université Paris XI, URA n°775 du CNRS, 91405 Orsay, France

Abstract. Photoelectron measurements have been carried out with three different sources of synchrotron radiation for studying photoexcitation and autoionization of hollow lithium atoms. The comparison of the three sets of data demonstrate the need of a third generation storage ring for a better understanding of atomic structures and dynamical photoionization processes of atomic systems. Partial cross sections have been measured with high spectral resolution at ALS, for hollow atoms formed by triple excitation of lithium atoms in the ground- or first optically excited states, providing the first definitive test of advanced ab initio calculations for this highly excited four-body atomic system. The existence of Rydberg series within the many observed hollow lithium states is demonstrated.

INTRODUCTION

A hollow atomic system is an atom or an ion in which at least one inner-shell is empty. Doubly-excited states of helium, first studied either by electron (1) or photon (2) impact and theoretically described by Fano (3) more than thirty years ago, are, in fact, the first hollow atomic states observed, even though the name given to these exotic states was invented later to describe multiply charged ions with empty inner-shells. Studies of these states in helium were continuously improved over recent years (4, 5), especially with the use of the third generation Advanced Light Source (ALS) storage ring in Berkeley (6, 7). Recently, very high resolution studies (1 meV at 60 eV) provided the best information on the correlated motion of a pair of electrons in the field of a nucleus (7). In a completely different domain, recent advances in collision experiments of highly charged ions with metallic surfaces (8-12) demonstrated the formation of highly excited hollow ions. In such ions, the K and L shells are empty and all of the electrons are in outer orbitals, resulting from the capture of many electrons by the highly charged ions approaching the surface. For instance, the capture of up to ten electrons in the M- and N-shells by Ar^{18+} ions, leaving completely empty the K- and L-shells, has been observed by studying the K- and L relaxation by x-ray fluorescence decay of the complex atomic system. The potential applications of such experiments, e.g., to create nanostructures on insulator surfaces, are now under heavy investigation.

Lithium, with one additional electron outside the $1s^2$ core, is the simplest open-shell many-electron system. Simultaneous excitation of all three electrons can create hollow atoms of the type $n\ell n'\ell' n''\ell''$, with n greater or equal to 2. The third electron makes the theoretical and experimental study of hollow lithium states an even greater challenge than that of helium. Such states provide the best opportunity for investigating the four-body Coulombic problem, because their formation by single photon - three electron excitations relies completely on electron-electron correlations. With an empty K shell surrounded by three electrons, the lithium system provides a most stringent test of the theoretical approximations aiming to describe many-electron interactions. The great advantage of experiments on isolated atoms as in the case of hollow lithium is that they can provide information on the basic physical processes involved in their decay, without perturbation by the external environment.

The first observation of triply excited $2\ell 2\ell' 2\ell''$ states in lithium has been reported in collision experiments (13, 14). Later on, these states were studied by electron (15) impact. However, only photon-excitation experiments provide the selectivity, sensitivity and resolution required to unravel the many hollow lithium states. The dipole selection rules ensure that only the $^2P^o$ final states can be photoexcited from the $1s^2 2s\ ^2S_{1/2}$ ground state of the lithium atom. The photoexcitation and decay dynamics of hollow lithium atoms has recently become the subject of intense experimental and theoretical interest. The first exploration of the lowest $2s^2 2p\ ^2P^o$ resonance at 142.3 eV energy was recently achieved in photoabsorption (16), with a Dual Laser Plasma (DLP) technique, allowing a first determination of the Fano parameters of this resonance. Soon after, two experiments (17, 18) at the HASYLAB and Photon Factory storage rings, using the total photoion yield technique, measured the autoionization of several hollow states and revealed the existence of numerous higher-lying resonances decaying into the Li^+ and Li^{2+} final charge states. Figure 1 shows, as a summary of these measurements, the results of the first absorption experiment at the excitation energy of the lowest energy $2s^2 2p\ ^2P^o$ hollow state (142.3 eV, left panel), and of one of the ionic yield measurements (17) extending up to 163 eV photon energy (right panel). Over this energy region, direct photoionization of atomic lithium into the continuum is competing with autoionization of the resonantly excited states, producing interference in some of the continuum channels and strongly enhancing partial cross sections in other channels. However, neither photoabsorption nor ion yield measurements can discriminate against transitions into the various continua of the Li^+ ion.

The energy level diagrams of Li, Li^+, and Li^{2+}, in Figure 2, shows the high number of transitions energetically allowed. Since the excited states of Li^+ can be reached either by direct photoionization into the continuum or by autoionization from the triply-excited states, one understands easily the need for using electron

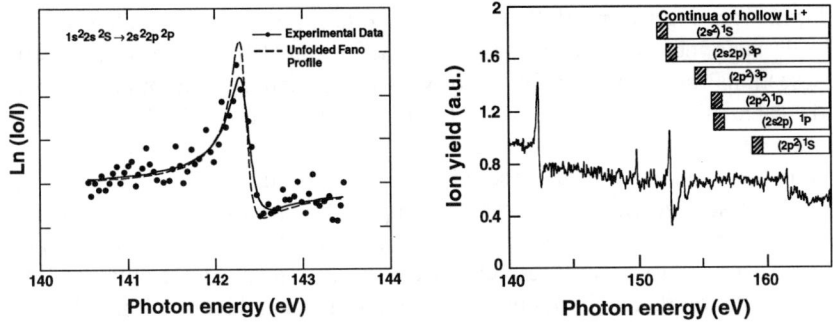

Figure 1. First measurements of the photoabsorption (left part) and photoion yield (right part) measurements over the energy range of the lowest energy hollow lithium state (from Refs. 16 and 17).

spectrometry in order to disentangle the various contributions to the Li$^+$ and Li^{++} ionic yields. By measuring the energies of the electrons emitted in the different transitions, photoelectron spectrometry has the unique capacity to analyze separately the many decay channels; partial photoionization cross sections can be measured and insight into the decay dynamics achieved. This technique was first

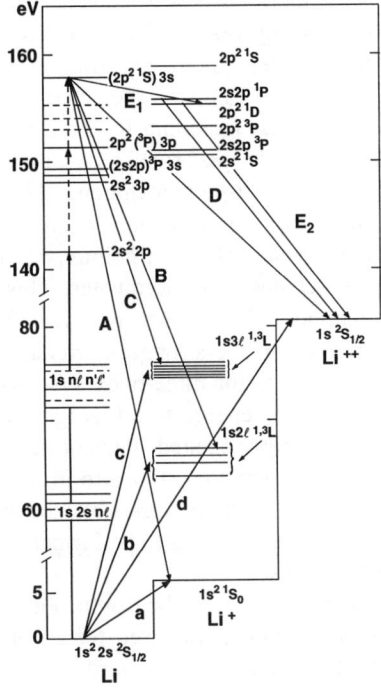

Figure 2. Energy level diagram of Li, Li$^+$, and Li^{2+}.

used with low spectral resolution to study the decay of the first hollow lithium $2s^22p$ $^2P^o$ state into the various continua of the Li$^+$ ion (19). The measured partial cross sections are compared favorably with cross sections determined from R-matrix calculations. However the high spectral band pass (0.5 eV) limited the critical nature of the comparison with theory. Subsequently, experiments were carried out with a slightly better resolution (0.2 eV) by using an undulator of the Super ACO storage ring. However, this improvement was not yet sufficient to achieve a precision test of the R-matrix calculations. Recently, much better spectrally resolved (19 to 39 meV resolution) and sensitive results were obtained using the higher photon flux and brightness available from an undulator of the Advanced Light Source in Berkeley (20). These experimental data provided the first definitive test of the *ab initio* R-matrix calculations resulting in an extremely good agreement between theory and experiment for the energy dependence of the continuum and resonant partial cross sections. The first measurements of even-parity hollow lithium states by triple photoexcitation of *laser-excited* lithium atoms were very recently achieved (21). Finally, the existence of Rydberg series among the hollow states of neutral lithium, first predicted by the R-matrix calculations but contested in previous experiments (18), has been clearly demonstrated by observing the Auger decay of doubly-excited states of Li$^+$ into the ground state of the Li^{2+} ion (22).

On the theoretical side, there have been considerable new developments (19-21, 23-33) during the past twenty years. When correlation effects are so dominant, it is not satisfactory to consider the strong correlations as a perturbation. The recent R-matrix calculations (19-21) have predicted the existence of a very high number of hollow atomic states, including Rydberg series. They have produced extensive photoionization partial crosss sections over the photon energy range between 140 eV and 167 eV. A saddle-point technique (27, 31, 32) has also been used to produce to date the most accurate resonance energies and widths for a number of resonances. For doubly excited states of helium, an alternative description has been developed using hyperspherical coordinates. This approach has led to the definition of new angular and radial quantum numbers (34) and has provided a full interpretation of the behavior of the strongly correlated electron motions. These calculations were used first (35) for three-body systems such as He and H$^-$, and helped to predict the resonance energies and assign them to series. Preliminary application of these ideas to triply excited states of lithium was made some years ago (28), but accurate enough potential curves to make some quantitative predictions regarding this four-body system were obtained very recently (33). These various theoretical methods are expected to be further refined in terms of treating multiply-excited configurations.

This paper presents a short review of the results obtained so far. Part II will describe the experimental setup which was used for all photoelectron spectrometry experiments and demonstrates the need of a third generation storage ring to provide data that can be useful for a valuable test of the theoretical approaches. Part III will present the data obtained for the lowest lying hollow states formed by triple excitation of lithium atoms in the ground and first excited states. Part IV will comment on the results obtained in measuring the autoionization of higher energy hollow states up to 154 eV photon energy. Finally, Part V describe the recent experiments which have succeeded in observing hollow-hollow autoionizing transitions, i. e., autoionization from triply excited states of neutral lithium atoms to helium-like doubly excited states of the Li^+ ion, followed by decay into the ground state of the Li^{2+} ion.

EXPERIMENTAL

Three different synchrotron radiation beam lines were used, with the same experimental setup, to energy analyze the electrons after photoexcitation of neutral lithium atoms over the energy range of the triply excited states. An electron cylindrical mirror analyzer (CMA) was used for angle-integrated measurement of the energies and intensities of electrons emitted in the autoionization of the photoexcited hollow states. Electrons produced in the interaction of the monochromatized photon beam with an atomic lithium beam in the source volume of the CMA were energy analyzed at the magic angle and detected with a channeltron. For the experiments on laser-excited lithium atoms, the beam of an Ar^+-ion pumped dye laser, tuned to 671 nanometers perpendicular to the lithium beam, was focused into the source volume of the CMA. The details of the experimental setup and procedure were similar to the ones already used in previous experiments involving inner-shell ionization in laser-excited atoms (36-38). We utilized sufficiently high densities of lithium atoms to cancel possible alignment induced by the linearly polarized laser beam. The integrated number of electrons of a given energy detected at the magic angle is thus proportional to the absolute values of the partial cross section.

The three different sources of synchrotron radiation used in these experiments were: the SA23 bending magnet beam line of the Super ACO storage ring, the SU7 undulator beam line of Super ACO, and the 9.0.1 undulator of the Advanced Light

Table I. Characteristics of the photon beam and CMA in the different experiments. $\Delta E/E$ is the resolution of the CMA at a pass energy E.

Beam line	Monochromator	Spectral resolution (140 - 160 eV)	Photon flux range	($\Delta E/E$)	Beam density
SA23	1.8 m TGM	0. 5 to 0.7 ev	10^9/s	1%	10^{13}/cm^3
SU7	10 m TGM	0. 23 eV	10^{11}/s	0.7%	10^{13}/cm^3
9.0.1.	10m SGM	0.019 to 0.039 eV	10^{12}/s	0.4%	10^{12}/cm^3

Source at Berkeley. In Table I, we summarize the characteristics of the photon beam delivered by these various sources and the parameters of the CMA used in the experiments.

The details of the experimental procedure are given in the original papers (19-21). Here, we compare in the left part of Figure 3, the photoelectron spectra obtained by direct photoionization into the continua of lithium atoms in the ground state and in the first $1s^2 2p\ ^2P_{3/2}$ excited state at the SA23 (three upper panels) and 9.0.1 experimental stations (lower panel) with 90 eV photons (at this energy, the spectral resolution at SA23 and 9.0.1 were 0.3 eV and 0.015 eV, respectively). The energy level diagram of Li, Li$^+$, and Li^{2+}, shown in Figure 2, helps the assignement of the different photoelectron lines, especially since photoionization with excitation of lithium atoms occurs, giving rise to intense correlation satellites. When lithium atoms in the ground state are photoionized in the 1s shell at this photon energy (top upper panel), according to $1s^2 2s\ ^2S_{1/2} + h\nu \rightarrow 1sn\ell\ ^{1,3}L + \varepsilon\ell'$, the residual Li$^+$ ion can be left in any of the following ionic states with the corresponding electron lines observed at the expected energy in the electron spectrum: $1s2s\ ^3S$ (at 64.41 eV bin-

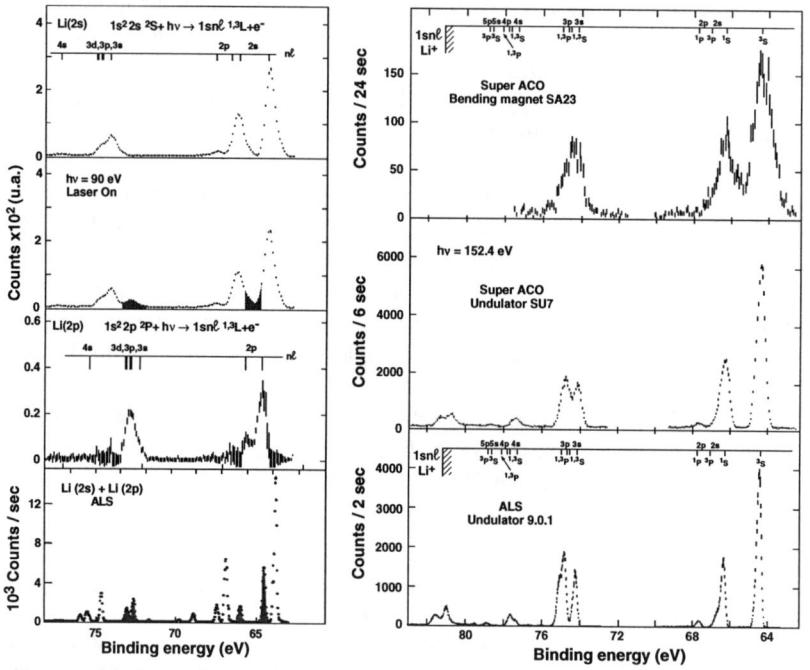

Figure 3. Left part: photoelectron spectra resulting from photoionization of lithium atoms in the ground state and in the first excited state taken at Super ACO (three upper panels) and at the ALS (lower panel) at 90 eV photon energy (see text for explanation); right part: photoelectron spectra resulting from photoionization of lithium atoms in the ground state at 152.4 eV eV photon energy with SR from Super ACO (bending magnet, upper panel; undulator, middle panel) and from the ALS (lower panel).

ding energy, BE, in Figure 3); $1s2s\,^1S$ (at 66.31 eV BE) and $1s2p\,^3P$ (at 66.67 eV BE), not resolved in the spectrum; $1s2p\,^1P$ (at 67.31 eV BE), barely seen in the spectrum; $1s3\ell\,^{1,3}L$ (not resolved here, between 75 eV and 76 eV BE), $1sn\ell\,^{1,3}L$, with n >3 at higher BE, not seen in the spectrum. Peaks $1s2s^{1,3}S$ are the so-called main lines, the other ones can be classified into various categories of correlation satellites. In the second upper panel, the electron spectrum includes photoelectron lines due to photoionization of lithium atoms in both the ground- and in the first $1s^22p\,^2P_{3/2}$ excited states. Since only 15% to 20% of the lithium atoms existing in the atomic beam are transferred into the first optically excited state, 75% to 80% of the atoms are still remaining in the ground state. Thus, the photoelectron lines arising from photoionization of laser-excited lithium atoms (the lines corresponding to photoionization of the $1s^22p\,^2P_{3/2}$ excited atom into the $1s2p\,^{1,3}P^o$ final ionic states should appear at 66.3 eV and 65.7 eV BE) strongly overlap with electron lines due to photoionization of lithium atoms in the ground state, because of the modest spectral and electron resolutions. One can only observe a broadening of the first two electron lines and the presence of additional lines at the binding energies (74 eV to 75 eV) corresponding to photoionization of the excited atoms into the $1s3\ell$ excited ionic states. The true photoelectron spectrum due to photoionization of the excited atoms can be obtained by subtracting the second spectrum from the first spectrum, after adequate normalization. The result of this substraction is shown in the third panel. The error bars are rather large and the shape of the electron lines is not well enough defined, because this spectrum is obtained from the subtraction of two high counting rates. However, when one uses the photon beam from the ALS 9.0.1 beam line, the brightness is so high that it is possible to reduce the widths of the CMA slits (with 2 mm slits, the electron resolution becomes 0.4% of the pass energy) and to resolve most of the electron lines, as shown in the lower panel. The lines arising from photoionization of excited atoms are marked as grey areas. The photoelectron spectra of lithium atoms in both the ground- and excited states are simultaneously observed without any overlap, allowing full analysis of all processes. Another interesting comparison is shown in the right part of Figure 3, where the photon energy, 152.4 eV, has been chosen to coincide with the excitation energy of the $2p^23p\,^2P^o$ hollow lithium state from lithium atoms in the ground state. From top to bottom, the three panels display the photoelectron spectra obtained at the SA23, SU7, and 9.0.1 beam lines. The use of an undulator at Super ACO (middle panel) evidently increases considerably the photon flux available as compared to the bending magnet beam line (upper panel), but the resolution is not yet sufficient to resolve most of the electron lines, while the spectrum measured at the Advanced Light Source (lower panel) shows clearly the dramatic increase in both the resolution and counting rates (see in particular the n = 3 correlation satellites). These results (19-21) demonstrate that a high-precision undulator installed on a third-generation storage ring in combination with excellent x-ray optics is able to provide at the same time high resolution and high intensity monochromatic photon beams, allowing a detailed test of the atomic theories.

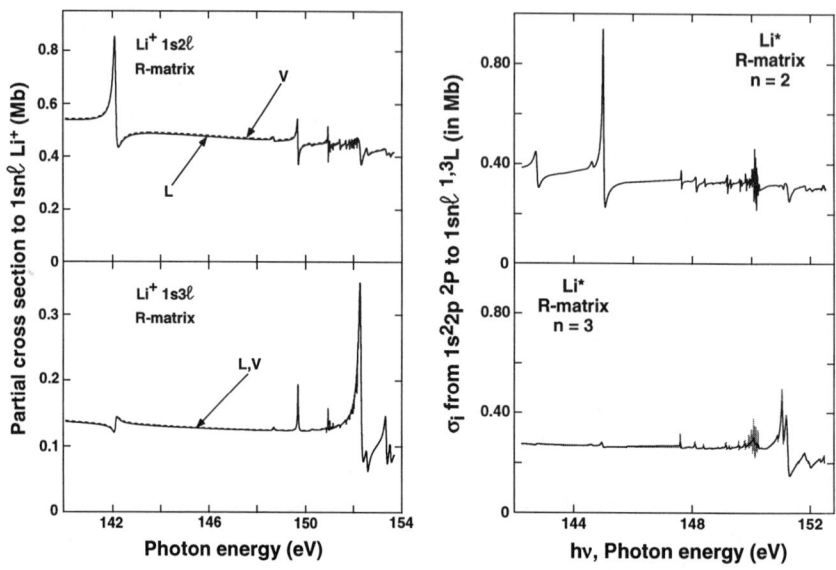

Figure 4. R-matrix results showing the variation of the partial photoionization cross sections of Li atoms in the ground (left) - and first excited (right) states to the 1s2ℓ (upper part) and to the 1s3ℓ (lower part) final Li+ ionic states (from Refs. 19 and 21).

Prior to the experiments at ALS, the results of the R-matrix calculations were available for ground- as well as for the first excited states, allowing the determination of the energy regions of interest. We show in Figure 4 the predictions (19-21) of these calculations for photoionization of lithium atoms in the ground (left) - and excited (right) states. In both cases, the upper panel presents the photoionization cross sections into the 1s2ℓ final ionic states, while the lower panel shows the same data for 1s3ℓ final ionic states (correlation states). It was clear from these calculations that the lowest energy hollow states ($2s^22p$ $^2P^o$ in the ground state, $2s2p^2$ $^2S^e$, $^2P^e$, $^2D^e$ in the excited state) autoionize mainly in the n = 2 final states while the hollow states at higher energy decay mostly to the higher energy ionic states.

THE LOWEST ENERGY HOLLOW STATES

The lowest lying hollow state which can be populated by triple excitation of lithium atoms in the ground state is the $2s^22p$ $^2P^o$ state at 142.3 eV photon energy. We have measured the partial photoionization cross sections to 1s2ℓ and 1s3ℓ ionic states of Li+. We show, in the upper panels of Figure 5, how the partial sections we have determined at the Advanced Light Source for photoionization of

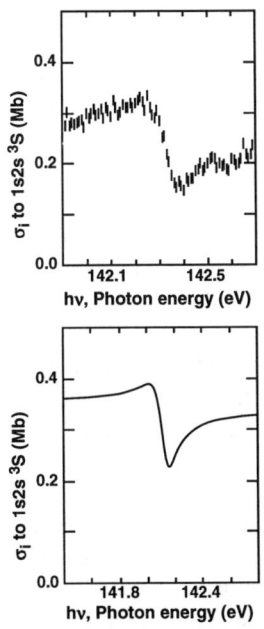
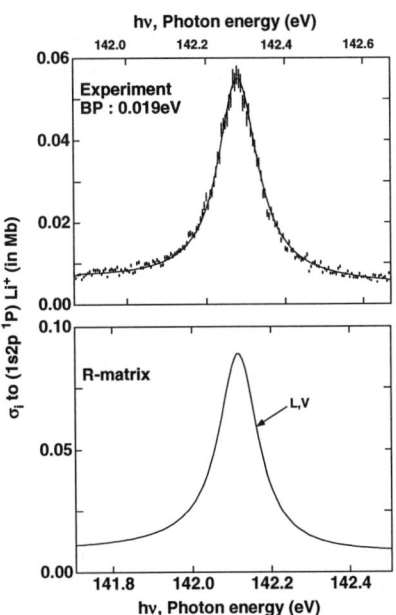

Figure 5. Photoionization cross sections of lithium atoms to the 1s2s ^{3}S (left part) and to the 1s2p ^{1}P (right part) final Li$^+$ ionic states over the photon energy region of the first triply excited resonance at 142.3 eV. The L- and V- curves are the results of the *ab initio* R-matrix calculations after convolution with the instrumental function (from Ref. 20).

lithium atoms, vary over the resonance energy region, into the 1s2s ^{3}S + εp (left part, further noted n = 2 ^{3}S) and the 1s2p ^{1}P + εs (right part, n = 2 ^{1}P) continuum channels, respectively. Both profiles show a quite different behavior: while in the (n = 2 ^{3}S) channel the shape of the cross section has a Fano-type profile, in the (n = 2 ^{1}P) channel one observes a symmetrical profile. This was expected, since the direct photoionization cross section to the 1s2s ^{3}S state is the highest of all partial cross sections, allowing interferences to occur over the energy range of the resonance, while the cross section to the 1s2p ^{1}P state is very weak at these photon energies (only 2% of the former one). In Fig. 5, the solid line was obtained from a fitting of the experimental data to a Fano-Starace (39) profile valid in the case of one resonance interfering with many continua. We show also in the lower parts of the figure the results of the *ab initio* theoretical R-matrix calculations after convolution with our instrumental function. The agreement in terms of the resonance width and profile and the relative amplitude of the cross-section is impressive, especially since there is no normalization of the experimental results to any theoretical parameter or *vice versa*. With regard to the absolute values of the cross sections, the precision of the experimental values is limited by the uncertainty

(± 25%) in the measured total photoabsorption cross-section (40) to which all partial cross sections are normalized. The measured absolute energy of the resonance differs from the R-matrix calculated value by 0.10 eV; an almost constant shift of 0.10 to 0.20 eV between experimental and R-matrix calculated energy values was observed throughout the whole energy range examined in this work. This shift comes mostly from the calculation of the ground state energy. In the (n = 2 ^{1}S and ^{3}P) channels, the conclusion is the same as for the ^{3}S and ^{1}P channels, respectively. For photoionization into the $1s^2$ ^{1}S final ionic state (ionization of the outer 2s electron), we observed a deep window resonance. For photoionization into the n = 3 channels, we did not measure any detectable variation of the cross section, within a few percents, over the resonance energy region.

Following deconvolution of the spectral profile (with a Gaussian width of 0.019eV), a FWHM of 0.118(3) eV is obtained for the $2s^22p$ ^{2}P resonance profile. In Table II, we show a comparison between the various experimental results with recent theoretical results. The new result provides the smallest to date measurement of the resonance width with much better error-bars implying an autoionization lifetime of approximately 10^{-13} sec for this hollow atomic state. It agrees within the experimental error-bars, for both the resonance energy and width, with the very recent high precision calculation using the saddle-point technique (31). The values we have determined from the profile of the partial cross sections for decay of the $2s^22p$ ^{2}P state into the $1s2s$ ^{3}S, $1s2p$ ^{3}P and $1s^2$ ^{1}S ionic states are 0.122(5)eV, 0.119(3) eV, and 0.118(3) eV, respectively. All these values determined in the photoelectron spectrometry work are in excellent agreement with each other. In Table II, we show a comparison between the various experi-

Table II. A comparison between the experimental results for the energy and lifetime of the $2s^22p$ ^{2}P^o state with a selection of theoretical results.

Experiment	Eo (eV)	Γ (eV)
Photoabsorption[16]	142.32(5)	0.20(6)
Photoion[17]	142.32(3)	0.14(3)
Photoion[18]	142.35(10)	0.15(2)*, 0.20(4)*
Photoelectron[19]	142.30(5)	0.20(4)
Photoelectron[20]	142.28(3)	0.118(3)
Theory	**Eo (eV)**	**Γ (eV)**
R-matrix[20]	142.18	0.13
Saddle point[31]	142.255	0.1175
MBPT[26]	142.661	0.13

* Photoion yield meas.   ** Li^{++} (Time of flight) meas.

mental results with the most recent theoretical results. The table shows immediately the advantage of the high spectral resolution experiments which succeeds in determining the profile width with an uncertainty sufficiently small to test the most recent theoretical calculations.

For Li-atoms in the $1s^22p$ ^2P excited state, the dipole selection rules allow the population of even-parity hollow states which are not accessible from the ground state. In addition, since the initial state is a ^2P state, three hollow states with different symmetries are accessible from the initial state: $2s2p^2$ $^2S^e$, $^2P^e$, and $^2D^e$. Only one experimental value was previously estimated in the electron-Li$^+$ collision experiment (15), the energy of a resonance tentatively attributed to the $2s2p^2$ ^2D state. We relied mainly on the results of the R-matrix calculations which predicted more accurately the energies of these hollow states and the partial cross sections for photoionization into various final ionic states of Li$^+$. As expected, the calculations predict the existence and provide the energy of three hollow states of even parity. Autoionization of these states into the Li$^+$ continua produces electron lines of various energies, according to, e.g., Li*** $2s2p^2$ $^2S^e$, $^2P^e$, $^2D^e$ → [$1s2p$ $^{1,3}P$ + εp] ^2S, ^2P, ^2D. We show the results of the R-matrix calculations for photoionization into the individual ^2P, ^2S, and ^2D channels of the $1s2p$ ^3P ionic state in the left part (three upper panels) of Figure 6, respectively (21). The theoretical energies of the three hollow states are listed in Table III. Adding the individual cross sections provides the partial cross section σ_i for photoionization into the $1s2p$ ^3P final ionic state as shown in the lower panel. The predicted profile of the three resonances is similar in the $1s2p$ ^1P channel, the ^2P resonance having the largest oscillator strength, followed by the ^2D. The ^2S resonance should appear only as a weak undulation

Table III. Energies and lifetimes of the $2s2p^2$ $^2S^e$, $^2P^e$, and $^2D^e$ hollow states of lithium

Experiment			Assignment	R-matrix[a]		Saddle-pointmethod[c]	
Energy (eV)	Lifetime[a] (meV)	q[a]		Energy (eV)	q	Energy (eV)	Lifetime (meV)
142.92(5)[a]	103(10)	-2.6(2)	$2s2p^2$ ^2D	142.76	-2.71	142.91	90
			$2s2p^2$ ^2S	144.63		144.63	64
145.08(5)[a]	47(5)	-2.1(2)	$2s2p^2$ ^2P	145.00	-2.09	145.07	48
145.5(1)[b]			$2s2p^2$ ^2D				

a) Ref. 21
b) Ref. 13
c) Ref. 32

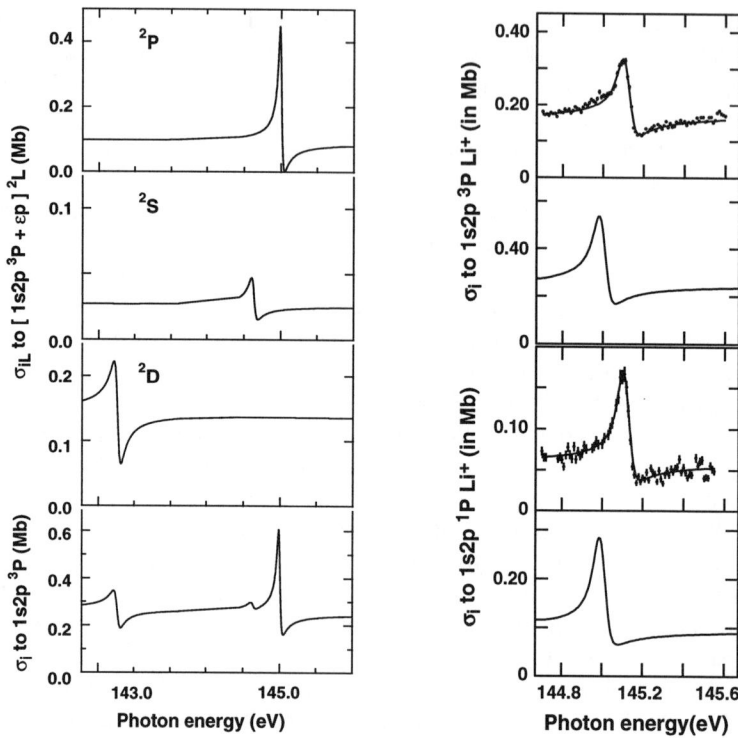

Figure 6. Left part: individual cross sections for photoionization of Li $1s^22p$ $^2P_{3/2}$ states into the $[1s2p\ ^3P + ep]^2P$, ^2S, and ^2D ionic states (three upper panels) and the sum (lowest panel), as obtained in the R-matrix calculations (from Ref. 21); right part: experimental and R-matrix results of cross sections for photoionization of the Li $1s^22p$ $^2P_{3/2}$ state into the $1s2p$ ^3P and ^1P states of the Li$^+$ ion at 145 eV (excitation energy of the $2s2p^2$ ^2P hollow states) (from Ref. 21).

in the σ_i partial cross section. We also show in Table III the experimental value of Müller et al. (15) and the results obtained by Chung and Gou (32) using the saddle-point method. Their results differ from the R-matrix results by about 0.1 eV. Photoion yield experiments on laser-excited lithium are also in progress (41).

After recording electron spectra outside the resonant energy range to determine the density of atoms laser-transferred into the $1s^22p$ $^2P_{3/2}$ excited state, we measured the partial cross sections for photoionization of the excited atoms into the $1s2p$ ^1P and ^3P final ionic states over the predicted energy range. The experimental results (21) for the autoionization of the ^2P resonances into the $1s2p$ $^{1,3}P$ final ionic states at 145 eV are shown in the right part (first and third panels) of Fig. 6. The error bars at each data point reflect only the magnitude of the statistical error, indicating the accuracy of the relative cross section measurements. To obtain the absolute cross section uncertainty, one has to add a 10% error for the determination of the density of excited atoms and a 25% error due to uncertainty in measuring the photoabsorption cross section (40) to which all cross sections are finally normalized. We show the results of the *ab initio* R-matrix calculations after convolution with the Gaussian monochromator function in the second and fourth panels. Experimental and theoretical results are in amazing agreement one with each other on a relative scale, providing confidence in the quality of the experimental data and R-matrix calculations. The calculated values for the hollow states differ on an absolute scale from the experimental values by 0.1 eV for the energies and by about 50% for the absolute values of the partial cross sections.

Like in Fig. 5, the solid lines in Fig. 6 are the results of our analytical fit. The expressions for partial photoionization cross sections established by Starace (39) in the case of one resonance interacting with several continua were convoluted with our Gaussian instrumental function. The resulting curves shown in Fig. 6 are the fits giving the lowest value of the χ^2 parameter. We were able from this fit to check the spectral resolution treated as a free parameter (39 ± 2 meV) and to determine the parameters of the resonance: Γ-width, E_0-energy and C_1- and C_2- Starace coefficients. By adding the partial cross sections into the 1s2p ^{3}P and ^{1}P final states and a flat cross section for photoionization into the n = 3 satellite states, we obtain the total photoabsorption cross section and determine the q-index of the resonances. Table III shows the present experimental values (21) together with results of both theoretical calculations. The measured energies of the 2s2p^2 ^{2}D^e and ^{2}P^e resonances are in excellent agreement with the values given by the saddle-point method (31, 32). The R-matrix values are slightly lower than the experimental values. The measured natural lifetimes of the 2s2p^2 ^{2}D and ^{2}P hollow states agree again very well with the results of the saddle-point method, as was also the case for the 2s^{2}2p ^{2}P state. The variation of the cross section induced by the existence of the 2s2p^2 ^{2}S state is too weak, as suggested by R-matrix calculations, to allow its observation in the present experiments, taking into account the limited beam time available.

THE HIGHER ENERGY HOLLOW STATES

The previous photoion experiments (17, 18) demonstrated a rich spectrum of multiply excited states in the 145-165 eV energy region, as is already clear in Fig. 1. The distribution of intensity among the observed resonances did not show any apparent regularity. However, due to cancellation effects particular resonances may sometimes be very weak in such experiments but be large if observed in a specific decay channel as in a photoelectron experiment; application of photoelectron spectroscopy to these higher lying resonances was therefore particularly advantageous. Even channels with very weak values of the resonant cross sections may show up strongly provided the corresponding background and the off-resonance signal are very low in this particular channel. In the upper panels of Figure 7, we show, as an example, the measured (20) cross sections for photoionization of neutral lithium atoms into the 1s3s ^{3}S (left part) and the 1s4ℓ ^{1,3}L channels (right part), respectively, in the energy range 149-154 eV range which encompasses many of the hollow atomic states previously observed in photoion experiments. The R-matrix calculations predict spectacular variations of the partial cross sections over this photon energy range. In the lower parts of Fig. 7, we show the *ab initio* theoretical calculations of partial crosss sections for photoionization into the (n = 3 ^{3}S) channel and in the (n = 4) channels (lower parts of the figure). The theoretical results have been convoluted with the instrumental function. Apart from an energy shift of 0.1 to 0.3 eV, the detailed agreement (on a relative cross-section basis) in the 1s3s ^{3}S ionic channel is again a tribute to both the quality of the experimental data and the R-matrix calculations. The profiles of the resonances are different in the different channels. For example, the 152.5 resonance predominantly decays into the n=3 ^{3}S channel whereas the 150 eV feature shows up more strongly in the n = 4 channels. The detailed structure in the neighborhood of the strong 152.5 resonance is accurately reflected in both experimental and theoretical curves, including the minimum at 152.55 eV. This pattern provides valuable clues to the principal configurations in the decaying resonances. However, there are additional weak and narrow resonances in the 1s2p ^{1,3}P channel (not shown here), which are not fully reproduced in the experimental data. The behavior of the par-

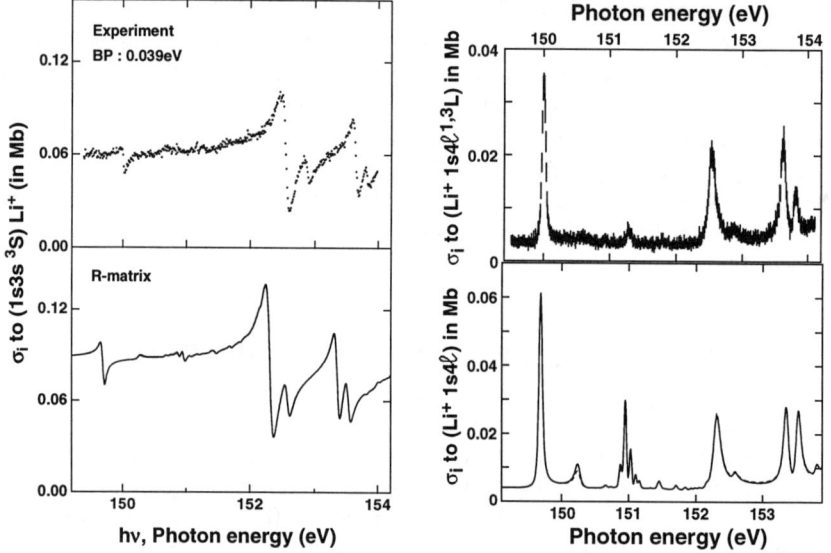

Figure 7. Experimental (upper parts) and R-matrix (lower parts) results of partial cross sections for photoionization of lithium atoms into the 1s3s ^{3}S (left) and 1s4l ^{1,3}L final ionic states (from Ref. 20).

tial cross section in the 1s4ℓ channels is quite different. At all resonances, the partial cross section is strongly enhanced, often with a quite symmetrical shape. This could be expected, since direct photoionization into the 1s4ℓ channels is very weak. Thus, almost no interference effects can occur, leaving a Lorentzian profile for the natural profile of the resonance. But the differences in the relative intensity of the partial cross section at the various resonances is something new, in particular the narrow and intense resonances near 151 eV excitation energy in the convoluted theoretical curve, while they are of weak intensity in the experimental results. This could be explained remembering that the number of Li$^+$ states included to represent the total wavefunction did not include any components of the 1s4ℓ type. In a way, the "inner box" used in the R-matrix calculations has too small a radius. The absolute cross-section values calculated by the R-matrix method are generally greater than the experimentally determined values. In Table IV, we give our energy measurements for the resonances observed in this energy range, taking into account our previously published data (20) and newly obtained results (22). The R-matrix calculated values of the various 2ℓ2ℓ' ionization thresholds of Li$^+$ are also given, in bold characters. One can note an overall agreement, within the error-bars, of our experimental results, with those determined in the high resolution (17) photoion yield experiment. The results obtained in the other photoion yield measurement (18) (capital letters A to H in the last column of the table) differ somewhat from these measurements, suggesting that the relative energy scale in this experimental work was not rigorously measured.

As already mentioned, a detail examination of partial cross sections for each resonance can provide valuable clues to the principal configurations in the decaying resonances. Care however needs to be taken in the analysis; the uniqueness of normal spectroscopic orbital labelling of the resonances (based on the independent-particle model) is questionable in view of the strongly correlated states involved in the description of the hollow atom resonances. In recent work on H$^-$ (35), it was shown that the autodetaching Feshbach resonances decay primarily into the nearest energetically accessible continuum channel, with similar angular and radial correlations. Here, we found that it is, indeed, the n = 4 channel in which the decay of the resonances is the strongest on a *relative* scale.

Fig. 4 already showed the existence of many more hollow states of lithium of the type nℓn'ℓ'n"ℓ". Our calculated values for the energies of these states are also given in Table IV, together with a proposed identification. They are compared to other previously determined theoretical and experimental values.

Our latest results strongly supports the identification of the main resonance at 152.5 eV as being mainly a 2p^{2}3p excited state, although other hollow states may also be present around this energy. This resonance (152.48 eV) is only a few hundredths of an eV above the 2s^2 ^{1}S threshold (152.43 eV) and can decay to this state of Li$^+$, decreasing the lifetime of the excited state and broadening the structure as it seen in Fig. 7. In the decay of this 2p^{2}3p excited state, the 3p electron can stay as a spectator, the autoionization process involving only the two 2p electrons in a

Table IV. Calculated energies and assignment of $n\ell n'\ell' n''\ell''$ ^2P hollow excited states in atomic lithium

Assignment	Calculated energy (eV)		Experimental (eV)
	R-matrix[a]	Saddle-point[b]	
$2s^2$ (^1S)$2p$ ^2P	142.10	142.25	142.28[a], 142.32[b], 142.34(A)[d]
($2p^2$ ^3P)$2p$	148.71		148.77[c], 148.7(B)[d]
$2s^2$ (^1S)$3p$	149.01	149.24	149.25[c]
($2s2p$ ^3P)$3s$	149.68	149.84	149.91[c], 149.79(C)[d]
($2s2p$ ^3P)$4p$	150.27		
($2s2p$ ^3P)$5p$	150.68		
($2s2p$ ^3P)$6p$	150.88		
($2s2p$ ^3P)$4s$	150.95		151.2[a], 151.20[c,e]
($2s2p$ ^3P)$5s$	151.45		
($2s2p$ ^3P)$6s$	151.71		
($2s2p$ ^3P)$7s$	151.85		
$2s^2$ ^1S			151.67[e]
($2s2p$ ^3P)$8s$	151.94		
$2s2p$ ^3P	152.19		152.43[a], 152.41[e]
($2s2p$ ^1P)$6s$	155.01		
($2s2p$ ^1P)$7s$	155.17		
$2p^2$ ^1D	155.16		155.33[e]
$2s2p$ ^1P	155.49		155.71[e]
($2p^2$ ^1S)$3p$	157.00		
($2p^2$ ^1S)$4p$	158.08		
($2p^2$ ^1S)$5p$	158.52		
($2p^2$ ^1S)$6p$	158.73		
($2p^2$ ^1S)$7p$	158.86		
$2p^2$ ^1S	159.19		159.17[e]

a) Refs. 19 - 21
b) Ref.16
c) Ref. 17
d) Ref. 18
e) Ref. 22

decay to the (n = 3 ^3P) channel. However, this kind of argument is limited in view of our earlier remarks on the differences in the orbitals for the different configurations.

Other interesting resonances have electronic configurations such as 2s2p3s, 2s2p3d, $2p^3$, $2p^2$3p. The $2p^3$ state (near 149 eV) is particularly interesting, because, although the cross section is small, it is expected to be one of the most highly correlated. However, its observation in photoelectron spectroscopy require photon beams with a much higher brightness and longer beam times. The profile of these resonances, including angle-resolved data, has been recently measured (22).

In order to make a comparison between our new results and the earlier photoion work, we have summed the different ^{2S+1}L decay spectra into the n = 2, 3 and 4 channels, as shown in the the left pannel of Figure 8. This provides a good esti-

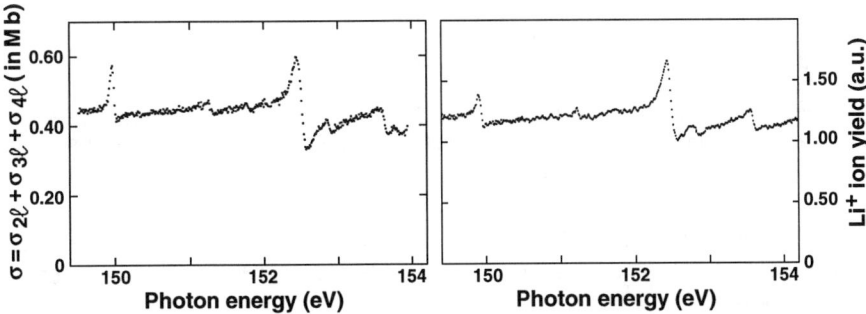

Figure 8. Photoionization cross section as determined in the electron spectrometry (left panel, from Ref. 20) and photoion yield Li+ ionic continua (right panel, from Ref. 17) experiments.

mate of the total photoabsorption cross-section in this spectral region as decay to higher channels is expected to be very weak. The ion yield results (17) are shown in the right panel of Fig. 8. Our total cross section is in good agreement with the measurements of the photoion yield in the Li+ continua, except in a small energy range above 152.5 eV.

For hollow atoms produced by triple excitation of laser-excited lithium atoms over the higher photon energy part of the explored energy range, we present an example of the experimental partial cross section into the 1s3p $^{1,3}P$ satellite states between 151 and 152 eV, where our R-matrix calculations predicted significant variation of this cross section. Our results are shown in the left panel of Figure 9, together with the results of the convoluted R-matrix calculations as shown in the right panel of the figure. Both sets of results are in excellent agreement one with each other on a relative scale. Our experimental data confirm the existence of two strong resonances followed immediately by a deep minimum in the cross section. The two resonances are so close in energy that they interact with each other, preventing experimental determination of accurate energy values. One can, however, say that they are observed in the predicted energy region. The assignment given to these resonances in the calculations is $2p^23d$ $^2P^e$ (at 151. 10 eV) and $2p^23d$ $^2D^e$ (at 151. 23 eV), respectively.

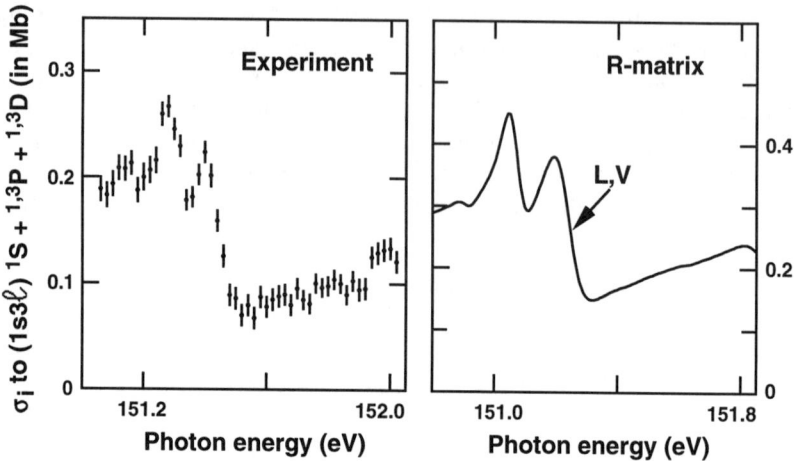

Figure 9. Measurements and R-matrix calculations of the cross section for photoionization of the Li $1s^22p\ ^2P_{3/2}$ state into the $1s3p\ ^{1,3}P$ states of the Li^+ ion near 151 eV excitation energy (from Ref. 21).

As already mentioned for atoms in the ground state, one should keep in mind that the 2s and 2p orbitals in the hollow states are quite different from the similarly labelled orbitals in the initial or final ionic states. For the highest resonances, the description of these autoionizing states using one-electron orbitals was considered as even more questionable, because every hollow state was described by the mixture of an increasing number of electronic configurations (17,18). Even though the one-electron labeling is kept throughout this paper, its purpose is to indicate the dominant configuration involved in the description of the hollow state. However, we will see in the next paragraph that some of the highly excited states can be assigned to Rydberg series in which the two core-excited electrons keep the same electronic configuration and symmetry, while the third electron is well described by a one-electron configuration with increasing values of the principal quantum number n within a Rydberg series.

RYDBERG SERIES OF HOLLOW STATES

Many excited states in Table IV can be classified according to several Rydberg series such as: $1s^22s\ ^2S_{1/2} + h\nu \rightarrow (2s^2\ ^1S)np$, $(2s2p\ ^3P)ns$, $(2p^2\ ^1S)np$, $(2s2p\ ^1P)ns$, $(2p^2\ ^3P)np$, and $(2s3s\ ^3S)np$. They are piling up to the respective thresholds, as was already suggested by preliminary results (20) obtained below 152 eV, even though the variation of the relative intensities within the first terms of each series can be modified by the strong perturbation introduced by the creation of two holes in the 1s shell. This is in contradiction with Azuma et al. (18, 41) who stated that the concept of Rydberg series totally disappears in the triply excited lithium spectrum. Three main series emerge from the R-matrix calculations, $(2s^2\ ^1S)np$, $(2s2p\ ^3P)ns$ and $(2p^2\ ^1S^o)np$, the first two having a 2s electron in their electronic configuration. In helium, there are also three series with largely different intensities. Azuma et al. (18) attributed the perturbation of the regular Rydberg series to a strong contraction of the outer orbitals in the hollow states. However this argument should be used with caution and only on a qualitative basis. Actual calculations of the dipole matrix elements with the *same* set of wavefunctions for the initial and final states, as in the R-matrix calculations, are necessary, before any quantitative conclusion can be drawn. In most recent, yet unpublished experiments (22), we have observed quite strong Rydberg series by measuring the non radiative decay of doubly excited states of the Li$^+$ ion into the ground state of the Li^{2+} ion, according to:

Li: $1s^2 2s\ ^2S \rightarrow$ Li*** $[(2p^2\ ^1S)np\ ^2P] \rightarrow$ Li$^+$: $2p^2\ ^1D$ (or $2s2p\ ^1P$) + e_1^-
Li$^+$: $2p^2\ ^1D$ ($2s2p\ ^1P$) $\rightarrow$ Li^{2+}: $1s\ ^2S + e_2^-$

The e_1^- electron has a very low kinetic energy and was not detected in the experiment. The e_2^- electron has a high kinetic energy and its intensity was measured as a function of the photon energy, reflecting the behavior of the cross section in the first step of the autoionization process, since the radiative decay of Li$^+$ excited states is negligible. The corresponding series is shown in Figure 10. Even though the crosss section is only in the kilobarn range, the very low background in this

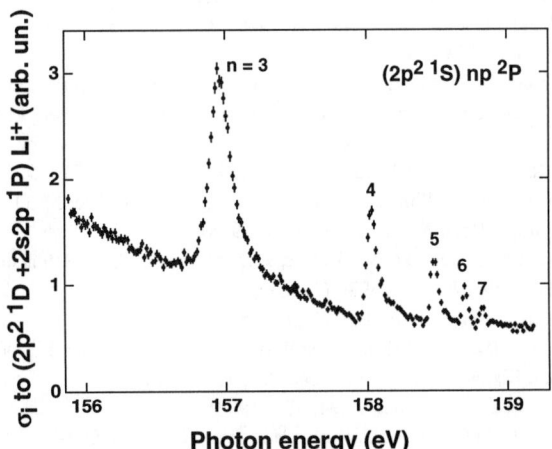

Figure 10. First measurements of the Rydberg series $[(2p^2\ ^1S)np]\ ^2P$ as observed in the decay of the $2p^2\ ^1D$ and $2s2p\ ^1P$ Li$^+$ doubly excited Li$^+$ states into the $1s\ ^2S$ Li^{2+} continuum.

channel allows us to clearly observ five terms of this Rydberg series. The measured energy of each term belonging to the $(2p^2\ {}^1S)np\ {}^2P$ series shows that all terms have the same quantum defect (22).

To conclude, we have experimentally observed and measured odd and even-parity hollow lithium states over an extended energy range. The saddle-point method proved to be very efficient in theoretically predicting the energy and the lifetime of the new hollow states observed in this work, while the R-matrix calculations give excellent agreement, on a relative scale, with the measured partial cross sections.

ACKNOWLEDGEMENTS

The support of the Centres d'Etudes Nucléaires de Limeil-Valenton (CEA-DAM), the staff and management of Advanced Light Source, and EU Human Capital and Mobility under contract 93-0361 is gratefully acknowledged. The authors are very grateful to E. T. Kennedy for helpful discussions and critical reading of the manuscript. They like also to thank their other collaborators helping in taking the data: N. Berrah, T. J. Morgan, L. Journel, B. Rouvellou, S. Al Moussalami, C. Blancard, J. Bozek, and A. S. Schlachter. They are also very grateful to L. VoKy and A. Hibbert for calculating, on request, the energy of many hollow states and partial photoionization cross sections.

REFERENCES

1. Silvermann, S., and Lassettre, E., *J. Chem. Phys.* **40**, 1265 (1964).
2. Madden, R. P., and Codling, K., *Phys. Rev. Lett.* **10**, 516 (1963).
3. Fano, U., *Phys. Rev.* **124**, 1866 (1961).
4. Domke, M., Xue, C., Puschmann, A., Mandel, T., Hudson, E., Shirley, D. A., Kaindl, G., Green, C. H., Sadeghpour, H. R., and Petersen, H., *Phys. Rev. Lett.* **66**, 1306 (1991).
5. Domke, M., Remmers, G., and Kaindl, G., *Phys. Rev. Lett.* **69**, 1171 (1992).
6. Menzel, A., Frigo, S. P., Whitfield, S. B., Caldwell, C. D., Krause, M. O., Tang, J. Z., and Shimamura, I. *Phys. Rev. Lett.* **75**, 1479 (1995).
7. Schulz, K., Kaindl, G., Domke, M., Bozek, J. D., Heimann, P. A., Schlachter, A. S., and Rost, J. M., *Phys. Rev. Lett.* **77**, 3086 (1996).
8. Donets, E. D., *Nucl. Instr. Meth. B* **9**, 522 (1985).
9. Briand, J. P., de Billy, L., Charles, P., Essabaa, S., Briand, P., Geller, R., Desclaux, J. P., Bliman, S., and Ristori, C., *Phys. Rev. Lett.* **65** , 159 (1990).
10. Schneider, D., et al. *Surf. Science* **294**, 403 (1993).
11. Aumayr, F., and Winter, H., *Com. Atom. Mol. Phys.* **29**, 275 (1994).
12. Morgenstern, R., and Daas, J., *Europhys. News* **25**, 3 (1994).
13. Bruch, R., Paul, G., Andra, J., and Lipsky, L., *Phys. Rev. A* **12**, 1808 (1975).
14. Rodbro, M., Bruch, R., and Bisgaard, P., *J. Phys. B* **12**, 2413 (1979).

15. Müller, A., Hoffmann, G., Weissbecker, B., Stenke, M., Tinschert, K., Wagner, M., and Salzborn, E. *Phys. Rev. Lett.* **63**, 758 (1989).
16. Kiernan, L. M., Kennedy, E. T., Mosnier, J.-P., Costello, J. T., and Sonntag, B. F., *Phys. Rev. Lett .* **72**, 2359 (1994).
17. Kiernan, L. M., Lee, M.-K., Sonntag, B. F., Sladeczek, P., Zimmermann, P., Kennedy, E.T., Mosnier, J.-P., and Costello, J. T., *J. Phys. B* **28**, L161 (1995).
18. Azuma, Y., Hasegawa, S., Koike, F., Kutluk, G., Nagata, T., Shigemasa, E., Yagishita, A., and Sellin, I., *Phys. Rev. Lett.* **74**, 3768 (1995).
19. Journel, L., Cubaynes, D., Bizau, J.-M., Al Moussalami, S., Rouvellou, B., Wuilleumier, F. J., VoKy, L., Faucher, P., and Hibbert, A., *Phys. Rev. Lett.* **76**, 30 (1996).
20. Diehl, S., Cubaynes, D., Bizau, J.-M., Journel, L., Rouvellou, B., Al Moussalami, S., Wuilleumier, F. J., Kennedy, E. T., Berrah, N., Blancard, C., Morgan, T. J., Bozek, J., Schlachter, A. S., VoKy, L., Faucher, P., and Hibbert, A., *Phys. Rev. Lett.* **76**, 3915 (1996).
21. Cubaynes, D., Diehl, S., Journel, L., Rouvellou, B., Bizau, J.-M., Al Moussalami, S., Wuilleumier, F. J., Berrah, N., VoKy, L., Faucher, P., Hibbert, A., Blancard, C., Kennedy, E. T., Morgan, T. J., Bozek, J., and Schlachter, A. S., *Phys. Rev. Lett.* **77**, 2194 (1996).
22. Diehl, S., et al. , to be published.
23. Ahmed, M., and Lipsky, L., *Phys. Rev. A* **12**, 1176 (1975).
24. Nocolaides, C. A., and Beck, D. R., *J. Chem. Phys.* **66**, 1982 (1977).
25. Safronova, U. I., and Senashenko, V. S., *J. Phys. B* **11**, 2672 (1978).
26. Simmons, R., Kelly, H. P., and Bruch, R., *Phys. Rev. A* **19**, 682 (1979).
27. Chung, K. T., *Phys.Rev. A* **23**, 1596 (1981); *ibid.* **24**, 1350 (1981); *ibid.* **25**, 1596 (1982).
28. Greene, C. H., and Clark, C. W., *Phys. Rev. A* **30**, 2161 (1984).
29. Conneely, M. J., Lipsky, L., and Russek, A., *Phys. Rev. A* **46**, 4012 (1992).
30. Nicolaides, C., *J. Phys. B* **25**, L91 (1992).
31. Chung, K. T., and Gou, B. C., *Phys.Rev. A* **52**, 3669 (1995).
32. K.T. Chung, K. T., and B.C. Gou, B. C., *Phys.Rev. A* **53**, 2189 (1996)
33. Yang, X., Bao, C. G., and Lin, C. D., *Phys. Rev. Lett.* **76**, 3096 (1996).
34. Herrick, D. R., and Sinagoglu, O., *Phys. Rev. A* **11**, 93 (1975).
35. Sadeghpour, H. R., and Shinagoglu, O., *Phys. Rev. A* **45**, 1587 (1992).
36. Bizau, J.-M., Wuilleumier, F. J., Ederer, D. L., Keller, J.-C., LeGouët, J.-L., Picqué, J. L., Carré, B., and Koch, P., *Phys. Rev. Lett.* **55**, 1281 (1985)
37. Cubaynes, D., Bizau, J.-M., Wuilleumier, F. J., Carré, B., and Gounand, F., *Phys. Rev. Lett.* **63**, 2460 (1989)
38. Rouvellou, B., Cubaynes, D., Journel, L., Bizau, J.-M., Pahler, M., Wuilleumier, F. J., VoKy, L., Faucher, P., Morgan, T. J., Marinelli, C., Berrah, N., and Schlachter, A. S., *Phys. Rev. Lett.* **75**, 33 (1995).
39. Starace, A., *Phys. Rev. A* **16**, 231 (1977).
40. Mehlman, G., Cooper, J. W., and Saloman, E. B., *Phys.Rev. A* **25**, 2113 (1982)
41. Azuma, Y., Hagasawa, S., Koike, F., Kutluk, G., Nagata, T., Shigemasa, E., Yagishita, A., and Sellin, I., in *Proceedings of the Oji Seminar on Atomic and Molecular Photoionization*, edited by A. Yagishita and T. Sasaki (Universal AC Press, Tokyo, Japan, 1996), pp. 253-262.

Extending Synchrotron–Based Atomic Physics Experiments into the Hard X-Ray Region

T. LeBrun

Physics Division, Argonne National Laboratory, Argonne, Illinois 60439, USA

Abstract. The high–brightness, hard x-ray beams available from third–generation synchrotron sources are opening new opportunities to study the deepest inner shells of atoms, an area where little work has been done and phenomena not observed in less tightly bound inner-shells are manifested. In addition scattering processes which are weak at lower energies become important, providing another tool to investigate atomic structure as well as an opportunity to study photon/atom interactions beyond photoabsorption. In this contribution we discuss some of the issues related to extending synchrotron–based atomic physics experiments into the hard x-ray region from the physical and the experimental point of view. We close with a discussion of a technique, resonant Raman scattering, that may prove invaluable in determining the spectra of the very highly–excited states resulting from the excitation of deep inner shells.

Introduction

Synchrotron sources are evolving rapidly and driving changes in x-ray optics and experimental techniques that are transforming the capabilities of synchrotron–based experiments. The very bright and energetic x-ray beams available at third-generation sources allow the wide array of measurements performed using VUV photoexcitation to be extended into the x-ray region where new phenomena are expected. In addition, the range of high photon energies includes a domain where scattering processes compete strongly with and even dominate photoabsorption, shifting focus for atomic physics experiments to include photon/atom scattering in the hard x-ray region. The photon beams from third generation sources such as the APS, the ESRF and Spring-8 are well suited to such studies because the high photon energies open the deepest inner-shells of the periodic table to investigation and the high brightness compensates for the small interaction cross sections while allowing more detailed investigations via highly differential measurements and coincidence techniques.

The structure and relaxation dynamics of the deepest inner shells has received much less attention than those shells accessible in the VUV, and this contribution discusses some of the physical and instrumental implications of the shift from low energy to high energy synchrotron-based atomic physics experiments. The author would like to take this opportunity to recognize Paul Cowan's contributions as a scientist, collaborator and a friend. All of the results discussed here were taken at beamline X24A at the NSLS, a facility whose great productivity testifies to Paul's ongoing contributions to many areas of synchrotron-based research.

Background

Valence shell atomic spectra have been actively studied for more than a century and have made essential contributions to such advances as the development of the quantum theory, quantum electrodynamics, and the understanding of atomic structure. While studies of inner-shell atomic spectra also began early, it was the introduction of tunable synchrotron-based light sources in the VUV and x-ray region of the electromagnetic spectrum that sparked development of the field.

The wide range and tunability of the available photon energies prompted a new generation of experiments probing the behavior of atoms and molecules at excitation energies above, and well beyond the ionization threshold. These synchrotron-based experiments were unique because the tunability and intensity of synchrotron radiation allowed quasi-discrete states near ionization thresholds to be selectively excited and their decay studied. These studies revolutionized our understanding of the structure and dynamics of atomic inner shells, illuminating such phenomena as the photoelectric effect, the interaction of discrete and continuum states (autoionization), giant resonances in scattering continua, and correlations in strongly interacting many body systems.

But only half of the subshells of neutral atoms have excitation energies less than one keV. Higher photon energies probe more tightly bound shells where the strong nuclear potential increases the importance of relativistic effects (1) and K-shell electrons may be sensitive to nuclear size effects or couple to nuclear magnetic moments — effects visible in x-ray spectra (2). When one also considers systems such as the rich variety of negative ions which have their transitions shifted to even higher energies, the importance of x-rays to atomic structure studies becomes evident. Thus far, however, the exploration of resonant excitation of atoms by hard x-rays remains in its infancy.

X-ray Interactions

We will begin by briefly discussing the characteristics and magnitudes of the x-ray/atom interaction processes which are dominant for photon energies up to 150 keV, and then turn to experimental implications. The upper limit of 150 keV has been selected here to include the absorption thresholds of the heaviest stable elements (e.g. U 1s = 115 keV), as well as higher energies where multiple core-hole states far above thresholds are found and where Compton and Rayleigh scattering can dominate the attenuation cross section. Note that this range of energies is now fully accessible at currently operating synchrotron sources by using wigglers and wavelength shifters. Indeed, Compton scattering measurements have already been performed at 1 MeV photon energy by Suortti and coworkers at the ESRF (3). For more information on these scattering processes the reader is referred to recent reviews by Crasemann (4) and Pratt et. al. (5).

The interaction of a photon with an atom is traditionally distinguished by whether the photon is absorbed (photoabsorption), elastically scattered (also called Rayleigh or coherent scattering), or inelastically scattered (Compton or incoherent scattering). While these are convenient descriptions, they are not unambiguous because the absorption and re-emission of a photon is indistinguishable from scattering in experiments which are not time-resolved. As a result some processes which involve resonant photoexcitation can fall under more than one of these classifications, as we will show later. At the photon energies of interest here, scattering from electrons dominates atom/x-ray interactions so processes which are important at higher photon energies such as pair production and nuclear scattering will be neglected in our discussion for the sake of brevity. For discussions of nuclear scattering see the contributions by Smirnov, and Kikuta in these proceedings.

Concentrating on electronic scattering, we then have the following processes to consider:

- Photoabsorption (followed by Auger or fluorescent decay)

$$\gamma + A_0 \to \begin{Bmatrix} A_i^* \\ A_{i'}^* + e_{ph}^- \end{Bmatrix} \to \begin{Bmatrix} A_f^* + e_A^- \\ A_{f'}^* + \gamma \end{Bmatrix} + (e_{ph}^-)$$

Here the first set of braces indicates the highly-excited intermediate state that follows photoexcitation: a quasi-discrete state for resonant excitation (top row), or a state with a continuum of energy eigenvalues for photoionization (bottom row). The second set of braces shows the two competing decay mechanisms for the intermediate state: Auger decay (top) or fluorescence decay (bottom).

Photoabsorption is the dominant attenuation process at low to moderate x-ray energies: it exceeds elastic scattering by 3-4 orders of magnitude near the K threshold of light atoms like carbon and is 1-2 orders of magnitude stronger than elastic scattering near the K thresholds of heavier elements like lead. It is also a clean and selective form of excitation because the energy and angular momentum transferred to the target are well known when the dipole approximation is satisfied, however the widely–used dipole approximation must be applied cautiously in the x-ray region. Recent experimental results (6, 7) have verified earlier predictions that show photoelectron angular distributions measured at high photon energy or near threshold to be sensitive to interference between dipole and higher-order multipole transition amplitudes. These studies are just beginning, and measurements of non-dipole effects are likely to be an active area of research at high energy synchrotron sources (8).

- Elastic = Rayleigh scattering

$$\gamma + A_0 \to A_0 + \gamma$$

Elastic scattering from atoms is also called Rayleigh scattering[1], or coherent scattering because the scattered radiation has the same wavelength as the incident

---

[1] Note that in contrast to Rayleigh scattering, Thompson scattering is elastic scattering for the idealized case of a single free charged particle, such as an electron or nucleus, when the recoil energy of the particle is ignored. Inclusion of the recoil kinematics leads to a description of Compton scattering which reduces to the Thompson formula at low photon energy where the recoil energy is small.

radiation as well as a definite phase shift. Interference effects such as diffraction are therefore observed in elastic scattering. Because diffraction is such an important technique for structure determination, cross sections and angular distributions for elastic scattering are important to applications in structural biology and materials science. The data for these applications must cover a wide range of chemical elements and photon energies, and tables of calculated values are widely used. Away from threshold, currently available theoretical results compare well with experiment and serve the needs of experiments using traditional fixed–energy sources. However the tunability of synchrotron radiation allows crystallographers to obtain more information using scattering near thresholds (e.g. information regarding the phase of the structure factor) — an area where measurements to test theory are still needed. Indeed even the most complete calculations frequently use the independent particle approximation (9), which is expected to be less reliable near thresholds. The need for testing theory near threshold is striking in light of the rapid evolution of structural biology at synchrotron sources, and provides a good opportunity for synchrotron-based atomic and molecular science to make an important contribution while demonstrating it's role as a source of fundamental data for a wide range of fields of research.

- Inelastic = Compton scattering + Raman scattering

$$\gamma + A_0 \rightarrow A_f^* + \gamma' + e^-$$

Compton scattering is photon scattering from a free electron, and in the range of photon energies we are considering it is the basic interaction process between an electron and a (real) photon. In the center-of-mass frame Compton scattering is just elastic scattering, but the transformation to laboratory coordinates gives rise to Doppler shifts and results in the well known characteristics of Compton scattering (10), for example the Compton shift of the *wavelength* of the scattered photon which depends only on the scattering angle and is independent of the incident photon energy.[2]

In scattering from atoms the electrons are of course not free and the effects of binding modifies Compton scattering in a number of ways. The cross section at low energy is reduced and the spectrum of scattered energies is broadened and slightly shifted. While simple kinematic approaches to binding effects serve well in many cases, the recent availability of second-order S matrix calculations (11) has raised the bar for synchrotron-based experiments seeking to study rigorously Compton scattering.

Atomic structure is also manifested in inelastic channels which can be described as Compton scattering. For example, photoionization followed by fluorescence decay can be considered an inelastic scattering process, indeed the two are indistinguishable in an experiment that can not resolve the core-hole lifetime (12). Because this process corresponds to inelastic scattering with a free electron in the final state, it can clearly be described as a form of Compton scattering, however it is also a form of x-ray Raman scattering — the term which

---

[2] The Compton shift at 90° is equal to 0.0243 Å, which is the wavelength of a photon whose energy equals mc^2. Note that the change in *energy* of the Compton-scattered photon is not independent of the incident photon energy because energy is inversely proportional to wavelength.

appears to be preeminent among the synchrotron research community. For more information on Compton scattering, as well as the relation between Compton and Raman scattering see Heitler (13), or the articles by Åberg (14) or Bergstrom (11).

Experimental Implications

As we have seen, photoabsorption dominates photon-atom interactions at energies near absorption thresholds. This implies that for light elements photoabsorption is most important at low energies (up to 20 keV for carbon) while it contributes strongly to attenuation in heavier elements over the whole range of photon energies considered here (1-150 keV). We henceforth concentrate on experiments that use the photoeffect and the subsequent decay processes to study inner-shell structure. To illustrate the experimental consequences of doing such experiments with harder x-rays we will compare cross sections and natural widths for different elements, referencing the values to the threshold energy instead of Z because most studies at synchrotrons are concerned with near-threshold excitation. Figure 1 illustrates the variation of cross sections at threshold for photoabsorption, Auger decay, and fluorescent decay as a function of the energy of the K and L_3 absorption edges of the elements.

The lower left figure shows the magnitude of the K shell photoabsorption cross section for the elements neon to thorium (Z = 10-90) at the K edge (15). The cross section for K-shell ionization at threshold decreases by two orders of magnitude in going from Ne to Th, and this highlights the need for very intense x-ray sources in probing the deepest shells of elements. If the photoabsorption cross

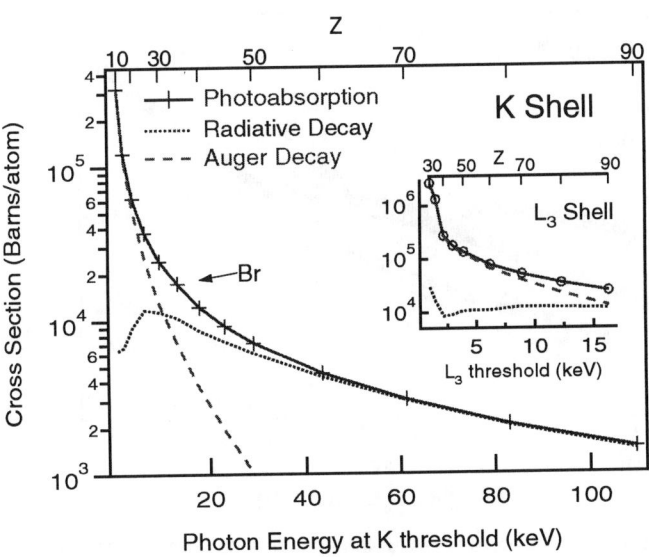

FIGURE 1. Cross sections at threshold for photoabsorption (solid line with crosses), Auger decay (dashes), and fluorescent decay (dots) for the K shell and L_3 shell (inset) of the elements for Z = 10 to 90.

sections are then multiplied by the radiative and nonradiative yields (16), one obtains the cross sections for fluorescence and Auger relaxation following K–shell ionization. The plot shows that the intensity of Auger decay from a K–hole falls much *more* rapidly than the photoabsorption cross section as Z is increased, while the intensity of K fluorescence varies *less* rapidly than the photoabsorption cross section. Therefore, though Auger decay is typically the dominant relaxation mechanism at low energy, radiative decay is stronger for K core-holes beyond 10 keV binding energy.

As a result, Auger spectroscopy of K–hole relaxation in heavy elements poses a challenge because of the declining cross sections and the experimental difficulties associated with the required high fields and accompanying low transmission of the electron spectrometers. While the small spot sizes available from third-generation synchrotron sources may help alleviate these problems, the strength of the radiative decay channels provides a complimentary approach which, combined with advances in high resolution x-ray spectroscopy, increases the potential for application of x-ray spectroscopy to atomic physics experiments at synchrotron facilities.

The fall–off of the Auger decay intensity is not as dramatic for other shells as shown in the inset where analogous data for the L_3 subshells is presented. Here Auger decay remains strong, but still decreases slightly faster than photoionization while fluorescence varies relatively little.

However, in determining the feasibility of experiments, intensity needs to be balanced with the required resolution. Some factors related to experimental resolution are summarized in figure 2 where the natural widths of core holes in different subshells of the elements sodium to uranium (17) are plotted versus the threshold energy of the subshell. The natural width of K holes varies from 0.3 eV

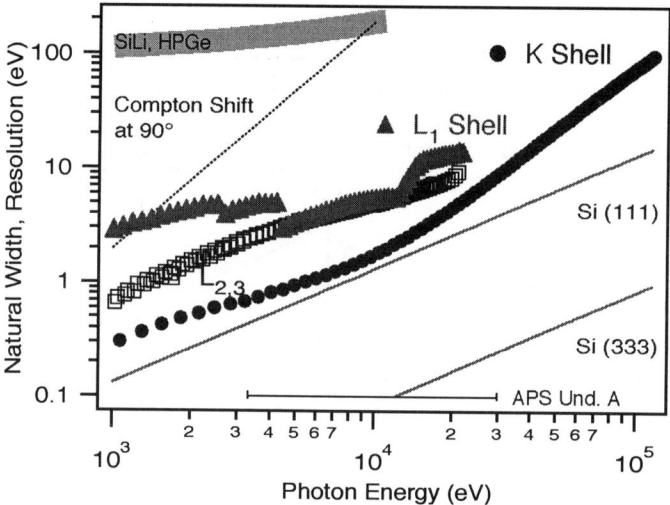

FIGURE 2. Natural widths of core holes for the elements sodium to uranium (Z=11 – 92) plotted versus the ionization energy of the core hole. The theoretical resolution of selected crystal reflections and detectors are included for comparison. The other quantities in the plot are explained in the text.

for Na to 96.1 eV for U, while the L_1 natural width of U is roughly 14 eV. When these widths are compared to the resolution required to study resonance or Rydberg structure near threshold (less than 1 eV) it is clear that most of the near-edge features will be washed out by the natural width. As we shall discuss in the last section, resonant Raman spectroscopy may provide a route to studying near-edge structure with an effective resolution that is narrower than the lifetime width, so we include that possibility in this discussion of instrumental requirements.

For experiments such as photoabsorption and nonresonant photoelectron spectroscopy, a bandpass that matches the core-hole lifetime width or is slightly narrower is optimal for most experiments. The workhorse of many double crystal monochromators is the 111 reflection of silicon, and it is fortunate that the theoretical resolution of Si(111) is reasonably well matched to the lifetime widths (see Fig. 2), although flux–limited experiments in the L shells or high-Z K shells may benefit from more reflective crystals such as Ge (111).

In order to use resonant Raman scattering to resolve structure near threshold, spectral widths narrower than lifetime widths will be required. Si(111) will serve for L shell measurements at intermediate energies, but if the technique lives up to its promise, higher-order reflections (intrinsic resolution $\Delta E/E = 10^{-5}$ or better) and high resolution post-monochromators will ultimately become necessary. These target values for the resolution also set tough standards for x-ray and electron spectrometers, where the transmission falls rapidly for high resolution and high analyzing energy. Fortunately the L shell threshold energies for high Z elements fall within the range of energies available from the undulator A at the Advanced Photon Source, indicating that insertion devices at third-generation sources can provide the higher intensities needed in the required range of photon energies.

Finally, in relation to scattering studies note that the Compton shift at 90° is generally large compared to the natural width for energies beyond a few keV, so high purity germanium or Si(Li) detectors can resolve Compton and Rayleigh scattering above approximately 10 keV. Such detectors can also be very useful in coincidence experiments which only need to distinguish relaxation channels between different subshells.

Resonant Raman Scattering

It is clear from the above considerations that lifetime broadening will limit efforts to study structure near absorption edges of deep inner-shells. However resonant Raman scattering has recently emerged as a promising technique to address this limitation. This stems from the fact that the natural width of the initial core hole does not explicitly contribute to the linewidths of fluorescence or Auger lines following resonant excitation, thus resonant peaks are narrower than peaks resulting from ionization. This is exploited in resonant inelastic x-ray scattering to study the short-lived intermediate states that result from core excitation. The energy loss of the scattered x-ray is characteristic of the final state, but the intensity of the emitted radiation also depends on the transition amplitude to the core-excited state. The intermediate density of states is in no way changed of course, but contributions from different intermediate states can be separated *if interference between different intermediate states is neglected*. This assumption is not generally satisfied, but if the manifestations of the interference can be well–understood the technique may prove uniquely powerful.

Resonant x-ray Raman scattering was first observed by Sparks (18) using a Cu Kα source to irradiate a variety of targets, and subsequently investigated using synchrotron radiation (19-22). The field benefited from early theoretical developments by Tulkki and Åberg (14, 23-27) who described the process in terms of the resonant scattering theory and also derived a simplified model which continues to provide valuable guidance to experimentalists. The Auger analog to this process was first observed by Brown et. al. (28, 29), work which has been significantly expanded upon recently (30-34).

The aspects of *resonant* Raman scattering that make it potentially unique for spectroscopy are the narrow spectral widths of the emitted radiation (photons or electrons) and the dispersion of the emitted energy with varying incident photon energy. The former characteristic effectively enhances the resolution while the latter allows discrete states to be distinguished from continuum states. These characteristics are simple consequences of energy conservation, and discussed elsewhere (35, 36). Here we simply note that the important point is whether one or two bodies share energy with the atom (ion) in the final state, and in this context the reasoning is the same whether the bodies are electrons or photons. The features of radiative and nonradiative Raman scattering are compared in Table 1.

These characteristics are best illustrated with a spectrum, for which we use an Auger Raman spectrum of the K-shell absorption edge in argon. To avoid the complicated forest of lines usually present in Auger spectra we select a single line that dominates the KLL spectrum, the L_2L_3 $^1D_2$ transition (37), and follow it's evolution as the photon energy is scanned across the K threshold.

Table 1. Characteristics of X-ray and Auger Resonant Raman Scattering

Resonant Excitation X-Ray Raman	Auger Raman
• Energy of emitted x-ray disperses nearly 1:1 with incident photon energy.	• Energy of emitted electron disperses nearly 1:1 with incident photon energy.
• Spectral width of emitted x-ray includes final state lifetime and incident photon bandpass.	• Spectral width of emitted electron includes final state lifetime and incident photon bandpass.
• Resonant peak maximum near characteristic fluorescence energy.	• Resonant peak maximum greater than diagram energy due to higher charge in final state and PCI.

Ionization X-Ray Raman	Auger Raman
• Well above threshold, energy of characteristic fluorescence is non-dispersive and the spectral width includes the natural widths of the intermediate and final states, but not the incident photon bandpass.	• Well above threshold, energy of the diagram line is constant and the spectral width includes the natural widths of the intermediate and final states, but not the incident photon bandpass.
• Near threshold, the fluorescence peak is truncated on the high-energy side due to conservation of energy.	• Near threshold, the diagram peak is broadened and asymmetric due to PCI, although E. cons. truncation effects should be present too.
• Centroid of fluorescence peak moves due to varying cut off from E. cons.	• Peak maximum shifted due to PCI, varying cut-off expected also.
• Observable well below threshold.	• May not contribute strongly below threshold due to PCI recapture.

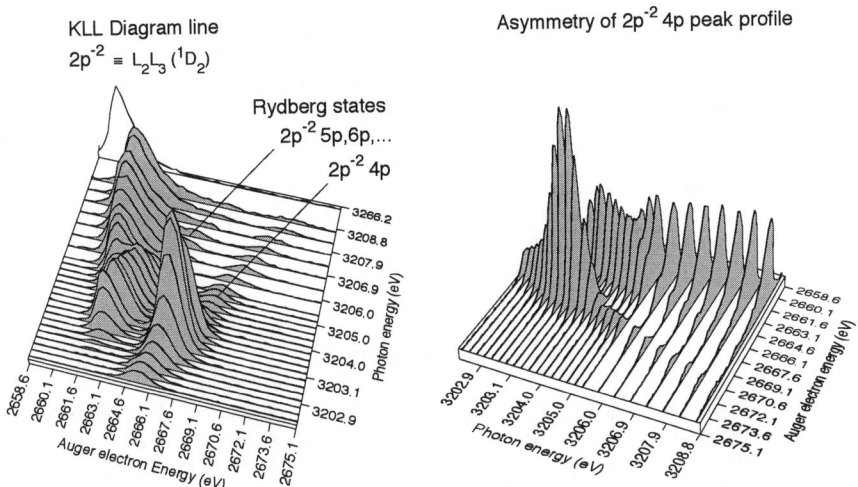

Figure 3. Auger resonant Raman spectrum of the K edge of argon.

The data presented in figure 3 is a series of electron spectra of the L_2L_3 $^1D_2$ peak in the KLL auger spectrum as the photon energy is scanned across the K absorption threshold at 3206.26 eV (38). The photon energy and electron energy scales have been set using the values of Asplund *et. al.* (37) and Breinig *et. al.* (38) respectively. The individual Auger spectra have been background subtracted and normalized to the total LMM electron yield. No smoothing is applied.

The $^1D_2$ diagram line measured 60 eV above threshold is shown at the rear of the left-hand spectrum. In comparison, the line observed just above threshold is broadened, asymmetric and shifted in energy due to post-collision interaction effects (PCI). Below threshold, we observe two peaks (indicated by straight lines) resulting from resonant excitation to Rydberg states instead of ionization. The 4p is well resolved and shows intensity well beyond threshold, while the higher Rydbergs are not well-resolved from each other and blend into the onset of the continuum line. The 4p peak is much more separated from the higher Rydbergs than in the absorption spectrum due to the resonant Auger energy shift which results from the change in nuclear charge between the intermediate and final states(39). This increases the effective resolution of resonant Auger Raman scattering relative to resonant x-ray Raman scattering.

A simple spectator model would assume that the initially excited electron stays in the same Rydberg orbital. If so, the resonant peaks in these spectra (which show how the excitation to the *final* states changes as a function of the incident photon energy) would also tell us how the excitation to the intermediate states varies — thus giving the intermediate density of states with high resolution. However, while the spectator model is very useful for interpretation, it is not suitable for quantitative analysis as can be seen in the right-hand spectrum in Fig. 3. Assuming spectator relaxation we would expect the profile of the 4p peak as a function of photon energy to be a symmetric Voigt shape. However the peak is strongly asymmetric due to contributions from intermediate states with the excited electron in a 5p or higher orbital. This effect has been calculated by Armen *et. al.* (40) and their results agree well with experiment, demonstrating the need to take interference effects into account when trying to extract the intermediate density of states from such experimental results.

Understanding the effects of interference is essential to applying this technique for high resolution spectroscopy, and several groups are already studying interference effects in resonant x-ray Raman spectra (40–44). In addition, early studies of the polarization and angular distribution of resonantly scattered x-rays proved very successful in probing the symmetries of molecular orbitals (45, 46), and point to many interesting future developments.

Summary

We have discussed some of the issues related to exploiting third-generation synchrotron sources to extend studies of inner-shell structure to the deepest shells of atoms. The tremendous progress in the VUV and with traditional x-ray sources has laid the ground work for such experiments, but the shift in range of photon energies also entails a shift to a broader perspective to include scattering processes as subjects of investigation. And while this range of energies presents opportunities to study new physics, it also presents new obstacles. We have endeavored to illustrate some of these challenges, as well as possible approaches to dealing with them. The author warmly acknowledges Steve Southworth and Mike MacDonald for collaboration on this and other projects. This work was supported by the U.S. Department of Energy, Office of Basic Energy Sciences, under Contract No. W-31-109-Eng-38.

References

1. Grant, I. P., in *Atomic and Molecular Physics Handbook,* edited by Drake, G. W. F., Woodbury, NY: AIP Press, 1996, ch. 22, pp. 258–286.
2. Hansen, P. G., Jonson, B., Borchert, G. L., and Schult, O. W. B., in *Atomic Inner-Shell Physics,* edited by Crasemann, B., New York: Plenum, 1983.
3. Suortti, P., in these proceedings.
4. Crasemann, B., in *Atomic and Molecular Physics Handbook,* edited by Drake, G. W. F., Woodbury, NY: AIP Press, 1996, ch. 60, pp. 701–711.
5. Pratt, R. H., Kissel, L., and Bergstrom, P. M., in *Resonant Anomalous X-Ray Scattering,* edited by G. Materlik, C. J. Sparks, K. Fischer, Amsterdam: Elsevier Science, 1994.
6. Hemmers, O., Fisher, G., Glans, P., Hansen, D.L., Wang, H., Whitfield, S.B., Lindle, D.W., Wehlitz, R., Levin, J.C., Sellin, I.A., Perera, R.C.C., Dias, E.W.B., Chakraborty, H.S., Deshmukh, P.C., and Manson, S.T., private communication.
7. Krässig, B., Jung, M., Gemmell, D. S., Kanter, E. P., LeBrun, T., Southworth, S. H., and Young, L., *Phys. Rev. Lett.* **5**, 4736 (1996).
8. Krässig, B., in these proceedings.
9. Kissel, L., Zhou, B., Roy, S. C., Gupta, S. K. S., and Pratt, R. H., *Acta Cryst.* **A51**, 271-288 (1995).
10. Evans, R. D., in *Handbuch der Physik,* edited by Flügge, S., Berlin: Springer Verlag, 1958, vol. 34, pp. 218–298.
11. Bergstrom, P. M., Suric, T., Pisk, K., and Pratt, R. H., *Phys. Rev. A* **48**, 1134 (1993).
12. Sakurai, J. J., *Advanced Quantum Mechanics,* New York: Addison–Wesley, 1967, ch. 2-6, pp. 56–57.

13. Heitler, W., *The Quantum Theory of Radiation*, London: Oxford University Press, 1954.
14. Åberg, T., "in *Proceedings of the conference on Raman Emission by X-rays Workshop*, edited by Ederer, D. L. and McGuire, J. H., 1996.
15. Chantler, C. T., *J. Phys. Chem. Ref. Data* **24**, 71 (1995).
16. Hubbell, J. H., *J. Phys. Chem. Ref. Data* **23**, 339 (1994).
17. Krause, M. O. and Oliver, J. H., *J. Phys. Chem. Ref. Data* **8**, 329-337 (1979).
18. Sparks, C. J., *Phys. Rev. Lett.* **33**, 262 (1974).
19. Eisenberger, P., Platzman, P. M., and Winick, H., *Phys. Rev. Lett.* **36**, 623 (1976).
20. Briand, J. P., Girard, D., Kostroun, V. O., Chevalier, P., Wohrer, K., and Mossé, J. P., *Phys. Rev. Lett.* **46**, 1625 (1981).
21. Hämäläinen, K., Siddons, D. P., Hastings, J. B., and Berman, L. E., *Phys. Rev. Lett.* **67**, 2850 (1991).
22. MacDonald, M. A., Southworth, S. H., Levin, J. C., Henins, A., Deslattes, R. D., T.LeBrun, Azuma, Y., Cowan, P. L., and Karlin, B. A., *Phys. Rev A* **51**, 3598 (1995).
23. Tulkki, J. and Åberg, T., *J. Phys. B* **15**, L435-L440 (1982).
24. Tulkki, J., *Phys. Rev. A* **27**, 3375 (1983).
25. Åberg, T. and Tulkki, J., in *Atomic Inner-Shell Physics*, edited by Crasemann, B., Plenum, 1985.
26. Åberg, T. and Crasemann, B., in *X-Ray Anomalous (Resonance) Scattering: Theory and Experiment*, edited by Fischer, K., Materlik, G., and Sparks, C. J., Amsterdam: Elsevier/North-Holland, 1994.
27. Tulkki, J., thesis, Helsinki University of Technology, 1985 (unpublished).
28. Brown, G. S., Chen, M. H., Craseman, B., and Ice, G., *Phys. Rev. Lett.* **45**, 1937-1940 (1980).
29. Armen, G. B., Åberg, T., Levin, J. C., Crasemann, B., Chen, M. H., Ice, G. E., and Brown, G. S., *Phys. Rev. Lett.* **54**, 1142 (1985).
30. Kivimäki, A., Brito, A. N. d., Aksela, S., Aksela, H., Sairanen, O. P., Ausmees, A., Osborne, S. J., Dantas, L. B., and Svensson, S., *Phys. Rev. Lett.* **71**, 4307 (1993).
31. Wang, H., Woicik, J. C., Åberg, T., Chen, M. H., Herrera-Gomez, A., Kendelewicz, T., Mäntykenttä, A., Miyano, K. E., Southworth, S., and Crasemann, B., *Phys. Rev. A* **50**, 1359 (1994).
32. Aksela, S., Kukk, E., Aksela, H., and Svensson, S., *Phys. Rev. Lett.* **74**, 2917 (1995).
33. Armen, G. B. and Wang, H., *Phys. Rev. A* **51**, 1241 (1995).
34. Drube, W., Treusch, R., and Materlik, *Phys. Rev. Lett.* **71**, 42 (1995).
35. Cowan, P. L., in *X-Ray Anomalous (Resonance) Scattering: Theory and Experiment*, edited by Materlik, G., Sparks, C. J., and Fischer, K., Amsterdam: Elsevier/North-Holland, 1994.
36. LeBrun, T., "Interpreting x-ray and Auger resonant Raman spectra," in *Proceedings of the conference on Raman Emission by X-rays Workshop*, edited by Ederer, D. L. and McGuire, J. H., 1996.
37. Asplund, L., Kelfve, P., Blomster, B., Siegbahn, H., and Siegbahn, K., *Physica Scripta* **16**, (1977).
38. Breinig, M., Chen, M. H., Ice, G. E., Parente, F., and Crasemann, B., *Phys. Rev. A* **22**, 520 (1980).
39. Eberhardt, W., Kalkoffen, G., and Kunz, C., *Phys. Rev. Lett.* **41**, 156 (1978).
40. Armen, G. B., Levin, J. C., and Sellin, I. A., *Phys. Rev. A* **53**, 772 (1996).
41. Neeb, M., Rubensson, J.-E., Biermann, M., Eberhardt, W., Randall, K. J., Feldhaus, J., Kilcoyne, A. L. D., Bradshaw, A. M., Xu, Z., Johnson, P. D., and Ma, Y., *Chem. Phys. Lett.* **212**, 205 (1993).
42. Carra, P., Fabrizio, M., and Thole, B. T., *Phys. Rev. Lett.* **74**, 3700 (1995).
43. Luo, Y., Agren, H., Guo, J. H., Skytt, P., Wassdahl, N., and Nordgren, J., *Phys. Rev. A* **52**, 3730–3736 (1995).
44. Gel´mukhanov, F. and Agren, H., *J. Phys. B* **29**, 2751–2762 (1996).
45. Lindle, D. W., Cowan, P. L., Jach, T., LaVilla, R. E., and Deslattes, R. D., *Phys. Rev. A* **43**, 2353–2366 (1991).
46. Southworth, S. H., Lindle, D. W., Mayer, R., and Cowan, P. L., *Phys. Rev. Lett.* **67**, 1098–1101 (1991).

Nondipolar photoelectron angular distributions

B. Krässig, M. Jung, D. S. Gemmell, E. P. Kanter, T. LeBrun, S. H. Southworth, and L. Young

Physics Division, Argonne National Laboratory, Argonne, IL 60439, USA

Abstract. The deviations of photoelectron angular distributions from the simple, highly symmetric shapes predicted within the electric-dipole approximation are investigated. The admixture of an electric-quadrupole component in the photon-atom interaction causes an asymmetry in the angular distribution with respect to the direction of photon propagation. The reported measurement of the angular distributions of argon $1s$, krypton $2s$, and krypton $2p$ photoemission within 2–3 keV above their respective thresholds reveal pronounced asymmetries which are present even at low electron kinetic energies. The measured asymmetry parameters are in good agreement with recent predictions from nonrelativistic calculations.

INTRODUCTION

The interaction of low-energy to soft x-ray photons with matter has largely been studied within the framework of the dipole approximation. This approximation is used when the photon's wavelength can be regarded large in comparison to the atomic dimensions. Consequently, the photon momentum, being proportional to the inverse of the wavelength, is considered small and the dependence on the photon momentum is neglected. The photoelectron angular distribution in the dipole approximation therefore remains unchanged if the direction of photon propagation is reversed. An extensive body of both theoretical and experimental work is concerned with the physical information that can be extracted from angular distributions in cases where the dipole approximation is valid (cf. the reviews [1,2]).

With increasing energy the forward-backward symmetry in the angular distributions disappears. The first measurements of photoelectron angular distributions in the 1920s, using high-energy x rays, displayed pronounced forward peaking of the distributions [3,4]. It was shown that this could be related to the momentum of the absorbed radiation, however not in such a way, as one might assume, that the emitted electrons are simply kicked forward by the photon momentum [5]. The dependence on the photon momentum is retained when the photon wave's exponential is approximated by the first two terms, $e^{i\mathbf{k}\cdot\mathbf{r}} \sim 1 + i\mathbf{k}\cdot\mathbf{r}$, rather than only by the unit term. This expansion of the exponential has a close correspondence to the multipole decomposition of the photon-atom interaction: the unit term leads to the long-wavelength limit of electric-dipole (E1) interaction, and the term linear in kr is related to magnetic-dipole (M1) and electric-quadrupole (E2) interactions. It is those additional contributions which are responsible for the observed forward-backward asymmetry in the angular distributions. In the early calculations, based on a hydrogenic model, this "retardation" effect was found to be proportional to v/c, in agreement with the experimental observations (cf. [6]).

Since the early papers relatively few theoretical and even fewer experimental studies have been reported on this subject. On the theoretical side, both relativistic and nonrelativistic calculations were performed for a variety of cases using a more refined model [7,8]. On the experimental side, however, progress in this field had been hampered by the restriction to the limited spectrum and intensity obtained from the x-ray sources used (cf. [9]; for a listing of experiments before 1978, see [10]). With the availability of intense and tunable x-ray radiation at high-energy synchrotron radiation facilities, renewed interest for the topic has emerged. Recently, theoretical predictions of nondipolar angular distributions have been reported which differ significantly from the simpler retardation result, particularly for low photoelectron energies [11–13]. Stimulated by these findings, we performed an experiment to measure the angular distributions of photoelectrons from the Ar K and Kr L shells within 2–3 keV of the respective thresholds. In this brief report we present a summary of the experiment and the results. For details on the experimental procedure and the data treatment the reader may refer to the recent publications [14,15]. Similar results for the Ne L shell have been obtained in a recent experiment [16].

PHOTOELECTRON ANGULAR DISTRIBUTIONS

The photoelectron angular distribution, described by the differential cross section $d\sigma/d\Omega$, is proportional to the square of the matrix element for photon-induced transitions between the initial state ψ_i and the final state ψ_f

$$\frac{d\sigma}{d\Omega} = f\,|\langle\psi_f\,|\,\exp(i\mathbf{k}\cdot\mathbf{r})\hat{\epsilon}\cdot\mathbf{p}\,|\,\psi_i\rangle|^2\,. \tag{1}$$

Here, $\hat{\epsilon}$ is the polarization vector of the photon, $\hbar\mathbf{k}$ the photon momentum, and $\mathbf{r}$ and $\mathbf{p}$ are the position and momentum operators of the electron. The quantity f represents the combined cofactors in this expression. For simplicity, the

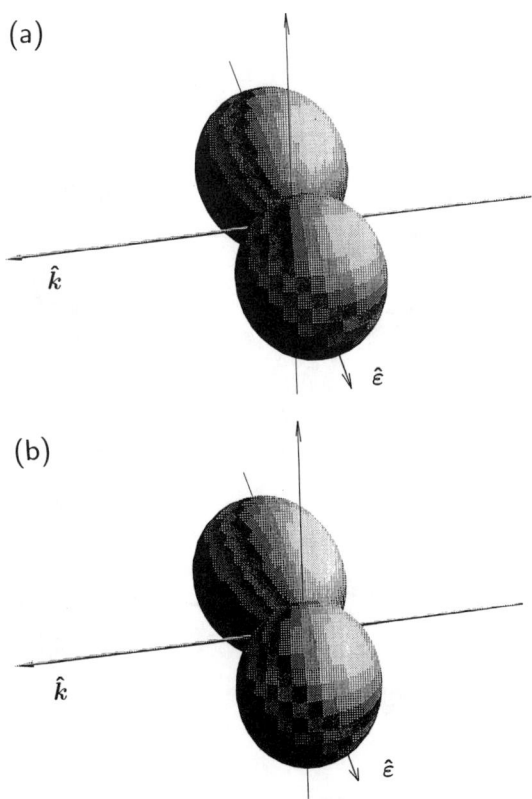

FIGURE 1. The angular distribution of photoelectrons from an s subshell. (a) dipole approximation (pure E1 interaction); (b) including the E1-E2 nondipolar contribution.

matrix element in the following will be abbreviated by the symbol $\langle \mathcal{O} \rangle$. Employing a decomposition of the interaction with the photon in terms of electric and magnetic multipoles (cf. [17]), the transition matrix element is replaced by a sum of individual multipole transition matrix elements, $\langle \mathcal{O} \rangle = \sum_{\pi,j} \langle \pi j \rangle$. Here, the multipole transition elements are characterized by their parity π and order j. In terms of the multipole decomposition, the differential cross section breaks down into a sum of individual multipole interactions $|\langle \pi j \rangle|^2$ and cross terms of combinations $\langle \pi j \rangle \langle \pi' j' \rangle^*$ where $\pi' j' \neq \pi j$. As a result of the angular properties of the multipole components, this sum, e.g. for unpolarized radiation, transforms according to[1]

---

[1] The corresponding expression for linearly polarized radiation contains for $L \geq 2$ additional terms $P_L^{(2)}(\cos\Theta)\cos 2\Phi$, weighted by factors B'_L, which are closely related to the parameters B_L. $P_L^{(2)}$ are second associated Legendre polynomials, and the azimuthal angle

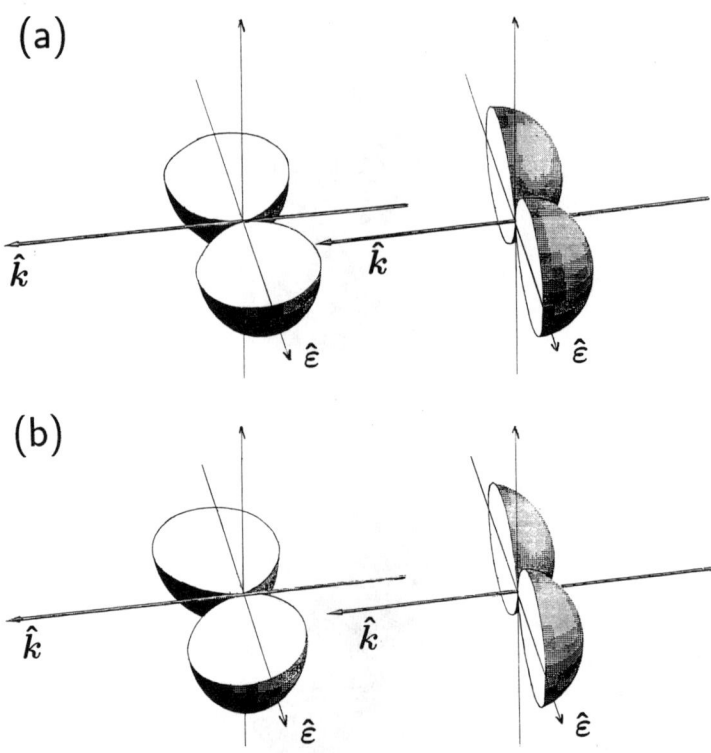

FIGURE 2. Vertical and horizontal cuts through the angular distributions depicted in Fig. 1. (a) dipole approximation (pure E1 interaction); (b) including the E1-E2 nondipolar contribution.

$$\frac{d\sigma}{d\Omega} = f \sum_{\pi,\pi',j,j'} \langle \pi j \rangle \langle \pi' j' \rangle^* = \frac{\sigma}{4\pi} \sum_L B_L P_L(\cos\Theta). \quad (2)$$

In this expression the angle Θ represents the emission angle of the photoelectron with respect to the photon beam. The angular integrations in a term $\langle \pi j \rangle \langle \pi' j' \rangle^*$ contribute Legendre Polynomials P_L of orders $|j-j'| \leq L \leq j+j'$, and these orders L are exclusively *even* for $\pi' = \pi$ and exclusively *odd* for $\pi' \neq \pi$. In the same manner, the corresponding radial integrations contribute to the respective coefficients B_L [11]. In Eq.(2) the B_L are normalized such that $B_0 = 1$ and $\sigma = f \sum_{\pi,j} |\langle \pi j \rangle|^2$.

Eq.(2) gives a convenient way of parameterizing the differential cross section with a set of angular distribution parameters B_L. The sum in L extends no further than to $2j$, with j being the highest contributing multipole order. In general the angular distribution is well described by a small number

---

Φ is measured from the direction of linear polarization (cf. [11]).

of terms, because the multipole amplitudes decrease rapidly with increasing order. The M1 and E2 interactions are smaller than the E1 interaction by a factor of $Z\alpha$, and higher multipoles are further suppressed by higher powers of $Z\alpha$. The M1 interaction acts only on the angular and spin part, but not on the spatial part of the electron's wavefunction, and thus, depending on the theoretical model, either vanishes or contributes very little. Consequently, the next-higher level of approximation to the dipole approximation includes the even-parity electric-quadrupole interaction up to terms of order $Z\alpha$ [7,12,13]. The parameterization of the angular distribution extends up to $L = 3$ and involves three angular distribution parameters B_1, B_2, B_3.

The angular distributions for pure E1 interaction and for E1 with additional E1-E2 interference are juxtaposed in Fig. 1 (a) and (b) for the case of ionization in an s subshell with linearly polarized x rays. It is clear that the E1 interaction is still the dominant feature in the angular distributions depicted in (b), which represents the strongest nondipolar asymmetry that has been observed in our experiment. The degree of the asymmetry can be better appreciated in the cuts through the distribution shown in Fig. 2. On the left, parts (a) and (b) each contain the cuts in the plane spanned by $\mathbf{k}$ and $\hat{\epsilon}$, and on the right the cuts in the plane perpendicular to $\mathbf{k}$ are shown. The nondipolar angular distribution is strongly asymmetric in the plane of the photon beam. In the plane perpendicular to the photon beam there is no difference between the dipolar and the nondipolar angular distributions.

EXPERIMENT

The idea pursued in the experiment was to probe the angular distribution by rotating an electron spectrometer on a circle around the polarization direction $\hat{\epsilon}$. Pure dipolar interaction results in an isotropic signal on this circle, and the nondipole effect causes an asymmetry between the forward and backward directed semicircles.

The experiment was performed using the monochromatized and highly linearly polarized x-ray beam from beamline X-24A at the National Synchrotron Light Source and an apparatus designed for angle-resolved electron spectrometry. A schematic of the experimental setup is shown in Fig. 3. The interaction region is defined by the intersection of the collimated x-ray beam and the target gas emanating from an effusive jet. A parallel-plate analyzer (PPA) is mounted such that it can be rotated on a cone with opening angle $\theta = 54.7°$. A stationary cylindrical mirror analyzer (CMA) and a downstream p-i-n diode (not shown in Fig. 3) were used to monitor the target density and the photon flux during the experiment. The photoelectron intensity was recorded with the PPA-angle setting varied in 15° increments over a full 360° range. The dwell time per angle was 60–120 s, and several such angular scans were added up for each x-ray energy.

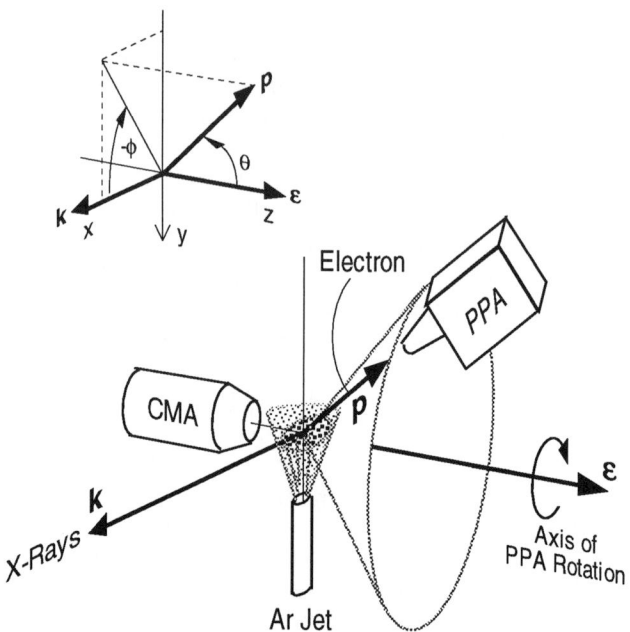

FIGURE 3. The setup of the experiment and the coordinate frame used in the representation of the angular distribution, Eq.(3). See text.

The angular distribution measured with this geometry is more conveniently represented in the system of coordinates shown in the inset of Fig. 3. In conjunction with this coordinate frame we employ an alternative parameterization to Eq.(2) and adopt the terminology for linear polarization used in Ref. [13]:

$$\frac{d\sigma}{d\Omega} = \frac{\sigma}{4\pi}\left(1 + \beta P_2(\cos\theta) + (\gamma \cos^2\theta + \delta)\sin\theta\cos\phi\right). \quad (3)$$

The parameters β, γ, δ and those used in the introduction, B_1, B_2, B_3, are connected by the relations

$$\beta = -2B_2; \quad \gamma = -5B_3; \quad \delta = B_1 + B_3. \quad (4)$$

The parameter β describes the angular anisotropy of the E1 interaction, and γ and δ govern the nondipolar part of the angular distribution. Positive/negative values of γ and δ signify a forward/backward-directed angular distribution.

The angle θ of the experimental setup was chosen to be the so-called "magic angle", $\theta_m = 54.7°$, which is the zero of $P_2(\cos\theta)$, to remove the influence of the dipolar anisotropy parameter β on the measurement. The photoelectron intensity as a function of the azimuthal angle ϕ can then be expressed as

$$I(\theta_m, \phi) = I_0 \left(1 + \sqrt{\frac{2}{27}}(\gamma + 3\delta) \cos \phi \right). \quad (5)$$

It is clear that one can only determine a combined quantity $\gamma + 3\delta$ with this experimental geometry.

There are two instrumental effects which cause the actually observed angular distribution to deviate from the form given by Eq.(5). In this brief report these will only be summarized; for a detailed description of these effects and their incorporation in the data evaluation procedure, see Ref. [15].

The first effect pertains to the inherent anisotropy of the setup depicted in Fig. 3. It is caused by the oblong source volume formed by the ~ 1 mm-diameter x-ray beam traversing the target gas. In order to assess this anisotropy, we measured the angular response of a variety of Auger electrons with different kinetic energies: Ar LMM, N KVV from N_2, O KVV from CO_2, Xe MNN, Ne KLL, Kr LMM, Ar KLL. Within the description of the two-step model, Auger electrons emitted in KLL transitions are emitted isotropically [18], and any nondipole terms related to the mixing of different parities, e.g. $\langle E1 \rangle \langle E2 \rangle$, vanish, rendering the remaining nondipole contributions negligible [19]. As a result, all of the measured Auger transitions should emit isotropically on the cone with opening angle equal to the magic angle. The recorded intensity variation of Auger electrons therefore represents a good measure of the instrumental anisotropy.

The second effect to cause a deviation from Eq.(5) is caused by noncomplete linear polarization of the x rays (here, $P_\perp \approx 0.95$) and by any misalignment of the experiment's rotation axis with respect to the polarization vector of the x rays (cf. [15,20]). Even a small tilt λ between the rotation axis and the polarization vector $\hat{\epsilon}$ (here, $\lambda \approx 1°$) creates an asymmetry between the upper ($0° < \phi \leq 180°$) and lower ($180° < \phi \leq 360°$) semicircles. The dependence on λ can be essentially removed by averaging data points at azimuthal angles ϕ and $-\phi$. This procedure gives the same result as would have been obtained for a measurement with perfect alignment of the rotation axis, but with a slightly reduced degree of linear polarization, $P' = P_\perp \cos 2\lambda$. The experimental angular distribution for partially linearly polarized x rays, after correcting for the instrumental anisotropy and averaging between the upper and lower semicircles, has the form

$$I_{P'}^{\text{av}}(\theta_m, \phi) = I_0 \left[1 + \sqrt{\frac{2}{27}}(\gamma + 3\delta) \cos \phi \right.$$
$$\left. - \frac{(1-P')\beta}{4} \cos 2\phi - \frac{(1-P')\gamma}{3\sqrt{6}} \cos \phi \cos 2\phi \right]. \quad (6)$$

When using Eq.(6) as fitting function, a reasonable choice of four fitting parameters is $I_0, [\gamma + 3\delta], [(1-P')\beta], [(1-P')\gamma]$, since their associated angular terms are distinctly different. In particular, the polarization-dependent terms in Eq.(6) vanish at the angles $\phi = 45°, 135°, 225°, 315°$. Just as in the case of complete linear polarization, a combined nondipole quantity $\gamma + 3\delta$ is readily obtained from such a fit without knowing the quantities P' or β. Furthermore, if the dipolar anisotropy parameter β is known, a fairly accurate determination of P' can be made. Conversely, however, the higher the degree of linear polarization and the smaller the tilt angle λ (i.e. the closer the quantity P' approaches unity), the less accurate become any evaluations of either β or γ from the fit parameters $[(1 - P')\beta]$ and $[(1 - P')\gamma]$, respectively.

For illustrative examples of raw data sets obtained in the angular scans, of the correction for the instrumental anisotropies, and of the corresponding fitting curves, see Refs. [14,15].

RESULTS

The collected results of nondipolar anisotropy parameters for Ar $1s$, Kr $2s$, and Kr $2p$ photoionization are displayed in Fig. 4. The experimental data points are plotted as open symbols with error bars, and theoretical predictions from Refs. [12] and [13] are given for comparison as dashed and solid lines, respectively. For the level of approximation used in these calculations, i.e. including terms $\langle E1 \rangle \langle E2 \rangle$, the quantity δ vanishes for ionization from an s subshell. The results for Ar $1s$ and Kr $2s$ are therefore given in terms of the nondipole anisotropy parameter γ, whereas the results of Kr $2p$ are given as the combined quantity $\gamma + 3\delta$.

The agreement between the theoretical nonrelativistic central-field calculations and the experimental data is very good in all three cases. The experiment confirms the prediction that the nondipole asymmetry neither approaches zero towards threshold nor is exclusively positive, as would be expected on the basis of the simple retardation picture [6]. This difference is caused by the mutual screening of the electrons and would be absent in a simple hydrogenic model (cf. [7,12]).

The energy dependences of γ differ considerably for the Ar $1s$ and Kr $2s$ cases (top and middle panels in Fig. 4). Ultimately, this difference is caused by the different shapes and nodal structure of the wavefunctions in the initial states. The nondipole asymmetry parameter γ for ionization from an ns subshell can be expressed as

$$\gamma = 3k \frac{Q(ns \to \epsilon d)}{D(ns \to \epsilon p)} \cos(\delta_{\epsilon d} - \delta_{\epsilon p}). \tag{7}$$

The quantities $Q(ns \to \epsilon d)$ and $D(ns \to \epsilon p)$ are the radial quadrupole and dipole matrix elements, and $\delta_{\epsilon d}$ and $\delta_{\epsilon p}$ are the phase shifts of the continuum partial waves for quadrupole and dipole transitions, respectively. The zeros in the energy dependences for Ar $1s$ and Kr $2s$ reflect the zeros of the quadrupole matrix element and of the cosine of the phase difference. From Eq.(7) it is also

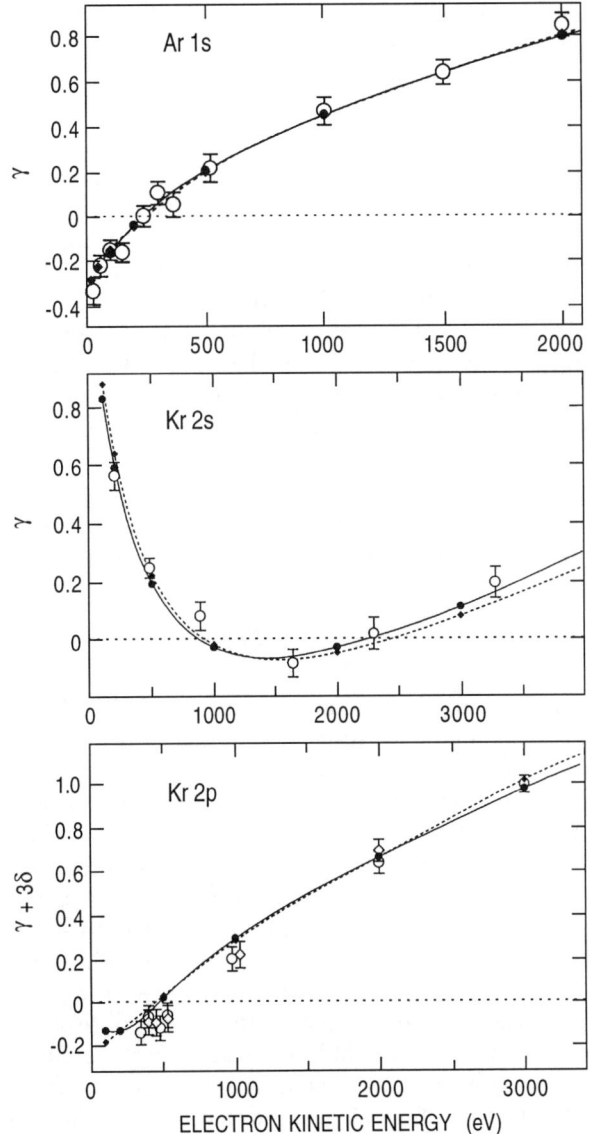

FIGURE 4. Energy dependence of the nondipole angular distribution parameter γ for Ar $1s$ (top), Kr $2s$ (middle), and of the combined quantity $\gamma + 3\delta$ for Kr $2p_j$ (bottom). Open circles/diamonds, experimental results; in the case of Kr $2p_j$ the circles and diamonds refer to the $j = 1/2$ and $j = 3/2$ fine structure components. Dashed and solid lines, theoretical predictions from [12] and [13], respectively.

clear that a zero in the dipole transition amplitude would create extremely enhanced nondipolar asymmetries.

The individual fine structure components, $j = 1/2, j = 3/2$, could be resolved in the experiment for Kr $2p$ for all but the highest energy point (bottom part of Fig. 4). No difference in the energy dependence of $\gamma+3\delta$ for the two fine structure components was detected. The agreement between prediction and experiment is not quite as good as for Ar $1s$ and Kr $2s$. At the lower energies the experimental data points are slightly, yet systematically lower than both of the theoretical predictions. For Kr $2p$, too, backward directed nondipolar asymmetries are detected towards threshold and steadily rising positive values for increasing energies. Expressions similar to Eq.(7) for both γ and δ are given in Ref. [13]. Many more transition elements and phase differences have to be taken into account for the partial waves occurring in conjunction with the ionization from p or higher-ℓ subshells. It is an interesting observation that the two theoretical predictions agree closely in their results for $\gamma + 3\delta$, even though they obtain somewhat different results for γ and δ. For further tests of the theory, future experiments will have to make provisions that enable separate determinations of all three angular distribution parameters β, γ, δ.

OUTLOOK

As intense tunable x-ray beams ranging from 1–100 keV in energy are currently becoming available at third-generation synchrotron radiation sources in Europe, the United States, and Japan, the study of nondipole effects and their inclusion in the interpretation of photoionization data will increasingly become part of data acquisition and analysis. The nondipolar asymmetries reported in this paper, particularly in the cases of ionization from s subshells, are representative of rather straight-forward physical systems and hence the validity of nonrelativistic central-field descriptions has to some extent been expected. Just as for determinations of the β parameter in the electric-dipole interaction, it is the less straight-forward situations, for example the nondipolar asymmetries in the threshold region, in the regions of resonances [7] and of Cooper minima [21], which represent interesting subjects for future experimental and theoretical investigations. The information derived from the electric-quadrupole interaction is complementary to that obtained from electric-dipole interaction, because the atomic wavefunction is probed in a different way. In addition, relativistic effects gain importance in studies with higher-Z elements and higher x-ray energies. Even at relatively low photon energies the effect of the electric-quadrupole interaction should be observable when, e.g., in a resonance, the electric-dipole amplitude is strongly suppressed or the electric-quadrupole amplitude strongly enhanced. Such experiments, be it at high photon energies where the photoionization cross section decreases or in cases where the dominant electric-dipole contribution is suppressed, will be faced with the problem of very low counting rates. Progress in this field will therefore strongly profit from instrumental developments which increase the detection efficiency in angular distribution measurements.

ACKNOWLEDGMENTS

We thank C. A. Kurtz and B. J. Zabransky for excellent technical support. We are indebted to P. M. Dehmer, J. L. Dehmer, and R. D. Deslattes for the loan of equipment. We are especially thankful to M. Peshkin for helpful suggestions and illuminating discussions. Measurements were carried out at the National Synchrotron Light Source, Brookhaven National Laboratory, which is supported by the U.S. Department of Energy, Division of Materials Sciences and Chemical Sciences. This work was supported by the U.S. Department of Energy Office of Basic Sciences under Contract W-31-109-Eng-38.

REFERENCES

1. A. F. Starace, in *Handbuch der Physik*, edited by W. Mehlhorn (Springer-Verlag, Berlin, 1982), Vol. XXXI, pp. 1–121.
2. V. Schmidt, Rep. Prog. Phys. **55**, 1483 (1992).
3. W. Bothe, Z. Phys. **26**, 59 (1924).
4. P. Auger and F. Perrin, Journal de Physique **6**, 93 (1927).
5. A. Sommerfeld and G. Schur, Ann. d. Phys. **4**, 309 (1930).
6. H. A. Bethe and E. E. Salpeter, *Quantum Mechanics of One- and Two-Electron Atoms* (Springer-Verlag, Berlin, 1957).
7. M. Ya. Amusia, P. U. Arifov, A. S. Baltenkov, A. A. Grinberg, and S. G. Shapiro, Phys. Lett. **47A**, 66 (1974); M. Y. Amusia and N. A. Cherepkov, in *Case Studies in Atomic Physics* (North-Holland, Amsterdam, 1975), Vol. 5, pp. 47–179; M. Ya. Amus'ya, V. K. Dolmatov, and V. K. Ivanov, Sov. Phys. Tech. Phys. **31**(1), 4 (1986), and references therein.
8. R. H. Pratt, A. Ron, H. K. Tseng, Rev. Mod. Phys. **45**, 273 (1973); A. Ron, R. H. Pratt, and H. K. Tseng, Chem. Phys. Lett. **47**, 377 (1977); Y. S. Kim, R. H. Pratt, A. Ron, H. K. Tseng, Phys. Rev. A **22**, 567 (1980).
9. M. O. Krause, Phys. Rev. **177**, 151 (1969); F. Wuilleumier and M. O. Krause, Phys. Rev. A **10**, 242 (1974).
10. H. K. Tseng, R. H. Pratt, S. Yu, and A. Ron, Phys. Rev. A **17**, 1061 (1978).
11. J. H. Scofield, Phys. Rev. A **40**, 3054 (1989); Phys. Scr. **41**, 59 (1990).
12. A. Bechler and R. H. Pratt, Phys. Rev. A **39**, 1774 (1989); Phys. Rev. A **42**, 6400 (1990).
13. J. W. Cooper, Phys. Rev. A **42**, 6942 (1990); Phys. Rev. A **45**, 3362 (1990); Phys. Rev. A **47**, 1841 (1993).
14. B. Krässig et al., Phys. Rev. Lett. **75**, 4736 (1995).
15. M. Jung et al., Phys. Rev. A **54**, 2127 (1996).

16. O. Hemmers *et al.*, 1996, to be published.
17. M. Peshkin, Adv. Chem. Phys. **18**, 1 (1970).
18. B. Cleff and W. Mehlhorn, J. Phys. B: Atom. Molec. Phys. **7**, 593 (1974); E. G. Berezhko and N. M. Kabachnik, J. Phys. B: Atom. Molec. Phys. **10**, 2467 (1977).
19. N. M. Kabachnik and I. P. Sazhina, 1996, to be published.
20. P.-S. Shaw, U. Arp, and S. Southworth, Phys. Rev. A **54**, 1463 (1996).
21. M. S. Wang, Y. S. Kim, R. H. Pratt, and A. Ron, Phys. Rev. A **25**, 857 (1982).

Inelastic X-ray Scattering Including Resonance Phenomena

K. Hämäläinen*, S. Manninen*, W. Caliebe[†],
C.-C. Kao[†] and J. B. Hastings[†]

*Department of Physics, POB 9, FIN-00014 University of Helsinki, Finland
[†]NSLS, Brookhaven National Laboratory, Upton New York 11973 USA

Abstract. The availability of intense synchrotron sources has recently made it possible to study weak interaction phenomena using inelastic x-ray scattering utilizing high resolution crystal spectrometers. The total resolution $\Delta E/E$ of the order of 10^{-4} (better than 1 eV at 8 keV) is rather easily achievable using backscattering geometry and the measured count rates, especially from the low-Z elements, have turned out to be reasonable. Moreover, the tunability of the incident energy makes it possible to study the resonance phenomena while scanning the incident photon energy in the vicinity of these intrinsic resonances. One of the breakthroughs has been the possibility to measure the evolution of fluorescence radiation through an absorption edge and get 2-dimensional information which reveals fine structures which are normally washed out by the core-hole lifetime. In this paper we will review several examples of resonant- and non-resonant inelastic x-ray scattering experiments accomplished at NSLS and ESRF. These include Fermi-surface studies using Compton scattering, magnetic Compton scattering, high resolution fluorescence and absorption spectroscopy including site- and local-spin selectivity.

INTRODUCTION

Inelastic x-ray scattering and absorption spectroscopies give information on the electronic and atomic structure of the target material. In these cases the photon is used as an indirect probe and therefore it is very important to understand the interaction phenomena between the electromagnetic field and the target material. The ultimate goal of an x-ray spectroscopic experiment is to gain some information on the material structure and behavior and therefore it is essential to understand the basic approximations applied in each particular case. The evolution of more powerful and brighter synchrotron sources has also contributed much better resolution and the generally assumed approximations have started to fail. This can lead to two opposite kinds of outcomes: (i) either we can gain new or more detailed information on the target material or (ii) the failure of fundamental approximations makes the direct and simple comparison between the experimental data and the desired physical quantities much less straightforward. In the following presentation we will make a brief review of the basic theory behind the inelastic scattering phenomena and discuss specific examples.

THEORY

The interaction between the radiation field (photons) and the electrons are described by the Hamiltonian

$$H_{int} = \frac{e^2}{2mc^2}\mathbf{A}^2 - \frac{e}{mc}\mathbf{p}\cdot\mathbf{A}, \qquad (1)$$

where $\mathbf{p}$ is the electron momentum and $\mathbf{A}$ the vector potential of the electromagnetic field. The inelastic scattering is usually related to the processes where the incident photon is loosing part of its energy due to some elementary excitation. The differential cross section is given by the Kramers-Heisenberg formula (1)

$$\frac{d^2\sigma}{d\omega_2 d\Omega} = r_0^2\left(\frac{\omega_2}{\omega_1}\right)\left|(\mathbf{e}_1\cdot\mathbf{e}_2)\langle b|e^{-i\mathbf{k}\cdot\mathbf{r}}|a\rangle + \frac{1}{m}\sum_i\left[\frac{\langle b|(\mathbf{e}_2\cdot\mathbf{p})e^{-i\mathbf{k}_2\cdot\mathbf{r}}|i\rangle\langle i|(\mathbf{e}_1\cdot\mathbf{p})e^{-i\mathbf{k}_1\cdot\mathbf{r}}|a\rangle}{E_a - E_i + \hbar\omega_1 + i\Gamma/2} + \right.\right.$$

$$\left.\left.\frac{\langle b|(\mathbf{e}_1\cdot\mathbf{p})e^{-i\mathbf{k}_1\cdot\mathbf{r}}|i\rangle\langle i|(\mathbf{e}_2\cdot\mathbf{p})e^{-i\mathbf{k}_2\cdot\mathbf{r}}|a\rangle}{E_a - E_i - \hbar\omega_2 + i\Gamma/2}\right]\right|^2 \delta\left(\frac{E_b - E_a}{\hbar} - \omega\right) \qquad (2)$$

where $(\hbar\omega_1,\mathbf{k}_1)$ and $(\hbar\omega_2,\mathbf{k}_2)$ are the incident and scattered photon energies and wave vectors, $\mathbf{k}$ the scattering vector and $\mathbf{e}_1$ and $\mathbf{e}_2$ the unit polarization vectors, respectively. The atomic states involved in the transition are denoted as $|a\rangle$ (initial state), $|i\rangle$ (intermediate state) and $|b\rangle$ (final state) where E_a, E_i and E_b are the corresponding energies. The lifetime Γ of the excited atomic state appears as a result of the time-dependent perturbation theory. The energy difference between the incident and scattered photon is denoted by $\hbar\omega$.

The first term is related to the $\mathbf{A}^2$-contribution of the interaction Hamiltonian (Eq. 1) and gives the Compton scattering cross section, for example. The second term is due to the $\mathbf{p}\cdot\mathbf{A}$-term in second-order perturbation theory and involves a resonant behavior. This term is responsible, for example, for fluorescence radiation and when the incident photon energy is below the absorption edge for so called resonant Raman scattering (RRS) first observed with x-rays by Sparks (2). There is no fundamental difference in the theoretical treatment of fluorescence and RRS, however, the terminology is somewhat ambiguous. Generally, this term is referred to as the resonant inelastic scattering term (3). The different type of scattering processes are discussed with some examples in this paper.

INSTRUMENTATION

Standard channel-cut monochromators are able in many cases to deliver sufficient incident photon energy resolution for the inelastic x-ray scattering studies. The bottleneck has been the analysis of the scattered radiation. When crystal optics are used for the energy analysis the role of the optical source size becomes important. The source size in most cases is dominated by the incident beam size at the sample but sometimes, especially at higher energies, the finite penetration depth can play a crucial role. This means that the incident beam has to be well focused but should also at the same time preserve the intrinsic monochromator resolution and should have high enough throughput i.e. collect as large as possible opening angle of the synchrotron beam. For the undulator sources this is no problem since the natural opening angle is very narrow and reasonable beam size can be reached even without any focusing element (at ID16 at ESRF, for example). Furthermore, the heat load is a much less severe problem than in the case of wiggler radiation. At the inelastic scattering beamline X21 at NSLS (4) the incident beam optimization is accomplished by using a single element cylindrically bent and asymmetric Si monochromator crystal (Fig. 1). This solution can overcome most of the previously mentioned problems. However, limited scanning range and non-fixed focal point create some practical limitations.

The source size contribution to the energy resolution can be overcome simply by using backscattering geometry. For the analyzer crystal the energy resolution $\Delta E/E = \cot\theta \Delta\theta$ where θ is the Bragg angle and normally the angular opening $\Delta\theta$ is dominated by the source size. It can be easily seen that close to backscattering ($\theta \approx 90°$) the energy resolution is least sensitive to the source size. However, this is also limiting the available energy range. For non-resonant scattering this is not a problem but the fluorescence lines have fixed energies and in some cases it is not possible to find a suitable analyzer crystal reflection close enough the backscattering geometry (5). The experimental set-up utilizing Rowland circle geometry used for several examples discussed in this text at the NSLS beamline X21 is shown in Figure 1.

Also the size of the analyzer crystal is an important factor. Larger analyzer area would give a larger solid angle and more intensity which could be utilized especially in the fluorescence detection since this type radiation is normally isotropic. However, the non-resonant inelastic scattering has normally a q-dependence where the momentum transfer q is determined by the scattering angle and the analyzer crystal size cannot be increased in order to preserve reasonable momentum resolution. Specifically, this is the limiting factor for the momentum resolution in Compton measurements at low energies (6). Increasing the size of bent crystal analyzers is a technological challenge where significant development has already been reached for ultra high resolution inelastic scattering spectrometers (7). One of the main breakthroughs has also been the focusing scheme utilizing so-called *dispersion compensation* which has enabled many pioneering experiments in the field of inelastic x-ray scattering (8). Dispersive optics using position-sensitive detectors has been an alternative technique especially at somewhat higher energies (9,10). This method has the advantage of parallel detection giving more stability but it is at the same time limited to a fixed energy range. On the contrary, the focusing optics allow both short and long range energy scans.

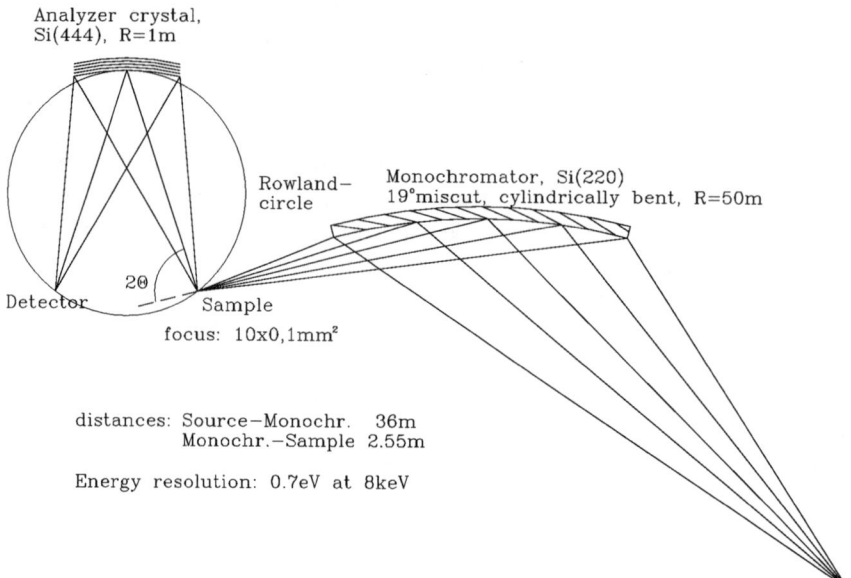

FIGURE 1. Experimental set-up for inelastic x-ray scattering experiments at NSLS beamline X21. Some relevant parameters are given in the figure.

NON-RESONANT INELASTIC SCATTERING

High Resolution Compton Scattering

Inelastic scattering of x-rays at large momentum transfers, usually called Compton scattering (11), gives a unique tool to study the *ground state properties* of outer electrons. Providing that the energy transfer to the Compton electron is much larger than the binding energy it can be shown that the scattering cross section is proportional to the so-called Compton profile, defined as

$$J(p_z) = \iint n(\mathbf{p}) dp_x dp_y . \qquad (3)$$

This relationship between the scattering cross section and the ground state momentum density $n(\mathbf{p})$ offers a direct link between the experiment and theory. Unlike in x-ray diffraction studies where information in the case of solids is limited to relatively large $\sin(\theta)/\lambda$ values, for Compton scattering this is not true. The measured energy spectrum, although it is a projection of the momentum density onto the scattering vector, involves the full valence electron contribution. By measuring directional Compton profiles the observed anisotropy of the momentum density can be analyzed theoretically using up-to-date band structure calculations. The main obstacle for the extensive use of the Compton scattering technique has been the momentum resolution. The inelastic scattering cross section is relatively low, especially for heavy elements, and therefore energy-dispersive spectrometers collecting the full scattered spectrum simultaneously have been extensively used. These are generally based on γ-ray sources and solid state detectors having a typical resolution of 350 eV at 50 keV. This corresponds to about 0.55 a.u. in momentum scale which is much worse than achieved using the angular correlation of positron annihilation (ACAR), an alternative way to study momentum distributions (12). The third generation synchrotron sources have offered an opportunity to solve this problem. High photon fluxes of hard x-rays together with well-collimated beams have made it possible to build spectrometers based on focusing crystal optics. Most of them use a curved analyzer crystal and the energy spectrum is measured using a position sensitive detector (9,10). The other approach is to use a focusing geometry, where using synchronous motions, the sample, analyzing crystal and detector stay on the focusing circle during the energy scan (13).

The importance of high resolution was shown in a momentum anisotropy study of FeAl (13). In the low resolution experiment (0.55 a.u.) only gross information can be obtained whereas the result measured using the high resolution crystal spectrometer (BL 25 at ESRF, resolution 0.15 a.u. at 50 keV) follows nicely the details predicted by a full-potential linearized augmented plane wave method

(FLAPW). Additionally, it has been shown that the anisotropic features could be divided into wave function and Fermi surface related parts (14).

To study the details of the Fermi surface even better experimental resolution is required. Because the Fermi momentum is typically of the order of 1 a.u. a resolution better than 0.1 a.u. is required. Briefly speaking the Fermi surface should be seen as a break in the Compton profile but the electron-electron correlation which smears the discontinuity complicates the analysis. Therefore, the resolution plays a crucial role in the separation of these two effects, especially because deconvolution methods fail due to the discontinuity. High resolution Compton scattering is useful for measuring the correlation effects directly.

Recently several studies have been made to determine the Fermi surface anisotropy and the effect of the correlation-induced reduction of the discontinuity in simple metals. Sakurai et al. (15) and Schülke et al. (16) have studied in detail the momentum and Fermi surface anisotropy and correlation effects in Li with a resolution of about 0.1 a.u. Hämäläinen et al. have studied Be (6), Li and Na using the latest high-resolution spectrometers at NSLS and ESRF. The strength of focusing optics is demonstrated in Figure 2 which shows the full Compton profile of Na. After scanning the full profile showing the gross features one can concentrate on the desired energy region and features. These can be non-resonant Raman-type edges or the Fermi break, both clearly visible in the figure.

The details of these types of instruments were described in the previous section. The momentum resolution can be as good as 0.01 a.u. at 8 keV, 5 times better than in prior high-resolution experiments. A closer analysis of the high resolution data has brought up new problems to be solved. The crystal optics within backscattering geometry works only at relatively low energies where the impulse approximation is not perfectly valid. In order to study the free-electron contribution and the correlation effects, scattering from the core electrons should be calculated very accurately in order to separate this contribution from the measured data. Various approaches have been developed for these calculations (17) but their accuracy is still somewhat uncertain.

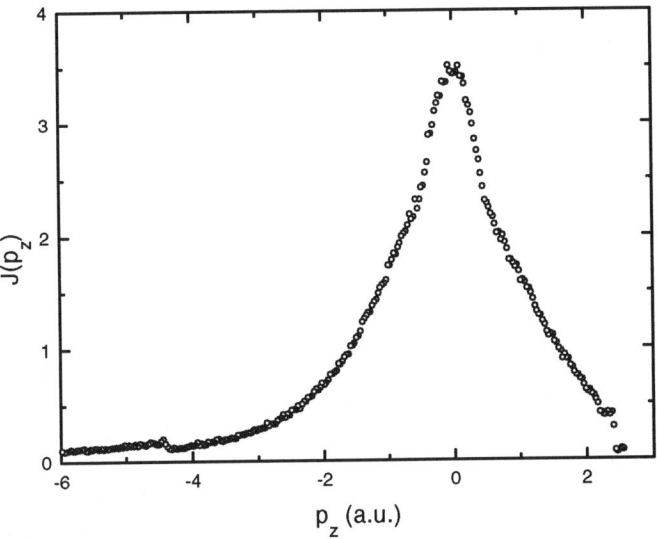

FIGURE 2. High resolution Compton profile of Na measured at ESRF beamline ID16. The momentum resolution is about 0.01 a.u.

Magnetic Compton scattering

Inclusion of the electron spin in the interaction between the electromagnetic field and the electron gives new views in the studies of magnetic materials. It turns out (18) that the $\mathbf{A}^2$-term in the interaction Hamiltonian should be replaced by $\mathbf{A} = \mathbf{C} + i\beta\,\mathbf{M}$, where the first term $\mathbf{C}$ is related to the pure charge scattering amplitude, and the second term to the pure magnetic scattering; β is a factor which includes the photon energy, scattering angle and the polarization of the incident radiation, $\mathbf{M}$ describes the magnetization of the sample and its direction gives information about the electron spin direction. In the first approximation the magnetic scattering is proportional to $\hbar\omega_1/mc^2$. Because the scattering cross section is proportional to $\mathbf{A}^2$ the pure magnetic term is therefore second order in both polarization and magnetization and very small at usual x-ray energies. The cross term $2i\beta\mathbf{C}\cdot\mathbf{M}$ is imaginary unless the photon polarization is complex, i.e. circular polarization is used. Under these circumstances this contribution is real, first order and its sign can be flipped either by changing the magnetization direction or the helicity of the photon polarization. Circularly polarized radiation can be obtained by extracting radiation above (or below) the orbital plane of the synchrotron or by using special magnetic arrays. Spin flipping using an external magnetic field is in practice easier than changing the helicity of the incident beam.

First obvious experiments in magnetic Compton scattering are on Fe, Co and Ni (19,20,21). The results were a critical test of the band structure calculations, especially of spin-polarized wave functions. Even in the case of iron, which is certainly the most intensively studied magnetic material, existing calculations could not fully explain the experimental observations (22). Simultaneously with these experiments the idea of separating the orbital and spin magnetization densities using the Compton scattering technique (22) was developed. Despite the first positive results on iron and cobalt (23) it turned out that within the experimental uncertainties the orbital contribution was zero and the magnetic Compton scattering therefore arises solely from the spin magnetization (24). Of course, if the total magnetization is known and the spin contribution is measured, the orbital part could then be calculated.

A particularly interesting material in these studies has been $HoFe_2$ which is a cubic ferrimagnet and the magnetic contribution comes from Fe-3d and Ho-4f electrons. The iron spins are anti-parallel to the holmium moments and the holmium spin and orbital moments are parallel. Additionally, the magnetization of the iron sub-lattice is very weakly temperature dependent whereas this is not true for holmium. By changing the temperature and measuring the magnetic Compton profile Cooper et al. (24) managed to determine the spin compensation temperature as well as the spin moment of iron and both spin and orbital moments of holmium.

In the latest experiments on magnetic Compton scattering efforts have been made to study the size of the magnetic effect when the incident photon energy exceeds the electron rest energy. Using the superconducting wavelength shifter at ESRF (BL 25) energies as high as 1 MeV have been used in magnetic Compton scattering experiments (25). The approximation made in the beginning of this section is no longer valid and a more sophisticated theory is required to explain the results.

Non-Resonant Raman Scattering and Valence Band Transitions

The A^2 contribution in Eq. (2) at low momentum transfer is usually written in terms of the dynamic structure factor S

$$\frac{d^2\sigma}{d\omega_2 d\Omega} = \left(\frac{d\sigma}{d\Omega}\right)_{Th} S(\mathbf{q},\omega). \qquad (4)$$

Depending on the momentum transfer q two different cases will be discussed: (i) *Raman scattering*, when $qa < 1$ and $\hbar\omega \approx E_B$ where a is the radius of an inner shell electron and E_B the binding energy of the electron (inner-shell excitation). Mizuno and Ohmura (26) derived a relationship between dynamic structure factor and the probability of absorption if the energy transfer $\hbar\omega$ is close to a soft x-ray absorption edge and the momentum transfer is still small enough, so that $qa < 1$ is still fulfilled. Inelastic x-ray scattering can therefore be used to measure soft x-ray absorption edges with hard x-rays. The energy of the absorption edge is only given by the energy transfer and not by the energy of the incident photon. The resolution is determined by the resolution of the spectrometer.

This method has been used to study the 1s absorption edges of carbon in diamond and graphite. The bonds in these two carbon modifications are completely different. In diamond the electron of the 2s orbital and the 3 electrons of the 2p orbitals form 4 sp^3-hybrid orbitals, which are all equal and form the bonds to the neighboring atoms. In graphite, however, two different kinds of bonds exist. In the plane, the 2s and 2 2p electrons form 3 sp^2-hybrid orbitals, which build the bonds in the graphite layers, while the last p-electron forms the bond to the next graphite layer. All these bonds have different energies, which is directly reflected in the corresponding spectra.

The experiment has been done with a single-crystal diamond and two different highly-oriented pyrolitic graphite crystals with the surface parallel and perpendicular to the graphite planes. The results are shown in Figure 3. The different energies and shapes for different bonds are clearly visible.

These data demonstrate that inelastic x-ray scattering can be used to measure soft x-ray absorption edges. The high resolution allows the different types of bonds in both carbon modifications to be distinguished. It can also be used to look at the different chemical valences. One improvement for the future will be an instrument that works at an energy of $\approx$ 40 keV with a resolution of about 100 meV, which would allow the sample to be kept in a stainless steel environment, so that chemical problems like catalysis can be measured with high energy resolution.

(ii) *Valence band transitions*, when $\mathbf{q}\cdot\mathbf{r}_c \approx 1$ and $\hbar\omega \approx \omega_p$ where $\mathbf{r}_c$ is the interparticle distance and $\hbar\omega_p$ the free electron plasma energy. In this case the fluctuation-dissipation theorem (27) connects the dynamical structure factor $S(\mathbf{q},\omega)$ and the dielectric function $\varepsilon(\mathbf{q},\omega)$ which describes the response of the electrons to an external electromagnetic field. The dynamical structure factor is therefore important in description of the optical and electronic properties.

Experiments have been done to determine the location and energy of the indirect band gap in diamond. The band gap was determined 30 years ago by optical absorption for the energy (28) and by neutron scattering for the momentum transfer (29). Inelastic x-ray scattering allows both values to be determined at the same time by measuring at different scattering angles the energy loss of the photons. The experimental results are shown in Figure 3. Three typical spectra for different scattering angles (2θ=16.9°, 34° and 42°, which correspond to 0.66ΓX, 1.33ΓX and 1.6ΓX) are shown for an energy loss between 2 and 10 eV. The band gap is clearly visible at 5.5 eV for a momentum transfer of 1.33ΓX but not for 0.66ΓX, which is expected from the band structure. All calculations treat the first and second Brillouin zone equally, so that some information of the symmetry gets lost. Absorption measurements can only determine the energy, because the momentum of the photon is negligible. The momentum is determined from the phonon dispersion curves, which were measured with inelastic neutron scattering. The energy of the neutrons is too low to provide the required energy transfer for the electron. Inelastic X-ray scattering is therefore an ideal technique to measure the energy and the momentum of the excited state at the same time, because both conditions have to be fulfilled, which gives more information.

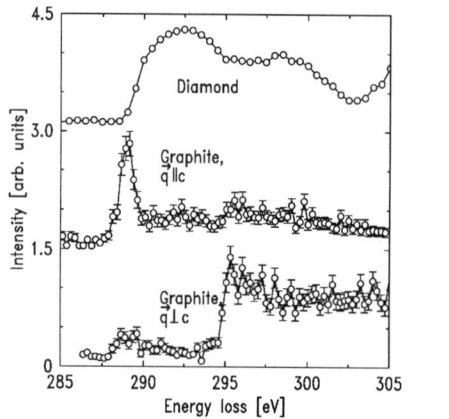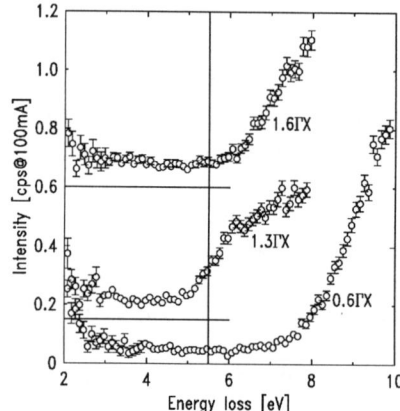

FIGURE 3. *Left:* Non-resonant Raman Scattering spectra of diamond and graphite in two different orientations. The data for diamond have been collected with a resolution of 0.7 eV, while the graphite data were collected with a resolution of 0.3 eV at a lower flux which explains the worse statistics. *Right:* Inelastic X-ray scattering spectra of diamond in the (001) direction for different scattering angles.

RESONANT INELASTIC SCATTERING

Absorption Spectroscopy

In x-ray absorption spectroscopy the incident photon energy is scanned through an absorption edge while monitoring the transmitted photon flux. In the absorption process the photon ejects an inner-shell electron creating a core hole. In the vicinity of the absorption edge the transition probability depends on the unoccupied density of states (XANES) and further above the edge is modified due to the electron backscattering from the neighboring atoms (EXAFS). Therefore both local and near-neighbor structural information can be achieved (30). It is important to notice that here the photoelectron is used as an indirect probe whose energy can be controlled only by the total energy conservation requirement. The significance of the core hole is only to give the element sensitivity or to choose certain type of transition and final state via selection rules. However, the core hole has a finite lifetime which is reflected in the energy broadening of the measured spectrum. Actually, in these cases one is always studying the system in an excited state but in case of deep core holes the interesting properties do not significantly deviate from the ground state system properties.

High Resolution Fluorescence Spectroscopy

The problems related to the lifetime broadening can be partially overcome by measuring the system absorption by fluorescence technique. Traditionally, the fluorescence XANES and EXAFS has been used as an alternative technique to measure the absorption in dilute systems in order to improve the signal-to-background ratio. This is based on the fact that the total fluorescence yield is directly proportional to the absorption coefficient. However, if one is able to measure the energy distribution of the fluorescence radiation with better resolution than its natural lifetime width supplementary information can be obtained. This can be seen from Eq. (2) where the fluorescence (and RRS) cross section is described by the second term. The extra information comes from the *final state* effect which modifies the absorption cross section. As already pointed out in the introduction this can either lead to more detailed information about the system or can unnecessarily complicate the data analysis.

The first attempt to use RRS to gain more detailed information on density of states was made by Suortti et al. (31) using a high resolution spectrometer and a conventional x-ray tube. In their novel experiment Hämäläinen et al. (32) demonstrated that by simply defining the fluorescence energy with better resolution than its natural lifetime width one can reveal fine structures in the absorption spectra which are normally smeared out. As later correctly pointed out by Carra et al. (33) one has to be very careful interpreting this kind of data in terms of the one-electron model. In a proper theoretical model one has to take account of multi-electron effects and include the effect of the core hole and the final state configuration. This is already well-known from soft x-ray absorption spectroscopy where the role of the core hole can be very important and has to be taken account in order to get true information on the density of states.

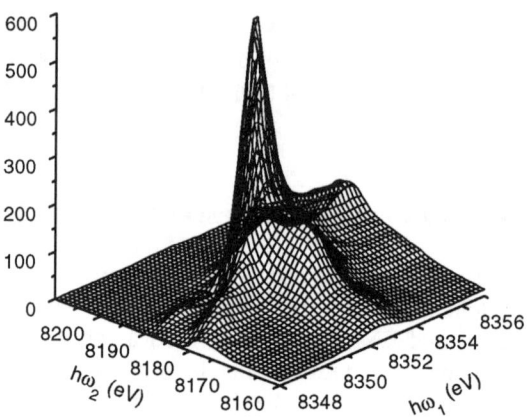

FIGURE 4. Evaluation of Er Lα_1 fluorescence in the vicinity of L$_{III}$ absorption edge in erbium iron garnet.

To get a full picture of the fluorescence process close to an absorption edge it is very important to make a 2-dimensional mapping i.e. to measure fluorescence spectra with fine enough incident energy grid. For atomic gas systems using relatively soft x-ray this was demonstrated by MacDonald et al. (34). For a rare-earth metal system an example using harder x-rays is shown in Figure 4. It can be clearly seen that very detailed fine structure is revealed compared with the conventional absorption spectrum. Especially, the weak fine structure related to quadruple transitions below the actual absorption edge is clearly seen. It has been pointed out by Krisch et al. (35) that different dispersion behaviors can be utilized to separate the transitions to bound states or to continuum. In the experiment using a Gd sample the extra information was used to study specifically the different transitions and give more insight to results from a MCD experiment (36). Recently, theoretical progress has been made to understand these more complicated experimental results. Tanaka et al. (37) have been able to numerically reproduce the observed fine structure for Dy (32) and Carra et al. (33) have been pointed out the importance of multi-electron effects and suggested making these experiments with constant energy transfer.

Local-Spin Selective Absorption Spectroscopy

High-resolution fluorescence detection also opens up possibilities to measure the satellite structure of the fluorescence lines which are related to the different final states of the system. One striking example is the Kβ fluorescence line of certain Mn compounds which involve a well-separated satellite structure related to a very strong final state effect (38). In the case of MnO and MnF$_2$, for example, all the spin-up $3d$ states are occupied while all the spin-down states are vacant. After Kβ emission the final state $3p$ hole is very strongly coupled with the $3d$ electrons and this coupling is related to the spin state. As a result the Kβ satellite structure can be identified to spin-up final state hole while the main line is almost purely spin-down final state. Since the spin-state is conserved in the fluorescence process this gives the possibility to choose the spin-state of the photoelectron. Therefore, by monitoring the different final states one can obtain the local-spin selective absorption spectrum and the spin-separated density of states. This was demonstrated by Hämäläinen et al. for MnO and MnF$_2$ (39) which are actually antiferromagnets making traditional MCD measurements impossible. This is due to the fact that the spin is observed in the local atomic system and no global long-range magnetic ordering is needed. Subsequent studies on MnP have also successfully made by de Groot et al. (40). However, it should be emphasized that this kind of spectroscopy is possible only if the nature of the final state is well understood and if the system in question posseses this kind of satellite structure at all.

Site-Selective Spectroscopy

The inner-shell electron levels and therefore also the x-ray fluorescence energies of certain element depend on the chemical environment i.e. ionization state, coordination etc. The variation is not large and typically can be less than the lifetime width. However, if one is able to separate the different fluorescence lines related to the corresponding oxidation states, for example, it is possible to accomplish site-selective absorption spectroscopy. This is highly important for the complicated protein structures which can posses multi-valence structures and one would like to separate the local environments of the different sites of the same base atom. In traditional EXAFS this is not possible as one is always averaging over all different sites of the same element.

However, as in the case of local-spin selective spectroscopy this demands well understood system final states. For this purpose a systematic study of Mn Kβ fluorescence lines was accomplished for an extended set of different chemical environments (38). This study made it possible to reconstruct the different components of fluorescence lines for mixed-valence compounds. The Mn compounds form biologically interesting proteins and play a crucial role in photosynthesis, for example. Once the behavior of the fluorescence lines were fully understood, site-selective XANES and EXAFS techniques were demonstrated by Grush et al. (41). The method was shown to work for concentrated samples but even for third generation synchrotron sources dilute protein systems present a demanding challenge. For this purpose design of multi-array analyzer crystals are in progress and partly in use (42).

Resonant Enhancement of the Charge Transfer Excitations

Limited by the incident photon flux, most of the x-ray RRS experiments performed up to now involved resonant Raman processes in which both the intermediate states and the final states correspond to the localized core electron excitations. High energy resolution studies of resonant Raman processes in which the final states correspond to the elementary excitations of the conduction electrons or the valence electrons in condensed matter has only become possible recently with the development of high flux insertion devices. An important class of experiment is the recent high energy resolution study of the excitation energy dependence in soft x-ray emission spectroscopy (43,44,45). For wide band solids, such as Si and diamond, strong excitation energy and orientation dependence of the valence band emission spectra were observed, and interpreted using the one-electron band structure of the material. On the other hand, for narrow band solids, such as transition metal and rare earth compounds, the excitation energy dependence is usually interpreted in terms of multi-electron satellites or correlation effects, and is believed to provide unique information on the electronic structure of these important materials (46).

Recently, anion-to-cation charge transfer excitations have been observed in a high resolution RRS study of NiO (47), one of the prototypical Mott-Hubbard systems. In the experiment, inelastic x-ray scattering spectra near the NiO valence band emission energy were measured as the incident photon energy was scanned through the Ni K absorption edge. A new spectral feature, which is not observed for incident energies far away from the absorption edge, appears in the inelastic scattering spectra in addition to the valence band emission spectrum. While the average energy of the valence band emission spectrum remains constant, the average energy of this new feature disperses linearly with the incident photon energy. Figure 5 shows the inelastic scattering spectrum obtained at the peak of the Ni K absorption spectrum. The average energy transfer of the new feature is about 6.5 eV, close to the recently reported anion-to-cation charge transfer energy of NiO (48,49,50). The inelastic scattering cross section shows a large resonant enhancement and a strong incident-energy dependence. These observations are interpreted using a configuration-interaction cluster model of NiO.

It should be noted that the charge transfer energy is usually indirectly deduced from the results of a combination of valence band photoemission and inverse photoemission measurements (50), or other core electron spectroscopies (48,49). Thus, RRS may provide a new and more direct way to study the charge transfer excitations and other types of elementary excitations in many highly correlated systems (51).

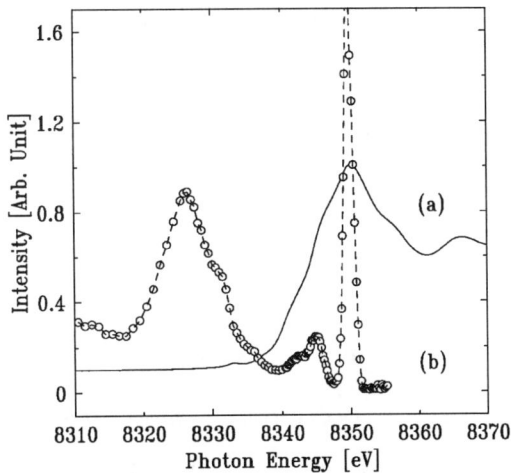

FIGURE 5. (a) Ni K absorption spectrum of NiO; (b) Resonant inelastic x-ray scattering spectrum of NiO obtained with the incident photon energy tuned to the main peak of the absorption spectrum (a).

SUMMARY

We have made a review and progress report of high resolution inelastic x-ray scattering spectroscopy covering both resonant and non-resonant scattering. The intense and naturally well-collimated synchrotron beams have made it possible to construct focusing spectrometers to study weaker interaction phenomena with better resolution. Moreover, the tunability of the incident energy combined with high-resolution fluorescence detection have opened up totally new possibilities compared with conventional absorption spectroscopy which is probably the most utilized technique within the synchrotron community. Future progress depends on the development of: (i) larger size analyzer crystals (multi-element) which are custom shaped for non-backscattering geometry, possibly with controlled mosaicity, (ii) high energy and position resolution multi-element detectors for dispersive optics, (iii) improved theoretical and computational methods to get proper information from the complicated experimental results, (iv) more incident photons for dilute systems and for weak non-resonant scattering.

ACKNOWLEDGEMENTS

All the work here is done in collaboration with lot of people whose names can be found in the referred papers. The authors would specially like to acknowledge M.J. Cooper, S.P. Cramer, P. Suortti and T. Åberg. This work is supported by the US department of Energy under Contract No. DE-AC02-76CH00016 and by the Academy of Finland (SA-8582).

REFERENCES

1. Kramers, H. A., and Heisenberg, W., *Z. Phys.* **31**, 681-708 (1925).
2. Sparks, C. J., *Phys. Rev. Lett.* **33**, 262-5 (1974).
3. Åberg, T., and Tulkki, J., edited by Crasemann, B., *Atomic Inner-Shell Physics*, New York; Plenum, 1985.
4. Kao, C.-C., Hämäläinen, K., Krisch, M., Siddons, D. P., Hastings, J. B., and Oversluizen, T., *Rev. Sci. Instr.* **66**, 1699-702 (1995).
5. Stojanoff, V., Hämäläinen, K., Siddons, D. P., Hastings, J. B., Berman, L. E., Cramer, S., and Smith, G., *Rev. Sci. Instr.* **63**, 1125-7 (1992).
6. Hämäläinen, K., Manninen, S., Kao, C.-C., Caliebe, W.,Hastings, J.B., Bansil, A., Kaprzyk, S., and Platzman, P., *Phys. Rev.* B **54**, 5453-9 (1996).
7. Sette, F., Ruocco, G., Krisch, M., Bergmann, U., Masciovecchio, C., Mazzacurati, V., Signorelli, G., and Verbeni, R., *Phys. Rev. Lett.* **75**, 850-53 (1995).
8. Schülke, W., and Nagasawa, H., *Nucl. Instr. Meth.* **222**, 203-6 (1984).
9. Loupias, G., and Petiau, J., *J. Physique* **41**, 265-71 (1980).
10. Sakurai, Y., Ito, M., Urai, T., Tanaka, Y., Sakai, N., Iwazumi, T., Kawata, H., Ando M., and Shiotani N., *Rev. Sci. Instr.* **63**, 1190-3 (1992).
11. Cooper, M. J., *Rep. Prog. Phys.* **48**, 415-81 (1985).
12. Peter, M., Jarlborg, T., Manuel, A. A., Barbiellini B., and Barnes, S. E., *Z. Naturforsch.* **48a** 390-7 (1991).
13. Manninen, S., Honkimäki, V., Hämäläinen, K., Laukkanen, J., Blaas, C., Redinger, J., McCarthy, J., and Suortti, P., *Phys. Rev.* B **53**, 7714-20 (1996).
14. Blaas, C., Redinger, J., Manninen, S., Honkimäki, V., Hämäläinen, K., and Suortti, P., *Phys. Rev. Lett.* **75**, 1984-7 (1995).
15. Sakurai, Y., Tanaka, Y., Bansil, A., Kaprzyk, S., Stewart, A.T., Nagashima, Y., Hyodo, T., Nanao, S., Kawata, H., and Shiotani, N., *Phys. Rev. Lett.* **74**, 2252-5 (1995).
16. Schülke, W., Stutz, G., Wohlert, F., and Kaprolat, A., *Phys. Rev.* B, in press.
17. Issolah, A., Garreau, Y., Levy, B., and Loupias, G., *Phys. Rev.* B **44**, 11029-34 (1991).
18. Platzman, P., and Tzoar, N., *Phys. Rev.* B **2**, 3556-9 (1970).
19. Cooper, M.J., Laundy, D., Cardwell, D.A., Timms, D., Holt, R.S., and Clark. G., *Phys. Rev.* B **34**, 5984-7 (1986).
20. Timms, D.N., Brahmia, A., Collins, S.P., Cooper, M.J., Holt, R.S., Kane, P.P., Clark, G., and Laundy, D., *J. Phys.* F **18**, L57-61 (1988).
21. Kubo, Y., and Asano, S., *Phys. Rev.* B **42**, 4431-46 (1990).
22. Collins, S.P., Cooper, M.J., Lovesey, S., and Laundy, D., *J. Phys.:Condens.Matter* **2**, 6439-49 (1990).
23. Cooper, M.J., Zukowski, E., Collins, S.P., Timms, D.N., Itoh, F., and Sakurai, H., *J. Phys.: Condens. Matter* **4**, L399-404 (1992).

24. Cooper, M.J., Zukowski, E., Timms, D.N., Armstrong, R., Itoh, F., Ito, M., Kawata, H., and Bateson, R., *Phys. Rev. Lett.* **71**, 1095-98 (1993).
25. McCarthy, J.M., Cooper, M.J., Lawson, P.K., Timms, D.N., Manninen, S.O., Hämäläinen, K., and Suortti, P., to be published.
26. Mizuno, Y., and Ohmura, Y., *J. Phys. Soc. Japan*, **22**, 2 445-449 (1967).
27. Pines, D., and Nozières, P., *The Theory of Quantum Liquids*, New York, Addison-Wesley, 1989.
28. Clark, C. D., Dean, P. J., and Harris, P. V., *Proc. Roy. Soc. London*, A **277**, 312-329 (1964).
29. Dean, P. J., Lightowlers, E. C., and Wright, D. R., *Phys. Rev.* **140**, A352-68 (1965).
30. Theo, B. K., *EXAFS: Basic Principles and Data Analysis*, Berlin, Spinger-Verlag, 1986.
31. Suortti, P., Eteläniemi, V., Hämäläinen, K., and Manninen, S., *J. Physique Coll.* **48** C9, 831-4 (1987).
32. Hämäläinen, K., Siddons, D. P., Hastings, J. B., and Berman, L. E., *Phys. Rev. Lett.* **67**, 2850-53 (1991).
33. Carra, P., Fabrizio, M., and Thole, B. T., *Phys. Rev. Lett.* **74**, 3700-3 (1995).
34. MacDonald, M. A., Southworth, S. H., Levin, J. C., Henins, A., Deslattes, R. D., LeBrun, T., Azuma, Y., Cowan, P. L., and Karlin, B. A., *Phys. Rev.* A **51**, 3598-603 (1995).
35. Krisch, M. H., Kao, C.-C., Sette, F., Caliebe, W. A., Hämäläinen, K., and Hastings, J. B., *Phys. Rev. Lett.* **74**, 4931-4 (1995).
36. Schütz, G., Knülle, M., Wienke, R., Wilhelm, W., Wagner, W., Kienle, P., and Frahm, R., *Z. Phys.* B **73** 67-76 (1988).
37. Tanaka, S., Ogasawara, H., Okada, K., and Kotani, A., *Jpn. J. Appl. Phys.* **32**, 101-3 (1992).
38. Peng, G., deGroot, F. M. F., Hämäläinen, K., Moore, J. A., Wang, X., Grush, M. M., Hastings, J. B., Siddons, D. P., Armstrong, W. H., Mullins, O.C., and Cramer, S.P., *J. Am. Chem. Soc.* **116**, 2914-20(1994).
39. Hämäläinen, K., Kao, C.-C., Hastings, J. B., Siddons, D. P., Berman, L. E., Stojanoff, V., and Cramer, S. P., *Phys. Rev.* B **46**, 14274-7 (1992).
40. deGroot, F. M. F., Pizzini, S., Fontaine, A., Hämäläinen, K., Kao, C.-C, and Hastings, J. B., *Phys. Rev.* B **51**, 1045-52 (1995).
41. Grush, M. M., Christou, G., Hämäläinen, K., and Cramer, S. P., *J. Am. Chem. Soc.* **117**, 5895-6 (1995).
42. Kao C.-C., and Cramer, S. P., private communication.
43. Rubensson J. E., Mueller, D., Shuker, R., Ederer, D. L., Zhang, C. H., Jia, J., and Callcott, T. A., *Phys. Rev. Lett.* **64**, 1047-50 (1990)
44. Ma, Y., Wassdahl, N., Skytt, P., Guo, J., Nordgren, J., Johnson, P. D., Rubensson, J.-E., Boske, T., Eberhardt, W., and Kwan, S. D., *Phys. Rev. Lett.* **69**, 2598-2601 (1992).
45. O'Brien, W. L., Jia, J., Dong, Q.-Y., Callcott, T. A., Miyano, K. E., Ederer, D. L., and Kao, C.-C., *Phys. Rev. Lett.* **70**, 238-41 (1993).
46. Tanaka, S., and Kotani, A., *J. Phys. Soc. Jpn.* **62**, 464 (1993).
47. Kao, C.-C., Caliebe, W. A., Hastings, J. B., and Gillet, J.-M., Phys. Rev. B in press.
48. Geunseop Lee, and Oh, S.-J., *Phys. Rev.* B **43**, 14674-82 (1991).
49. van Elp, J., Eskes, H., Kuiper, P., Sawatzky, G. A., *Phys. Rev.* B **45**, 1612-22 (1992).
50. Sawatzky, G. A., and Allen, J. W., *Phys. Rev. Lett.* **53**, 2339-41(1984).
51. Luo, J., Ph.D. Thesis, Rice University (1994).

Angular Correlations in Auger and Fluorescence Cascades

Nicolai M. Kabachnik[1]

*Fakultät für Physik, Universität Bielefeld,
D-33501 Bielefeld, Germany*

Abstract. Angular correlations in cascades of Auger electrons and X-rays (fluorescence) are discussed. These correlations can be studied by angle-resolved coincidence measurements of two sequential emissions of Auger electrons and/or fluorescence photons. A general expression for the angular correlation function for two sequential radiation is obtained using the density matrix and statistical tensor formalism. As an example, the cascades in inner-shell atomic photoionization and photoexcitation are considered. Spectroscopic and dynamic information which can be gained from angular correlation measurement in cascade transitions is discussed. Finally, the alignment transfer and non-coincidence measurements of the angular distribution of any radiation from a cascade are discussed.

INTRODUCTION

When a vacancy is produced in a deep inner shell of a many-electron atom, the following rearrangement of the highly excited electronic system usually proceeds through a stepwise series of radiative and non-radiative decay processes known as a vacancy cascade [1]. This complex process results in the emission of X-ray (fluorescence) photons and Auger electrons and often leads to a highly charged residual ion. A multitude of decay pathways and variety of the processes which can occur in a highly correlated many-electron system contribute to the complexity of the Auger electron and fluorescence spectra and lead to a broad ion charge distribution. To overcome the difficulties in the analysis of the vacancy cascade and the resulting spectra, various coincidence techniques were exploited such as Auger-electron – photoion coincidence measurements [2] or the threshold photoelectron – residual ion coincidence spectroscopy [3]. The advantage of the coincidence technique is that it selects a subset of all transitions, and the resulting spectra or ion charge

---

[1]) On leave of absence from Institute of Nuclear Physics, Moscow State University, Moscow 119899, Russia

distribution is greatly simplified. Even more informative are the coincidence measurements of any two emitted particles (photons) from the cascade. The first electron – electron coincidence studies of the Auger cascade of core-excited nobel gas atoms [4] and the electron – fluorescence photon coincidence studies of photoexcited Ca atoms [5] show great promise as a way of unraveling the complex spectra and investigating in detail the cascade dynamics.

Recently it was suggested to study the angular correlation between two Auger electrons from a cascade detected in coincidence [6]. The first experiments including those presented at this conference [7–9] confirm high potential of this method. The aim of this report is to discuss the theoretical background of angular correlation measurements in the cascade processes. A short overview of possible applications of the method in inner-shell physics is given.

THEORY OF ANGULAR CORRELATIONS IN CASCADES

The theory of angular correlations in collision and decay processes is a well developed part of quantum mechanics. This theory was first applied to cascade transitions in nuclear physics. The most elegant formulation of the theory is based on the formalism of density matrices and statistical tensors. Excellent reviews and books are published (see, for example [10,11]) in which the angular correlations in cascade processes are considered in detail.

General Expression for Directional Correlation of Two Successive Radiation

Consider the simplest case of two successive radiations emitted from an excited atom (ion). Here and below "radiation" means either Auger (autoionization) electron or X-ray (fluorescence) photon. We assume that the atomic transitions occur in steps and the atomic states involved at each stage have well-defined total angular momentum and parity.

The initial state with the total angular momentum J_0 may have a non-statistical population of magnetic substates; in other words it may be oriented or/and aligned. It is convenient to describe the anisotropy of the state by statistical tensors [10,11] $\rho_{k_0 q_0}(J_0, J_0)$. In general, all components of the statistical tensors with $k_0 \leq 2J_0$, $-k_0 \leq q_0 \leq k_0$ can be non-zero. If the initial state is unpolarized, the statistical tensor with $k_0 = q_0 = 0$ is the only non-zero one. If the initial state is produced by photoionization or photoexcitation of an inner-shell electron with unpolarized or linearly polarized light, then the vacancy state is in general aligned and the degree of alignment is described by the second rank statistical tensor ($k_0 = 2$).

We shall further assume that the detectors are insensitive to the polarization (spin) state of the radiation. Then the general expression for the angular correlation function of two successive radiations from the polarized initial state can be presented in the following form:

$$W(\mathbf{n}_1, \mathbf{n}_2) = 4\pi \hat{J}_0 \sum_{k_0 q_0 k_1 k_2} G_{k_1 k_2 k_0} \rho_{k_0 q_0}(J_0, J_0) \{Y_{k_1}(\mathbf{n}_1) \otimes Y_{k_2}(\mathbf{n}_2)\}_{k_0 q_0}. \quad (1)$$

Here $\mathbf{n}_1$ ($\mathbf{n}_2$) are unit vectors which determine the direction of the first (second) radiation, and we use a standard notation $\hat{J} = \sqrt{2J+1}$. The tensorial product of spherical functions is defined as

$$\{Y_{k_1}(\mathbf{n}_1) \otimes Y_{k_2}(\mathbf{n}_2)\}_{k_0 q_0} = \sum_{q_1 q_2} (k_1 q_1, k_2 q_2 \,|\, k_0 q_0) \, Y_{k_1 q_1}(\vartheta_1, \varphi_1) \, Y_{k_2 q_2}(\vartheta_2, \varphi_2) \quad (2)$$

where $(k_1 q_1, k_2 q_2 \,|\, k_0 q_0)$ is a Clebsch-Gordan coefficient.

The generalized angular correlation coefficients $G_{k_1 k_2 k_0}$ contain angular momentum coupling coefficients and amplitudes of the decay processes. Due to the assumption of the stepwise character of the process, each of the terms $G_{k_1 k_2 k_0}$ may be split into two factors one of which depends entirely on the first transition and the other on the second:

$$G_{k_1 k_2 k_0} = B_{k_1 k_2 k_0}(J_0, J_1) A_{k_2}(J_1, J_2). \quad (3)$$

(J_1 and J_2 are the total angular momenta of the intermediate and the final atomic states, respectively.) These factors are given by

$$B_{k_1 k_2 k_0}(J_0, J_1) = (-1)^{k_1 + k_2 - k_0} \hat{J}_0 \hat{J}_1 \sum_{L_1 L_1'} c_{k_1 0}^*(L_1, L_1')$$

$$\times \begin{Bmatrix} J_1 & L_1 & J_0 \\ J_1 & L_1' & J_0 \\ k_2 & k_1 & k_0 \end{Bmatrix} \langle J_1, L_1 \,||\, \hat{O} \,||\, J_0 \rangle \, \langle J_1, L_1' \,||\, \hat{O} \,||\, J_0 \rangle^* \quad (4)$$

$$A_{k_2}(J_1, J_2) = (-1)^{J_1 + J_2 + k_2} \hat{J}_1 \sum_{L_2 L_2'} c_{k_2 0}^*(L_2, L_2')$$

$$\times (-1)^{L_2'} \begin{Bmatrix} J_1 & L_2 & J_2 \\ L_2' & J_1 & k_2 \end{Bmatrix} \langle J_2, L_2 \,||\, \hat{O} \,||\, J_1 \rangle \, \langle J_2, L_2' \,||\, \hat{O} \,||\, J_1 \rangle^*. \quad (5)$$

Here we use the standard notations for the Wigner 6j and 9j coefficients; $\langle J_b, L \,||\, \hat{O} \,||\, J_a \rangle$ denotes the decay amplitude which describes a transition from the state J_a to the state J_b by emission of the radiation with the total angular momentum L. $c_{k0}(L, L')$ are the radiation parameters. Explicit expressions for these parameters for any radiations are given in [10,11]. For example, for Auger electrons the radiation parameters are

$$c_{k0}(L, L') = (-1)^{L'+l+\frac{1}{2}} (4\pi)^{-1} \hat{l}\hat{l}'\hat{L}\hat{L}'(l0, l'0 \mid k0) \begin{Bmatrix} l & L & \frac{1}{2} \\ L' & l' & k \end{Bmatrix} \tag{6}$$

where L and l are the total and orbital angular momenta of the Auger electron, respectively. (In this case the summations in (4) and (5) include also sums over l and l'.) For the dipole fluorescence (the detector is not sensitive to polarization) the radiation parameters are

$$c_{k0}(1, 1) = \frac{3}{8\pi} (11, 1-1 \mid k0). \tag{7}$$

Since the atomic states involved are assumed to have well-defined parities and parity is conserved in the decay processes, it is easy to show that only even values of k_1 and k_2 can occur in (1). In addition the summation in (1) is limited by the triangle rule $|k_0 - k_1| \leq k_2 \leq k_0 + k_1$.

Expression (1) shows that each of the terms under summation consists of two parts. The first factor, the angular correlation coefficient, is determined by the dynamics of particular transitions, while the second factor is purely kinematic – it is entirely determined by the geometry of the experiment and the initial conditions. The second factor is common for any radiations. The main goal of a correlation experiment is to obtain the angular correlation coefficients and as a final result to determine the amplitudes of the process. However, analysis of the kinematic factor in (1) is necessary in order to choose the experimental conditions.

The Case of Unpolarized Initial State

If the initial atomic state is unpolarized the general expression (1) simplifies. This case can be realized, for example, in the ionization of an inner s-shell. Then $k_0 = q_0 = 0$ and from equation (2) it follows that $k_1 = k_2$ and expression (1) takes the form:

$$W(\bar{\vartheta}) = \sum_{k=even} \overline{A}_k(J_0, J_1) A_k(J_1, J_2) P_k(\cos\bar{\vartheta}) \tag{8}$$

where $P_k(\cos\bar{\vartheta})$ are Legendre polynomials, factors $A_k(J_1, J_2)$ are defined by (5), while for $\overline{A}_k(J_0, J_1)$ one obtains

$$\overline{A}_k(J_0, J_1) = (-1)^{J_1+J_0} \hat{J}_1 \sum_{L_1 L_1'} c_{k0}^*(L_1, L_1')$$

$$\times (-1)^{L_1} \begin{Bmatrix} J_1 & L_1 & J_0 \\ L_1' & J_1 & k \end{Bmatrix} \langle J_1, L_1 \| \hat{O} \| J_0 \rangle \langle J_1, L_1' \| \hat{O} \| J_0 \rangle^*. \tag{9}$$

Now the angular correlation function depends only on one angle $\overline{\vartheta}$, the relative angle between two radiations. This fact has a simple physical explanation. The initial state is unpolarized, therefore the first radiation is emitted isotropically. However, if we fix the direction of its emission and if it takes a definite angular momentum from the atom, then the intermediate atomic state will be aligned along the direction of the first radiation. The second radiation can be anisotropic, but its angular distribution will be axially symmetrical about the direction of the first radiation. k is even, therefore the distribution is also symmetric with respect to reflection through the $\overline{\vartheta} = 90°$ plane. The complexity of the angular correlation is determined by the condition $k \leq min(2L_1, 2L_2, 2J_1)$ where L_1 and L_2 are total angular momenta of the two radiations. From this result it is clear that if any of the radiation involved or the intermediate state have angular momenta equal to 0 or $\frac{1}{2}$ the angular correlation disappears, $W = const$.

If one of the radiations is a dipole photon, then $k \leq 2$ and the angular correlation takes a familiar form [12]:

$$W(\overline{\vartheta}) \sim 1 + \alpha_2^{J_1 \rightarrow J_2} \mathcal{A}_{20}(J_1) P_2(\cos \overline{\vartheta}) \qquad (10)$$

where $\alpha_2^{J_1 \rightarrow J_2} = A_2(J_1, J_2)/A_0(J_1, J_2)$ is an intrinsic anisotropy parameter [13,14] and $\mathcal{A}_{20}(J_1) = \overline{A}_2(J_0, J_1)/\overline{A}_0(J_0, J_1)$ is the parameter characterizing the alignment of the intermediate state along the direction of the first radiation.

In complete analogy with the angular distribution of a single radiation, one can show that if both decays are single-channeled, (either only one multipole contribute to the radiative transition or only one partial wave is possible for the Auger electron) then the angular correlation coefficients become model-independent, they are determined only by the angular momentum coupling coefficients.

Examples of Particular Experimental Conditions

A choice of some particular geometry of the experiment can simplify the angular correlation function and makes the analysis of experimental data easier. Since in the majority of recent studies of vacancy cascades the synchrotron radiation was used for creating the initial vacancy, we consider in the following photoionization or photoexcitation of the core electron by linearly polarized light. We choose the z-axis along the direction of linear polarization. The vacancy or core-excited initial state is in general aligned along this direction. If one of the cascade radiations is detected along the polarization direction, the experimental conditions for the second radiation turns out to be axially symmetrical with respect to this direction. Therefore the angular correlation will depend only on one polar angle ϑ. This result can be easily obtained from the general expression (1). In the considered case the angular correlation can be expressed as

$$W(\vartheta) = \sum_{k=even} a_k P_k(\cos\vartheta), \qquad (11)$$

where

$$a_k = \hat{J}_0 \sum_{k_0 k_1} \hat{k}_0 \hat{k}_1 (k_1 0, k0 \mid k_0 0) \rho_{k_0 0}(J_0, J_0) G_{k_1 k k_0}. \qquad (12)$$

Another particular case is the geometry used practically in all experiments where both detectors are placed in a plane perpendicular to the photon beam direction. In this case the angular correlation function can be expressed in terms of associated Legendre polynomials:

$$W(\vartheta_1,\vartheta_2) = 2\hat{J}_0 \sum_{k_0 k_1 k_2} \sum_{q=0,2...} G_{k_1 k_2 k_0} \rho_{k_0 0}(J_0, J_0)(2-\delta_{0q})$$
$$\times (k_1 q, k_2 - q \mid k_0 0) \tilde{P}_{k_1}^q(\cos\vartheta_1)(-1)^{q/2} \tilde{P}_{k_2}^q(\cos\vartheta_2). \qquad (13)$$

Several comments can be made concerning the angular correlation of different types of radiation. If both radiations are Auger or autoionization electrons (e-e correlations, like in experiments [4,7,9]) the complexity of the correlation function depends on their orbital angular momenta. In particular, if the initial vacancy is unpolarized the maximal rank of the Legendre polynomials in (8) is limited additionally by the maximal orbital momentum of the Auger electrons in each of the transitions, $k \leq min\{2l_{1max}, 2l_{2max}\}$.

In the case of γ-γ correlations of two succesive photon emissions, the correlation coefficients are especially simple. If the dipole approximation is valid the transition amplitudes do not affect the angular correlation. The coefficients $G_{k_1 k_2 k_0}$ are determined only by the values of angular momenta of the states involved.

The case in which the first radiation is a photon and the second is an Auger electron (γ-e correlations) has been studied experimentally in [8] where a K_α photon from Ar was detected in coincidence with the L-MM Auger transitions. In this case the initial $1s$ vacancy is isotropic and the first radiation is the dipole K-L transition. The L_3 vacancy can be aligned along the direction of the photon emission, and the angular distribution of the subsequent L_3-MM Auger electrons has exactly the same form (10) as in the direct photoionization of the $2p_{3/2}$ subshell. However, as was noticed by Cooper [8], the alignment of the L_3 vacancy in the cascade transition is larger than in the direct photoionization. In the latter case, far from the threshold, the transition $2p \to \epsilon d$ dominates and the alignment parameter is small, $\mathcal{A}_{20}(3/2) \sim 0.05$ [12]. When the L_3 vacancy is produced in a cascade via the K-L_3 transition, its alignment can be obtained using expressions (9,10):

$$\mathcal{A}_{20}(J_1) = \overline{A}_2(J_0, J_1)/\overline{A}_0(J_0, J_1)$$
$$= (-1)^{J_1+J_0+1} \hat{J}_1 \sqrt{\frac{3}{2}} \begin{Bmatrix} J_1 & 1 & J_0 \\ 1 & J_1 & 2 \end{Bmatrix} \quad (14)$$

and in the considered case $\mathcal{A}_{20}(3/2) = 0.5$. This formal result can be easily understood. In the non-relativistic approximation alignment of the $2p_{3/2}$ vacancy is determined by the alignment of its orbital angular momentum:

$$\mathcal{A}_{20}(2p_{3/2}) = \frac{1}{\sqrt{2}} \mathcal{A}_{20}(l=1) = \frac{w_1 - w_0}{2w_1 + w_0} \quad (15)$$

where w_m ($m = 0, 1$) are populations of the magnetic substates. When the photon is emitted from the s-state only $m = \pm 1$ substates are populated (the quantization axis is along the emission direction), therefore $\mathcal{A}_{20}(3/2) = 0.5$.

The case of e-γ correlations in which the first radiation is an autoionization electron and the second is a fluorescence photon was studied experimentally in [5]. In addition to angular correlations the polarization of the emitted photon was measured. The theory of such measurements was developed in [15].

WHAT CAN BE LEARNED FROM THE CASCADE ANGULAR CORRELATION MEASUREMENTS?

Spectroscopic Aspect

As mentioned in the introduction, the coincidence technique was originally used in studying the vacancy cascades with the aim of selecting a particular pathway in the cascade and therefore simplifying the measured spectrum. In this way a better understanding of the cascade is achieved. For example, coincidence measurements by simultaneously detecting two Auger electrons emitted upon the two-step decay of the inner-shell resonances in noble gases [4] permitted the unraveling of the complex Auger spectrum and the identification of particular lines, which is especially difficult in the regime of low kinetic energies where first and second-step Auger lines strongly overlap. These measurements were angle integrated. The angle-resolved coincidence measurements [7,9] provide more information on the studied transitions. The angular correlation patterns are very characteristic for particular momenta of the atomic states involved. Experience in the angular correlation measurements of nuclear cascades shows that such measurements can be used effectively for determining the angular momenta and parities of resonances.

Decay Dynamics

Over the last several years extensive study of the angular distribution of Auger electrons and especially of resonant Auger electrons has revealed high

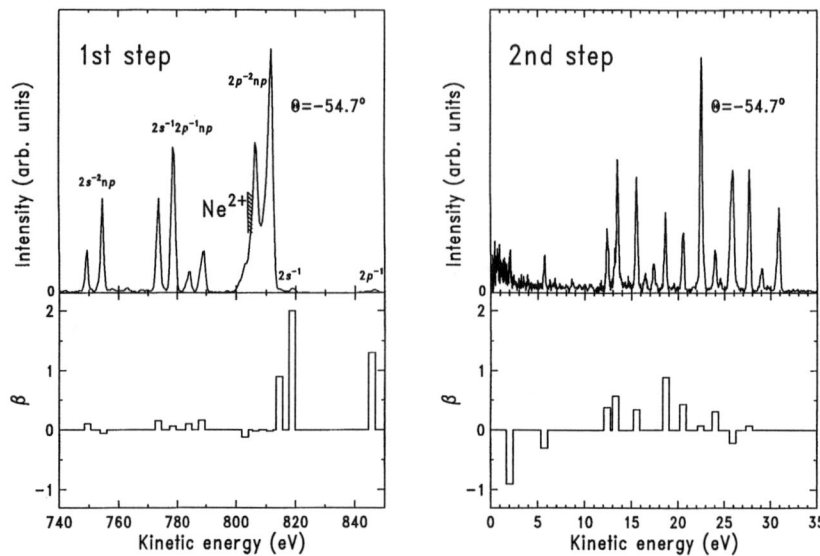

FIGURE 1. Spectra and anisotropy parameters $\beta = \alpha_2 \mathcal{A}_{20}$ for cascade Auger electrons folowing the $1s \to 3p$ photoexcitation of Ne measured in non-coincident experiment [7]. Left panel: first step $K\text{-}LL$ Auger electrons, right panel: second step Auger electrons.

sensitivity of the anisotropy parameters to such dynamical effects as initial and final state configuration interaction, exchange and relaxation effects [16–19]. Dynamics of the decay process affect the anisotropy coefficients through the values of the decay amplitudes and to a lesser extend through their phases. As it was demonstrated recently the anisotropy coefficients may depend crucially on the mixing of configurations [20,21]. Similar sensitivity to dynamical effects can be expected for the angular correlation in cascades since the correlation coefficients (3–5) contain, in general, the transition amplitudes.

As an example, consider recent experimental data on the angular distributions of Auger electrons from the resonant $1s \to 3p$ photoexcitation of Ne atom [7]. It is known that this resonance decays predominantly via $K\text{-}LL$ Auger transitions while the $3p$ electron plays a role of a spectator. The residual ion with a vacancy in the L_1 or L_2 subshells can decay further to a doubly ionized Ne with two vacancies in the valence L_3 subshell. (We do not discuss here more complicated modes of the Auger cascade.) In the discussed experiment the angular distributions of both steps in the cascade was measured both in a coincident and non-coincident manner. The striking feature of the results of these measurements is that the first step Auger electrons are emitted almost isotropically whereas the second step electrons reveal strong anisotropy of the angular distribution (see fig.1). Qualitative explanation of this phenomena is purely dynamical. The resonantly excited $1s^{-1}3p\,^1P_1$ state of Ne is strongly aligned along the direction of photon polarization. However, in the pure spectator model [22] the K-LL decay of the $1s$ vacancy in the presence of the $3p$ spectator is isotropic because the $1s$ hole itself cannot be aligned. One can say that the alignment of the $1s^{-1}3p$ state is entirely determined by the alignment of the spectator electron. Therefore the resulting intermediate $(2s2p)^{-2}3p$ state should be again strongly aligned due to alignment of the spectator electron. In the second step the excited $3p$ electron participates in the transition and therefore the angular distribution of the second Auger electron is expected to be strongly anisotropic as it is seen in the experiment. Small but non-zero anisotropy of the first step Auger transitions is due to configuration interaction. A similar effect was recently observed in resonant photoexcitation of Mg [21].

Post-Collision Interaction

Another dynamical effect which may be very important for the angular correlation in Auger cascades, especially near the ionization threshold, is post-collision interaction (PCI). In a stepwise description of the Auger cascade the Coulomb interaction between the photoelectron and the Auger electrons is ignored. It is well known, however, that in a single Auger decay this interaction leads to remarkable effects in the Auger electron spectrum: the lineshape becomes asymmetric and broadened, and its maximum is shifted in energy. The angular distribution of the Auger or autoionization electrons is also distorted by PCI [23,24]. One can expect therefore PCI effects in the cascade Auger emission as well. The influence of PCI in Auger cascades was indeed observed in the angle-integrated measurements of the photoion yields [2,25,3]. The theory of PCI in the Auger cascade induced by near-threshold photoionization was developed by Koike [26]. However, it does not include the angle-dependent PCI effects.

From the theoretical point of view inclusion of PCI means that the simple stepwise picture of the Auger cascade is no more valid. However, the general expression for the angular correlation function (1) is still valid. It may be considered as an expansion of the multiple ionization cross section in terms of bipolar harmonics for the case when two of the electrons are detected in an angle-resolved experiment. The breakdown of the step model is reflected in the fact that the coefficients of the expansion cannot be expressed in the form of (3). Also the values of k_1 and k_2 can be either even or odd.

Recently effects of PCI in the cascade Auger processes have been theoretically considered by Sheinerman [27] in the angle-dependent formulation. He considered the case when a photoelectron and two Auger electrons from a cascade have comparable and not small velocities. Using the eikonal approximation he obtained an analytical expression for the cross section differential in

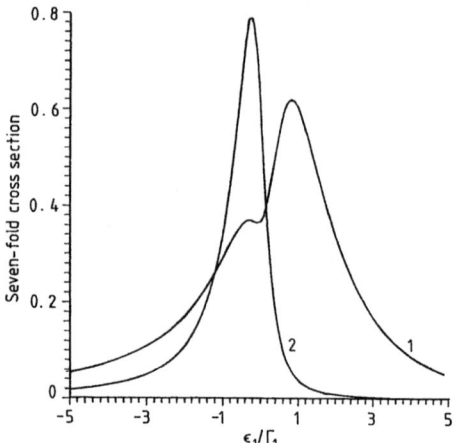

FIGURE 2. Line profiles of the photoelectrons calculated taking into account the PCI with two cascade Auger electrons [27]. Two curves correspond to the following emission angles (in rad): 1 : $\theta_1 = 0, \theta_2 = 0.3, \theta_3 = 0.1$; 2 : $\theta_1 = 0, \theta_2 = 0.6, \theta_3 = 1.4$. Other parameters of the calculation can be found in [27].

energy and angle of the emitted electrons. As it was expected the lineshapes in the electron spectrum depends on the relative velocities and angles of electron emission. Very interesting structure of the photoelectron line is predicted in conditions when both Auger electrons are emitted at small relative angle and hence are strongly interacting (see fig.2). This structure is not associated with an interference effect but rather with the addition of two PCI distortion factors due to the successive Auger decay of two quasistationary atomic states.

ALIGNMENT TRANSFER AND ANGULAR DISTRIBUTION OF CASCADE RADIATION IN NON-COINCIDENCE MEASUREMENTS

In the majority of the experiments so far the angular distributions of the Auger or autoionization electrons are measured in a non-coincident arrangement. Thus it is important to predict the angular distribution for any radiation in the cascade in conditions when the preceeding radiations are not detected. Consider again the simplest case of two sequential radiations of a photoexcited or photoionized atom. If the first radiation is not observed, one can obtain the angular distribution of the second radiation by averaging (1) over all possible directions of $\mathbf{n}_1$. Naturally, the angular distribution can be presented in a standard form [14] (z-axis is along the photon linear polarization):

$$W(\vartheta_2) = \frac{W_0}{4\pi} \left\{ 1 + \sum_{k=even} \alpha_k^{J_1 \to J_2} \mathcal{A}_{k0}(J_1) P_k(\cos \vartheta_2) \right\} \quad (16)$$

where $\alpha_k^{J_1 \to J_2}$ is the usual anisotropy parameter [13,14], and for the alignment tensor one obtains

$$\mathcal{A}_{k0}(J_1) = \rho_{k0}(J_1, J_1)/\rho_{00}(J_1, J_1)$$

$$= \mathcal{A}_{k_00}(J_0)\delta_{k_0 k} \left(\sum_{L_1} |\langle J_1, L_1 || \hat{O} || J_0 \rangle|^2 \right)^{-1}$$

$$\times \left(\sum_{L_1} \hat{J}_0 \hat{J}_1 (-1)^{L_1+J_0+J_1+k_0} \begin{Bmatrix} J_0 & J_1 & L_1 \\ J_1 & J_0 & k \end{Bmatrix} |\langle J_1, L_1 || \hat{O} || J_0 \rangle|^2 \right) \quad (17)$$

where $\mathcal{A}_{k_00}(J_0) = \rho_{k_00}(J_0, J_0)/\rho_{00}(J_0, J_0)$ is the alignment tensor of the initial state J_0 of the cascade. Equation (17) illustrates the general property, namely, if the radiation is not detected, the statistical tensors of the ion after the first decay can have only those components which were present in the initial state. In other words the symmetry of the ion is preserved in the cascade until the radiation is detected. For example the initial alignment induced by photoexcitation or photoionization may be transferred to the lower states in the cascade, but no other tensor components can appear. The efficiency of the alignment transfer in the cascade is determined by the quantum numbers of the states involved and in the general case by the ratios of the decay matrix elements. In a particular case of a single channel decay, for example if the first step is dipole photon emission, matrix elements cancel out and only angular coupling coefficients determine the alignment transfer.

Notice that only squares of the amplitudes of the first decay enter into (17), therefore different decay channels enter incoherently. In addition k is even having a maximum value $k \leq min(2J_0, 2J_1, 2L_2)$.

Evolution of the statistical tensors in an arbitrary cascade of radiative and Auger transitions was considered recently on the basis of Pauli kinetic equation in [28]. In this way expression (1) can be generalized to describe the directional correlation of any two radiations from an arbitrary cascade where the intermediate radiations are not observed.

ACKNOWLEDGMENTS

This paper was completed at Bielefeld University. The hospitality of Bielefeld University and financial support of Sonderforschungsbereich 216 is gratefully acknowledged. I am very greatful to U. Arp, T. LeBrun, S. Southworth, R. Wehlitz, U. Hergenhahn, J. Viefhaus, and U. Becker for sending the results of their experiments prior to publication. Discussions with V. Schmidt and R. Wehlitz were especially useful for formulating the concept of the report. I am thankful to E. Sidky for his help in preparation of the manuscript.

REFERENCES

1. Carlson T.A., and Krause M.O., *Phys. Rev.* **137**, A1655 (1965).
2. Levin J.C., Biedermann C., Keller N., Liljebi L., O C.-S., Short R.T., Sellin I.A., and Lindle D.W., *Phys. Rev. Lett.* **65**, 988 (1990).
3. Hayaishi T., Murakami E., Morioka Y., Shigemasa E., Yagishita A., and Koike F., *J. Phys. B* **27**, L115 (1994).
4. Von Raven E., Meyer M., Pahler M., and Sonntag B., *J. Electr. Spectr. Relat. Phenom.* **52**, 677 (1990).
5. Beyer H.-J., West J.B., Ross K.J., Ueda K., Kabachnik N.M., Hamdy H., and Kleinpoppen H., *J. Phys. B* **28**, L47 (1995).
6. Kabachnik N.M., in *Proceedings of the International Workshop on Photoionization 1992*, Eds. Becker U. and Heinzmann U., New York: AMS Press, Inc., 1993, pp. 20-23.
7. Viefhaus J., Avaldi L., Hentges R., Wiedenhöft M., Wieliczek K., and Becker U., *17th Intern. Conf. X-ray and Inner-Shell Processes, Hamburg, Germany, 1996. Book of Abstracts*, p. 220.
8. Arp U., LeBrun T., Southworth S.H., Cooper J.W., MacDonald M.A., and Jung M., *17th Intern. Conf. X-ray and Inner-Shell Processes, Hamburg, Germany, 1996. Book of Abstracts*, p. 192.
9. Wehlitz R., private communication, 1996.
10. Devons S., and Goldfarb L.J.B., *Handbuch der Physik*, Ed. Flügge S., Berlin: Springer-Verlag, 1957, v. 42, p. 362.
11. Ferguson A.J., *Angular Correlation Methods in Gamma-Ray Spectroscopy*, Amsterdam: North-Holland, 1965.
12. Berezhko E.G., Kabachnik N.M., and Rostovsky V.S., *J. Phys. B* **11**, 1749 (1978).
13. Berezhko E.G., and Kabachnik N.M., *J. Phys. B* **10**, 2467 (1977).
14. Kabachnik N.M., and Sazhina I.P., *J. Phys. B* **17**, 1335 (1984).
15. Kabachnik N.M., and Ueda K., *J. Phys. B* **28**, 5013 (1995).
16. Tulkki J., Kabachnik N.M., and Aksela H., *Phys. Rev. A* **48**, 1277 (1993).
17. Tulkki J., Aksela H., and Kabachnik N.M., *Phys. Rev. A* **50**, 2366 (1994).
18. Chen M.H., *Phys. Rev. A* **47**, 3733 (1993).
19. Aksela H., Jauhiainen J., Kukk E., Nommiste E., Aksela S., and Tulkki J., *Phys. Rev. A* **53**, 290 (1996).
20. Hergenhahn U., Lohmann B., Kabachnik N.M., and Becker U., *J. Phys. B* **26**, L117 (1993).
21. Whitfield S.B., Hergenhahn U., Kabachnik N.M., Langer B., Tulkki J., and Becker U., *Phys. Rev. A* **50**, R3569 (1994).
22. Hergenhahn U., Kabachnik N.M., and Lohmann B., *J. Phys. B* **24**, 4759 (1991).
23. Cordrey I.L., and Macek J.H., *Phys. Rev. A* **48**, 1264 (1993).
24. Kuchiev M.Ju., and Sheinerman S.A., *J. Phys. B* **27**, 2943 (1994).
25. Ueda K., Shigemasa E., Sato Y., Yagishita A., Ukai M., Maezawa H., Hayaishi T., and Sasaki T., *J. Phys. B* **24**, 605 (1991).
26. Koike F., *Phys. Lett.* **193A**, 173 (1994).

27. Sheinerman S.A., *J. Phys. B* **27**, L571 (1994).
28. Balashov V.V., Bodrenko I.V., Dolinov V.K., and Strakhova S.I., *Opt. i Spectrosc.* **77**, 891 (1994).

High Resolution Core-Level Electron Spectroscopy on Free Molecules and Atoms

S. Svensson

Department of Physics, Uppsala University, Box 530, S-751 21 Uppsala, SWEDEN

Abstract. Electron spectroscopy on free atoms and molecules using synchrotron radiation from an undulator is outlined. The high brilliance of the synchrotron radiation in combination with high efficiency monochromators and modern high resolution electron spectrometers make it possible to record electron spectra from core levels with a resolution of the order of 10-100 meV, even at high kinetic energies. New aspects of the spectroscopy are discussed, Auger resonant Raman spectra are presented, both in the case of atoms and molecules, and also angular resolved results at sub-natural resolution are reviewed. Particular attention is paid to the dissociation of core excited molecules in connection with resonant Auger decay. Vibrational resolution in core photoelectron spectra is also discussed using propene as an example, and the relevance of high precision chemical binding energy shifts is discussed.

INTRODUCTION

Recently, undulator beamlines for the study of free atoms and molecules have become operative. One important research field at these new beamlines is high resolution core electron spectroscopy, which is the subject of the present paper. We use here the terminology "high resolution", although its meaning changes with time. If we look in the literature concerning core electron spectroscopy, from the classical ESCA experiments in the fifties and onwards, one finds that the FWHM (full width at half maximum) that different groups have referred to as "high resolution" has decreased exponentially with the years. See Fig. 1.

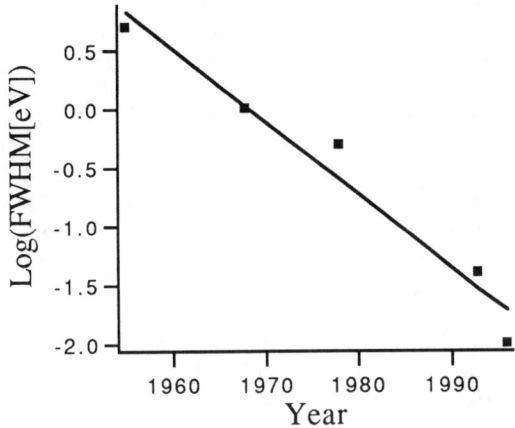

FIGURE 1. The development of what has been referred to as "high resolution" in core photoelectron spectroscopy as a function of time.

© 1997 American Institute of Physics

This type of exponential behaviour is very often presented in rapidly growing research fields. Another often quoted example is the curve for the typical brilliance of x-ray sources, as a function of time. The driving force behind the curve in Fig. 1 is improvements in technology. The early pioneers of electron spectroscopy in the fifties were working with standard x-ray anodes, there were substantial difficulties with the electron spectrometers, the electronic equipment was mostly based on drifting analogue electronics, voltages were often even hand controlled, the detectors were very simple etc. In the seventies rotating anode x- ray sources were developed, and new detectors, such as the channel-plate detector, became commercially available. Also the experiments could be computerized. The development from the middle of the eighties up to now has been dominated by the implementation of synchrotron radiation, where the latest improvements, a resolution giving a measured FWHM of 10-100 meV of the lines in the spectra, to a great extent relies on the use of undulators giving very high brilliance.

Are all these efforts to improve the resolution compellingly motivated? One might answer this question by a simple "yes", meaning that it always pays off to sharpen the scientific tools, but it is also possible to qualify the answer. Core electron spectroscopy is a very effective scientific tool to study the electronic structure of atoms and molecules and most often we use the sharpest, i.e. the outermost, core levels of the atoms. This means the C1s level for the elements up to Ne, the 2p level for the elements up to Ar etc. These core levels are of course broadened by the inherent lifetime; for the levels of interest in core electron spectroscopy the lifetime width varies in the range 30-300 meV and only from this consideration the latest efforts towards "high resolution" are motivated.

Another motivation comes from the study of vibrational sublevels, which have a typical splitting of a few tenths of an meV up to about 400 meV. Also the study of molecular field splitting motivates the efforts for increasing the resolution.

Tuneable synchrotron radiation sources make it possible to study autoionization from core excited states [1]. The exciting photon energy is tuned to one of the core excited states that exist in the vicinity of the ionization threshold and the following decay is observed by electron spectroscopy. In such experiments the core excited state acts as an intermediate scattering state [2]. One interesting consequence of this is that the lifetime width of the core excited state does not limit the width of the Auger lines. If the monochromator band-pass is decreased and the electron spectrometer has sufficient resolution, the observed line width is limited only by the lifetime width of the final state. This is referred to as the Auger resonant Raman effect (ARRE) [3]. From the experimentalists' point of view this means that it is meaningful to improve the total resolution all the way down to a few meV. It must be noticed that in the case of Auger electrons the kinetic energy is typically high. One needs to get a resolution up to a few meV at kinetic energies of several hundred eV, i.e. the resolving power $E/\Delta E$ must be $\approx 10\,000$ or larger both for the electron spectrometer and for the monochromator in order to observe the ARRE for the interesting core states, as discussed above.

In this paper we will discuss high resolution electron spectroscopy, using as example the experimental set-up at the beamline for high energy studies of atoms and molecules at the MAX laboratory, and we will present a review of some recent results.

THE EXPERIMENTAL SET-UP

BL 51 at the MAX laboratory has been operative since 1993 [4]. This beamline is situated on a straight section of the MAX I storage ring, which is a second generation synchrotron radiation facility. The energy of the stored electrons is relatively low, 550 MeV. The straight section is only about one meter long and very special precautions have been taken to enable the operation of a high energy undulator; a very small period has been used (35 mm) and the minimum gap is as low as 8 mm. The undulator harmonics are about 9 eV broad and monochromatization of the radiation is achieved by a plane grating monochromator of the well known SX-700 type from Zeiss which has been modified by using a plane-elliptical instead of an ellipsoidal refocussing mirror [5]. The beam line does not use windows to separate the experimental section from the monochromator. Instead a differential pumping/refocussing stage has been constructed, which reduces the pressure by 5 orders of magnitude [4]. The beamline has demonstrated a photon resolution of the order of 10 000 [6].

As pointed out above, an electron spectrometer must be designed with a resolution that matches the monochromator. We have chosen to work with truncated spherical or hemispherical electron analyzers. This choice is based upon the fact that the most highly resolved electron spectra have been obtained using such devices [7, 8]. Modern electron spectrometers utilize an electrostatic lens to retard the electrons before entering into the analyzer. The resolution of the analyzer depends on the size of the spherical electrodes. In principle it would be possible to achieve high resolution at high primary kinetic energies of the electrons using a very strong retardation. However, in practice the transmission of the spectrometer decreases very fast for a large retardation ratio, and therefore we have chosen to work with fairly big analyzers at BL 51, although this implies increased costs, mechanical and engineering problems.

So far two electron spectrometers have been used. The first one is a 144 mm truncated spherical sector analyser that was built by our group in Uppsala [9]. This instrument, which only allows the study of gas phase samples, is mounted at a fixed angle of 54.7 ° to the polarisation plane of the radiation. Assuming that the undulator radiation is linearly polarised (see below), this angle corresponds to the so called magic angle [10], where branching ratios are directly measured. This instrument was also constructed to allow transportation: the weight of the frame and of the pumping system were kept as small as possible. This spectrometer has also been used at other beamlines at the MAX laboratory.

In 1995 a new end station was placed into operation at beamline 51 . This end station is designed to make possible not only gas phase studies. An introduction chamber and a preparation chamber are connected to the system and the vacuum system is constructed so as to enable studies of liquid surfaces and thin films of polymers and molecular crystals.

The main component in the system is a 200 mm spherical electron analyser (SES-200) from the Scienta company in Uppsala [8]. Gas cells have been constructed for the study of free molecules and atoms and these cells have been provided with compensation electrodes in order to correct for the plasma potentials [7], which is necessary in order to get optimum resolution and stable operation.

It is a considerable engineering effort to make such a large instrument rotatable around the incoming x-ray beam. In order to achieve high resolution it is, as discussed above, advantageous to use a large radius of the hemispheres. On the other hand a large instrument implies a large weight. In order to measure angular distributions of the electrons, the instrument must be rotated around the photon beam. The rotation axis must be kept stable to a high precision, in practice one has to keep the axis centred within a few tenths of a mm. This rotational movement is performed on an instrument that includes the vacuum tank and pump. It weighs about 200 kg!

The solution to this mechanical problem is dynamical balances [11]. The instrument is attached to a constant force spring, and the force on the bearings can be kept very small. Rotary, differentially pumped seals were used in the design of the vacuum system. In Fig 2. we show a sketch of the new end station which has been described in detail in Ref. [12].

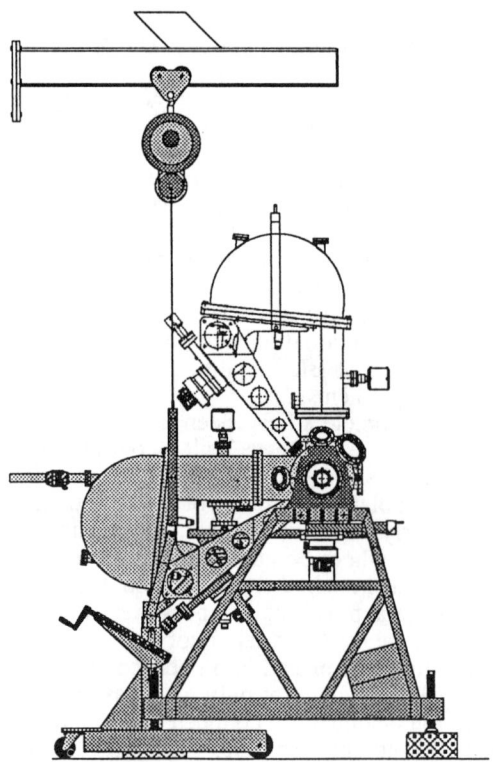

FIGURE 2. The new end station for the study of gases, liquids and solid films at BL 51 at the MAX laboratory.

Several methods can be used to measure the resolution of a monochromator at a synchrotron radiation facility. One method is to study a yield spectrum of a core level. If the monochromator has a photon band-pass that is substantially smaller than the typical lifetime widths of core levels of interest, one then has to apply deconvolution methods to determine the monochromator resolution. Another possibility is to use the He photoabsorption spectrum which has features that are very sensitive to the resolution. A third alternative, which is relevant when considering high resolution electron spectroscopy, is to use a valence photoelectron line. Such lines can be found with a negligible lifetime width. In Fig. 3 we show a valence photoelectron spectrum of Xe recorded with the new end station. The exiting photons had an energy of 65.1 eV, there is a Doppler contribution of 8 meV to the line width at this kinetic energy. The monochromator and electron spectrometer contributions to the FWHM were about 8 meV and 12 meV, respectively.

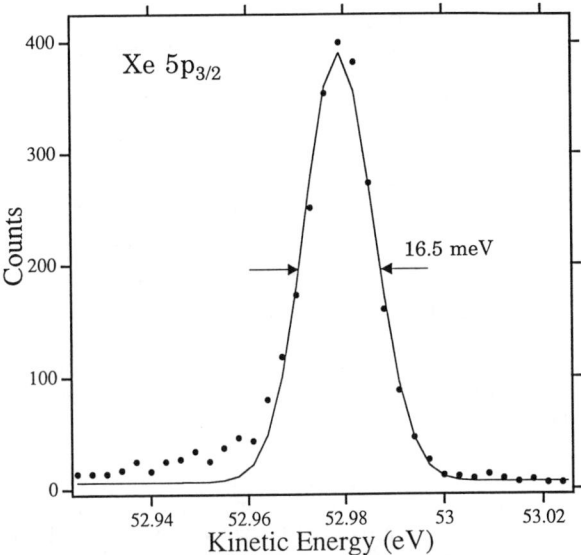

FIGURE 3. The Xe $5p_{3/2}$ photoelectron spectrum obtained at an excitation energy of 65.1 meV. The total measured FWHM is 16.5 meV (From Ref. [12]).

The MAX II third generation storage ring was taken into operation 1995. During 1997 the equipment at BL 51, including the new end station, will be relocated to an undulator beamline at this facility.

RECENT EXPERIMENTAL RESULTS

Angle-Resolved Auger Resonant Raman Spectroscopy on Atoms

Using synchrotron radiation, it is possible to create excited states with an inner shell vacancy. Such a state may decay via an Auger transition giving a singly ionized atom and an escaping electron in the final state. This process is often referred to as a Resonance Auger process and one may compare it with the Normal Auger process (Fig. 4).

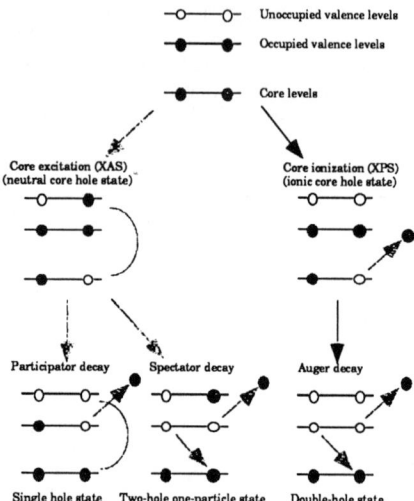

FIGURE 4. Photoionization, photoexcitation and the main Auger deexcitation processes. Resonant Auger deexcitation follows core excitation (left) and normal Auger decay follows core ionization (right).

In order to see how the Auger resonant Raman effect works, one should notice the fundamental difference between the normal Auger and the resonant Auger process. In the latter case there are only two particles in the final state, the outgoing Auger electron and the cation. The energy conservation law then gives a unique kinetic energy E_k to the outgoing electron:

$$E_k = h\nu - (E_f - E_i) \qquad (1)$$

Here, $h\nu$ is the photon energy and E_f and E_i are the final state and initial state energies, respectively. One should notice that in eq. (1) the intermediate core excited state is not present. This state only acts as an intermediate scattering state that blows up the cross section.

In fact, the theoretical line profile $I(E_k)$ of the resonant Auger line can be written as a product (note not a convolution) of two distributions [13-15]:

$$I(E_k) = \Phi(h\nu) L(h\nu - h\nu_0, \Gamma) \qquad (2)$$

where $\Phi(h\nu)$ gives the photon energy distribution from the monochromator and $L(h\nu - h\nu_0, \Gamma)$ is the absorption probability for a resonance centered around the nominal resonance energy $h\nu_0$. In order to get the observed electron line profile one also has to convolute by the electron spectrometer function.

From eq. (2) it can be seen that for a broad photon bandwidth, the absorption profile $L(h\nu - h\nu_0, \Gamma)$ dominates. This is for atomic core levels generally a Lorentzian distribution, and for a low resolution monochromator one will observe a Lorentzian like lineshape, provided that the electron spectrometer contribution is negligible. In this case the width of the line is almost equal to the inherent lifetime width of the core excited state. An interesting consequence of eq. (2) is that one observes a pure Lorentzian only in the case of an infinite photon band-pass, i.e. when white light excitation is used.

If $\Phi(h\nu)$ is the narrowest of the two distributions, the lineshape will be determined by the monochromator band-pass. In the latter situation, one observes subnatural line widths in the spectra, this is a consequence of the Auger resonant Raman effect.

The most obvious use of ARRE is of course that the resolution in resonant Auger spectroscopy can be pushed to the limits set by the lifetime width of the final state, by the monochromator band-pass, and by the electron spectrometer resolution. For gas phase studies the Doppler broadening, which depends on the molecular weight, the temperature and the kinetic energy, also contributes to the line width. For a heavy atom like Xe this contribution may be substantial, about 8 meV at a kinetic energy around 50 eV.

The use of the ARRE thus opens up the possibility to obtain very highly resolved spectra. Several studies of rare gas atoms have been performed recently at BL 51 at MAX laboratory and a comparison with very detailed calculations has given new insights, especially concerning relativistic and correlation effects [16-19].

Resonant Auger spectra exhibit very strong *angular dependence*, and it has been theoretically found that the so called angular anisotropy parameter for the Auger emission is a very sensitive test of theoretical models [20].

In Fig. 5 we show the $Kr\ 3d_{5/2}^{-1} 5p \rightarrow 4p^{-2} 5p$ resonant Auger spectrum, obtained in Ref. [12] using ARRE. The FWHM of the lines in this spectrum is about 33 meV. Two spectra are shown, one obtained at 0° and one at 90° angle versus the electric polarization vector. The angular dependence of the photoionization differential cross section $\frac{d\sigma_{if}}{d\Omega}$ is described by [10]

$$\frac{d\sigma_{if}}{d\Omega} = \frac{\sigma_{if}}{4\pi}\left[1 + \frac{\beta_{if}}{4}(1 + 3P_1 \cos 2\theta)\right] \quad (3)$$

here β_{if} denotes the angular anisotropy parameter ("the β parameter") ranging from $\beta_{if} = 2$ (emission along the electric vector axis) to $\beta_{if} = -1$ (emission perpendicular to this axis). P_1 is the degree of linear polarization. The 4s line in Fig. 5 is almost completely due to direct photoionization and its angular anisotropy parameter is 2. One can therefore determine P_1 directly from the spectra, the estimate is over 99%.

One can also notice the strong Auger angular dependence, with very different angular anisotropy of the decay. These spectra have been analysed in Ref. [21],

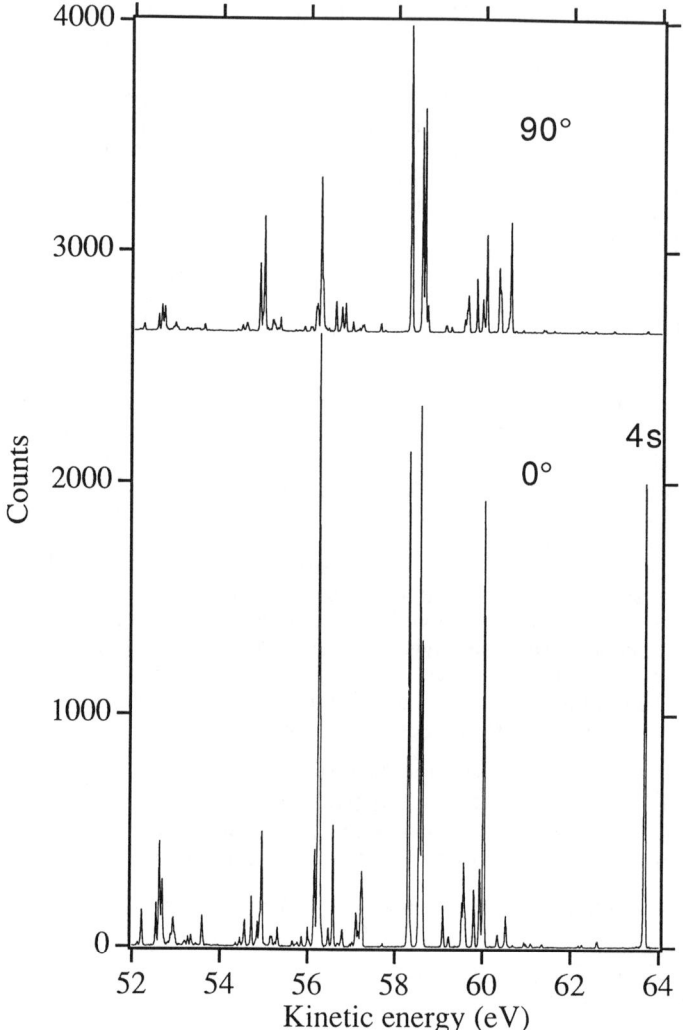

FIGURE 5. The $Kr\ 3d_{5/2}^{-1} 5p \rightarrow 4p^{-2} 5p$ angle-resolved resonant Auger spectrum obtained using ARRE. The FWHM of the electron lines is 33 meV, i.e. substantially smaller than the inherent lifetime width of the core excited state (about 83 meV). The 4s line is almost totally due to direct photoionization. It is hardly visible in the 90° spectrum, indicating over 99% polarization. The acquisition time for a spectrum was 9 minutes.

where also the $Kr\ 3d_{3/2}^{-1}5p \rightarrow 4p^{-2}5p$ Auger resonant Raman spectra were presented. The resolution in the spectra allowed a very detailed analysis using the most recent theoretical calculations on the angular anisotropy of the Auger decay [20,21]. Earlier only group β_{if} values for non-resolved lines had been obtained experimentally. Using ARRE the individual transitions to all but a few of the sublevels of the $4p^{-2}5p$ configuration can be identified in the spectra. In Fig. 6 we show a detail of the decay spectrum.

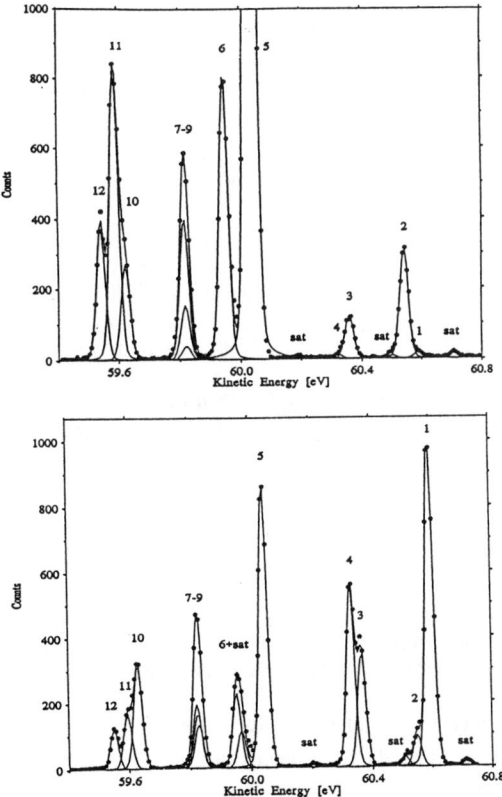

FIGURE 6. A detail of Fig. 5. One can e.g. notice the positive $\beta=1.07$ for line 5 in comparison to $\beta=-0.97$ for line 1 (From Ref. [21]). Upper spectrum was recorded at 0° and lower spectrum at 90° versus polarization plane of beam.

Recently, angle-resolved ARRE results for Xe, obtained at the Advanced Light Source in Berkeley, have also been reported [22].

Radiationless Raman Scattering from Dissociating Core Excited States

X-ray radiationless resonant scattering from molecules exhibits special features. In this case one also has to take into consideration the molecular dynamics associated with the core excitation and the Auger decay. The situation can be quite complicated since both the excited intermediate state and the final state can be bound or dissociative. The characteristic dissociation time scale can be comparable to the typical time scale for the decay. It was first observed by Morin and Nenner [23] that the resonant Auger spectrum of HBr, when exciting the core electron to the anti bonding σ^* level, revealed lines that could be explained as originating from a decay in the core excited Br atom. Thus the ionization process was considered to proceed in three steps:

$$h\nu + HBr \rightarrow (HBr)^* \rightarrow H + Br^* \rightarrow H + Br^+ + e^- \qquad (4)$$

It is interesting to see how ARRE is affected in such a case. The first observation of ARRE for a molecule in the VUV region was made by Liu et al. in a study of the HBr molecule [24]. A study of the HCl molecule at higher resolution performed at the MAX laboratory has been presented by Kukk et al. [25] and we will here shortly review this study. The different situations of resonant and non-resonant ionization of HCl are illustrated in Fig. 7.

The spectra in this figure were obtained with monochromator band-passes of 440 meV and 100 meV. Panel A in the figure displays an Auger line obtained at the Cl $2p_{3/2} \rightarrow \sigma^*$ resonance at 201.0 eV. Since the core excited state in this case is strongly dissociative, the transition is atomic like and can be explained as an "atomic" $2p_{3/2} \rightarrow 3p^{-2}$ (^1D) Auger transition. In this case dissociation precedes the decay. From eq. (4) there are three particles in the final state, which share the available energy. In this respect the situation is similar to a normal Auger process, where there also are three particles in the final state, the doubly ionized molecule, the photoelectron and an Auger electron. In the case shown in panel A in Fig. 7, no Raman narrowing is observed, since the excess energy can be taken up by the escaping hydrogen atom.

In panel B we display an Auger line obtained at 204.4 eV, corresponding to the $2p_{3/2} \rightarrow 4s\sigma$ excitation. The $4s\sigma$ is a non bonding Rydberg orbital and the core excited state is bonding. The Auger transition is therefore molecular, no dissociation is involved, and we have two particles in the final state; the molecular cation and the Auger electron. As in the case of an atom, the energy is uniquely distributed between the cation and the Auger electron. Consequently a clear Raman narrowing is observed in the line profile.

Finally, in panel C the HCl $2\pi^{-1}$ photoelectron line, obtained at 204.4 eV is displayed. The participator contribution to this line is small, the inherent width of the final state is negligible, and therefore we observe the normal behaviour for non resonant photoionization, namely that the width of the line is determined by the monochromator and electron spectrometer resolution (and Doppler broadening).

When dissociation is involved, the concepts of "atomic decay" and "molecular decay" do not give an adequate description of the process.

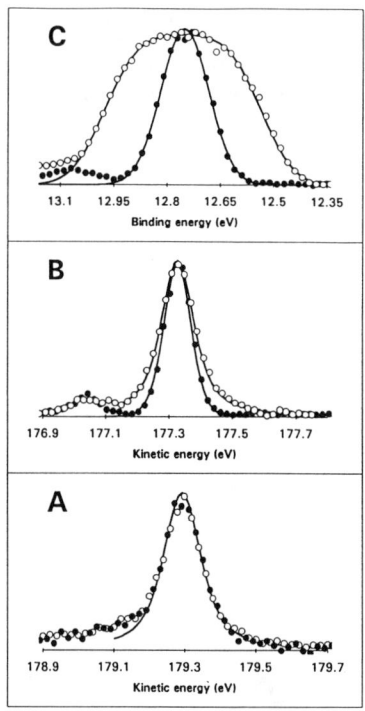

FIGURE 7. Electron spectra of HCl taken at A) the 2p→ σ* resonance where dissociation takes place. B) The 2p→ 4sσ resonance. The intermediate state is bonding. C) Shows the 2π photoelectron line at 204.4 eV photon energy. Note the Raman behaviour in B, which is absent in A (From Ref. [25]).

In reality one has both "molecular decay lines" (as broad structures) and also "atomic like Auger lines" as narrow peaks in the spectrum. The relative strength of these processes depends on the time scales for Auger decay and dissociation. A theoretical treatment of this complex behaviour for the HCl case has been obtained by a semiclassical method in Ref. [26]. In this report the dissociation was simulated using a molecular dynamics approach and the decay was assumed to follow an exponential behaviour. Thus, the Auger spectrum was simulated as an addition of spectra over a large time interval, covering the decay time scale.

In a recent theoretical review [27] Gel'mukhanov and Ågren have given the full quantum mechanical treatment of x-ray inelastic scattering involving dissociative states. This paper also covers the case of ARRE. Many interesting features are predicted from the full quantum mechanical treatment. For example, one should be able to directly measure the vibrational wave function and an oscillatory behaviour is also predicted in parts of the spectra.

Here we will discuss the resonant Auger process involving dissociative states using a simple model, based on the concept of the "core hole clock" [28]. In this approach one calculates a typical dissociation time for the core excited molecule by comparing the total "molecular" and "atomic" contributions to the spectra.

In Fig. 8 we show the resonant Auger spectrum of H$_2$S obtained at 164.35 eV. The spectrum, taken at BL 51 at MAX, has been reported in Ref. [29]. At this

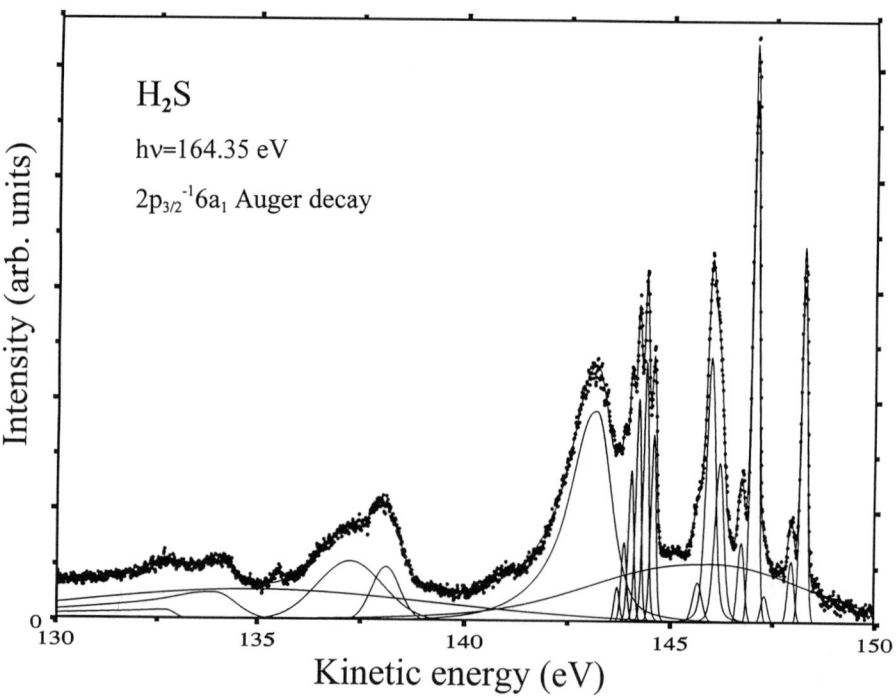

FIGURE 8. Auger decay spectrum of H$_2$S after resonant excitation to the $S\ 2p_{3/2}^{-1}6a_1$ state which dissociates rapidly. The narrow lines correspond to Auger transitions in the HS fragment, whereas the broad features of the background are due to Auger transitions in the non dissociated molecule. By comparing the intensities the dissociation time can be calculated, using the "core hole clock" approach (From Ref. [29]).

energy one reaches the $S\ 2p_{3/2}^{-1}6a_1$ core excited state, which rapidly dissociates into a core excited HS* fragment and a hydrogen atom. In a study at lower resolution, the narrow decay lines were interpreted as originating from the Auger decay of the core excited fragment to HS$^+$ by Aksela *et al.* [30].

One can see two types of lines in Fig. 8. The lines corresponding to transitions in the dissociated HS* fragment are very narrow, one can observe vibrational substructure, corresponding both to vibrationally excited intermediate states ("hot bands") and to transitions to vibrationally excited levels in the final HS$^+$ cation. Resonant Auger spectroscopy gives us a remarkable opportunity to study the free HS radical.

One should notice in Fig. 8 the very broad structures that form the background. These structures are several eV broad and are due to transitions in the molecule, during the very short time when the core excited molecule is still dissociating. In Ref. [29] one introduced the ratio:

$$n(decays(H-S^*H)) = \frac{\text{"Molecular" Auger Intensity}}{\text{Total Auger Intensity}} \quad (5)$$

Assuming that the dissociation time, t_D is short compared to the decay time τ, one gets, by integration of the exponential decay equation, a simple formula for the dissociation time:

$$t_D = -\tau \ln(1 - n(\text{decays H-S*H})) \quad (6)$$

An integration of the spectrum in Ref. [29] gave n(decays H-S*H)=0.43. The lifetime width of the S 2p core hole is 70 meV giving a value of $\tau = 9.4$ fs. The dissociation time according to eq. (6) is then 5.3 fs. The lifetime of the core hole has been used as a time standard to calculate the dissociation time, thus a "core hole clock" measures time intervals in the femtosecond region.

Dissociation of Neutral Doubly Excited Core Hole States

The first observations of the neutral dissociation phenomenon in resonant Auger spectroscopy of molecules were made for singly core excited states, i.e. the excitation energies were below the ionization edge and the observed fragment spectra were associated with resonant Auger spectra.

There are several non resonant mechanisms leading to normal Auger satellites. The core ionization can be accompanied by a shakeup or shakeoff process giving an excited intermediate state. A valence electron may shake up or shake off during the Auger decay, and finally, for excitation energies near to the threshold, we may observe postcollision recapture.

The first observations of structures associated with a normal Auger spectrum that clearly resonate were made at BL 51 at MAX [31]. In Fig. 9 we show the LVV Auger electron spectrum of H$_2$S excited at several photon energies ranging from 171 eV to 180.5 eV. One can observe several weak structures at higher kinetic energies than that of the main line. The structures resonate at an energy around 176 eV and the structures do not disperse with the photon energy.

In the absorption spectrum there exists a several eV broad structure around 176 eV photon energy which has been assigned to multielectron processes (Neutral Doubly Excited Core Hole States). The resonant structures in Fig. 9 cannot be due to molecular Auger decay. The bandwidth of the photons used to excite the spectra was substantially smaller than the width of the structure around 176 eV. This width is determined by dynamical effects in the molecule. Therefore, radiationless Raman Scattering from these states in the molecule would give lines that disperse with photon energy. Deexcitation in the molecule would lead to two particles in the final state and energy conservation would lead to a dispersion of the electron kinetic energy with the photon energy.

The weak structures in Fig. 9 must therefore be explained in terms of a rapid dissociation.

The core doubly excited molecule dissociates before the Auger decay that consequently must take place in a core excited HS* radical. Similar results have later been obtained for the nitrogen molecule by Neeb et al. [32]. One can foresee new studies in these directions in the future.

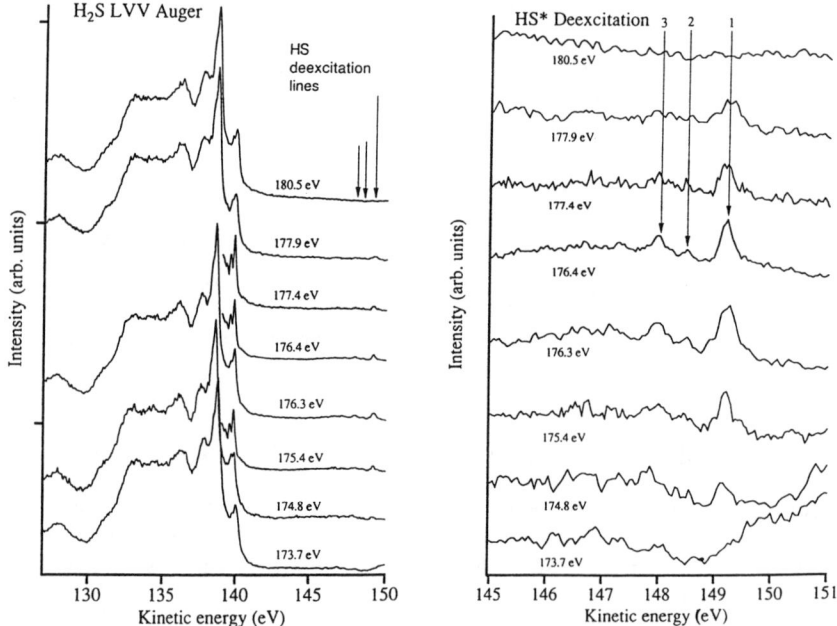

FIGURE 9. Resonant structures associated to the normal LVV Auger spectrum of H_2S. Core doubly excited states above the ionization threshold dissociate before the decay, giving rise to the small resonant structures at 148 eV, 148.5 eV and 149.2 eV. (From Ref. [31]).

Lifetime Vibrational Interference in Molecular Resonant Auger Spectra

So far dissociative intermediate states in the resonant Auger process have been discussed. In the case of bound states one has to consider the lifetime vibrational interference effect [33]. Spectra have now been obtained using ARRE and it has been possible to achieve vibrational resolution, both in the excitation and in the deexcitation. In a recent study of Osborne et al. [34] the resonant Auger decay following C1s-π* excitation in the CO molecule was studied. In Fig. 10 we show a result from this report.

The Lorentzian tails of the resonances from different vibrational sublevels in the intermediate core excited states overlap, and consequently one will observe coherent interference between the decay from these sublevels to a particular final state. High

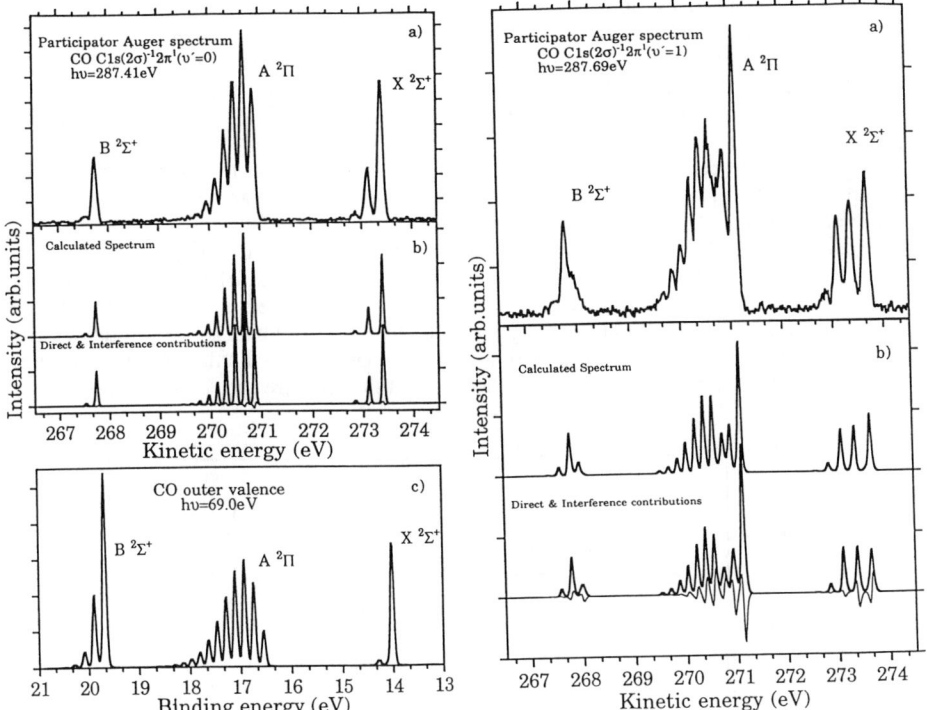

FIGURE 10 The participator Auger spectrum of CO. To the left (upper part) the spectrum excited at 287.41 eV, corresponding to an excitation of the v'=0 level in the intermediate core excited state. In the lower part the non resonant photoelectron spectrum is shown. In the left part the spectrum recorded at an energy of 287.69 eV (C1s-π* excitation) is shown. The v'=1 vibrational state has been reached. However, the lifetime tail of the v'=0 state falls under the same excitation and a strong interference contribution is seen in the spectrum.

resolution implies that this process can be studied in detail. In particular, one was able in Ref. [34] to measure the very small lifetime vibrational interference energy shift.

Molecular Core Photoelectron Spectroscopy

Core photoelectron spectroscopy has been developed since the early pioneering experiments in the middle of the fifties. The development of the field was illustrated in Fig. 1 where the concept of "high resolution" is seen to be time dependent. In this section we will discuss the present level of experiments and the objectives to spend money and effort in the field of core photoelectron spectroscopy, which by now is over 30 years old.

One of the most important aspects of core photoelectron spectroscopy is the determination of chemical binding energy shifts. With a modern undulator beamline it is possible to obtain a resolution well below 100 meV for the narrow core levels that are used to determine the chemical shifts. As a matter of fact, the lifetime widths are now in many cases the limiting factor. At this resolution level it is possible to unveil not only the vibrational structure of the core photoelectron lines, but it has recently also been demonstrated that molecular field effects can be observed also for deeper core levels [35].

Binding energy shifts can be correlated with various chemical properties of molecules and in the context of free molecules this is one of the most important motivations for such studies. A discussion has been given e.g. in Ref. [36]. If one looks at a typical chemical property such as the activation energy one finds that this quantity is typically measured with a precision of about 1 kcal/mole, i.e. of about 43 meV/molecule. In connection to chemistry it is therefore interesting to obtain the

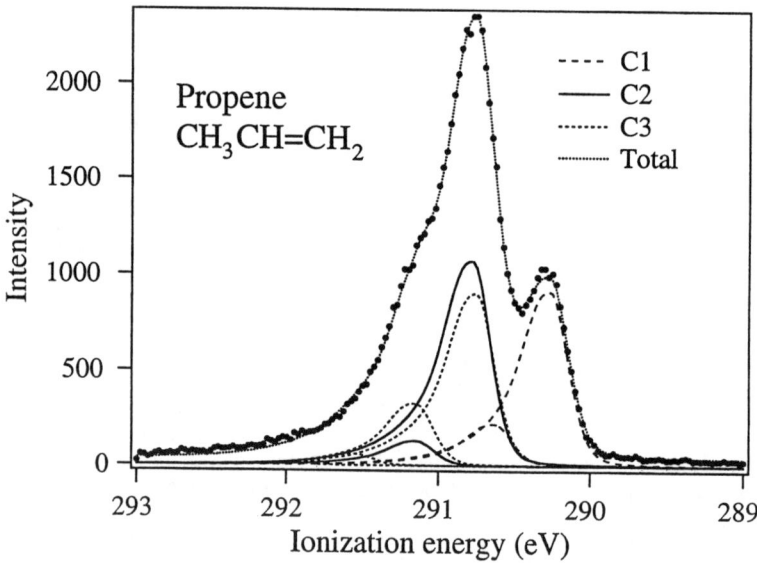

FIGURE 11. The C1s core photoelectron spectrum of propene. The molecule contains three carbon atoms. However, the vibrational substructure associated with the respective chemically shifted electron lines differs. With the resolution at an undulator beam line it is possible to deconvolute the spectrum properly.

chemical binding energy shift with a precision at this level or better. Molecular field splitting and vibrational splitting of core levels typically fall in the range of 10-400 meV. In order to obtain high precision shifts it is therefore desirable to unveil such substructure of the electron lines. In Fig. 11 we show a result that is typical for what can be obtained at a high energy undulator beam-line [37].

The C1s photoelectron of the propene molecule, $C(1)H_3 - C(2)H = C(3)H_2$ is shown in this figure. The spectrum shows three distinct features, and it would be tempting to assign these structures to the three carbon atoms. However, careful studies of similar molecules made in Refs. [37, 38] have shown that the vibrational substructure associated with carbon atoms with different numbers of attached hydrogen atoms gives rise to distinct vibrational profiles. It is interesting to notice that the shoulder on the high binding energy side of the highest structure in the spectrum of Fig. 11 is not due to a chemically shifted line, but must be associated to the v'=1 level of the C(1) 1s core electron line. The chemical shift between the C(1) and C(2) carbon atoms is therefore very small. The binding energy shift between the three atomic positions can be determined to a precision of at least 20 meV, which enables a comparison of the data with chemical parameters of interest.

CONCLUSIONS

Recent results from high resolution core electron spectroscopy have been discussed. Experimental techniques have been reviewed and results concerning different aspects of core excitation and core photoionization have been presented. Experimental observations of the Auger resonant Raman effect for free atoms and molecules have been reviewed and special attention has been paid to the situation where a molecule undergoes dissociation in connection with core excitation. It has been shown that the resolution presently obtainable at undulator facilities enables the regular detection of molecular field effects and vibrational sublevels in core photoelectron spectra, and the relevance of this for the determination of chemical shifts has been discussed.

ACKNOWLEDGMENTS

The author want to thank the research group in Uppsala: Dr. Andrus Ausmees and the research students Stuart Osborne and Stefan Sundin. Special gratitude is due to Ing. Jan-Olof Forsell who was responsible for the design of the electron spectrometers described in this report.

Our colleagues in Oulu, Finland, Profs. S. Aksela and H. Aksela and their group are thanked for a long and successful collaboration at BL 51 at MAX laboratory ("the Finnish beamline").

REFERENCES

1. W. Eberhardt, G. Kalkoffen, C. Kunz, *Phys. Rev. Lett.* **41**, 156 (1978).
2. T. Åberg, B. Crasemann in *Anomalous (resonant) x-ray scattering*, edited by K. Fischer, G. Materlik and C. Sparks (Elsevier, Amsterdam, 1994)

3. G.S. Brown, M. H. Chen, B. Crasemann, G.E. Ice, *Phys. Rev. Lett.* **45**, 1937 (1980).
4. S. Aksela, A. Kivimäki, A. Naves de Brito, O.-P. Sairanen, S. Svensson, J. Väyrynen, *Rev. Sci. Instrum.* **65**, 831 (1994).
5. R. Nyholm, S. Svensson, J. Nordgren, A. Flodström, *Nucl. Instr. Meth.* **A246**, 267 (1986).
6. S. Aksela, A. Kivimäki, O.-P. Sairanen, A. Naves de Brito, E. Nõmmiste, S. Svensson, *Rev. Sci. Instrum.* **66**, 1 (1995).
7. P. Baltzer, L. Karlsson, M. Lundqvist, B. Wannberg, *Rev. Sci. Instrum.* **64**, 2179 (1993).
8. N. Mårtensson, P. Baltzer, P.A. Brühweiler, J.-O. Forsell, A. Nilsson, A. Stenborg, *J. Electron Spectrosc. Relat.Phenom.* **70**, 117 (1994).
9. S.J. Osborne, A. Ausmees, J.-O. Forsell, B. Wannberg, G. Bray, L.B. Dantas, S. Svensson, A. Naves de Brito, A. Kivimäki, S. Aksela, *Synch. Rad. News* **7**, 25 (1994).
10. V. Schmidt, *Rep. Prog. Phys.* **55**, 1483 (1992).
11. J. Guo, N. Wassdahl, P. Skytt, S.M. Butorin, S.-C. Duda, C.J. Englund, J. Nordgren, *Rev. Sci. Instrum.* **66**, 1561 (1995).
12. S. Svensson *et al.*, *Rev. Sci. Instrum.* **67**, 2149 (1996).
13. F. Gel'mukhanov, H. Ågren, *Phys. Rev. A* **49**, 4378 (1994).
14. G. Armen, H. Wang, *Phys. Rev. A* **51**, 1241 (1995).
15. S. Aksela, E. Kukk, H. Aksela, S. Svensson, *Phys. Rev. Lett.* **74**, 2917 (1995).
16. O.-P. Sairanen, H. Aksela, S. Aksela, J. Mursu, A. Kivimäki, A. Naves de Brito, E. Nõmmiste, S.J. Osborne, A. Ausmees, S. Svensson, *J. Phys. B: At. Mol. Opt. Phys.* **28**, 4509 (1995).
17. A. Kivimäki, A. Naves de Brito, S. Aksela, H. Aksela, A. Ausmees, S. J. Osborne, L.B. Dantas, S. Svensson, *Phys. Rev. Lett.* **71**, 4307 (1993).
18. H. Aksela, S. Aksela, O.-P. Sairanen, A. Kivimäki, A. Naves de Brito, E. Nõmmiste, J. Tulkki, S. Svensson, A. Ausmees, S.J. Osborne, *Phys. Rev. A* **49**, R4269 (1994).
19. H. Aksela, O.-P. Sairanen, S. Aksela, A. Kivimäki, A. Naves de Brito, E. Nõmmiste, J. Tulkki, A. Ausmees, S.J. Osborne, S. Svensson, *Phys. Rev. A* **51**, 1291 (1995).
20. J. Tulkki, H. Aksela, N.M. Kabachnik, *Phys. Rev. A* **50**, 2366 (1994).
21. H. Aksela, J. Jauhiainen, E. Nõmmiste, S. Aksela, S. Sundin, A. Ausmees, S. Svensson, *Phys. Rev. A* **54**, 605 (1996).
22. B. Langer, N. Berrah, A. Farhat, O. Hemmers, J.D. Bozek, *Phys. Rev. A* **53**, R1946 (1996).
23. P. Morin, I. Nenner, *Phys. Rev. Lett.* **56**, 1913 (1986).
24. Z.F. Liu, G.M. Bancroft, K.H. Tan, M. Schachter, *Phys. Rev. Lett.* **72**, 621 (1994).
25. E. Kukk, H. Aksela, S. Aksela, F. Gel'mukhanov, H. Ågren, S. Svensson, *Phys. Rev. Lett.* **76**, 3100 (1996).
26. E. Kukk, H. Aksela, O.-P. Sairanen, S. Aksela, A. Kivimäki, E. Nõmmiste, A. Ausmees, A. Kikas, S.J. Osborne, S. Svensson, *J. Chem. Phys.* **104**, 4475 (1996).
27. F. Gel'mukhanov, H. Ågren, *Phys. Rev. A* **54**, 379 (1996).
28. O. Björneholm, A. Nilsson, A. Sandell, B. Hernäs, N. Mårtensson, *Phys. Rev. Lett.* **68**, 1892 (1992).
29. A. Naves de Brito, Al. Naves de Brito, O.Björneholm, J.S. Neto, A.B. Machado, S. Svensson, A. Ausmees, S.J. Osborne, L. J. Saethre, H. Aksela, O.-P. Sairanen, A. Kivimäki, E. Nõmmiste, S. Aksela, T.C.P.I.p. 1996, *Theor. Chem. Phys.* In press (1996).
30. S. Aksela, H. Aksela, A. Naves de Brito, G.M. Bancroft, K.H. Tan, *Phys. Rev. A* **45**, 7948 (1992).
31. S. Svensson, H. Aksela, A. Kivimäki, O.-P. Sairanen, A. Ausmees, S.J. Osborne, A. Naves de Brito, E. Nõmmiste, G. Bray, S. Aksela, *J. Phys. B : At. Mol. Opt. Phys.* **28**, L325 (1995).
32. M. Neeb, A. Kivimäki, B. Kempgens, H.M. Köppe, J. Feldhaus, A.M. Bradshaw, *Phys. Rev. Lett.* **76**, 2250 (1996).
33. F. Gel'mukhanov, L. Mazalov, A. Kontratenko, *Chem. Phys. Lett.* **46**, 133 (1977).
34. S.J. Osborne, A. Ausmees, S. Svensson, A. Kivimäki, O.-P. Sairanen, A. Naves de Brito, H. Aksela, S. Aksela, *J. Chem. Phys.* **102**, 7317 (1995).
35. S. Svensson *et al.*, *Phys. Rev. Lett.* **72**, 3021 (1994).
36. L.J. Saethre, T.D. Thomas, *J. Phys. Org. Chem.* **4**, 629 (1991).

37. L.J. Saethre, O. Sværen, S. Svensson, S. J. Osborne, T.D. Thomas, J. Jauhiainen, S. Aksela, Submitted to *Phys. Rev. A* (1996).
38. S.J. Osborne, S. Sundin, A. Ausmees, S. Svensson, L.J. Saethre, O. Sværen, S.L. Sorensen, J. Végh, J. Karvonen, S. Aksela, A. Kikas, Submitted to *J. Chem. Phys* . (1996).

Selection Rules in Resonant X-Ray Emission of Free Molecules

P. Glans*, P. Skytt, K. Gunnelin, J.-H. Guo, and J. Nordgren

Department of Physics, Uppsala University, Box 530, S-75121 Uppsala, Sweden
*Atomic Physics, Stockholm University, Frescativägen 24, S-10405 Stockholm, Sweden

Abstract. X-ray emission spectra obtained using energy-selective synchrotron-radiation excitation for small molecules such as CO, N_2, and CO_2 are shown. The spectra illustrate how the intensities are governed by the dipole selection rules associated with the absorption-emission process. It is shown, in particular, how the spectra can provide symmetry information of both occupied and unoccupied valence orbitals. The importance of the vibrational motion of the molecules is also stressed.

INTRODUCTION

The progress in synchrotron instrumentation has led to a resurgence in core-level spectroscopies. Soft x-ray emission spectroscopy which was previously limited to the usage of broad-band electron-gun and x-ray tube sources can now be performed with tunable narrow-bandpass photon-beam excitation at synchrotron facilities. The possibility to make selective resonant excitations adds a whole new dimension to x-ray emission studies, and a wealth of information, previously inaccessible, can now be obtained.

In this paper resonant and non-resonant x-ray emission spectra of small molecules such as CO, N_2, and CO_2 are presented. The resonant spectra are obtained by tuning the excitation energies at or near one (or a set of) resonance(s). The intermediate states of the absorption-emission (scattering) processes are thus core-excited states in the neutral molecule. In the fluorescent decay of these states the core hole is filled by a valence electron. We may distinguish between participator and spectator decay depending on whether the excited electron fills the hole or remains as a spectator while another valence electron fills the hole. For participator transitions the final states correspond to the electronic ground states, but not necessarily the vibrational ground states. The final states of the spectator transitions, on the other hand, correspond to one hole – one particle states. The non-resonant x-ray emission spectra are obtained by tuning the photon energies above the ionization threshold. The decay, in this case, is between core-hole and valence-hole states in the singly

© 1997 American Institute of Physics

ionized molecule. One of the main advantages of exciting with monochromatized synchrotron radiation instead of conventional sources is that it is possible to suppress shake-up and shake-off transitions in the primary excitation step by choosing suitable excitation energies. Non-resonant spectra more or less free from satellite lines due to multiple excitations can thus be obtained.

Absorption and emission of soft x-rays are governed by dipole selection rules. The aim of this paper is to illustrate how some of these rules manifest themselves in resonant and non-resonant x-ray emission spectra of small molecules.

EXPERIMENTAL

For shallow core holes the fluorescence yields are typically very low, normally a few parts per thousand, and for gas-phase samples the particle densities are inherently low. These circumstances make soft x-ray emission spectroscopy of free molecules demanding and, in order to get high enough count rates, it is important to combine a high-brightness source with a well-designed gas cell and an efficient spectrometer.

The experiments presented here were performed at beamline 7.0 of the Advanced Light Source (ALS) of the Lawrence Berkeley National Laboratory (LBNL) in Berkeley, California. The beamline is comprised of a 99 period, 5.0 m undulator and a spherical grating monochromator (1) . It provides a well-focused, linearly-polarized beam in the interaction region.

In these experiments the photon beam entered a gas cell through a silicon-nitride window with a thickness of about a 1000 Å and a diameter of about 250 μm. The emitted x-rays were viewed at a right angle through an aluminium-nitride-coated polyimide window, which also was about 1000 Å thick. The target gas pressure in the cell was typically in the 0.5 - 2.0 Torr range.

A grazing-incidence grating spectrometer (2) was used for detection of the emitted x-rays. It consists primarily of three fixed spherical gratings sharing a common adjustable entrance slit and a two-dimensional detector. The detector is moved to the focal position using three translation stages. The spectrometer was mounted so that its entrance slit was close to the gas cell and parallel to the direction of the exciting photon beam.

The chamber housing the spectrometer and gas cell can be rotated around the beam axis (3) allowing for angle-resolved measurements. The spectra presented here were recorded either parallel ($\theta = 0°$) or perpendicular ($\theta = 90°$) with respect to the electric vector of the exciting photon beam. An electrode mounted inside the gas cell was used to record absorption spectra. These spectra were used for calibration of the the excitation energies and, in some cases, for deducing the bandpass of the monochromator function, i.e. the energy width of the exciting photon beam.

RESULTS

Local Selection Rule

In an atomic system, one-electron dipole transitions are allowed if $\Delta l = \pm 1$. In a molecule (or solid) this selection rule can be used to estimate the strength of different core-valence transitions by employing the so-called one-center approximation (4). This approximation leads to a "local selection rule" where the spectral intensities for a molecule (or solid) reflect the local, atomic character of the valence levels at the core-hole site.

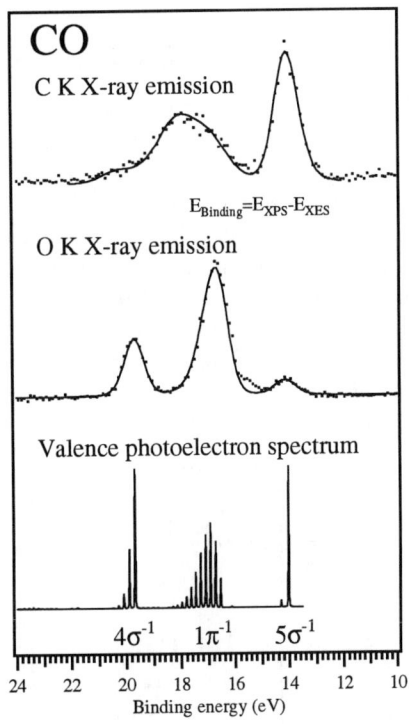

Figure 1. Non-resonant x-ray emission spectra of CO recorded by tuning the excitation energies a few eV above the ionization thresholds. Also, included is a valence photoelectron spectrum (5).

The local selection rule is exemplified in Figure 1, which shows non-resonant carbon and oxygen K emission spectra of CO. Because the core holes are of s-type ($C1s^{-1}$ and $O1s^{-1}$) the spectra reflect the p-type character of the valence orbitals. From the spectra we can deduce that the outermost, 5σ orbital is mainly of $C2p$ character with only a small amount of $O2p$ character. The 4σ orbital, on the other hand, has almost no $C2p$ character at all. The only peak which is strong in both spectra is due to the 1π orbital. That the 1π orbital has both substantial $C2p$ and $O2p$ character explains the strong bonding properties of this orbital. The removal of an electron from this orbital also causes a dramatic increase in bond length and, therefore, a large number of vibrational levels are populated.

The spectral intensities can often be very well predicted by employing the one-center approximation. However, "two-center" terms and/or relaxation and correlation may affect the intensities substantially in many cases. For small molecules these effects can be accounted for in *ab initio* calculations. The advantage with energy-selective excitation is, as mentioned earlier, that satellite peaks can be suppressed which enables a more accurate determination of experimental relative intensities. X-ray emission spectra of small molecules provide a critical test of theoretical calculations at various levels of sophistication and of the applicability of the one-center approximation.

Lifetime Interference

The traditional two-step model where the excitation and decay are treated as two independent steps is in general not correct in describing resonant as well as non-resonant x-ray emission. If there is more than one pathway to reach the same final state, that is if there are close-lying intermediate states that overlap energetically and which can decay to the same final state, it is necessary to account for the interference between the different pathways. This can be done by treating the excitation and decay as a one-step scattering process. In molecules the vibrational spacing is typically of the same order of magnitude as the lifetime width of the core-hole states and, therefore, lifetime-vibrational interference effects (6) commonly affect the vibrational band profiles associated with the various electronic transitions.

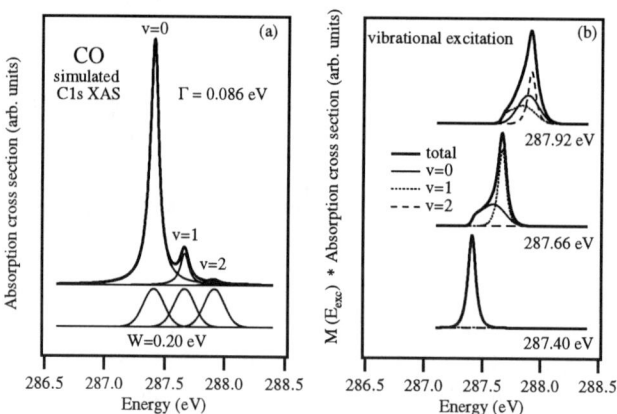

Figure 2. (left panel) Simulated $C1s \rightarrow 2\pi$ (π^*) x-ray absorption spectrum of CO. Resonant x-ray emission spectra were recorded using the monochromator photon energy distributions, $M(E_{exc})$, shown below the absorption spectrum. (right panel) Products of the photon energy distributions and the absorption cross section.

Figure 3. Resonant x-ray emission spectra obtained by tuning the excitation energies to the resonant energies of the $v = 0 - 2$ vibrational levels. Dashed and solid curves correspond to simulated vibrational band profiles obtained using two-step and one-step models, respectively.

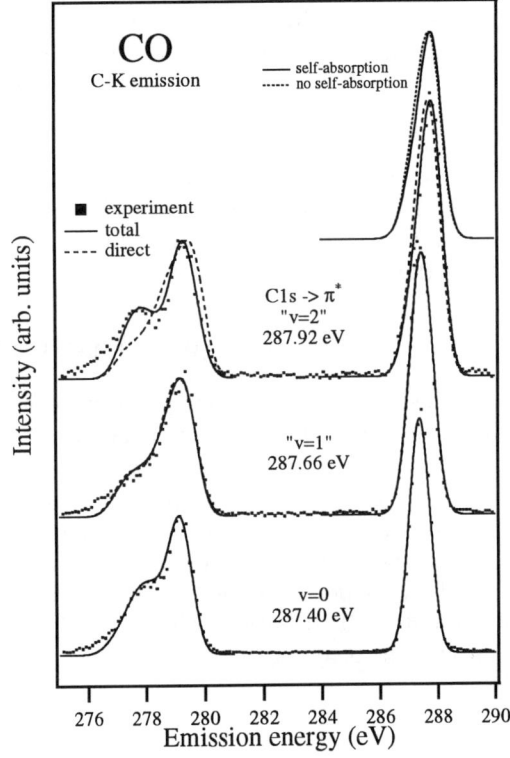

In Figure 2, a simulated $C1s \to \pi^*$ x-ray absorption spectrum of CO is shown (7). The spectrum is dominated by the $v = 0$ vibrational level, since the bond length of the core-excited state is very close to that of the ground state. In Figure 3, resonant x-ray emission spectra of CO obtained by tuning the excitation energy to the first three vibrational peaks in the absorption spectrum are displayed (7). The dashed and solid curves are simulated vibrational band profiles using potential curves from the literature. The dashed curves are obtained using a two-step model and the solid curves are obtained using a one-step model, which accounts for interference between the different vibrational levels of the intermediate state. When the excitation energy is tuned to the strong $v = 0$ level the likelihood of exciting any of the higher, much weaker, vibrational levels is negligible. Since there is only one intermediate state no interference effects are possible and both models therefore give the same result. At the $v = 1$ resonant energy, the tail of the $v = 0$ level gives an appreciable contribution. However, the interference effects are rather small and both models give similar results. In contrast, when the excitation energy is tuned to the $v = 2$ resonance, strong interference effects are observed and the two-step model fails to reproduce the experimental spectrum. The tails of the $v = 0$ and $v = 1$ levels are almost as strong as the $v = 2$ level, even on resonance, which explains the observed interference effects.

In the above example, the interference only affects the vibrational part of the transitions but not the electronic part and, hence, a two-step model can be used to calculate the intensities of the electronic transitions. However, in other cases, lifetime-interference effects may be significant among different close-lying intermediate electronic states and a one-step treatment has to be used for the electronic part as well.

Angular Distributions

It has been shown experimentally in some of the first resonant x-ray emission studies of free molecules, that the polarization (8) and angular (9) distributions are in general anisotropic. The reason for the anisotropies is the fact that the absorption cross section depends on the orientation of the molecule with respect to the electric vector of the exciting photon beam, and because of that an aligned ensemble of core-excited molecules is produced. The angular distributions depend on the symmetries of the final states of the emission process and the distributions can be used to gain symmetry information about the orbitals involved in the process.

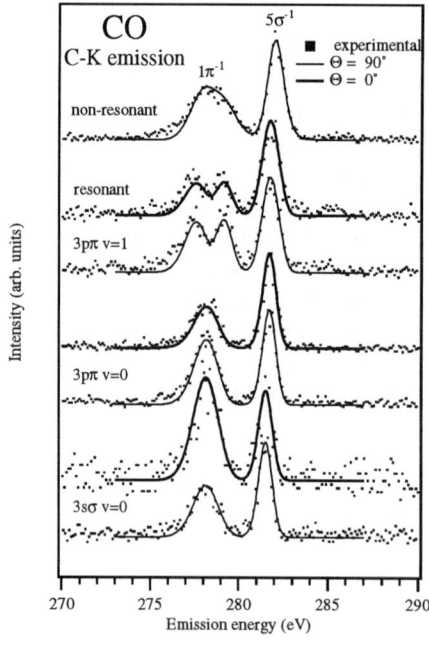

Figure 4. $C\text{-}K$ x-ray emission spectra of CO obtained by resonant excitation to the $3s\sigma$ ("v=0") and $3p\pi$ ("v=0" and "v=1") Rydberg resonances. Spectra recorded both at $\theta = 0°$ and $\theta = 90°$ are presented. A non-resonant spectrum is included for comparison. All the spectra were normalized to have the same peak height for the 5σ band. The solid lines show simulated vibrational band profiles.

To illustrate the angular dependence of resonant x-ray emission we show spectra of CO, recorded parallel ($\theta = 0°$) and perpendicular ($\theta = 90°$) to the electric vector of the synchrotron beam[7], in Figure 4. The spectra were obtained by tuning the excitation energies to the $C1s \rightarrow 3s\sigma$ and $C1s \rightarrow 3p\pi$ Rydberg resonances, respectively. Two vibrationally broadened peaks are observed in all the spectra. These peaks correspond to 1π and 5σ electrons filling the core holes, respectively. In the $3s\sigma$ resonantly excited x-ray emission spectra the 1π peak is enhanced at $\theta = 0°$ and reduced at $\theta = 90°$ in comparison to the 5σ peak. The opposite trend is observed in the $3p\pi$ spectra. The angular distributions of the emission peaks, in other words, depend on whether the core electron is promoted to a σ or π orbital. This implies that the spatial, σ or π, symmetry of the unoccupied orbital can be obtained from angle-resolved resonant x-ray emission spectroscopy.

It should be noted that there are three possible final states that can contribute to the 1π peak when exciting to a π orbital, since a $\pi^{-1}\pi^1$ configuration can be coupled to three singlet states: $^1\Sigma^+$, $^1\Sigma^-$, and $^1\Delta$. The angular distributions for these final states are different and the "total" angular distribution of the 1π peak, therefore, depends critically on the relative intensities of the transitions to those states. For example, the 1π and 5σ intensity ratio is almost identical at $\theta = 0°$ and $\theta = 90°$ in the π^* resonant x-ray emission spectra of CO. The reason, in this case, is that only the $^1\Sigma^-$ and $^1\Delta$ final states contribute to the main 1π peak, and the angular anisotropy becomes much smaller than in the $3p\pi$ resonant x-ray emission spectra (10).

Parity Selection Rule

Diatomic Molecules

For molecules with inversion symmetry the parity must change in an electric dipole transition, i.e. when a photon is absorbed or emitted. Since resonant x-ray emission is a two-photon process the parity should be conserved in the full scattering process. This means that if a core electron is promoted to an orbital of gerade (ungerade) symmetry only electrons from gerade (ungerade) orbitals are allowed to fill the core hole.

This parity selection rule for resonant soft x-ray emission was recently verified experimentally, from studies of O_2 (11), to be strictly obeyed for homonuclear diatomic molecules. In Figure 5 x-ray emission spectra of N_2 are shown (12). Nitrogen is a very good molecule to use for illustrating the parity selection rule because it has a fairly simple electronic structure and the peaks in the absorption as well as in the emission spectra are well separated. In the non-resonant spectrum, the topmost spectrum in Figure 5, there are three vibrationally broadened peaks, due to transitions to the $2\sigma_u^{-1}$, $1\pi_u^{-1}$, and $3\sigma_g^{-1}$

final states. The resonant spectra show that when the core electron is promoted to a gerade orbital (x_g) the $2\sigma_u^{-1}x_g$ and $1\pi_u^{-1}x_g$ peaks vanish and only one spectator transition, to the $3\sigma_g^{-1}x_g$ final state, remains. In contrast, exciting to an ungerade orbital (x_u) allows only electrons from ungerade orbitals to fill the core hole and the two $2\sigma_u^{-1}x_u$ and $1\pi_u^{-1}x_u$ spectator peaks are observed.

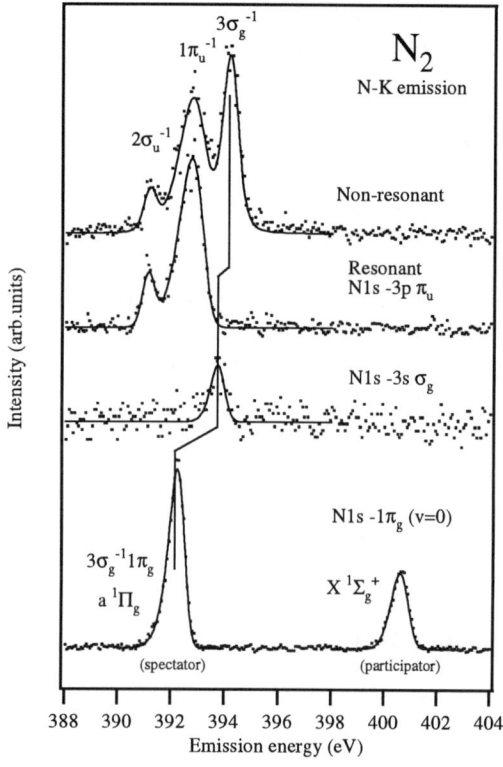

Figure 5. Non-resonant and resonant x-ray emission spectra of N_2. The resonant spectra were obtained by selective excitation to the $1\pi_g$ ("$v = 0$"), $3s\sigma_g$, and $3p\pi_u$ orbitals, respectively. The solid lines are simulated vibrational band profiles obtained using a one-step, lifetime-vibrational interference model.

If the excitation energy is tuned above the ionization threshold the photoelectron can have either gerade or ungerade parity and all three peaks are allowed. However, if the probability is higher for producing photoelectrons of a certain parity, it will be reflected in the relative intensities of the emission peaks. For instance, if the excitation energy is tuned to the σ_u^* shape resonance of N_2, the intensity of the $3\sigma_g^{-1}$ peak is greatly reduced compared to the $2\sigma_u^{-1}$ and $1\pi_u^{-1}$ peaks (12,13) .

Other noticeable effects in the spectra, in Figure 5, are the large energy shift of the spectator peak and the strong participator peak which arise when the core excitation is to the π^* ($1\pi_g$) orbital. These effects reflect the penetrating and screening properties of the π^* orbital (10). Excitations to the more diffuse Rydberg orbitals, on the other hand, lead to emission spectra with small energy shifts of the spectator peaks and very weak participator peaks.

Polyatomic Molecules

An O-K non-resonant x-ray emission spectrum of CO_2 (14,15) is shown at the top of Figure 6. The high-energy peak in this spectrum is associated with decay to the $1\pi_g^{-1}$ final state, and the low-energy peak is primarily due to the $1\pi_u^{-1}$ and $3\sigma_u^{-1}$ final states with some additional contribution from the $4\sigma_g^{-1}$ final state. According to the parity selection rule we only expect transitions involving the $1\pi_u$ and $3\sigma_u$ electrons to be allowed following $O1s \rightarrow \pi^*$ $(2\pi_u)$ excitation, which means that the high-energy peak should be absent in π^* resonant x-ray emission spectra. However, this peak is not only observable, it is, in fact, as strong as the low-energy peak in the spectrum recorded at the π^* resonance ($h\nu = 535.0$ eV).

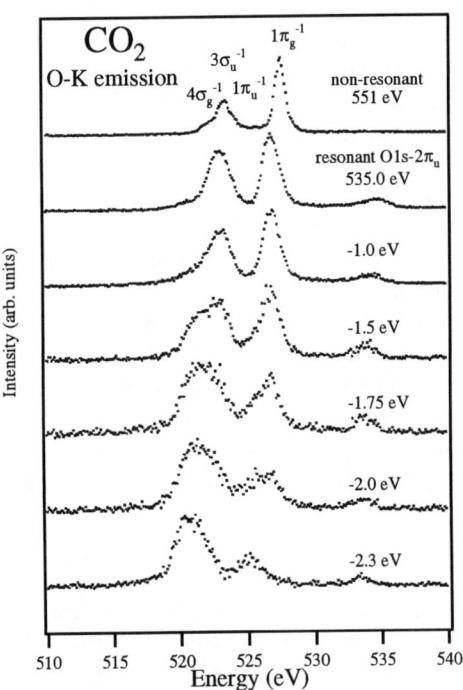

Figure 6. (top) O-K non-resonant x-ray emission spectrum of CO_2. Below, π^* resonantly excited x-ray emission spectra are displayed. The detuning energy from the maximum of the π^* absorption peak (535.0 eV) is indicated next to each spectrum.

The reason that the parity selection rule is no longer strict is that the vibrational motion along the non-totally symmetric modes can break the inversion symmetry of the molecule (16). COF is the equivalent molecule to the

$O1s \to \pi^*$ core-excited CO_2 molecule in the $Z+1$ approximation. The C-F equilibrium bond length for this molecule is substantially longer than the C-O bond length and this implies that the molecule will vibrate along the asymmetrical stretch mode, thereby breaking the inversion symmetry of the CO_2 molecule (17).

Initially the oxygen atoms in the CO_2 molecules are in equivalent sites. In the core-excited state the two oxygen sites will gradually become inequivalent as the inversion symmetry of the molecule is broken by the dynamical vibrational motion. The intensities of the parity "allowed" versus "forbidden" transitions will be affected by how rapid the symmetry breaking is in comparison to the decay of the excited state. The fact that the relative intensity of the low-energy peak is larger in the $h\nu = 535.0$ eV spectrum than in the non-resonant spectrum demonstrates that the symmetry breaking is not complete and that there is a propensity for the parity allowed transitions in the resonant spectrum.

The lower spectra in Figure 6 were obtained by lowering the excitation energy away from the resonant energy of 535.0 eV. As the excitation energy is detuned from the $O1s \to \pi^*$ resonance the forbidden high-energy peak gradually becomes weaker and the spectra become more dominated by the allowed transitions. The observed "symmetry purification" can be understood, for instance, in a time-dependent picture by recalling that the tails of the Lorentzian-broadened intermediate vibronic states are formed at an earlier time than the peaks. As the excitation energy is detuned from the absorption peak, and primarily the tails of the intermediate vibronic states are excited, the time that the molecules spend in the core-excited states is reduced. The larger the detuning, the more sudden the decays become and there is less time for vibrational motion and breaking of the molecular inversion symmetry, and therefore the propensity for the parity allowed transitions increases (18).

In summary, the CO_2 spectra illustrate that the parity selection rule is not strict for polyatomic molecules due to the non-totally symmetric vibrational modes breaking the inversion symmetry of the molecules. Furthermore, the spectra show that there still is a propensity for the parity allowed transitions and that it is possible to enhance this propensity by detuning the excitation energy to the tail of the absorption peak.

Symmetry Assignment

Molecular x-ray absorption spectra can be very complex and they often contain several unidentified features. The symmetry information contained in resonant x-ray emission spectra provides knowledge that can be used to make assignments of unidentified features in the absorption spectra. In the lower part of Figure 7 an O-K x-ray absorption spectrum of CO_2 is displayed. Three absorption features (A, B, and C) are observed below the ionization threshold. The strong peak, absorption feature A, is primarily due to $O1s \to$

π^* ($2\pi_u$) excitation, whereas the origin of the other two features has been less clear. In a forthcoming paper we show that it is possible to deduce the predominant symmetry of the unoccupied orbitals giving rise to features B and C from resonant x-ray emission spectra (19). The O-K x-ray emission spectra consist mainly of a "high-energy" and a "low-energy" peak as was shown in the previous section. In the upper part of Figure 7 we show the intensity ratio of these peaks from spectra recorded at various excitation energies at two angles, $\theta = 0°$ and $\theta = 90°$.

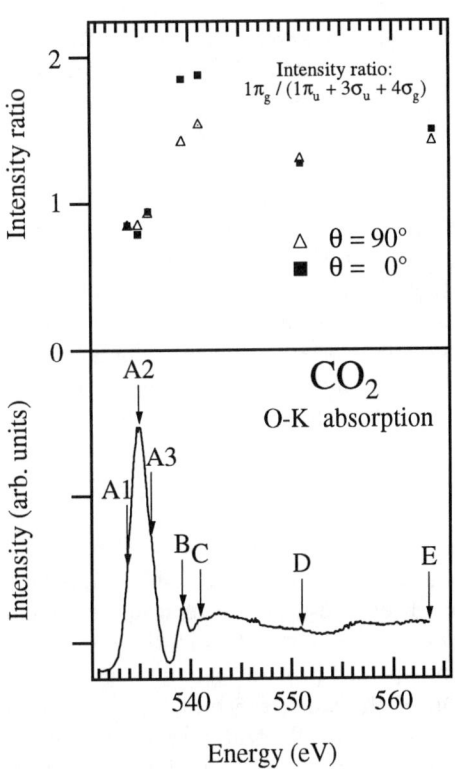

Figure 7. The lower part shows an O-K x-ray absorption spectrum of CO_2 with arrows indicating where emission spectra were recorded. The upper part displays the intensity ratio of the high-energy ($1\pi_g$) to the low-energy ($1\pi_u$, $3\sigma_u$, and $4\sigma_g$) peak recorded parallel ($\theta = 0°$) and perpendicular ($\theta = 90°$.) to the polarization of the exciting beam.

At absorption features B and C the measured emission intensity ratios are different at the two angles. In both cases the intensity ratio is larger in the $\theta = 0°$ spectrum, and this tells us that the unoccupied orbitals giving rise to features B and C are mainly of σ character. This assignment is in agreement with what was concluded from a recent angular-resolved ion-yield work (20). In the non-resonant spectra we do not expect any strong angular anisotropy, and as expected the intensity ratio is about the same at both angles. There is also no strong angular dependence when exciting to absorption peak A. This is probably due to a similar effect as was found for CO, namely that not all $\pi^{-1}\pi^*$ terms contribute significantly.

While the spatial "σ/π" character can be determined also with other angular-resolved emission techniques, x-ray emission spectroscopy provides a unique technique for determining whether the resonant features are due predominantly to unoccupied orbitals of gerade or ungerade parity. The low-energy peak in the O-K x-ray emission spectra of CO_2 is primarily due to decay from the $1\pi_u$ and $3\sigma_u$ orbitals, whereas the high-energy peak is due to decay from the $1\pi_g$ orbital. Because of the propensity for parity allowed transitions we expect a reduced intensity ratio if we excite to an orbital of ungerade parity and an increased ratio if the excitation is to a gerade orbital. At peak A the intensity ratio is lower than in the non-resonant spectra, while the ratio is higher at features B and C. The π^* orbital has ungerade parity and a lower intensity ratio is therefore expected at peak A. From the increased intensity ratio at features B and C we conclude that these features are mainly due to transitions to unoccupied orbitals of gerade parity.

From our x-ray emission spectra we find that the features B and C are primarily due to excitations to unoccupied orbitals of σ_g character. A plausible explanation for the dominance of excitations to σ_g orbitals just below the ionization threshold is Rydberg-valence mixing, i.e. Rydberg orbitals of σ_g symmetry borrow intensity from the unoccupied $5\sigma_g$ orbital. This mixing also depletes the expected $5\sigma_g$ shape resonance, anticipated just above the ionization threshold, and it explains why there is no strong σ_g shape resonance observed in the absorption spectrum.

Conclusions

X-ray emission spectroscopy results, obtained by employing excitation with energy-selective synchrotron radiation, have been shown. The presented spectra have been chosen to illustrate how some of the dipole selection rules influence the relative intensities in the emission spectra of small molecules. The spectroscopy can provide element specific information about the chemical composition of the valence orbitals because of the local atomic character of the core holes, as was illustrated by comparing C-K and O-K non-resonant x-ray emission spectra of CO. Angular-resolved resonant x-ray emission spectra of CO were shown to illustrate that the intensities of the emission peaks are anisotropic. The parity selection rule was demonstrated by spectra of N_2 and CO_2. In the diatomic case the parity selection rule is strict, whereas in the polyatomic case it becomes a propensity rule because of dynamical vibronic coupling breaking the inversion symmetry of the molecule. In addition, it was shown that the anisotropic angular distribution and the parity selection rule of x-ray emission can be utilized for spectral assignments of x-ray absorption features. The importance of lifetime-interference effects was also briefly discussed, and experimental and simulated vibrational band profiles in resonant x-ray emission spectra of CO were compared.

The examples given above demonstrate some of the promising prospects of x-ray emission spectroscopy, utilizing energy-selective excitation, for studies of free molecules. Spectra of small molecules are ideal for tests of theoretical models and for estimating the importance of various effects. The knowledge gained from studies of small molecules may also be of use when larger systems are studied and the spectra become more complex.

ACKNOWLEDGMENTS

The theoretical input provided by H. Ågren, F.Kh. Gel'mukhanov, Y. Luo, and A. Cesar is greatly appreciated. We would also like to thank C.-J. Englund, N. Wassdahl, C. Såthe, A. Langereis, T. Wiell, E. Rotenberg, J. Denlinger, and T. Warwick for their contribution to the success of the experiments. In addition, P.G. would like to thank S.B. Whitfield, H. Wang, O. Hemmers, and D.W. Lindle for many stimulating discussions related to the material presented herein. This work was supported by Göran Gustafssons Foundation for Research in Natural Sciences and Medicine and by the Swedish Natural Science Research Council (NFR). The experiments were performed at the ALS of Lawrence Berkeley National Laboratory operated by DOE under contract No. DE-AC03-76SF00098.

1. Warwick T., Heimann P., Mossessain D., MacKinney W., and Padmore H., *Rev. Sci. Instrum.* **66**, 2037 (1995).

2. Nordgren J., Bray G., Cramm S., Nyholm R., Rubensson J.-E., and Wassdahl N., *Rev. Sci. Instrum.* **60**, 1690 (1989).

3. Guo J.-H., Wassdahl N., Skytt P., Duda L.C., Butorin S., Englund C.-J., and Nordgren J., *Rev. Sci. Instrum.* **66**, 1561 (1995).

4. Manne R., *J. Chem. Phys.* **52**, 5733 (1970).

5. Wannberg B., Nordfors D., Tan K.L., Karlsson L., and Mattsson L., *J. Electr. Spectr. Relat. Phenom.* **47**, 147 (1988).

6. Gel'mukhanov F.Kh., Mazalov L.N., and Kondratenko A.V., *Chem. Phys. Lett.* **46**, 133 (1977).

7. Skytt P., Glans P., Gunnelin K., Guo J.-H., and Nordgren J., accepted for publication in *Phys. Rev. A*.

8. Lindle D.W., Cowan P.L., LaVilla R.E., Jach T., Deslattes R.D., Karlin B., Sheehy J.A., Gil T.J., and Langhoff P.W., *Phys. Rev. Lett.* **60**, 1010 (1988).

9. Southworth S.H., Lindle D.W., Meyer R., and Cowan P.L., *Phys. Rev. Lett.* **67**, 1098 (1991).

10. Skytt P., Glans P., Gunnelin K., Guo J.-H., Nordgren J., Lou Y., and Ågren H., accepted for publication in *Phys. Rev. A*.

11. Glans P., Gunnelin K., Skytt P., Guo J.-H., Wassdahl N., Nordgren J., Ågren H., Gel'mukhanov F.Kh., Warwick T., and Rotenberg Eli, *Phys. Rev. Lett.* **76**, 2448 (1996).

12. Glans P., Skytt P., Gunnelin K., Guo J.-H., and Nordgren J., accepted for publication in *J. Electr. Spectr. Relat. Phenom.*

13. Ågren H. and Gel'mukhanov F.Kh., *Phys. Rev. A* **54**, 379 (1996).

14. Skytt P., Glans P., Guo J.-H., Gunnelin K., Nordgren J., Gel'mukhanov F.Kh., Cesar A., and Ågren H., submitted to *Phys. Rev. Lett.*

15. Cesar A., Gel'mukhanov F.Kh., Lou Y., Ågren H., Skytt P., Glans P., Guo J.-H., Gunnelin K., and Nordgren J., submitted to *J. Chem. Phys.*

16. Cederbaum L.S., *J. Chem. Phys.* **103**, 562 (1995).

17. The promotion of an electron to the π^* orbital may also induce some vibrations along the bending modes because of the Renner-Teller effect. However, the symmetry breaking caused by the bending modes is minor compared to that caused by the asymmetrical stretch mode.

18. As the detuning is increased other intermediate states than the $O1s^{-1}\pi^*$ state will gradually become more important. This means that there is a limit to how far it is possible to detune and still observe the "symmetry purification" effect.

19. Skytt P., Glans P., Gunnelin K., Guo J.-H., Nordgren J., and Ågren H., submitted to *Phys. Rev. A*

20. Bozek J.D., Saito N., and Suzuki I.H., *Phys. Rev. A* **51**, 4563 (1995).

FREE CLUSTERS STUDIED BY X-RAY SPECTROSCOPY

A.V.Soldatov and T. Ivanchenko
Department of Physics, Rostov University, Sorge str.5, Rostov-Don, 344090, Russia

Abstract. x-ray spectroscopic investigations of free clusters as well as new data on rare gas clusters are presented. X-ray absorption K-edge spectra of free neon clusters covering the whole size range from the atom to the solid are reported. Absorption into Rydberg states as well as XANES of neon clusters is investigated with high spectral resolution. The x-ray absorption fine structure is shown to be sensitive to the size of cluster. For theoretical analysis of the experimental spectra full multiple-scattering method has been applied. In the framework of our approach we also succeeded to analyze the cluster size dependence of the XANES data. In order to study the sensitivity of XANES on cluster symmetry, theoretical simulations have been performed for both face-centered cubic and icosahedral structures of rare gas clusters. The two-electron excitations, involving monopole $2p \rightarrow ep$ transitions, have been identified in the energy interval about 30 eV above the K-edge of solid neon.

INTRODUCTION

Free atomic clusters are very interesting physical objects because they present a 'bridge' between atoms and molecules from one side and solids from another side. Their studies support a link between atomic physics and the physics of condensed matter (1). Of particular interest is the evolution of electronic energy levels with size and their relation to the geometrical structure of the clusters, which are found to be formed into the fragments containing "magic" numbers of atoms (2). In this context rare gas clusters play an important role, as their size can easily be varied from a few atoms to rather large microcrystals (3). They are model systems for insulators and possess a simple electronic structure without the problems associated with directed bonds or strong electronic correlation as in metals. Fluorescence spectroscopy of valence shell excitations of rare gas clusters has already provided a wealth of information. Most notably surface and bulk sites have been studied as a function of cluster size (4). However, despite the simple electronic structure of closed shells, these excitations are often difficult to analyze. In the case of $L_{2,3}$ levels excitation the resulting spectra are complicated by spin orbit splitting and the possibility of transitions into s- and d-symmetry states. However, since the absorption coefficient at L-edges is considerably larger than at the K-edge which usually provide much simpler excitations (s>p), several studies have been performed at L-edges, e.g.

studies on small Ar clusters excitonic states (5,6), studies in the EXAFS region of Ar clusters (7), zero-kinetic-energy photoelectron spectroscopy (8), and photoelectron-photoion-photoion coincidence spectrocsopy of atomic and molecular clusters. Furthermore, both orbitals involved in a transition are usually modified by clustering, making it difficult to separate the different contributions. This turns out to be impossible for higher excitations that form a broad continuum of overlapping bands (10).

What makes rare gases especially interesting are tightly bound electronically excited states - excitons of the solid and Rydberg states of atoms and molecules - which are to a certain extent equivalent. The radius r_e of these excitations (0.2 nm - 10 nm, in principle up to infinity) covers the range of the radius of clusters containing up to 10^5 atoms. Therefore, strong variations of the observed spectral features with the cluster size are expected. Decay process are discussed in more detail elsewhere (11,12). The photoabsorption process in clusters can be treated within two energy intervals: low energy excitonic region, where one can onbain information on the excited electronic states, and more extended continuum region of absorption (XANES/EXAFS) where one can obtain the details of the local structure of the cluster. We will treat these two regimes separately and present some results, obtained from the study of free rare gas clusters by x-ray absorption spectroscopy.

Since clusters-with a large number of nonequivalent atomic sites- also possess a far lower symmetry than the solid, there is an urgent need to simplify the task if possible. So, we perform theoretical calculations only for two types of atomic sites: "central" and "surface", and the final comparison of theoretical results with experimental data has been done for solid Ne XANES.

X-ray absorption spectra of 'infinite' rare gas solids were measured several times, but mostly in the region of transitions from 2p ($L_{2,3}$), 3p ($M_{2,3}$) or 3d ($M_{4,5}$) levels (13-15). There one must take into account strong correlation effects in these well - localized electronic shells (16) and the spectra have nearly atomic-like character. The K-edge XANES spectra of solid phase heavy rare gases have been measured, see for example ref. 17, but because of the large core hole energy width these spectra exhibit very broad features in the XANES region. Recently the K-edge x-ray absorption near edge structure (XANES) spectrum of solid Ne (18) as well as of Ne (19) and Ar (20) clusters has been measured.

DETAILS OF EXPERIMENT AND COMPUTATION

In figure 1 we show the experimental set-up used for XANES measurements of free rare gas clusters. The clusters were produced in an adiabatic expansion through a liquid He-cooled conical nozzle (diameter = 100 μm, opening angle $2\theta = 30°$) and crossed with monochromated synchrotron radiation after passing through a skimmer into the main vacuum chamber. The average cluster size was varied by choosing appropriate expansion parameters with stagnation pressures ranging from 1 to 3 bars and temperatures down to 37 K. Details of the cluster source have been published elsewhere (21). The average cluster size in the beam was calculated from the parameters of the expansion using scaling laws (22). The experiments were performed at the new SX700 undulator beam line BW3 at Hasylab (23). The resolution used at the neon K edge was typically 300 meV, which is comparable to the natural linewidth of the atomic resonance lines of about 250 meV. As a result of this broadening only the 1s - 3p and 1s - 4p transitions are evident in the spectra. The resolution used is close to the best values reported so far.

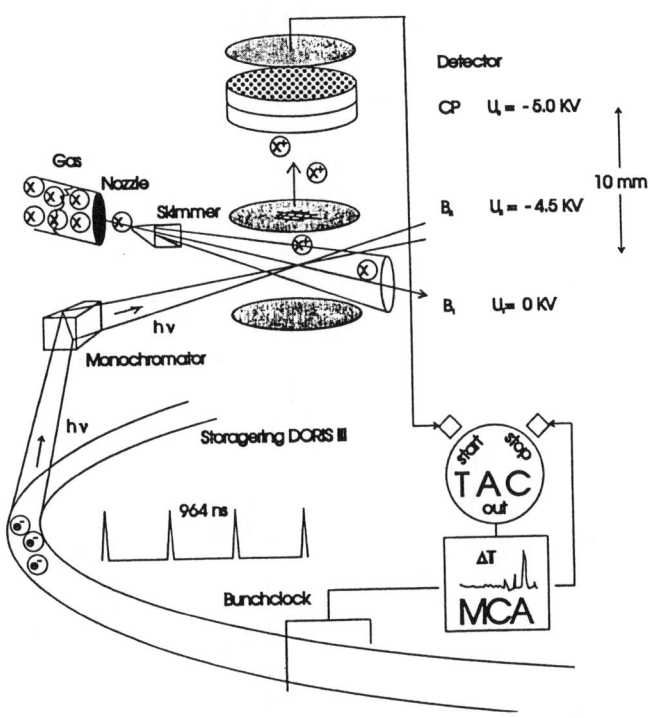

FIGURE 1. Experimental setup for producing free rare-gas clusters and x-ray spectroscopic measurements.

The full multiple-scattering calculation scheme we used has been discussed previously (24). For phase shifts calculation we have obtained crystal muffin-tin (MT) potential with touching MT spheres. We treated Ne clusters as both f.c.c. symmetry with lattice parameter equal to 4.426 Å and icosahedral symmetry with first shell radius equal to 3.1316 Å. For calculation we have used up to 5 shells cluster. Our procedure of MT potential construction and MT radii choosing was described elsewhere (25). In our calculation we included phase shifts with orbital momentum up to 4, but even for l =3 there are almost no changes in the spectra. We take into account all main factors that cause the broadening of the spectra (i.e. final lifetime of core hole, limitations of mean free path of photoelectron and, also, experimental resolution) and treated all these factors contributing to the imaginary part of the complex potential we have used (for details see (25)). But in order to obtain more resolved XANES features we included all three factors only for the calculation of the spectrum that was compared with the experimental one in Fig. 4. In other cases we used a constant imaginary part of potential equal to -0.002 Ry.

RESULTS AND DISCUSSION

Excitons region

One of the main questions in cluster physics concerns the evolution of energy levels with size, e.g. the transition from atomic and molecular Rydberg states to the excitons of the solid. In both cases the excited electron moves around the

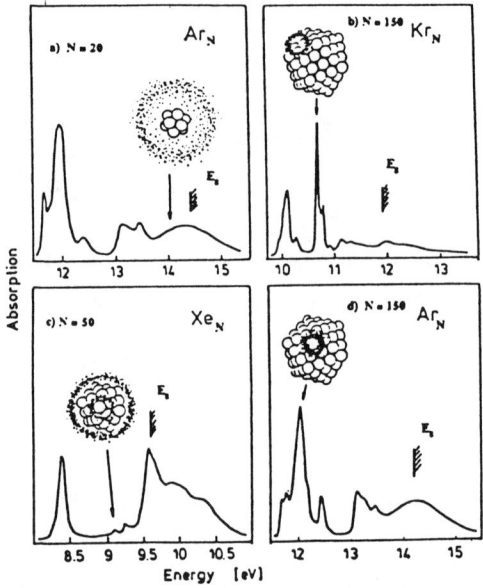

FIGURE 2. Absorption spectra of rare gas clusters and schematic illustration of the character of electronic excitations.

ref. 18. As one can see the spectra are rather sensitive to the size of the cluster and their shape became more sharp when the size of cluster increases. As it has been predicted from theoretical calculations for infinite solid Ne fragments with f.c.c. symmetry (28) a very large change must take place for small clusters, namely of 13, 19, 43, 55 atoms.

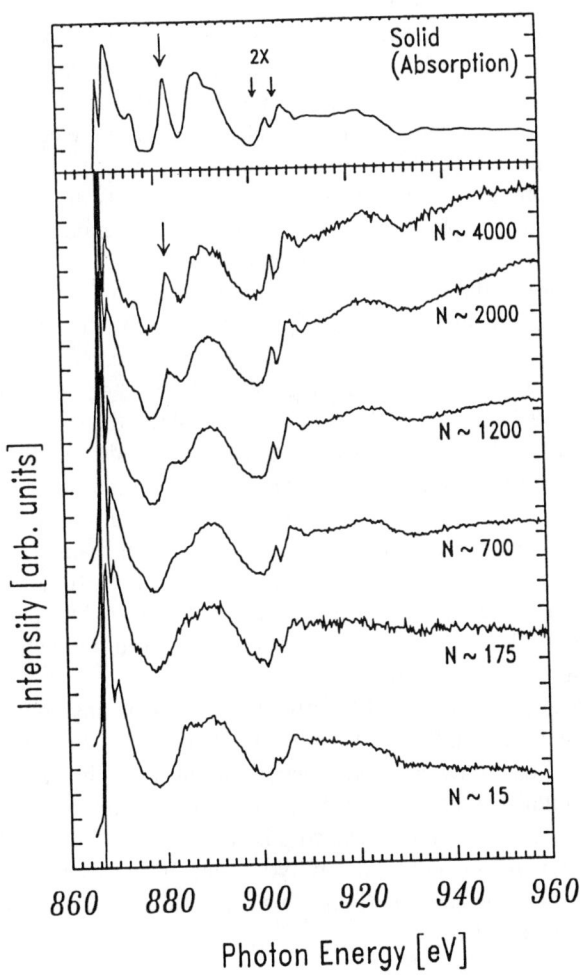

FIGURE 3. Ne X-ray absorption spectra above K-edge in free clusters as a function of cluster size.

positive charge. However, in the solid these bound electron hole states form bands classified according to their principal quantum number n and the orientation of the polarization (transverse, longitudinal) relative to the direction of propagation of the light (26). There are three limiting situations according to the relationship between the radius r_e of the electronic excitation and the cluster radius R: (regime 1) $r_e < R$, (regime 2) $r_e \approx R$ and (regime 3) $r_e > R$. Since the radius of the excited orbital depends strongly on the principal quantum number n, the different regimes might be present in clusters of a given size. In the following, we like to give an overview of the different kinds of excitations which are observed in pure rare gas clusters. As an example, absorption profiles of small Ar_N, Kr_N and Xe_N clusters are presented in figure 2. Energetically below the ionization energy E_b strong absorption bands are observed. Some of them rather sharp. The character of the electronically excited state is schematically visualized in the figure. The assignment of the different absorption bands comes from a comparison with the bands in rare gas solids and from that observed in gas phase experiments. In addition, model calculations have been performed which support the assignment (10).

Energetically close to the ionization limit a broad continuum is observed which is assigned to molecular Rydberg states (figure 2a)). Here, the excited electron is weakly bound in a state characterized by large main quantum numbers and an extended orbital with a radius much larger than the cluster radius R (regime 3). Therefore, the excited electron is located mainly outside the cluster. The onset of absorption is dominated by several strong absorption bands which are assigned in accordance with corresponding bands in the solid either to tightly bound bulk (figure 2d) or surface (figure 2b) excitons. Depending on the 'radius' of the orbital they are classified as n=1 Frenkel or intermediate type, or for larger main quantum numbers, as Wannier type excitons. In case of surface excitons, the excitations takes place mainly in the first and second monolayer of the cluster depending on the size of the orbital (4,27). Excitonic absorption bands are only observed if the cluster radius R is considerable larger than the exciton radius.

In addition, small clusters (N<500) show cluster specific absorption bands which are not observed in the solid. Some of them are rather sharp with widths significantly smaller than that of excitonic absorption hands (figure 2c). They are assigned to so called 'cluster excitons' where the 'radius' of the excitation is comparable to the cluster radius R (regime 2). While the hole is located in well defined shells of the cluster, the electron has a high probability to stay inside as well as outside the cluster. Consequently these excitations have a character in between molecular Rydberg states and excitons.

Continuum region

Another very important difference of free cluster XANES from the model cluster calculation corresponding to the infinite solid is that in free clusters there are a many unequivalent atomic positions, while in the infinite solid one can treat all atoms as equivalent. Therefore in order to study the difference in partial contribution from these different types of atoms into total XANES of the whole cluster we simulate XANES for two sites : central atom in the cluster and surface atom of the cluster (Fig. 4). As one can see the spectra of these two types of atoms are rather different and one can conclude that it is necessary to calculate partial spectra of all type of unequivalent atoms and take their weighted normalized sum in order to simulate the spectrum of a free cluster.

Another problem is the symmetry of a free cluster. While clusters - fragments of an infinite solid - one can treat as having f.c.c symmetry like a whole crystal, free cluster may have different symmetry. Because there is evidence for icosahedral symmetry of free rare gas clusters (as it was found from electron diffraction data of Ar (29) in contrary to the metallic cluster case (30))we have simulated the XANES of icosahedral clustera in comparison

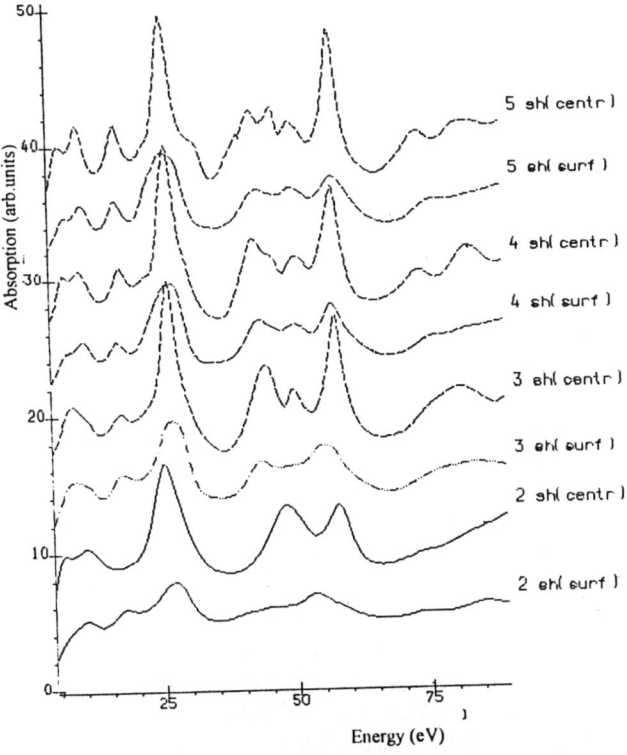

FIGURE.4. Theoretical X-ray absorption coefficient above K-edge for Ne cluster with *f.c.c.* symmetry within clusters of different size calculated for two types of atomic site : "central" and "surface". The origin of energy scale corresponds to the muffin-tin zero.

with f.c.c. cluster (see Fig.5). It was found that even for very small 1 shell cluster (13 atoms) there is a large difference in partial XANES contribution from surface atoms of these two kinds of clusters.

In the discussion we assumed Ne clusters to be solid (In spite of that very interesting results can be obtained for the case of liquid phase clusters like He ones (31)). While this has been shown for argon clusters produced in an adiabatic expansion (32), the situation is not so clear for neon due to the lower mass and increased zero point vibrational energy. Recent investigations hint to a phase change from liquid to solid for Xe-doped neon clusters (33).

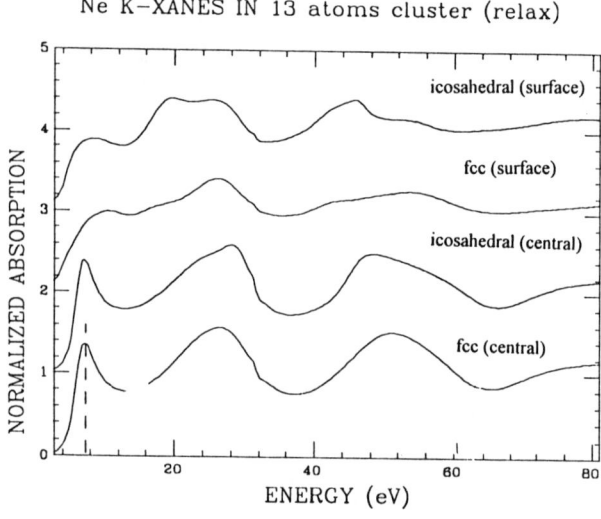

FIGURE.5. Theoretical X-ray absorption coefficient above K-edge for 13 atomic Ne cluster with both f.c.c. and icosahedral symmetry.

Structure related information can be extracted from the intensity variations of the 1s ionization continuum in the near edge region which is known as XAFS (XANES/EXAFS). This has been demonstrated in principle for argon clusters, where the unselected ion yield of relatively large clusters has been used in an EXAFS analysis (20). In our spectra only slight undulations are observed for small clusters while more structure continuously emerges for larger systems, where the agreement with data collected from solid is better.

The task of total XANES calculation from large free cluster is very difficult, because, as it was mentioned above, one needs to calculate partial XANES from all unequivalent types of atoms and makes appropriate average procedure. In order to verify the theoretical simulation results without full calculations we compare the results for the simple case of solid Ne where all atoms are equivalent. We take into account that in the process of X-ray absorption the

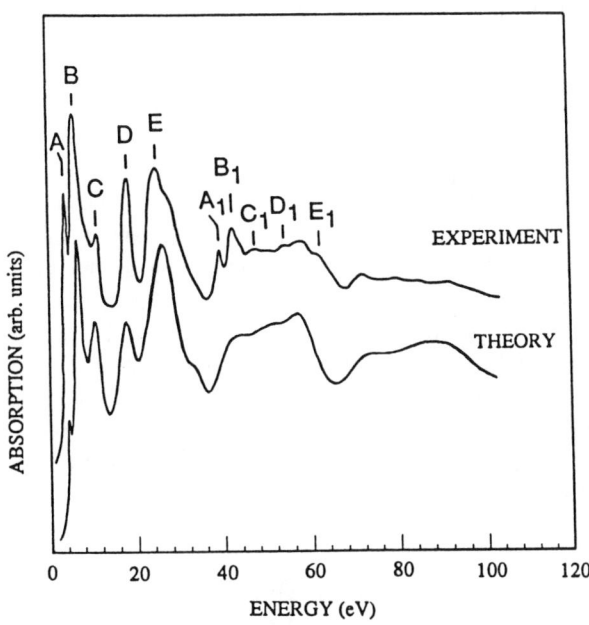

FIGURE.6. Comparison of experimental K-XANES in solid Ne (9) with the theoretical results for the large cluster including effects of core hole and full broaderning (see text).

excited photoelectron moves in the presence of the core hole (28). In the case of an insulating material (like solid rare gases) it leads to the appearance of so-called core excitons in the lowest energy part of XANES (34). We have assumed that the electronic system of solid Ne is fully relaxed in the field of 1s (K) core hole and treated this situation in the frame of Z+1 approximation (35), i.e. while constructing MT potential we have used (for the central atom in the cluster) the atomic electron density of Na atom. Now we can compare our theoretical spectrum with experimental data for solid neon from (18) (see Fig.6). As it was just mentioned above we take into account all the main factors that result in the broadening of the spectrum: we include a constant term of 0.4 eV according to the core hole lifetime and experimental resolution, and the energy dependent term corresponds to the mean free path of the photoelectron in solid phase. Because we did not know the last function directly for solid Ne, we have used the averaged function for metals (36) that is expected to be an approximation. In spite of these approximations one can see that the agreement between the theoretical and experimental results (in the energy region from the main edge up to about 40 eV) is rather good. But starting from the energy of about 35 eV and beyond there are sharp structures in the experimental spectrum that are not reproduce in the theoretical spectrum. According to its similarity with the first sharp peaks and according to the energy separation from them these structures we have associated them with two-electron excitations in the central atom corresponding to $1s^2 2s^2 2p^6$ to $1s^1 2s^2 2p^5 ep^2$ states. One can see in Fig.6 that each one-electron excitation multiple-scattering maximum (labelled as A, B, C, D and E) corresponding to $1s^2 2s^2 2p^6 \rightarrow 1s^1 2s^2 2p^6 ep^1$ transitions has its own two-electron excitation "replica" (labelled as A1, B1, C1, D1 and E1) corresponding to $1s^2 2s^2 2p^6 \rightarrow 1s^1 2s^2 2p^5 ep^2$ transitions. Such interpretation is confirmed by the sharp character of this structure in comparison with the bandwidth of the one-electron excitation multiple scattering resonances at the energy about 30 eV above the main edge (that is strongly broadened according to the small value of the photoelectron mean free path for this energy). In the XANES spectra of more heavy rare gas solids (for example in solid Kr XANES) this double excitation structure is not visible in the XANES spectrum because of decreasing of relative probability of two- and one-electron excitations with increasing of atomic number(37). But in the case of the absence of XANES modulations i.e. in gas phase one can see such a structure in the X-ray absorption spectrum of Kr (38) at the energy about 20 eV above the main edge .

AKNOWLEDGMENTS

The authors are grateful to T. Möller (Hamburg) and his group for providing experimental results. One of the authores (A.V.S.) would like to aknowledge NATO grant CRG930305 support during his work at laboratory of A.Bianconi (Rome University), where a part of this work has been done and thanks A. Hitchcock for the supply with the articles in this field.

REFERENCES

1. *Clusters of Atoms and Molecules I*, edited by H.Haberland, Springer Series in Chemical Physics vol 52 (Springer, Berlin, 1994).
2. Echt O., Sattler K. and Recknagel E., *Phys. Rev. Lett*. **47**, 1121-1124 (1981).
3. Jortner J.,*Z. Phys. D* **24**, 247 (1992).
4. Stapelfeldt J., Wörmer J. and Möller T., *Phys. Rev. Lett.* **62**, 98 (1989).
5. Rühl E., Jochims H.W., Schmale C., Biller E., Hitchcock A. and Baumgartel H., *Chem Phys. Lett*. **178**, 558-564 (1991).
6. Rühl E., Heinzel C., Baumgartel H., Drube W. and Hitchcock A., *Jpn. J. Appl. Phys.* **32** 791-793 (1993).
7. Rühl E., Heinzel C., Hitchcock A and Baumgartel H., J. Chem. Phys. 98, 2653-2663 (1993).
8. Knop A., Jochims H.W., Kilcoyne A.L.D., Hitchcock A.P. and Rühl E., *Chem. Phys. Lett*. **223**, 553-560 (1994)
9. Rühl E., Hitchcock A.P., Morin P. and Lavollee M., J. Chem. Phys. 92, 521-540 (1995).

10. Wörmer J., Joppien M., Zimmerer G. and Möller T., *Phys. Rev. Lett.* **67**, 2053 (1991).
11. Lengen M., Joppien M., Wörmer J. and . Möller T., *Phys. Rev. Lett.* **68** *2362 (1992).*
12. Joppien M., Groteluschen F., Lengen M Muller R., Wörmer J. and . Möller T.,Proceedings of the International Conference "Synchrotron Radiation and Dynamic Phenomena, Grenoble (1991).
13. Haensel R., Keitel G., Schreiber P. and Kunz C., *Phys. Rev.* **188**, 1375 (1969).
14. Haensel R., Keitel G., Koch E.E., Skibowski M. and Schreiber P., *Phys. Rev. Lett.* **23**, 1160 (1969).
15. Haensel R., Keitel G., Kunz C. and Schreiber P., *Phys. Rev. Lett.* **25**, 208 (1970)
16. Yavna V.A., Hopersky A.N., Nanolinsky A.M. and Popov V.A., Proc.2nd European Conf. on progress in X-ray Synchrotron Radiation Research, (SIF, Bologna, 1990) p. 101.
17. Polian A., Itie J.P., DartygeE., Fontain A. and Tourillon G., *Phys. Rev.* **B 39**, 3369 (1989).
18. Hiraya A., Fukui K., Tseng P-K., Murata T. and Watanabe M., *J. Phys. Soc. Jpn.* **60**, 1824-1828 (1991).
19. Federmann F., Björneholm O., Beutler A. and . Möller T., *Phys. Rev. Lett*, **73**, 1549-1552 (1994).
20. Rühl E., Heinzel C., Hitchcock A, Schmelz H., Reynaud C., Baumgartel H.,. Drube W. and Frahm R., *J. Chem. Phys.* **98**, 6820-6826 (1993).
21. Kambach R., Joppien M., Stapefeldt I, Wörmer J. and . Möller T., *Rev. Sci. Instrum*,. **64**, 2838 (1993).
22. Hagena O.F., Z. Phys. D 4 291 (1987). The average cluster size is obtained with the additional empiric relation $N = 38 \, (\Gamma/1000)^2$.
23. Larsson C.U.S., Beutler A., Björneholm O., Federmann F., Hahn U., Rieck A., Verbin S. and Möller T., *Nucl.Instrum. Methods Phys. Res. A* **337**, 603 (1994).
24. Durham P., in X-ray Absorption : Principle, Applications, Techniques of EXAFS, SEXAFS, XANES (Ref.1), p. 53.
25. Della Longa S., Soldatov A.V., Pompa P., and Bianconi A., *Comput. Materials Science* **4**, 199-210 (1995).
26. Schwentner N., Koch E.E,and Jortner J., Electronic excitations in Condensed Rare Gases (Springer, Berlin, 1985).
27. Wormer J. and Moller T., *Z. Phys. D.* **20**, 39 (1991).
28. Soldatov A.V., Ivanchenko T.S., Della Longa S., and Bianconi A., *Phys. Rev. B* **47**, 16155-16162 (1993).
29. Farges J., DeFerandy M.F., Raoult B., and Torchet G., *Ber. Bunsen-Ges Phys. Chem.* **88**, 211 (1984).
30. Montano P.A., Shenoy G.K. and Alp E.E., *Phys. Rev. Lett.* **56**, 2076-2079 (1986).
31. Joppien M., Karnbach R. and . Möller T., *Phys. Rev. Lett*, **71**, 2654-2657 (1993).
32. Farges J., de Feraudy M.F., Raoult B. and Torchet G., *J. Chem. Phys.* **84**, 3491 (1986).
33. von Pietrowski R., Diplomarbeit Universitat Hamburg, 1993 and R. von Pietrowski, M. Rutzen, K. von Haeften, S. Kakar, and
T. Möller, Z. Phys. D, in press
34. Pudewill D., Himpsel F.J., Saile V., Schwentner N., Skibowski M. and.Koch E.E., *Phys. Status Solidi B* **74**, 485 (1976).
35. Bianconi A., in X-ray Absorption : Principle, Applications, Techniques of EXAFS, SEXAFS, XANES (Ref.1), p. 573.
36. Muller J.E., Jepsen O. and Wilkins J., *Solid State Commun.* **42**, 365 (1982).
37. Krause M.O., *J. Phys. (Paris)* **32**, C4-67 (1971).
38. Hayaishi T., Murakami E., Morioka Y., Shigemasa E., Yagishita A., Aksela H. and Aksela S., *Photon Factory Reports* **8**, 244 (1990).

Soft X-Ray Emission Excited Resonantly and Nonresonantly by Synchrotron Radiation

D.L. Ederer[1], J.A. Carlisle[2a], J. Jimenez[1b], J.J. Jia[3c], Ling Zhou[3], T.A. Callcott[3], R.C.C. Perera[4], A. Moewes[5], L.J. Terminello[2], E. Shirley[6], A. Asfaw[2d], J.van Ek[1], E. Morikawa[5], and F.J. Himpsel[7]

[1]*Tulane University, New Orleans, LA 70118*
[2]*Lawrence Livermore National Laboratories Livermore, CA 94551*
[3]*University of Tennessee, Knoxville, TN 37996*
[4] *Advanced Light Source, Lawrence Berkeley Laboratory, Berkeley, CA 94720*
[5] *Center for Advanced Microstructures and Devices, Baton Rouge, LA 70888*
[6]*National Institute of Standards and Technology, Gaithersburg, MD 20899*
[7]*Physics Dept. University of Wisconsin, Madison, WI 65500*

Abstract. This paper is a summary of some of the recent activities of our soft x-ray spectroscopy group. We are using soft x-ray emission spectroscopy to probe the electronic properties of matter, emphasizing atoms in the bulk and at interfaces. In particular we have used incoherent photon excitation to obtain a basic understanding of the electronic properties of a wide variety of materials such as yttrium oxide and titanium diboride, we have used the penetrating power of x-rays to study multilayers of silicon and iron and have shown that iron silicide is present in the silicon layer and provides evidence that in the antiferromagnetic Fe/Si multilayer system the $FeSi_2$ layer is conducting rather than insulating. The ubiquitous presence of Raman scattering has been used to elucidate the electronic band structure of materials including graphite and hexagonal boron nitride. Such scattering can produce dramatic changes in the emission spectrum that can further the basic understanding of the electronic band structure. We have made a systematic study of Raman scattering in several transition metal compounds, including their borides, oxides, and sulfides. Photon-in photon-out soft x-ray spectroscopy has many applications, and is adding a new dimension to soft x-ray spectroscopy by providing many opportunities for exciting research, especially at third generation synchrotron light sources.

INTRODUCTION

Soft x-ray emission spectroscopy has been applied extensively toward the study of materials since the discovery of x-rays(1-4). The topics covered in the X-96 conference addressed a broad spectrum of topics concerning the production and use of x-rays as a scientific tool to understand the electronic properties of matter. This report is a summary of some of the activities of our group, which has focused its interest on photon-in photon-out spectroscopy in the spectral photon energy range below 1 keV. We began our studies in the 50-300 eV photon energy range about a decade ago at the National Synchrotron Light Source.(5) Similar work was begun in Sweden by a team at Uppsala University led by Nordgren and his co-workers.(6) Since then several third generation light sources have come on line where the high brightness available at these sources has been put into service studying many different materials and compounds by a number of groups.(7-9)

Until recent times, the absorption of a photon and the relaxation of the excited state of the atom by reemission has been considered as two independent processes. The process of photon absorption and photon emission is driven by the usual first order process involving the inner product of the electron momentum, **p**, and the vector potential of the photon field, **A**. Weak second order two photon processes involving two photons involve the elastic scattering of photons (Raleigh Scattering) and the inelastic scattering of photons (Raman and Compton scattering). Raman scattering was first observed in the x-ray spectra region two decades ago(10). Since the initial measurements in the keV region of the x-ray spectrum(11,12) the importance of these coupled scattering processes has become much more evident. Last year a workshop was held to survey the field and to bring theorists and experimentalists together to discuss various aspects of x-ray Raman scattering(13). This workshop followed the publication of a survey of x-ray scattering(14) and was a precursor of a satellite conference dealing with x-ray scattering(15)connected with X-96. The emergence of several workshops on this topic, over a period of a few years are an indication of the dynamic growth and activity in this field of research.

Soft x-rays are especially powerful as a probe to study bulk electronic structure, because they are atomic site specific, penetrate tens to hundreds of atomic layers, and produce less damage than electrons or ions.(16) The large penetration depth of x-rays make them an important tool to probe multilayers and atoms at buried interfaces.(17) The soft x-rays can excite elementally specific shallow core levels that have a natural width that is about a factor of ten smaller than more deeply bound levels, thus the valence band fluorescence spectrum is not unduly broadened by the width of the core level. In addition the fluorescent photons are not affected by electric or magnetic fields, so sample charging does not distort the data, allowing insulators and magnetic materials to be effectively studied. The dipole selection rule allows one to obtain a selected localized density of states of one of the elements in the compound under investigation, thus, one obtains complimentary information to that obtained by photoelectron spectroscopy.

This paper is a brief summary of some of our recent work that has been carried out at the Advanced Light Source (ALS) and at the Center for Advanced Microstructures and Devices (CAMD). The use of these two x-ray facilities has been very efficacious. The ALS is a third generation source where we have limited beam time, while CAMD is a second generation light source where we have full time use of a beam line. At CAMD we have the opportunity to try out ideas and conduct surveys to determine the experiments that have highest priority for the limited time available at the ALS.

In the following sections of this paper we present recent work invoking conventional incoherent x-ray excitation and show its use in the interpretation of band structure and the determination of the electronic properties of relatively complex solids. We have also used the penetrating power of x-rays to study multilayers of iron and silicon and will present some results that prove iron silicide is present in the silicon layers and is metallic rather than insulating. We shall provide some illustrations showing how Raman scattering can be used to elucidate the electronic properties of graphite (18) and h-BN. (19) Finally we will also show a few examples from a systematic study of Raman scattering in several transition metal compounds, their borides, oxides, and sulfides. Our work has been concerned with solids, however, there has also been a great deal of work carried out in gases(20) which will be summarized during this X-96 Conference.

INSTRUMENTATION

Until recently the exploitation of photon excited soft x-ray emission spectroscopy in the spectral range (50 eV to 1000 eV) as an extremely powerful, and useful technique to study the electronic structure of atoms in solids and molecules, has been limited by the small fluorescence yields (10^{-3}-10^{-4}) of materials with a low atomic number and by inefficient soft x-ray spectrometers. To counter the deficiency of low yield, and low spectrometer efficiency, improved soft x-ray grating spectrometers(5,21) with sensitivities 10^3-10^4 times greater than conventional spectrometers have been coupled with third generation synchrotron radiation sources. For the first time a narrow band of x-ray radiation ($E/\Delta E \geq 1000$) can be used to excite soft x-ray emission in the entire soft x-ray spectra range. This combination allows high quality spectra to be obtained from some elemental samples (22) in about thirty seconds.

At present we employ a spectrometer mounted on a beam line at CAMD (23) at Louisiana State University and another spectrometer end station on an undulator at the ALS (22) at Lawrence Berkeley Laboratory. At these sources, an electron beam, white light, or monochromatic radiation from the storage ring can be used to excite the sample.

When white light is used for excitation, the reflected beam from the sample can be detected by the fluorescence spectrometer, and provide a measure of the absorption coefficient of the sample. In the soft x-ray region, light emitted (or reflected) from the sample is energy analyzed by a grating spectrometer and a computer-interfaced multichannel detector(5). The spectrometer may be placed horizontally or vertically relative to the synchrotron radiation, and in some cases the spectrometer rotates about the photon beam. The ability to rotate the spectrometer allows the experimenter to study the angular distribution of the fluorescence radiation that has been excited by either S or P polarized x-rays.(24) Typically, the energy resolution employed while measuring the spectra presented in the figures was ~0.1-0.4 eV.

RESULTS

Description of Photon-in Photon-out Processes

X-ray absorption and emission have generally been thought of as independent processes to produce and destroy the core hole shown schematically in Figure 1a. X-ray absorption spectroscopy is universally used to probe the unoccupied states above the threshold to produce a core hole, while x-ray emission spectroscopy probes a dipole selected group of valence states. Far above threshold the emission and absorption process are not coupled and are incoherent, however, inelastic scattering, a second order process involving two photons can be an especially important process near threshold because it can be resonantly enhanced by six orders of magnitude. In this case the absorbed photon of frequency, v_1, is coupled to the emitted photon of frequency, v_2, through one or more intermediate states. This process has been studied for some time (10-12,25-28) for K shell excitations, and is a more general interaction of photons with matter, where an atom in an initial state I absorbs a photon and passes through intermediate states η and winds up in a final state F. The scattering can be classified as:

i) resonant fluorescence, an elastic scattering process where the input photon energy equals that of a dipole allowed transition and,

ii) inelastic or Raman scattering, where the input photon interacts with the excited electron and is emitted at a lower energy hv_2.

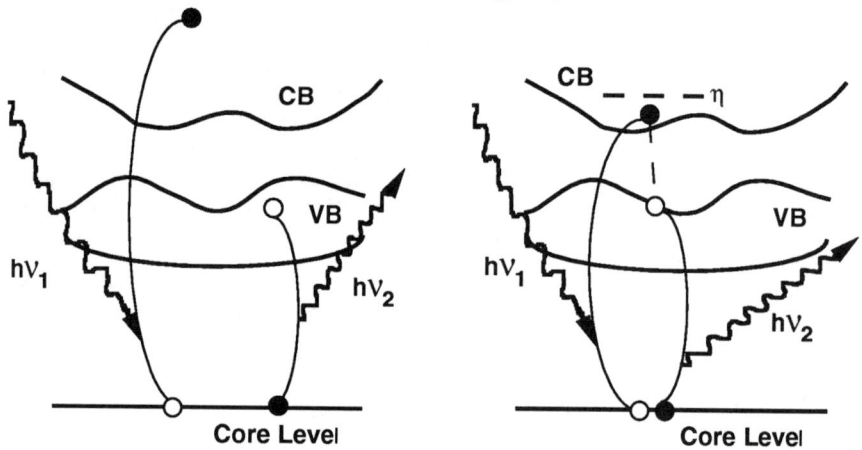

a) Incoherent Emission   **b) Coherent Emission**

FIGURE 1. A schematic representation of the absorption emission processes. The panel on the left (a) illustrates the independent photon-in photon out process. Photons of energy hv_1 produce a core hole, which decays by emitting a fluorescence photon of energy hv_2. The panel on the right (b) illustrates the coupled coherent process where a photon of energy hv_1 promotes a core electron to an intermediate state η which is coupled to the final state through the emission of a photon of energy hv_2.

The processes i) involves the $\mathbf{A \cdot A}$ term in the electromagnetic interaction while ii) involves the square of the $\mathbf{p \cdot A}$ term. Recall the quantity $\mathbf{p}$ is the electron momentum and $\mathbf{A}$ is the electromagnetic vector potential.

In their review, Åberg and Craseman(29) presented the cross section for the inelastic scattering process ii) which consists of the Compton term, the nonresonant, and the resonant terms. The cross section, derived from second order time dependent perturbation theory is given by the following equation as:

$$\frac{d\sigma(v_1, v_2)}{d\Omega} = r_e^2 \frac{v_2}{v_1} \sum_F |\langle F|e^{i(k_1 - k_2) \cdot r}|I\rangle + \frac{1}{m_e} \sum_\eta \frac{\langle F|p \cdot \varepsilon_2|\eta\rangle\langle\eta|p \cdot \varepsilon_1|I\rangle}{E_\eta - E_I - hv_1 - i\Gamma_\eta/2} +$$
$$\frac{1}{m_e} \sum_\eta \frac{\langle F|p \cdot \varepsilon_1|\eta\rangle\langle\eta|p \cdot \varepsilon_2|I\rangle}{E_\eta - E_I + hv_1}|^2 \times \delta(E_F - E_I + hv_1 - hv_2) \quad (1)$$

This is the generalized Kramers-Heisenberg (K-H) equation. In this equation the first term is the Compton scattering, the second is resonant and the third term is non resonant, and does not contribute significantly to the scattering amplitude. The scaling factor for the inelastic scattering cross section is the square of the classical radius of the electron, r_e, which has a magnitude of $7.0 \times 10^{-24} cm^2$. In the spectral energy region of interest, this cross section is small compared to the first order photoionization processes and only the resonant term dominates the scattering. In many other cases the resonant scattering is characterized by a single intermediate state.(26,30)

The delta function in the K-H equation insures the conservation of energy in the scattering process. The indices, I and F, represent the initial and final states respectively. The energy of the input photon is $h\nu_1$. The quantities e_1 and e_2 are the polarization directions and k_1 and k_2 are the propagation directions of the input and output photons respectively. The energies E_η and E_I are the energy of the intermediate state η and the energy of the initial state respectively. The amplitudes are summed over the intermediate states, η, and the cross section is summed over the final states, F.

Because this is a two photon process the angular momentum selection rule is $\Delta l = 0, \pm 2$, the parity of the final state must be the same as the initial state. Under the constraint of these selection rules Gel'mukhanov, and Ågren(31) developed a formalism of the dependence of the inelastic scattering cross section on the molecular symmetry. The angular dependence of the scattered radiation has been described by Åberg and Tullki.(26) This scattering is not limited to atomic levels but has been observed to modify the band emission of solids near threshold. The phenomena was explained by Ma et al. (28) as a process that involved not only energy conservation, i.e.: $h\nu_1 - h\nu_2 = E_F - E_I$ but crystal momentum conservation as well so that two equations apply in solids, namely:

$$h\nu_1 - h\nu_2 = E_c(\kappa_c) - E_v(\kappa_v), \qquad (2)$$

$$q_1 + \kappa_v = \kappa_c + q_2 = G. \qquad (3)$$

In these expressions the quantities $E_c(\kappa_c)$ and $E_v(\kappa_v)$ are the conduction band and valence band binding energies indexed to the crystal momentum vectors κ_c and κ_v of the conduction band and the valence band electron, respectively. Thus, Raman scattering can involve states localized in real or momentum space. Raman scattering from states localized in momentum space can occur either below(19,32) or above the threshold(18,28,33-35), while Raman scattering involving electrons localized in real space occur when the energy of the incident photon is close to but less than the energy for a localized core excitation.

The photons emitted through the coherent or incoherent excitation process must reach a detector. The fluorescent counting rate observed from a sample of thickness, d, at frequency, ν_2, for coherent and incoherent emission is modified by absorption of the incident phonons having energy $h\nu_1$, and by absorption of the fluorescent photons having energy $h\nu_2$. Following Eisebitt et al (36) and Jaklevic et al (37) the expression for the fluorescence counting rate is given by:

$$I_D(\nu_1, \nu_2) = I_0(\nu_1) \left(\frac{\Omega}{4\pi}\right) \varepsilon \\ \frac{\sigma_x(\nu_1, \nu_2) \sec(\theta_1) \left[1 - \exp(\mu_T(h\nu_1)\sec(\theta_1) + \mu_T(h\nu_2)\sec(\theta_2))d\right]}{\mu_T(h\nu_1)\sec(\theta_1) + \mu_T(h\nu_2)\sec(\theta_2)} \qquad (4)$$

In this expression $I_0(v_1)$ is the incidence flux; $\Omega/4\pi$ and ϵ is the solid angle subtended by the detector and the detector efficiency respectively. The quantities θ_1 and θ_2 are the angles with of the incident and exit photon beam with respect to the normal to the sample surface, respectively. The quantity $\mu_T(E)$ is the absorption coefficient of all the species in the specimen at the photon energy E. The quantity $\sigma_x(hv_1)$ is the cross section for producing the fluorescent process, and is given by:

$$\sigma_x(v_1,v_2) = \sigma_{scatt}(v_1,v_2) + \sigma_{in}(v_1)*Y(v_2). \quad (5)$$

In Eq(5) the first term on the right hand side of the expression is the sum of Rayleigh, Compton, and Raman scattering from Eq(1). The second term is the product of absorption produced by the dipole absorption cross section and the fluorescence yield, $Y(v_2)$. The scattering cross section (of magnitude 7.0×10^{-24} cm^2) is only significant for low Z materials near thresholds where either a high density of states in the conduction band may produce electron scattering in the Brillouin zone or the scattering may occur through localized resonances. Near intermediate resonant states the scattering cross sections typically increase by a factor of 10^6.

There are three interesting limiting conditions that apply to Eq(4).

1) If the sample is thin and concentrated, then $I_d(v_1) \approx \sigma_x(v_1)$;

2) if the sample is thick and dilute, then $I_d(v_1) \approx \sigma_x(v_1)$, and finally,

3) if the sample is thick and concentrated, then $I_d(v_1)$ may or may not be proportional to $\sigma_x(hv_1)$.

For the first two cases the detected flux is proportional to the absorption cross section to produce fluorescence. However, the third condition prevails under most circumstances, and in this case, the observed spectral distribution is distorted by the various absorption cross sections. This self absorption must be considered when analyzing the frequency dependence (v_1) of the fluorescence emission (v_2).

Incoherent Fluorescent X-Ray Emission

When the absorption takes place at energies far removed from threshold and resonant excitations where coherent scattering can be appreciable or even dominate the x-ray emission, the absorption and emission of the photons is decoupled. As described in the introduction, x-ray emission under these circumstance dominated the literature for many decades because the spectra, due to transitions of the valence band electrons into core hole states gave an angular momentum selected, site specific picture of the valence band density of states. A great deal of information can be obtained from incoherent x-ray emission.(38) As our first illustration, the observed and calculated density of states are presented for yttrium, and yttria.

Fluorescence X-ray emission form Yttrium and Yttria

The X-rays emitted during electronic transitions between the N_{III} (4p$_{3/2}$) and the M_V (3d$_{5/2}$) core states and from the valence band into Y M_V and oxygen K level vacancies have been energy analyzed. These X-ray energy distributions were interpreted using yttrium p-type and oxygen p-type partial density of states information generated from a one-electron density functional calculation (39). The measurements and the calculated density of states for yttrium and for yttria are shown in Figure 2. The calculation of the electronic structure of yttria was based on the Linear Muffin Tin Orbital (LMTO) method within the atomic sphere approximation (40).

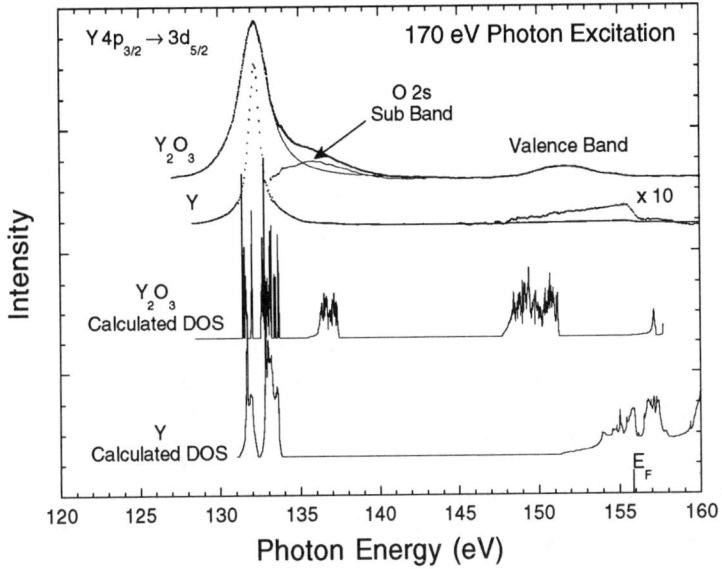

FIGURE 2. Soft X-ray emission spectra for transitions into yttrium 3d vacancies in yttrium metal and in yttrium oxide shown as dotted curves in the upper half of the figure. For clarity the yttrium oxide data has been offset along the vertical axis as indicated. The yttrium metal data has been multiplied by a factor of ten in the valence region. The calculated total density of states for yttria and yttrium metal are shown as solid lines in the lower half of the figure. The doublet structure at photon energies of 131.5 eV and 134 eV is the DOS associated with the Y-4p$_{1/2}$ and 4p$_{3/2}$ levels, respectively. Core-core 4p$_{1/2}$→3d$_{3/2}$ transitions were not observed but shown in the calculation. States at energies greater than the Fermi energy, E$_F$, are unoccupied.(From Ref. 39).

The x-ray emission data show a subband in the vicinity of the oxygen 2s core level, about 15 eV below the valence band. The 2s origin of this spectral feature is confirmed by our calculations of the yttrium X-ray emission spectrum in the region of the 3d$_{5/2}$ → 4p$_{3/2}$ transition. There is significant overlap of the Y-4p wave functions at the oxygen site, thereby hybridizing with the oxygen 2s sub band. In view of the bixbyite structure for Y$_2$O$_3$ this points toward a strong interaction between Y-atoms and the four oxygen atoms that are closest. In addition the observations indicate a broadening of the 3d$_{5/2}$ → 4p$_{3/2}$ transition in these materials with respect to the atomic levels. The broadening is verified by the calculation which shows band-like features extending to the shallow 4p core state. Furthermore, we observe additional broadening of 3d$_{5/2}$ → 4p$_{3/2}$ core-core transition in the oxide with respect to its profile for yttrium metal. The calculations show this additional broadening occurs because the yttrium in the oxide is distributed over two non-equivalent sites which results in different 4p binding energies (about 0.5 eV) for the yttrium atoms at the two sites in the oxide. Finally, we did not observe the $3d_{3/2} \rightarrow 4p_{1/2}$ core-core transition.

The calculated total densities of states for the 4p shallow core states and the valence bands of the metal and the oxide is shown in the lower half of Figure 2. These calculations are compared to the soft x-ray emission between the valence band and the $3d_{5/2}$ and the $4p_{3/2}$ and the $3d_{5/2}$ core state for yttria and yttrium metal. The yttrium oxide x-ray spectrum exhibits a valence emission band centered near a transition energy of 152 eV, while the valence emission band for the metal shows a sharp discontinuity at the Fermi energy in both the experiment and in the calculation. The yttrium and oxygen valence spectra are very similar in shape and also conform closely to the photoelectron spectrum. The similarity of the experimental spectra at the oxygen and yttrium sites indicates that the band structure is indeed pervasive, i.e. Y-5p states participate in the formation of the valence band of Y_2O_3.

These results suggest that a state of the art one-electron band structure calculation gives good results even for a compound like yttrium oxide which has 28 atoms per unit cell.

Incoherent X-ray emission for Titanium Diboride at the Boron Site

Our group has made a study of a large number of transition metal diborides and hexaborides. As a second example we would like to consider one of them. This class of compounds are of technical interest because they are quite hard, metallic, and chemically stable, and have high melting points.(41) In addition to the technical importance of the borides, our group is interested in compounds of this type because of the very strong 1s → π* resonance that has been observed and associated with the boron atom (19, 42-44).

Several calculations have been carried out for the hexaborides and the diborides.(45-47). Titanium diboride has been chosen for the example to be presented here. As $h\nu_1$ was scanned between 188 eV and 197 eV, resonant Raman or spectator emission as it sometimes called (42,43) was observed at the 1s → π* core-exciton resonance. At a photon energy of about 197 eV the emission spectrum is representative of the valence band density of states having type symmetry. The crystal structure of TiB_2 is designated as C32 and is characterized by alternating layers of titanium and boron atoms. The boron is arranged in hexagonal nets with a titanium atom centered above each hexagon (48,49). In this compound the three B valence electrons occupy sp-σ and p-π orbitals. Strong σ bonding occurs between the B atoms in the net whereas the π orbitals overlap with d_{z^2} or d_{xz}, and d_{yz} orbits on the Ti atom. Within the Ti layer σ type orbitals are formed from $d_{x^2-y^2}$ and d_{xy} atomic orbitals. These orbitals are unoccupied and nonbonding with respect to the boron layer (48,49), but form a resonance low in the conduction band. This is in contrast to the D^4_{6h} symmetry of the lattice of h-BN.(50). In h-BN the bonds between the B and N atoms in a layer are of σ symmetry. The interlayer bonding is produced by orbitals of π symmetry. The difference in bonding between h-BN and TiB_2 is undoubtedly the reason why the emission by boron in h-BN is so different than that of boron in TiB_2, even though both compounds have a hexagonal crystal structure, the point group for the B site is D_{3h} in both compounds. The emission spectrum of oriented h-BN also shows a strong asymmetry in the valence band x-ray emission as a function of the fluorescence emission angle relative to the c

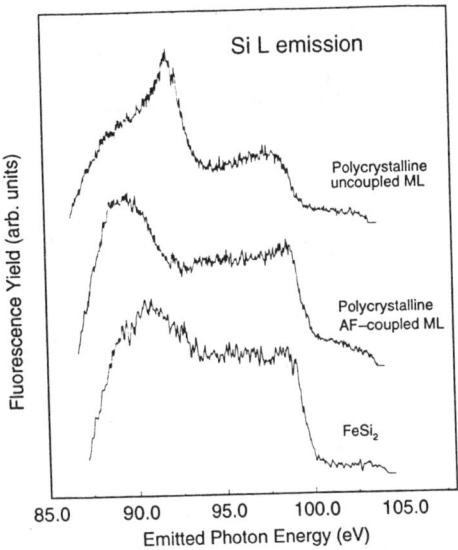

FIGURE 4. Soft x-ray fluorescence Si L-emission for an FeSi$_2$ reference sample and for the two polycrystalline Fe/Si multilayers. The incident photon energy was 132 eV. The data labeled uncoupled ML is from the (Fe30Å/Si20Å)x50 multilayer grown on glass. The data labeled "AF-coupled ML" is from the antiferromagnetically coupled (Fe30Å/Si14Å)x50 multilayer grown on glass.

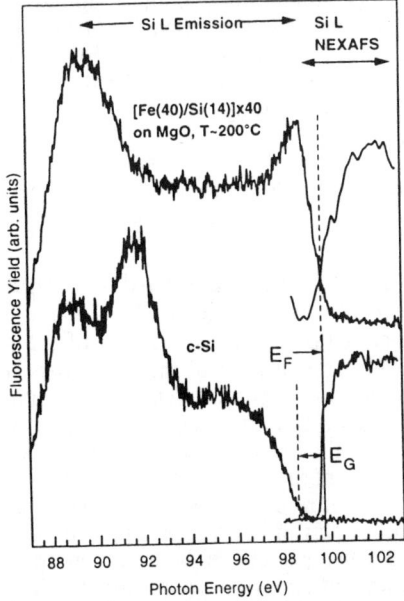

FIGURE 5. The Soft x-ray florescence and NEXAFS spectra from the antiferromagnetically coupled (Fe30Å/Si14Å)x50 multilayer grown on glass (curves in the upper part of the figure), and from crystalline Si (curves in the lower part of the figure). The Fermi energy is labeled E_F and the gap energy is labeled E_G.

759

comparable to those obtained by total yield at least near the threshold for absorption into the conduction band.(62) In these spectra the silicon emission spectra were excited at photon energies twenty eV or so greater than the absorption threshold to avoid modification of the spectrum by coherent inelastic scattering. The unambiguous nature of these results on thin buried silicide layers is an excellent illustration of the power of SXE as a bulk sensitive spectroscopy to study the buried interface.

Coherent X-Ray Emission: Raman and Elastic Scattering

In this section we will describe our results for hexagonal boron nitride, graphite, and several titanium compounds. These examples give especially revealing insights into the X-ray Raman process in solids. For these examples we have observed elastic scattering involving an excited discrete state and resonant Raman emission or "Spectator Emission", as it is known, due to the promotion of the core electron to a discrete localized state. Coherent emission also occurs which involves momentum localized scattering in the conduction band *above* threshold that modifies the valence band fluorescence. In these cases (19, 27, 32, 42, 63) the emission spectra depend sensitively on the screening role and coherent coupling of the spectator electron in the localized orbitals. In addition to this behavior, we have observed Raman Scattering in h-BN and titanium compounds that involve a scattering of electrons in the valence band to the conduction band state below the ionization threshold which tracks the photon excitation energy.

In the case of silicon(33-35), diamond(28), and graphite(18), the process for Raman scattering in the valence band observed *above* threshold has been described by Ma *et al* (28) and involves scattering of electrons localized in momentum space and delocalized in real space. The material is left in delocalized valence-excited states. Through a study of these scattering processes, it is possible map the electronic bands, as will be shown by our measurements in graphite.(18) The Raman scattering below threshold that involves a localized discrete state and momentum constrained excitations are illustrated by our measurements in h-BN(19) and titanium compounds (32).

Raman and Elastic Scattering in Graphite

Photons that excite core electrons to states in the conduction band of a particular symmetry can scatter the electrons to states of the same symmetry in the valence band through the resonant term of the K-H cross section as depicted by Equation 1. In solids the conservation of crystal momentum (Eq. 3) places an additional constraint on the scattering cross section. This **k**-conservation model is one of the factors that produce a variation of the intensity of emission as a function of photon excitation energy above threshold. Our measurements for graphite(18) are shown in Figure 9 where we observed crystal momentum conserving resonant absorption-emission processes that involve the excitation of core electrons to unoccupied states just above the Fermi energy. As Ma *et al.* (28) have suggested the absorption and emission events are coupled. In the figure the excitation spectrum above threshold is shown. The spectral range involving the excitation of the π^* resonance in this layered compound resonance is omitted because the spectra have a similar appearance to that of h-BN described in the

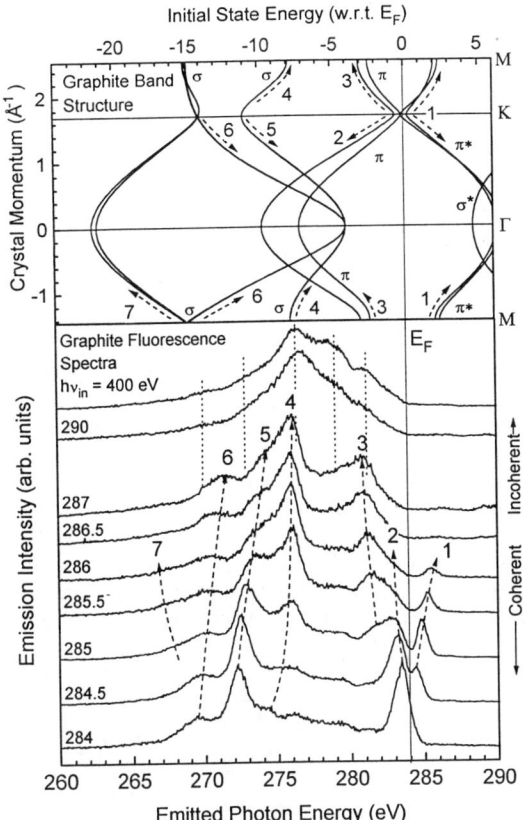

FIGURE 6. Resonant fluorescence and nonresonant fluorescence spectra from highly oriented pyrolytic graphite at incident energies $h\nu_1$. The upper panel displays the graphite band structure with the energy axis matched to the photon energy axis of the fluorescence data. The dispersive features labeled 1-7 in the resonant spectra are associated with the photons 1-7 of the graphite band structure indicated by dashed lines and arrows in the upper panel (From Ref. 18).

next section. In this example we wish to emphasize that the interpretation of this inelastic scattering process can result in new information about band structure as noted by Ma et al. (28) and by Carlisle et al.. (18). Excitation near resonance permits one to observe competing processes taking place: fluorescent decay of the exciton and fluorescence due to transitions between the valence band and the core hole with the exciton as a spectator, similar to the condition when the exciton was excited resonantly, and incoherent fluorescence where the coherent excitation is decoupled by the phonon interactions for example.

The uppermost spectrum in Figure 6 is characteristic for the nonresonant case where the excitation takes place far above threshold. Spectra such as these resemble those obtained by electron excitation. Below this spectrum are those obtained just above the Fermi energy using photon excitation energies between 284 eV and 290 eV. Most of these spectra have a quite different line shape compared to the nonresonant one. In these spectra various emission features disperse as is shown by the numbered spectral features. The energy of the emitted photon changes as the input photon energy changes. A dramatic change occurs when the input photon energy changes by just 0.5 eV, from 284.5 eV to 285 eV.

These energy-dependent emission features are the result of transitions from states with well defined crystal momentum. The band structure of graphite along the high symmetry directions is presented in the upper panel of Figure 6. This band structure was derived from a tight-binding parameterization of quasiparticle calculations (64). The π or σ bands give rise to the bands labeled in the top half of Figure 9 and correspond to the different type of bonding that arises between the carbon atoms in the sheets.

Raman and Elastic Scattering in h-BN

Our observations in h-BN (19) are shown in Figure 7. The photon energy of the output photon is plotted as the abscissa, and a spectrum for each excitation energy is shown displaced along the ordinate. At a photon energy of 189 eV considerably below threshold (about 195 eV) very weak Raman peaks occur at 173 eV and 177 eV along with the elastic scattered photons at 189 eV which can be seen on a scale reduced by a factor of 100. These two features labeled A and B have a Raman loss of 16 eV and 12 eV respectively. As the excitation photon energy increases approaching the localized 1s → π* resonance at 193 eV, the Raman scattering involving electrons in the valence band increases in intensity and track the excitation photon energy. When resonance is almost realized the

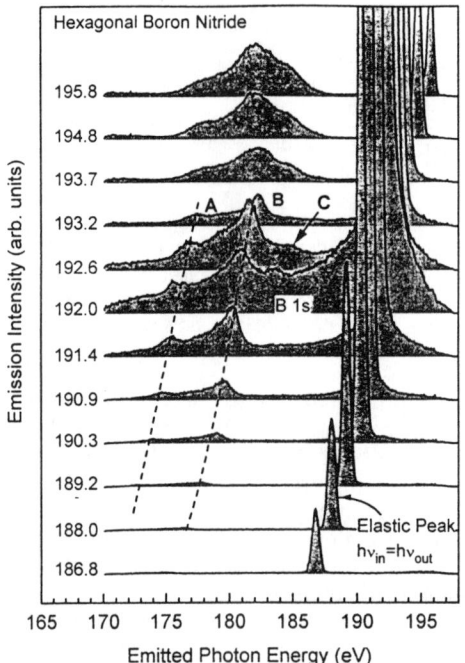

FIGURE 7. The boron K-edge soft x-ray fluorescence spectra of h-BN for excitation near the K absorption threshold. Labeled features are discussed in the text. (From Ref. 19).

valence band emission becomes more complicated due to the screening effects of the electron excited to the localized π^* state. In this case the spectator electron dynamically screens the final states and is mixed with elastic and Raman scattering at resonance. The relative intensity of the resonantly scattered photon depends sensitively on the crystal structure (32) and the atomic environment which influences the bonding in the vicinity of the boron. Boron nitride is a layered structure similar to graphite with sp^2 σ bonds in the layer and π bonds perpendicular to the layer. The band structure was calculated(19) and the quasiparticle band structure is shown in Figure 8. The transitions proposed for the peaks A and B exhibiting Raman scattering are indicated in the diagram as valence band to π^* transitions. The simulated spectrum, which does not include excitonic band-modifying effects, is shown in Figure 9. The fluorescence band emission was simulated in an independent, quasiparticle model, using a detailed description of the quasiparticle wave functions and energies. The generalized K-H formula (Eq. 1) for the cross section was used to calculate the scattered photon intensity. In Eq. 1 the matrix elements between the initial, intermediate, and the final states were approximated as one-electron matrix elements. The energies of the excited states were based on quasiparticle energies. We have not included excitonic or other complicated effects in this model. Similar below threshold Raman scattering has been observed in graphite which is another layered compound (63).

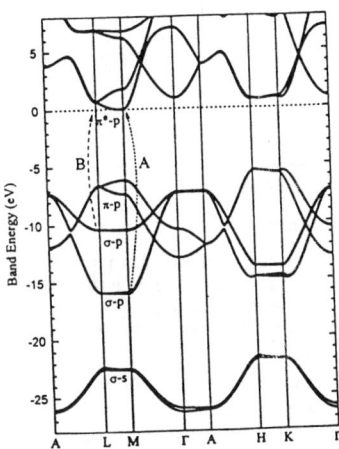
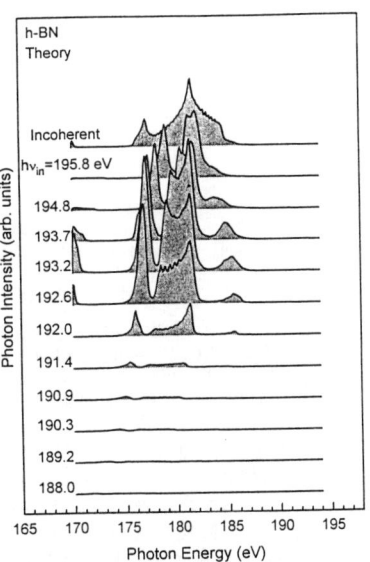

FIGURE 8. The h-BN quasiparticle band structure. Transitions proposed for peaks A and B in Figure 6 are indicated. (From Ref. 19).

FIGURE 9. The simulated h-BN fluorescence for several incident photon energies as labeled. Incoherent emission is not included. (From Ref. 19).

Raman Scattering in Transition Metal Compounds: Titanium and Titanium Compounds

The transition metal compounds form a very interesting and important set of materials. The diversity arises from the many valence states the transition elements may take when forming compounds. This variety provides ample opportunity for a large class of materials to have a vast range of electronic and magnetic properties (65). The x-ray spectroscopy of the transition elements is especially interesting because they have unfilled localized atomic-like states that are at the bottom of the conduction (47, 48, 66). Our group embarked on the systematic study of transition metal sulfides (67) and oxides (32). As an example of the type of spectra observed in some of these compounds we have chosen to showcase the $L_{II,III}$ emission and Raman scattering in some titanium compounds obtained by photon excitation.

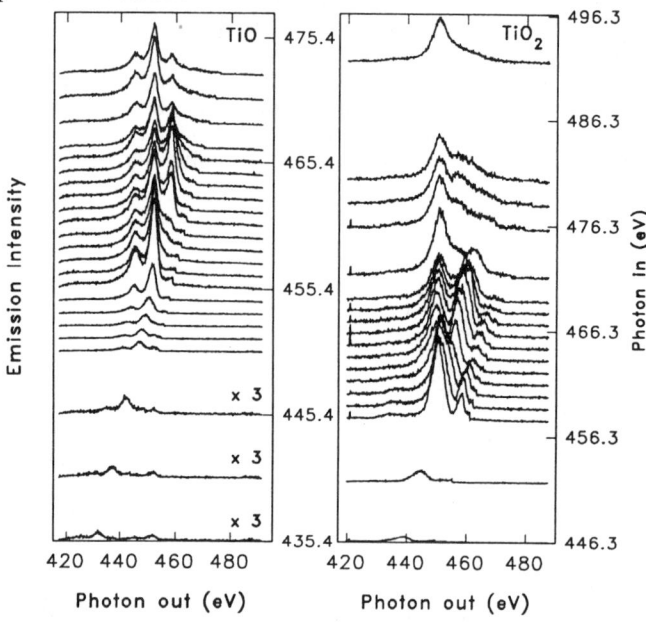

FIGURE 10. Evolution of the TiO (left panel) and TiO_2 (right panel) $L_{II,III}$ emission spectra. All curves show the normalized x-ray emission as a function of the emission energy $h\nu_2$ plotted as the abscissa. The scale on the right give the excitation energy $h\nu_1$. The intensity of the lower three spectra in TiO was multiplied by a factor of three. (From Ref. 32)

As in h-BN and graphite the exciton levels provide a means to promote Raman scattering with these excitonic states acting as the intermediate state, and the electrons in the valence band providing many final states. Figure 10 is an example of Raman scattering in TiO and TiO_2, where the $L_{II,III}$ emission is plotted on the abscissa for different input photon energies as the ordinate. Each spectrum is displaced according to the value of the excitation photon energy shown on the ordinate. At input photon energies much greater than the $L_{II,III}$ absorption threshold ($\approx$455 eV), the spectra resemble those obtained by Fischer and Baun (55) obtained by electron beam excitation. In these examples the

Raman scattering follows the classic pattern given in Eq(1) as described by Åberg and Tulkki (26), and reviewed by Åberg and Crasemann. (29) The energy positions of the $L_{II,III}$ emission versus the input energy is shown in Figure 11.

Just as in h-BN the *entire band* participates in the scattering constrained by the localization in momentum space defined by Eq(2) and Eq(3). The Raman scattering observed in the titanium compounds is an example of the momentum localized Raman scattering observed below threshold in h-BN reviewed here.(19) It is somewhat surprising that in some compounds the Raman scattering is visible up to *30 eV* below the ionization threshold, as can be seen clearly in Figure 11.

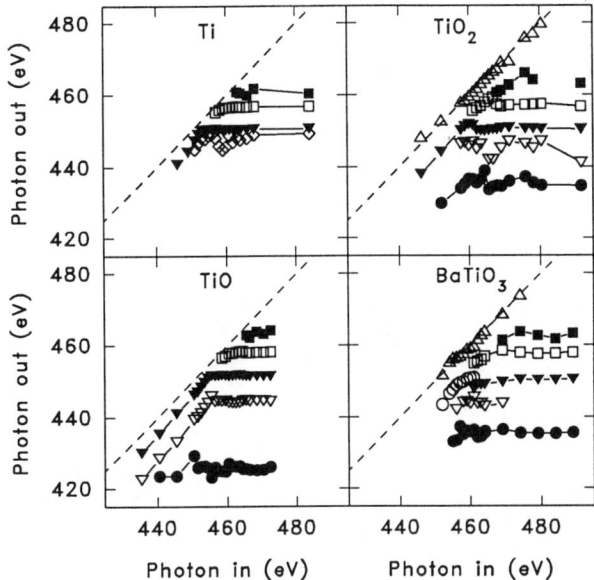

FIGURE 11. Peak position as a function of the excitation energy for all Ti compounds studied. The symbols are grouped as follows. Solid circle: L_{III} valence band (2s subband), open triangle down: L_{III} valence band, solid triangle down: L_{III} valence band, open square: L_{II} valence band, open triangle up: elastic peak, open circle: Raman peak, diamond: extra peak needed in the fitting routine. The exciting photon energy is indicated by a dotted line. (From Ref. 32)

Table 1: Raman Energy Loss

Peak	Ti	TiO	TiO$_2$	BaTiO$_3$
L_{III} (O 2s)		21(1)*	23.1(8)	22.5(8)
L_{III} (VB)		11(1)	14.4(8)	14.9(8)
L_{III} (VB)	3.8(9)	4.2(9)	7.9(8)	14.1(6)
L_{II} (V)	2.1(8)	2.2(1)	6.2(8)	6.8(8)
MS†	2.1(5)	2.4(5)	7.4(6)	7.6(6)
Raman¶				7.8(8)

* Estimated uncertainty in the last digit.
† Spectral feature probably produced by multiple scattering.
¶ Extra peak required by the fitting process between the L_{III} and the L_{II} edges.

The Raman loss energies for the titanium compounds studied are shown in the Table 1 labeled according to the appropriate peak associated with one of the 2p core states used in the algorithm that was applied to fit the spectrum to a model. These Raman losses correspond to energies required to excite a valence exciton, and should be of particular value when making computations that include the valence-exciton state.

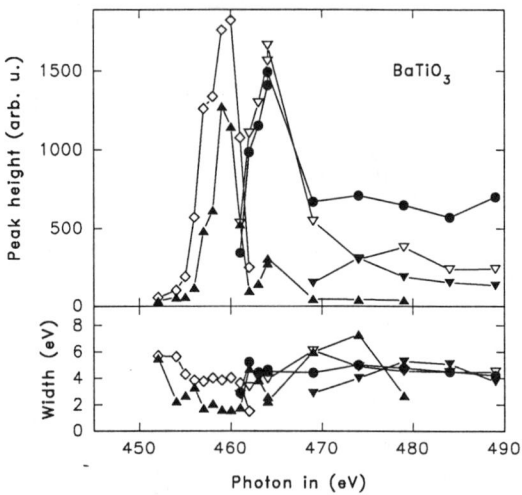

FIGURE 12. Peak heights (upper panel) and peak widths (lower panel) for BaTiO3. The symbols are grouped as follows. Solid circle: L_{III} valence band, open triangle down: L_{II} valence band, solid triangle down: MS (multiple scattering excitation), solid triangle up: elastic peak, diamond: Raman peak. (From Ref. 32)

By modeling the data it is a simple task to study the variation of the peak intensities as a function of the incoming photon energy. The intensities, corrected for self-absorption according to Eq. 4 are shown in Figure 12 for BaTiO$_3$. The fluctuation of the partial excitation cross sections for increasing input photon energy is consistent with the idea that the core electron is excited into a long lived (relative to the valence-core life time) localized state just at the ionization threshold of the 2p core states. Notice in Figure 12 the elastic scattering (▲) resonates at the L_{III} and the L_{II} thresholds and is observable over a large energy range. The variation in intensity of the partial cross sections suggests that the resonance width is approximately 1.2(5) eV.

Resonant Raman Scattering in Titanium Diboride

Our group has studied several hexaborides and diboride compounds. We illustrate their typical behavior by showing the boron K excitation spectra in Figure 13 and the titanium $L_{II,III}$ in Figure 14 for TiB$_2$.

The boron K absorption spectrum of TiB$_2$ shows the prominent π^* resonance at about 193 eV and a shape resonance just above the ionization threshold. Both compounds have a weak absorption feature at about 192 eV. The absorption spectrum of all the diborides and hexaborides have similar spectra and they all resemble closely the spectrum of gaseous BF$_3$, which suggests that these features are of a localized nature. While the emission spectra of boron and titanium discussed in the previous sections show strong resonant scattering, neither Ti L nor B K emission show strong Raman scattering below threshold. There is little Raman emission below the excitation threshold, but as the photon energy is scanned through the π^* resonance, one notes a slight change in the boron K

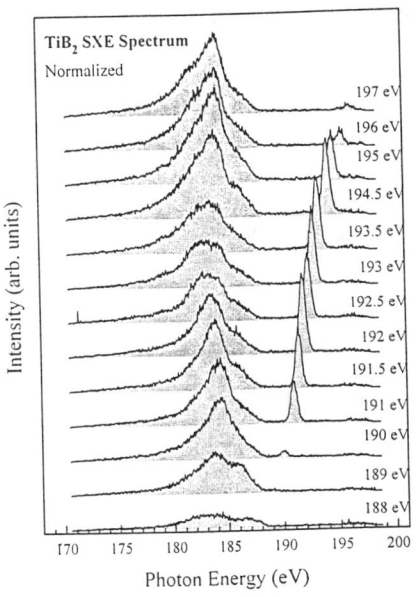

FIGURE 13. Evolution of the B K emission spectra in TiB_2. All curves show the normalized x-ray emission as a function of the emission energy $h\nu_2$ plotted as the abscissa. The scale on the right gives the excitation energy $h\nu_1$.

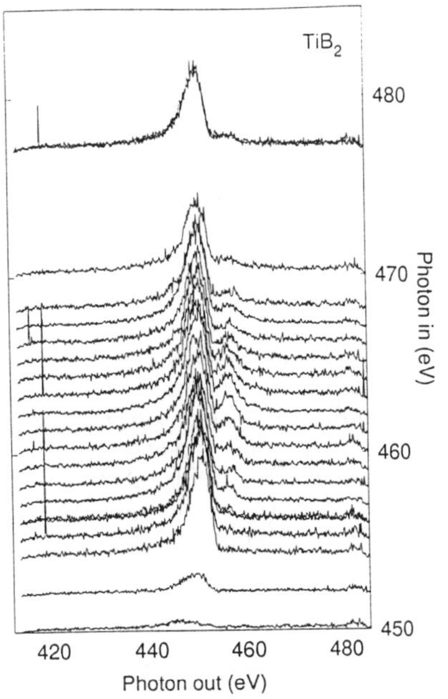

FIGURE 14. Evolution of the Ti $L_{II,III}$ emission spectra. All curves show the normalized x-ray emission as a function of the emission energy $h\nu_2$ plotted as the abscissa. The scale on the right gives the excitation energy $h\nu_1$.

emission spectrum in the energy range of the π^* excition due to changes in the valence band screening produced by the excited electron, which provides a clue toward an understanding of the bulk electronic structure of these borides. The energy dependent L emission spectrum shows almost no coherent excitation and the spectra are completely dominated by incoherent fluorescence emission even at threshold. Only a weak resonance excitation is seen at the L_{II} edge. Furthermore, the L_{II} spectra excited far above threshold is very weak suggesting suppression by Coster Kronig radiationless decay of the L_{II} core hole. These fast competing processes probably partly account for the lack of coherent processes observed in this system. The valence band emission from these compounds shows none of the crystal momentum conserving resonant absorption-emission processes that are so common in h-BN and other titanium compounds, where the resonant excitation of the localized excited state acts as a spectator for the valence emission.

CONCLUSIONS

We have given a number of examples to illustrate the power of photon stimulated x-ray fluorescence spectroscopy to act as a probe to understand the electronic properties of materials in greater detail. We have seen how state of the art one electron band structure calculations can provide a reasonable picture of the observed valence band emission in many materials. We have demonstrated how synchrotron radiation excitation and fluorescence can be used to determine the conducting or insulating nature of an interface layer. Our group has also demonstrated how x-ray Raman scattering, an almost ubiquitous process can be used to study the details of the density of states in different regions of the Brillouin zone. Finally we have seen how this coherent scattering may provide information about the dynamical processes in the material.

ACKNOWLEDGMENTS

The authors are especially grateful to Mr. R. Winarski, Mr. S. Stadler, and Ms. Tao Shu, for their assistance in taking and reducing some of the data. The authors would like to acknowledge support from NSF grant DMR-9017997, the University of Tennessee Science Alliance and a DoE-EPSCoR cluster research grant DoE-LEQSF (1993-95)-03. The Advanced Light Source and the National Synchrotron Light Source are supported by Department of Energy grants DE-AC03-76SF00098 and DE-AC02-76CH00016, respectively.

REFERENCES

[a] Present address: Physics Depart., Virginia Commonwealth University, Richmond VA 23284
[b] On leave from the Instituto de Ciencias Nucleares, 04510 Mexico, D.F., Mexico
[c] Present address: IMP, Inc. 2830 N. First St., San Jose, CA 95134
[d] Present address: Physics Department, University of Addis Abbaba, Addis Abbaba, Ethiopia.

1) Röntgen,W.C., (Trans. by A. Stanton) Nature **53**, 1896, 274.
2) Compton,A.H., and Allison, S.K., *X-Rays in Theory and Experiment*, New York: 1935, D. Van Nostrand Co
3) Parratt, L.G., Rev. Mod. Phys. **31**, 1959, 616-729.
4) Tomboulian, D.H., *Handbuch der Physik*, S.Flügge/Marburg, editors, Berlin: Springer-Verlag, 1957, Vol. **XXX**, pp. 246-304.
5) Callcott, T.A., Tsang, K.L., Zhang, C.H., Ederer, D.L., and Arakawa, E.T., Rev. Sci. Inst. **57**, 1986, 2680-2688.
6) Rubensson, J.-E., Wassdahl, N., Bray, G., Rindstedt, J., Nyholm, R., Cramm, S., Mårtensson, N., and Nordgren, J., Phys. Rev. Letts. **60**, 1988, 1759-1763.
7) Ederer, D. L., Callcott, T. A., and Perera, R.C.C., Synchrotron Radiation News, **7**, 1994, No. 4, 29-33.
8) Ma,Y., Synchrotron Radiation News, **8**, 1995, No. 1, 26-30.
9) Shin, S., Synchrotron Radiation News, **8**, 1995, No. 4, 16-22.
10) Sparks, C.J., Phys. Rev. Lett. **33**, 1974, 262-266.
11) Eisenberger, P., Platzman, P.M., and Winick, H., Phys. Rev. Letts. **36**, 1976, 623-627.
12) Briand, J.P., Girard, D., Kostroum, V.O., Chevalier, P., Wohrer, K., and Mossé, J.P., Phys. Rev. Lett. **46**, 1981, 1625-1629.
13) Proceedings of the Workshop: *Raman Emission by X-Rays*, Ederer, D.L. and McGuire, J.H., Editors (to be published by World Scientific Publishing, 1996).
14) *Resonant Anomalous X-Ray Scattering: Theory and Applications,* Materlik, G., Sparks, C.J., and Fischer, K., Editors, Amsterdam: North Holland, 1994.

15) *Proceedings of the International Workshop on Resonant Inelastic Soft X-Ray Scattering*, Walberberg Germany, Sept. 1996, to be published.
16) Callcott, T.A., Jia, J.J., Zhou, L., Ederer, D.L., Terminello, JL.J., Carlisle, A., Perera, R.C.C., Samant, M.G., Himpsel, F.J., and Arakawa, E.T., *Proceedings of the Materials Research Society* San Fransicso April 1996, to be published.
17) Ederer, D.L., Carlisle, J.A., Jimenez-Mier, J., Jia, J.J., Osborn, K., Callcott, T.A., Perera, R.C.C., Underwood, J.H., Terminello, L.E., Asfaw, A., and Himsel, F.J., J. Vac. Sci. Technol. A **14**, 1996, 859-866..
18) Carlisle, J.A., Shirley, E.L., Hudson, E.A., Terminello, L.E., Callcott, T.A., Jia, J.J., Ederer, D.L., Perera, R.C.C., and Himsel, F.J., Phys. Rev. Letters **74**, 1995, 1234-1238.
19) Jia, J.J., Callcott, T.A., Shirley, E.L., Carlisle, J.A., Terminello, L.J., Asfaw, A., Ederer, D.L., Himpsel, F.J., and Perera, R.C.C., Phys. Rev. Lett., **76**, 1996, 4054-4059.
20) Skytt, P., Guo, J., Wassdahl, N., and Nordgren, J., Phys. Rev. A **52**, 1995, 3572.
21) Nordgren, J., Bray, G., Cramm, S., Nyholm, R., Rubensson, J.-E.and Wassdahl, N., Rev. Sci. Instrum. **60**, 1989, 1690-1695.
22) Jia, J.J., Callcott, T.A., Yurkas, J., Himsel, F.J., Samant, M.C, Stöhr, J., Ederer, D.L., Carlisle, J.A., Hudson, E.A., Terminello, L.J., Perera, R.C.C., and, Shuh, D. K., Rev. Sci. Instrum. **66**, 1995, 1394-1398.
23) Asfaw, A., Ederer, D.L., Zhou, L., Lin, L., Osborn, K., Callcott, T.A., Miyano, K.E., Morikawa, E., Rev. Sci. Instrum. **66** (2), 1995, 1627-1629.
24) Guo, G.-H., Wassdahl, N., Skytt, P., Butorin, S.M., Duda, L.-C., Englund, C.J., and Nordgren, J. , Rev. Sci. Instrum. **66**, 1995, 1561-1567.
25) Cowan, P.L., Phys. Scr. **T31**, 1990, 112-118.
26) Åberg, T. and Tulkki, J., *Atomic Inner Shell Physics*, Bernd Crasemann Editor , New York: Plenum Publishing Corp., 1985. pp. 419-462.
27) O'Brien, W.L., Jia, J.J., Dong, Q.-Y., Callcott, T.A., Miyano, K.E., Ederer, D.L., Mueller, D., and Kao, C.-C., Phys. Rev. Lett. **70**, 1993, 238-242.
28) Ma, Y., Wasshahl, N., Skytt, P., Guo, J., Nørdgren, J., Johnson, P.D., Rubensson, .J.-E., Boske, T., Eberhardt, W., and Kevan, S.D., Phys. Rev. Lett. **69**, 1992, 2598-2602.
29) Åberg, T., and Crasemann, B., *Resonant Anomalous X-ray Scattering: Theory and Applications*, ed: G. Materlik, C.J. sparks, and K Fischer, editors, Amsterdam: Elsevier Science B.V., 1994, pp. 431-448.
30) Cowan, P, *Resonant Anomalous X-ray Scattering: Theory and Applications*, G. Materlik, C.J. sparks, and K Fischer, editors, Amsterdam: Elsevier Science B.V., 1994, pp. 449-472.
31) Gel'mukhanov, F., and Ågren, H., Phys. Rev. A **49**, 1994, 4378-4384.
32) Jimenez-Mier, J., Ederer, D.L., Diebold, U., Moewes, A., Callcott, T.A., Zhou, L., Jia, J.J., Carlisle, J. Hudson, E., Terminello, L.E., Himsel, F.J., and Perera, R.C.C., Proceedings of the Workshop: *Raman Emission by X-Rays* Ederer, D.L., and McGuire, J.H., Editors, to be published by World Scientific Publishing (1996).
33) Miyano, K.E., Ederer, D.L., Callcott, T.A., O'Brien, W.L., Jia, J.J., Zhou, L., Dong, Q.-Y., Ma, Y., Woicik, J.C., and Mueller, D.R., Phys. Rev. **B48**, 1993-I, 1918-1922.
34) Rubensson, J.E., Mueller, D.R., Shuker, R., Ederer, D.L., Zhang, C.H., Jia, J.J., and Callcott, T.A, .Phys. Rev. Lett. **64**, 1990, 1047-1051.
35) Shin,S., Agui, A., Watanabe, M., Fujisawa, M. ,Tezuka Y., and, Ishii, T., Phys. Rev. B **53**, 1996, 15660-15665.
36) Eisebitt, S., Böske, T., Rbensson, J.-E.and, Eberhardt, W., Phys. Rev. B **47**, 1993, 14103-14111.
37) Jaklevic, J., Kirby, J.A., Klein, M.P., Robertson, A.S., Brown, G., and Eisenberger, P., Solid State Comm. **23**, 1977, 679-674.
38) Meisel, A., Leonjardt, G., and Szargan, R., *X-Ray Spectra and Chemical Binding* , V.I. Goldanskii, F.P. Schäfer, and J.P. Toennies editors, Berlin: Springer-Verlag, 1989.
39) Mueller, D.R., Ederer, D.L., van Ek, J., O'Brien, W.L., Dong, Q.-Y., Jia, J.J., and Callcott, T.A., accepted for publication in Phys. Rev. B (1996).
40) Andersen, O.K., Phys. Rev. B. **12**, 1975, 3060-3068.
41) Tian, D.-C, and Wang, X-B., J. Phys: Condens Matter **4**, 1992, 8765-8771.
42) Muramatsu, Y., Oshima, M., and Kato, H., Phys. Rev. Lett. **71**, 1993, 448-452.
43) O'Brien, W.L., Jia, J.J., Dong, Q.-Y., Callcott, T.A., Miyano, K.E., Ederer, D.L., Mueller, D.R., and Kao, C.-C., Phys. Rev. Lett. **70**, 1993, 238-242.

44) Shin, S., Agui, A., Fujisawa, M. T., Yezuka, Y., Ishii, T., Minagawa, Y., Suda, Y,. Ebina, A., Mishima, O., and Erqa, K., Phys Rev. B **52**, 1995, 11853-11857.
45) Perkins, P.G., Armstrong, D.R., and Breeze, A., J. Physics C:Solid State Physics, **8**, 1975, 3558-3564.
46) Ihara, H., Hirabayashi, M., Nakagawa, H., Phys. Rev. B **16**, 1977, 726-730.
47) Wang, X-B., Tian, D.-C., and Wang, L.-L., J. Phys.: Condens. Matter **6**, 1994, 10185-10190.
48) Vvedensky, D.D., Eberhart, M.E., Christodoulou, L., and MacLaren, J.M., Materials Science and Engineering, **A126**, 1990, 33-38.
49) Burdett, J.K., Canadell, E., and Miller, G., J. Am. Chem. Soc. **108**, 1986, 6561-6566.
50) Robertson, J., Phys. Rev. B **29**, 1984, 2131-2142.
51) Tegeler, E., Kosuch, N., Wiech, G., and Faessler, A., Phys. Stat. Sol. (b) **84**, 1977, 561-573.
52) Mansour, A., Schnatterly, S.E., Phys. Rev. B **35**, 1987, 9234-9237.
53) Holliday, J. E., Adv. X-ray Analysis **9**, 1966, 365-374.
54) Lyakhovskaya, I.I., Zimkina, T.M., and Fomichov, V.A., Soviet Physics-Solid State **12**, 1970, 138-143.
55) Fischer, D.W., and Baun, D.W., J Appl. Phys. **39**, 1968, 4757-4763.
56) Carlisle, J.A., Chaiken, A., Michel, R.P., Terminello, L.J., Jia, J.J., Callcott, T.A., and Ederer, D.L., submitted to Phys. Rev. B Rapid Comms. (1996).
57) Carlisle, J.A., Perera, R.C.C., Underwood, J. H., Terminello, L.E., Hudson, E.A., Callcott, T.A., Jia, J.J., Ederer, D.L., Himsel, F.J., and Samant, M.G., Appl. Phys. Lett. **67**, 1995, 34-38.
58) Perrera, R.C.C., Zhang, C.H., Callcott, T.A., and Ederer, D.L., J. Appl. Phys. **66**, 1989, 3676-3681.
59) Fuyllerton, E.E., Mattson, J.E., Lee, S.R., Sowers, C.H., Huang, Y.Y., Felcher, G. Bader, S.D., and Parker, F.T., J. Appl. Phys. **73**, 1993, 6335-6341.
60) Mattson, J.E., Kumar, S., Fullerton, E.E., Lee, S.R., Sowers, C.H., Grimsditch, M., Bader, S.D., and Parker, F.T., Phys. Rev. Lett. **71**, 1993, 185-189.
61) Jia, J.J., Callcott, T.A., O'Brien, W.L., Dong, Q.-Y., Mueller, D.R., Ederer, D.L., Tan, Z., and Budnick, J.I., Phys. Rev. **B46**,1992, 9446-9451.
62) Stöhn, J., *NEXAFS Spectroscopy*, New York: Springer Verlag, 1992.
63) Carlisle, J.A., Terminello, L.E., Jia, J.J., Callcott, T.A., Ederer, D.L., Perera, R.C.C., and Himsel, F.J., Proceedings of the Workshop: *Raman Emission by X-Rays* (Eds: D.L. Ederer and J.H. McGuire) to be published by World Scientific Publishing (1996).
64) McGovern, I.T., Eberhardt, W., Plummer, E.W., and Fischer, J.E., Physica (Amsterdam) **99B**, 1980, 415-422.
65) Sharp, D.W.A., *Transition Metals*, London Butterworths, Baltimore University Park Press, 1972.
66) de Groot, F.M.F., Fuggle, J.C., Thole, B.T., and Swatzky, G.A., Phys. Rev. B**42**, 1990, 5459-5458.
67) Zhou, L., Callcott, T.A., Ederer, D.L., Perera, R.C.C., submitted to Phys. Rev. B ,1996.

Electronic Structure of Advanced Materials Studied by X-Ray Emission Spectroscopy

E. Z. Kurmaev, V. R. Galakhov, Yu. M. Yarmoshenko,
V. A. Trofimova, S. N. Shamin, V. M. Cherkashenko,
A. I. Poteryaev, and V. I. Anisimov

*Institute of Metal Physics, Russian Academy of Sciences-Ural Division,
620219 Yekaterinburg GSP-170, Russia*

Abstract. High resolution soft x-ray emission spectroscopy with high spatial resolution is used to study of the electronic structure and characterize advanced materials: high-T_C superconductors, transition metal compounds, porous silicon, solid-solid buried interfaces and hard materials. In high-T_C, the main attention is focused on the analysis of oxygen-cation interactions and the determination of the location of impurity atoms. In transition metal compounds the participation of different electronic states of constitute atoms in the valence band is analyzed and correctness of LDA band structure calculations is estimated. For $CuFeO_2$ an unusual mutual position of the $Cu3d$ and $Fe3d$ bands was found which is attributed to strong electron-electron correlations. In porous silicon the local structure of silicon atoms is found to depend on the type of doping of the initial Si wafer. Solid-solid buried interfaces in thin semiconducting films irradiated by eximer laser are investigated. For the hard materials boron-carbonitride a structure consisting of hexagonal lattice planes of carbon and boron nitride is proposed.

1. INTRODUCTION

X-ray emission spectroscopy is a powerful technique for the study of the electronic structure of solids and the characterization of materials (1). This method is particularly sensitive to chemical bonding since the wave function of a core electron is quite localized and its angular momentum symmetry is well defined. As a result the valence states involved in the electronic transition to the core level are projected to the selected atomic site and their symmetries are restricted by the dipole selection rules. This enables the site-projected and symmetry-restricted partial density of states curves to be measured and compared directly with LDA band structure calculations (see, for instance, Ref. (2,3)). The wave functions of core electrons are localized to the vicinity of the emitted atom which means that only the atoms in the first coordination sphere around the excited atom will primarily influence the x-ray emission spectra (as confirmed in Refs. (4,5)). In this respect, x-ray emission spectroscopy is a local structure sensitive method (as are NMR, EELS, EXAFS and Mössbauer effect) and can also be used to obtain structural information.

X-ray emission spectroscopy with variable energy electron excitation allows different layers of the material under investigation to be probed because the x-ray yield depth varies as a function of the energy of the incident electrons (6). This is used usually in EPMA (electron-probe microanalysis) for estimation of depth profiles of elements in solids. It was shown recently that the combination of x-ray emission spectroscopy with high energy resolution and the variation of electron exciting energies opens up unique possibilities for studies on solid-phase reactions in buried solid-solid interfaces and the reconstruction of depth distributions of different phases (7,8,9). The great improvement in lateral resolution represents possibly the greatest recent achievement in x-ray emission spectroscopy. Up to beginning of 80's, two different techniques: X-ray emission spectroscopy (with high energy resolution but poor spatial resolution) and EPMA (with good lateral resolution and poor energy resolution) were developed separately. The first technique was used mostly for study of electronic structure and chemical bonding of different substances and the second for the determination of the chemical composition in microvolumes. It was a dream to combine both techniques to be able to study the electronic structure in microvolumes. The first promising results were obtained by Dr. A. I. Kozlenkov (10,11,12) who successfully developed new optical schemes for small-spot x-ray spectrometers with diffraction gratings and achieved sufficient intensity while maintaining high energy resolution. Based on these developments the small-spot x-ray spectrometers with diffraction gratings were constructed simultaneously in Russia and Japan (the characteristics of these spectrometers are given in Ref. (13,14)). The first measurements with these spectrometers showed that x-ray emission spectroscopy with high spectral energy and lateral resolution can be used to detect small clusters of silicide phases in a transition metal/silicon system (15). This development opens up new possibilities for characterizing interfaces on an atomic scale.

In the present paper we review the new research opportunities that are being offered by x-ray emission spectroscopy using laboratory excitation sources for characterizing advanced materials. Readers who are interested in applications of soft x-ray fluorescence excited with tunable synchrotron radiation are refereed to review articles (16,17,18,19).

2. HIGH-T_c SUPERCONDUCTORS

It was shown in Ref. (20) that the shape and energy position of the $OK\alpha$ spectra ($2p \rightarrow 1s$ transition) of various cuprates may be described by superpositions of the spectra of oxygen atoms participating in M^{3+}-O, M^{2+}-O and Cu-O bonds (M^{3+}=Y, La, Bi; M^{2+}=Ca, Sr, Ba). However, the question remains open as to whether the O-M and O-Cu bonds in cuprates remain the same as in the constituent oxides. It was shown in Ref. (20) that the experimental $OK\alpha$ spectrum of the mixed oxide La_2CuO_4 can be well represented by the superposition of the $OK\alpha$

spectra of constituent simple oxides (La_2O_3 and CuO) in a ratio of 1:3, which corresponds to the relative number of O-Cu and O-La bonds. This approach was less successful when applied to the reconstruction of the $OK\alpha$ spectra of cuprate compounds where an alkali earth metal oxide participates in the bonding. The energy position of the reconstructed $OK\alpha$ band of $SrCuO_2$ obtained by the superposition of $OK\alpha$ spectra of CuO and SrO is considerably different from the energy position obtained from experiment (21). This may be due to the fact that O-M^{2+} bond, unlike O-M^{3+}, is deformed in the perovskite-like structure of the cuprate, as compared to the constituent simple M^{2+}O oxide. Although the selectively excited spectra from O-M^{2+} bonds are not yet available, some approximate information can be obtained by subtracting the $OK\alpha$ spectrum of CuO from those of $M^{2+}CuO_2$. Since the compound $CaCuO_2$ does not exist, we used the spectra of $Ca_{0.85}Sr_{0.15}CuO_2$ in order to extract the contribution related to the O-Ca bond. It is found that the $OK\alpha$ spectra from M^{2+}-O bonds in cuprates which have been extracted in such way are really different from the emission spectra of simple oxides. The extracted spectra are shifted towards higher energies by 0.5, 0.6 and 1.0 eV for O-Ca, O-Sr and O-Ba, correspondingly, and are closer in energy position to the $OK\alpha$ spectra from O-M^{3+} bonds (21).

In Figure 1, the $OK\alpha$ spectrum of the high-T_c superconductor $Bi_2Sr_2CaCu_2O_{8+\delta}$ is reconstructed using the above scheme and compared to the experimental data. Taking into account the aspects discussed above, we estimated for $Bi_2Sr_2CaCu_2O_{8+\delta}$ the contribution from the O-Cu bond based both on the $OK\alpha$ spectrum of CuO and $OK\alpha$ spectrum of Bi2212 under the conditions of selective excitation at E=528.5 eV (22). The latter spectrum probes only O-Cu bond in the planes of Bi2212 compound. The contribution from O-Bi^{3+} bonds by the spectrum of Bi_2O_3 (with the satellite subtracted), and the contributions from O-Sr and O-Ca bonds are taken from the difference spectra. The weights of the different contributions were based on the nominal composition of the compound in question. It is shown that the additive curves provide a reasonably good approximation to the measured $OK\alpha$ spectrum which, as in the case of La_2CuO_4, justifies this approach for describing $OK\alpha$ spectra in complex oxides. Similar results were obtained from the reconstruction of $OK\alpha$ spectra of $YBa_2CuO_{6.9}$ and $Tl_2Ba_2CaCu_2O_{8+\delta}$ compounds (21). It follows from the observed trends in the variety of compounds that the electrons populating the O-M bonds dominate the shape and energy position of the $OK\alpha$ x-ray emission bands in cuprates. It follows from the analysis of $OK\alpha$ spectra that the covalent contribution is enhanced in the O-M^{2+} bonds and reduced in the Cu-O bonds, reversing the trend shown by the constituent oxides. Whereas the O-Cu bonds in cuprates remain more covalent than O-M^{2+} bonds, the difference in the degree of covalency between them becomes less pronounced than it is in CuO and M^{2+}O oxides. Seemingly it manifests the redistribution of the electron density between O-M and O-Cu bonds as a result of the donor-acceptor interaction between cations where M-elements play the role of donors and Cu is acceptor. Therefore, X-ray spectroscopy data indicate that O-M^{3+} bonds are only slightly deformed on the cuprate lattice, whereas O-M^{2+} bonds become more covalent.

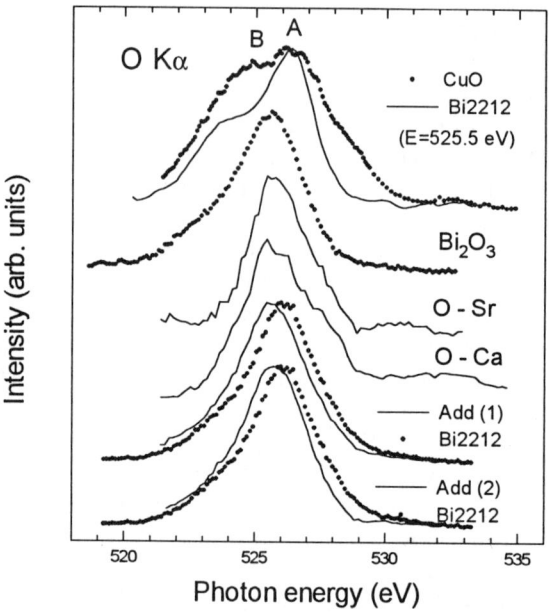

FIGURE 1. $OK\alpha$ spectrum of $Bi_2Sr_2CaCu_2O_{8+\delta}$ reconstructed by the superposition of spectra accounting for O-Cu, O-Bi, O-Sr and O-Ca bonds. The contribution from the O-Cu bonds is taken from the $OK\alpha$ spectrum of CuO and the $OK\alpha$ spectrum of Bi2212 measured by selective excitation at E=528.5 eV (22). The additive spectra Add(1) and Add(2) correspond to these both cases respectively.

One of the essential features of high-T_c superconductors is the very strong sensitivity of their superconducting properties to substitution of components by different elements. It is well known that one can form many such compounds with partial or full substitution of Y by RE elements for $YBa_2Cu_3O_7$. Also Ba and Cu can be replaced by various metallic ions. At the same time, the attempts to substitute oxygen by other chalcogens were not so successful. In this connection the determination of the local positions of impurity atoms in HTSCs seems to be very important for understanding their superconducting properties. We have applied x-ray emission spectroscopy to study the local structure of light dopants (B (23), F (24,25,26,27), S (28,29,30,31), P (29)) in high-T_c superconductors.

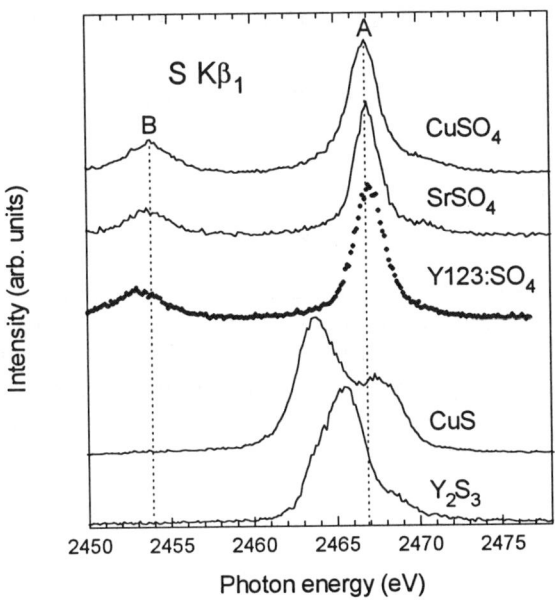

FIGURE 2. $SK\beta_{1,3}$ XES of sulphates, sulphides and Y123:S.

In Fig. 2, the results of the study of the sulphur doped Y123 compound are shown. It can be seen that the $SK\beta_{1,3}$ XES of Y123:S is very similar to that of the $CuSO_4$ and $SrSO_4$ sulphates and reveal two sub-bands, A and B. According to the theoretical calculations of the electronic structure of oxyanions (see, for instance, Ref. (32) the intense A sub-band is connected with S3p states of the central atom mixing with O2p states, and the B sub-band is due to S3p-O2s hybridization. The energy position and fine structure of these sub-bands depend on the degree of oxidation, local surrounding and the coordination number of the anion. On the other hand it can be seen that the $SK\beta_{1,3}$ XES of Y123:S is different from that of CuS and Y_2S_3. This suggests that Cu and Y atoms are absent in the first coordination sphere of the S atoms and it can be concluded that S atoms in doped YI23 substitute Cu-atoms in the chains and have the same environment as in SO_4-groups. Given the great similarity of the $SK\beta_{1,3}$ of Y123:S and other sulphates, it is necessary to point out that the A sub-band of Y123:S has a narrower width, which indicates the isolated character of SO_4-groups in this compound.

3. TRANSITION METAL COMPOUNDS

In this section we consider the most interesting applications of x-ray emission spectroscopy for the analysis of the electronic structure of transition metal compounds that require the development of new ideas and stimulated a new theoretical approach.

CuFeO$_2$

It has delafosite-type structure that consists of hexagonal layers of Cu, O and Fe stacked sequence along the c-axis to form a layered triangular lattice antiferromagnet. The triangular lattices of magnetic Fe^{3+}-ions (with a magnetic moment of about 4.4 μ_B) are separated by nonmagnetic ion layers of Cu^{1+} and O^{2-}. According to opto-electronic and electrical properties CuFeO$_2$ is a semiconductor with bandgap about 1.15 eV (33,34,35).

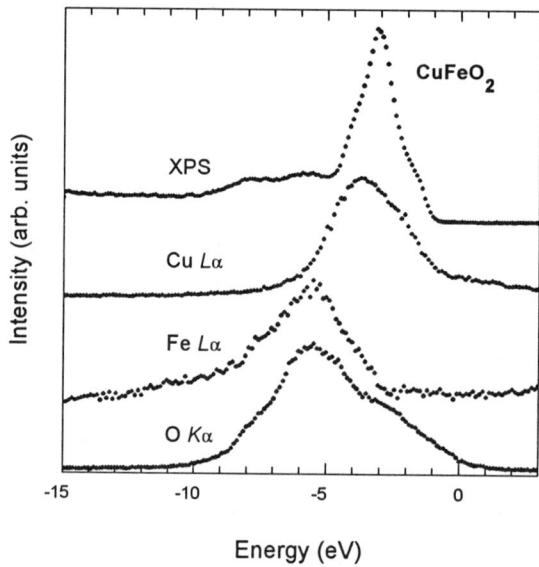

FIGURE 3. X-ray photoelectron spectrum of the valence band and Cu$L\alpha$, Fe$L\alpha$ and O$K\alpha$ x-ray emission spectra of CuFeO$_2$. X-ray emission spectra are brought in the common energy scale using the core level electron binding energies.

A full set of x-ray emission valence spectra (Cu$L\alpha$ ($3d4s \rightarrow 2p$ transition), Fe$L\alpha$ ($3d4s \rightarrow 2p$ transition) and O$K\alpha$ ($2p \rightarrow 1s$ transition) was measured for this compound (36) which (in accordance with dipole selection rules) probe Cu$3d4s$, Fe$3d4s$ and O$2p$ DOS in the valence band, respectively. The x-ray emission spectra were brought to the scale of the binding energies with respect to the Fermi level using the binding energies of relevant initial (core level) states of the x-ray transitions as measured by XPS. Corresponding binding energies are: $E_b(Cu2p_{3/2})$=932.4 eV, $E_b(Fe2p_{3/2})$=711.5 eV, and $E_b(O1s)$=530.05 eV. These spectra are compared in Fig. 3 with the x-ray photoemission valence band spectrum of $CuFeO_2$ (36) which probes (apart from differences in the atomic cross-sections of constituent atoms) the total DOS in the valence band. The maximum of intensity of $CuL\alpha$ emission band lies at about -3.6 eV relative to the Fermi energy, E_F (0.0 eV in Fig. 3) and this value agrees well with the main peak of the XPS valence band spectrum situated near the Fermi level at -3 eV. The intensity maxima of $FeL\alpha$ and $OK\alpha$ emission bands have the same binding energies and are located at -5.5 eV relative to E_F. From the joint analysis of experimental XPS and XES spectra we can conclude that Fe3d-states have higher binding energies than the Cu3d-states and are more strongly mixed with O2p-states than Cu3d-states.

This conclusion is in obvious contradiction with results of LSDA (local spin density approximation) band structure calculations of $CuFeO_2$ (36) according to which the position of Cu3d and Fe3d-bands are reversed (the Cu3d band is located 1 eV lower than the Fe3d one).

According to the LSDA calculation $CuFeO_2$ is a metal (not a semiconductor) where Fe ions are in the low-spin state and the magnetic moment is equal to 0.96 μ_B (which is 4.5 times less than experimental value) (36). So, one can conclude that the LSDA calculations are wrong and do not reproduce the experimental data. It was supposed (36) that in LSDA wrong results arise from failing to take into account the one-site Coulomb repulsion. The direct way to do it is to use the many-body perturbation theory and to estimate the exchange-correlation self-energy (GW approximation (37)), but due its complexity the application of this technique to real compounds is quite problematic. At the same time, more simplified methods can be used to improve LSDA. One of them is to accept the atomic point of view on correlations between localized electrons and include them in the same form as for free atoms through renormalized parameters of electron-electron interaction that can be estimated in the framework of LSDA. Such a new (LSDA+U) approach was suggested in Ref. (38) and successfully used for the analysis of the electronic structure of 3d-oxidic systems (39,40).

Fig. 4 shows the partial densities of states for $CuFeO_2$ calculated using the LSDA+U approach (36). As a result of this calculation it was found that $CuFeO_2$ is a semiconductor with the bandgap of 2.0 eV. The O2p band is located 5 eV below the Fermi level. The Cu3d completely filled band is located at the same energy region. The Fe3d↑ (t^3_{2g} and e^2_g) states are approximately 9 eV below zero. The Fe3d↓ band is empty and thus Fe ions are in the majority spin state. The magnetic moment is equal to 3.76 μ_B (in accordance to μ_{eff} = 4.65 μ_B determined from $\mu_{eff} = 2\sqrt{S(S+1)}$). It is also seen that the Cu3d band is placed above 5 eV the Fe3d one. Using LSDA+U quantitative agreement with the experimental XPS and XES spectra was obtained. It should denote that in this case the XES measurements stimulated a new theoretical treatment of the electronic structure of $CuFeO_2$.

3d-Metal Silicides

According to band structure calculations (41) upon formation of silicides tran-

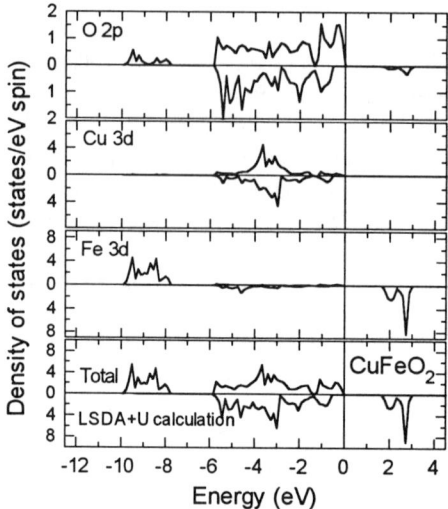

FIGURE 4. Calculated partial DOS of CuFeO$_2$ using the LDA+U approximation.

sition metal-silicon bonds replace Si-Si bond and *sp*-hybridization. As a result directed Si-Si covalent bonds do not occur in silicides. In general the Si3p electrons form chemical bonds with the transition metal yielding bonding and antibonding states whereas the majority of the transition metal d-states form a dominant d-bond between the bonding and antibonding M3d-Si3p orbitals. The location of the metal d-states below the Fermi level is a characteristic feature of silicide formation. These predictions of theory are in a good agreement with results of XPS and XES measurements.

In Fig. 5 a comparison of XPS and x-ray emission valence spectra (Fe$L\alpha$ (3d4$s\rightarrow$2p transition), Fe$K\beta_5$ (4$p\rightarrow$1s transition), Si$L_{2,3}$ (3s3$d\rightarrow$2p transition), Si$K\beta_{1,3}$ (3$p\rightarrow$1s transition)) with the results of band structure calculations of FeSi are given (2). The experimental Fe$L\alpha$ x-ray emission spectrum reproduces fairly well the position of the centre of gravity of the calculated Fe3d DOS distribution in the valence band of FeSi, but not the splitting of the Fe3d band. This is due to the large broadening of the Fe$L\alpha$ x-ray emission spectrum by the instrumental resolution (about 0.4 eV) and the width of the inner (core) Fe2p level (about 0.8-1.0 eV) which is determined by the lifetime of the core-level hole.

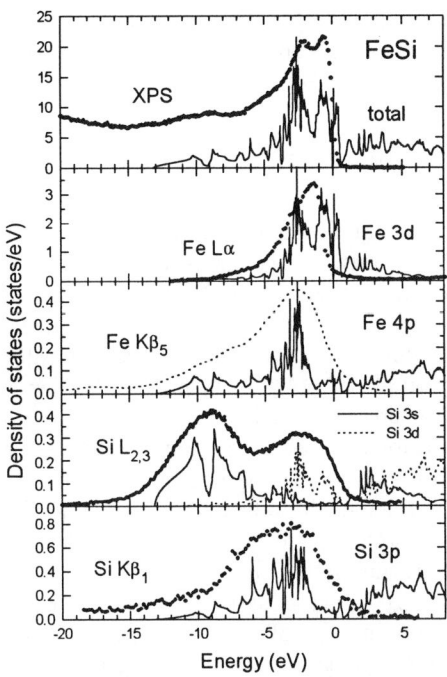

FIGURE 5. Measured XPS spectrum (dots; top panel) and x-ray emission spectra (dots for Fe$L\alpha$, Si $L_{2,3}$, Si$K\beta_{1,3}$ and broken curve for Fe$K\beta_5$ of FeSi) as compared with total (per unit cell) and partial (per atomic sphere) densities of states calculated using the LMTO method.

The experimental Fe$K\beta_5$ spectrum shows two maxima located at binding energies of approximately 3.5 and 9.0 eV, which is in accordance with the theoretical distribution of the Fe$4p$ partial DOS. It is also seen from both the theoretical and experimental spectra that Fe$4p$ states hybridize mostly with Si$3p$ and Si$3d$ states. The energy position and fine structure of the Si$K\beta_{1,3}$ x-ray emission spectrum is in good accord with the Si$3p$ partial DOS distribution. It is clearly seen that the two-peak structure in the Si$L_{2,3}$ XES cannot be explained only by the distribution of Si$3s$ states because the Si$3d$-states contribution is noticeable in the energy region close to the Fermi level. This is very important conclusion because usually it is considered (see Ref. (41) that the Si$3d$ states do not take part in the chemical bonding in $3d$ transition metal silicides. At first pointed out in Ref. (13) the high-energy sub-band of the Si$L_{2,3}$ x-ray emission spectra of the transition metal silicides FeSi, MnSi and NiSi, as well as those of the disilicides FeSi$_2$, MnSi$_2$ and NiSi$_2$ cannot be understood without the assumption that Si$3d$ states contribute to the chemical banding. Subsequently, the same conclusion was drawn for Pt-silicides (Pt$_2$Si, PtSi) in (42) based on an analysis of the Si$L_{2,3}$ x-ray emission spectra of these compounds.

4. POROUS SILICON

The discovery of photoluminescence in the visible region in porous (P-Si) silicon recently kindled considerable interests in this material for potential applications in microelectronic devices (43). For a better understanding of the origins of photoluminescence of P-Si it might be useful to study the local structure of the materials. Two porous Si layers were studied recently by means of x-ray emission spectroscopy in Ref. (44). Sample P-Si(2) was prepared by anodic etching of an n^--Si(100) substrate at 80 mA/cm^2 in a solution of hydrofluoric acid under white light exposure from a 300 W tungsten lamp for 0.5 min under the conditions described in Ref. (45). The specimen was exposed to air for about two years prior to the measurements described here. Sample P-Si (1) was prepared by anodizing p^+-Si at 20 mA/cm^2 for 5 min in darkness. This sample was manufactured 6 months before measuring the XE spectra.

The x-ray emission Si$L_{2,3}$ spectra for P-Si(1), and P-Si(2) along with c-Si and SiO$_2$ measured at electron energies of 2, 4, and 6 keV are depicted in Figs. 6 ((a) and (b)). It can be seen that the spectra for P-Si(1) prepared from p^+-Si are similar to those for c-Si (though it had been stored in air for 6 months). This effect is typical for the bulk of the sample as the spectra do not change for different accelerating voltages on the x-ray tube. In contrast to this result, Si$L_{2,3}$ XES for porous silicon (P-Si(2)) obtained from n^- Si and exposed to air for about 2 years were found to be very close to those of SiO$_2$ (Fig.6 (b)). It is necessary to point out that the intensity minimum between two spectral features for Si$L_{2,3}$ XES of P-Si(1) at photon energies of 89.3 and 92.0 eV is not as distinct than that for c-Si (see Fig. 6 (a)). This smearing-out can probably be attributed to the contribution of the Si$L_{2,3}$ XES spectrum of some a-Si or SiO$_x$ (x<1) whose spectra have flat extended maxima in this energy range. The best agrrement of the model spectrum of P-Si(1) to the experimental one is found for the superposition of Si$L_{2,3}$ XES of c-Si on a-Si and SiO$_{0.3}$ in the ratio of 0.65:0.25:0.10 (44). The contribution of a-Si to the spectra of P-Si(1) is prominent. We therefore conclude that the local structure of porous silicon prepared from p^+-Si can be characterized by short range order that has features typical for crystalline and amorphous silicon. This conclusion is in accordance with high-resolution transmission electron microscopy which shows that P-Si consists of small branches and particles that are crystalline in the centre surrounded by an amorphous layer (see Ref. (46)). Some amount of the SiO$_x$ phase in the forming of the P-Si(1) sample is found too.

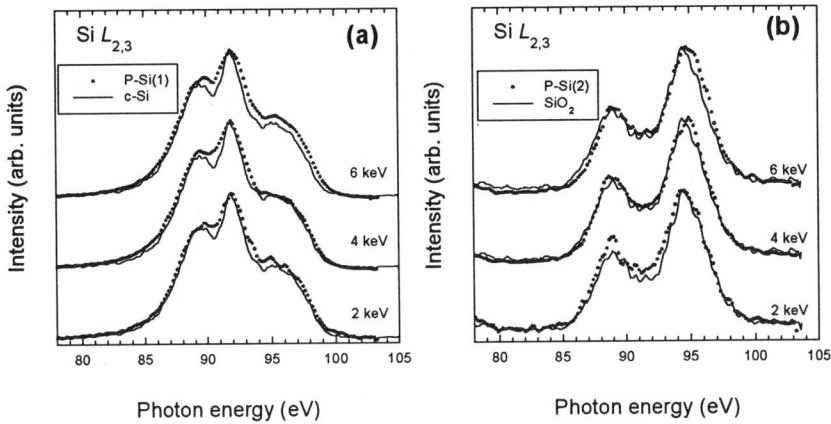

FIGURE 6. (a) Si $L_{2,3}$ X-ray emission spectra of P-Si(1) and c-Si measured using electron excitation energies of E=2, 4 and 6 keV; (b) Si $L_{2,3}$ X-ray emission spectra of P-Si(2) and SiO_2 measured using electron excitation energies of E=2, 4 and 6 keV.

The difference in the degree of oxidation of the P-Si(1) and P-Si(2) samples is most probably due to the difference in substrates rather than the time since preparation. A similar effect was found for the SiO_2/Si system (47). After irradiation by electrons of the energy of 11-12 MeV, the effect of oxidation was found on the n^--Si substrate but not on the p^+-Si substrate.

5. INTERFACES

Buried solid-solid interfaces are of great interest because of their practical applications in semiconductor devices (48), catalysts (49) and multilayer optical coatings (50). Characterization of such "interface region" between two solids is very important for the correct description of the formation of the interfaces. The main difficulty in such studies is due to the fact that interfaces are "buried" within the solid matter. Up to now the most investigations have been done using surface sensitive methods that can characterize the structure and electron states only of the top layers of the specimens (51,52). As mentioned in (53) "progress in new materials, interfaces and eventually device depends a great deal on continuing advances in diagnostic tools to help us characterize interfaces structurally, chemically and electronically on an atomic scale". The recent development of soft x-ray emission spectroscopy both with variable electron excitation (13), (14) and tunable monochromatic synchrotron radiation excitations (17) open new possibilities for the characterization of such materials. The main advantage of this technique is probing of bulk material and depth profiling without damage of specimens (7). According to measurements of x-ray emission spectra from Fe/Si system annealed at 900-950 °C the formation of α-FeSi$_2$ and β-FeSi$_2$ phase was found (15). In this case the grains of β-FeSi$_2$ were too small to be detected by x-ray diffraction. X-ray emission spectroscopy as a very powerful method to clarify the existence of small clusters located at interfaces. It means that this technique can be used to characterize the interfaces on an atomic scale. The combination of high lateral (about 1 μm) and energy (about 0.3 eV) resolution allows to characterize not only model systems but also commercial semiconducting devices.

Si$_3$N$_4$ films prepared on (100) GaAs substrate by the CVD method and irradiated by Kr$_2$ excimer laser were investigated in Ref. (9). After one laser pulse the sample was taken out of the laser apparatus and examined on an x-ray spectrometer at two points on the surface: a laser-irradiated area and a nonirradiated area. It was found that the Si$L_{2,3}$ XES of the irradiated area of Si$_3$N$_4$-film was different to the nonirradiated one and practically identical to the spectrum of crystalline silicon (c-Si) (9). Under excimer laser irradiation photochemical breaking of Si-N bonds takes place and c-Si precipitation occurs.

The effect of oxide layer of thickness 3, 7 and 10 nm grown on p^+-Si(100) on the formation of CoSi$_2$ was investigated with help of Si$L_{2,3}$ XES in Ref. (54). In each case 20 nm thick Co films were deposited onto the oxide by DC sputtering and were annealed at 900-1000 °C for 10–45 sec. It was found that the formation of SiO$_2$ started at annealing temperatures above 900 °C. A SiO$_2$ layer about of 3 nm thick is optimal for the formation of CoSi$_2$ at T=1100°C for 10 sec. It was found that the increasing the annealing time from 10 to 45 sec leads to the formation of CoSi$_2$ for thicker layers of SiO$_2$ (about 10 nm).

6. HARD MATERIALS

Because graphite and boron nitride have similar hexagonal structures, it may be possible to synthesize related compounds from their solid solutions. A possible starting material is hexagonal boron carbon nitride which may undergo a phase change to a cubic structure having a high degree of hardness (55). This assumption was the basis for many attempts to synthesize BCN. The main problem in the preparation of these materials is connected with the identification and characterization of ternary phases. According to Ref. (56) the lattice parameters of hexagonal boron carbonitrides and those of graphite and BN are very similar. In this situation there is no direct structural method of distinguishing between h-BCN, h-BN and graphite. In the Ref. (57) we applied x-ray emission spectroscopy to characterize the local structure of BCN compounds. The BCN samples were prepared by nitridation of boron carbide B$_4$C powder in a graphite element resistance furnace with continuous nitrogen gas flow near atmospheric pressure.

Each heat treatment (T=1200, 1800 and 2400 °C) was performed for 24 hrs. According to the XES measurements of the BCN materials and the comparison of the spectra obtained with those of the reference compounds we come to the following conclusions about the local structure of the constituent atoms:

1. Boron atoms are chemically bonded only to nitrogen atoms. These bonds are the same as in h-BN.
2. The local structure of nitrogen atoms is the same as that in *h*-BN.
3. Carbon atoms form only carbon-carbon bonds that are similar to the corresponding bonds in hexagonal graphite.

The results allow us to estimate the validity of the structural models of the BCN materials. All existing structural models of BCN can be subdivided into two major groups. The first one is a substitution solid solution of all three components in a honeycomb network with various in-layer arrangements depending on the overall composition. According to the second model the ternary system is a mixture of separate BN and graphite layers either in random or in regular intercalation. We conclude that the mixture model initially proposed for the B-C-N compounds synthesized by by CVD (58) and by pyrolysis of organic precursors is in full agreement with the XES data for the compounds obtained by nitridation of boron carbide. The substitution model cannot be accepted because we did not found any traces of the B-C and C-N bonds. The intercalation of BN and graphite layers is most probably random since the diffraction data (56) did not reveal any particular ordering.

REFERENCES

1. Kurmaev, E.Z., Cherkashenko, V.M., and Finkelstein, L.D., *X-Ray Spectra of Solids*, Moscow, Nauka, 1988, 175 p.
2. Galakhov, V.R., Kurmaev, E.Z., Cherkashenko, V.M., Yarmoshenko, Yu.M., Shamin, S.N. Postnikov, A.V., Uhlenbrock, St., Neumann, M., Lu, Z.W., Klein, B.M., and Shi, Z-P., *J. Phys.: Condens. Matter* 7, 5529 (1995).
3. Lu, Z.W., Klein, B.M., Kurmaev, E.Z., Cherkashenko, V.M., Shamin, S.N., Yarmoshenko, Yu.M., Trofimova, V.A., Uhlenbrock, St., Neumann, M., Furubayashi, T., Hagino, T., and Nagata, S., *Phys. Rev. B* 53, 9626 (1996).
4. Cherkashenko, V.M., and Kurmaev, E.Z., *Russian Journal of Structural Chemistry*, 22, 18 (1981).
5. Cherkashenko, V.M., Galakhov, V.R., and Kurmaev, E.Z., *ibid.* 27, 185 (1986).
6. Feldman, C., *Phys. Rev.* 117, 455 (1960).
7. Galakhov, V.R., Kurmaev, E.Z., Shamin, S.N., Elokhina, L.V., Yarmoshenko, Yu.M., Bukharaev, A.A., *Appl. Surf. Sci.* 72, 73 (1993).
8. Kurmaev, E.Z., Shamin, S.N., Galakhov, V.R., Wiech, G., Majkova, E., and Luby, S., *J. Mater. Res.* 10, 907 (1995).
9. Kurmaev, E.Z., Shamin, S.N., Dolgih, V.E., Kurosawa, K., Nakamae, K., Takigawa, Y., Kameyama, A., Yokotani, A., and Sasaki, W., *Jpn. J. Appl. Phys.* 33, L1549 (1994).
10. Kozlenkov, A.I., and Belov, Yu.I., *Optics and Spectroscopy* (*USSR*) 42, 567 (1977).
11. Kozlenkov, A.I., *ibid.* 46, 579 (1979); 48, 390 (1980).

12. Kozlenkov, A.I., Bogdanov, V.G., and Belov, Yu.I., *Pribory i Technika Eksperimenta* (*USSR*) No.5, 261 (1976).
13. Kurmaev, E.Z., Fedorenko, V.V., Shamin, S.N., Postnikov, A.V., Wiech, G. and Kim, Y., *Physica Scipta* T41, 288 (1992).
14. Iwami, M., Hirai, M., Kusaka, M., Kubota, M., Yamamoto, S., Nakamura, H., Watabe, H., Kawai, M., and Soezima, H., *Jpn. J. Appl. Phys.* 29, 1353 (1990).
15. Kasaya, M., Yamauchi, S., Hirai, M., Kusaka, M., Iwami, M., Nakamura, H., and Watabe, H., *Appl. Surf. Sci.* 75, 110 (1994).
16. Nordgren, E.J., *Physica Scripta*, T61, 32 (1996).
17. Ederer, D.L., Callcott, T.A., and Perera, R.C., *SRN*, 7 (1994).
18. Shin, S., *SRN*, 8, 16 (1995).
19. Ma, Y., *SRN*, 8, 26 (1995).
20. Finkelstein, L.D., Galakhov, V.R., Fedorenko, V.V., Elokhina, L.V., Samokhvalov, A.A., Kurmaev, E.Z., Butorin, S.M., Nordgren, J., Slobodin, B.V., Teterin, Yu.A., and Sosulnikov, M.I., *Solid State Communs.* 90, 769 (1994).
21. Finkelstein, L.D., Galakhov, V.R., Fedorenko, V.V., Elokhina, L.V., Samokhvalov, A.A., Kurmaev, E.Z., Butorin, S.M., Nordgren, J., Slobodin, B.V., Teterin, Yu.A., and Sosulnikov, M.I., *J. Electr. Spectr. Relat. Phenom.* 68, 431 (1994).
22. Butorin, S., Guo, J.-H., Wassdahl, N., Skytt, P., Nordgren, J., Ma, Y., Strom, C., Johansson, G., and Qvarford, M., *Phys.Rev. B* 51, 11915 (1995).
23. Kurmaev, E.Z., Shamin, S.N., Slater, P.R., and Greaves, C., *Physica C* 227, 309 (1994).
24. Fedorenko, V.V., Galakhov, V.R., Elokhina, L.V., Finkelstein, L.D., Naish, V.E., Kurmaev, E.Z., Butorin, S.M., Nordgren, E.J., Tyagi, A.K., Rao, U.R.K., and Iyer, R.M., *Physica C* 221, 71 (1994).
25. Kurmaev, E.Z., Fedorenko, V.V., Elokhina, L.V., Finkelstein, L.D., and Popova, T.B., *Physica C* 226, 58 (1994).
26. Kurmaev, E.Z., Fedorenko, V.V., Elokhina, L.V., Vinogradova, A.S., and Gao, X-H., *Solid State Commun.* 96, 967 (1995).
27. Kurmaev, E.Z., Elokhina, L.V., Fedorenko, V.V., Bartkowski, S., Neumann, M., Greaves, C., Edwards, P.P., Slater, P.K., and Francesconi, M.G., *J. Phys.: Condensed Matter* 8, 4847 (1996).
28. Yarmoshenko, Yu.M., Trofimova, V.A., Elokhina, L.V., Kurmaev, E.Z., Butorin, S., Cloots, R., and Ausloos, M., *Physica C* 211, 29 (1993).
29. Yarmoshenko, Yu.M., Trofimova, V.A., Kurmaev, E.Z., Slater, P.R., and Greaves, C., *Physica C* 224, 317 (1994).
30. Yarmoshenko, Yu.M., Trofimova, V.A., Dolgih, V.E., Korotin, M.A., Kurmaev, E.Z., Aguiar, J. Albino, Ferreira, J.M., and Pavao, A.C. *J.Phys.: Condens. Matter* 7, 213 (1995).
31. Yarmoshenko, Yu.M., Korotin, M.A., Trofimova, V.A., Galakhov, V.R., Elokhina, L.V., Kurmaev, E.Z., Uhlenbrock, St., Neumann, M., Slater, P.R., and Greaves, C., *Phys. Rev. B* 52, 11830 (1995).
32. Karlson G., and Manne, R., *Phys. Scripta* 4, 11 (1971).
33. Benko F.A., and Koffyberg, F.P., *J. Phys. Chem. Solids* 45, 57 (1984); *Can. J. Phys.*, 63, 1306 (1985); *Phys. Status Solidi* 94, 231 (1986).
34. Benko F.A., and Koffyberg, F.P., *J. Phys. Chem. Solids* 48, 431 (1987).
35. Dordor, P., Chaminade, J.P., Wichainchai, A., Marquestaut, E., Doumerc, J.P., Pouchard, M., Hagenmuller, P., and Ammar, A., *J. Solid State Chem.*, 75, 105 (1988).
36. Galakhov, V.R., Poteryaev, A.V., Kurmaev, E.Z., Anisimov, V.I., Bartkowski, St., Neumann, M., Lu, Z.W., Klein, B.M., and Zhao, T.-R. (*submitted to Phys. Rev. B*).
37. Hedin, L. *Phys. Rev.* 139, A796 (1965).
38. Anisimov, V.I., Zaanen, J., and Andersen, O.K., *Phys. Rev. B* 44, 943 (1991).

39. Anisimov, V.I., Solovyev, I.V., Korotin, M.A., Szyzyk, M.T., and Sawatzky, G.A., *Phys. Rev. B* 48, 16929 (1993).
40. Solovyev, I.V., Hamada, N., and Terakura, K., *Phys. Rev. B* 53, 7158 (1996).
41. Weaver, J.H., Franciosi, A., and Moruzzi, V.L., *Phys. Rev. B* 29, 3293 (1984).
42. Yamauchi, S., Hirai, M., Kusaka, M., Iwami, M., Nakamura, H., Ohshima, H. and Hattori, T., *Jpn. J. Appl. Phys.* 33, L1012 (1994).
43. Canham, L.T., *Appl. Phys. Lett.* 57, 1046 (1990).
44. Kurmaev, E.Z., Shamin, S.N., Galakhov, V.R., Sokolov, V.I., Ludwig, M.H., and Hummel, R.E., (*submitted to J. Phys.: Condens. Matter*).
45. Inoue, K., Maekashi, K., and Nakashima, H., *Jpn. J. Appl. Phys.* 32, L361 (1993).
46. Bao Xi-mao, Yan Feng, Liu Cheng-en and Zheng Xiang-qin, Proceeding of the 21st International Conference on "The Physics of Semiconductors" (eds. Ping Jiang and Hou-Zhi Zheng), Bejing, China (August 10-14, 1992), vol.2, World Scientific, p.1447.
47. Kurmaev, E.Z., Shamin, S.N., Galakhov, V.R., Makhnev, A.A., Kirillova, M.M., Kurennykh, T.E., Vykhodets, V.B., and Kashieva, S. (*submitted to NIMB*).
48. Poate, J.M., Tu, K.N., and Mayer, J.W., Eds., *Thin Films-Interdiffusion and Reactions*, Wiley, New York, 1978.
49. Deviney, M.L. and Gland, J.L., Eds., *Catalyst Characterization Science: Surface and Solid State Chemistry*, ACS Symposium, Ser. No. 288, American Chemical Society, Washington, DC, 1985.
50. Christiensen, F.E., Ed., "X-Ray Multilayers for Diffractometers, Monochromators and Spectrometers", *SPIE Proceedings, vol. 984, The International Society for Optical Engineering, Bellingham*, WA, 1988.
51. Callandra C., Bisi O., Ottaviani G., *Surf. Sci. Rep.* 4, 271, 1985.
52. Rossi G., *Surf. Sci. Rep.* 7, 1, 1987.
53. Ludeke R., "Formation of Semiconductor Interfaces", in *Proceedings of the 4th International Conference*, Eds., Bruno Lengeler, Hans Luth, Winfried Monch and Johannes Pollmann, p. 755, June 14-18, 1993.
54. Kurmaev, E.Z., Shamin, S.N., Galakhov, V.R., and Kasko I. (*submitted to Thin Solid Films*).
55. Riedel, R., *Adv. Mater.*, 4, 759 (1992).
56. Andreev, Yu.G., and Lundstrom, T., *J. Alloys Comp.*, 210, 311, (1994).
57. Kurmaev, E.Z., Ezhov, A.V., Shamin, S.N., Cherkashenko, V.M., Andreev, Yu.G., Lundstrom, T. (*submitted to J. Alloys Comp.*).
58. Derre, A., Filipozzi, L., and Peron, F., *J. Phys.* IV, Suppl., C3, 3, 195 (1993).

Author Index

A

Ahlers, D., 521
Amusia, M. Ya., 415
Ando, M., 351
Anisimov, V. I., 771
Asfaw, A., 749
Attenkofer, K., 521

B

Backe, H., 57
Beiersdorfer, P., 121
Bell, F., 369
Bianconi, A., 585
Bizau, J.-M., 625
Bourgeois, D., 267
Brenzinger, K.-H., 57
Burgdörfer, J., 475
Buskirk, F., 57

C

Caliebe, W., 671
Callcott, T. A., 749
Campbell, J. L., 431
Canney, S. A., 383
Carlisle, J. A., 749
Cherkashenko, V. M., 771
Crasemann, B., 3
Cubaynes, D., 625

D

Dambach, S., 57
de Groot, F., 497
Dev, B. N., 249
Diehl, S., 625
Doerk, Th., 57
Dräger, G., 557

E

Ederer, D. L., 749
Eftekhari, N., 57
Elleaume, P., 43
Euteneuer, H., 57

F

Fadley, C. S., 295
Fischer, P., 521
Freudenberger, J., 73

G

Galakhov, V. R., 771
Gemmell, D. S., 659
Gerchikov, L. G., 447
Glans, P., 723
Görgen, F., 57
Gunnelin, K., 723
Guo, J.-H., 723
Guo, X., 383

H

Hagenbuck, F., 57
Hagmann, S., 169
Hämäläinen, K., 671
Harami, T., 351
Hastings, J. B., 671
Herberg, C., 57
Himpsel, F. J., 749

I

Igarashi, H., 351
Ivanchenko, T., 737
Izumi, K., 351

J

Jia, J. J., 749
Jiminez, J., 749
Johann, K., 57
Jung, M., 659

K

Kabachnik, N. M., 689
Kaiser, K.-H., 57
Kanter, E. P., 659
Kao, C.-C., 671
Karnatak, R. C., 569
Kettig, O., 57
Kheifets, A. S., 369, 383
Khemliche, H., 137
Kikuta, S., 351
Knies, G., 57
Koizumi, A., 399
Kollmus, H., 153
Korol, A. V., 447
Koyama, I., 351
Krässig, B., 659
Kube, G., 57
Kunimune, Y., 351
Kurmaev, E. Z., 771
Kurp, F. F., 369

L

Lanzara, A., 585
Lauth, W., 57
LeBrun, T., 647, 659
Len, P. M., 295
Lienert, U., 175
Liesen, D., 93
Limburg, B., 57
Lind, J., 57
Lyalin, A. G., 447

M

Manninen, S., 671
Materlik, G., 295
McCarthy, I. E., 383
McGuire, J. H., 475

Mitsui, T., 351
Miyamoto, N., 399
Moewes, A., 749
Moffat, K., 267
Morgenstern, R., 603
Morikawa, E., 749
Moshammer, R., 153
Mourou, G., 267

N

Naylor, G., 267
Nordgren, J., 723

O

Olson, R. E., 153

P

Padmore, H. A., 193
Papp, T., 431
Perera, R. C. C., 749
Pisk, K., 465
Poteryaev, A. I., 771
Pratt, R. H., 465
Prior, M. H., 137

Q

Qiu, Y., 475

S

Saini, N. L., 585
Sakai, N., 399
Schmidt-Böcking, H., 153
Schmitt, W., 153
Schneider, J. R., 369
Schöpe, H., 57
Schotte, F., 267
Schulte-Schrepping, H., 369
Schulze, C., 175
Schütz, G., 521
Seto, M., 351

Shamin, S. N., 771
Shimizu, T., 351
Shirley, E., 749
Skytt, P., 723
Smirnov, G. V., 323
Soldatov, A. V., 737
Solov'yov, A. V., 447
Southworth, S. H., 659
Stephan, G., 57
Stern, E. A., 535
Streli, C., 233
Suortti, P., 175
Surić, T., 465
Svensson, S., 703

T

Tanaka, Y., 399
Terminello, L. J., 749
Tonn, Th., 57
Trofimova, V. A., 771
Tschentscher, Th., 369

U

Ullrich, J., 153
Unverzagt, M., 153

V

van Ek, J., 749
Varga, D., 431

Vos, M., 369, 383
Voss, J., 209

W

Walcher, Th., 57
Walker, R. P., 21
Wang, J., 475
Weigold, E., 369, 383
Wobrauschek, P., 233
Wuilleumier, F. J., 625
Wulff, M., 267

Y

Yacoby, Y., 535
Yarmoshenko, Yu. M., 771
Yoda, Y., 351
Young, L., 659

Z

Zahn, R., 57
Zhang, X., 351
Zhou, Ling, 749

AIP Conference Proceedings

	Title	L.C. Number	ISBN
No. 374	High Energy Solar Physics (Greenbelt, MD 1995)	96-84513	1-56396-542-9
No. 375	Chaotic, Fractal, and Nonlinear Signal Processing (Mystic, CT 1995)	96-85356	1-56396-443-0
No. 376	Chaos and the Changing Nature of Science and Medicine: An Introduction (Mobile, AL 1995)	96-85220	1-56396-442-2
No. 377	Space Charge Dominated Beams and Applications of High Brightness Beams (Bloomington, IN 1995)	96-85165	1-56396-625-7
No. 378	Surfaces, Vacuum, and Their Applications (Cancun, Mexico 1994)	96-85594	1-56396-418-X
No. 379	Physical Origin of Homochirality in Life (Santa Monica, CA 1995)	96-86631	1-56396-507-0
No. 380	Production and Neutralization of Negative Ions and Beams / Production and Application of Light Negative Ions (Upton, NY 1995)	96-86435	1-56396-565-8
No. 381	Atomic Processes in Plasmas (San Francisco, CA 1996)	96-86304	1-56396-552-6
No. 382	Solar Wind Eight (Dana Point, CA 1995)	96-86447	1-56396-551-8
No. 383	Workshop on the Earth's Trapped Particle Environment (Taos, NM 1994)	96-86619	1-56396-540-2
No. 384	Gamma-Ray Bursts (Huntsville, AL 1995)	96-79458	1-56396-685-9
No. 385	Robotic Exploration Close to the Sun: Scientific Basis (Marlboro, MA 1996)	96-79560	1-56396-618-2
No. 386	Spectral Line Shapes, Volume 9 13th ICSLS (Firenze, Italy 1996)		1-56396-656-5
No. 387	Space Technology and Applications International Forum (Albuquerque, NM 1997)	96-80254	1-56396-679-4 (Case set) 1-56396-691-3 (Paper set)
No. 388	Resonance Ionization Spectroscopy 1996 Eighth International Symposium (State College, PA 1996)	96-80324	1-56396-611-5
No. 389	X-Ray and Inner-Shell Processes 17th International Conference (Hamburg, Germany 1996)	96-80388	1-56396-563-1